Handbook Factory Planning and Design

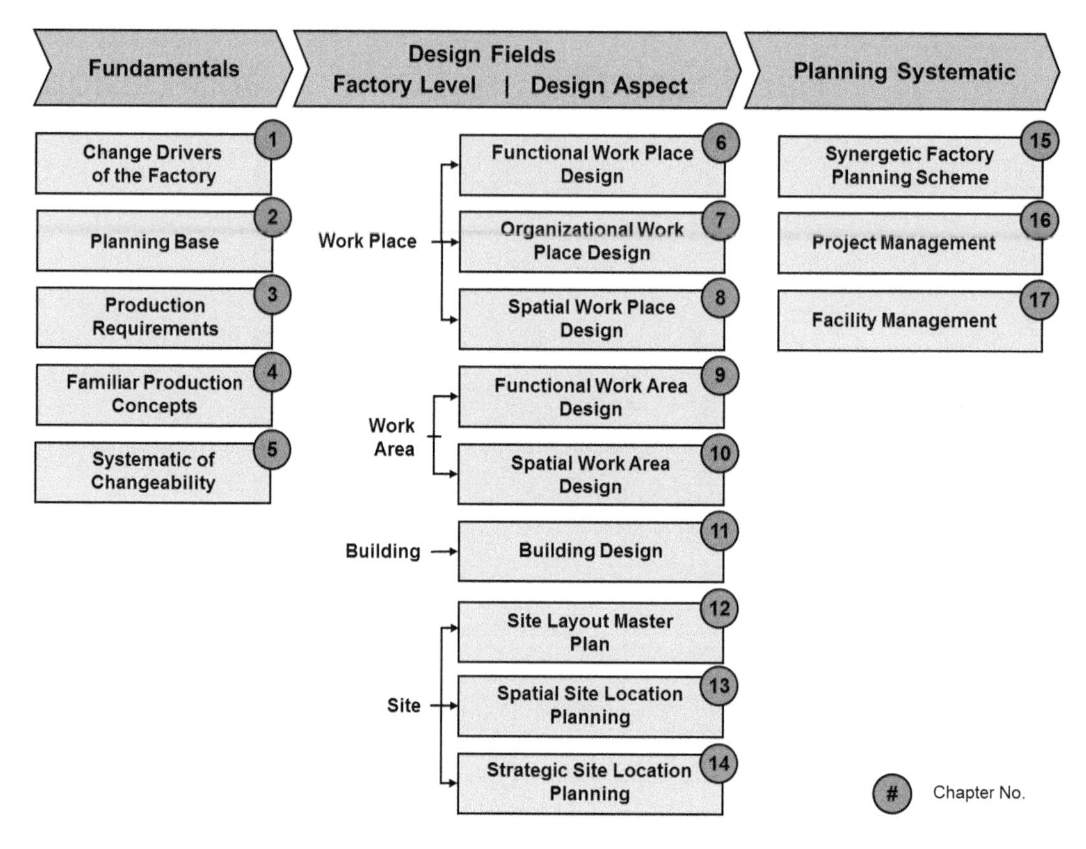

Structure of the Book

Hans-Peter Wiendahl
Jürgen Reichardt · Peter Nyhuis

Handbook Factory Planning and Design

Hans-Peter Wiendahl
Garbsen
Germany

Jürgen Reichardt
Fachbereich Baukonstruktion und Industriebau
Fachhochschule Münster
Münster
Germany

Peter Nyhuis
IFA Institut für Fabrikanlagen und Logistik
Leibniz Universität Hannover
Garbsen
Germany

Translated by Rett Rossi

Additional material to this book can be downloaded from http://extras.springer.com.

ISBN 978-3-662-46390-1 ISBN 978-3-662-46391-8 (eBook)
DOI 10.1007/978-3-662-46391-8

Library of Congress Control Number: 2015933365

Springer Heidelberg New York Dordrecht London

Printed on acid-free paper

Springer-Verlag GmbH Berlin Heidelberg is part of Springer Science+Business Media
(www.springer.com)

Preface

The 1990s saw the rapid development of both the Internet and business logistics. Less than two decades later, the globalized economy was a reality. Nowadays, sustainability and resource efficiency are guiding principles to run a factory. The digital communication of orders, processes, and resources is the next foreseeable development step in manufacturing.

Enterprises now frequently distribute their productions over several sites in a number of countries, and their productions are usually subject to strong fluctuations. Individual sites thus have to be highly reactive and changeable. This in turn necessitates a paradigm change; generally speaking, we need to invert the way we have traditionally considered a factory. Whereas previously, the primary task of a parent company was seen as developing a product, producing it and processing orders, while procuring and distributing finished goods to customers were secondary, today's priority is reliably supplying globally distributed markets from the most advantageous sites. Instead of central factories with a broad manufacturing depth, transformable or even temporary production sites located near the individual markets are now essential.

With this in mind, we realized that a critical look at factory planning up until now had to be undertaken. In gathering information from numerous research projects and industrial-based projects conducted in various branches, it became clear that in addition to the customary primary goal of being as efficient as possible, additional demands have arisen:

- Depending on the impulse for change, a factory needs to be able to adjust itself within a suitable time period with regard to both production technology and spatial demands on each of the impacted factory levels.
- Manufacturing and assembly systems need to take into consideration local perspectives concerning know-how, wage costs, and required value-adding (i.e., local content).
- Production facilities and buildings need to be designed so that they conserve resources and are energy efficient.
- The external appearance of the factory needs to represent the corporate identity of the enterprise, while the internal appearance needs to meet the claim of the product.
- The spatial design of production sites needs to provide comfortable workplaces, thereby expressing the company's high regard for its employees.

In consideration of all this and over a number of years, we have developed the tri-fold structure of this book. It is based on the second edition of the German "Handbuch Fabrikplanung" (Handbook Factory Planning), published in 2014 by Hanser Verlag Munich.

The first part of the book consists of five chapters and begins by developing a deeper understanding of the drivers behind factory changes and the resulting planning basis including future demands. Following that, we review existing production concepts and conclude by deriving various characteristics of what we refer to as a 'site's changeability'.

In the second part of the book, we describe the planning and design process of a production site from the level of individual workstations to the level of various sections, up to the levels of the building and location itself. Depending on the level, we discuss strategic and functional planning aspects as well as aspects pertaining to the actual organization of work—all with a special emphasis on changeability. Describing the spatial specifications of these levels plays a central role in directing the factory planner's view to the notion that form not only follows function, but also follows the performance of the buildings and the building services they are equipped with.

With three chapters in the third part of the book, we focus on the systematic factory planning process with respect to these new requirements. The center of our discussion is the synergetic factory planning model. In seven stages, it describes the creative interplay between production planning and spatial planning based on a continuous 3-D-modeling starting with the goal-setting right up to the ramp-up. The second chapter takes a look at project management, including the aspects of forming a project team, the responsibilities or team tasks, as well as a brief overview of digital tools for planning a factory. In view of the frequent changes of use, it becomes all the more important to efficiently use real estate properties; the last chapter of the book is therefore dedicated to facility management.

Our goal with this handbook is first and foremost to provide a comprehensive, methodical, and practical support for the management of production enterprises as well as for planners and designers of production sites. The same applies to architects and construction planners who design and realize industrial buildings. Moreover, this handbook is also intended for those studying production technology and industrial logistics from the perspectives of both engineering and management, and for architecture and building construction students.

Before delving into our subject matter, we would like to thank first of all Mrs. Rett Rossi, our most valued translator, who went deep into the complex subject and delivered a perfect performance. Next to thank is Jens Lübkemann from the IFA Institute of Production Systems and Logistics Leibniz University, Hannover, for coordinating the work between the authors, our reviewers, and Mrs. Rossi as well as the preparation of the correct format of text and figures. Mr. Gerhard Hoffmann, CEO of IFES GmbH in Cologne, has contributed Sect. 11.3 and Detlef Gerst Chap. 7; to both, we have to express our sincere thanks. In addition, we are much indebted to Indranil Bhattacharya, from the architectural firm Reichardt–Maas and Associates (Essen/Bangalore), for energetically supporting Chaps. 11–14 on spatial planning especially with regard to adapting it to international aspects as well as adding British building norms and quoting of English standard literature sources. Many thanks go further to our colleagues Prof. Hoda and Waguih ElMaraghy, University of Windsor Canada, and Prof. Neil Duffie, Madison Wisconsin University, for carefully reviewing several chapters. Last but not least we would like to thank the members of the Scientific Publishing Services in Chennai, India for the excellent preparation of the final book lay out. This concerns mainly Mr. Udhaya Kumar P. and Ms. Shilpa Soundararajan.

Garbsen, March 2015 Hans-Peter Wiendahl
Essen Jürgen Reichardt
Garbsen Peter Nyhuis

Contents

About the Authors

Hans-Peter Wiendahl born in 1938, studied Mechanical Engineering at the RWTH Aachen and at MIT USA and worked after that as junior researcher for 4 years at the WZL Laboratory for Machine Tools and Production Engineering in University of Aachen. There, he received a Doctorate in Engineering in 1970, and in 1972 he graduated as lecturer (Dr.-Ing habil.). From 1972 to 1979, he moved to a large machine building company working in different management positions. From 1979 until 2003, Wiendahl was appointed Full-time Professor and Director of the Institute of Production Systems and Logistics at the Leibniz Universität Hannover (www.ifa.uni-hannover.de). Prof. Wiendahl is full member of the German Academic Society for Production Engineering (WGP) and Emeritus Member of the International Academy for Production Engineering CIRP. He has authored more than 10 books and over 300 articles in the field of Factory Planning, Assembly, and production logistics. He received three doctorates of honor from the Universities of Magdeburg, ETH Zurich, and Dortmund and the SME Golden Medal.

Jürgen Reichardt born in 1956, studied Architecture at the University of Karlsruhe and the Technische Universität Braunschweig. From 1988 to 1995, he worked as project manager for design and implementation of complex industrial buildings at Agiplan AG in Mülheim/Ruhr. In 1992, he founded his office Reichardt Architects with a focus on planning of industrial plants and logistics centers at home and abroad. Since 1996, he is Professor in Industrial Construction at the Muenster school of architecture (https://www.fh-muenster.de). In 2006, he founded the office Bhattacharya Reichardt Architects & Engineers in Bangalore (BRAE), India, and operates since 2008 its German office as RMA Reichardt–Maas–Associated Architects GmbH & Co. KG in Essen.

Peter Nyhuis born in 1957, studied Mechanical Engineering at the Leibniz Universität Hannover. He received his Ph.D. in 1991 after his time as junior researcher at the Institute for Production Systems and Logistics. In 1999, he was graduated as lecturer (Dr.-Ing. habil.). From 1999 to 2003, Professor Nyhuis worked at Siemens AG in SPLS Supply Chain Consulting, an in-house consulting entity. In 2003, he was appointed as Full-time Professor and Managing Director of the Institute for Production Systems and Logistics at the Leibniz Universität Hannover for the topics factory planning, production logistics, assembly planning, and Industrial Engineering. Since 2008, he is also Managing Director of the Institute of Integrated Production Hannover (IPH). Prof. Nyhuis is full member of the German Academic Society for Production Engineering (WGP), Associate Member of the International Academy for Production Engineering (CIRP), and member of the International Federation for Information Processing (IFIP) (Working Group 5.7: Production Control). He is author of several books and book chapters on production planning and control, logistic curves, factory planning, and procurement logistics.

1 Factory Change Drivers

1.1 Introduction

Due to the variety and speed at which factors influencing a factory change, a factory can quickly lose its competitiveness. The main reason for this is the factory's inability to adapt its facilities and organization fast enough. When strategically planning a factory for long term use it is therefore essential to bear in mind the change drivers that have impacted factories in the past and present and will impact them in the future. In this first chapter we will consider the symptoms of a change resistant factory, describe the basic stages of developing a modern factory and outline the first approaches of a competitive manufacturing enterprise.

1.1.1 Stagnant Factories

Since the beginning of the 1990s the role and significance of production has been intensely discussed in Germany both in research as well as in the practice. Computer Integrated Manufacturing (CIM), developed in the 1980s, had failed to provide the anticipated success in countering high labor costs world-wide. Moreover, the illusory outlook following German reunification belied the increasingly obvious weaknesses of Germany as a location for productions. A study conducted by the Massachusetts Institute of Technology (M.I.T.) about the automobile industry in Japan, the United States and Europe was the first to suddenly make it clear that German industrial enterprises in particular were on the verge of losing their competitiveness with regards to productivity, delivery times and quality [Wom90]. The main cause of this was identified as the enterprises' insufficient ability to innovate and adjust to the massively increasing dynamism of markets and technology. We refer to this weakness, which is primarily due to poor management, as a *stagnant factory*, the characteristics of which are outlined according to four main criteria in Fig. 1.1.

In a stagnant factory, a *complex organizational structure* has been developed over the course of a long business tradition. Numerous sections are strictly structured in five to seven hierarchical levels with precisely defined tasks and competencies. Employee participation is not desired and remuneration is based on output rather than on results. The emphasis is on functionally optimizing marketing, design and production processes. As a result, decisions cannot be made quickly and the responsibility to customers with respect to the processing of orders is widely spread.

The lack of proximity to customers is closely related to the *lacking market orientation*. Consequently, the functional organization is not centered on customers and fulfilling their wishes, but rather focuses on operational goals such as high utilization of machinery or manufacturing in 'economical lots'. However, in order to successfully act in a market an enterprise has to

H.-P. Wiendahl et al., *Handbook Factory Planning and Design*,
DOI 10.1007/978-3-662-46391-8_1, © Springer-Verlag Berlin Heidelberg 2015

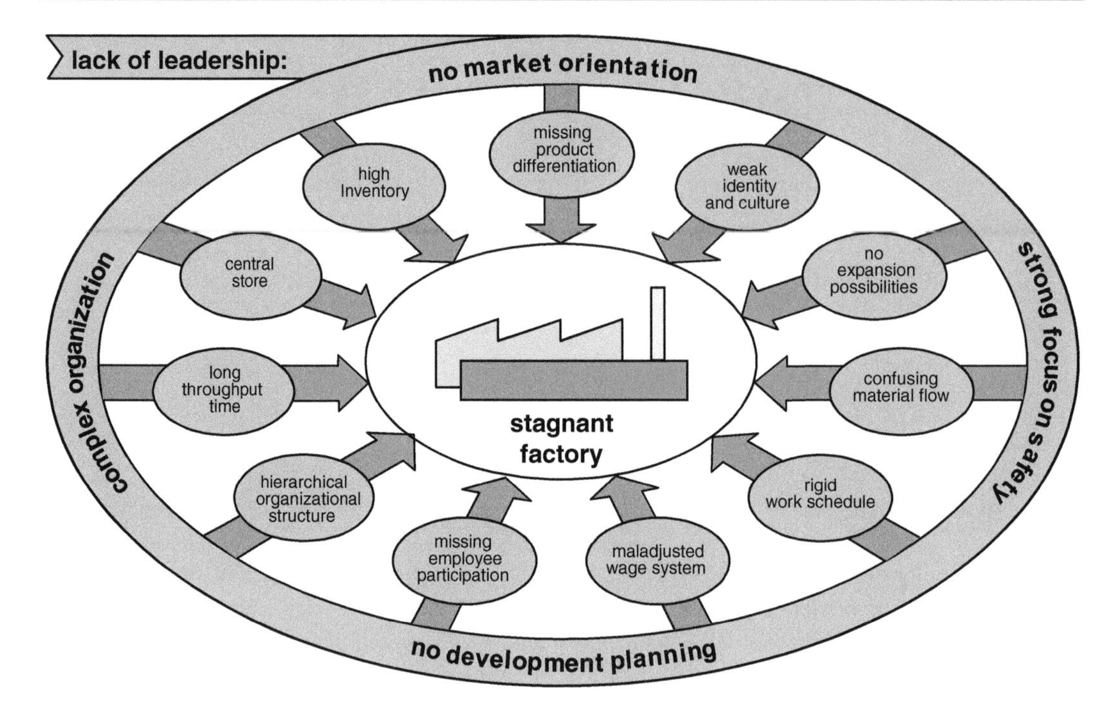

Fig. 1.1 Characteristics of a stagnant factory. © IFA G6181SW_Wd_B

follow the principle that everything that does not serve the customer is waste. Stagnant enterprises lack this orientation. They frequently fail to differentiate internally about their offerings according to customer groups or markets. Long throughput times, high inventories and central warehouses are visible signs of an enterprise that is wrongly oriented.

Moreover, a firm such as this frequently fails to have a guiding vision clearly mediating the business' fundamental operational goal to every employee. One of the dangerous results of this is the disappearance of the corporation's identity and culture. No longer able to identify with the enterprise and its products, employees see themselves as tiny pinions in a big gear box. To some extent they 'internally resign'; struggling day-to-day through the complicated organization, no energy remains for new ideas. Customers also sense this and rightfully complain about their partners lack of engagement.

Without such an overall goal, the *development of the enterprise* can also not be planned. The entrenched structures are mirrored in the unsystematic construction of the buildings, causing a disorderly material flow and long transportation paths. Considerable effort is required in order to quickly adjust sections (e.g., due to increasing production demand) because there are no possibilities for expansion nor are any planned. Unsightly buildings, unorganized spatially-scattered storage areas with raw materials, unfinished parts and accumulations of junk along with dirty, poorly lit workshops that impede a positive work attitude strengthen the cultural decay. No one wants to give customers a tour of the facilities, because the discrepancy between the production's claims and the appearance of the factory are too obvious.

Ultimately, the developments described here lead to a false sense of safety. Large inventories of raw materials, purchased parts, semi-finished products and end products feign a capability to react which the structure itself can no longer manage. Should non-routine orders be placed, long delivery times, rush orders and schedule delays arise. Furthermore, it is almost inevitable that ecological aspects such as conserving resources and protecting the environment completely fall into the background.

Figure 1.2 shows a typical example of stagnancy found in the industry. In the depicted production area, the strongly non-directional material flow is immediately obvious. Products manufactured here cover a distance of more than 1 km while being processed. This is one of the reasons that there are throughput times longer than four weeks when the actual processing time is only two days. The order throughput is also decelerated by long setup times and a high percentage of reworking. This structure was first questioned because of the need to integrate a new product into the production, for which there was an area deficit of 1400 m^2. A study showed that by rigorously orienting on three product groups (runners, repeaters and rarities), standardizing the work processes and introducing the pull-principle for controlling orders, it would be possible to reduce throughput times by 50 % and floor space by 40 %.

1.1.2 Previous Methods of Corporate Management

The developments outlined above have made it clear that previously successful maxims for managing industrial firms are no longer effective in view of an increasingly unpredictable environment. In particular, according to [Lut96, AbRe11] the following maxims have been followed:

- Plan and optimize all operational processes as much as possible especially in production. This was typically characterized by a large amount of work preparation and strong emphasis on time management.
- Clearly demarcate departments, specialized competencies and hierarchical responsibilities based on the division of labor. This was frequently documented by extensive organizational handbooks with precise descriptions of positions and processes.
- Equate specialized competencies with hierarchical positions. This traditional career path inevitably led to increasing hierarchy instead of decreasing.
- In-house solutions above all else. Supposed or actual company specific know-how was reluctantly passed on (i.e., in the form of outsourcing), resulting in an increasing number of parts and variants.
- Maximize the use of economies of scale. This typically, resulted in large lots being generated, orders being started too early or stock

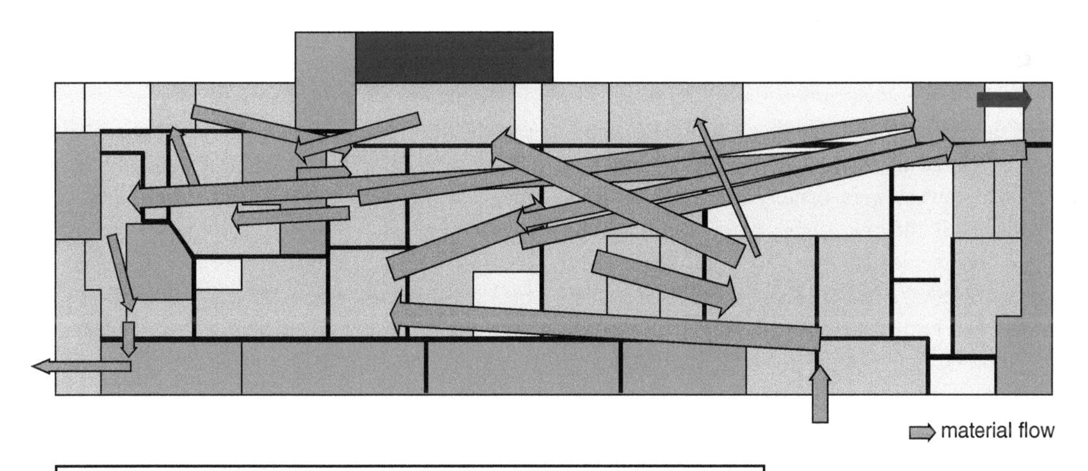

analysis results:
- strongly non-directional material flow
- production distances1300 m to 1500 m
- setup times up to 16 hours
- rework share 20%
- lead time app. 38 working days
- area deficit 1400 m ²

Fig. 1.2 Actual state of a production area. © IFA G3207SW_Wd_B

orders being released without concrete customer orders.
- Sustain a market position with incremental product innovation in the form of gradual improvements to an existing product. Subsequent to a dominant base product (frequently the invention of the company owner) strong customer loyalty could thus be attained over a long time.
- Develop new products as so-called 'breakthrough' innovations only occasionally and in order to exploit new markets. These innovations were rarely the result of researching customer needs (market pull), but rather developed from the potential of the enterprise's technology (technology push). In the favorable case scenario the product met existing customer need or triggered it.
- Focus investments and innovations on saving work force. Since the market was not yet saturated, the aim was to compensate for wages and ancillary wage costs along with continually greater overhead costs by disproportionally rationalizing the production process.
- When rationalizing, externalize as many costs and charges as possible. In particular this included costs related to the environmental impact and specific social costs e.g., operation related terminations or lay-offs.

The success of these principles was linked to relatively stable environmental conditions which, since the 1990s, have only been limitedly applicable when at all. Thus for example, changes in the sales market were usually predictable far in advance and the mid-range corporate planning range was typically three to five years. The number of competitors in these markets was also limited and both their strengths and weaknesses were known. Moreover, investment capital and resources could be raised at minimal cost and environmental impact played a subordinate role in the enterprise's success, just as the stock market price of the enterprise itself. Finally, highly motivated and well qualified workers were available everywhere [Lut96].

Since the start of the 1980s, these conditions have changed at a speed never seen before. The most significant challenge was the globalization of the goods and information streams, driven by the inventions of logistics and the internet. A wealth of products surged onto the world market from young aggressive industrial nations. Consequently, changes in the market became more and more difficult to plan.

Starting with Warnecke [War93] and Westkämper [West99] the term "turbulent environments" was coined to describe these phenomena. According to it, all of the parameters relevant to the production such as the product structure, competition, sales figures and available technology can vary quite quickly and suddenly. The predictability of changes in the industrial environment thus strongly diminishes. Indications of a turbulent environment include continually shorter lifecycles of products from their entry onto the market up to their discontinuation as well as the replacement of products with an ever greater number of variants.

An example of this is the growing number of niche vehicles—a typical lifestyle product. Figure 1.3 depicts the trend for these during the last five decades. Whereas there were only three categories (limousine, coupé and convertible/roadster) in the 1960s, in 2006 fourteen different segments were already known. Moreover, renowned automobile makers have already announced an increase of more than 40–50 models in the next decade.

Together with the variety of the product, comes the fast permeation of technological developments. Whether in the form of materials, manufacturing methods, information and communication technology, the internet or RFID (radio frequency identification devices) and virtual reality, they open up new possibilities for both the design engineer and factory planner.

A further, more structural development regards the diverging lifecycles of the technical factory elements i.e., processes, buildings and sites in comparison to the product. In Fig. 1.4, cited according to [ScWi04, p. 106], Wirth clearly illustrates these challenges. The *product lifecycle* (A) grows shorter and shorter last but not least due to the self-generated diversity of variants. In order to meet this development, the product is frequently divided into base modules

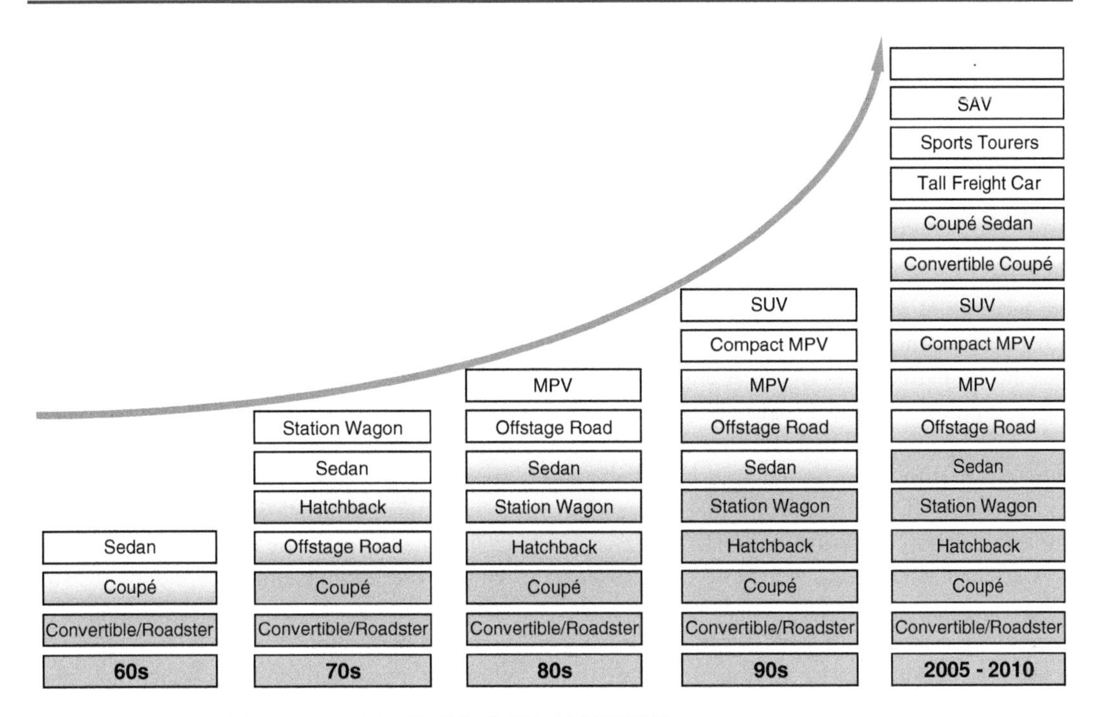

Fig. 1.3 Market trend for niche vehicles [Pol06]. © IFA 14.051SW_B

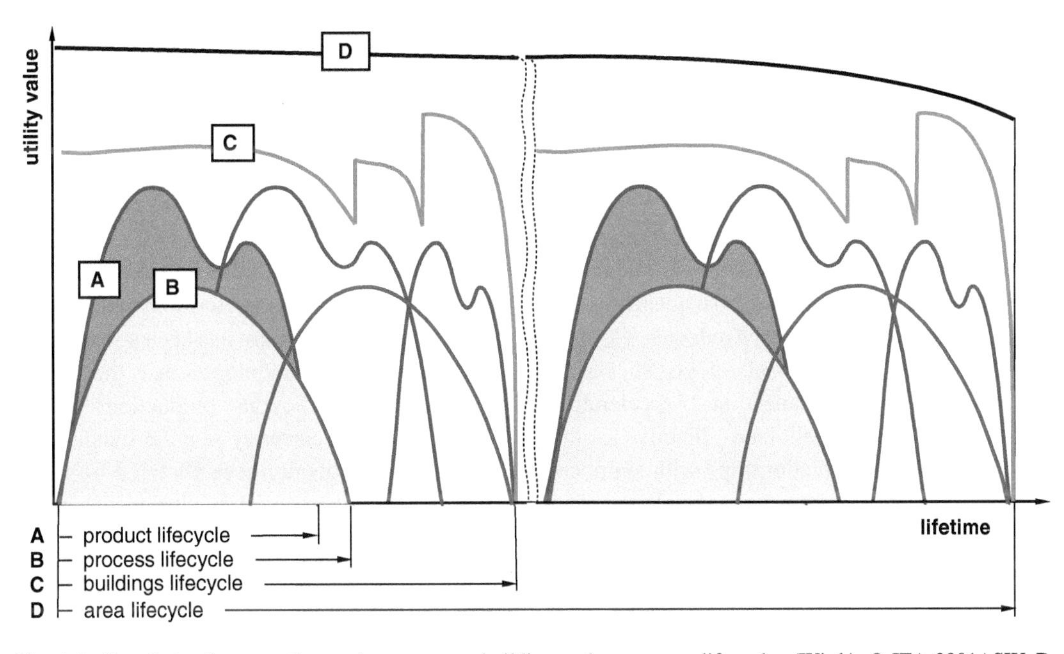

Fig. 1.4 Correlation between the product, process, building and area usage lifecycles (Wirth). © IFA 9901ASW_B

and variant dependent components. The base modules outlast a number of product cycles while the variant dependent components constitute the product's claimed innovation e.g., an additional function or a new design. The *process lifecycle* (B) is determined by changes in technology and their efficiency. Generally, the process lifecycle is longer than the product lifecycle and is applied to a number of product generations, if for no other reason than their depreciation value. With the *building lifecycle* (C) the structure of the building, which can last 30–50 years, has to be distinguished from the technical building services, which can be used for perhaps 10–15 years. Usually both cycles are a number of times longer than both the process and product cycles. Finally, the *area usage lifecycle* (D) is dependent on the location of the property and the related building rights. It is generally in the magnitude of decades and is longer than the use of the building. From this, Wirth deduces that sub-systems should be designed to be adaptable and harmonized temporally into the lifecycle of the entire factory [ScWi04, p. 107].

Despite the resulting, frequently interlinked, decision-making and execution processes when developing a product, introducing it to the market and processing orders, the time available for enterprises to react to changes in the environment decreases. The first basic reaction in response to this development seemed to be to reduce complexity. Driven by the concepts of lean production [Wom90] and business re-engineering [Ham93] five approaches arose:

- Products and production programs were broken down into components, modules and subsystems, and key competencies were concentrated on. In-house manufacturing was then drastically reduced by shifting required items to external suppliers. Consequently, the workforce was considerably reduced.
- The entire procurement logistics were restructured, differentiated and accelerated: components were delivered directly to the assembly line and relationships with suppliers for modules and systems were developed. The suppliers assumed responsibility for everything from the design up to integrating the modules/systems into the end product. A further example here is allocating the entire spectrum of C-parts (i.e., articles that are only worth 5–10 % of the value of a product, but make up 50–80 % of the volume) to a logistics service provider.
- The direct value-adding area of the manufacturing and assembly was fundamentally restructured into segments and decentralized. Based on the 1960s group technology [Mit60] and the 1970s/1980s manufacturing cells, the concept of modular factories [Wild98] and fractal factories [War93] were created. The general idea is to form groups of parts or components requiring similar manufacturing or assembly technology for a market segment with specific demands regarding delivery times and delivery reliability and to produce them in one segment. A certain part or component is then triggered by call-off orders and manufactured ready-to-install after their quality is tested and found to be 100 % faultless. All of the indirect functions such as the material and tool planning, scheduling, servicing and maintenance up to and including planning personnel and capacities are integrated into the segment. They thus appear as an in-house supplier.
- An alternative to relocating, which is receiving greater attention, is the integration into a network of enterprises. Here, companies join together into a virtual enterprise, appearing from the outside like a large enterprise and offering all of the services from 'one place'. In particular it allows small and mid-size enterprises to successfully bid and develop large projects with low overhead costs.
- In addition to these structural changes in the value adding chain, there has been an increased orientation on methodology since the 1990s. Based on the Toyota production system [Ohn88], which currently sets the standard for an efficient production (see Sect. 2.3.6), many enterprises have recognized that they have to orient all of their processes on preventing waste. The concept of lean production, brought into the forefront by Womack and Jones was initially understood as an instrument for downsizing personnel. However, since the start of the 21st century, it has been reassessed and has inspired the development of numerous 'holistic production systems' (HPS) [Spa03].

Within this context, value stream mapping, introduced by Rother and Shook [RoSh03], is a pragmatic approach to quickly analyzing wasted time, inventory, production areas and movements of material and staff. This approach has since lead to the concept of value stream factories [Erl10].

If we summarize the evolutionary stages of a factory together under the aforementioned aspects since the 1960s and strongly simplify them, four principle forms can be identified as shown in Fig. 1.5.

With a stable and predictable market, the *functional factory* was oriented on increasing efficiency by bundling know-how. The accompanying workshop principle with the workforce divided accordingly ensured that resources were highly flexible, nevertheless, at the price of high inventories and long throughput times. The need to more closely orient themselves on the markets and corresponding products led to the described *modular, fractal or segmented factories*. Order processing was then noticeably accelerated; facilities, however, were occasionally under-utilized. Personnel could only be fully utilized when they were cross-trained and flexible shift models were implemented. As the products and markets diversified more and more, the complexity also grew, so that with the aid of the described measures for reducing the manufacturing depth, especially in the automobile industry, strategic *supply networks* (also known as *supply chains*) were created. The enterprise responsible for supplying the end customer, now concentrates on their key competencies, in the extreme case only on the product design, final assembly and sales. Cost potentials are increased by resolutely outsourcing processes for everything from procurement, manufacturing and distribution up to and including development. Networks such as this are usually limited to the lifecycle of a product i.e., typically 3–5 years.

With increasing market turbulence and simultaneous demand for a faster and broader scope of goods and services, regional and national *production networks* visibly develop. They form production clusters which configure themselves very quickly with a high degree of innovation based on orders and dissolve just as quickly once the good or service has been yielded.

All of the factory forms outlined here have one thing in common—they all presume immobile resources (buildings, equipment and infrastructure) and sites. In Chap. 2, we will discuss the extent to which they satisfy already existing and foreseeable future demands. With the described concepts, production enterprises have quite successfully managed to take a first step in increasing

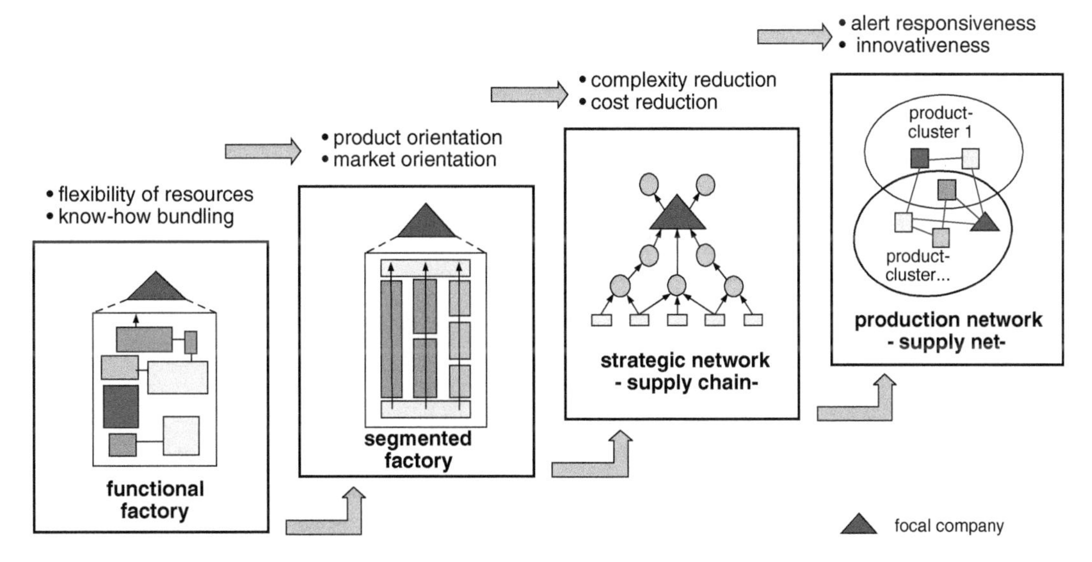

Fig. 1.5 From a functional factory to the location within a production network. © IFA G8147SW_Wd_B

their operational efficiency and responsiveness in order to match the challenges of a globalizing market. In doing so, the essential unique selling points have proven to be superior product functionality, punctual supply and high quality.

Since the 1990s, the relatively new field of *product integrated services* has also developed. These services encompass the entire lifecycle of the supplied product, beginning with supporting the customer in the planning and design phase and including everything from the assembly and ramp-up, to providing internet supported teleservices and replacement parts as well as return services. This approach has been further developed in *BOT models* (build-operate-transfer models). Here, the equipment or facility remains the property of the company that produces it and the customer only pays for the products actually generated. BOT models also represent an important contribution to sustainable development. The aim is to minimize the consumption of resources such as raw materials and energy by extensively reusing and recycling the product and thus keeping the impact on air, water and soil as low as possible.

Nevertheless, many enterprises also considered *relocating* part of their production to 'low wage countries' a solution because supposedly more favorable manufacturing conditions with regards to labor costs and work hours were to be found there. The Fraunhofer Institute for Systems and Innovation Research (ISI) conducts systematic investigations in respect to this in the German industry, the latest results of which are depicted in Fig. 1.6 for the two period's mid-2004 to mid-2006 and mid-2007 to mid-2009 [Kin12]. Generally labor costs, proximity to key customers and access to new markets are still the

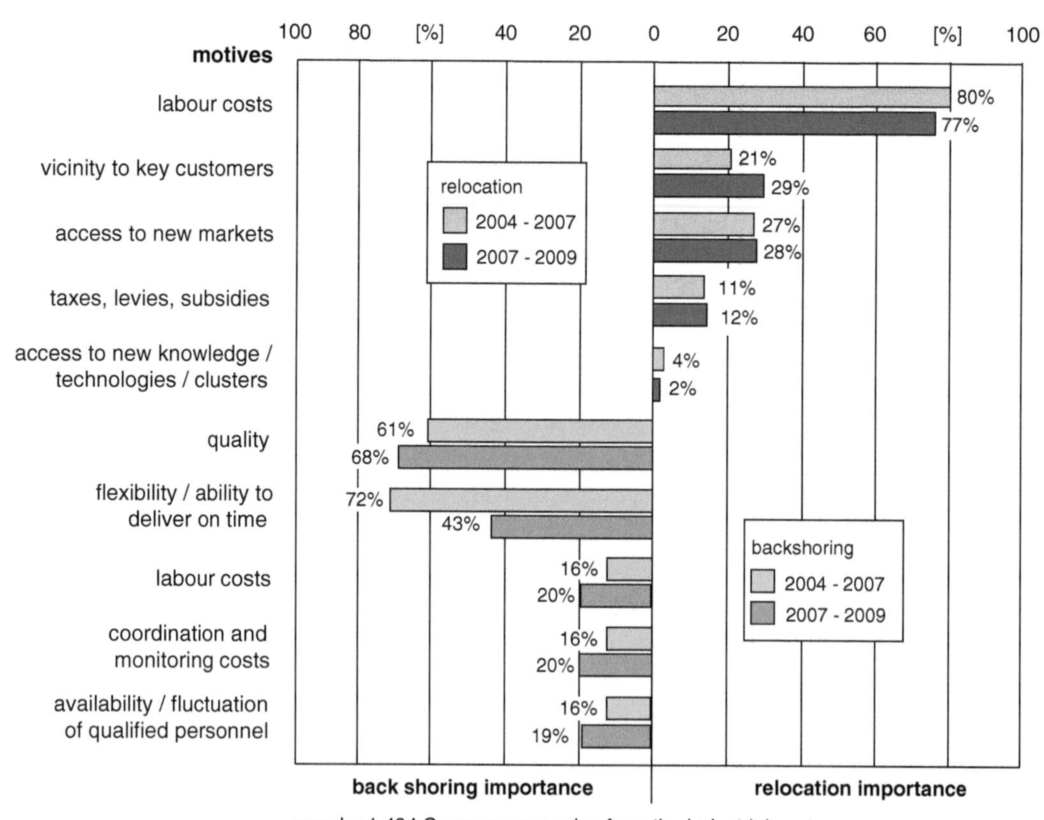

Fig. 1.6 Relocation and Backshoring motives in the German metal and electrical industry (per Kinkel). © IFA 14.663SW_B

dominating motives for relocating productions. The new Eastern European Community Countries, China and Asia are the main target regions.

Nevertheless, approximately one third of the firms that relocated have returned. The main motives for backshoring are insufficient quality, lacking logistic flexibility, rising labor costs and the problem of losing and/or finding qualified personnel.

In former studies ISI lists the basic reasons for the lack of success as follows:

- a lack of coherence between strategies and evaluation criteria,
- inadequately considering the potential for internal optimization,
- not evaluating network requirements at the respective site,
- gauging the site evaluation statically instead of dynamically,
- not weighing the ratings of individual site factors for the complete result,
- underestimating the start-up time required to ensure the process certainty, quality and productivity, and
- underestimating the costs of managing the foreign site.

On the one hand, it is indisputable that direct investments in other countries had a positive impact on employment in Germany. On the other hand, the study provides important impulses for more extensive approaches to improving the competitiveness of small enterprises in particular and to protect them from making rash decisions.

1.1.3 Competitive Factors of Superior Enterprises

Nevertheless, the efforts made up until now have been insufficient since the strategy of reducing complexity is oriented more at cushioning market turbulence and does not continually impact the entire value-adding chain. In particular, there is a threat of losing the ability to react. Against the background of high educational standards, a stable social system, excellent infrastructure and a robust currency one of the promising future strategies that the internal strength of German enterprises has is the considerable potential for *controlling complexity*. After all, turbulent markets also offer the opportunity to capture additional market shares with an offensive strategy. This of course requires the enterprise to be able to react not only to external developments, but also to be able to enter the market proactively. This also includes being able to generate turbulence, for example by suddenly cutting delivery times in half, offering new products for a specific market segment in an unexpectedly high frequency or taking a quality offense by doubling the length of a guarantee.

A strategy such as this however requires more than controlling costs, quality and time in order to obtain customer satisfaction (see Fig. 1.7). First of all, a strong *innovative drive* needs to be developed and promoted. This means permanently questioning products, services, processes and behaviors, not only through continual optimization but also through innovative leaps. This in turn entails the company culture to be oriented on communication, with employees clearly participating and a stronger focus on results instead of on performance.

The second key property of enterprises able to benefit from turbulence is their capacity to quickly utilize something new i.e., organizationally they are *quick learners*. The most predominant characteristic of such enterprises is the ability to develop common visions and goals for bundling energy and knowledge. This includes continual qualification measures with the primary purpose of conveying methods and social skills, a high degree of informal communication and pronounced self-organization in flat hierarchies with autonomous organizational units [Gau04].

The third, generally 'new' property is changeability [West99, Rein00, Wien99, Wien07]. This describes the ability of a factory to realize structural changes on all levels with minimal expenditures in response to internal or external triggers. The planning and realization of this adaptation process has to occur at a specific speed set by the market. This changeability differs from related concepts such as responsiveness, reconfigurability, adaptability, flexibility and agility—as is more extensively clarified in

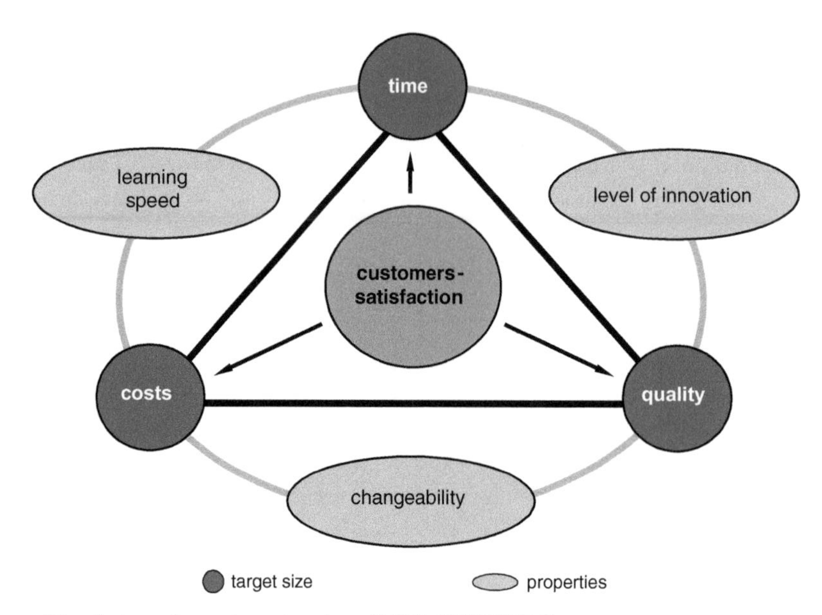

Fig. 1.7 Competitive factors of superior enterprises. © IFA G5990SW_B

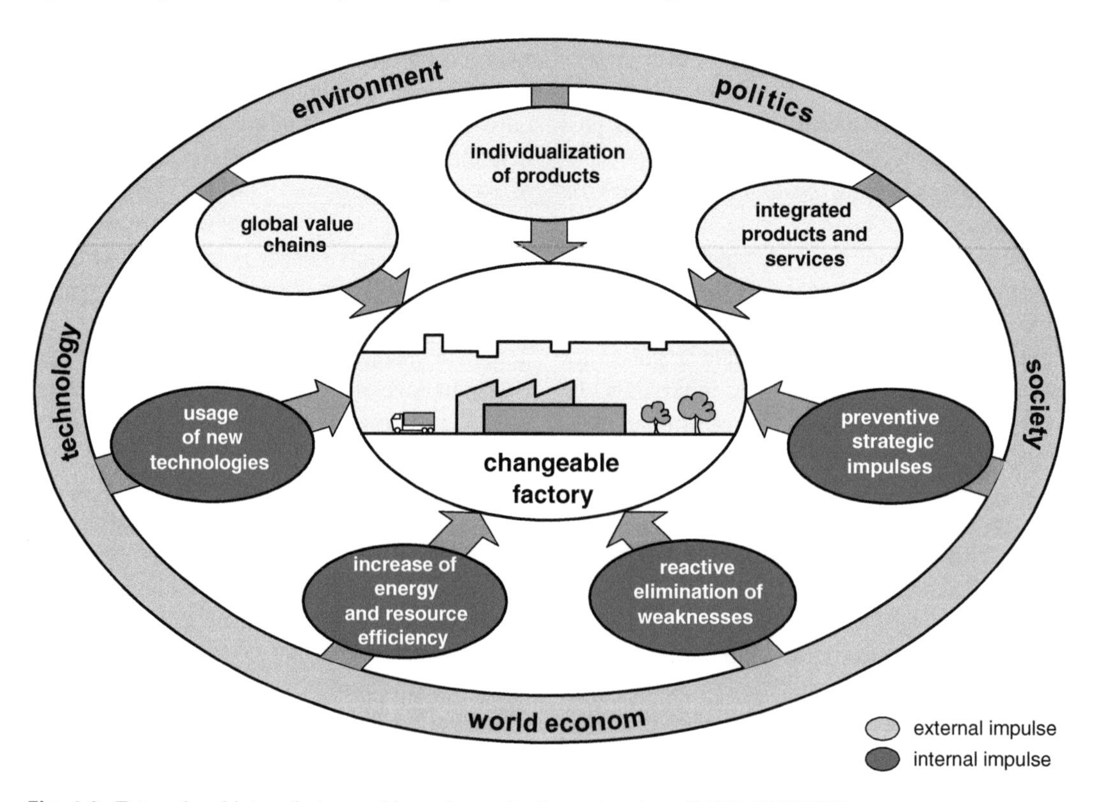

Fig. 1.8 External and internal change drivers for production enterprises. © IFA G8776SW

Chap. 5. We would suggest that changeability is the key concept that allows an enterprise to be successful in turbulent surroundings.

Before we move on to develop the new requirements, strategies and design fields of changeable factories in the next chapter, Fig. 1.8

provides a summary of the change drivers. Further information can be found in [Jov08, AbRe11] and under Horizon 2020—The Framework Program for Research and Innovation of the European Union [http://www.manufuture.de/COM-2011-Horizon2020.pdf].

Surrounding conditions that directly impact enterprises include the world economy, environment, politics, society and technology. These lead to change drivers that have an indirect impact and that can be differentiated according to whether they are external or internal impulses. Globalization, technology and society result in a growing individualization of products with shorter product lifecycles and an expansion of the market performance up to and including services across the entire lifecycle. Delivery times thus sink further, the demand for delivery reliability increases and all of this occurs alongside strongly fluctuating consumption and turbulence. Moreover, enterprises still have to face continuous pressure regarding costs and quality. Products and services increasingly are offered out of global networks, whether from in-house, joint or external enterprises.

Strong internal impulses come from preventative, strategic considerations such as entering a new market, expanding available products or a fundamental restructuring triggered by a change in management or ownership. In comparison, reactive internal impulses are created by eliminating noticeable weakness in technological or logistical performances, developing new work models for an aging workforce or realigning the production volume between domestic and international locations due to currency related risks. Finally, it is about taking up new challenges regarding energy and resource efficiency, but also about using the potential of new technologies.

1.1.4 Summary

Many manufacturing companies have imperceptibly lost their competitiveness in a globalized environment. Typical symptoms include large inventories, long throughput time, unclear material flows and complex organization. Generally, business goals such as the utilization of machinery and optimal lot sizes are emphasized rather than orienting on the customers' needs.

Since the 1990s, the entry of younger industrial nations into the market, the related dramatic increase in product variants as well as the demand for quicker and more punctual deliveries have forced production companies to rethink their practices. This is evident in the modularization of products, the decreased manufacturing depth, the new orientation of procurement logistics, the production segmentation as well as the prevention of all types of waste. These measures are subject to the primacy of the customers' absolute satisfaction.

In addition to traditional objectives (time, costs and quality), outstanding features of a competitive factory which prove to be crucial for survival include: aligning their level of changeability with the market, being highly innovative and being quick learners.

Bibliography

[AbRe11] Abele, E., Reinhart, G.: Zukunft der Produktion. Herausforderungenb, Forschungsfelder, Chancen. (Future of Produktion. Challenges, Reserch Fields, Chances). Hanser Verlag, Munich (2011)

[Erl10] Erlach, K.: Wertstromdesign. Der Weg zur schlanken Fabrik (Value stream design. Path to the lean factory), 2nd edn. Springer, Berlin (2012)

[Gau04] Gausemeier, J., Hahn, A., Kespohl, H.D., Seifert, L.: Vernetzte Produktentwicklung. Der erfolgreiche Weg zum Global Engineering Networking (Cross linked product development. The successful path to global engineering networking). Hanser, Munich (2004)

[Ham93] Hammer, M., Champy, J.: Reengineering the corporation: a manifesto for business revolution, 1st edn. New York (1993)

[Jov08] Jovane, F., et al.: The incoming global technological and industrial revolution towards competitive sustainable manufacturing. CIRP Ann. Manuf. Technol. **57**, 641–659 (2008)

[Kin12] Kinkel, S. et al.: Wandlungsfähigkeit der deutschen Hightech-Industrie (Changeability of the hightech Industry) Fraunhofer ISI. Karlsruhe (2012)

[Lut96] Lutz, B., et al. (eds.): Produzieren im 21. Jahrhundert: Herausforderungen für die deutsche Industrie. Ergebnisse des Expertenkreises Zukunftsstrategien (Manufacture in the 21st century: challenges for the German industry. Results of the expert group strategies for the future), vol. 1. Frankfurt/M. (1996)

[Manu07] www.manufuture.de

[Mit60] Mitrofanow, S.P.: Wissenschaftliche Grundlagen der Gruppentechnologie (Scientific base of group technology), 2nd edn. VEB Verlag Technik, Berlin (1960)

[Ohn88] Ohno, T.: Toyota Production System. Beyond Large-Scale Production. Productivity Inc., Portland (1988)

[Pol06] Polk, R.L.: Marktentwicklung Nischenfahrzeuge (Market development of niche vehicles). Studie, Essen (2006)

[Rein00] Reinhart, G.: Im Denken und Handeln wandeln (Change in thinking and action). In: Reinhart, G., Hoffmann, H. (ed.): Nur der Wandel bleibt. Wege jenseits der Flexibilität (Only change last, paths beyond flexibility), pp. 19–40. Utz Verlag, Munich (2000)

[RoSh03] Rother, M., Shook, J.: Learning to See: Value Stream Mapping to Add Value and Eliminate MUDA. The Lean Enterprise Institute, Cambridge (2003)

[ScWi04] Schenk, M., Wirth, S.: Fabrikplanung und Fabrikbetrieb. Methoden für die wandlungsfähige und vernetzte Fabrik (Factory planning and factory operation. Methods for the changeable and cross linked factory). Springer, Heidelberg (2004)

[Sch10] Schenk, M., Wirth, S., Müller, E.: Factory Planning Manual. Situation-Driven Production Facility Planning. Springer, Heidelberg (2010)

[Spa03] Spath, D. (ed.): Ganzheitlich produzieren (Holistic production). Log_X Verlag, Stuttgart (2003)

[War93] Warnecke, H.-J.: Revolution der Unternehmenskultur – Das Fraktale Unternehmen (Revolution of the corporate culture—the fractal enterprise), 2nd edn. Springer, Heidelberg (1993)

[West99] Westkämper, E.: Wandlungsfähigkeit der industriellen Produktion (Changeability of industrial production). TCW-Verlag, Munich (1999)

[Wien99] Wiendahl, H.-P., Hernández, R.: Bausteine der Wandlungsfähigkeit zur Planung Wettbewerbsfähiger Fabrikstrukturen (Components of changeability for planning of competitive factories). 2. Dt. Fachkonferenz Fabrikplanung, Fabrik 2000+ am, Stuttgart 26/27 Oct 1999

[Wien07] Wiendahl, H.-P., et al.: Changeable manufacturing—classification, design and operation. Ann. CIRP **56**(2), 783–809 (2007)

[Wild98] Wildemann, H.: Die modulare Fabrik – Kundennahe Produktion durch Fertigungssegmentierung (The modular factory—production near to the customer by manufacturing segmentation), 5th edn. TCW-Verlag, Munich (1998)

[Wom90] Womack, J.P., Jones, D.T., Roos, D.: The Machine that Changed The World. New York (1990)

2 Planning Basis

This chapter describes a strategically justified planning basis which can be regarded as a guideline for the planning team. First, based on the factors that impact a plant and that were outlined in Chap. 1, the production strategy needs to be determined. This step, which is the task of corporate planning, includes identifying products and business processes that are to be manufactured or executed in the factory. Within a factory, generally speaking, the order fulfillment with its sub-processes stands in the foreground while the products and business processes typically determine the location and factory areas that are to be designed. The other major components of the planning basis include decisions about the type of factory from the customers' perspective, the position of the factory in the supply chain and possibly the integration of the factory into a production network.

2.1 Production Strategies

A factory is not operated for its own sake, rather it is one of a number of instruments a production enterprise utilizes to realize their business strategy. Up until the 1970s, the necessity of operating the factory was never a question; rather the priority was safeguarding employment. Nowadays, discussions primarily focus on which role in-house production should play in the competition for markets and how the enterprise's financial resources should be allocated. The possibilities of global procurement and co-operations as well as developments in technology and logistics have created new degrees of freedom for configuring and positioning the own production. These should be utilized against the background of a well thought out competitive strategy for the entire company, thus ensuring its long term economic viability.

According to M.E. Porter, a competitive strategy includes in particular:

- concentrating on selected market segments,
- differentiating the products and services in comparison to the competition as well as
- gaining a comprehensive cost leadership [Por98].

The competitive strengths and determinants that should be analyzed and evaluated in this respect are briefly summarized into five so-called 'forces' in Fig. 2.1.

The starting point is the number of competitors and the intensity of the rivalry among the existing firms in the branch. The latter is determined for example by over capacities, brand identity and exit barriers. Following that new entrants and their entry barriers in the branch are analyzed. The third force concerns buyers and their bargaining power and sensitivity to prices, while the fourth takes into consideration possible substitute products or services and the danger of one's own product being replaced. Finally, the fifth force focuses on the bargaining power of suppliers.

An important approach within this context involves evaluating effectiveness ("doing the right thing") and efficiency ("doing things right")

H.-P. Wiendahl et al., *Handbook Factory Planning and Design*,
DOI 10.1007/978-3-662-46391-8_2,

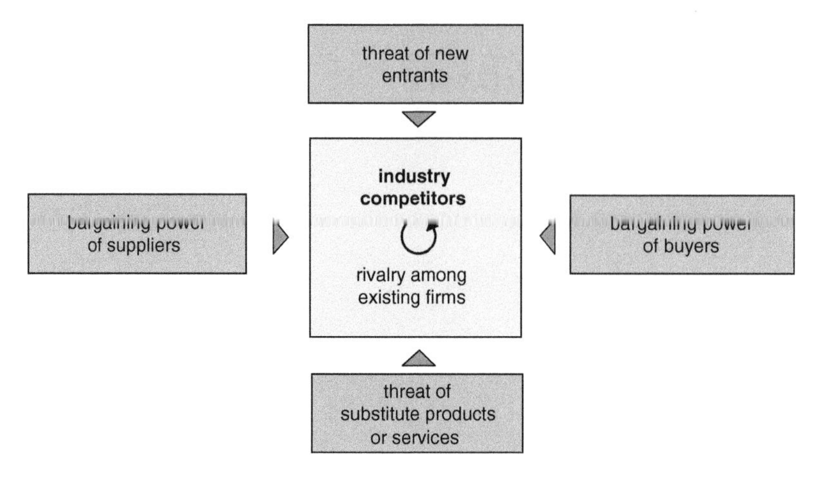

Fig. 2.1 Forces driving industry competition (per M.E. Porter). © IFA D3436_Wd_B

with the aid of the 'Balanced Scorecard'. The Balanced Scorecard supports the multi-dimensional strategic planning and control of an enterprise or division. Based on a suggestion from Kaplan and Norton [Kap96], starting with a superordinate vision and strategy four perspectives are developed (see Fig. 2.2). Strategic goals are to be formulated for each perspective, from which operative targets and actions are derived; compliance with these is then monitored based on specific parameters.

The *financial perspective* examines whether a selected or implemented strategy improves the business results or not from the perspective of the

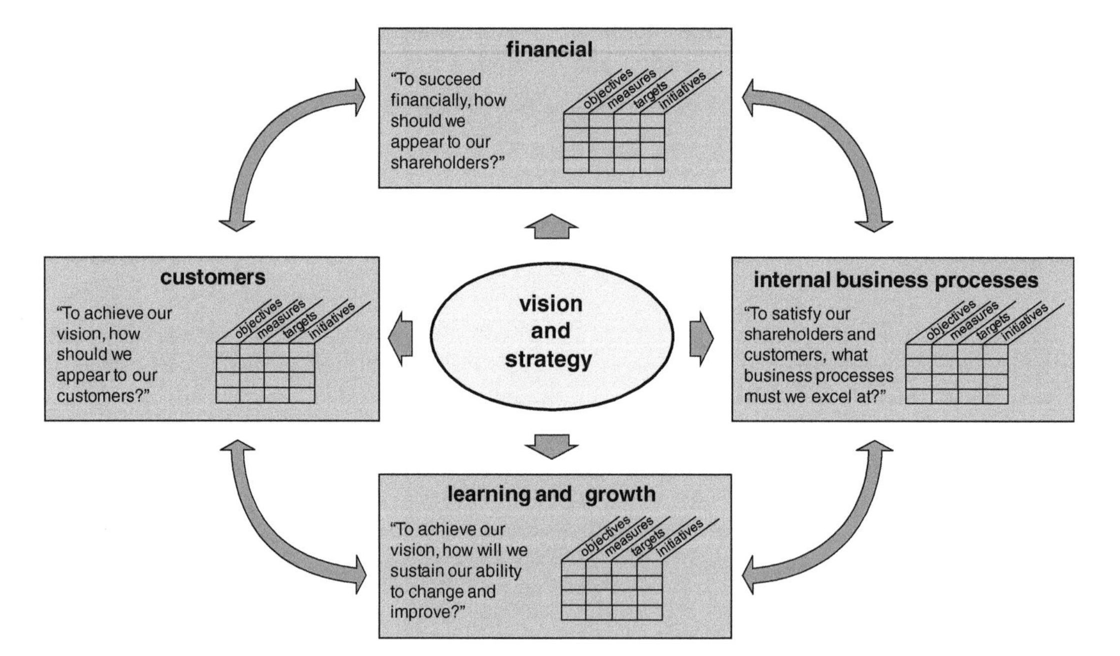

Fig. 2.2 Basic concept of the balanced scorecard (per Kaplan and Norton). © IFA G8889SW_B

shareholders. Based on the analysis objectives, variables, targets and measures can be derived. If we consider the production for example, this would concern the share of in-house production, the resources that should be implemented and the location.

In consideration of the factory itself, the *customer perspective* poses the question of whether or not the factory fulfills the service characteristics that the market demands e.g., delivery time, delivery reliability and the quality of the product. However, general objectives such as customer satisfaction and customer loyalty are also taken into consideration here as drivers for the company's success. Corresponding measures might then include focused sub-factories close to the customer or re-designing the corporate identity from the bottom-up.

The *perspective from the internal business processes* prioritizes the structures and processes that play a decisive role in satisfying customers' wishes and whose improvement is perceived by customers. From the factory view this could for example include, internal throughput times, the possibility of making late decisions about variants or implementing a product quality concept that saves customers from having to inspect goods upon receipt.

With the *learning and growth perspective*, the importance of the ongoing and progressive development of products and methods is emphasized. From the view of production this concerns for example, continually improving production technology, introducing team work or developing uninterrupted logistic chains from the in-house production up to and including the customer.

One of the noteworthy aspects about the Balanced Scorecard is that in comparison to traditional methods, such as the Return on Investment (ROI) concept and Shareholder Value approach, it does not draw one-sidedly on financial and to some degree strongly historical based parameters for making decisions. Rather, the view is oriented more equally on the customer, competition and internal factors that are not only difficult to measure (e.g., the ability to innovate and learn) but also increasingly significant for the success of a business in a turbulent market. The concept thus offers a flexible framework for developing each of the enterprise specific strategies, which is in turn indispensable especially with regards to the future role of the factory.

2.2 Factory Strategies

When planning a factory, it is essential to have knowledge about the part of the business strategy pertaining to the market and the production; without this information the orientation on cost related aspects dominates too easily. Figure 2.3 depicts the key strategic elements of the planning basis for a factory, which are subject to three premises: First, they have to be *sustainable* in an economical, ecological and societal respect and thus not aimed at short term success. Second, the call to be *innovative* arises from the dynamic environment and applies not only to products, but also and especially to the production and administrative processes. Third and lastly, changeability is imperative—not just for the factory, but certainly for it in particular.

Business areas developed from visions and models form the core of the strategic basis. They designate a distinct external market which has clearly demarcated competitors and is closely aligned with the enterprise's philosophy, values and culture. Every business area is defined by a market offer and a market segment, described by the types of customers, distribution channels or geographical regions [Gau99]. Determining the sales region according to revenue and the regional market share is particularly important for strategically positioning the factory. The sales volume on the one hand and the local competitive environment on the other hand, result from there and provide the starting point for decisions about where to locate the factory and the scale of production.

2.3 Market Offer

The products and services available in market segments are defined for every business sector and are summed up together under the term

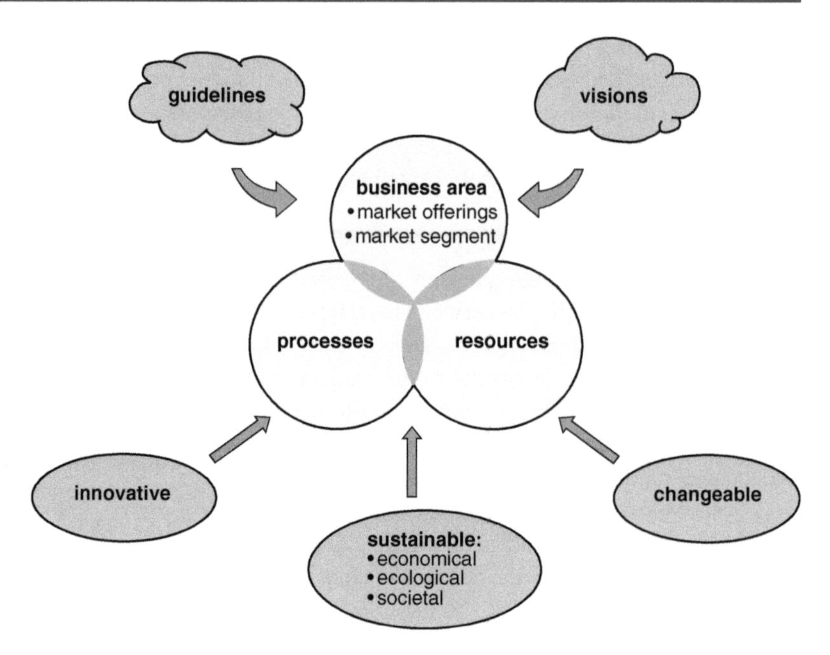

Fig. 2.3 Strategic basis for planning and designing a factory. © IFA G8891SW_B

market offer [Gau99]. This market offer requires *processes* that the enterprise's potential should yield. These are then generally divided into management, business and support processes. The business processes—to the extent that they concern the factory—are value-adding and consist of production engineering processes, material flow processes as well as information and communication processes. They require *resources* which basically consist of people, equipment and capital.

The market offer yielded by the enterprise can be considered from the perspective of logistics as well as according to the type of market service. With regards to logistics, the classification system developed by Siemens for their market offer is practical for this purpose; it defines four types of businesses according to the point in time at which the final product is defined and where the value is added (see Fig. 2.4) [Faß00].

The business types include consumer oriented products, systems for industry equipment, large projects related to plant construction and after-sale services. Each poses clearly different demands on the factory and its logistics.

Products are ready-to-use consumer goods usually meant for the end user such as household appliances, entertainment electronics, communication technology etc., which, for the most part, are self-produced. They are developed independent of specific orders and factors for success include, extremely short delivery times and a high service level due to good inventory management as well as an efficient, frequently world-wide distribution system.

Systems consist of custom designed configurations of—as much as possible standardized—hardware and software elements, whose function-defining modules are manufactured in-house and completed with purchased system components. The development and logistics cycles are thus only partially decoupled. Here, factors for success include the ability to quickly configure standard and procured components, managing order-specific supplies with high delivery reliability, directly supplying complete systems that have already been tested and then immediately installing and putting them into operation.

Plant related businesses mainly consist of engineering i.e., the technical design and planning of custom-made large plants, such as steel mills, paper mills or power stations. Products here cannot be made-to-stock due to the unique character of each; engineering and logistics cycles are therefore coupled for each order. Since components produced in-house play a small role,

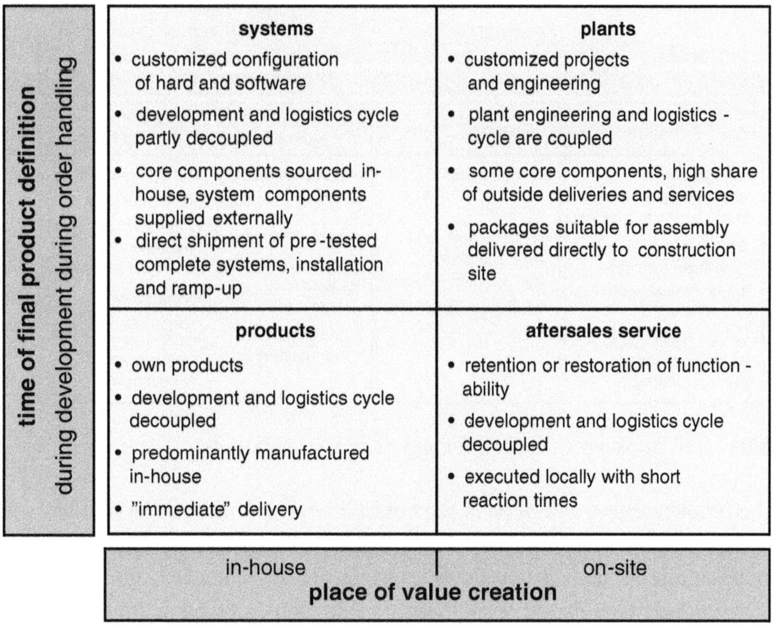

Fig. 2.4 Logistical business types (Siemens). © IFA G8908SW_Wd_B

time of final product definition / during development during order handling	in-house	on-site
	systems • customized configuration of hard and software • development and logistics cycle partly decoupled • core components sourced in-house, system components supplied externally • direct shipment of pre-tested complete systems, installation and ramp-up	**plants** • customized projects and engineering • plant engineering and logistics - cycle are coupled • some core components, high share of outside deliveries and services • packages suitable for assembly delivered directly to construction site
	products • own products • development and logistics cycle decoupled • predominantly manufactured in-house • "immediate" delivery	**aftersales service** • retention or restoration of function - ability • development and logistics cycle decoupled • executed locally with short reaction times
	place of value creation	

success factors include professional project management, controlling and coordinating the numerous customized deliveries and services that are predominantly supplied externally as well as punctually delivering the assembly-ready packages to the construction site which represents the location of the greatest value-adding processes.

The fourth type of business, *services*, refers only to the after-sales services of a product, system or plant. They serve to maintain their functionality (e.g., via regular inspection and maintenance) or in the case of disruptions, to restore it. The development of these services in the form of maintenance plans, repair kits, spare parts etc. is temporally decoupled from their provision in the logistics cycle. Factors for success here include the ability to react quickly and provide fast information while maintaining a minimal store of spare parts as well deploying and supervising technicians with a high rate of jobs completed in only one visit.

With regards to how markets are served, there are two known extremes of market offers, which apply mainly for products. On the one hand, there are mass programs, with which a cost leadership is pursued by increasing quantities of standardized outputs i.e., so-called 'economies of scale'. On the other hand, there is the 'economies of scope' strategy which strives to maximize the utility by focusing on multiple products for special groups of customers. In between these two there are strategies aimed at individualized mass production based on modular systems and flexible manufacturing methods. Customers then receive products that are to a large extent finely tuned to meet their needs, comprised of different quantities and variants of standardized parts and components that are quickly and flexibly assembled.

In addition, the strategy of supplementing products with services and selling the product utility instead of just the product is being pursued increasingly, especially in highly industrialized countries. It has been shown that customized products with value-adding services open up very promising potential in regards to global competition.

In Fig. 2.5, the basic dimensions of a competitive market offering are outlined. The key idea is that the market offer is oriented on the customer's value-adding chain. Based on a product which has a large benefit for a customer, the enterprise considers how they can attain long term customer loyalty by integrating services into

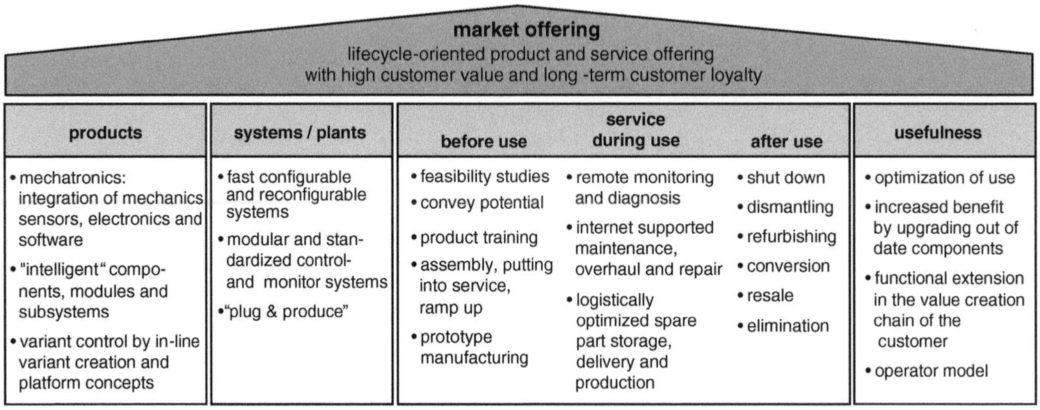

market offering
lifecycle-oriented product and service offering
with high customer value and long -term customer loyalty

products	systems / plants	service			usefulness
		before use	during use	after use	
• mechatronics: integration of mechanics sensors, electronics and software • "intelligent" components, modules and subsystems • variant control by in-line variant creation and platform concepts	• fast configurable and reconfigurable systems • modular and standardized control- and monitor systems •"plug & produce"	• feasibility studies • convey potential • product training • assembly, putting into service, ramp up • prototype manufacturing	• remote monitoring and diagnosis • internet supported maintenance, overhaul and repair • logistically optimized spare part storage, delivery and production	• shut down • dismantling • refurbishing • conversion • resale • elimination	• optimization of use • increased benefit by upgrading out of date components • functional extension in the value creation chain of the customer • operator model

Fig. 2.5 Dimensions of market offerings. © IFA G8899SW_B

the product above and beyond the entire lifecycle of the product. This leads to four categories of market offers, namely products, system/plants, services and usefulness, which can now be considered from the view of design and production.

With *products* it can be seen that 'mechatronics', i.e., combining mechanical parts and electronic components together with integrated software, is emerging increasingly. The latter ranges from sensor technology (for recognizing operating states) to electronics for everything from product application and control up to and including corresponding software. In the meantime, in many engineering products the portion of production costs due to mechanics, electronics and software is equally large.

In order to quickly configure and reconfigure the products to the changing needs in the lifecycle of the customer products, efforts are being made to develop so-called 'intelligent components', modules and sub-systems that are equipped with sensor and control technologies and can communicate with other devices. Not only do they considerably decrease the scope of the higher level control, but they also monitor themselves, allow their functionality to be tested during production before they are integrated and thus greatly reduce the effort involved in the final assembly. Module and platform concepts such as this allow a wide diversity of variants especially when variants can be formed by configuring software.

Finally, it can be determined that for many products, especially in the capital goods industry, there is a trend towards businesses for *systems and plants*. The customer frequently wants to receive so-called 'plug and produce' systems (e.g., a manufacturing system) or plants (e.g., packing plant) and expects a service packet that includes everything from engineering, delivery, commissioning and staff training up until the ensured yield is attained as well as optimization during its use. Depending on their value and complexity, the system or plant requires an extensive range of additional services to be operated. More and more often, the users are no longer able to perform these services themselves because they no longer have the skills required. Whereas previously, factories had their own planning, maintenance and repair departments, nowadays these tasks have been largely transferred to specialized service providers. Their services are then divided into three phases: before, during and after use of the system.

The *service* begins in the pre-use phase with feasibility studies and is supplemented with offers to illustrate the potential of the proposed investment. The latter can include for example, sample parts, delivering so-called 'pilot series' or educating design engineers in using a new technology. Thus, for example, a known producer of machines for processing sheet metal offers a workshop in which the technical and economic advantages of sheet metal construction in comparison to welding and casting are methodically conveyed to the designers based on sample parts. Afterwards, the participants have the opportunity to design and produce a prototype of a customized part and subsequently to economically evaluate it. Further traditional services in the pre-use phase concern the product training of later users, assembly, commissioning and start-up, especially of systems and plants up to the agreed upon capacity.

The second type of service refers to the operating phase and is shaped by the quickly developing possibilities of information and communication technology. The previously mentioned intelligent product components and systems meanwhile allow products to be remotely monitored and diagnosed by suppliers, whether in regular maintenance cycles or when there are malfunctions. In many cases an internet based maintenance and repair service can be developed from this, allowing the customer to remotely access services from the manufacturers. This could for example be special repair manuals, linked with digital disassembly and reassembly drawings.

It is also possible for the product supplier to design their on-site customer service more quickly and more productively by giving their service employees access to in-house product data and repair instructions remotely. Finally, for the spare parts service, new possibilities arise by using the internet to optimize their storage, delivery and production.

With the quicker follow-up of products, the final phase of the product life for products, systems and plants gains significance. Previously their shutting down, dismantling and disposal was a rather burdensome side-issue. Increased awareness about environmental protection and stricter legislation requires this phase to be professionally considered as well in the sense of waste recycling and management. Thus offers are being developed to properly dismantle plants down again to a green field and refurbish them for the purpose of reuse and resale. When this is not possible or economical, it then comes down to recycling and/or disposing of them harmlessly (see Sect. 2.7).

Another approach goes even farther than just offering services during the three use phases to the extent that it is not even concerned with producing a defined material or immaterial output. Rather it focuses on selling the purpose of the output and thus on making the *utilization* of a product, system or plant a product itself. With that a particularly close, almost symbiotic customer relationship is achieved. This fourth dimension of the market offer is generally yielded by the manufacturers, to limit the financial risk, frequently in the form of a spin-off firm.

One possibility is to combine single services into a service package with the aim of ensuring or increasing the utilization of the product for the customer's value-adding process. This can, for example, affect the availability of a 24-hour server center, the yield of a production machine per shift or the operating costs of a pumping station. This service is created by extending the remote maintenance offer by optimizing operating parameters and the spare-parts stock. Yet another service might be increasing the utilization of a system by exchanging outdated components, e.g., electronic control panels. Finally, it is also possible to extend the functionality of a supplied product in the value-adding chain of the customer e.g., by installing an automatic loading system for a production plant as a replacement for a manual solution.

The most extensive example of this utilization-oriented market offer is the so-called 'BOT model' (Build-Operate-Transfer model). In this case, the manufacturer of the plant or an external service provider operates the production plant and only delivers finished products to a consumer, frequently directly to the site of the customer's plant.

The factory for manufacturing the Smart Car in Hambach, France is a known example of this. There, 15 suppliers contribute 80 % of the value-adding to the finished product (see also Fig. 2.14). One of the suppliers, Eisenmann, supplies and operates the painting plant there. They are paid for every car body that is painted [Bar98].

It was anticipated that BOT models would be very important in the future because they reduce the complexity of the user's production as well as decrease their investment risk and costs. At the same time, they open up a long-term customer relationship for the producer who has the know-how. Nevertheless, the operator is also directly

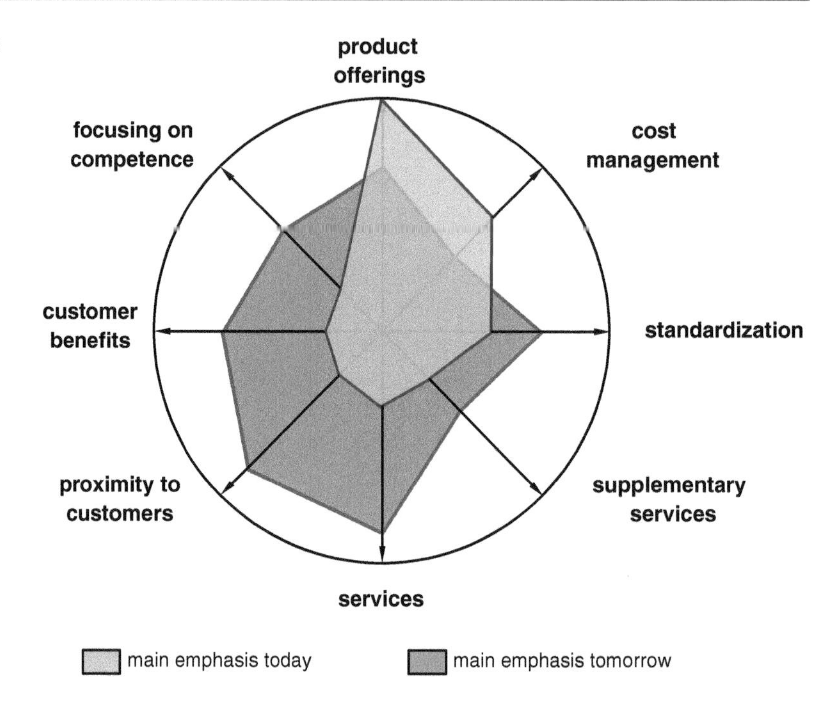

Fig. 2.6 Future emphases for production enterprises (per Boutellier, Schuh, Seghezzi). © IFA G8915SW_B

impacted by their success or lack of success in the market and thus correspondingly takes on the risks. Usually, a special core business firm is established in order to financially decouple them. Against expectation the model has not wide spread to industry mainly because of the financial risk for the plant deliverer.

If we are to summarize the information about market offers, it can be seen that the dominating model for all of the business' activities is clearly oriented on the customer. Moreover, it is dependent on offering customers individual solutions in their value-adding chain, taking on calculable customer risks in one's own value-adding chain and including the customer in designing and creating the solution [Bou97].

With that (from the perspective of the market offer) a number of future emphases for production enterprises can be identified (see Fig. 2.6).

In order to escape the dilemma between pricing pressure and increasing customer wishes, paid product-integrated services should be extended and traditional product outputs reduced to core components. This requires standardizing individual services and products, developing more intensive value-adding supplementary services and focusing on key competencies. In doing so the pressure on managing costs decreases since the new market offers are paid for. As a result, the customer benefit and proximity to the customer increases on the whole [Bou97].

2.4 Business Processes

As already mentioned, the offerings defined in the business sectors are to be yielded through *processes*. As Gausemeier wrote: "A process is a number of activities aimed at yielding a result which is of value to the customer" [Gau99]. With this, Gausemeier expressed the break from functional organization (characterized by splitting the work down into continually smaller units). Processes are linked to process chains which can be identified as either main business processes or supplementary processes.

Figure 2.7 depicts a breakdown of the business processes which are well-established for production enterprises. The *main business processes* follow the lifecycle of the market offering. The 'market opening' process is responsible for

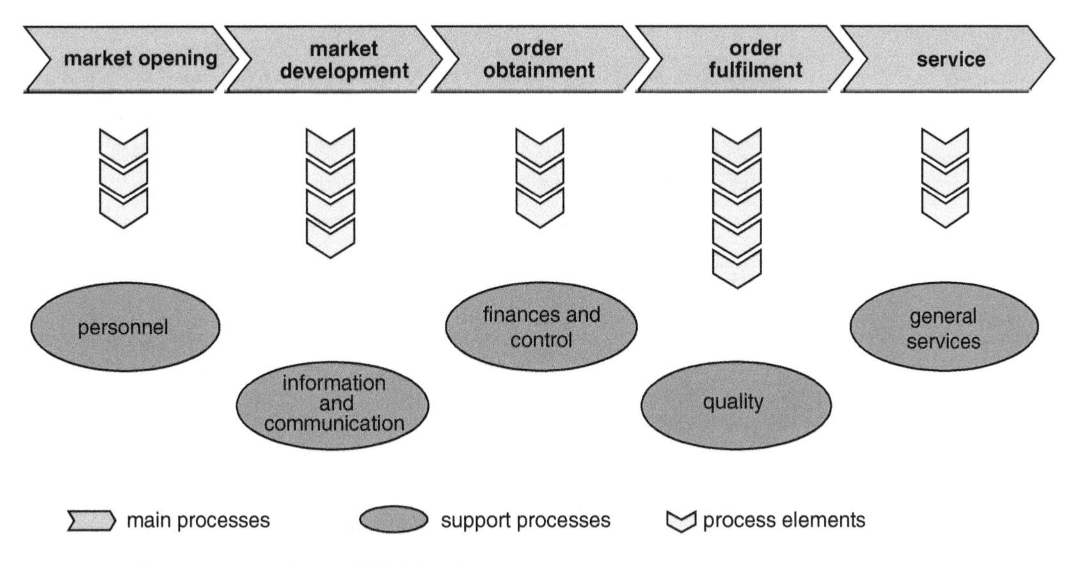

Fig. 2.7 Business processes. © IFA G8902SW_B

defining the market offer in form of a specification based on the business sector strategy. From that, a functional product, suitable for series, emerges in the 'market development' process. This product is then offered and sold to customers during the 'order obtainment' phase, whereby the technical, logistical and economic feasibility has to be ensured when signing the contract. The 'order fulfillment' phase summarizes the processes from order confirmation to shipment including the necessary procurement procedures. Once the customer has begun using the product, the 'service' phase (as described above) begins. One of the key attributes of these main processes is that one person (whether that be a supervisor, manager or a team leader) is completely responsible for the results and resources.

The *support processes* 'personnel', 'finances and control', 'quality management' (including planning, monitoring and testing), 'information and communication' as well as 'general services' (which ranges from the building maintenance up to site security) serve to supplement the main processes. They have to sell their services to the owner of the main processes at the agreed upon price and are thus competing against external service providers.

2.5 Aspects of Factory Design

The main business process that is essential to the factory is the *order fulfillment*. The sub-processes that need to be yielded here include the order input, product design (as far as the order specifications require it), job prep, sourcing of raw materials and purchased parts, part manufacturing, assembly, testing, packaging and shipping as well as the related quality checks and job control. These sub-processes are to be generated by factory resources, which are summarized together under the headings technology, organization and employees in Fig. 2.8. They form, so to speak, the pillars of the factory that are built upon a site and its buildings. A convincing market offer, however, is not just created from material and human resources, but rather is also determined by aspects of the enterprises culture and sustainability, which result both from the enterprise's comprehensive vision and from the local conditions.

Figure 2.8 identifies the key aspects of factory design that we will address in this book with regards to how they can be structured and dimensioned particularly with respect to changeability. Cost and feasibility considerations

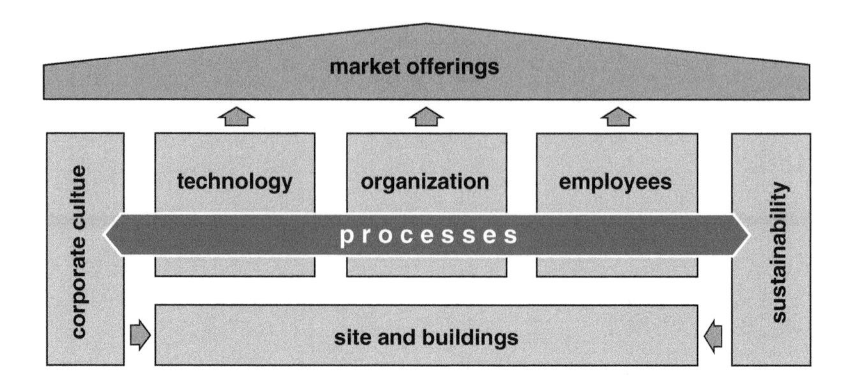

Fig. 2.8 Aspects of factory design. © IFA G8900SW_B

are components of the planning process and are discussed there. The order processing influences the factory layout to a far greater degree than any of the other remaining main processes or support processes mentioned in Fig. 2.7. These other processes mainly require resources such as office space, personnel and infrastructure, which are organized during the general planning.

All of the enterprise's processes and functions have to be oriented though on customer demands, market offers and a guiding vision that is developed in consideration of changeability.

2.6 Manufacturing Location and Factory

In view of our discussion about developing a production concept there is still the question in which scope and with what strategic orientation the individual enterprise wants to produce its products. The decision about the geographic location of the production is then made accordingly. When doing so, two different perspectives need to be distinguished; an external and an internal (see Fig. 2.9).

The term *manufacturing location* represents the external perspective. Within the scope of developing business sectors, market offers and necessary processes, a suitable manufacturing location has to be selected from a global perspective. These provide a market segment with specific goods and services related to the selected business sector in view of its economic and logistics criteria. In a second step, most often on a closed site, the factory is designed in the sense of an internal perspective of the manufacturing location. The *factory* is thus a local bundling of the primary production factors (personnel, resources, buildings and materials) as well as the necessary knowledge, qualifications and capital.

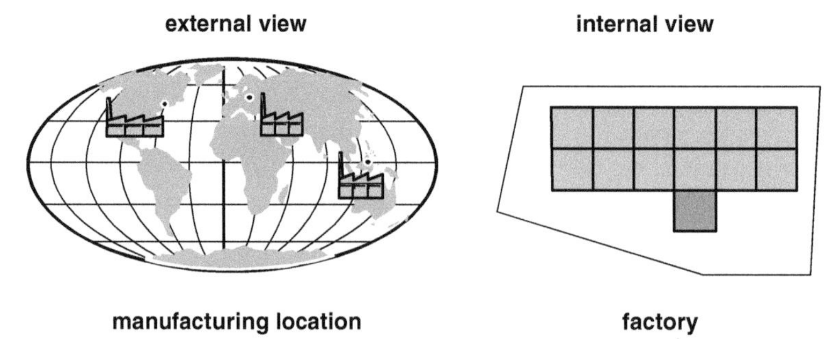

Fig. 2.9 Comparison of manufacturing location and factory. © IFA G9630SW_B

In the form of processes, these factors realize the parts of the value chain necessary for delivering the goods required by the manufacturing location. The term 'value chain' in comparison to value-adding chain also includes the activities such as storage, transportation, testing etc., that are inevitable due to the selected manufacturing principle but that do not add value.

Within the factory, a number of products for different business sectors and different proportions of the value chain can be manufactured. As already mentioned, due to the clear responsibilities for costs, quality and delivery capability the aim here is to operate sub-factories (often referred to as 'mini-factories', 'business units' etc.) which are demarcated as much as possible both spatially as well as organizationally. These sub-factories then only use a common infrastructure with regards to their energy supply, data processing, social facilities etc.

2.7 Morphology of Factory Types

Based on these considerations, we can now develop a morphology of factory types that combines four attribute levels. These originate from specific views of a factory and are primarily determined by the production strategy (see Fig. 2.10).

The first perspective is concerned with the *position of the enterprise* in the supply chain between the raw material suppliers and the end consumer (see Fig. 2.11). An extreme case is an enterprise that manufactures the raw goods required for its product itself as well as the end product with all of its interim stages before delivering it directly to the end user. This was the case at the start of the industrial age in North American automobile manufacturers. As a result of continually greater differentiation and specialization this is no longer economically and logistically feasible. Thus in the meantime, suppliers for raw materials, parts, components, modules, sub-systems and end products have developed. Each of these covers a stage of interim products and delivers them to a customer. Here, a customer can be a company that further processes it, a middleman or an end customer.

The second perspective is concerned with *how the customer perceives the factory*, that is, what its most predominant strategic attribute is in the sense of positioning themselves amongst their competitors. Six different forms can be identified here (see Fig. 2.12).

The *high tech factory* is characterized by products that are on the leading edge of technology in the world market e.g., with regards to their function, level of performance, lifecycle costs, availability etc. The manufacturing and assembly processes are operating close to their

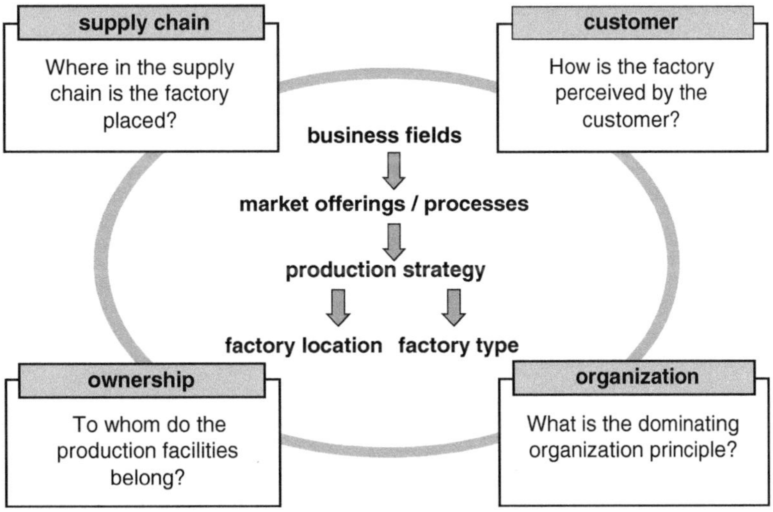

Fig. 2.10 Perspectives for developing factory types. © IFA G9637SW_B

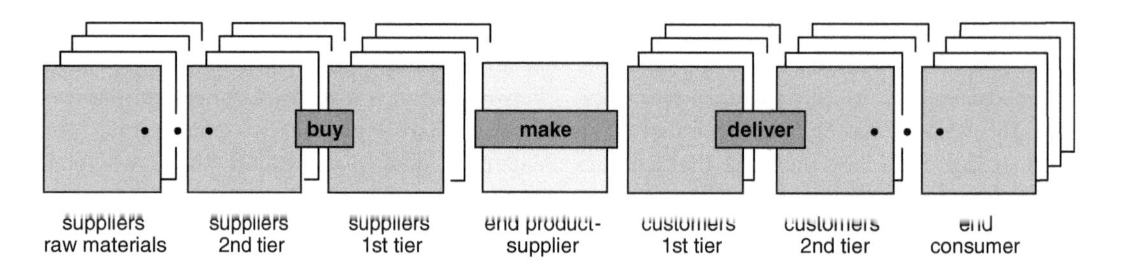

Fig. 2.11 Components of a supply chain. © IFA G9638SW_B

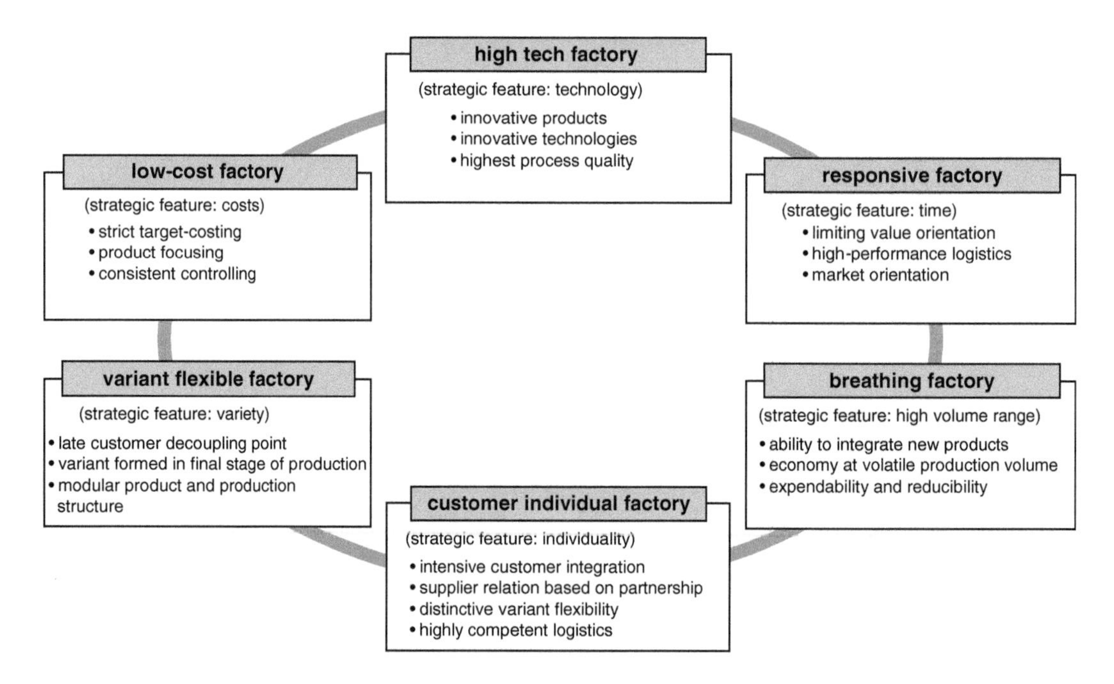

Fig. 2.12 Types of factories from a customer's perspective. © IFA G8629SW_B

natural limits (see Sect. 3.3) usually with self-developed technologies and with the highest process quality. The factory, both inside and outside, reflects the extraordinary demands set by the company on high-end technology. Since these are the innovators, premium prices are targeted and costs, delivery times and controlling variants do not play a large role.

The *responsive factory* focuses on the time factor. It is characterized by highly efficient logistics, which is also oriented on limits—in this case on throughput times. Since the products do not lay claim to any leading technology, their competitive edge lies in products being quickly available to customers. Orders are often inserted into the production directly by the customer or distributor.

In the *breathing factory* the focus is on economically manufacturing products with seasonally dependent sales fluctuations (e.g., household appliance and sporting goods industries) with largely varying production quantities. This is achieved with a comparably low degree of automation, very flexible work hours, and cross-trained employees. Consequently, new products can be quickly integrated and the factory output can be rapidly increased or decreased.

If the product spectrum is marked by a large number of variants, in the sense of a customized market supply, the *variant-flexible factory* should be pursued. It is characterized by modular structures as well as manufacturing technologies that allow variants to be generated as late as possible in the production process.

The further developed form of variant-flexible factories is the *customer specific factory*. It pursues the idea of mass customization, which will be discussed further in Sect. 4.10. In this situation, every order is different from the next with regards to technical specifications, quantity and due date. In the extreme case, the customer can configure the product themself via the internet, order it direct from the factory and follow its production over the internet as well. One of the conditions for this is that all business processes are mastered from the customer's order specification up to supplying the product to the customer.

If products are in the mature stage and thus subject to strong price pressure due to numerous competitors, the *low cost factory* is aimed at continually decreasing self-costs by strictly managing target costs, focusing on few products with large production quantities and consistently avoiding any kind of waste. This requires strict monitoring of the performance figures.

The described types of factories from the customer perspective will not appear in their pure form, since in real factories just about all of the strategic characteristics have to be taken into consideration with varying emphasis. In Fig. 2.13 the qualitative value of the competitive factors developed in Fig. 1.7 are depicted for the six factory types. It can be seen that the customer specific factory fulfills the most competitive factors, followed by the variant flexible factory.

The next dimension to be discussed in developing a morphology of factory types according to Fig. 2.10 is that of the *dominating organizational principle*. Here, factories can be differentiated as functional, segmented, networked or virtual.

The *functional factory* is organized into areas using the same technology through which a number of different products are routed e.g., mechanical processing, electronic manufacturing and assembly. This proves to be advantageous with regards to the utilization and flexibility of resources and bundling of know-how. Nevertheless, it is also associated with long through put times and large inventories resulting in stagnancy.

	factory type ▷	high tech factory	responsive factory	breathing factory	variant-flexible factory	low cost factory	customer specific factory
	feature ▷	technology	speed	volume range	variety	costs	customer request
competitive factors	costs	○	○	⊖	⊖	●	●
	time	○	●	⊖	⊖	○	●
	quality	●	○	○	○	○	●
	innovativeness	●	○	⊖	⊖	○	⊖
	learning speed	⊖	○	○	●	⊖	⊖
	changeability	⊖	⊖	●	●	○	●

characteristic: ○ weak ⊖ middle ● strong

Fig. 2.13 Characterization of factory types from customer's perspective. © IFA G9586SW_B

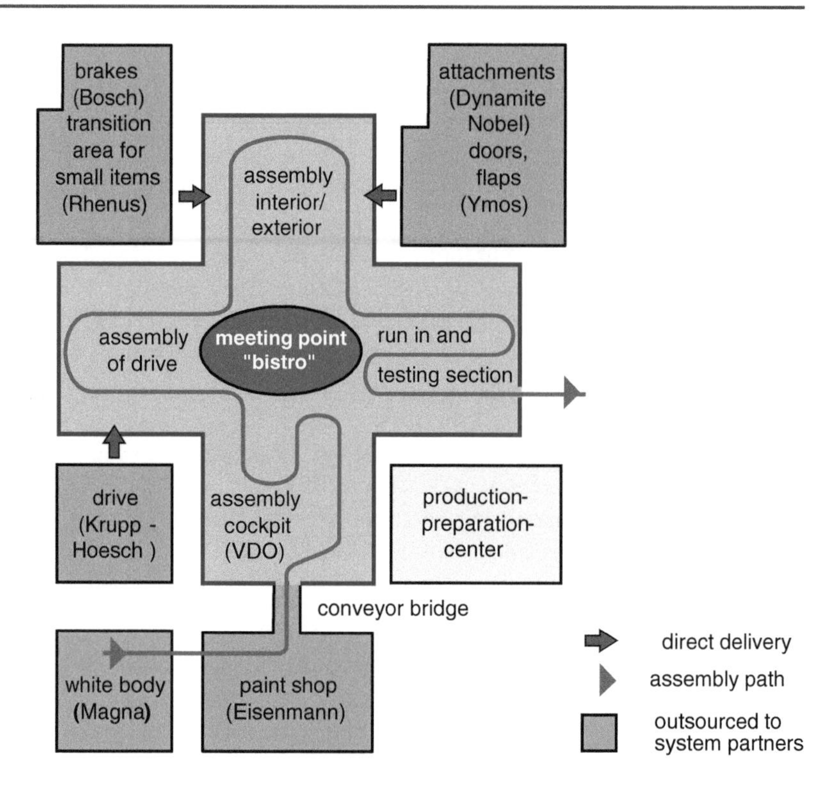

Fig. 2.14 BOT model example (based on MCC). © IFA G5341SW_Wd_B

If greater flexibility and more responsiveness are required, the *segmented factory* arises; comprised of powerful, small production units, clearly oriented on the product and market, it is also fully responsible for financial results. These units, depending on the quantity of the production and the number of variants, are then organized according to either the line principle, segment principle or workshop principle.

If the number of products and their variants is continually growing, reducing the complexity by drastically decreasing the in-house part manufacturing and suppliers is indispensable for preventing a collapse. Thus a *networked factory* emerges with several tiers of suppliers for subsystems, modules, components and parts. These are coordinated by intermediary logistics service providers.

In order to quickly seize opportunities for a complex product or system, a number of factories can temporarily join together for a project and bundle their processes and resources. Such cooperations are also plausible among competitors when it concerns utilizing very expensive equipment. When the enterprise that has the direct contact with the customer does not participate in the production itself, the term *virtual enterprise* is used. In extreme cases, these only look after marketing and processing orders.

In the fourth dimension of the factory morphology according to Fig. 2.10, the *ownership of property,* expresses itself in the production means. Starting with the objects that have the greatest risks many enterprises search for a release from the permanent tie to production plants through renting or leasing. There has thus been great interest in the abovementioned BOT models [Sche04, p. 441ff]. In these cases, either the plant manufacturer or an external service provider operates the production plant on the factory site or in the immediate near to it and delivers systems and components ready to be integrated in the final assembly. Figure 2.14 illustrates the BOT model based on the example of an automobile manufacturer which produces the compact Smart Car [Bar98].

In this case, 15 suppliers yield 80 % of the car's value-adding on the premises of the Micro Compact Car Company (MCC) located in Hambach, France. By skillfully bundling orders

views ▽	characteristic forms					
perception of the market	minimal price	variant flexible	volume flexible	reactive	high tech	customer specific
position in the supply chain	parts supplier	compo-nents supplier	module supplier	subsystem supplier	end product supplier	
orientation of the organization	functional	segment	networked	virtual		
ownership of production facilities	property	rent	leasing	cooperation	operator model	

Fig. 2.15 Morphology of factory types. © IFA G9585SW_B

across approximately 80 contractors 1000 positions in total are allotted as compared to 6000–8000 positions found in traditional automobile manufacturers. In this case, Magna operates the body shop and is paid for every automotive body it produces. The same applies to the paint shop operated by Eisenmann.

The plant operators thus act as their customer's problems solver, extending their competency as a plant manufacturer to include permanently controlling the production processes. This is not possible without their participation in developing both the end product and the entire factory; a long-term customer relationship is thus created. BOT models relieve the end product manufacturer from intensive capital investments and allow them to concentrate on the core processes of marketing and distribution, product development, final assembly and service. When there is a slump in the market, both partners are in the same boat and carry the corresponding risk. Nevertheless, the producer demands that the operator ensures the agreed upon quality, delivery reliability and price.

A further form of property ownership is co-operations in which two or more enterprises erect a production plant and use it for different products.

Based on the four dimensions described here, a morphology system for factory types results (see Fig. 2.15).

A factory—albeit still idealized—can then be described by combining one of each of the attribute forms from the four perspectives. A description such as this is particularly suitable as a basis for discussions about strategies when constructing or re-constructing a factory because it prevents the factory planning from becoming too focused on the subject of optimizing the layout and material flow.

2.8 Summary

The planning basis of a factory starts with the competitive factors found within the concerned industry. These include new entrants and substitutable products as well as the bargaining power of customers and suppliers. As part of the business planning a vision and strategy has to be developed, taking into account the perspectives of the customers as well as owners, the business processes and the ability to learn and grow.

The developed strategy describes the business sector, the market offer and the market segment. In doing so, sustainability in the economical, environmental and societal sense is vital. It is the strategy that first determines the design areas of the factory. These areas are concerned with the production facilities, organization and employees who interact within buildings on a site. The strategic orientation of the factory is still determined by the desired perception of the customer (function, price, high tech), the position of the products in the supply chain between the suppliers and customers (parts, modules, end product), the dominant principle of organization (functional, networked) and the owners of the means of production (ownership, rental, leasing, operating model). The resulting plant types can be characterized as high tech, responsive, flexible, breathing, variant, low-cost or customer-specific. In practice they tend to appear in mixed forms, since they usually serve several markets and customers with different products.

Bibliography

[Bar98] Barth, H., Gross, W.: Fabrik mit Modellcharakter—Neue Zielhierarchien bei der Fabrikplanung (Factory with Model Character—New Target Hierarchies when Planning a Factory) Zeitschr. f. Wirtsch. Fertigg., 93 (1998) 1–2, pp. 15–17

[Bou97] Boutellier, R., Schuh, G., Seghezzi, H. D.: Industrielle Produktion und Kundennähe—Ein Widerspruch? (Industrial production and customer service—a contradiction?) In: Schuh, G., Wiendahl, H.-P. (eds.) Komplexität und Agilität. Steckt die Produktion in der Sackgasse? (Complexity and Agility. Has Production Reached the Deadend?). Springer, Heidelberg (1997)

[Faß00] Faßnacht, W., Frühwald, Ch.: Controlling von Logistikleistung und –kosten. (Controlling Logistic Performance and Costs). In: Baumgarten, H., Wiendahl, H.-P., Zentes, J. (eds.) Expertensystem Logistik, Beitrag 5.03.03, pp. 1–16. Springer, Heidelberg (2000)

[Gau99] Gausemeier, J., Fink, A.: Führung im Wandel (Leadership in Change). Hanser Munich Vienna (1999)

[Kap96] Kaplan, R.S., Norton, D.P.: The Balanced Scorecard: Translating Strategy into Action. Harvard Business School Press (1996)

[Por98] Porter, M.E.: Competitive Strategy: Techniques for Analyzing Industries and Competitors with a New Introduction. The Free Press, New York (1998)

[Sche04] Schenk, M., Wirth, S.: Fabrikplanung und Fabrikbetrieb. Methoden für die wandlungsfähige und vernetzte Fabrik (Factory Planning and Management. Methods for the Changeable and Networked Factory). Springer, Heidelberg (2004)

3 Production Requirements

3.1 Introduction

The change drivers described in Chap. 1 pose a variety of requirements for future productions, the fundamental aspects of which are outlined in Fig. 3.1.

Our discussion begins with the *customer's requirements* from the turbulent markets. As can be seen, we have once again reduced these to three key concepts, summarized as follows: (1) Functionally superior products and services with (2) long-term benefit for the customer have to be (3) quickly available. The *market offerings* derived from these consist of products, systems and plants discussed in Sect. 2.3 supplemented here by the services yielded before, during and after the utilization phase.

These generate certain requirements for the production that can be developed from four perspectives organized according to two levels: inner/outer and rational/emotional. The rational outer view reflects the supplier's behavior perceived by the customer and is related to the concepts of responsiveness and quantity/variant flexibility. From there, inner requirements are derived according to the orientation of the processes on so-called natural limits, followed by (largely participative-designed) self-organization and a cooperative network with external value-adding partners.

The rational view is supplemented by an emotional view. Externally this is shaped by a specific brand identity and product image, while internally it is evident in the transparency of processes and aesthetically suitable appearance of the factory. The value view spans across all of these fields. Here the concepts of sustainable development and the commitment towards a corporate culture influence the product design during the lifecycle of the product and the process design during the lifecycle of the facilities.

Based on a wide range of publications and our own findings, we will now discuss each of these in enough depth to be able to develop initial ideas about visions, models and types of robust future factories.

3.2 Responsiveness

For the competitive factory, market oriented *responsiveness* can be considered the crucial factor. This means the factory has to be able to deliver the customer's desired product in the desired quantity at the desired time and with the desired quality. It is thus obvious that the customer's wishes have to be clearly defined when an order is confirmed. If this is not possible with more complex products—for example, with production systems—interim dates should be

H.-P. Wiendahl et al., *Handbook Factory Planning and Design*,
DOI 10.1007/978-3-662-46391-8_3, © Springer-Verlag Berlin Heidelberg 2015

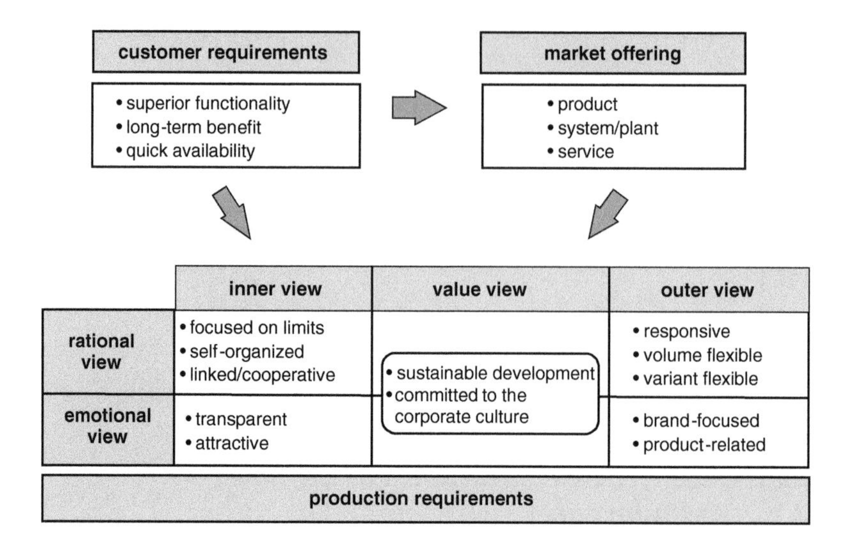

Fig. 3.1 Development of production requirements. © IFA G8903SW_B

agreed upon by which the remaining specifications have to be set.

Market oriented responsiveness implies the following: only that which has been sold will be produced. This concept is referred to as *production on demand.* It means not only that the manufacturing only starts after the order has entered the system, but also that the required material is ordered explicitly for each order. For this approach it is necessary that the sum of the procurement times and the internal delivery times are shorter than the desired delivery time. Despite tremendous effort this is not always possible. In cases such as this a *customer decoupling point* can be introduced. This refers to the point in the operational logistic chain (with its sub-functions procurement, manufacturing, assembly, and shipping) after which the orders are allocated to specific customers [Hoe92]. Orders are processed without specific customers based on sales forecasts up until the customer decoupling point.

Depending on the relationship between the demanded delivery times and internal throughput times for the four sections of the logistic chain, there are four order or supply strategies (see Fig. 3.2).

With a *make-to-stock* production the customer receives the order product directly from the finished goods warehouse. The products are procured, manufactured, assembled and stored based on a production plan. With an increasing number of variants this is more and more problematic because the tied-up capital is too great, the predictability of the individual variants is strongly reduced and, with that, the service level decreases.

In cases such as this, the aim is to pre-manufacture and temporarily store standardized parts, components or sub-systems based on a platform concept. These are then *assembled to order* specifically for a customer after an order has been placed. Numerous mechanical, traffic or electrical engineered products as well as electronics can be quickly delivered in this way—increasingly within 24 to 72 h.

It can however be technically impossible or uneconomical to pre-manufacture the components for all of the plausible customer wishes, whether it is because they have to be dimensioned according to the customer's requirements or because making them to stock is too expensive. In this situation, a *make-to-order* production might be pursued in which only the initial material and procured components are prestocked for key production components based on sales forecasts. The remainder consists of standard components assembled together with customized manufactured components into a customer product.

The fourth case is a *custom-specific one-of production.* It requires a complete new product

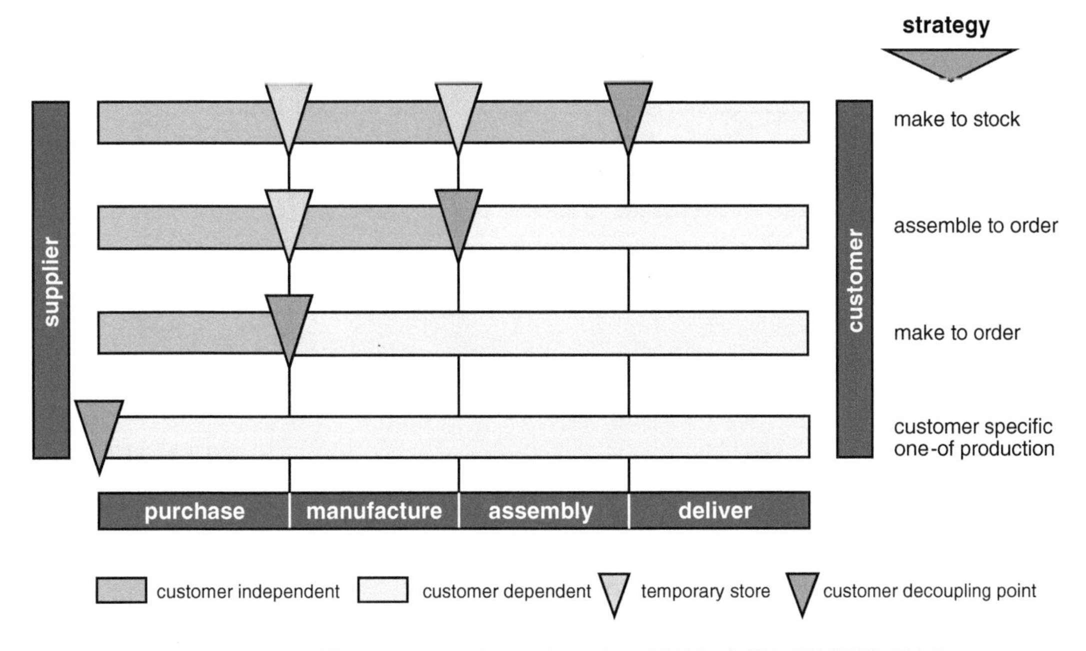

Fig. 3.2 Order strategies with different customer decoupling points (Eid95). © IFA G0268SW_Wd_B

construction in which components are only procured after the product has been designed and the parts dimensioned.

From a logistics perspective, customer decoupling points are represented as an interim store with a defined stock of the customer-neutral components, from which the subsequent segments in the direction of the flow are served as required, usually according to the super market principle. Once a specific quantity has been withdrawn, a replenishment order is automatically generated. The remaining storage points in Fig. 3.2 are usually designed as dynamic interim buffers, frequently in the form of mobile shelves with a fixed storage capacity. These help to balance out the queuing times between the workstations that arise due to different processing times and lot sizes. Furthermore, it should be kept in mind that during their lifecycle products clearly can be manufactured according to different order strategies depending on the desired relationship between the delivery times and throughput times, the size of the production volume per time unit and how large the number of variants is. Moreover, usually an enterprise would offer more than one product on a market. Managing these constant changes poses the key challenge in planning and controlling a factory.

3.3 Quantity and Variant Flexibility

In addition to a limited reaction time, one of the predominant characteristics of production in a turbulent market is strong demand fluctuations and simultaneous increase in the number of variants and their components. Whereas up until now it was possible to at least partially counter the variant problem with a skillful modular construction, the increasing quantity fluctuations pose a dilemma for enterprises. On the one hand, it is no longer possible to maintain stores of all the variants, and on the other hand, as outlined in Fig. 3.3, automated production concepts reach their limits in two respects.

Generally speaking, it is anticipated that there is a pronounced fluctuation of customer demands over the course of time. This range of fluctuations is also referred to as the *quantity variance* and expresses the maximum quantity sold during a period (e.g., one year) as a multiple of the

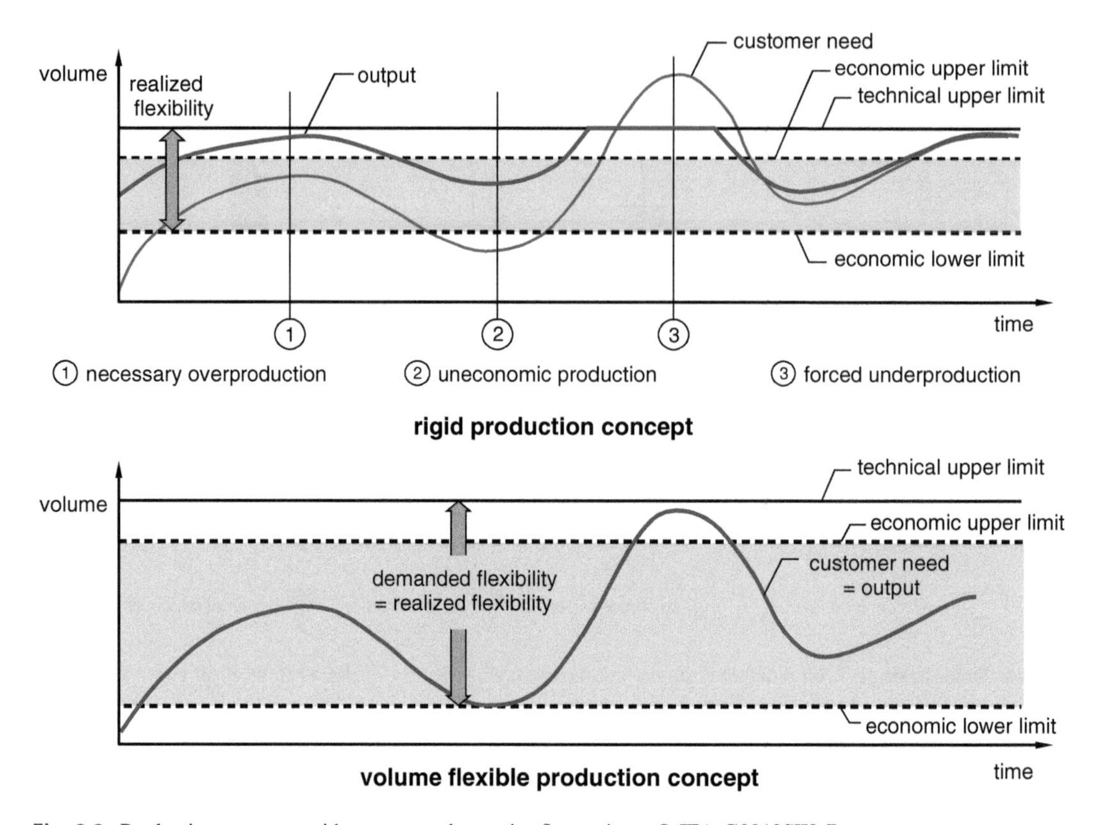

Fig. 3.3 Production concepts with pronounced quantity fluctuations. © IFA G8913SW_B

minimum quantity. Consumer goods with long lives and seasonal dependence such as washing machines, which are subject to a quantity variance of 1:6 in the course of a year, are typical of this.

A rigid production concept, characterized by extensively automated individual processes, linked workstations, long setup times and a small workforce usually operated in 2 or 3 shifts is defined by two limits in the output quantity (see Fig. 3.3 upper part). On the one hand, there is an economic upper limit. Due to economic reasons this limit is below the technical upper limit, defined by the maximum number of shifts and the cycle time. Typically the economic limit is 80 to 90 % of the technical upper limit. In the case of high demand above the economic or even technical upper limit, this usually requires overproduction in advance that then enters interim stores (Situation 1 in Fig. 3.3), or results in a temporary increase in delivery time (Situation 3 in Fig. 3.3).

On the other hand, the economic lower limit is set by a system's fixed costs. By nature, automated systems have high fixed costs (depreciation, interest, maintenance, repairs etc.) and comparably low variable costs (personnel, energy, operating supplies etc.). If the required production quantity is below the economic lower limit, losses arise (Situation 2 in Fig. 3.3).

The aim of a flexible volume production concept is to cover the volume fluctuations in the market as well as possible by first extending the economic upper and lower limits (see Fig. 3.3 lower part). By doing so, an economic production is even then possible when the sales volume is small—most likely due to an adjustable degree of automation. Moreover, it aims to quickly adjust the technical upper limit, e.g., through modular workstations.

This method is met with considerable reservations in the practice due to the dominating principle of cost effective analyses. It assumes there is a largely constant production quantity and the lifecycle of the production plant is longer than the product lifecycle, i.e., no significant adjustments to the plant are thus required. Both of these conditions, however, are not met in quantity/variant flexible productions and thus inevitably lead to the consideration of *lifecycle costs*. Here all of the costs that are directly or indirectly generated or predictable in processes during the system's phases of life (planning, construction, production, procurement, commissioning, start-up, operation and de-commissioning) are summarized [Far11]. In doing so, planned operating costs should be differentiated from unplanned costs that result from standstills (e.g., due to technical or organizational disruptions), rejects or the need to rework/reconstruct items.

Based on the example of a robot welding line for car bodies it can be seen that of the total (100 %) costs during the lifecycle of a body shop, only 25 % occur in the phases leading up to the start of operations, whereas 44 % arise from planned costs and 31 % from unplanned costs (see Fig. 3.4, [Perl98]). With that, the last cost block is greater than all of the costs related to the initial investment. The operating company estimates that costs can be reduced in future plants by approximately 30 % when improving the plant planning and procurement, orienting the product development on the production and changing the organization of the production and, in particular, by reducing the unplanned subsequent costs by 85 %. These can mainly be reduced with a plant concept that allows the product, product quantity and production technology to be quickly changed and the plant brought back into operation faster. Recommendations for this are found in [ElM09].

Approaches to making production more flexible with regards to quantities differ significantly between the manufacturing section and the assembly. Since the introduction of numerical control in the 1950s, the development of manufacturing technology has been shaped by the increasing connection of workstations to automated changes of work-pieces and tools up to flexible manufacturing systems. Nevertheless, with the increasing number of part variants with smaller lots and shorter delivery times, this proved to be not flexible enough. Rejecting automated manufacturing technologies in favor of organizational solutions such as lean production, business process orientation, total quality management or the introduction of group work also turned out to be equally lacking. It then became evident that combining new manufacturing technology concepts together with flexibly organized work and extensive operational instrumentation for measurement and control brought progress. The goal of such adaptive manufacturing systems is not to reduce automation, but rather to continually re-design the physical concept of the system, the Planning and

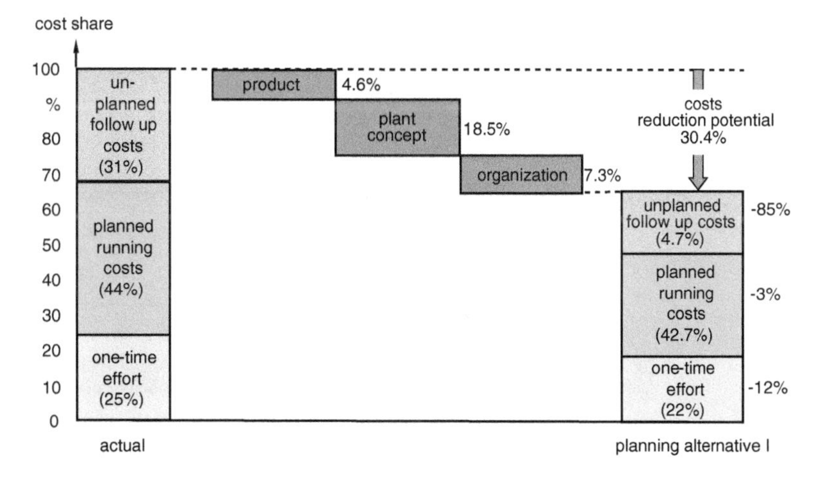

Fig. 3.4 Potential for reducing lifecycle costs (Perlewitz, BMW). © IFA G8918SW_B

Control system, sensor technology and human/machine interfaces with the goal of greater variant and process flexibility.

'Intelligent' universal modules (i.e., modules equipped with their own sensors, actuators and controls) that can be configured into mobile manufacturing units are promoted as the solution to strive for. Due to their extensive independence from a special processing task they are sustainable, changeable and therefore a safe investment. Together with their intuitive operation based on standardized human/machine interfaces, their ability to be easily reconfigured within the frame of a defined range of flexibility allows fast learning curves and specific user abilities to be developed [Abe06]. Another approach involves combining a number of manufacturing processes in one machine and with that allowing a part to be completely manufactured in one setup. In addition to the greater manufacturing precision resulting from the elimination of a number of clampings, there are also considerable gains with regards to time and flexibility.

In comparison to manufacturing, assembly is characterized by fitting together of many parts using various joining processes. Besides the joining processes, a substantial portion of the assembly costs are due to supplying, feeding and positioning parts as well as transporting them between assembly stations. Whereas, for quality purposes, the joining processes are predominantly automated or at least the testing, parts are still frequently handled manually. Only with a large number of pieces, as is typical for consumer goods or in the electronic and automobile branches, are the automated joining and testing stations linked with conveyor belts upon which objects to be joined are identified, positioned, saved and transported with the aid of workpiece carriers (see e.g. Fig. 6.38).

The question of the quantity and variant flexibility is most important for automated assembly plants that, similar to manufacturing plants, can become uneconomical when operated in two shifts and utilization is less than 90 %. The problem is amplified by short product life-cycles, which for some electronic products such as mobile phones span less than a year.

Here too, the perspective of considerations about economic viability has to be broadened. As an example, Fig. 3.5 depicts the results of an analysis of the yearly operating costs of an

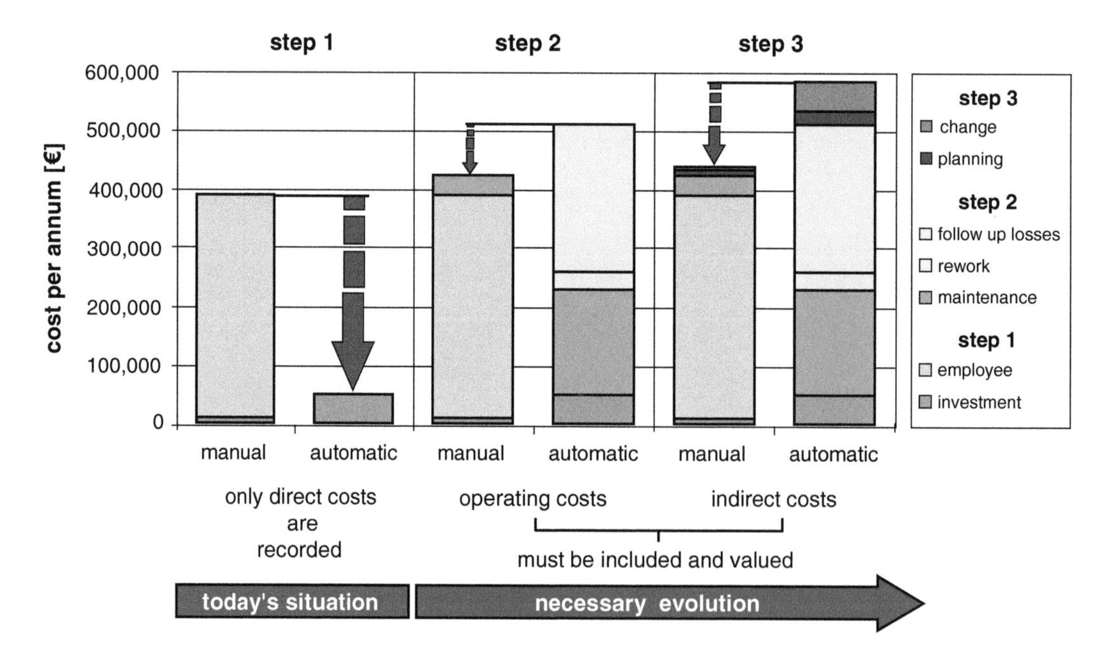

Fig. 3.5 Extended feasibility studies for two assembly stations. © IFA G6732SW_B

assembly station in the final assembly of commercial vehicles.

The objects of comparison were two assembly stations with similar tasks, of which one was operated for the most part manually and the other was for the most part automated. Step 1 shows the yearly costs arising from workplace and employee related costs in accordance with standard cost accounting. Despite distinctly lower costs for the workstation, the manual assembly station is clearly inferior to the automated station due to the high costs for personnel.

The manual station becomes more favorable when, in Step 2, the costs for maintenance, reworking and so called 'Follow up Losses' are also included. Follow up losses refer to downtimes not caused by the station itself, but rather by disruptions before or after the station. Regardless of the typical buffers, these disruptions either lead to waiting times, because no workpiece carriers are entering the station, or to blockades, because the carriers cannot flow to the next station.

The comparison between manual and automated assemblies favors the manual solution all the more in Step 3 where the costs for planning and changing the station as well as the production downtimes resulting from changes in the car model are included in the overall consideration of the lifecycle costs.

In view of the demanded flexibility of the assembly it is clear on one hand that highly automated solutions can be unfeasible and inappropriate due to hidden costs and losses. On the other hand, the solution cannot be to unconditionally replacing automated systems with manual systems. Rather, it is recommended that the degree of automation be quickly adjustable to changes in the quantities or products by adding, removing or replacing automated/manual process and transportation modules. Such solutions have already been realized as so-called 'hybrid assembly systems' [Lot12]. This, however, requires that operators and lower management be involved in the configuration process and can only be realized when all employees are qualified accordingly and motivated.

3.4 Focusing on Limits

By focusing on limits, the factory development can get valuable impulses especially with regards to responsiveness. The term 'limit' stems from mathematics and refers to the value of a numerical sequence, towards which it converges. In environmental and industrial health and safety standards, limits refer to permissible values of certain measurands (e.g., sound or the percentage of contaminants in air, earth or water) that are not to be exceeded.

For the factory area, the value-adding steps that reach the theoretical possible limits for minimizing costs for the organization and conducting production under stable conditions can be referred to as limit optimized processes. This idea was investigated within the framework of a joint research project from the perspectives of manufacturing technology, machining technology and logistics [Doe00]. The value-adding steps considered are based on the key factory processes, i.e., procurement, part manufacturing, assembly, transportation and storage. The input factors in these processes include materials, energy, information, space, employees and capital. Technical processes such as discrete-part manufacturing and assembly as well as logical organizational procedures for processing orders are relevant as basic limit objects.

Another critical differentiation concerns the limit level. Figure 3.6 depicts three levels with their prerequisites. The individual processes are evaluated according to the targets on the right side of the diagram in order to obtain an objective yardstick for the limit.

The first target, given the primacy of economic efficiency, is the costs. Following that is time a universal goal and evaluation parameter because a large number of subsequent parameters, such as the stock tied to the process or the required area, are derived from it. Quality is also an important target for all processes; after all, the reliability of the process' yield is dependent on it. In addition to these three traditional targets, controlling the variety of processes, parts and products is gaining significance. Finally, the term

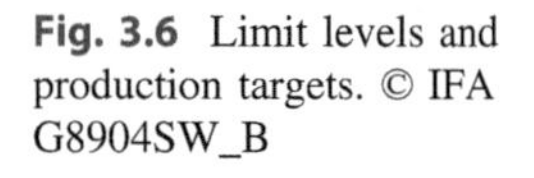

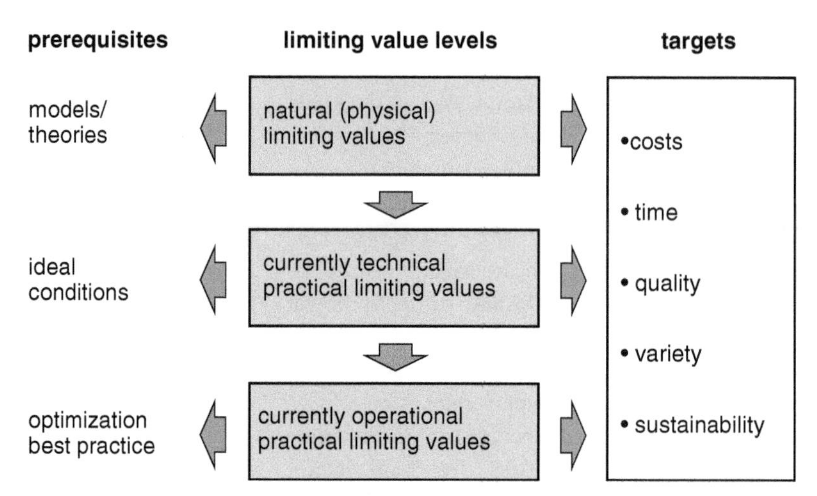

Fig. 3.6 Limit levels and production targets. © IFA G8904SW_B

'sustainability' is meant to emphasize the production processes' conservation of resources and environmental protection.

On the first level, the operationally-possible limit attained by optimizing existing systems and processes is assumed. Farther-reaching goals can already be defined in contrast to the actual state by comparing processes internally or across the industry in the form of benchmarking and by implementing continuous improvement or waste avoidance methods [Suz93]. The second limit level requires ideal conditions; these values can be technically realized, however, only attained in practice under laboratory conditions. Due to this, implementing this level is not economically feasible; nevertheless, it can be used to formulate goals that are attainable within a foreseeable time. Finally, the third limit level is focused on theoretical models of the processes. Based on these, it defines so-called 'natural physical limitations'. This view requires a long planning horizon and visionary thinking.

Based on the example "use of coolant fluid for metal grinding", the three limit levels lead to the following conclusions. The current operationally attainable value is 10 l/min of fluid per millimeter (mm) of grinding wheel width. With so-called 'minimum lubrication and cooling', 0.0001 l/min per mm of grinding wheel width can be attained in the lab, and from a purely physical perspective it would seem that it is theoretically possible to grind steel without any fluid at all [Doe00].

In a very simplified form Fig. 3.7 depicts how such natural limits might look when considered individually for the targets of the process elements shown.

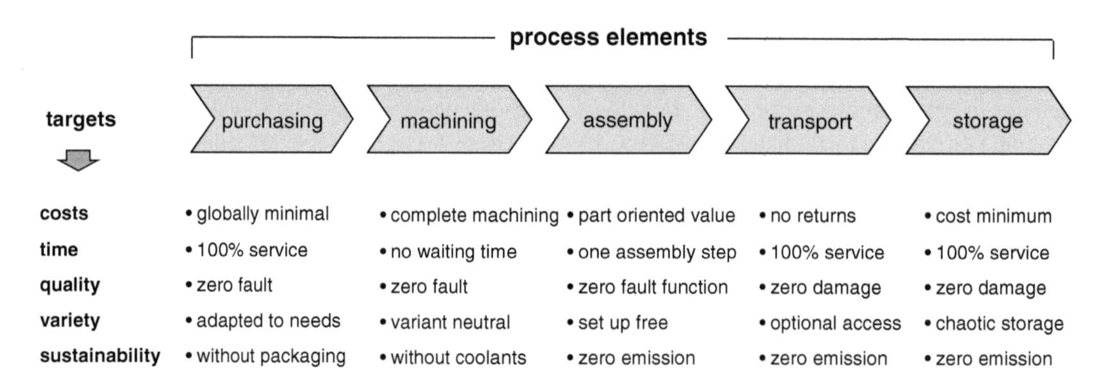

Fig. 3.7 Example of limit approaches in production. © IFA G8901SW_B

For example, with the purchasing costs we can imagine a scenario in which: (a) the cost price is the global minimum, (b) all delivered articles arrive punctually and in the ordered quantities at the site of consumption, (c) all articles are set up fault-free, (d) there are enough variants to cover demand and (e) there is no packaging that requires disposal. With machining these might include: (a) completely processing a workpiece with a number of technologies within one setup, (b) throughput times consist only of value-adding processes, (c) all articles are set up fault-free, (d) variation is controlled by non-wearing, programmable tools (e.g., laser beams) and (e) processes are not requiring any coolant or lubricant. Similar considerations are also indicated for the remaining processes.

The problem with this simplified perspective is that changes in the process parameters almost always impact a number of targets. Therefore, it is recommendable to use an approach based on a combination of looking back as well as forward for determining the limit (see Fig. 3.8).

First, the range of the analysis has to be set and the actual state determined. Targets are then defined depending on the chosen object of observation (e.g., the entire factory, a single process or sub-process). It is important that only one target is initially focused with respect to a limit. Doing so not only reduces the complexity arising from the dependency of the targets but also ensures that information can be clearly and simply communicated. Part of defining the goal is also identifying the deficits and the influences which cannot be initially changed.

Subsequently, it is critical to determine the process parameter that influences the chosen target the most. From a management perspective it is the cost drivers for example, logistic-wise it is the system inventory, whereas with machining it is the cutting speed, etc. Concrete values, which are not trivial but rather logically comprehensible, then need to be set for this significant parameter. For example, setting a throughput time of zero for the manufacturing of a workpiece would be an illogical limit; comparatively, a more practical time limit is the sum of the process times, which would mean that work-pieces never have to wait.

Occasionally it is also possible to actually calculate the limit. Thus, for example, the operating limit of a traffic lane in number of vehicles per hour can be calculated as a function of the vehicle length, the emergency break constant,

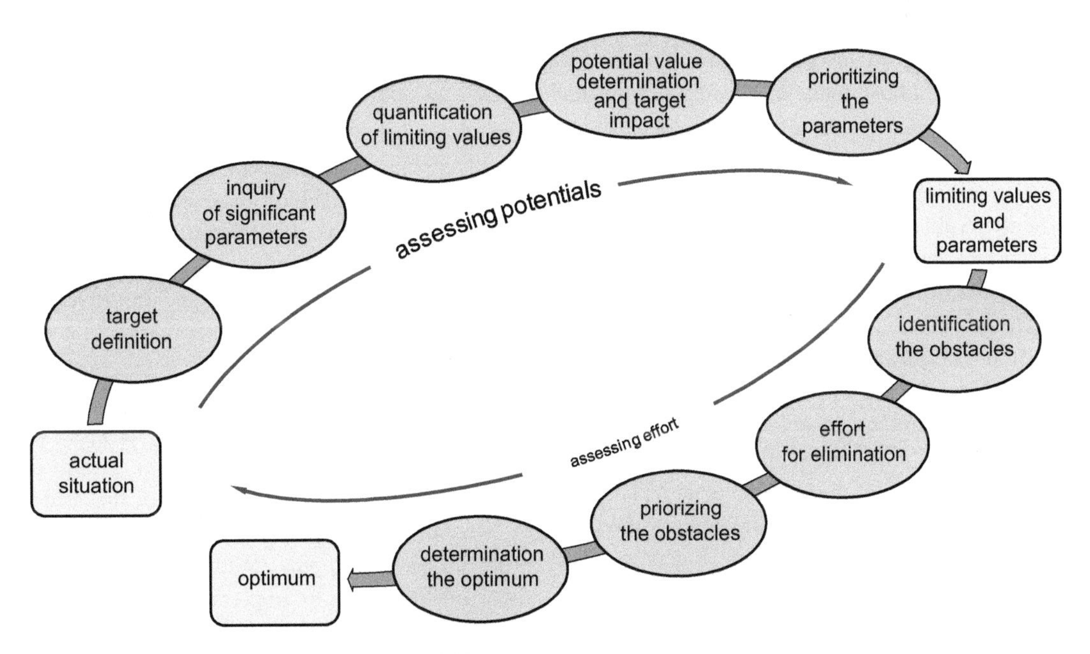

Fig. 3.8 Limiting value cycle. © IFA G8890SW_B

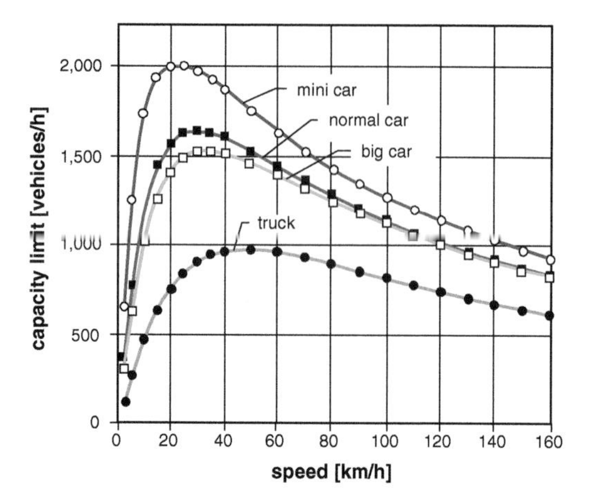

Fig. 3.9 Operating limit of a traffic lane (Gudehus). © IFA G8917SW_B

a reaction time and the vehicle speed. Figure 3.9 depicts the results of such a calculation based on the example of four vehicle classes [Gud05]. It is clear that for each type an (in this case surprisingly low) optimal limit exists for the vehicle speed, which when exceeded or fallen below leads to a decreased capacity limit. It needs to be emphasized here that this type of specific operating limit values are implemented to control traffic capacity, but in no way represent the cost minimum for the transportation volume itself.

Comparing the value of the determined limit with the actual value reveals the range that is available for adjustments and indicates how far the existing process is already technically or logistically pushed to its limit.

The next step is concerned with determining the potential value and influence of various process parameters on the selected target value. The greater the distance between the actual value and the limit value, and the greater the impact of changing the parameter on the target value is, the greater the potential. Nevertheless, due to system dependent correlations, parameter changes will work in opposition to one another or parallel to one another. Understanding these interactions is the key challenge of orienting on limits and generally requires an analytical approach together with expert knowledge. In contrast to FMEA (Failure Mode and Effects Analysis), the aim is not to find failures and their effects. Rather in the sense of an impact analysis, the effects of the changes to the process parameters on the developed potential are determined. The parameters, which can then be differentiated based on their potentials, now have to be prioritized, whereby small changes can develop large potential e.g., the potential yield of an automated assembly plant by increasing the availability of a bottleneck station.

Once the parameters have been set with their limits and dependencies, the question of implementation costs has to be addressed. In a type of predictive-backwards consideration, obstacles that need to be overcome have to be identified. The seven types of waste formulated by Suzaki can be helpful here: large inventories, over production, waiting times, transportation, production errors, movement and work processes [Suz93]. Subsequently, solutions for eliminating these obstacles need to be developed and the related costs determined. Depending on the limit level and scope of the object, methods typical for the respective fields can be applied e.g., design methodology [Pah07] or the logistic analysis of bottlenecks [Nyh09]. Finally the obstacles should be evaluated with regards to their cost-benefit relationships and prioritized accordingly.

In conclusion, Fig. 3.10 compares the underlying strategic idea of developing improvement potential focused on limits with traditional procedures. Usually, the target state, defined either by benchmarking or customer requirements, is strived for reactively with the aid of individual improvement steps; further potential remains hidden due to the lack of an objective yardstick. In comparison, the limit approach selects a clearly more aggressive starting position which leads to the economic optimum through the described steps.

For factory planning it seems more practical to focus on limits on an aggregated level including the procurement process, the processing chain for manufacturing a part group or assembly of a product group and the order throughput from the entry of the customer's order up to its shipping. Within this range, the focus on limits can be implemented on an operative level with regards to technical, logistical, or organizational aspects.

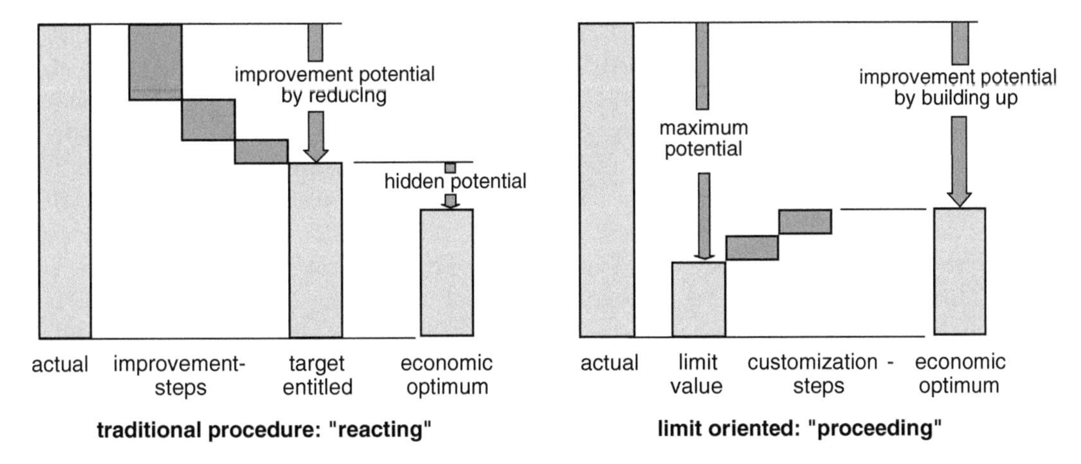

Fig. 3.10 Traditional and limit-oriented improvement (Hartung, Mc Kinsey). © IFA G8919SW_B

As an example, we will consider the question of generating variants. Frequently, product variants are determined by the different characteristics of individual parts. This requires details to be set very early on in the order throughput. Should the customers change their mind during this time, the part is no longer useable for this order and the product may have to be disassembled or even discarded. However, the limit can now be defined so that the part variants are differentiated in the last stage of the assembly. This approach to a solution consists of shifting the sub-processes that determine variants into final assembly and thus to some degree dissolving the traditional boundary between manufacturing and assembly. The problem that has to be overcome here is the controlling of this process in an industrial assembly setting without damaging parts that have already been integrated. In order to implement this method, the processing has to be separated into a variant neutral pre-manufacturing and a variant specific finishing that is incorporated into assembly as a station. The specifics are discussed in more detail in Sect. 4.11 with regard to production stage concepts.

If we summarize the focus on limits in a very simplified manner according to Fig. 3.11 i.e., from an external market view and an internal process view, the following features are indicated.

Fig. 3.11 Features of a limit oriented factory. © IFA G8898SW_B

market offering:

- 100% customer desire fulfilment (function, price, delivery date)
- attention to the complete product life cycle
- production for customer orders only

processes:

costs	⇨	only value adding activities
time	⇨	no waiting times of material and resources
quality	⇨	self-controlled processes and fault-free products
variety	⇨	variant formation in the final stage of production
sustainability	⇨	complete life cycle consideration of resources and processes

Based on fulfilling the customer's desires by 100 % with regards to functionality, price and due date, the enterprise strives to accompany the customer during the use of the supplied product over its entire lifecycle. It does not produce any orders to stock. For the processes, it follows that only activities which increase the value of the product to be delivered and its components should be conducted. Moreover, there should be no waiting times in the process—not for the materials, nor for the resources and employees. From the quality view, all processes should only deliver fault-free products to the internal or external customers. The demand for a maximum diversity of variants should be met by the possibility of determining the final design in the last stage of production, the so called postponement. Last but not least, sustainability requires the lifecycle to be considered in view of the re-use and recycling of both the product delivered to the customer as well as the resources and processes implemented in the enterprise itself. The latter are subject to guidelines for minimizing energy consumption up to the physical limit and to having zero impact on the environment. There is unmistakably a strong similarity between the concept of orienting on limits and that of lean production; these are addressed more extensively in Sect. 4.6.

3.5 Self-organization and Participation

The organization of work in the last decades has been characterized by tasks being more and more differentiated according to specializations. The strict division between manual and mental work has led to a number of specialized direct activities with correspondingly special job profiles (e.g., lathe operator or welder) as well as equally specialized planning, control and monitoring jobs such as production engineers, production schedulers, machine setters, quality controllers etc. The goal has been to plan as far ahead as possible using standards with as much mechanization and automation as possible. This has resulted in the hierarchical, bureaucratic, change-resistant organization we described at the start of this book.

Three developments have caused this concept to be called into question: First, the growing level of automation continually decreased the percentage of standardizable and manual manufacturing tasks. The remaining, increasingly demanding activities are precisely the ones resistant to planning and control. Second, as a result of a general social shift in values, strong hierarchies with minimal room to maneuver and management based on orders/obedience coupled with strict control over behavior and performance are no longer acceptable. Third and finally, in addition to current professional expertise the turbulent production conditions described above also require more methodological expertise preferably for finding and evaluating solutions for unexpected problems as well as social skills for resolving conflicts and abilities for working together as a team.

There is extensive agreement both in research and in operational practice that a new organizational concept based less on discipline and subordination and more on self-control, engagement, self-initiative as well as a willingness to communicate and cooperate, opens up new opportunities for controlling a complex and turbulent environment. Frequently this approach is referred to as self-organization and is closely linked to the concept of participation.

In this context, *self-organization* means taking responsibility for defined processes with a clear relation to results [Brö00]. Greater room for employees to maneuver with regards to actions and decisions is required, not only in terms of dealing with the daily work program (horizontal participation), but also in designing and changing the workstations layout and processes (vertical participation). By involving employees in demanding tasks, it is anticipated that they will be more engaged in process innovations and in improving productivity and quality [Grei89]. How strongly participants can be tied-in, depends on the degree of decentralization as well as the lifecycle phase of the production system (see Fig. 3.12 and [Menz00]).

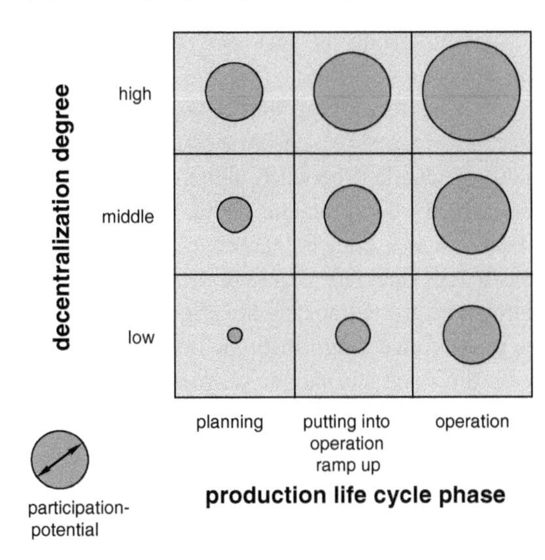

Fig. 3.12 Potential of employee participation. © IFA G8895SW_B

If we break the production system down into the different phases of planning, start-up (including the ramp-up to the target output) and steady continuous operation, then, even with a low degree of decentralization employees during the operating phase can have the greatest impact in designing the workstations and processes. Their potential for having an impact is less during the other two phases. Very good results have been obtained with skilled workers as plant operators when they are involved right from the beginning in building the facilities, testing them piece-by-piece and starting them up. In comparison, the participation in long-term, business relevant decisions during the planning phase can be restricted to informing and surveying employees. Nevertheless, with strongly networked and decentralized enterprises it is also possible for employees to participate continually in the planning, especially when restructuring the system due to changes in the product and technology.

This type of self-organization and participation fundamentally changes the relationship of employees to the enterprise. Instead of extensive targets and controls, flexible, result oriented control instruments arise—among which is the so-called 'management by objectives'—and new responsibilities and roles result for employees (see Fig. 3.13).

The previous dominating focus on functions resulted in highly regulated work for the individual employees structured according to their profession. The responsibility for processes and results decreased the closer the employee was to the actual product. They were paid according to output and presence. With an increasing focus on results, new roles arise for employees in an organization structured according to teams. An individual becomes a group member and

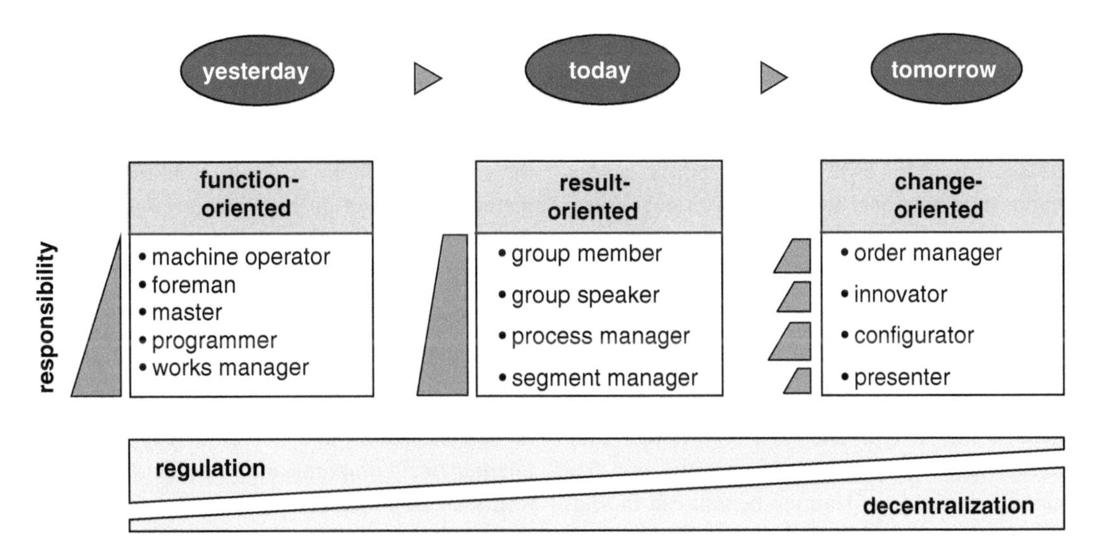

Fig. 3.13 Changes in the roles of employees. © IFA G8894SW_B

possibly a group representative. A process or segment leader coordinates a number of such groups. The responsibility is shifted lower, individual profiles regarding abilities, experiences and responsibilities are created. The extent to which additional forms of self-organization develop remains to be seen. Plausible roles include order managers, who act on behalf of a specific order spectrum such as an enterprise, supported by innovators who develop new products and processes as well as configurators who arrange customized products and moderators who accompany change processes [Brö00].

A further recommendation developed by Wirth is referred to as competence cell [Wir00]. A competence cell is a factory's smallest, changeable, value-adding unit. It consists of a multi-lateral collaboration of people equipped with resources and skills who offer their services as an enterprise within a skill network (cited according to [Sche10], p. 365ff).

The extent to which such roles are taken on and successfully implemented is strongly dependent on the willingness and ability of an enterprise to change. However, despite the fact that it will barely be possible to plan and operate production facilities and highly productive/changeable plants in the future without intensive employee participation, with increasing pressure to succeed, open and hidden opposition is to be expected. This resistance results from uncertainties, fears and suspicions about losing income, power and prestige. The promise of more autonomy and sovereignty stands face to face with the uncertainty of work relationships and job security. Moreover, in decentralized, autonomous organizational structures traditional career paths such as skilled laborers or operations engineers are called into question. Finally, it also has to be kept in mind that there is the loss of a 'professional home', so to speak, for the individuals. They are often the only specialist for a specific method or technique on their team, thus there is the danger that their knowledge may quickly grow outdated. If the teams are frequently changed, no learning benefit can unfold, and there is the additional threat of losing an 'organizational home'; the sense of belonging to a specific group is lost and social contacts languish.

In order to develop and obtain the indisputable advantages of self-organization within the field of conflict between dynamic and stable production environments and to overcome obstacles and dangers, courage, determination and a few rules are required: Goals and procedures need to be transparent, agreements need to be made concerning the employment of the staff after the restructuring, employees need to participate not only in operational considerations but also in management's strategic deliberations, communication needs to be 'top-down' as well as 'bottom-up' and results need to be transparent communicated [Brö00]. Above all a strongly established 'culture of trust' is required among all those involved—this usually develops gradually over time and needs to prove itself.

Establishing a model of participation is an imperative component of a changeable factory. The process for developing this is frequently underestimated with regards to duration and staff intensity. Moreover, it generally entails many more conflicts than creating a new manufacturing concept or new procurement logistics. Integrating employees early on in the factory planning process is therefore strongly recommended.

3.6 Communication

Participation, however, should not be limited to planning processes; rather it must continue with the factory operation as well. Routine tasks are increasingly taken on by machines and computers nowadays. Employees focus now more on special tasks and continuous improvements. These tasks are characterized by their open results and strong contextual dependency. Successfully addressing the lack of clarity with respect to input and output requires a high degree of coordination and communication. Here, interpersonal communication is superior to other forms of communication. Because the participants share a common perceptual space, the content is supplemented with additional

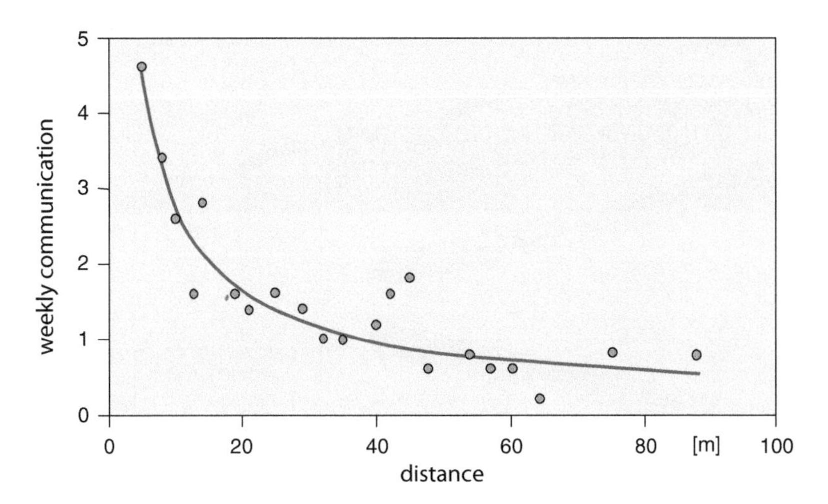

Fig. 3.14 Probability of communication as function of the distance between workplaces (Allen). © IFA 14.781_B

information and linked with individual know-how, whereby, their physical presence can prevent misunderstandings and allow more specific decisions to be made.

This creates a series of organizational requirements for production. The starting point is a communication concept that defines all of the basic communication forms within the factory. There are two approaches to this: First, the company organization can determine the form or flow of communication in the factory through the flow of processes and the structural organization of partners e.g., favoring flat hierarchies and the integration of indirect work content in the communication. Second, factory planning can determine the physical possibilities of the communication. Thus, shorter distances between communication partners accelerate and support the creation of a comfortable work environment, just as establishing places for communication and speeds up communication processes. In doing so the spatial development of a building influences very much the communicational behavior of those involved. In particular, research has shown how spatial proximity is correlated to the frequency of communication. Figure 3.14 depicts the probability of communication between two people as a function of the spatial distance [All07].

Whereas the layout of the processes can make a factory conducive to communication, the architecture can support this foundation by structural means. Thus for example, spatially integrating indirect areas in the production facility (frequently in the form of a gallery concept) is considered state of the art as is arranging information stands, meeting points and conference rooms within the factory. These aspects will be discussed in more detail in Chaps. 9 and 10.

3.7 Networking and Cooperation

It has become clear in the preceding sections that in addition to technical/logistical considerations soft aspects such as the organization and future roles of employees are critical factors for the success of a changeable factory. This is strengthened by the fact that, since the early 1990s in reaction to increasing diversity and rate of change, the organization of production has gradually become more decentralized (see Fig. 3.15, [Win01, p. 11]).

Based on the strongly hierarchical form of the company organization, lean production helped to develop small, increasingly independent profit and cost centers, supported by group work and

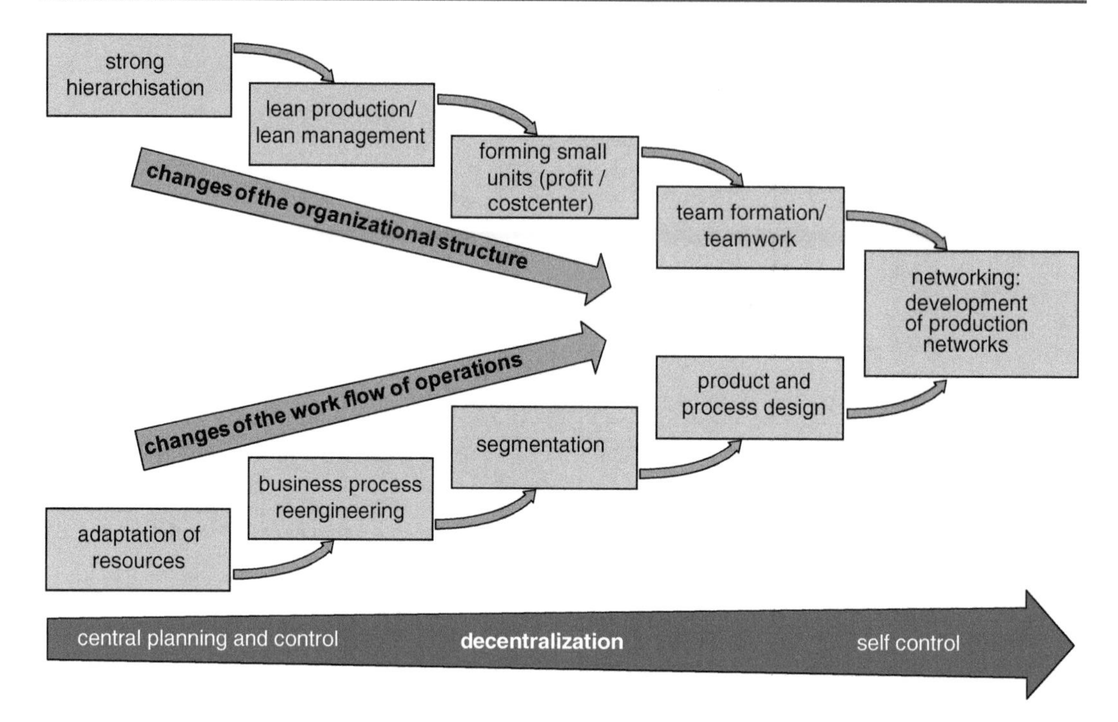

Fig. 3.15 Phases in decentralizing production (Windt). © IFA D5007ASW_Wd_B

team building. Parallel to this workflow management was simplified. Starting with adapting the resources primarily in view of reducing overhead costs, business processes were radically reorganized along the value adding chain. Largely autonomous mini-factories were created within the factory from a number of product/ market combinations. Consequently, products and processes were also frequently redesigned to be more modular.

With this gradual development of independent business divisions, while concentrating at the same time on key skills and shifting the remaining tasks to external suppliers and service providers, it was a natural step to grow co-operation between enterprises, whether temporary or long-term. These collaborations are no longer limited just to production; rather, more and more alliances for purchasing, supplying and development can be observed. Beyond the pure logistic chains in the sense of traditional customer/supplier relationships, stable network arrangements are formed and, in a further step, changeable production networks as well. With changeable production networks we are referring to business alliances that temporarily and dynamically configure themselves [Wien96]. Suppliers and customers become value-adding partners who are already tied into the development of the products and processes.

It is now obvious that traditional production planning and control (PPC) has to change in order to find decentralized solutions. This is supported by Internet technologies that are constructed according to the principle of decentralized processing power, data transport and data storage. The possibility that all objects in the real world are a part of the Internet was investigated as a vision within a project called the "Internet of things". This is realized when objects are permanently connected to the digital world using microcomputers and RFID tags (also referred to as smart tags). The objects are then able to find their way by means of agent technology [Bull07]. In Germany this concept has been defined by the government under the label Industry 4.0 as part of the so called high tech strategy.

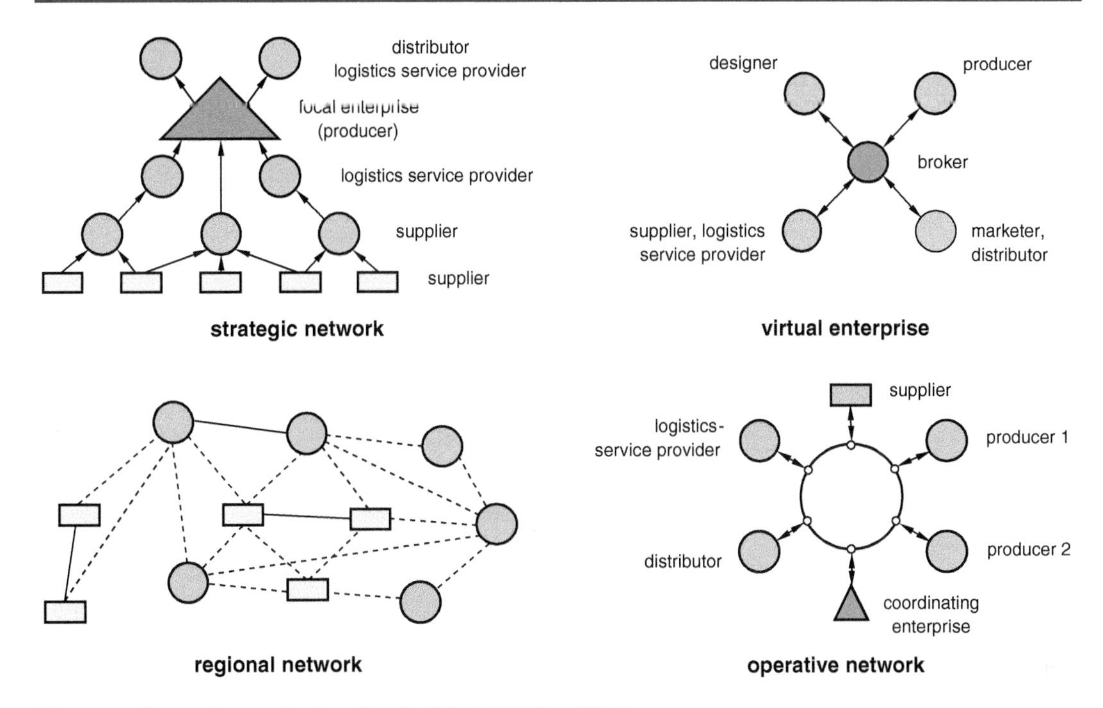

Fig. 3.16 Types of networks (acc. Pfohl). IFA D5087SW_Wd_B

One of the fundamental characteristics of changeable production networks are consciously maintained redundancies, so that a number of partners in a network can deliver the same output. In order to eliminate capacity bottlenecks or to avoid doubling large capital investments, it is possible to share resources among certain network partners. Moreover, it is also typical of such partnerships that functions are divided up among them. This can mean that individual partners concentrate on key skills or on bundling functions, for example in managing purchasing. Finally, it is possible to be tied into a number of different networks [Win01]. These production networks form themselves for different reasons and according to Pfohl can be categorized into four types (see Fig. 3.16, according to [Pfo04]).

Strategic networks are led by an enterprise which is usually the center of the network—usually an end product manufacturer or commercial enterprise with a close proximity to customers. These networks have been pioneered by automobile enterprises which closely connect their suppliers contractually. There is no 'real' partnership however as the one-sided advantages and dependencies are too great. *Regional networks* bundle specialized small and mid-sized firms that activate their relationship case-wise, yet still clearly compete against one another. Due to relationships with local firms, there are competitive advantages compared to competitors farther away. From the outside such a local alliance appears as one large enterprise. In co-*operative networks* partners use a network-wide information system to access services provided by the other partners, whereby the focus is on manufacturing and logistic capacities. Frequently, a number of partners are in the position to execute the same process without them being competitors; the product spectrum is usually different. Finally, *virtual enterprises* collaborate temporarily and in a project-like manner based on a common understanding of business in order to benefit from an opportunity. They remain independent, but appear to the customer as a single unit. Products with short lifecycles such as fashion and toys, but also software and electronic products are examples of where this might be applied.

For a factory, being integrated into a production network, provides a further critical influence

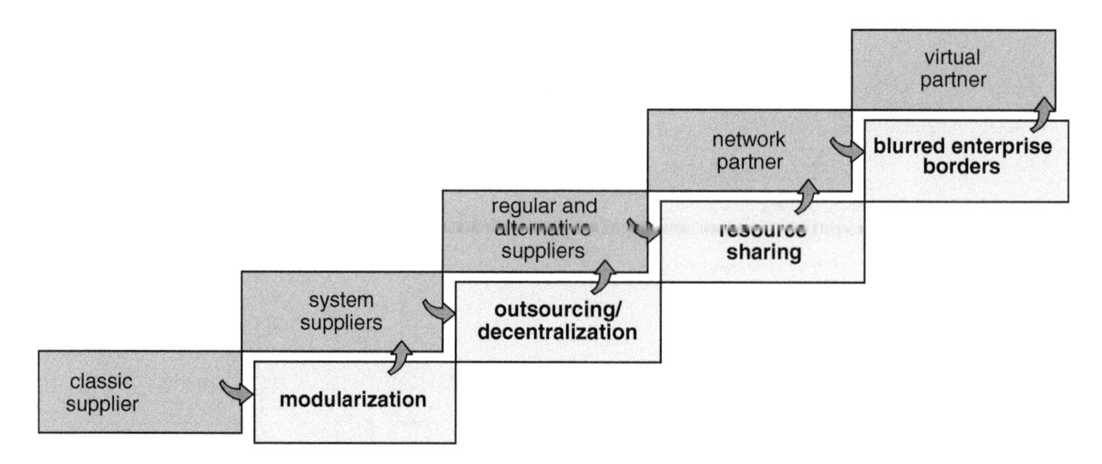

Fig. 3.17 Stages of co-operations (Windt). © IFA G3676ASW_Wd_B

on its design as it may be necessary to change more quickly than in the case of just changing a product or a process. There is also another problem from the perspective of co-operations: Networks are after all not formal organizations with standardized processes and regulations, but rather comparably loosely coupled communication structures. Sociology considers networks to be independent social structures established between enterprises and markets, which through their flexibility and openness allow participants to form an image of their own way of operating and to align expectations and intentions with one another accordingly. Uncertainties, risks and information deficits are thus more controllable than in formal organizations.

The character of networks such as these is determined by the degree to which participants give up their autonomy. In view of these aspects, typical stages of co-operations can be found in industry (see Fig. 3.17 [Win01, p. 1]).

Based on the tradition of suppliers as purchasers of finished and semi-finished products, system suppliers with close contractual obligations arose as a consequence of product modularization. With increasing sub-contracting and decentralization, suppliers were divided into regular suppliers and alternative ones. A generally new step is the common planning and use of resources in a network of equal partners. With growing transparency and intensive collaborations the enterprise alliances become blurred.

Due to the different organization's cultures and established procedures, surprising situations occasionally arise for the participants in such networks. They represent a new type of challenge for enterprises, which, due to the limited possibilities of formalizing the partnership or regulating it contractually, can only be conquered by gradually developing trust; trusts binds without the obligations in a formal sense. It is developed by investing time and personnel as well as through personal contact. Stable, long-term co-operative relationships are vital.

However, the trusted, longstanding partner also does not have to produce the most innovative and productive solution in a product network designed around a project-like, short-term collaboration. The challenge then is to balance the conflict between the desire for reliable relationships and innovative solutions through new partners, who also bring with them new risks.

Generally, it can be assumed that in the product networks discussed here: (a) the reciprocity of interests serves as a means of coordination, (b) discursive negotiation processes are a form of coordination, (c) and they are based on mutual interests. Moreover, conflicts are resolved in a negotiation process based on the partners' individual power of influence and the network is controlled and regulated with a focus on earnings (which are then divided according to the agreed upon rules).

In the organizational layout of the factory, the aspects briefly outlined here play a key role in shaping procurement as well as in dimensioning, planning and controlling capacities.

3.8 Demographic Development

An important factor in designing production is also aging populations. In Germany, the structure of the population is regularly calculated by the Federal Office of Statistics. The numbers calculated in 2006 for the next four decades are depicted in Fig. 3.18. They predict a continual decrease in the population and above all a clear change in the age structure. According to it, the age group of 20–60 that is available for working in production will decrease from the peak value of 46.3 million to 35.4 million people. This also means that the mean employed age will climb in the long run. Thus, for example, Volkswagen determined that the employees who built the Golf IV in 1998 had a mean age of 38.9 years. Ten years later this had climbed to 42.2 and without any countermeasures will climb even further to 47.1 in 2018. The *Zukunftsreport demographischer Wandel* (Future Report on Demographic Changes) presents important research that sheds light in particular on the consequences with regards to capacities for innovation [Pac00].

Since many enterprises fear their employees will be less productive due to this development in age, they are faced with the challenge of developing strategies for an aging workforce that take into consideration the deterioration of physical abilities and skills. There are three known models for this [Ger07]. The *deficit model* assumes an unavoidable physical and mental decline and reacts with early retirement and age oriented workplaces. The *skill model*, developed in the 1990s, recognizes the decrease in physical abilities, but emphasizes the rich experiences and abilities for problem solving. Finally, the *difference model* separates the calendar age from the biological age and considers the biological age as resulting from the individual talents, education and physical constitution and above all work histories.

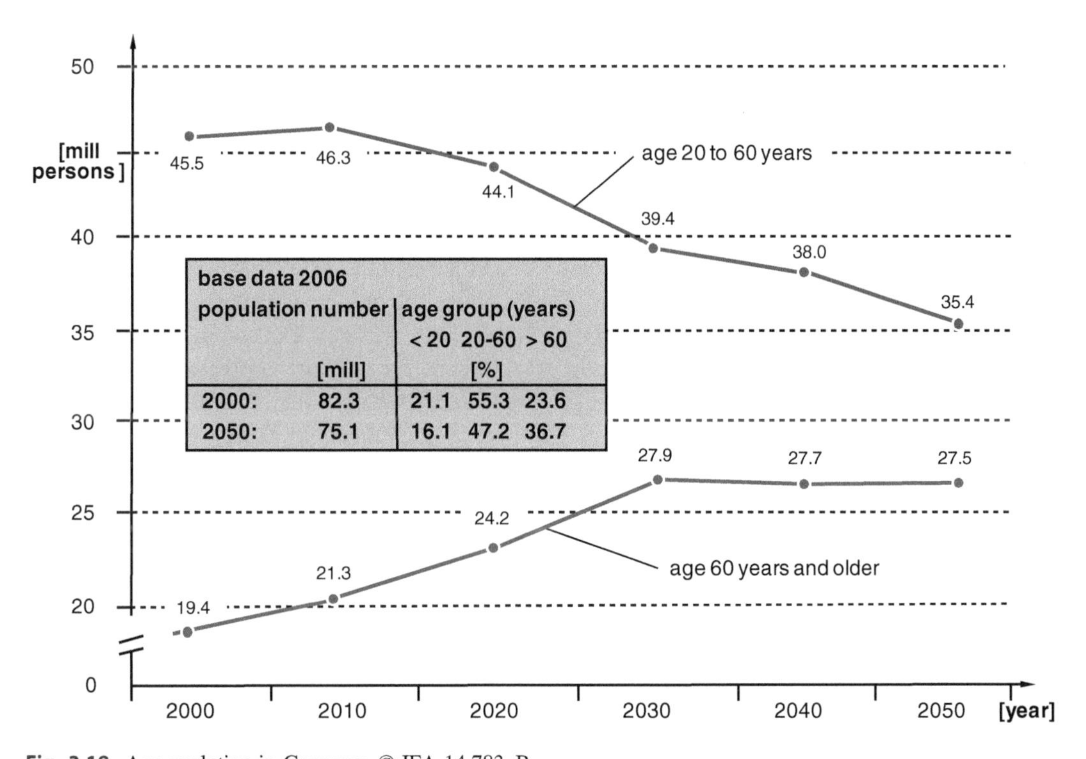

Fig. 3.18 Age evolution in Germany. © IFA 14.783_B

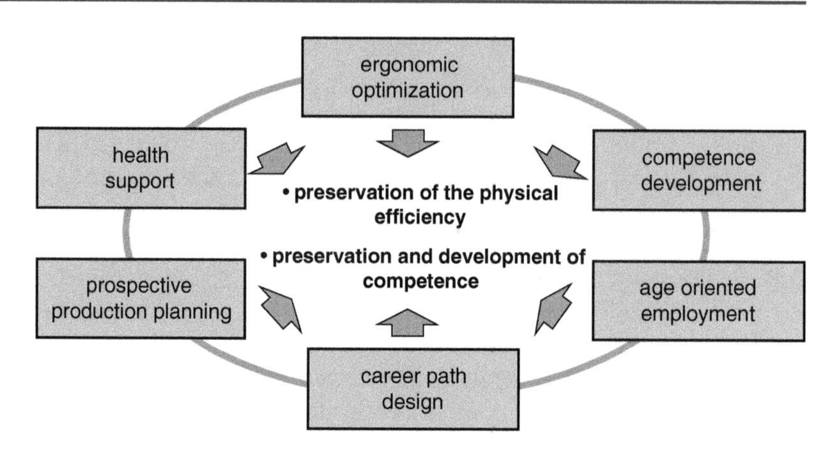

Fig. 3.19 Aims and Measures for age-oriented work design. © IFA 14.932SW_B

Based on the latter model, a workgroup recommended the measures outlined in Fig. 3.19 for age-oriented work design. However, in order to maintain abilities and skills these should be implemented much earlier than when employees reach the age of 50 [Ger07].

Ergonomic optimization focuses on preventing bad physical postures, while developing skills is aimed at continually upgrading qualifications and providing an environment conducive to learning (e.g., through group work). Job rotation and ergonomic workplaces suited for elder workers support the age-oriented management. By focusing on careers, a holistic approach is pursued. This facilitates an extensive change in stress and demands over an entire career. The age-oriented approach to production planning is particularly relevant for factory planners. It meets the challenges of the foreseeable change in age structure with adaptable workplaces and work content. Finally, preventatively promoting health should be taken into consideration, with sport programs, back training workshops, nutrition seminars etc. providing the incentive to actively focus on health and aging. More details will be explored in Sect. 7.8.

3.9 Corporate Culture

3.9.1 Organizational View

The preceding sections have shown that internally oriented self-organization and participation as well as externally oriented networking and cooperations are imperative prerequisites for managing a complex and dynamic environment. However, whether or not the enterprise is able to successfully change from a hierarchic, bureaucratic structure to a more open, spontaneous and venturesome behavior depends less on organizational structure and management systems than on the corporate culture.

Very simply put, corporate culture refers to all of the values, goals, perceptions, concepts, symbols, visions, models, myths, ways of thinking and behaving accepted by people in an enterprise as a common basis for their actions [Blei96]. Accordingly, corporate culture evolves and shapes employee attitudes towards their responsibilities, the product, their colleagues, management and the enterprise. According to Schein, it is realized on three levels (see Fig. 3.20, [Schei84]).

On the artifacts level, behaviors, presentation forms (e.g., clothing, architecture, offices etc.), rituals and symbols are communicated, which sometimes can only be interpreted in view of the level below. This level is comprised of all the underlying values and norms that usually unconsciously guide the behavior and actions of those in the organization. On the lowest level are the unquestioned, fundamental assumptions about the business environment as well as the actions and relationships between those connected with the enterprise. The way in which corporate culture can be reflected in various forms of expression is depicted on the basis of a few examples in Fig. 3.21 [Blei96].

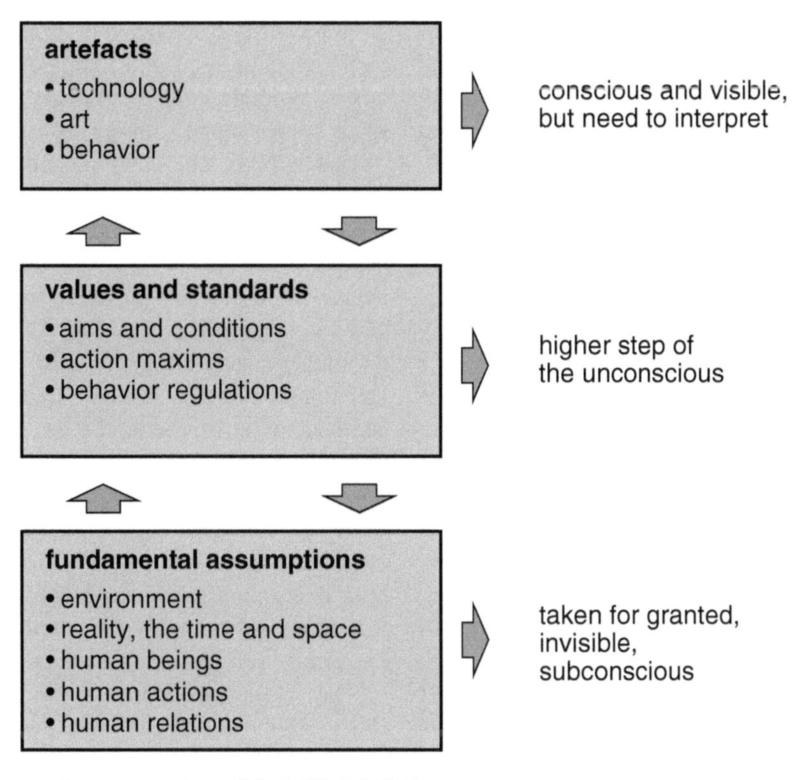

Fig. 3.20 Levels of corporate culture. © IFA G8905SW_B

Fig. 3.21 Expressions of corporate culture. © IFA G8907SW_B

aspect	characteristics		
communication:	formal	⟺	informal
dealing with criticism:	searching for faults	⟺	fault as a chance
cooperation:	lone fighter	⟺	team
personal relationships:	hierarchical and title emphatic	⟺	uncomplicated fact oriented
promotion method:	unclear	⟺	transparent methods
customer evaluation:	letting down disparaging	⟺	respect oriented
information policy:	gossip factory	⟺	fast, relevant information
identification:	malicious jokes dissociation	⟺	proud of the company

The culture is usually convincingly shaped by the personality of the founders and works to create a corporate identity both internally and externally. Its further development is dependent on history and the example set by management. Corporate culture however does not stand alone; rather it is also part of a national and branch specific culture and thus contains a number of sub-cultures. What seems to be essential is that the business policy and strategy are aligned with the values and norms of the corporate culture [Blei96].

Corporate culture thus represents a type of filter for the perception of internal and external demands and is as far as that goes restrictive, because changes may, if at all, be recognized too late or reacted to wrongly: The typical response of a traditional corporate culture is 'we've been successful for a century, why do we have to call everything into question now?'

The *expression* of corporate culture varies extraordinarily and is very dependent on how long those involved have been working together and how homogenous they initially were; it is thus always unique. Nevertheless, attempts at finding a structural typology can be found in the extensive publications on this topic. Bleicher formulates these in a number of dimensions [Blei96]: These concern (a) the openness (external/internal, change resistant/change friendly), (b) diversity (cutting edge/basic, unified/diversified), (c) influence of management on the culture (instrumental/development oriented, cost/benefit oriented) as well as (d) the impact of the workers on the culture (member/actor, collective/individual).

From there, Bleicher describes both the opportunistic and the obligatory corporate cultures. The first is characterized by a minimally reflective, tradition based, insular management style, which forces employees into a regimented fulfillment of tasks by quantifying all operations and by pursuing a technocratic, cost-oriented approach. The staff turns to pursuing opportunistic values; objective and social needs take second place behind urgencies. In comparison, the objective and socially obligated corporate culture proves to be open and ready for changes: it reacts sensibly to environmental changes, tolerates diverse subsystems with a basic focus and puts benefit before costs.

Against the background of changeability, corporate culture becomes extremely important. It is long lasting; however it is also subject to change—especially in critical situations. It seems pragmatic for a culturally-aware management to consciously nudge a necessary change in corporate culture, but to not expect that every step along the way can be planned [Blei96].

In order to objectivize this strongly emotional topic, it can be helpful to compare the actual and target corporate culture, whereby the corporate focus is considered with regard to the customer/market, strategy and future, products, innovation, technology, enterprise, results/output and costs. From this, Gausemeier derives a portfolio that compares stable, old and new cultural components with respect to their future significance and current characteristics (see Fig. 3.22 [Gau99]).

Whereas stable components are important for the future and are already established, new components still need to be developed. Moreover, old components though still strongly pronounced lose their significance in the future. Irrelevant components, like pure cost focus, do not contribute to future success. The process of both developing new components and eliminating old ones is primarily the job of management, frequently with the aid of external consultants.

The closer the strategy for formulating changeability is aligned with the organization that would implement it, and the greater the corporate culture is aligned with the inner agreement of the participants, the more likely the enterprise's changeability can be increased. Only after an enterprise reflects about whether or not an organization of their kind is in the position to want changes, to make those changes and to maintain them, can the desired values and attitudes truly be attained. Here, it is important to explore cultural questions in order to find a common accepted understanding of certain values. These values have to be lived, supported by incentive and sanction systems.

3.9.2 Architectural View

An important possibility for expressing corporate cultures exists in aligning the factory's appearance with the corporate culture through the architectural master plan. A visitor's first impression already unconsciously shapes their attitude and deepens when they enter into the buildings, factory halls and offices. For the factory planner, these 'soft' sides of planning need

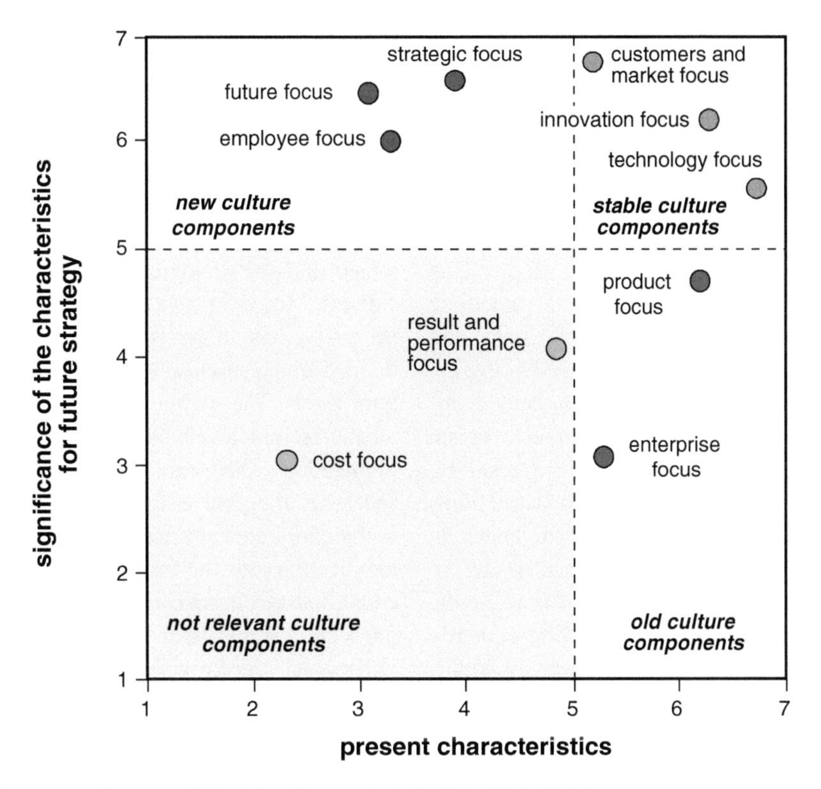

Fig. 3.22 Corporate culture portfolio (after Gausemeier). © IFA G8916SW_B

to be taken into consideration with respect to distinguishing the enterprise from their competitors. We would like to introduce this topic next. Later, in Sect. 11.5, we will discuss it further with respect to impressions and aesthetics.

The design of a factory building cannot be derived from the production requirements alone, rather it grows from a creative process in the context of the site, climate, society and people. Beyond the pure functional suitability, a practical building structure provides a positive force for motivation and communication [Rei05].

Unfortunately, the inhospitableness of industrial and commercial areas shapes often the appearance of our cities and landscapes. Confusion between the economic goals 'cheap' and 'cost-effective' justifies anonymity, banality and ugliness. The 'appearance' of many enterprises across the country seemingly has been patched together from Do-it-Yourself-stores and mistakes 'simple' for 'simple-minded'. The architectural critic C. Hackelsberger accurately refers to these areas as "commercial steppe". No one willingly stays here longer than the paid work time. The buildings and the space leftover between them are accepted without complaint as a social 'no-man's-land'.

Industrialists 'ruin' their own chances for the future by making short-sighted, strategically unwise building decisions, because the next change in production often requires relocating the enterprise. What remains are ecologically damaged, fallow areas which permanently scar the city and landscape.

Small budgets, tight construction schedules and sequential rather than cooperative planning prevent the natural development of a good project. At the same time, industrial construction in particular is a special domain of architecture, which far from the fashions of traditional building projects has maintained an inner freedom. It is open to new technologies, constructions and materials, and represents an extremely interesting task in the search for new concepts.

A key question from an architectural perspective as well as the assessment of changeability is regarding the relationship between the form and function of a building. Following [Ben78], architectural theory developed two seemingly diametrical approaches to finding form. Based on the American architect and theorist Louise Sullivan, at the end of the 19th century, the slogan "form follows function" marked functional necessity as the reason and expression of formal building design. During the peak of the Bauhaus movement, modern architects tried to overcome the ties of eclecticism with this slogan. In reaction to the resulting aesthetical banality of later box constructions, many architects promised greater diversity in architecture and a formal dominance of design with the motto "function follows form" in the second half of the 20th century. Consequently programs and processes were 'designed to fit' predefined geometries.

Neither of these strategies is particularly productive for the questions we are dealing with here in regards to constructing changeable factories. Each of them only considers one dimension of one criterion of the complex correlation between the environment, humans, functions, and form. Often during a project, a question is raised as to which of the current functions and forms will last in the long run. A snapshot of a temporary production program or the fashion of a short-lived, aesthetic zeitgeist is not well suited for determining a robust design. Rather, what is needed are holistic solutions developed equally from a process (function) point of view as well as a spatial (form) point of view.

Thus, it is necessary to find a consciously positive bundle of traits that provide many, preferably mutually complementary partial answers to complex questions. Following in the footsteps of the American engineer, Buckminster Fuller [Kra99], the result of this approach can be characterized by the concept of "performance". The "form follows performance" strategy [Rei05] derived from this is aimed at a comprehensive answer to finding a form in response to a holistically composed problem.

The actual formal impression of the building is not set, but rather results from the spatial solutions to the specified performance issues. Then for example, based on each of the underlying visions: (a) new construction technologies which are supportive of project aims should be used, (b) energy consumption should be optimized and (c) ecological issues should be considered. Moreover, flexibility, which is a known necessity, has to be ensured in the form of a defined changeability across all of the architectural levels. The spatial design and furnishings of all the factory levels should also promote communication between personnel. Generally speaking, the goal is to recognizably contribute to the corporate culture and to building a sense of identity through the factory construction. These collaborative efforts can result in highly efficient production shops, with well-proportioned spaces, interesting construction and comfortable workplaces.

3.10 Sustainability

3.10.1 The Term and Concept

In 1972 the book "Limits of Growth" first drew attention to the foreseeable exhaustion of natural resources on earth and triggered a debate that continues to this day [Mea72]. Thirty years later, its predictions have for the most part been confirmed [Mea04]. The authors' findings have triggered world-wide efforts not only to reduce the consumption of energy, but also to protect the environment as a whole and to establish rules for the responsible use of its resources. In 1987 this subject was addressed in a report from the UN World Commission on Environment and Development entitled "Our Common Future" [UN87]. Within this document is what is recognized today as the fundamental definition of sustainability: "Humanity has the ability to make development sustainable to ensure that it meets the needs of the present without compromising the ability of future generations to meet their own needs".

The report lead to the UN Conference "On Environment and Development" in Rio de Janeiro in 1992 and the program for action referred to as "Agenda 21". Then in 1997, the World Climate Summit held in Kyoto was aimed at setting binding targets for greenhouse gas emissions in industrialized countries. Finally in 2002, the World Summit on Sustainable Development (WSSD) was held in Johannesburg. However, in the eyes of many observers, it failed to create significant concrete governmental policies regarding sustainability targets (http://www.worldsummit2002.org/).

Non-Government Organizations (NGOs) focus on increasing awareness about the impending impact of declining resources of all kinds. This included organizations such as the World Commission on Environment and Development (WCED), World Wide Fund for Nature (WWF) and the Global Footprint Network (GFN). In Germany, there is the German Environmental Agency (DBU) (http://www.dbu.de/), the German Advisory Council on the Environment (SRU) established in 1971 (www.umweltrat.de) and since 1992, the German Advisory Council on Global Change (WBGU) (www.wbgu.de).

On the one hand, these efforts have resulted in management rules for the sustainable use of renewable and non-renewable natural goods as well as for release of substances and energy [SRU96]. On the other hand, measureable targets have been developed in relation to the material intensity of the economy and resource productivity, e.g., the MIPS factor (material input per unit service) [Sch10], TMR (Total material requirement) (http://scp.eionet.europa.eu/definitions) or TMC (Total Material Consumption) (OECD Glossary of Statistical Terms: http://stats.oecd.org/glossary). A large number of basic environmental protection principles are already legally binding in many countries.

There can only be a change in global thinking when governments and corporations take up the challenge of sustainably managing the earth's natural resources. In particular, economic enterprises that are more focused on profits have to incorporate ethics for responsibly dealing with resources into their managerial principles. The following ten principles developed by the UNGC (United Nations Global Compact) describe this approach in concrete terms and set environmental protection within the larger context of responsible management. They start by addressing human rights and humane working conditions in principles 1 through 6, followed by environmental protection in principles 7 through 9, and end with a call to prevent corruption of any kind [http://www.unglobalcompact.org/] (Fig. 3.23).

An important parameter within this framework is the so-called 'ecological footprint'. "The ecological footprint is an indicator of human pressure on nature. It measures how much land and water people need to produce the resources they consume (like food and timber), provide land for infrastructure, and absorb the CO_2 they generate and then compares this to the biocapacity, i.e., nature's ability to meet this demand" [GFN10].

In a global overview of debtors and creditors (in terms of ecological balance) Fig. 3.24 illustrates that the industrial nations as well as a number of countries in the Near East and Africa live clearly above their means, because their footprint is to some extent dramatically larger than their biocapacity [GFN10]. It should also be kept in mind that the consumption of natural resources has doubled in the last 40 years [Pol10]. For over 20 years now, humans have consumed 25–33 % more per year that the earth can regenerate.

Figure 3.25 clarifies this fact by placing the nations within a coordinate system. The horizontal axis is the United Nations Human Development Index (HDI) and the vertical axis is the ecological footprint [GFN10, WWF10]. As the Human Development Index HDI serves a weighted mean of sub-indices for life expectancy at birth, mean years of schooling and expected years of schooling and the GNI (Gross National Income) per capita. A high value is considered to be approximately 0.67, while a very high value is 0.79. The ecological footprint is measured in hectares per person. Whereas the world mean for this value was 4.5 hectares/person in 1961, due to the population growth it had sunk by 60 % to 1.8 ha/person by 2008.

• **Human Rights**

Principle 1: Business should support and respect the protection of internationally proclaimed human rights; and

Principle 2: make sure that they are not complicit in human rights abuses.

• **Labour**

Principle 3: Business should uphold the freedom of association and the effective recognition of the right to collective bargaining;

Principle 4: the elimination of all forced and compulsory labour;

Principle 5: the effective abolition of child labour; and

Principle 6: the elimination of discrimination in respect of employment and occupation.

• **Environment**

Principle 7: Business should support a precautionary approach to environmental challenges;

Principle 8: undertake initiatives to promote greater environmental responsibility; and

Principle 9: encourage the development and diffusion of environmental friendly technologies.

• **Anti-Corruption**

Principle 10: Business should work against corruption in all its forms, including extortion and bribery

Fig. 3.23 UN global compact's ten principles

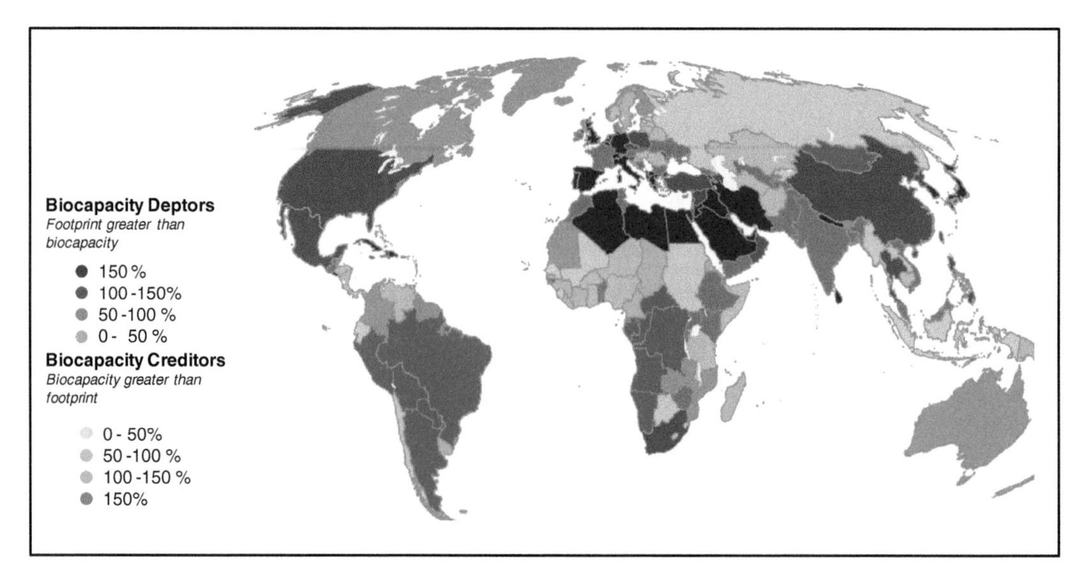

Fig. 3.24 The Ecological wealth of nations. http://www.footprintnetwork.org

There are four HDI thresholds: low, medium, high and very high. The graph clearly shows that no country with a very high HDI and only few countries with a high HDI meet the minimum criteria for sustainability in the sense that it does not exceed the world average available hectares per person. The countries are divided into two groups: One of the groups, primarily Europe, the USA and a few Asian countries, consumes far too many resources. The other group, primarily

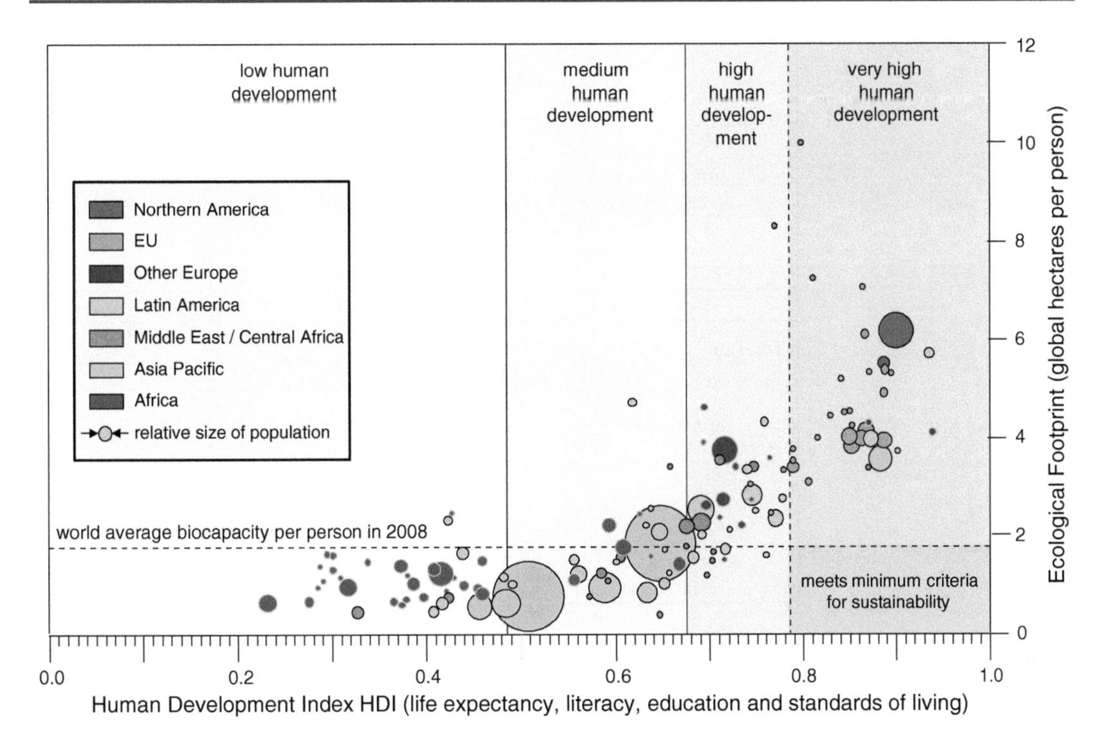

Fig. 3.25 Ecological footprint and human development (Global Footprint Report 2010)

African, Asian and Latin American countries, lives below a desirable standard of living and consumes comparably little resources.

In 2011 it was estimated that the world population had reached 7 billion and by 2050 it is expected to be 9.5 billion. If, given current production methods and consumer behavior, the quality of living for these people, especially in countries with transitional economies (e.g., China, Brazil, and India), is to be maintained or reached, every responsible measure of consumption will be exceeded [WWF10]. The Living Planet Report 2014 states that "The Ecological Footprint shows that 1.5 Earths would be required to meet the demands humanity makes on nature each year" [WWF14]. Only by embracing sustainability from an economic, ecological and social perspective as a future principle can the quality of living be improved and the natural resources and ecosystem be preserved for future generations [Jov08].

The worldwide increase in energy consumption is just as dramatic as the decreased biocapacity per person. Based on statistics gathered by British Petroleum BP, Fig. 3.26 depicts the consumption of primary energy in 2000 and 2010 for 6 regions of the world and the top ten countries. Of these, the first four countries are responsible for almost exactly 50 % of the world energy consumption [BP11].

During the 10 years, the Asia Pacific region recorded the greatest increase in consumption with 42 %, and with 38 % of the total it was the largest consumer in 2010. China alone was responsible for 20 % of world's consumption, and thus slightly exceeded the USA which was responsible for 19 % of the world's consumption. It is worth noting the rate of increase in consumption for China (57 %), India (45 %), Brazil (27 %) and South Korea (26 %). These countries obviously have changed from so-called 'developing countries' to 'transition countries'.

Fig. 3.26 Consumption of primary energy (figures by BP)

	consumption [mill tonnes oil equivalent]		increase [%]	share [%]	
World	**2000**	**2010**		**2000**	**2010**
North America	2.757,2	2.771,5	0,5	29,4	23,1
South and Central America	464,4	611,9	24,1	4,9	5,1
Europe and Eurasia	2.821,4	2.971,5	5,1	30,1	24,8
Middle East	416,2	701,1	40,6	4,4	5,8
Africa	272,1	372,6	27,0	2,9	3,1
Asia Pcific	2.651,2	4.573,8	42,0	28,3	38,1
Total	**9.382,4**	**12.002,4**	**21,8**	**100,0**	**100,0**
Top Ten					
China	1.038,2	2432,2	57,3	11,1	20,3
United States	2.313,7	2285,7	- 1,2	24,7	19,0
Russian Federation	620,4	690,9	10,2	6,6	5,8
India	295,8	542,2	45,4	3,2	4,5
Japan	514,1	500,9	- 2,6	5,5	4,2
Germany	332,3	319,5	- 4,0	3,5	2,7
South Korea	188,9	255,0	25,9	2,0	2,1
Brazil	185,2	253,9	27,1	2,0	2,1
France	254,2	252,4	- 0,7	2,7	2,1
United Kingdom	224,1	209,1	- 7,2	2,4	1,7
Total	**5.966,9**	**7.741,8**	**22,9**	**63,6**	**64,5**

Note: consumption of primary energy comprises commercially traded fuels, including modern renewables used to generate electricity.
Figures from BP Statistical Review of World Energy June 2011

3.10.2 Consequences for Factory Planning

The concept of sustainability has an economical, ecological and social dimension. Economically it is sensible not to maximize short-term profits at the expense of the environment because otherwise one's own market will wither. Ecologically it is advisable not to consume more than can be renewed or substituted by other material. Finally, from a social perspective it is completely against humanitarian principles to exploit workers by overloading them and providing poor working conditions. "Sustainable development occurs when all humans can have fulfilling lives without degrading the planet" [GFN10].

Production management together with factory planners therefore have a large responsibility because they fundamentally determine productivity as well as working conditions and consumption of energy [Jov08]. Manufacturing methods, material flow, work organization and buildings are design fields that considerably influence the sustainability. Due to the energy problem discussed here, the subject of energy efficiency has become highly significant.

If we first consider the share of the gross production value that energy costs represent in Germany's processing industry, the importance of energy consumption at 2.4 % does not seem significant from a purely economic perspective. In the production plants addressed in this book (i.e., machine building and electrical engineering as well as the automobile industry) this is even as low as 0.8–1.0 %. However, when we take the gross value that has been yielded by the company itself as a basis—that is without costs for material and purchased items—the share of energy costs is on average 5.2 %. For the previously mentioned plants the corresponding value is 2.0 % for machine building, 2.0 % for electrical equipment and 1.3 % for the automobile industry [Stat11]. If savings in the range of 15–20 % were to be targeted, then these would have a clear impact on the cost structure (Fig. 3.27).

If we then consider the total value-adding chain or production network that lies before the manufacturing, the savings possibility is multiplied a number of times and it can represent almost 50 % of the raw material costs.

Because all production energy consumption also pollutes the air by emitting greenhouse

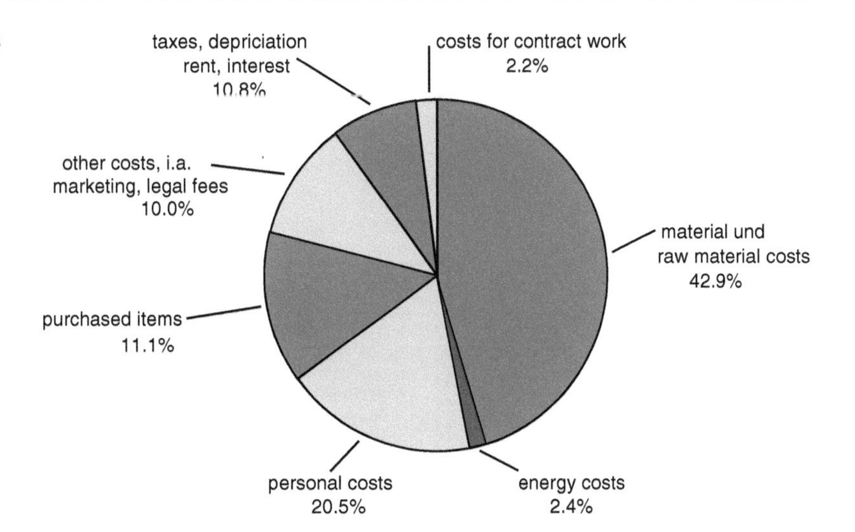

Fig. 3.27 Shares of gross production value in Germany's manufacturing industry 2009 (after [Stat11])

gases, its influence on the climate is also a critical subject. An indicator of this is the so-called carbon footprint. In the words of the European Platform on Life Cycle Assessment: "Carbon footprint (CF)—also named Carbon profile—is the overall amount of carbon dioxide (CO_2) and other greenhouse gas (GHG) emissions (e.g. methane, laughing gas, etc.) associated with a product, along its supply-chain and sometimes including from use and end-of-life recovery and disposal. Causes of these emissions are, for example, electricity production in power plants, heating with fossil fuels, transportation operations and other industrial and agricultural processes" [EU09].

A low carbon footprint has therefore become a strongly considered sign of quality for intelligently planned and resource-efficient construction projects. It is thus important to increasingly consider corresponding aspects when planning new production facilities or re-organizing existing ones.

In order to do so, the following management rules apply: Over the long term

- the rate of use of renewable natural goods must be greater than their regeneration rate.
- the use of non-renewable natural goods must be greater than the substitution of their functions.
- the release of materials and energy must be greater than the adaptability of the environment.

In order to apply these rules, a generic model for the flow of energy in a factory is required. Figure 3.28 depicts the definition of a factory system in view of the interactions between the production facilities and the factory building from this perspective [Her10]. It supplements the model of the material, information and value flow that will be discussed in greater detail in Sect. 15.4.2.

The starting point is the production equipment that, on the one hand, requires primary energy as well as compressed air, vapor and cold water. A portion of this is fed back into the energy recovery system but the majority of it consists of energy that escapes into the environment. On the other hand, some processes require special room conditions (e.g., temperature, humidity and cleanliness) that have to be supplied by building services. Moreover, in addition to the production facilities' requirements, the building itself has to meet needs in relation to the local climate and finally, the building must have an atmosphere that supports the health of the factory workers.

In the sense of energy efficiency a few design principles can be derived:

- The production processes and machinery represent the starting point. Here, in addition to targeting highly efficient processes the aim is to decrease idle time and prevent load peaks [Her13].

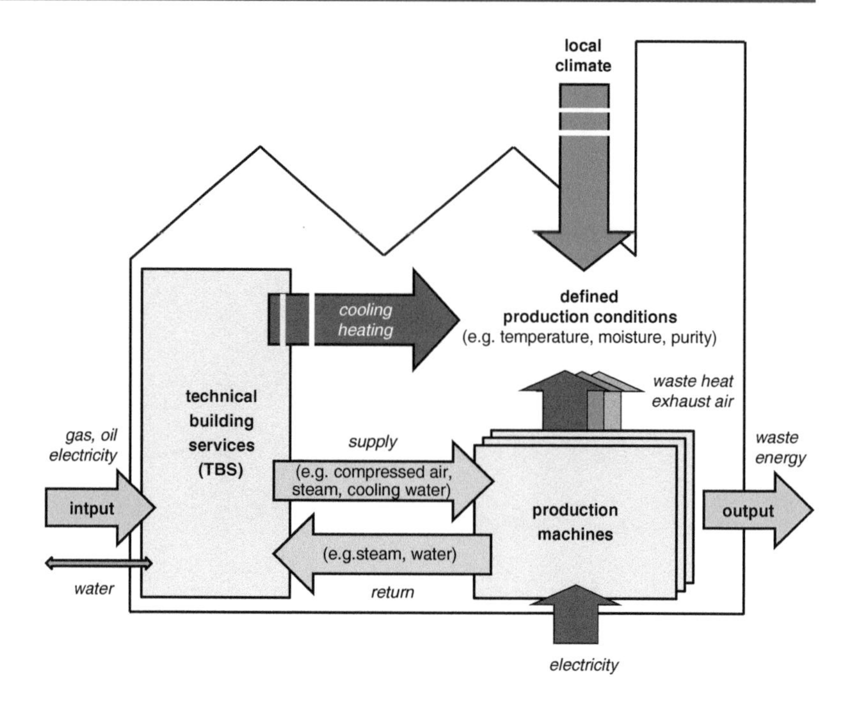

Fig. 3.28 Energy flow model of a factory (Herrmann)

- The building and building services should be energy neutral; i.e., it should be a 'zero energy' or 'zero net energy' building. The quality and conditioning of the space is then determined by the requirements for the processes and workplaces.
- The necessary minimum required energy for the building and its building services should if possible be covered 100 % from energy losses from processes and from regenerative energy earnings.

Currently, comprehensive certification systems in the sense of Green Building Standards for buildings and plants are being developed. This will be discussed in considerable detail in Sect. 15.7.

With all of the efforts being made to improve energy efficiency, in practice it is often obvious that exploiting the potential to reduce on-going energy costs both in newly constructed as well as renovated production facilities is generally related to comparably high investments, which in turn require long amortization times. With company sights on short amortization periods (e.g., 2 years) and limited financial means the necessary investments often are not made. Instead, companies seek measures requiring little investment; these, however, not only have little sustainable impact, but also fail to compensate for increasing costs of energy.

3.10.3 Recycling Economy

Our discussion up until now has focused on saving energy by designing the energy cycle to be as closed as possible. This applies both to the raw materials as well as to the implemented production means. However, the products themselves should also be designed so that they consume as few as possible resources that are detrimental to the environment during their use. Moreover, the components and materials contained in them should be reused as much as possible or recycled.

The term 'recycling economy' was coined for this approach. The various basic stations in the recycling economy are depicted in Fig. 3.29 [Sel97].

As can be seen, a number of cycles are involved in the recycling economy. In the first cycle, the goods used by the end customer after they have been developed, produced and distributed are consumed directly by further users up

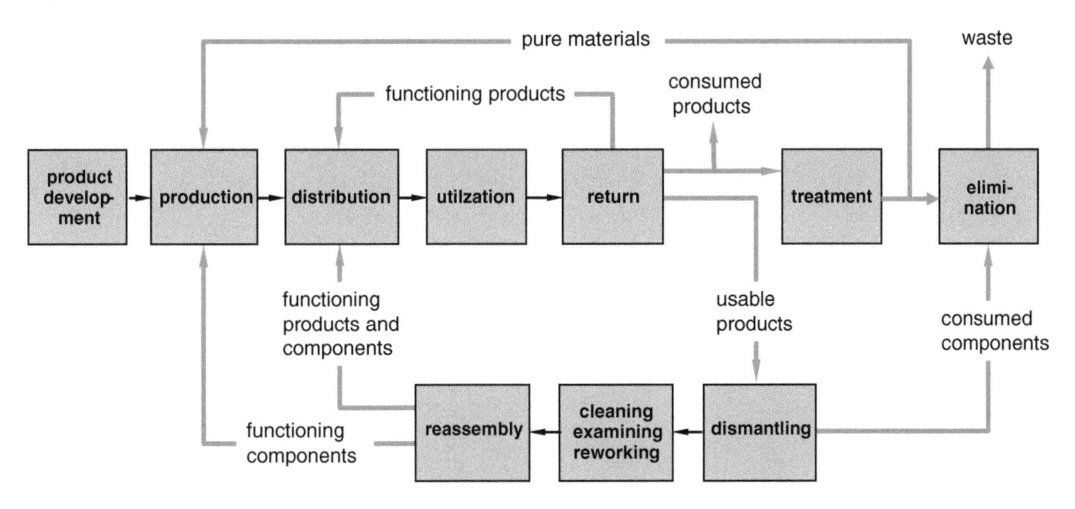

Fig. 3.29 Stations of the recycling economy. © IFA G8892SW_B

until the end of their useful life. Even when the product as a whole no longer functions, not all of its components have reached the end of their useful life, thus disassembling them, re-working them and re-assembling them can create products or components that are once more functional and that can again be placed in use. When it can no longer be reused, the product or component is then broken down with the aim of recovering pure types of raw materials that can in turn be reused for new purposes (e.g., copper or other metals) or safely eliminated (e.g. chemicals).

It is easy to see that with this approach, new streams of goods arise, that in turn further impact the environment. It is thus critical to develop a so-called 'material stream management' strategy based upon ecologically and economically balanced and optimized logistics.

The Recycling and Waste Management Act that was passed in Germany in 1994 and went into effect in 1996 comprehensively addresses this approach [KrW96]. Its essential points include:

- waste prevention is to be prioritized before recycling, and recycling before disposal
- resources should be recycled within the plant
- producers and distributors are responsible for products during their entire lifecycle
- products should be designed so that components and pure raw materials can be recovered for reuse and dangerous materials can be separated and safely disposed of.

Moreover, strategies for making products more durable, extending the life of their usefulness and intensifying their use are encouraged.

These and other laws, which impact a company's obligation to accept the return of consumed products as well as the workers and environmental protection, result in further conditions and concerns that need to be considered when designing a factory. On one hand, all of the factory's internal processes need to be re-thought in view of the internal recycling system. In particular, attention should be paid to manufacturing waste, e.g., metal chips as well as the ancillary and operating materials related to them such as emulsions, lubricants, grease, acids, alkaline solutions, etc.

On the other hand, the lifecycle of the equipment implemented in the factory needs to be considered, because the German Recycling Act also applies to them. Thus, to ensure that they are properly maintained and repaired and that components can be exchanged in order to extend their life, the equipment has to be selected accordingly. Moreover, the possibilities for maintenance and repair, disassembling the equipment and retrofitting also need to be ensured. Finally, the factory's structural framework, building shell, media supply, ventilation or air conditioning and lighting should be designed in view of their lifecycles.

For factory planning, these guidelines directly impact the design of the production processes.

responsive:	surpass customary market delivery times and delivery performance.
volume and variant flexible:	master volume fluctuations and product variants economic.
limiting value focused:	overcome known limits and make natural physical and logical limits the goal.
self organized:	carry out necessary structure and work flow changes at all levels initiatively and participatively.
network integrated:	link core competences across the enterprise dynamically and temporarily.
culture-conscious:	develop jointly accepted moral concepts and behaviors and transfer them convincing outside and inside the company.
sustainable:	design and operate products, production processes and production facilities during the complete life cycle obeying the principle of operating in an energy, resources and environmentally harmless way.

Fig. 3.30 Guiding principles for a future production. © IFA 15.052_B

By implementing an internal recycling system in the plant, resources such as raw materials, energy and the natural environment should be saved and protected in alignment with the abovementioned guidelines. Moreover, products should be designed so that they consume as few resources as possible, are not detrimental to the environment, and so that their components and materials can be reused or recycled as extensively as possible.

Furthermore, the increasing amount of used goods flowing back to producers raises the question of developing disassembly plants that could yield a service. Whether or not a disassembly and product recycling industry will develop in the future is not yet foreseeable. Initial experiences have indicated though that for a number of reasons integrating new production with industrial refurbishing in a single factory is not practical. Accordingly, it is anticipated that the maintenance, repair and customer specific refurbishing will develop further in the sense of a service and should be considered when planning new factories as well as when renovating existing ones. As preliminary result of our discussion, future demands on production can be expressed by the properties and guiding principles listed in Fig. 3.30.

3.11 Summary

Future design of factories has first of all to consider abrupt changes in the market, which in turn need high responsiveness and flexibility with respect to volume and variants. In order to enable the enterprise to meet this challenges it needs to focus on limits, involvement and empowering of employees, fostering personal communication and networking inside and across the firm. The demographic change has to be answered by designing age-oriented workplaces. All measures have to be aligned with corporate culture, as well as focused on resource and energy efficiency.

Our discussion here has shown that in a real factory, not all of these challenges can be equally met. In concrete cases, it is thus important to be able to recognize the reasons that speak for change. Such reasons are for instance insufficient delivery abilities and reliability, sinking market share and unsatisfying returns on capital. Forces for change that are induced by the enterprise itself are mainly due to changes in ownership or strategies, as well as basic changes in products and processes.

Before we develop concrete design aspects from these principles, we will introduce a few important factory concepts, which have developed over time and attempt to meet the challenges presented here in various ways.

Bibliography

[Abe06] Abele, E., Versace, A., Wörn, A.: Reconfigurable machining systems (RMS) for machining of case and similar parts in machine building. In: Dashchenko, A.I. (Hrsg.): Reconfigurable Manufacturing Systems and Tranformable Factories, S. 327–339. Springer, Berlin (2006)

[All07] Allen, T.J., Henn, G.W.: The Organization and Architecture of Innovation. Managing the Flow of Technology. Elsevier, Amsterdam (2007)

[Ben78] Benevolo, L.: Die Geschichte der Architektur des 19. und 20. Jahrhunderts. (History of architecture of the 19th and 20th century) Deutscher Taschenbuch Verlag München (1978)

[Blei96] Bleicher, K., Müller-Stewens, G.: Unternehmenskultur (Corporate Culture). In: Eversheim, W. und Schuh G. (Hrsg.): Betriebshütte, Produktion und Management, 7. Aufl., S. 2–38 bis 2–50. Springer, Berlin (1996)

[BP11] BP Statistical Review of World Energy. British Petroleum, London (June 2011)

[Brö00] Brödner, P., Kötter, W. (Hrsg.): Frischer Wind in der Fabrik. Spielregeln und Leitbilder von Veränderungsprozessen (Fresh wind in the factory. Rules and models of change processes). Springer, Berlin (2000)

[Bull07] Bullinger, H.-J., ten Hompel, M. (eds.): Internet der Dinge (Internet of things). Springer, Berlin (2007)

[Doe00] Doege, E., et al. (ed.): Potentiale der Grenzwertorientierung von Fertigungstechnologien und Abläufen (GreFA). Abschlußbericht zur Vordringlichen Aktion 6 des BMBF (Potential of the limit orientation of manufacturing technologies and processes (GreFA). Final Report to the first priority action 6 of the BMBF). Published by: WZL, RWTH Aachen (March 2000)

[Eid95] Eidenmüller, B.: Die Produktion als Wettbewerbsfaktor. Das Potenzial der Mitarbeiter nutzen – Herausforderungen an das Produktionsmanagement. (Production as competitive factor. Using the potential of our employees —challenges for the production management), 3rd edn. TÜV Media, Köln (1995)

[ElM09] ElMaraghy, H. (ed.): Changeable and Reconfigurable Manufacturing Systems. Springer, Heidelberg (2009)

[EU09] European Platform on Life Cycle Assessment: Carbon Footprint—what it is and how to measure it. Joint Research Centre Institute for Environment and Sustainability, Ispra, Italy (2009)

[Far11] Farr, J.F.: Systems Life Cycle Costing: Economic Analysis, Estimation, and Management. CRC Press, Boca Raton (2011)

[Gau99] Gausemeier, J., Fink, A.: Führung im Wandel (Leadership in Change). Hanser, München Wien (1999)

[Ger07] Gerst, D., Hattesohl, S., Plettke, M.: Wie leistungsfähig sind ältere Arbeitnehmer? In: Unimaganzin Hannover. Schwerpunktheft Demographischer Wandel. Forschung für eine zukunftsfähige Gesellschaft (How effective are older workers? In: Unimagazin Hannover. Special Issue on demographic change. Research for a Sustainable Society). Iss. 3/4, pp. 24–26 (2007)

[GFN10] Footprintnetwork. Annual Report (2010). http://www.footprintnetwork.org/

[Grei89] Greifenstein, R., Jansen, P., Kißler, L.: Sachzwang Partizipation? Mitbestimmung am Arbeitsplatz und neue Technologien. In: Aichholzer/Schienstock (Hrsg.): Arbeitsbeziehungen im Wandel (Participation—a constraint? Participation in the workplace and new technologies). In: Aichholzer/Schienstock (ed.): Industrial relations in transition), Berlin (1989)

[Gud05] Gudehus, T.: Logistik: Grundlagen, Strategien, Anwendungen (Logsitics: Basics, strategies, application), 3rd edn. Springer, Berlin (2005)

[Her10] Herrmann, C.H.: Ganzheitliches Life Cycle Management. Nachhaltigkeit und Lebenszyklusorientierung in Unternehmen (Comprehensive life cycle management: sustainability and life cycle orientation in enterprises). Springer, Heidelberg (2010)

[Her13] Herrmann, C.H., Posselt, G., Thiede, S. (eds.): Energie- und hilfsstoffoptimierte Produktion. (Energy and auxiliary optimized production). Springer, Heidelberg (2013)

[Hoe92] Hoeckstra, S., Romme, J.: Integral Logistic Structures. Industrial Press, New York (1992)

[Jov08] Jovane, F., Yoshikawa, H. et al.: The incoming global technological and industrial revolution towards competitive sustainable manufacturing. CIRP Annals—Manufacturing Technology **57**: 641–659 (2008)

[Kra99] Krause, J., Lichtenstein, C.: Your private sky. R. *Buckminster Fuller. Lars Müller Baden* (1999)

[KrW96] Gesetz zur Förderung der Kreislaufwirtschaft und Sicherung der umweltverträglichen Beseitigung von Abfällen (Kreislaufwirtschafts- und Abfallgesetz (Law to promote circular economy and ensure environmentally sound disposal of waste. Recycling and Waste Management Act)—KrW/AbfG 12.9.1996. BGBI, p. 1354 ff

[Lot12] Lotter, E.: Hybride montagesysteme. In: Lotter, B., Wiendahl, H.-P. (Hrsg.): Montage in der industriellen Produktion - Ein Handbuch für die Praxis. (Hybrid assembly systems). In: Lotter, B., Wiendahl, H.-P. (eds.) Assembly in Industrial Production—A Handbook for Practice), 2nd edn. Springer, Berlin (2012)

[Menz00] Menzel, W.: Partizipative Fabrikplanung - Grundlagen und Anwendung. (Participative factory planning—Principles and application. PhD thesis, Univ. Hannover. Publ. in: Fortschritt-Berichte VDI, Reihe 2, Nr. 389. VDI Düsseldorf (2000)

[Mea72] Meadows, D., et al.: Limits to growth. A Report for the Club of Rome's project on the predicament of mankind). Potomac Associates (1974)

[Mea04] Meadows, D., et al.: Limits to growth. The 30-years update. Chelsea Green Publishing Company, United States (2004)

[Nyh09] Nyhuis, P., Wiendahl, H.-P.: Fundamentals of Production Logistics. Theory, Tools and Applications. Springer, Berlin (2009)

[Pac00] Pack, J., Buck, H., Kistler, E., Mendius, H.G., Morschhäuser, M., Wolff, H.: Zukunftsreport demographischer Wandel. Innovationsfähigkeit in einer alternden Gesellschaft. Veröffentlichung aus dem Förderschwerpunkt ("Demographischer Wandel" Future Report demogaphic Change. Innovation ability in an aging Society. Publication of the Research Priority Program "Demographicals Change"). Bonn (2000)

[Pah07] Pahl, G., et al.: Engineering Design: A Systematic Approach, 3rd edn. Springer, Berlin (2007)

[Perl98] Perlewitz, U.: Konzept zur lebenszyklusorientierten Verbesserung der Effektivität von Produktionseinrichtungen (Lifecycle approach to improve the efficiency of production facilities). PhD thesis, TU Berlin (1998)

[Pfo04] Pfohl, H.-Ch.: Logistikmanagement. Konzeption und Funktionen (Logistics Management. Conception and function), 2nd edn. Springer, Berlin (2004)

[Pol10] Pollard, D. (ed.): Living Planet Report 2010. Global Footprint Network, Oakland, California (2010)

[Rei05] Reichardt, J.: Form follows performance. In: Licht Architektur Technik + Büro, 3/2005, S. 1

[Sch10] Schmidt-Bleek, F.: Das MIPS-Konzept: weniger Naturverbrauch - mehr Lebensqualität durch Faktor 10 (The MIPS concept: less consumption of nature, more quality of life). Droemer, München (1998)

[Schei84] Schein, E.: Coming to a new awareness of organizational culture. Sloan Manag. Rev. **24**, 3–16 (1984)

[Sche10] Schenk, M., et al.: Factory Planning Manual: Situation-Driven Production Facility Planning. Springer, Berlin (2010)

[Sel97] Seliger, G., Müller, K., Perlewitz, H.: Nachhaltiges Wirtschaften eröffnet neue Geschäftsfelder (Sustainable business opens new business fields. Zeitschrift für wirtschaftlichen Fabrikbetrieb. ZWF **92**(6), 299–302 (1997)

[SRU96] Sachverständigenrat für Umweltfragen (Hrsg): SRU Umweltgutachten 1996, S. 50 ff, (Expert Panel for Enviromental Topics (ed.): Environmental Report) Deutscher Bundestag Drucksache 13/4108, Berlin (1996)

[Stat11] Destatis Statistisches Bundesamt: Statistical Year Book 2011, Wiesbaden (2011)

[Suz93] Suzaki, K.: New Shop Floor Management: Empowering People for Continuous Improvement. Free Press, New York (1993)

[UN87] United Nations: Our Common Future. Report of the World Comission on Enviroment and Development. New York (1987)

[Win01] Windt, K.: Engpassorientierte Fremdvergabe in Produktionsnetzwerken (Bottleneck oriented outsourcing in production networks). Ph.D. thesis, Univ. Hannover 2000. In: Fortschritt-Berichte VDI, Reihe 2 Nr. 579, VDI-Verlag Düsseldorf, 2001

[Wien96] Wiendahl, H.-P. et al: Optimierung und Betrieb wandelbarer Produktionsnetze (Optimization and Operation of changeable Production Networks). In: Kirsten, U., Dangelmaier, W. (eds.): Endbericht zum BMBF-Projekt Vision Logistik—Wandelbare Produktionsnetze zur Auflösung ökonomisch-ökologischer Zielbereiche. Wiss. Berichte Nr. 181, Forschungszentrum Karlsruhe (1996)

[Wir00] Wirth, S., Enderlein, H., Peterman, J.: Kompetenznetze der Produktion. In: Vortragsband IBF-Fachtagung "Vernetzt planen und produzieren" (Competence nets in production. In: Proceedings IBF Conference "Networked plan and produce", pp. 17–27. TU Chemnitz, Germany (2000)

[WWF10] Living Planet Report 2010. Biodiversity, biocapacity and development. World Wide Fund for Nature WWF (Formerly World Wildlife Fund), Gland, Switzerland (2010)

[WWF14] WWF International (ed.): Living Planet Report 2014. Species and spaces, people and places. Gland, Switzerland (2014)

4 Known Production Concepts

4.1 Introduction

During the evolution of industrial production, the factory has continually undergone changes. To some degree this has been due to the demand for an increasingly net-worked economy, but it has also been in response to new technological possibilities. Breaking work down into small standardized steps was first suggested by Taylor in 1911 and put into practice by Ford. Together with the introduction of the assembly line, this approach allowed consumer goods to be mass produced by semi-skilled workers. This basic principle remained in use up until the 1950s in order to cover high post-war demands. In the 1960s, product variants began to grow and productions focused more on the customer. As a result group work, manufacturing cells and segments were introduced. At the same time, numerical control for machines tools and robots established themselves in industry, leading to machining centers and flexible manufacturing systems. Consequently, certain part groups could then be manufactured automatically in a random sequence and to some extent at night without supervision. Cost related pressures and in particular competition from Japan in the 1980s required new efforts to avoid waste, thus giving rise to Lean Production. In Germany, fractal and modular factories posed further impulses. With the resounding globalization since the 1990s and the ever growing uncertainty in the market, the concept of flexible and changeable factories has come into the foreground. Current discussions focus on digitally networked factories, in which individually customized products should move self-controlled through factories. Alongside this, there is the demand for workplaces that are adapted to an aging workforce and for factory operations to manage resources more efficiently. In this chapter, we will provide a brief overview of the roughly outlined stages marking the evolution of the factory.

4.2 F.W. Taylor

As the 'father of scientific management' the American, Frederic Winslow Taylor (1856–1915), was the first to fundamentally examine rationally designing manufacturing processes. His findings are summarized in his book "The Principles of Scientific Management" [Tay11]. Taylor outlined the main responsibility of management as ensuring the maximum prosperity of both the enterprise and its employees. In doing so he emphasized four principles:

- developing a truly scientific approach to management,
- scientifically selecting workers,
- scientifically training and developing employees, and
- trusting and close collaboration between managers and employees.

Through his experience in various enterprises, Taylor became convinced that an enormous waste was caused by applying rules of thumb and

H.-P. Wiendahl et al., *Handbook Factory Planning and Design*,
DOI 10.1007/978-3-662-46391-8_4, © Springer-Verlag Berlin Heidelberg 2015

implementing conventional work methods without really thinking about them. Based on thorough time and motion studies, he broke work processes down into their core steps and developed standardized procedures; whereby preparatory planning was strictly separated from performing actual tasks.

Taylor identified the following aspects as vital to this new approach:

- time studies using appropriate methods and equipment,
- 'function supervisors' responsible for educating employees and planning work,
- standardizing all tools, processes and movements,
- developing a planning room or department,
- managing by exception,
- using slide-rulers and other time-saving tools,
- having work instructions for employees,
- defining targets linked to a large bonus when successfully met,
- differential wages,
- using classification systems for products and operating equipment,
- having production control systems, and
- using modern accounting.

Taylor's thoughts influenced concepts about organizing work especially in large batch and mass production up until after World War II. 'Taylorism', however, was criticized early-on. Its limits became clearer especially with increasing automation, the explosion of the number of products and their variants, increasing demands on the quality of the product and processes as well as an evolving understanding of role distribution between management and workers. Nevertheless, Taylor's basic ideas are still worth considering; Gaugler summarizes these in six points, which we briefly outline here [Gaug96, p. 44ff]:

- a market economy with permanent pressure to decrease piece costs is conditional,
- aligning the business interests of the employers and capital investors with those of the employees,
- systematically developing best practices instead of rules of thumb,
- decisive roles for middle and lower management,
- intensive collaboration between management and workforce, and
- emphasizing extrinsic motivations (promoted externally) without ignoring intrinsic (personal reasoned) motivations.

Taylor's methods could be considered the first comprehensive system for scientifically operating a business. In addition to Taylor, Frank Bunker Gilbreth (1868–1924) and his wife Lillian (time and motion studies) as well as Henry Lawrence Gantt (inventor of the Gantt-Chart) systematically observed production processes, improved and implemented them in new factory processes.

Henry Ford (1863–1947) was known for resolutely applying Taylor's lessons. In his factories, he developed and integrated standardization, typing and interchangeability of parts along with the division of work and precision work into his mass production. This in turn made the automobile affordable for a broad sector of the population [Spur00].

In Germany, Taylor's ideas were taken up and further developed by scientists such as Adolf Wallichs in Aachen and Georg Schlesinger in Berlin [Spur00]. Germany adapted many of the American production methods under the concept of rationalization. Dedicated in particular to time and movement studies, REFA, Germany's leading organization in work design, industrial organization and company development, publishes a comprehensive methodology for designing, planning and controlling office and production work that is regularly updated (http://www.REFA.de).

4.3 Group Work

With the rapid development of the world economy following World War II, a pure seller's market arose offering comparably standardized products. These were produced in large numbers extensively based on principles set out by Taylor and Ford.

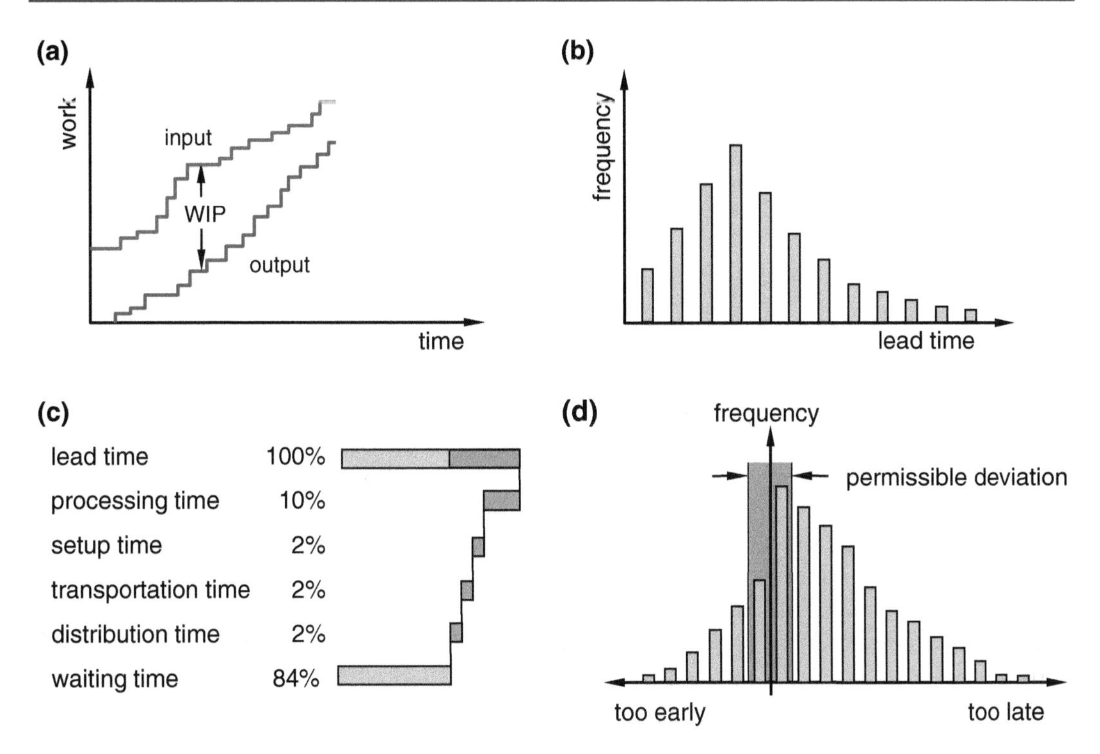

Fig. 4.1 Characteristic properties of an uncontrollable production. **a** Fluctuating trend of work-in-process. **b** Wide lead time distribution. **c** High share of waiting time. **d** Poor due date performance. © IFA G9125SW_B

As products became further differentiated and more and more markets were tapped into, lot sizes decreased while increasingly diverse orders made factories progressively complex. The amount of work-in-process (WIP) grew, throughput times became longer and schedule reliability decreased. Figure 4.1 depicts the characteristic behavior of a workshop that has become uncontrollable in this way.

In the Throughput Diagram (which shows the cumulated input and output at a workstation over time) it is obvious that in case of strongly fluctuating input delays in adjusting the capacities result in strong fluctuations in the output. Consequently, WIP levels are high and fluctuating strongly (Fig. 4.1a). Since WIP levels directly influence lead times, broad leftward skewed frequency distributions result with a few quick orders, while there is a broad midrange with normal orders and a few but to some extent extremely slow orders (Fig. 4.1b). If we now consider the breakdown of lead times (Fig. 4.1c) in a job shop production, the actual share of value adding processing time rarely reaches 10–15 %. Theoretical and practical research have shown that with this type of production and a target utilization of 96–98 %, generally waiting times of no less than 70–80 % of the whole throughput time can be achieved [Nyh09]. Thus in the end the schedule reliability is utterly inadequate (Fig. 4.1d); some orders are completed too early, the majority though too late. With a tolerance zone of for example ±2 work days for permissible lateness, frequently only 30 % of the orders are on-time.

Numerous strategic, organizational and engineering based approaches arose for solving the resulting problems. From a *strategic perspective*, the recommendation was to form business units which then track individual targets for each of the markets. As a result, autonomous factories, each of which focused on a different product family, developed within a plant.

Personnel and organizational methodology focused on employees. The idea here was that the progressive diversity of products and their variants could be better controlled by returning planning and control tasks from the offices back to the shop floor, reducing hierarchies and rules and fostering communication between personnel. At the same time the employees' growing need for autonomy and self-realization could also be addressed.

The forms of work which developed from there can be summarized under the concept of group work. A group takes on an overall task (e.g., manufacturing a group of similar workpieces or assembling an unit completely or partially) and autonomously distributes the required tasks among the group members; only the end results are controlled.

Based on the example of an assembly task and beginning with the traditional distribution of work (i.e., job sharing with strictly separated tasks) the three different forms of group work are clarified in Fig. 4.2.

With *job rotation*, those employees entrusted with the actual assembly are able to take on different assembly tasks because they have appropriate qualifications. All of the remaining tasks such as machine setup, quality checks, repairs, packing and transportation are excluded from this. Nevertheless, the job rotation already creates a break in the monotony of the usually short cyclical work.

In comparison, *job enlargement* is aimed at combining a number of different tasks on the same qualification level into a new, expanded job. The group members' activities can already be decoupled from the cycle. Tasks that do not add value though are still conducted at different stations. In addition, it is hoped that workers will develop a stronger identification with 'their' product which will in turn improve job satisfaction.

Job enrichment goes a step further in the direction of so-called 'part-autonomous work groups'. Here, not only the immediately productive tasks but also the remaining indirectly productive tasks such as quality checks and set-ups as well as other jobs like ordering materials, maintenance and repair etc. are transferred to a workgroup. Higher level planning and control tasks such as planning the personnel, orders and work hours are conducted by a foreman or supervisor.

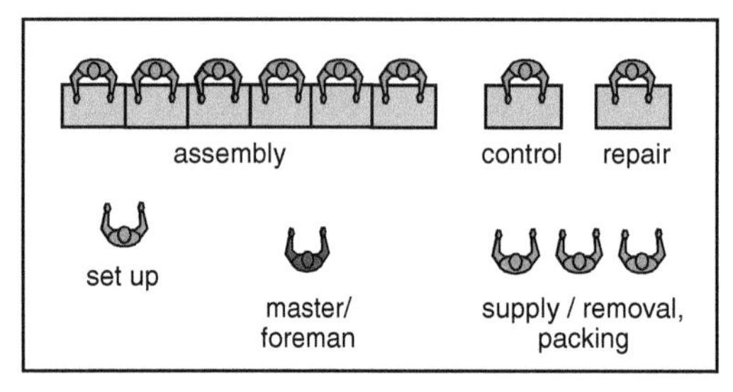

assembly
control
repair
set up
master/
foreman
supply / removal,
packing
job rotation

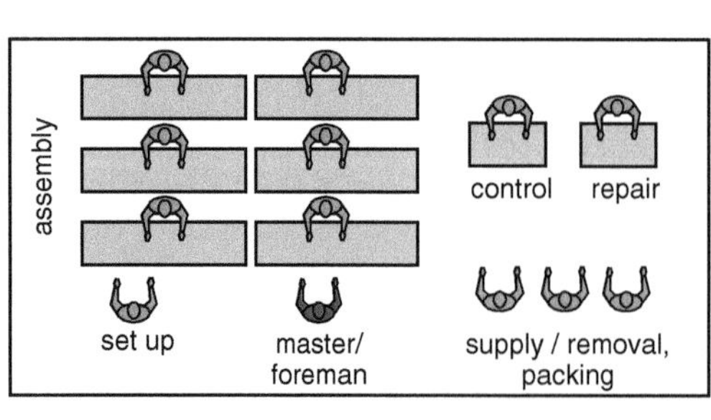

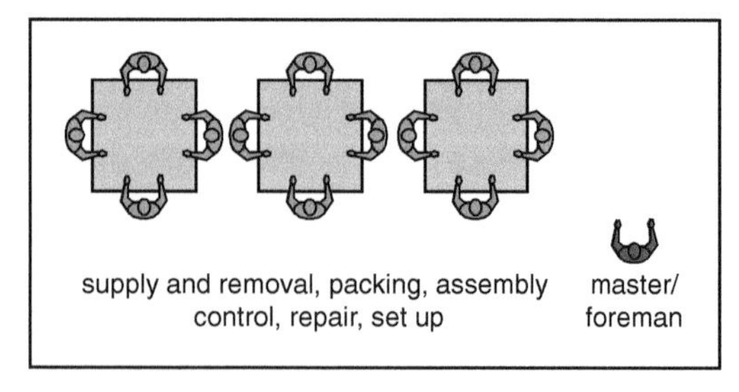

Fig. 4.2 Forms of work structures (Bullinger). © IFA G2949ASW_Wd_B

If these indirectly productive tasks are also transferred to the team, autonomous workgroup are created. These groups organize their jobs by themselves and are responsible for delivering products in the demanded quality and quantity on the appointed date. The contact person both internally and externally is usually the spokesman of the group and is an active member of the team.

Partially autonomous and full autonomous workgroups accommodate the workers' needs for self-realization the most. However, they also place the highest demands on employees not only on a professional level, but also on a personal level. Critical points here include regulating work hours and the wage system as well as on-going education and improvement.

4.4 Manufacturing Cells

The strategic and organizational approaches outlined here are supplemented by engineering solutions developed from research as an answer to the continually growing diversity of parts in enterprises. In addition to the work of Lange and Roßberg [Lan54], Mitrovanow's group technology approach [Mit60] formed the basis for these studies in Germany. Further work has shown that in every apparently still heterogeneous part spectrum groups of similar parts exist. They represent only approximately 20–30 % of the production costs for the corresponding products, but make-up almost 70 % of all the parts [Arn75]. In order to find these 'part families', diverse classification systems were implemented —in Europe, the Opitz coding system for single mechanical engineering parts was the most widespread of these [Opi66].

For part families such as shafts, gears, levers etc., it first seems that using variant designs and standard work-plans would be practical. Part families such as these generally have a similar work flow and thus are routed through a sequence of the same machines. The different lots only require small change-overs at the different work places thus setup times can be saved. However, this approach has failed to establish itself: Since it continued to be based on the principle of workshop production, the organizational costs and efforts for planning and control of these part families was considerable.

The necessity of shorter throughput times and lower WIP levels required renouncing the primate to fully utilize all workstations. In further developing the idea of part families, 'manufacturing cells', which combine all of the operating equipment and employees for producing a group of similar products spatially and organizationally, were created in the 1970s. The fact that not all of the machines would be fully utilized was accepted. By implementing group work, the manufacturing cells could be operated to a wide extent independently, which in particular also included the material supply, finite scheduling and sequencing orders.

Figure 4.3 depicts the basic structure of a manufacturing cell. Raw pieces are withdrawn from bins before the cell. In comparison to workshop productions though, the corresponding operation is not completed for the entire lot before it is transported further, rather once the operation of a part of the lot is completed it is immediately sent to the next workplace, either by manual handling of the operator or on a conveyor belt. Nowadays, this is commonly referred to as "one-piece-flow" or "single-piece-flow" manufacturing.

Manufacturing cells are characterized by low WIP levels along with short throughput times. Waiting times for parts between the individual operations are avoided because machines are arranged according to the workflow and lots are manufactured overlapping. Figure 4.4 depicts a simple example of the enormous WIP reductions and time savings for a manufacturing cell in comparison to a workshop production (according to [Suz87]). In this case, the total throughput time could be reduced from 20 to 8 min. In practice, throughput time reductions ranging from a number of weeks to a single day have been proven.

However, due to the most extreme differences in the processing times of individual operations it is not possible to fully utilize all of a manufacturing cell's resources. This can lead to

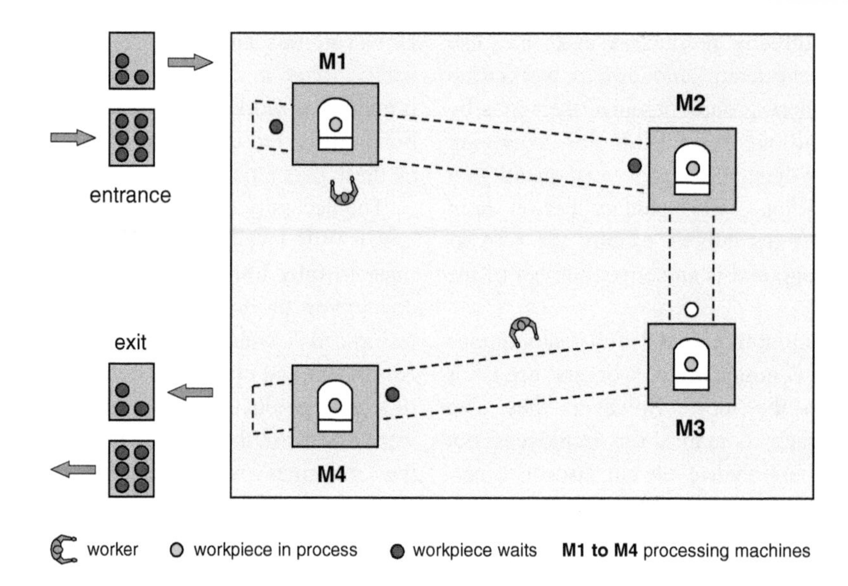

Fig. 4.3 Principle of a manufacturing cell. © IFA G9121SW_B

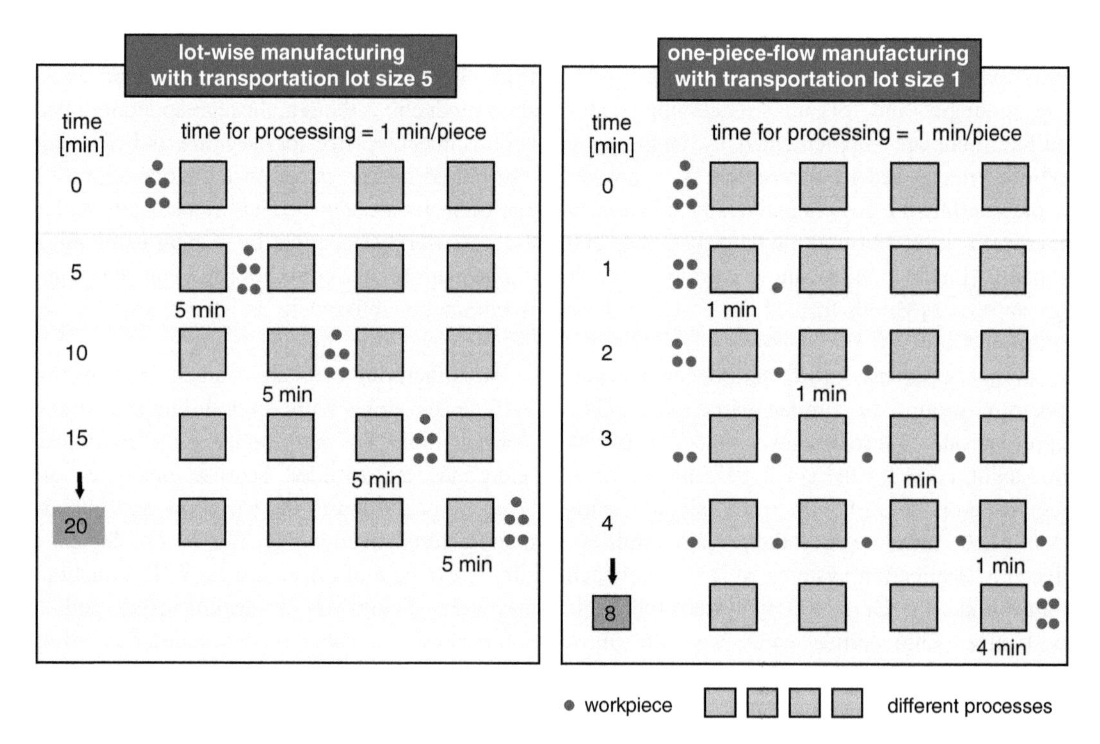

Fig. 4.4 Lot-wise and one-piece-flow production (after Suzaki). © IFA G8979SW_B

unjustifiable idle capacity costs. Furthermore, with an increasing number of part variants being produced in a manufacturing cell, the setup times can become problematic. Thus, even in the future, traditional workshop productions will continue to exist.

4.5 Flexible Manufacturing Systems

The development of manufacturing cells outlined here initially leaves open the question of the degree of automation and the way in which cells are linked. With the breakthrough of numerical control for machine tools in the 1960s and thus the possibility of automatically executing complete operations for different parts immediately following one another, new options for manufacturing part families developed.

One of the first important steps was establishing *machining centers*. These allow a workpiece to be processed in one setup from different sides with different technologies such as milling, drilling, thread-cutting etc. Automated tool changes are prerequisite for this concept.

If a number of these processing centers are then linked together along with additionally required workstations for washing and testing the workpieces, a *flexible manufacturing system* is created. Figure 6.30 in Chap. 6 depicts the layout of such a concept.

In addition to processing and support stations, central stores for workpieces and tools are integrated—each equipped with their own transport system. The individual stations also have local buffers for workpieces and tools. Components are controlled in a hierarchical system with a higher level master computer. The transport system for the workpieces and tools—usually clamped onto work piece pallets—frequently consists of transport wagons guided by rails with a conveyor device. Automated guided vehicles (AGV) and programmable handling devices, such as industrial robots, are however also employed.

Flexible manufacturing systems thus allow the automated, unsynchronized and non-directional manufacturing of a defined group of similar parts and thus in a way represent an automated manufacturing cell. Initially, these were commonly implemented in the 1980s but lost their economic feasibility due to: (a) the no longer controllable diversity of parts induced by the market, (b) enormous costs for storage devices and systems for connecting them, as well as (c) labor-intense programming and controlling. These costs were considerably reduced through simpler connections (e.g., with the aid of robots and standardized buffer devices) so that manufacturing systems have once again found their place today, especially for pilot lots and spare parts in the automobile industry and their suppliers.

4.6 Manufacturing Segments

Due to the history of their origins and use, both manufacturing cells and flexible manufacturing systems were initially not tied to the parts of a specific product. It was first the continually stronger orientation of the production on the market needs and the accompanying decentralization that suggested not to limit to just the value-adding processes in manufacturing. Rather other value-adding stages such as assembly, packaging and shipping could be combined as integrated organization units. Moreover, these units could each concentrate on one product so that a specific competitive strategy could be pursued. Wildemann referred to these units as *manufacturing segments* and expanded them into the concept of *modular factories* [Wild98].

Figure 4.5 itemizes the features of manufacturing segments. With regards to the market and target orientation, manufacturing segments serve specific market-product combinations, which each apply different strategies e.g., the highest quality or the shortest throughput times. The product orientation reduces coordination and control costs while minimizing the interdependency between segments, thus resulting in greater production depth.

Integrating a number of steps of a logistic chain into a manufacturing segment makes it possible to spatially combine the manufacturing and assembly, therefore eliminating the traditional separation of these two areas. The resulting technical problems related to the close proximity of incompatible technologies (e.g., emission intensive manufacturing processes next to precise

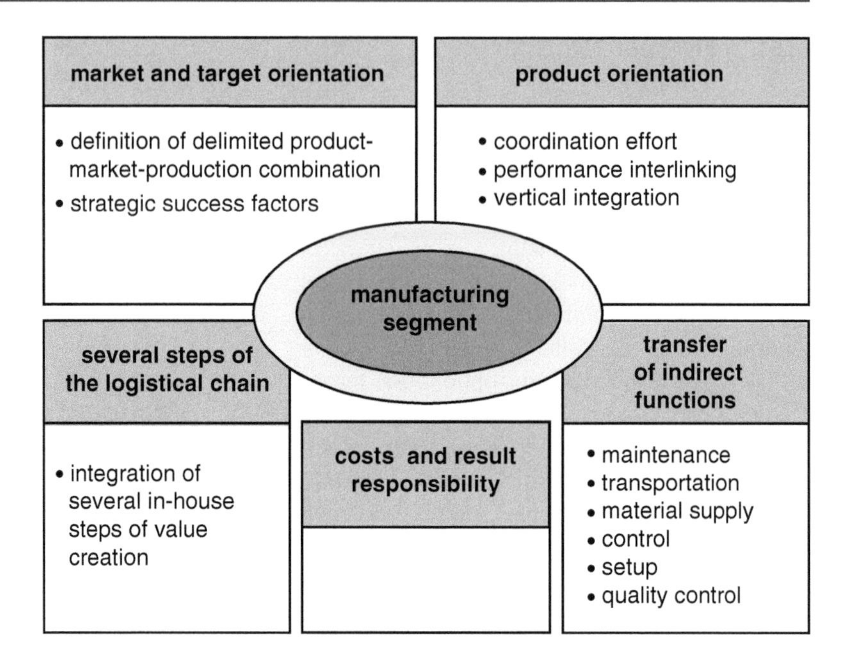

Fig. 4.5 Features of manufacturing segments (Wildemann). © IFA G9122SW_B

and clean assembly processes) need to be resolved within the scope of factory planning with suitable measures for workplaces and areas (e.g., appropriately housing of machinery).

By *transferring indirect functions* to employees in the manufacturing segment the end results are holistically influenced as is the case with manufacturing cells. Frequently, an incentive system can also provide impetus for continually improving the processes with the goal of avoiding all possible waste.

Finally, the *responsibility for costs and results* differs according to the type of customer. If the segment's product is passed on within the enterprise, it is organized as a cost or service center. If, however, the segment delivers a final product at market prices, it is designed as a profit center [Wild98].

Wildemann mentions a number of principles for designing such segments. These include

- optimizing the flow,
- providing smaller and possibly duplicate capacities,
- spatial concentration of equipment with a variable layout,
- self-regulating control loops,
- processing parts and groups completely,
- workers testing parts or components by themselves,
- temporally decoupling the manual work from the operating times of the machines and
- team orientation.

Manufacturing cells and manufacturing segments could be considered the most important development in production organization during the 1990s. As is summarized in Fig. 4.6, they have caused demonstrable, significant improvements in the market relevant objectives 'quality' and 'costs' [Wild00].

A further development in the segmentation approach led to 'indirect segmentation', in which business processes for indirect functions are also handled with responsibility for the results [Wild00].

By also integrating the distribution, marketing and product development into the thus defined direct and indirect segments, so-called 'product units' or 'business units' are created. These serve a market segment during the entire product lifecycle. The danger of losing know-how in the individual product units is countered by forming *support or function centers* e.g., for CAD technology, specific manufacturing methods or procurement procedures.

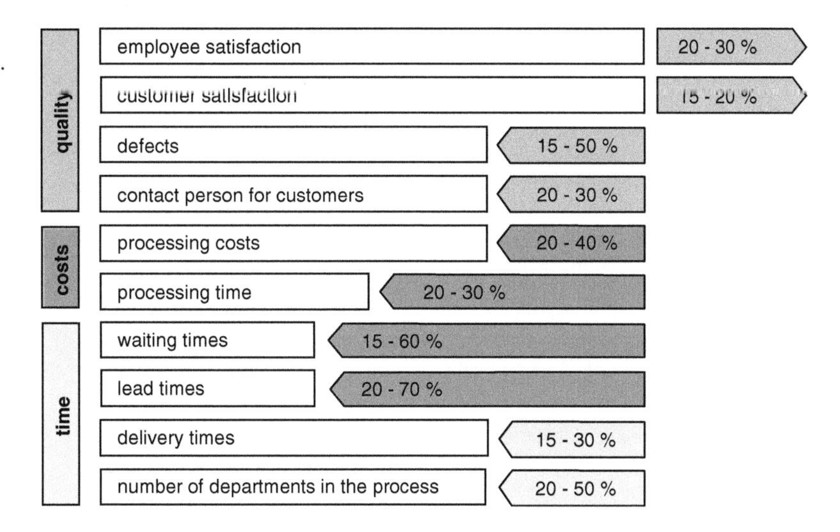

Fig. 4.6 Impact of segmentation (Wildemann). © IFA G9126SW_B

4.7 Lean Production and the Toyota Production System

The concept of lean production can be traced back to a five-year, worldwide study by the Massachusetts Institute of Technology of the final assembly conducted in approximately 100 automobile factories in Japan, North America and Europe [Wom90]. In studying four groups of factory characteristics (performance, layout, workforce and automation) in three geographical regions (America, Europe and Asia), severe differences were observed. Based on their research the authors concluded that, on average, Japanese plants had much better results in all of the categories than their American and European competitors (see Fig. 4.7). Nevertheless, significant differences were also visible within the groups.

Moreover, considerable differences could also be seen in the performance for the product development. This was true for both the costs and time required for developing a new model as well as the suppliers' contribution to the development, the time required till normal productivity and the quality achieved once production began.

The study released in the USA in 1990 and published in German in 1991 was met with great interest in Germany and triggered an enduring debate about the competitiveness of the local automobile industry and Germany as a production location. Soon it became clear that the deficit could no longer be eliminated through individual measures such as overhead value cost analyses, computer integrated manufacturing, shorter setup times and new PPC systems; instead it was obvious that a holistic approach with an "eye towards perfection" was required i.e., continually lowering prices, zero defects, no inventories and variable product diversity [Wom90]. This in turn meant that design and production processes—including those of the suppliers—had to be continually reviewed along with the customers' wishes.

With unprecedented effort the German industry was able to significantly improve both the products as well as the processes as an answer to this threatening challenge. However, this was also related to a tremendous climb in unemployment figures and thus to considerable critique of the so called lean production concept. For the first time since World War II, the number of employed in Germany did not grow with increasing production. Securing the workforce despite constantly decreasing work volume requires new products in new markets along with developing services and merging enterprises.

Lean production is not a closed, theoretically based business concept; rather it is the quintessence of an analysis of successful enterprises. It is generally based on Toyota Motor Company's

key figure	Japanese plants in Japan	Japanese plants in North America	American plants in North America	all European plants
• **performance**				
productivity (hours/car)	16.8	21.2	25.1	36.2
quality (assembly defects/10 cars)	60.0	65.0	82.3	97.0
• **layout**				
area (sqm/car/year)	0.5	0.8	0.7	0.7
size of the repair area (% of assembly area)	4.1	4.9	12.9	14.4
stock level (days for 8 chosen parts)	0.2	1.6	2.9	2.0
• **workers**				
% of the workers in teams	69.3	71.3	17.3	0.6
job rotation (0 = none, 4 = frequently)	3.0	2.7	0.9	1.9
number of suggestions per employee	61.6	1.4	0.4	0.4
number of wage groups	11.9	8.7	67.1	14.8
education of new production workers (hrs)	380.3	370.0	46.4	173.3
absence (%)	5.0	4.8	11.7	12.1
• **automation**				
welding (% of the operations)	86.2	85.0	76.2	76.6
painting (% of the operations)	54.6	40.7	33.6	38.2
assembly (% of the operations)	1.7	1.1	1.2	3.1

Fig. 4.7 Characteristics of high volume automobile manufacturers 1989 (Womack et al.). © IFA G9128SW_B

continually developed Toyota Production System (TPS), which aims at the highest possible product quality at the lowest possible costs with the shortest possible delivery time [Ohn88]. The goals of the production derived from this include:

- *productivity* by eliminating every type of waste,
- *quality* through reliable processes which facilitate high quality products,
- *flexibility* through responsive workplaces and employees, and
- *humanity* by including the knowledge of workers as much as possible.

The core of the concept is an organizational and human centered production model based on motivation, creativity and abilities of employees. In addition to intensifying customer relations, goals for the production include: decreasing levels of hierarchy, shortening decision making processes, shifting tasks to the executional level and increasing the collaboration of suppliers already during the development phase of one's own products.

Based on these, five inter-related elements of Toyota Production System were developed; a summary of these is found in Fig. 4.8 according to Oeltjensbruns' depiction [Oelt00].

The basic guide line of TPS is *avoiding waste*. This is in itself an orientation towards limits, aimed at the necessity of value-adding for a product along with the absolute fewest resources, materials, parts, space and work hours possible [Suz87].

While resolutely complying with the customer-supplier relationship *waste and reworking* is prevented by only sending parts and components that have been verified as 100 % faultless from one workstation to the next operation or station. This also applies to external suppliers.

Overproduction results from processing 'optimal lot sizes', saving setup times and processing fill-orders in order to avoid machine standstills as well as minimizing the material waste. It leads to unnecessary inventories in warehouses and production. The pull-principle combined with overlapped manufacturing according to the one-piece-flow principle prevents this.

Unnecessary waiting times for materials, people and machines are also closely related to the topic of inventories. They arise due to waiting times following an operation, queuing times, waiting times while workplaces are setup and waiting of parts in lots. Important approaches to avoiding these types of waste include:

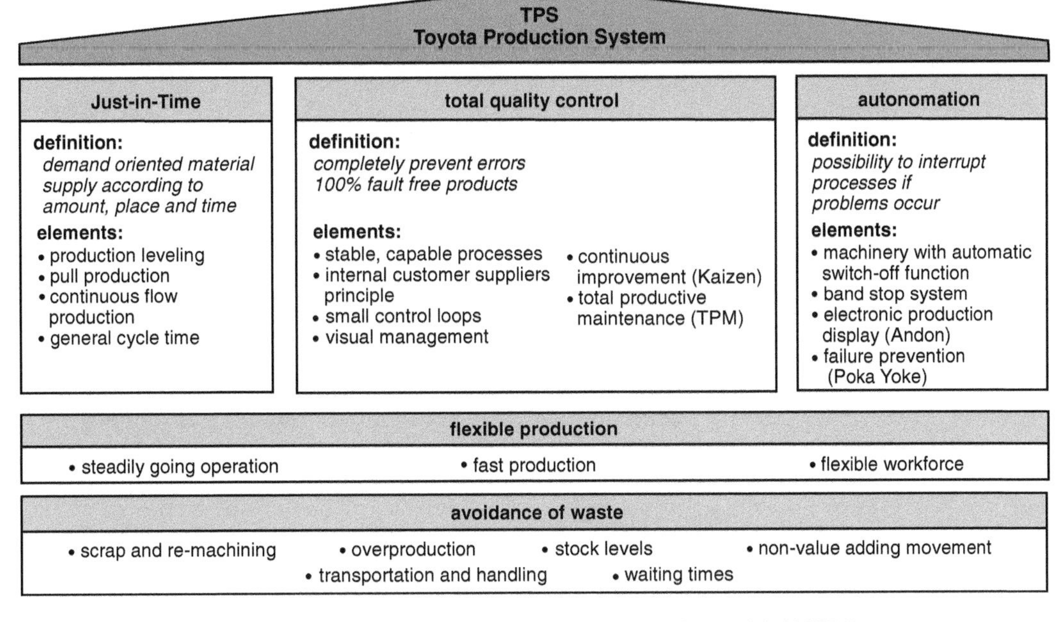

Fig. 4.8 Elements of the Toyota production system (after Oeltjensbrun). © IFA G9129SW_B

- manufacturing segments with workstations arranged in the shape of an 'U',
- extremely short setup times for the machinery, and
- limiting the WIP in a system by defining the number of permissible parts before or in a system.

Unnecessary transportation and handling of parts is caused by multiple interim storages and commissioning on the way from the supplier to the assembly place in the factory. An ideal delivery and supply concept prevents this through an integrated supply chain with which parts from the supplier's last workstation arrive directly at the site where they are to be used by the consumer in the requested amounts, without packaging and on suitable reusable workpiece carriers. This does not always happen. In this case it is the responsibility of the cross-plant logistics to ensure the demand driven delivery and the job of the internal transport system to ensure the places of consumption are punctually supplied. The handling of parts within the segment is then the worker's responsibility.

The second fundamental principle of the Toyota Production System consists of creating a *flexible production* that reacts quickly to changes in the product quantities and variants sold as well as the methods and processes implemented. This is accomplished by:

- distributing the work content as equally as possible based on a balanced sequence of product variants with large and small work contents (i.e., leveling),
- fast reactions to faults with the goal of permanently eliminating them and
- broadly qualified workers who, depending on demand, control one or more work operations.

The group work described previously is a prerequisite for this.

With the third element, *total quality control*, TPS strives to prevent faults entirely so that customers are able to unpack products which are completely defect-free. This requires all business processes to be continually monitored from marketing over to product development and sales up to the order processing and servicing. Stable, *capable* processes designed according to the

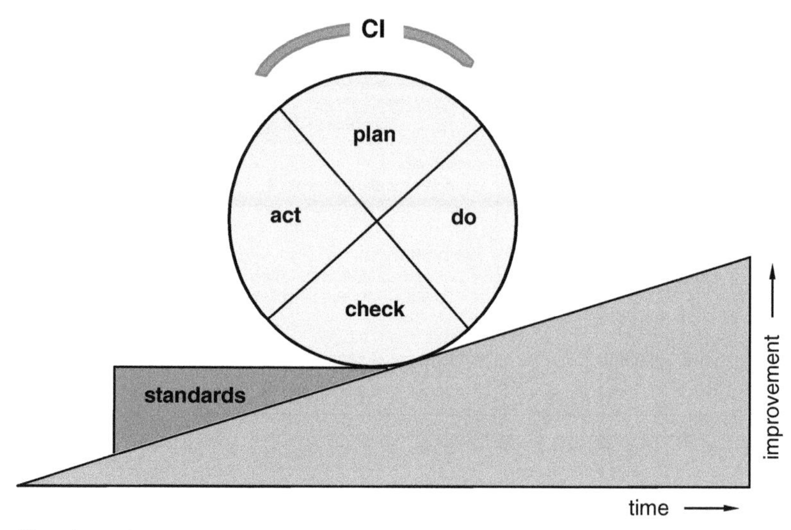

Fig. 4.9 Continuous improvement with the PDCA cycle (after Deming). © IFA G9130SW_B

principles of quality management along with the *internal customer-supplier principle* (only sending zero defect goods to the next workstation) and *small control loops* allow errors to be discovered as early as possible. This in turn allows eliminating them at the root without having to implement additional quality checks. Besides the self-testing conducted by a worker once they have completed a work step (supplier), the worker conducting the next step (customer) also tests the goods received from the previous station thus ensuring that even more rare faults are discovered. Should a defect be found, the worker refuses to accept the piece and to process it further; instead the faulty piece is returned to the responsible originator.

Another important instrument is transparently presenting information about the work process (e.g., production numbers, defect rates, degree of utilization, material consumption etc.) for all of the employees. With this '*visual management*' it is possible to quickly identify both the progression of key performance figures in comparison to the target values as well as unusual states. It is for this reason that visual management also includes the clear identification of materials and tools along with their storage location in addition to detailed information about the work processes and operating states of machines and systems.

Continuous improvement processes (Japanese *kaizen*: change for the better) still play a significant role. The goal here is to achieve a constant increase in productivity and quality through continuous small improvements, suggested and implemented by employees. The Plan-Do-Check-Act Cycle (see Fig. 4.9) is a proven method for the systemizing this process.

Also called the Deming Cycle after its founder Edwards Deming, this sequence of activities begins with selecting the subject, analyzing the situation, determining a method to solve it and developing an improvement plan (plan), followed by implementing the plan (do) and verifying the results (check). When the results are positive, the method is standardized and visualized to ensure its immediate application (act).

Moreover, a clean, safe and organized workplace is essential and can be achieved with the so-called '5S campaign'. The 5S include [PP96]:

- Seiri (Sort): Everything not used for the job should be removed from the workplace (materials, tools, papers etc.).
- Seiton (Set in Order): Everything should have its own place and be in its place when not being used.
- Seiso (Shine): The workplace should be cleaned and kept clean (machinery, resources, floor).

- Seiketsu (Standardize): The first three steps should be standardized and made part of a daily work routine.
- Shitsuke (Sustain): Continue to adhere to these steps and improve the pristine nature of the work place.

Continuous improvement (CI) cannot replace innovations in the sense of a radical new design; rather it is aimed at utilizing the potential of a process as quickly as possible and preventing a drop in productivity. Continuous improvement requires organized and detailed business goals down to the employee level. These goals then have to be processed continually in quality circles or temporarily in small groups either tied into a specific division or inter-divisionary. It is essential that employees identify with these goals.

In considering quality from a holistic approach, *Total Productive Maintenance* (TPM) is aimed at plant efficiency and is the product of availability, performance and quality. The availability is determined by machinery disruptions and setup times, while the performance factor results from missing orders and a loss of speed. Finally, the quality factor reflects the defective parts and start-up losses. With the aid of preventive maintenance and repair services during the plant's entire phase of use and workers taking on simple maintenance work, the threat of possible breakdowns can be identified early-on. When combined with the continuous improvement process and its techniques, lifecycle costs can be minimized and the plants yield maximized.

4.8 Just-in-Time

The fourth basic element of TPS—commonly known as just-in-time (JIT) as coined by Taiichi Ohno—aims to supply all of the factors needed for production exactly as needed [Ohn88]. This supports the logistic objectives 'low inventories, 'short throughput times' and 'high punctuality'. JIT considers the entire value-adding chain from the suppliers to the enterprise's production up to the delivery to the customer.

The JIT elements named in Fig. 4.8 are directed at an even flow of orders through the production. The production leveling is first aimed at smoothing the work content of the most irregular incoming orders, while the pull-principle emphasizes the warehouse principle. The ideal here is a continuous flow production with a constant cycle time across all of the value-adding stages.

For the *procurement process* JIT means delivering purchased parts synchronized with the production—preferably without an incoming goods store—directly to the place of consumption. In case of supplying variants in the sequence they are assembled (e.g., in model variants of automobiles) the phrase just-in-sequence procurement (JIS) is common. This type of procurement is suitable for parts with a high to medium consumption value (A and B parts) and a high to medium predictability of consumption (X and Y parts). For this type of procurement, with which there is sometimes only 2–4 h of stock maintained at the place of consumption, it is necessary to clearly align the supplier and consumer while ensuring high delivery reliability and controlled processes. Moreover, quantity fluctuations should not be too strong i.e., in the range of 20 to 30%. For the remaining part groups other supply concepts are pursued such as 'C part management' in which the entire organization and controlling of C parts is outsourced to an external service provider or article groups are delivered by suppliers. Generally, the aim here is to reduce the number of supplier positions, whether by bundling articles or by forming development partnerships with later suppliers of components and systems.

Two basic conditions are required for JIT delivery: (a) a supplier with highly reliable processes and (b) a reliable delivery. The first is met by introducing a supplier qualification phase and regular auditing. In the case where there is sufficient supply volume, the second condition is met by settling suppliers in 'supplier parks' located close to the consumer. We will take a

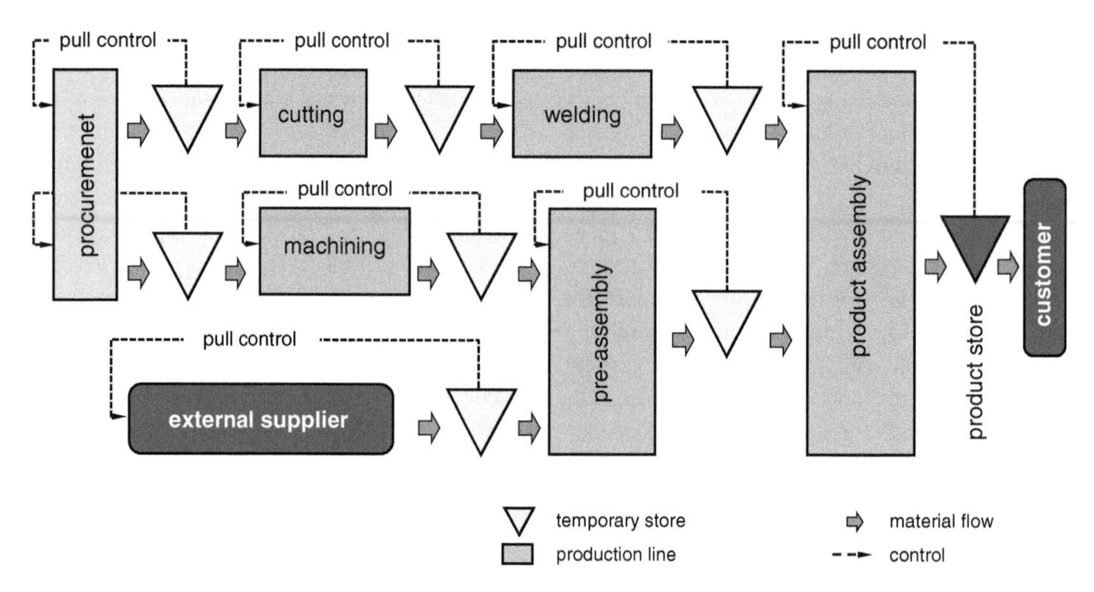

Fig. 4.10 Example of a production controlled according to the pull principle. © IFA G1278SW_Wd_B

look at some of the other delivery concepts that are constantly being developed later when we discuss designing procurement logistics (see Sect. 9.5.1).

JIT is applied in a production via the *pull principle*. In comparison to the push principle, in which individual orders are pushed from the workstation in the sequence according to the routing plan, orders are pulled out of the production according to the super market principle beginning with the last workstation. Figure 4.10 depicts an example of a production controlled according to the pull principle.

Starting with a stored final product the consumer (here, sales) withdraws the product according to an existing concrete customer order and delivers the desired quantity immediately. If the remaining stock falls below a specific level, information is transferred to the product final assembly to deliver this product variant in a defined quantity and in an agreed upon time. This information is transferred on a card which defines the identification number of the item, production area, consumption area and the quantity of the item to be manufactured. This document is referred to as a Kanban, which in Japanese means card or sign. The product assembly serves itself from the buffer preceding it, in which welded parts or pre-assembly groups ready to be mounted are stored; these in turn are re-manufactured with the aid of additional Kanbans. These control loops continue back to the cutting of sheet metal parts or to the mechanical shop. In this concrete case the external suppliers for the sheet metal and purchased components (hydraulic parts) are also tied-in via Kanbans communicated by fax or increasingly, electronically transferred protocols.

The pull principle thus extensively realizes a continual flow production; however, it is also dependent on certain conditions:

- each section of the production has to deliver 100 % faultless parts,
- the number of variants has to be limited,
- the capacity flexibility has to be sufficient for managing fluctuations in quantities, and
- delivery time of each Kanban has to be strictly adhered to.

When these conditions are met, not only can the production control be simple and decentralized, but inventories can also be controlled.

Should these conditions not be met, other methods such as Constant Work-in-Process (CON-WIP) [Hop11] or Load Oriented Order Release (LOOR) [Wie95] come into play. *CON-WIP* is unique in that it refrains from variant specific storage buffers; instead, further production runs are triggered from the preceding stage directly at the start of the production chain. It thus works according to a WIP controlled pull-production following the flow principle. In situations where orders are very different and processed according to the workshop principle (push), a *Load Oriented Order Release* can be applied. LOOR works by combining a WIP control with a throughput time control. Orders are only released once it is ensured that in future a previously defined WIP level on each of the workstations involved will not be exceeded. This method protects the workshop from being overloaded and thus from an uncontrolled increase or decrease in WIP. In addition to the already mentioned reliable production processes, just-in-time requires dependable transportation, storage and supply processes from the arrival of goods up to and including shipping and distribution.

From the perspective of factory planning it is necessary to be able to easily position workpieces well on the load carrier. Furthermore, mobile interim stores (where possible with flow racks), easy possibilities for transferring load carriers from transportation vehicles to interim stores or to the place of consumption as well as a reliable transport system with very frequent deliveries are recommendable.

The Toyota Production System does not provide any information with regards to *distributing* the goods created in the production. Nevertheless, the flow principle is also applicable here along with keeping stock levels as low as possible and shipping times as short as possible. The distribution logistics are responsible for distributing goods to the market and end user, usually via interim stores in which articles are bundled and shipped to the retrieval location near the customer. The basic technical functions that arise here include handling, storage, order picking and transportation. Since distribution only impacts the factory planning with regards to the output point on the factory site, we will not go into this subject in detail.

With the increased interweaving of the global goods stream, the object of consideration spreads from a single enterprise to logistic chains and networks. Not only is the flow of goods upstream to the supplier of the supplier considered, but also downstream to the customer of the customer. This is referred to as a 'value-adding chain' or more commonly as a '*supply chain*'. The Supply Chain Operations Reference Model (SCOR-Model 10.0), which is continually being revised and updated (http://www.supply-chain.org) identifies the key tasks for each of the links in the chain as source, make, deliver, and return (e.g., for complaints). *Supply Chain Management* (SCM) designs, plans and controls the affected materials, information and value flows in the networks with the aim of achieving high customer satisfaction (price, quality, delivery reliability) as well as reducing costs (inventories, interfaces) and quickly adjusting to markets. Thus, it is no longer a matter of individual enterprises competing with one another, but rather entire value-adding chains. Supply chains such as these are found especially in the automobile industry.

The last main element of the Toyota Production System is referred to with the artificially coined word *autonomation* paraphrasing the ability of an automated system to stop (either independently or through the intervention of an operator) when problems arise in the form of machine malfunctions, quality problems, or assembly faults. Technically, this usually occurs through internal sensors in the production machines or via a 'rip cord' along the manufacturing or assembly line. A malfunction is indicated by a signal lamp on the affected station as well as on a clearly visible electronic indicator board, quickly triggering specialists from quality control and maintenance and repair services [Oelt00].

In short, it can be said that Lean Production and the underlying Toyota Production System comprises principles for successfully designing procurement and production processes. A large number of enterprises have taken up lean production and since the 1990s established it in the form of so-called '*production systems*'.

Spath extended this approach to what he refers to as 'holistic production system' (HPS):

"Holistic production systems are methodical sets of rules and manuals for producing goods. They present a type of operating instructions for the production especially in consideration of organizational, personnel and economic aspects" [Spa03].

The principles of Lean Production are applicable not only for the automobile industry, but also for many other branches of mechanical, electrical and electronic products. Once again the automobile industry was the pioneer for this approach; however mid-sized enterprises are also increasingly applying it.

Since the focus is more on the actual production systems of the manufacturing and assembly, existing publications on lean production and holistic production systems tend not to address the implications of the concept for factory planning and changeable factories. There is tremendous potential though, in transferring the basic ideas to the factory as a system. This will be shown in the next chapters as we develop it step-by-step into the concept of a changeable factory.

4.9 Fractal Enterprises

Warnecke's concept of a fractal factory [War92] represents an attempt at combining recognizable developments in production organization since the 1970s into a holistic approach. Later he also extended this to include the entire enterprise [War93].

In a relatively new branch of mathematics—fractal geometry—a fractal refers to a geometric form which is self-similar and allows complex structures to be built with a few simple laws. If we apply this to an enterprise, a fractal enterprise is an open system consisting of autonomously acting and similarly oriented units—the fractals—which can clearly be described with regards to their objectives and performance. Through dynamic organizational structures they form a vital organism that reacts to external triggers by changing their structure and behavior [War93].

The concept emphasizes four fundamental organizational principles:

- self-organization through self-responsibility and functional integration,
- self-optimization through a continually developing enterprise,
- orientation on objectives through a holistic, market-oriented system of business objectives, and
- dynamic restructuring measured against the enterprise's individual fractals' rate of obtaining objectives.

Although fractals are in some ways similar to segments, fractal enterprises go beyond segmented and modular factories. Firstly, self-organization is emphasized based on the common orientation of objectives in the sense of an internal market economy. As a result it is possible to make self-initiated changes to a certain extent not only within the fractal, but also with regards to changing the job scope and relation to other fractals from the bottom to the top. Secondly, the employees within the fractals should become the drivers for improvements and changes because they are able to react considerably quicker to turbulent market movements than a bureaucratic organization with centralized planning.

An approach such as this requires the traditional view of processes, material flow, information, efficiency and finances to be extended to soft factors such as informal social relationships, strategic aspects and corporate culture [War95]. It is generally emphasized that there is no universal project plan that leads to a fractal enterprise. Instead, readiness for transformation and for changes in the corporate culture is a key element of a dynamic enterprise's pragmatic approach.

This concept has significantly contributed to a holistic view of production, emphasizing the appreciation of the fractal as permanent value-adding units that collaborate in a network of internal and shared potentials.

4.10 Agility Oriented Competition

In the USA, a business concept known as "Agile Manufacturing" has been developed based on an extensive industrial study; once again the increased responsiveness (agility) to all of the customers' wishes is emphasized [Gold91]. In doing so, the process of developing a new product or new service is prioritized above the manufacturing process so that the time between the first idea and the first sales revenue (concept to cash flow time) is kept as short as possible.

In order to open up new groups of customers, enterprises position themselves within a "competitive space" comprised of four dimensions:

- increasing customer value,
- co-operation,
- organization, and
- people.

This thus takes into account the reality of the constant and unpredictable changes enterprises are exposed to. Within this competitive space the authors recommend enterprises implement a holistic design that is broken down into six levels in order to equip enterprises for this challenge. Figure 4.11 summarizes these along with corresponding methods.

Based on the articulated or anticipated customer wishes marketing is responsible for defining combinations of products and services that provide the maximum benefit for customers. In addition, it is the responsibility of the production to supply these in the desired lot sizes at the required time. The entire design process is then holistically oriented on these tasks, whereby the relationship with the supplier is linked to the production processes in consideration of the customer relationship. Once delivered, the product use and removal phases should be accompanied in the sense of servicing the product throughout its lifecycle. The organizational aspect is concerned with new combinations of technology and highly specialized knowledge within a network consisting of both internal and external members including competitors. Management is then responsible for dispensing of a centralized command and control system and instead leading the way in the sense of a living example of the enterprise's corporate culture,

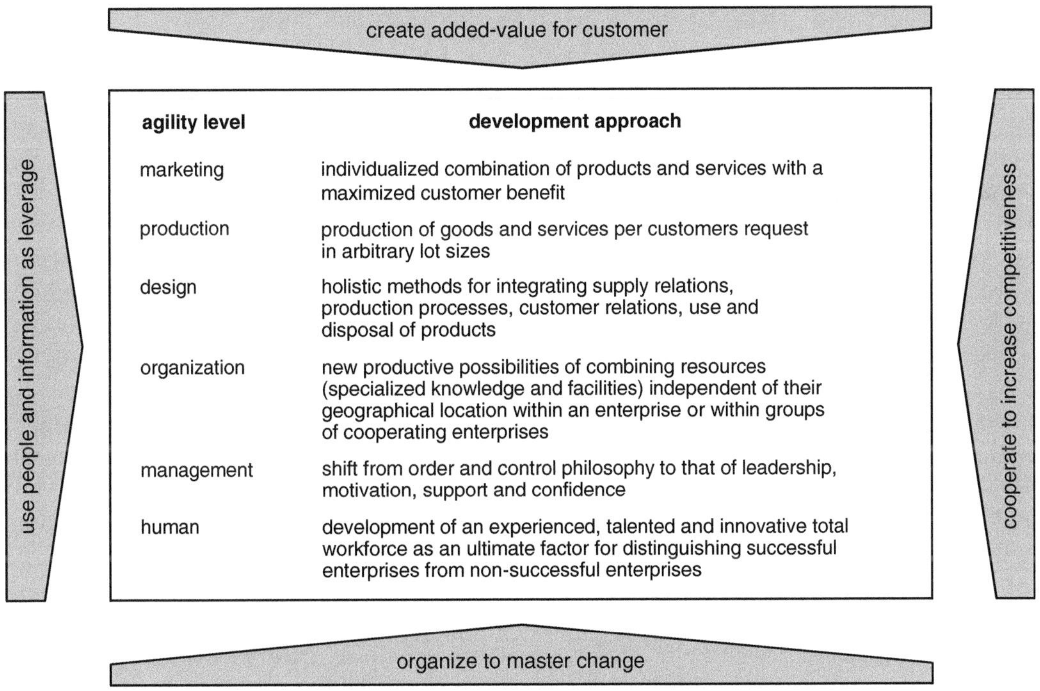

create added-value for customer

use people and information as leverage

cooperate to increase competitiveness

agility level	development approach
marketing	individualized combination of products and services with a maximized customer benefit
production	production of goods and services per customers request in arbitrary lot sizes
design	holistic methods for integrating supply relations, production processes, customer relations, use and disposal of products
organization	new productive possibilities of combining resources (specialized knowledge and facilities) independent of their geographical location within an enterprise or within groups of cooperating enterprises
management	shift from order and control philosophy to that of leadership, motivation, support and confidence
human	development of an experienced, talented and innovative total workforce as an ultimate factor for distinguishing successful enterprises from non-successful enterprises

organize to master change

Fig. 4.11 Characteristics of agile enterprises in a four-dimensional competition (per Goldman et al.). © IFA 10.030SW_B based on trust and supporting employees in a way that motivates them instead of patronizes them. An innovative, talented and experienced workforce is considered the ultimate factor for success and as such needs to be developed and promoted.

Generally, the authors of agility oriented competition do not provide any detailed recommendations or prescriptions for designing processes or factories, but rather guide their readers to consider a living customer relationship and underline the importance of employees and their knowledge, abilities and creativity. In doing so, agile enterprises understand transformation as an opportunity to offensively use new business possibilities. For factory planners, an approach such as this means designing a plant that allows the processes, product orientation and sequences to be quickly changed.

4.11 Mass Customization

In the course of further differentiating themselves from their competitors, many enterprises feel forced to develop continually new variants which meet their customers' specific wishes. In the preceding discussion on agility oriented competition this necessity was already suggested and was considered in view of the entire enterprise. However, this approach leads to continually greater effort and costs for the design, production and logistics departments, which can no longer be covered by adequate prices. Mass customization is an attempted way out of this dilemma (sometimes referred to as a complexity or variant trap).

The goal of this approach, first formulated by Davis, is to apply methods from large scale production to the production of customized products and services [Dav87]. Almost 10 years later Pine, a member of IBM's business consultation division, extended Davis' concept to a series of articles and later in 1993 published them in what can be considered the standard on this subject—his book "Mass Customization" [Pin93]. He moves away from the competition model developed by Porter which, based on a division of product strategies, either assumes the leadership of costs or differentiation into superior product functionality, quality and delivery time. In comparison to it, the goal of mass customization is explained as attacking traditional large scale producers by specifically occupying niche markets and gradually broadening them. Conditions for doing so include a highly flexible production, a networked collaboration of small producers and using information technology to closely link customers, suppliers and producers while developing new products. In the individual enterprises this requires flat hierarchies with largely autonomous teams.

In Germany, it was primarily Piller who promoted 'customized production' [Pil10]; based on numerous examples he describes mass customization as a logical development of the 1980s 'variant production' (see Fig. 4.12).

Whereas in the mass production of the 1960s, the market's demand for lower prices had to be weighed against the demand for a highly efficient production, in the 1970s the market requirements broadened to include the level of quality. This was met in turn by production firms introducing comprehensive quality concepts. With growing diversity on the product side during the 1980s, it became clear that greater flexibility in the production was required. An important approach here was computer integrated manufacturing (CIM) which however, never achieved its anticipated success because organizations failed to sufficiently adjust to it. The required flexibility is only ensured when all of the business processes are reorganized and the production is broken down into customer or product oriented segments. As products and services became more customized, it became necessary to tie-in not only customers, but also suppliers.

Based on this approach, Piller developed the following definition: "Mass customization is the production of goods and services for a (relatively) large sales market, which meets the different needs of each of those requiring the products. In doing so, the products are offered at prices corresponding to the amount purchasers are prepared to pay for comparable standardized mass products" [Pil06].

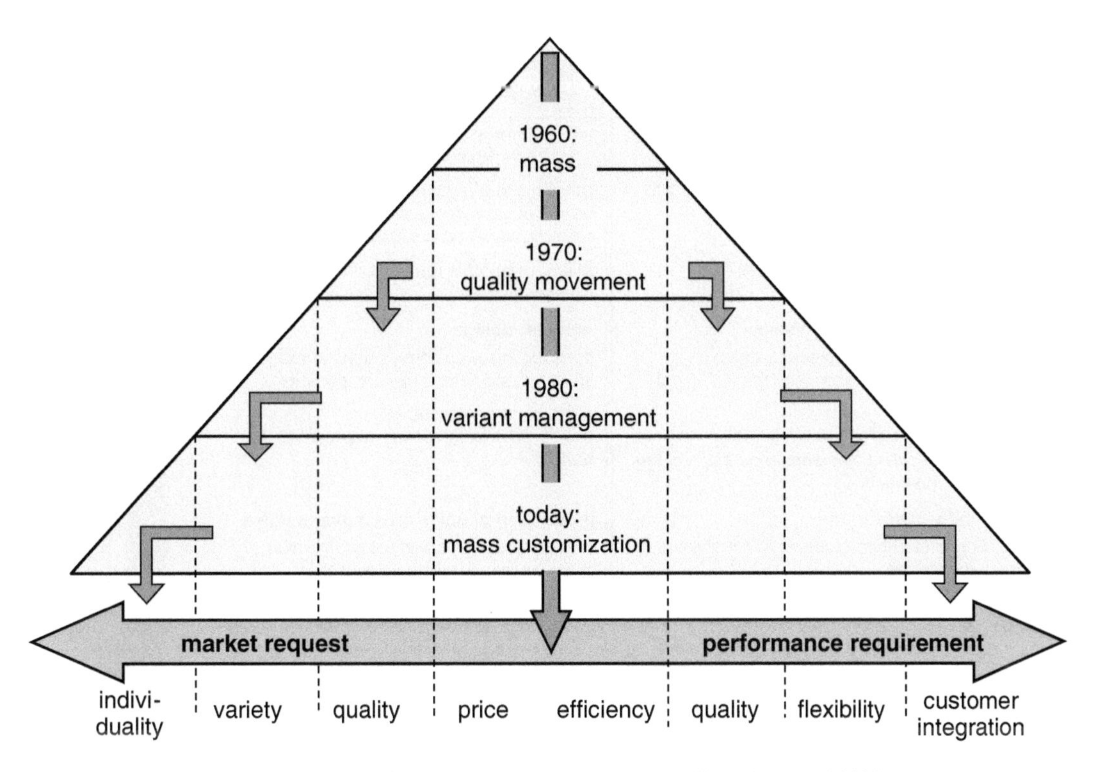

Fig. 4.12 Stages of developing the production for mass customization (Piller). © IFA 10.023SW_B

Mass customization is considered an important instrument for increasing customer loyalty. It can be implemented according to two concepts which differ from one another with respect to the point in time at which the product is individualized (see Fig. 4.13). With so-called 'soft customization' the product is either adjusted by the customers themselves, in the store, or within the frame of servicing. In comparison, with 'hard customization' the individualization occurs in the factory either through corresponding manufacturing operations, assembly of modules or via standardized processes which can be setup for single pieces. Figure 4.13 includes examples for each of the cases, clarifying their individual application strengths. A survey of 200 manufacturing plants in 8 countries applying the concept of Mass customization can be found in [Sal09].

For factory planning, mass customization according to the principle of hard customization means further developing the enterprises organization into product and market specific segments. Here, personnel from product design, production and logistics work together in close proximity, directly oriented on specific orders. Moreover, there is an extensive exchange of information and personal communication along the value-adding chain, production planning and control is more demanding and suppliers also have to be prepared to deliver small quantities quickly. Finally, exceptionally skilled logistics are essential for rapidly delivering finished products to the customer, who usually is an end user.

The next evolution step is towards personalized products and open-innovation; a broad overview of which is given in [Pil10].

4.12 Production Stages Concept

The concept of customized production fosters the continual creation of new product variants. Up until now, the strongly market driven approach primarily targeted consumer goods (e.g., shoes, clothing, furniture, etc.). Nevertheless, its spread

soft customization *no intervention into manufacturing, individualization occurs outside enterprise*	**hard customization** *variety based on manufacturing activities, change of internal functions necessary*	
self individualization design and manufacturing of standardized products with built-in flexibility, being adjusted by the customers themselves. *Bosch: self-designable instrument panel in the car* *Lutron: programming of light controls*	**individual final / pre-production with standardized rest production** either the first (material processing) or the last value-adding steps (assembly, refining) are customized, all others standardized *Mattel: adaptable Barbie doll* *Dolzer: custom-tailored men's suits*	**scope of customer individu-alized value creating steps**
individual finishing in trade/sales delivery of a uniform raw product, which is completed in the store according to customerís wishes *Paris Miki: individual eye glasses design* *Smart: customization of interior and design of the small car at the dealer*	**modular design** construction of custom-designed products from standardized compatible components *Dell: modular computers* *Krone: adaptable commercial vehicles and trailers*	
service individualization completion of standard products with individual secondary services *ChemStation: stock management for cleaning stores* *Zoots: profile administration at a chemical cleaning*	**production of unica on a massive scale** individualization across the entire value chain via standardized processes *Küche-Direkt: fitted kitchens* *My Twinn: dolls based on models* *NBIC: bicycles with individual frames*	

Fig. 4.13 Concepts of mass customization (Piller). © IFA 14.784_B

to capital goods cannot be prevented. The general approach is to generate customized variants by combining customer neutral components (made-to-stock) and custom made parts (made to order), whereby the latter are only manufactured after an order has been placed. The number of variants in the early stages of the manufacturing stages can thus be reduced. Figure 4.14 clarifies the effects that can thus be targeted [Wie04, p. 13].

When variants are built early-on in the production, the outer variance perceived by the customer causes high inventories of semi-finished goods and long throughput times while at the same time requiring a great deal of control effort. If we then question the traditional separation of manufacturing and assembly, it is possible to postpone the manufacturing processes which determine the variants into the assembly and thus reduce the inner variance and thereby the complexity of the entire production. In doing so, the customer neutral product basis is then formed by as few standardized components as possible.

Production is therefore no longer broken down into manufacturing and assembly, but rather into a variant neutral pre-production stage and a variant determining end-production stage. This is what is then referred to as the 'production stages concept'. It was developed and practically tested within the framework of a project conducted together with industrial partners [Wie04]. Figure 4.15 depicts the principle behind it.

Traditional discrete part manufacturing is thus replaced by pre-production processes in which both variant neutral parts and variant neutral components are made. The customized product is then completed in the end-production stage, decoupled by a buffer. This last stage includes manufacturing the remaining variant neutral parts and assembling the remaining variant neutral components before assembling and testing the final product. In the idealized end-production stage the finishing processes which determine the variants are conducted immediately before the variant part is integrated [Wie04].

Based on experiences on the shop floor, the following basic conditions for implementing production stages have been determined:

- The product should have technical variant features which are as uniquely definable as possible.

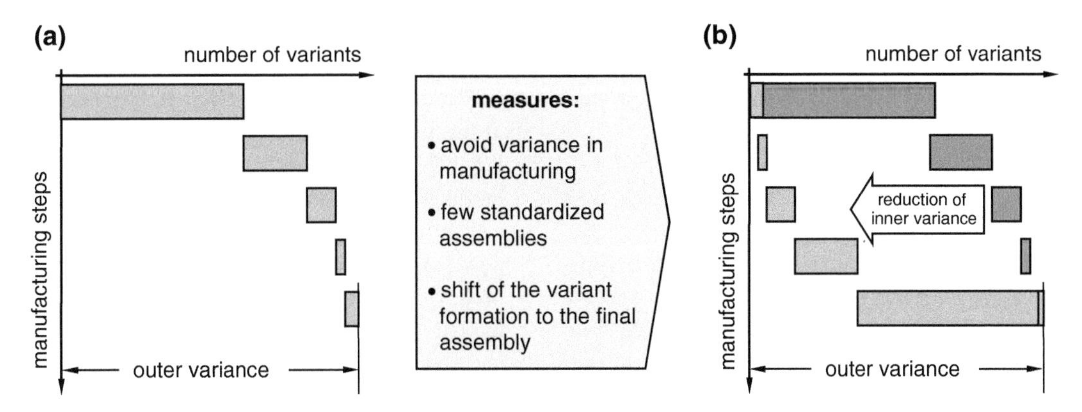

Fig. 4.14 Reducing complexity by creating variants late in the production. **a** Product structure with early variant formation. **b** Product structure with late variant formation. © IFA 12.082SW_B

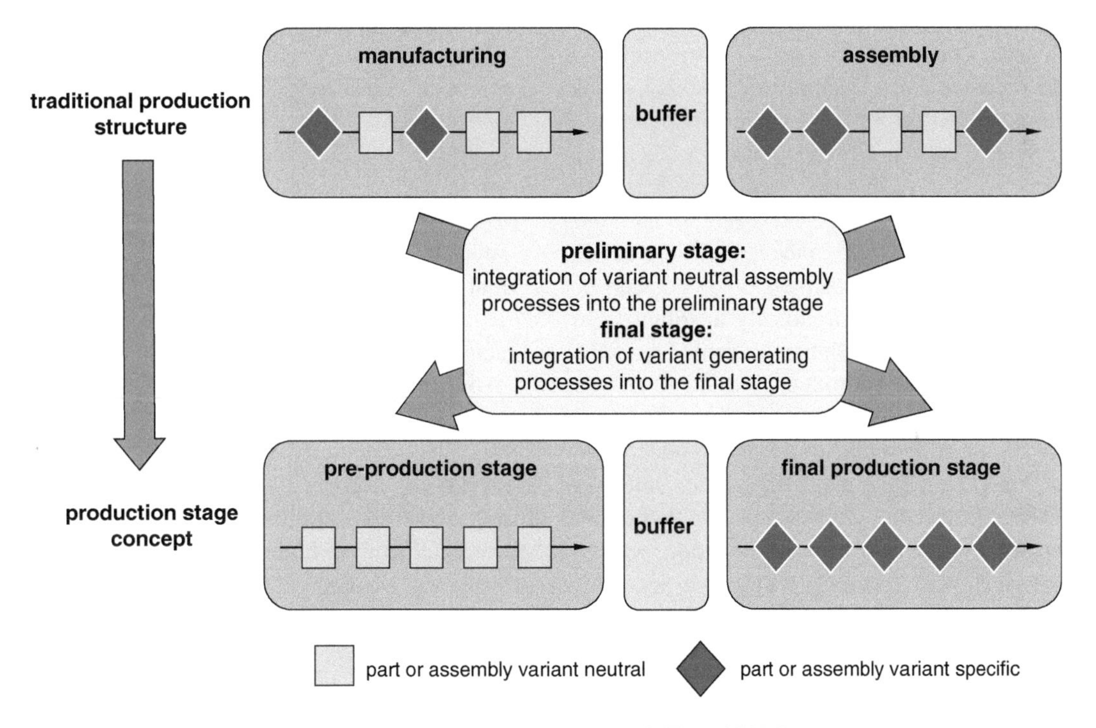

Fig. 4.15 Structure and elements of the production stages concept. © IFA 14.785_B

- It has to be technically possible (manufacturing-wise) to generate these variant features within the frame of the final assembly and to integrate these manufacturing processes into the assembly sequence.
- The product should be a final product or at least have a modular character in order to ensure a strong customer orientation.
- In view of the implemented assembly, manufacturing and testing resources, the end-production stage has to economically cover a large range of quantities without requiring different setups for each variant.
- The part supply is controlled by consumption in the pre-production stage and by demand in the end production stage.

- Employees working in the assembly have to become skilled assembly workers, leaving behind highly repetitive activities for product oriented skills.

This concept was tested in a number of case studies and documented in [Wie04]. It was shown that the product design has to be systematically adapted in relation to the clear separation of variant neutral and variant specific features. Among others, Schuh [Schu05] provides important methods for accomplishing this. In addition, it is necessary to have modular resources that allow pre-finished, variant-neutral parts to be locally processed without a decrease in quality and that can be integrated in the assembly processes as far as the cycle time and environmental conditions are concerned. Finally, the procurement and supply logistics as well as the quality control have to meet the challenges presented by the new requirements. Generally speaking, the production stage concept is technologically demanding, however, it can also mean a new advantage in the global competition.

In a further collaborative project, this concept was also extended to a globally distributed production. With the Global Variant Production System (GVP) the product is divided into the production stages: procurement, competence driven in-house manufacturing and product completion close to the market [Nyh08]. Since this division determines the site specific scope of a factory within a production network, we will extensively discuss the GVP System in Sect. 14.6.

For the factory planning, the production stage concept means special requirements regarding flexible and re-configurable production systems on the section level.

4.13 Research Approaches

- IMS

The most important research program for developing future production systems is the international framework "Intelligent Manufacturing System" (IMS) initiated by Japan. Underneath its umbrella, participants from Australia, Canada, the European Union, the European Free Trade Association, Japan and the USA initially examined the following five areas (www.ims.org):

- lifecycles of future products and production facilities including general models, communication networks, sustainability, recycling and new feasibility studies
- processes in view of sustainability, technological innovation as well as flexible and autonomous production modules
- strategic, planning and development tools for supporting the re-organization and development of strategies
- people, organization and social aspects for improving the reputation of the production, developing workforces, operating autonomous relocated factories, improved knowledge management and suitable performance indicators
- virtual, interconnected enterprises with regards to information and logistics in supply chains, supporting design co-operations and concurrent engineering processes as well as allocating costs, responsibilities and results to the participants of the networked production.

After ten years the first IMS project was completed. It was then re-started, this time with Japan, the Republic of Korea, Switzerland, the USA and the European Union. Currently, it primarily serves as a framework for industrial and research facilities to find world-wide partners for projects on managing 21st century production and organization problems. Five key areas have been defined:

- Sustainable Manufacturing, Products and Services
- Energy Efficient Manufacturing
- Key Technologies
- Standardization
- Innovation, Competence Development and Education.

The common vision from 2011 for the projects is headed "IMS 2020" and can be summarized through three main statements (www.ims.org, IMS2020 Brochure KAT1-5 Roadmap on Sustainable Manufacturing, Energy Efficient Manufacturing and Key Technologies):

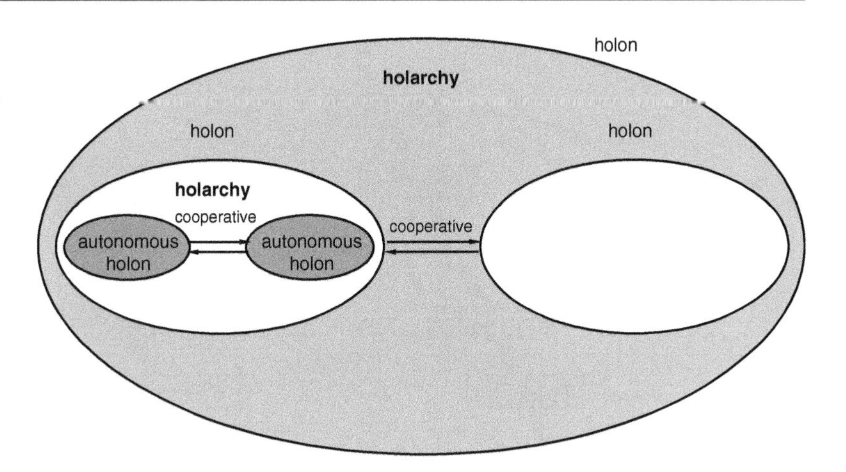

Fig. 4.16 Holonic system with cooperating autonomous holons. © IFA G9507SW_B

1. Rapid and adaptive user-centered manufacturing, which leads to customized and 'eternal' life cycle solutions
2. Highly flexible and self-organizing value chains, which enable different ways of organizing production systems, including infrastructures, and which reduce the time between engaging with end users and delivering a solution
3. Sustainable manufacturing possible due to cultural change of individuals and corporations supported by the enforcement of rules and a regulatory framework co-designed between governments, industries and societies.

One of the results from the first completed IMS project, the Holonic Manufacturing System (HMS), contains important stimuli for changeable factories; we would thus like to briefly introduce it here.

The key idea behind Holonic Manufacturing is the 'holon'—a term coined by Arthur Koestler in 1967 in his book "The Ghost in the Machine" [Koe67]. With 'holon' Koestler describes an autonomous structure within a social or biological system, which not only consists of smaller units but is also part of a larger unit. The word combines the Greek word *holos* (for whole) and the suffix 'on', which approximately means 'part'. Accordingly, a holon is both a whole and a part of a larger whole.

Within the IMS program mentioned above, this concept was transferred to the production world: A holon was thus defined as an autonomous and collaborative structural unit within a production system which converts, transports, stores and/or validates information and physical objects [Sei94].

Similar to a fractal, a holon consists of subunits that are strongly related to one another. The holon co-operates as a whole and in turn with other holons in an organized system referred to as a 'holarchy'. It is both autonomous and cooperative. Figure 4.16 schematically depicts this correlation [Thar96].

If one applies this scheme to a holonic production system, it integrates all of the necessary activities from the entry of an order to the design and manufacturing up to and including marketing in order to realize the agile production enterprise [Sei94]. In doing so, the holarchy defines the rules of the collaboration and the limits of the holons' autonomy. The individual holons can be a physical object (e.g., machinery or a facility) or it can represent information (e.g., a blueprint or a work plan). People are, to the extent it is practical, parts of holons. However, their roles and significance for an enterprise's success is not as strongly emphasized as it is with the approach to agility oriented competition.

The HMS consortium was established in order to implement this still very conceptual idea. Within the frame of this consortium reconfigurable machines that can independently and quickly adapt to changes in the product or quantity and react to disruptions with an

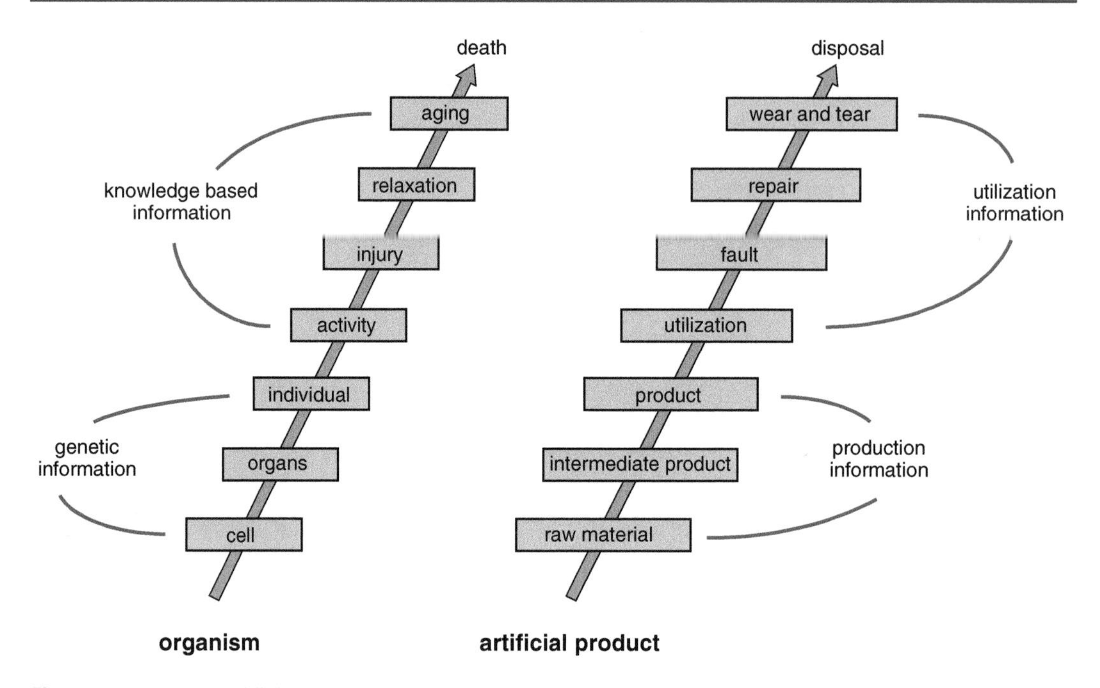

Fig. 4.17 Lifecycles of living organisms and man-made products (Ueda). © IFA G9506SW_B

automated restart or by operating in an emergency mode were developed. However, new control strategies as well as intelligent clamping concepts also belong to the project. For those who may be interested, a concise literature survey is presented by Babiceanu and Chen [Rad06] and an overview of the conceptual frame work is given by van Brussel et al. [Bru99]. Deen's book *Agent Based Manufacturing* [Deen08] describes a number of sub-factors as well as how holons differ from fractals and bionic systems. Meanwhile phase II of the HMS project has been defined and will be followed on [Gru03].

Generally, a holonic manufacturing system aims at being highly responsive to changes in the market and surroundings. As such it should adapt dynamically, continually controlling its plans and strategies.

- Bionic Manufacturing

Another noteworthy approach is the continually evolving Bionic or Biological Manufacturing project, which is supported by Japanese research funds. The approach promoted by Okino [Oki89] and Ueda is based on considerations about the analogy between the lifecycle of living organisms and industrially created products (see Fig. 4.17) [Ued95].

Based on genetic information stored in its DNA (deoxyribonucleic acid) an organism first builds organs from cells and from there a viable entity. It controls itself based on the inherited information and information learned through experiences with its environment (referred to as BN i.e., brains and neurons). The manmade product is similarly created in a production process from raw materials and intermediate products which arise as parts, subunits, and components. The information required to produce them consists of blueprints, parts lists, work plans and employee knowledge. In the course of a product's use, information about its use such as operating hours, malfunctions, and repairs is generated. Currently all of this information is stored externally i.e., not in the product itself.

Bionic Manufacturing aims to transfer knowledge gained from biology about the creation, growth and decay of biological life forms to the production, use, repair and disposal of industrial products. Accordingly, a Bionic Manufacturing System (BMS) arises from autonomous units, which, like organic cells, communicate with their surroundings. Cells are coordinated with the aid of enzymes which correspond to the production planning and control functions in a production system. Just as an organism is hierarchically structured in cells, organs and entities, a BMS also starts its hierarchical structure with the smallest autonomous units which then form groups and ultimately a complete production system.

Similar to the holonic approach a core element is defined with the name 'modelon' in a Bionic Manufacturing System [Oki89]. A modelon consists of a hierarchy of child modelons, operators (who act as enzymes) and an environment for saving information that can be exchanged between the modelons. The entire structure should act as an integrated and harmonious structure, but at the same time consist of autonomous units that are responsible for making their own local decisions.

One of the long term objectives of a Bionic Manufacturing System is for workpieces and products to inherently carry the information required to produce themselves as well as the knowledge about the tools and machines needed. If the product is then to be reproduced, it would transfer this information to the production units and elements and thus trigger the actual production processes.

This approach is being developed in so-called "gentelligent components" at the Leibniz University of Hannover Germany within the frame of the Collaborative Research Centre 653: "Gentelligent Components in Their Lifecycle" (http://www.sfb653.uni-hannover.de). The coined term 'gentelligent' expresses the genetic and intelligent character of the innovative components. A gentelligent part inherently carries the information necessary to uniquely identify and reproduce it as well as information about how it was created. It is thus possible to pass information on to subsequent generations of components in the sense of its lifelong learning during its use phase. New possibilities for use then arise in production and manufacturing planning, manufacturing, maintenance, repair as well as recycling and counterfeit protection. For those interested in reading a case study in which the basic idea was applied we recommend reading [Den10].

This research does not bare any direct consequences for factory planning. However, it is obvious that the general approach is to react quickly and flexibly to changes. This is achieved via dispersed, autonomous and collaborative structures with decentralized control, high flexibility and the abilities not only to learn but also to evolve and reproduce. This will impact production logistics over the long-term in the sense of the manufacturing object controlling itself.

- Manu*future*

The EU has also taken up the subject of future manufacturing within a framework program entitled "Manu*future*" which aims at a European production system comprised of all the components required for competitive and sustainable development (www.manufuture.org) [Jov09]. Figure 4.18 depicts the basic components that should be developed according to its call for applications [Wes07].

On the right side of the figure, the objectives emphasize the conditions of the production such as the corporate culture, quality, environment and social standards; those on the left aim at how the production is shaped with regards to its flexibility within a networked environment as well as its ability to deliver strong performances (both technically as well as logistically) adapted to the changing conditions in the production. These fields are extended to include a sustainable economic, ecological and social management as well as an approach towards an organization that is continually learning.

The initiative has led to the Manu*future* Technology Platform which developed a strategic research agenda based on five objectives (see Fig. 4.19).

The objectives are defined as

- competitiveness in manufacturing industries,
- leadership in manufacturing technologies,
- eco-efficient products and manufacturing, and
- leadership in products and processes, as well as in cultural, ethical and social values.

A comprehensive overview is given in [Jov09]. In the US the reshoring initiative launched in 2010, strives to bring back manufacturing jobs to the US. (http://www.reshorenow.org/).

sustainable management
changeable structures
European social standards
synergetic network formation
methodologies for competitive and sustainable development
European environmental standards
adaptive and high efficiency
innovation culture
technical intelligence
European quality standards
learning enterprise

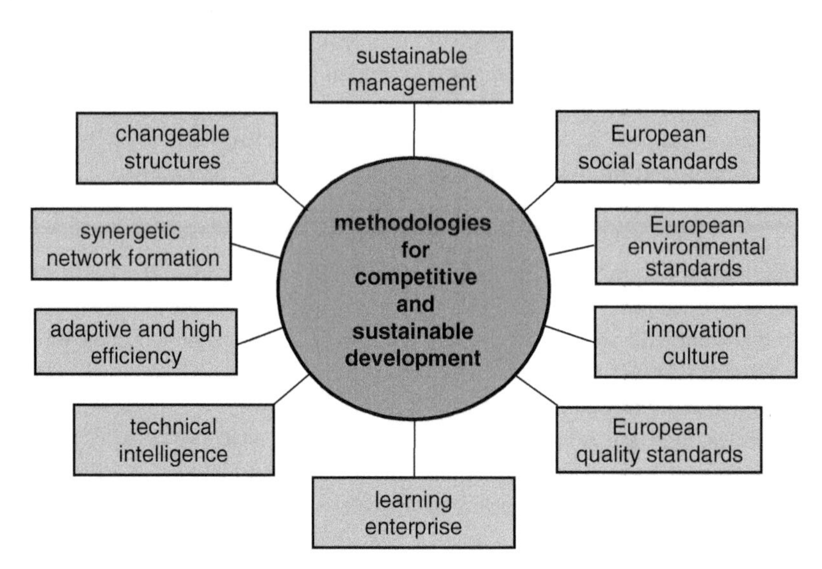

Fig. 4.18 European production system—Manufuture (Westkämper). © IFA 14.787_B

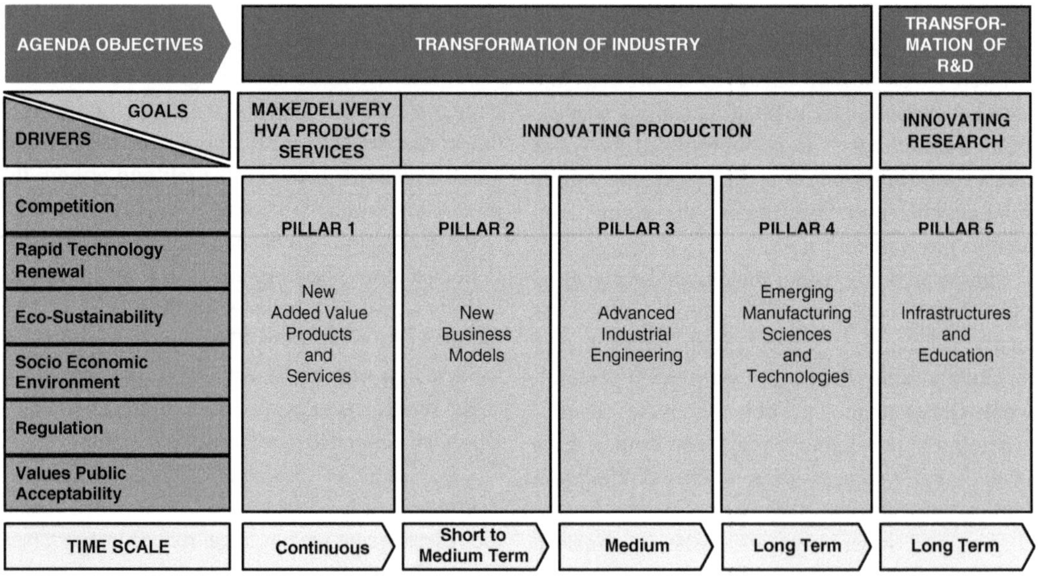

AGENDA OBJECTIVES
TRANSFORMATION OF INDUSTRY
TRANSFORMATION OF R&D
GOALS
DRIVERS
MAKE/DELIVERY HVA PRODUCTS SERVICES
INNOVATING PRODUCTION
INNOVATING RESEARCH
Competition
Rapid Technology Renewal
Eco-Sustainability
Socio Economic Environment
Regulation
Values Public Acceptability
PILLAR 1
New Added Value Products and Services
PILLAR 2
New Business Models
PILLAR 3
Advanced Industrial Engineering
PILLAR 4
Emerging Manufacturing Sciences and Technologies
PILLAR 5
Infrastructures and Education
TIME SCALE
Continuous
Short to Medium Term
Medium
Long Term
Long Term

HVA High Value Added

Fig. 4.19 EU manufuture strategic research agenda

4.14 Summary

All of the discussed production concepts can be summarized with the following characteristics:

1. Modular units with 'local intelligence' with integrated sensors, actuators and information processing/storage, for machining, assembling, storing and transportation.
2. Production systems that adjust quickly to changes in products, their variants and quantities because they are reconfigurable, scalable, mobile and have standardized interfaces.

3. These units can be easily configured into a process chain with which high quality parts, components and products can be produced in very small quantities.
4. Production units are able to be easily networked into cooperative productions.
5. Planning, control and monitoring the technical and logistic variables for the product and processes of the production modules is decentralized. Moreover, self-monitoring/controlling is supported.
6. Employee responsibilities are shifted from 'optimally' executing pre-planned production processes to designing and monitoring tasks with a clear responsibility for results related to the customer's defined product in the demanded quantity, quality, and delivery time.
7. Sustainable development is taken into consideration in relation to economics, the environment and social standards.

Consequently, it is clear that there is a need for changeable factories that can adjust to new challenges on one or more levels depending on what impulse triggers the change. In order to be able to asses this concept, we first have to clarify how it differs from similar concepts such as change-over ability, flexibility, re-configurability, transformability, and agility. The next chapter is dedicated to this.

Bibliography

[Arn75] Arn, E.A.: Group Technology: An Integrated Planning and Implementation Concept for Small and Medium Batch Production. Springer, Berlin (1975)

[Bru99] van Brussel, H., et al.: A conceptional framework for holonic manufacturing: identification of manufacturing holons. J. Manuf. Syst. **18** (1), 35–52 (1999)

[Dav87] Davis, S.M.: Future Perfect. Addison Wesley, Massachusetts (1987)

[Deen08] Deen, S.M. (ed.): Agent Based Manufacturing. Advances in the Holonic Approach (Advanced Information Processing). Springer, Berlin (2008)

[Den10] Denkena, B., Henning, H., Lorenzen, L.-E.: Genetics and intelligence: new approaches in production engineering. Prod. Eng. Res. Dev. (WGP), **4**(1):S. 65–73 (2010)

[Gaug96] Gaugler, E.: The Principles of Scientific Management" – Bedeutung und Nachwirkungen (Significance and Aftermath). In: Grüske, K.-D., Hax, H., Heertje, A., Schefold, B. (eds.) Vademecum zu dem Klassiker der wissenschaftlichen Betriebsführung (Vademecum to the Classics of Scientific Management), pp. 25–47. Verlag Wirtschaft und Finanzen, Düsseldorf (1996)

[Gold91] Goldman, S.L., Preiss, K. (eds.): 21st Century Manufacturing Enterprise Strategy: An Industry-Led View, vol. 2. Iacocca Institute at Lehigh University (1991)

[Gru03] Gruver, W.A., Kotak, D.B., Leeuwen, E.H., van Norrie, D.: Holonic manufacturing systems: phase II. Holonic and multi-agent systems for manufacturing, vol. 2744. Lecture Notes in Computer Science, pp. 1–14 (2003)

[Hop11] Hopp, W, Spearman, M.L.: Factory Physics. 3rd edn. Waveland Press, Long Grove (2011)

[Jov09] Jovane, F., Westkämper, E., Williams, D.: The ManuFuture Road. Towards Competitive and Sustainable High-Adding-Value Manufacturing. Springer, Berlin (2009)

[Koe67] Koestler, A.: *The* Ghost in the Machine. Arcana Books, London (1967)

[Lan54] Lange, P., Roßberg, W: Wege zur wirtschaftlichen Fertigung im Arbeitsmaschinenbau (Approaches for Efficient Manufacturing in Machine Production). Girardet, Essen (1954)

[Mit60] Mitrofanow, S.P.: Wissenschaftliche Grundlagen der Gruppentechnologie (Scientific Principles of Group Technology). VEB Verlag Technik, Berlin (Ost) (1960)

[Nyh09] Nyhuis, P., Wiendahl, H.-P.: Fundamentals of Production Logistics. Theory, Tools and Application. Springer, Berlin (2009)

[Nyh08] Nyhuis, P., Nickel, R.u., Tullius, K. (eds.): Globales Varianten Produktionssystem. Globalisierung mit System (Global Variant Production System). Verlag PZH Produktionstechnisches Zentrum GmbH, Garbsen (2008)

[Oelt00] Oeltjensbruns, H: Organisation der Produktion nach dem Vorbild Toyotas. Analyse, Vorteile und detaillierte Voraussetzungen sowie die Vorgehensweise zur erfolgreichen Einführung am Beispiel eines globalen Automobilkonzerns (Organization of Production Following the Example Toyotas. Analysis, Benefits, and Detailed Preconditions and Procedures for the Successful Introduction of the Example of a Global Automotive Group). Ph.D. thesis. TU Clausthal, Shaker Verlag, Aachen (2000)

[Ohn88] Ohno, T.: Toyota Produktion System. Beyond Large-Scale Production-Productivity Inc., Portland (1988)

[Oki89] Okino, N.: Bionic manufacturing systems—modelon based approach. In: Proceedings International Conference on Object oriented

Manufacturing Systems. University of Galgary, Galgary, pp. 297–302 (1993)
[Opi66] Opitz, H.: Werkstückbeschreibendes Klassifizierungssystem. Verschlüsselungsrichtlinien und Definitionen zum werkstück-beschreibenden Klassifizierungssystem (Workpiece Classification System. Encryption Policies and Definitions). Girardet, Essen (1966)
[Pil06] Piller, F.T.: Kundenindividuelle Massenproduktion. Die Wettbewerbsstrategie der Zukunft. (Customized mass production. The Competitive Strategy of the Future), 4th edn. Hanser, Munich (2004)
[Pil10] Piller, F.T., Tseng, M.M.: Handbook of Research in Mass Customization and Personalization, vol. 2. World Scientific, Singapore (2010)
[Pin93] Pine II, B.J.: Mass Customization. The New Frontier in Business Competition. Harvard Business School Press, Boston (1993)
[PP96] 5S for Operators: 5 Pillars of the Visual Workplace. Created by The Productivity Press Development Team. Adapted from Hiroyuki Hirano, 5 Pillars of the Visual Workplace. The Sourcebook for 5S-Implementation. English Edition 1995 by Productivity Press (based on 5S shido manyuaru, 1990 by Nikkan Kogyo Shimbun Ltd. Tokyo). Productivity Press, Portland (1996)
[Rad06] Babiceanu, R.F., Chen, F.F.: Development and applications of holonic manufacturing systems: a survey. J. Intell. Manuf. **17**(1):111–113 (2006)
[Sal09] Salvador, F., De Holan, P.M., Piller, F.: Cracking the code of mass customization. MIT Sloan Manage. Rev. **50**(3), 71–78 (2009)
[Schu05] Schuh, G.: Produktkomplexität managen. Strategien - Methoden – Tools (Managing Product Complexity. Strategies—Methods—Tools). Hanser, Munich (2005)
[Spa03] Spath, D. (ed.): Ganzheitlich produzieren. Innovative Organisation und Führung (Holistic Production. Innovative Organization and Leadership). LOG_X Verlag, Stuttgart (2003)
[Spur00] Spur, G., Fischer, W.: Georg Schlesinger und die Wissenschaft vom Fabrikbetrieb (George Schlesinger and the Science of Factory Operation). Hanser, Munich (2000)
[Sei94] Seidel, D., Mey, M.: Holonic manufacturing systems: glossary of terms. In: Seidel, D., Mey, M. (eds.) IMS—Holonic Manufacturing Systems: Strategies, vol. 1. IFW, University of Hannover, Hannover (1994)
[Suz87] Suzaki, K.: New Manufacturing Challenge: Techniques for Continuous Improvement. The Free Press, New York (1987)
[Tay11] Taylor, F.W.: The Principles of Scientific Management. Harper, New York (1911)
[Thar96] Tharumarajah, A., Wells, A.J., Nemes, L.: Comparison of the bionic, fractal and holonic manufacturing system concepts. Int. J. Comput. Integr. Manuf. **9**(3), 217–226 (1996)
[Ued95] Ueda, O.: A biological approach to complexity in manufacturing systems. Ann. CIRP (1995)
[War92] Warnecke, H.-J.: Die Fraktale Fabrik – Revolution der Unternehmenskultur (The Fractal Factory—Revolution of the Corporate Culture). Springer, Berlin (1992)
[War93] Warnecke, H.-J.: Revolution der Unternehmenskultur – Das Fraktale Unternehmen (Revolution in Corporate Culture—The Fractal Company), 2nd edn. Springer, Berlin (1993)
[War95] Warnecke, H.-J. (ed.): Aufbruch zum fraktalen Unternehmen – Praxisbeispiele für neues Denken und Handeln (Departure to the Fractal Company—Practical Examples of New Thinking and Action). Springer, Berlin (1995)
[Wes07] Westkämper, E.: Die Strategische Forschungsagenda Deutschland – Ergebnisse des Roadmapping-Prozesses MANUFUTURE D (The Strategic Research Agenda Germany—Results of the Road Mapping Process MANUFUTURE D. Manufuture Germany Platform (2007)
[Wie95] Wiendahl, H.-P.: Load Oriented Manufacturing Control. Springer, Berlin (1995)
[Wie04] Wiendahl, H.-P., Gerst, D., Keunecke, L. (Ed.): Variantenbeherrschung in der Montage. Konzept und Praxis der flexiblen Produktionsendstufe (Mastering Variants in the Assembly. Concept and Practice of Flexible Production Stage). Springer, Berlin (2004)
[Wild98] Wildemann, H.: Die modulare Fabrik: kundennahe Produktion durch Fertigungssegmentierung. (The Modular Factory: Customer-Oriented Production of Manufacturing Segmentation). 5th edn. TCW, München (1998)
[Wild00] Wildemann, H.: Innovative Fertigungsstrategien auf der Basis modularer Produktionsstrukturen (Innovative Production Strategies on the Basis of Modular Production Structures). In: Baumgarten, H., Wiendahl, H.-P., Zentes, J. (Hrsg.): Logistik-Management, Kap. 7 Feb 2001, pp. 1–36. Springer, Berlin (2000)
[Wom90] Womack, D.T., Jones, J.P., Roos, D.: The machine that changed the world : based on the Massachusetts Institute of Technology 5-million-dollar 5-year study on the future of the automobile. Rawson Associates, New York (1990)

5 Systematics of Changeability

5.1 Introduction

Generally, due to the globalization of the goods and service market since the 1990s, the extent and speed with which these changes have to be implemented, has significantly increased. These concern both the market offerings and the business processes, which are introduced in Sects. 2.3 and 2.4 as the strategic basis of a factory. We refer to the overall ability to undertake these adjustments as changeability. However, a countless number of related terms are to be found in publications and in the practice e.g., flexibility, reconfigurability, adaptability, agility, transformability and dynamic. Therefore, in the following, we will more closely consider the terms fundamental to production and factory planning and organize them systematically.

5.2 Flexibility

Flexibility of production is the most frequently discussed concept in this context. Extensive meta-analyses (e.g., [Ton98] which is based on 120 publications about this topic), have shown that flexibility is either static or dynamic. *Static flexibility* describes the ability to steadily operate within a defined range of products, processes and their quantities with regards to quality, costs and delivery time. In comparison, *dynamic flexibility* describes the ability to change the production system with regards to its capacity, structure and processes quickly and without any substantial costs. Flexibility can either refer to the entire value-adding chain from the supplier to the customer (horizontal classification) or to different layers of the production, from individual workstations to sections up to sites and production networks (vertical classification). Moreover the *time aspect of the flexibility*—which can also be described as the speed of response—should be considered. Here, flexibility can be categorized as short, medium or long term (also known as operative, tactic and strategic flexibility, respectively). Finally, the object which the flexibility of the production's performance is concerned is also pertinent. This then addresses on the one hand, the volume and mix of the product spectrum and on the other hand, the items it contains with their different base materials, manufacturing methods and work sequence.

What proves to be problematic is *measuring the flexibility* and the costs that are associated with it. Currently, there are still no generally accepted methods or approaches for determining this. The most important classes according to Toni and Tonchia are 'direct', 'indirect' and 'synthetic' aggregated indicators. The first, *direct*, analyzes how the flexibility of the observed system behaves in different situations based on possible options or measures, whereas *indirect* indicators examine the character of the flexibility (technological, organizational) or the costs and/or effort connected to the flexibility. With *synthetic* indicators the aim is to set the (internal) system flexibility in relation to the strived for (external)

H.-P. Wiendahl et al., *Handbook Factory Planning and Design*,
DOI 10.1007/978-3-662-46391-8_5, © Springer-Verlag Berlin Heidelberg 2015

objective and from that to calculate a type of fulfillment rate. In the end, the flexibility cannot be precisely measured; rather, it is comparable to the abilities of a person or organization to react to disruptions in their environment within an appropriate time and with a suitable amount of effort without endangering themselves.

Flexibility is also increasingly considered a strategic approach to turbulent environments and as such is further differentiated. In an extensive and thorough meta-analysis of 70 published sources, Rakesh Narain, R.C. Yadav et al. demonstrate the lack of guidelines for determining the flexibility an organization requires [Rak00]. The authors suggest that there are three different types of flexibility (necessary, sufficient and competitive) to each of which they allocate certain classes of problems and approaches to solutions (see Fig. 5.1).

Necessary flexibility is required to be able to quickly react to operative problems, which sporadically and unpredictably arise in the form of product changes, machine malfunctions, absent personnel, supplier problems and demand fluctuations. They directly impact the technological, logistic, and personnel related resources that participate in processing orders. Solutions in this area are aimed at ensuring they are sufficiently elastic and easily converted. Medium term, tactical flexibility—referred to by the authors as *sufficient flexibility*—ensures that processes have the ability and certainty required for today's business with regard to product quality, delivery time and delivery reliability as well as production costs. Manufacturing processes thus have to allow different parts to be finished with different materials and without higher costs. This requires: (a) machines and measuring tools that are easily converted, (b) flexible handling and supply of parts and (c) employees to be trained accordingly. Finally, the strategically based *competitive flexibility*, which works over the long term, aims at controlling product changes as well as the supplier's and market behavior. The entire production is considered here, whereby the solutions cited focus on the machinery level and their

feature	expression		
flexibility type	necessary	sufficient	competitive
focus	operative short-term	tactical medium-term	strategic long-term
problem class	A unforeseeable/ sporadic problems	B product quality, cost and time	C product and environmental changes
flexibility elements	• machine • product • workers • materials management • operation sequence • volume	• process • operations • program • material	• production • expansion • market
solution trials	• universal machines • layout • modular PPC system • universal fixtures • NC control	• machine flexibility • universal machining centers • tool and fixtures handling • material supply • cross qualified employee • design for manufacturing	• convertible machines • new manufacturing technologies • alternate routings • flexible material handling systems • modular, flexible machine cells • flexible layout • factory information and control system

Fig. 5.1 Characterization of types of production flexibility (per Rakesh Narain a.o.). © IFA 9897SW_B

handling devices as well as on the layout and control system level.

In keeping with the character of the publications, no concrete suggestions are made for the three types of flexibility. Moreover, the analysis does not consider the relation to logistics, buildings or their equipment and does not include the production site and the development of it. Nevertheless, this classification provides a valuable basis for systematically addressing how to design flexibility.

In German publications, Kaluza, among others, intensively examined the concept of flexibility based on his own extensive work as well as an evaluation of numerous publications [Kal05]. He defines a broad notion of flexibility which should include the fundamental operational aspects:

"Flexibility is the ability of a system to allow proactive or reactive as well as targeted changes in the system's configuration in order to fulfill the altering conditions of its surroundings" [Kal05, p. 9].

With regards to the notion of production flexibility, which is of particular interest for us here, Kaluza distinguishes between a 'real' and 'dispositive' flexibility [Kal95]. *Real flexibility* describes the ability of the personnel, technology and materials to adjust, whereby the first two factors are primarily of interest. Their flexibility is further categorized into qualitative, quantitative or structural. Figure 5.2 depicts the resulting system of flexibility types as well as the selected instruments or flexibility measures that are allocated to them [Kal89].

Whereas *qualitative flexibility* characterizes the basic ability of the personnel and technological resources to complete various tasks, *quantitative flexibility* describes the range of each of the performance indicators quantity, time and intensity wise. *Structural flexibility* pertains to both personnel and products. On the personnel side, it depends on how successfully the borders between planning, execution and control tasks can be removed with measures for expanding the work areas. In comparison on the production side, structural flexibility is determined by the type of layout and control and is described by the routing freedom, redundancy of production facilities and storage capacity.

In addition to this real flexibility, which could also be interpreted as potential flexibility, Kaluza posits the previously mentioned *dispositive flexibility*. Here, he distinguishes between two types:

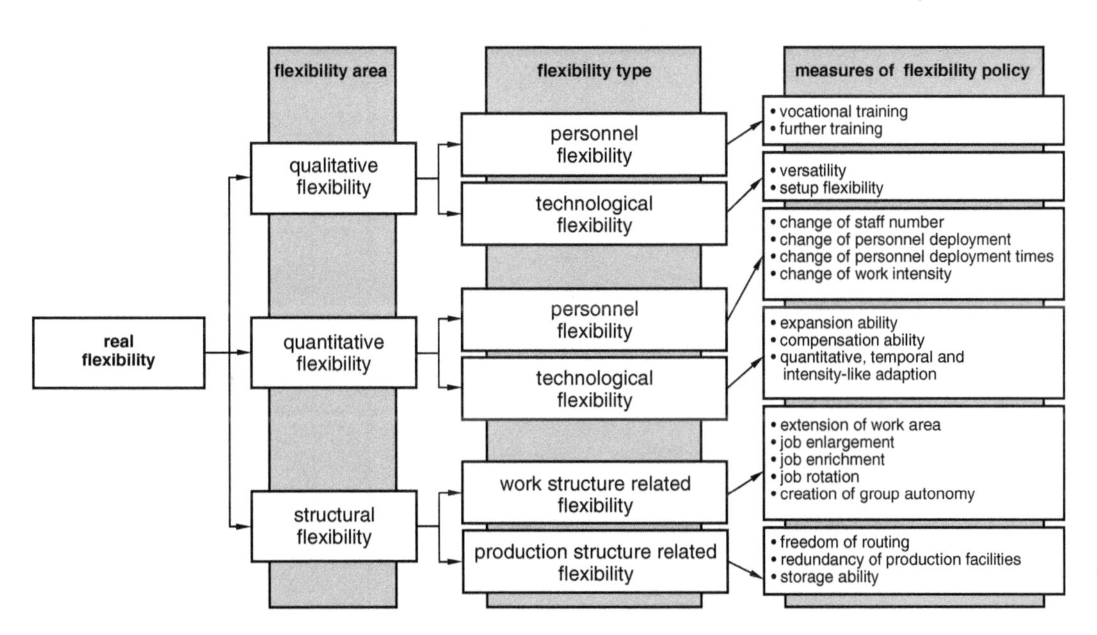

Fig. 5.2 Scope and types of real flexibilities (per Kaluza). © IFA 10.000SW_B

flexibility of the production planning and flexibility of the production control. Supportive measures for the first type of flexibility include measures for structuring the product and systems for planning production, whereas the second type of flexibility can be supported by production control methods and improved communication.

With these types of flexibility, Kaluza already addresses fundamental elements which clearly need to be considered when planning a changeable factory. These include the work organization, production facilities and logistical planning and controlling of the order processing.

Further important impetuses for considering a production's flexibility arise from the work on part family manufacturing and group technology that has been taking place since the 1960s and is aimed at overcoming the disadvantages associated with job shop productions i.e., high WIP levels and long throughput times. Manufacturing cells, segments, flexible production systems and even lean production are a result of it (see Sects. 4.4–4.7). All of these have to be flexible on the one hand and on the other hand allow machines to be utilized economically.

In 1981, the Institute of Production Systems and Logistics at the University of Hannover already attempted to describe how production flexibility is structured by breaking it down into three sub-concepts [Wie81]. Figure 5.3 outlines this suggestion according to areas and types of flexibility, supplemented with examples.

Technological flexibility describes the possibility of implementing different manufacturing processes in one machine (versatility). This allows different workpieces of a basic form e.g., rotational parts or cubic parts, to be completely manufactured as far as possible in one setup within the workspace of one machine. In comparison, *setup flexibility* means being able to execute different manufacturing tasks with an economically feasible degree of effort. *Structural flexibility*, also referred to as routing freedom, allows an order with different operation sequences to be guided through a manufacturing system. It is generally determined by the more or less strict orientation of the layout on the processing sequence of the operations. Finally, *capacitive flexibility* describes the quantitative reserve of a production system (expansion

flexibility area	flexibility type	examples
technological flexibility	versatility	turning drilling milling
	setup flexibility	tool work piece
structural flexibility	freedom of routing	tied to operations sequence; lead time; sequence orientation layout
capacitive flexibility	expansion ability	cap.; time
	compensational ability	vol.; A B C products; time
	storage capacity	vol.; sales; production; time

Fig. 5.3 Structure of production flexibility. © IFA 10.001SW_B

potential), the possibility for shifts within the production program (compensation ability), and the possibility to balance differences in sales and capacity trends by storing semi-finished or finished products (storage capability). These definitions which only refer to part manufacturing represent a further building block on the way to the concept of a changeable factory.

Similar considerations as those for part manufacturing were developed by Eversheim in the early 1980s for assembly systems [Eve83]. In order to design the necessary modular elements, the types of assembly flexibility had to be defined (see Fig. 5.4).

Whereas, *date oriented flexibility* refers to the assembly processes running on each of the individual stations and either allows switch-overs for individual workpieces or re-routing, *time period oriented flexibility* concerns the setup or conversion of an entire assembly system to another variant or product. *Disruption flexibility*, which plays a particular role in assembly systems due to the short cycle times, is event oriented and generally concerns failsafe strategies that are to be implemented when functions are unpredictably disrupted due to malfunctions. Solutions for such assembly systems are increasingly available on the market. Manual or automated stations can be swapped out quickly in an assembly system, in order to adjust to different products or fluctuations in the number of pieces [Lot06].

While traditional German studies on factory planning such as those by Kettner and Aggteleky already explored the notion of flexibility on the factory level quite early-on (see Fig. 5.5 and [Her03]), Anglo-American literature has yet to emphasize this topic (see e.g. [Mut89, Her06, Tom10]).

Kettner recommends planning as far ahead as possible during the factory planning phase, which allows certain flexibility in the schedule as a reaction to changes during the planning and maintains a reserve in the sense of over-dimensioning. The factory itself should be easily expandable and have capacities with a specific degree of flexibility [Ket84].

Aggteleky already distinguishes more concretely between the flexibility of the structure and that of the layout [Agg87]. The first ensures against deviating operating conditions through universal facilities and the fault elasticity of the production, whereas the layout flexibility

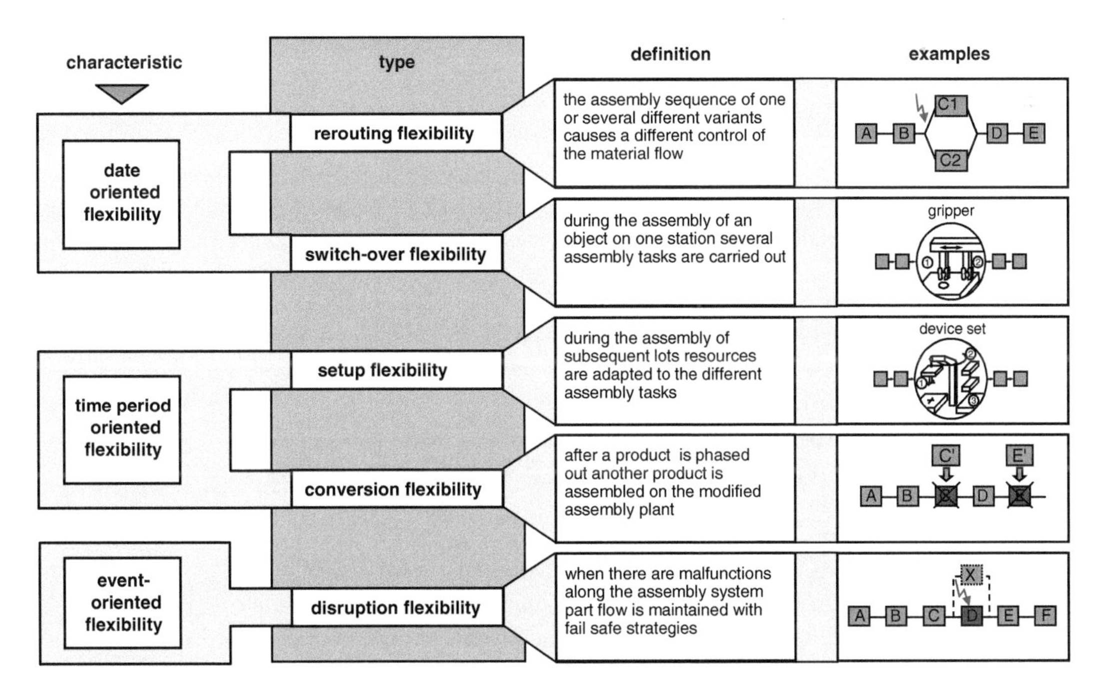

Fig. 5.4 Types of flexibility in assembly systems (Eversheim). © IFA G1115SW_B

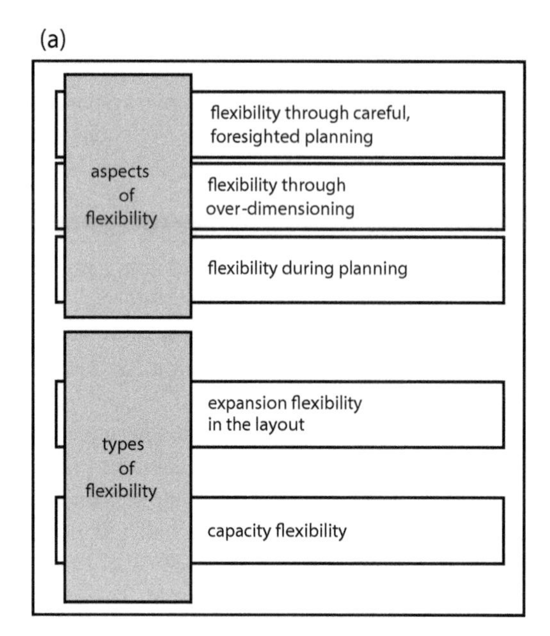

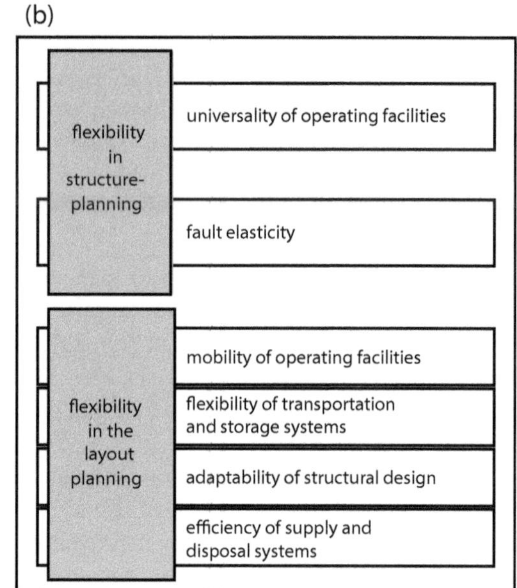

Fig. 5.5 Approaches to flexibility in factory planning. © IFA 10.125_B. **a** Approach according to Kettner, **b** approach according to Aggteleky

addresses the basic factory elements: manufacturing and assembly, storage and transport systems, buildings and technical infrastructure.

Both these approaches clearly show that due to the comparably stable market conditions, the scope of changes typical for today could not yet be considered back then and that the concept of changeability was not required.

5.3 Reconfigurability

From the perspective of manufacturing technology, solutions that make machine tools and production facilities more flexible technology wise need to be emphasized. Since the 1990s they have been discussed under the notion of reconfigurability. Here, the focus is on dividing manufacturing equipment into functional components and thus making it possible to quickly reconfigure machines e.g., by inserting a movement axis or a spindle. After being mechanically coupled, they are recognized by a higher level control and are productive once a control program starts. The first implementation of this technology was introduced in the USA by Koren [Kor01]. In Germany, within the frame of a public funded research project referred to as METEOR (http://www.meteor2010.de), solutions for reconfigurable machine tools and manufacturing systems were developed together with the machine tool industry [Abe06]. Whereas, reconfigurable assembly systems could be considered state of the art, reconfigurable manufacturing systems tend to still be in the research and development phase. A comprehensive overview of the current state of research on flexible and reconfigurable production systems is provided by [Wie07].

5.4 Changeability and Change Enablers

From the perspective of factory planning, the question of which flexibility is required for the entire factory has been discussed since the end of the 1990s under the concept of *changeability*.

Already in 1997, Reinhart referred to changeability as a new dimension of flexibility [Rein97]. He defined the concept as a combination of flexibility and responsiveness [Rein00], whereby flexibility is understood as the "possibility for change within the provided dimensions and scenarios" and "responsiveness [is] a potential for being able to act beyond expected dimensions and corridors". Later, Reinhart explained the concept of changeability more precisely:

"Changeability is understood as the potential which makes it possible to quickly adapt also beyond given corridors in relation to organization and technology without having to extensively invest" [Rein08].

Figure 5.6 provides a visual depiction of this (see also [Nyh08, p. 25]). Accordingly, when there are change drivers that do not exceed a certain degree, the "built-in" flexibility of the system comes into play. The required change thus takes place within the system without having to convert it and reconvert it. If the requirement of a change driver exceeds the thus defined flexibility corridor, the system has to be changed. A solution space within which the system can be modified is foreseen for such situations. This space allows almost any configuration of resources, however is still limited for example, with regard to the size and precision of products. If a change driver arises necessitating a modification, (e.g. a considerable increase in the number of pieces), a structural change, which can however be built-back, is required.

Westkämper [West99] also contributes important impulses for the changeability of the entire production enterprise. As depicted in Fig. 5.7, he differentiates the changeability of company structures according to elements (real estate, mobile property, information processing and personnel) as well as time horizons (short/mid/long term).

From there, Westkämper derives the technological innovation necessary for allowing the production to be continuously re-planned and reconfigured. He then also recommends concrete approaches for accomplishing this [West00]. Here too, flexibility is distinguished from changeability:

"A system is referred to as flexible when it is reversibly adaptable to changed circumstances within the frame of a generally anticipated span of features and expressions."

Moreover:

"A system is referred to as changeable when its processes, structures and behavior inherently possess a specific, implementable variability. Changeable systems are capable of not only adapting in reaction but also able to intervene in

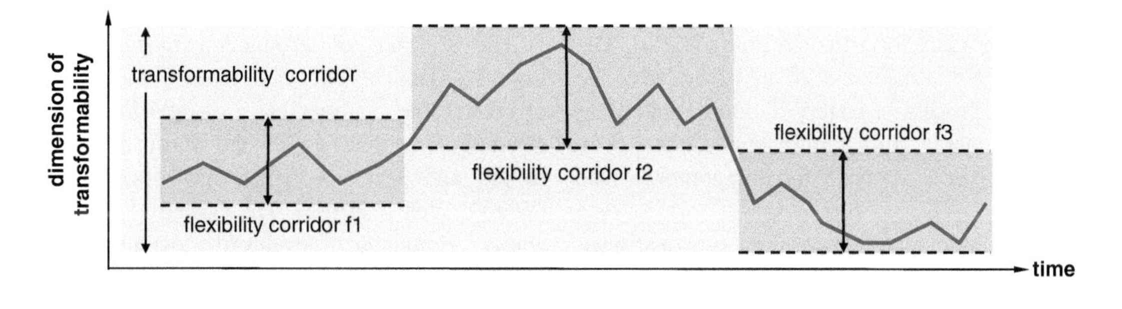

dimension	transformability	flexibility
▪ volume / variants	▪ preplanned solution space	▪ provided capability range
▪ costs / time	▪ transformation if necessary	▪ scalable within set corridors
▪ process quality	▪ dismantling option as basic attribute	▪ dismantling not planned

Fig. 5.6 Comparison of flexibility and changeability (Zäh, Reinhart). © IFA 14.788_B

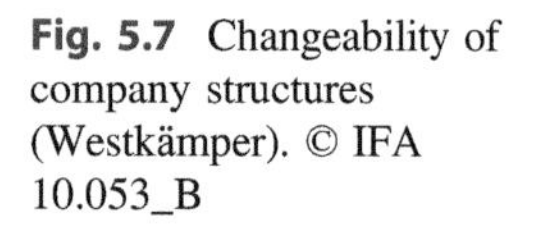

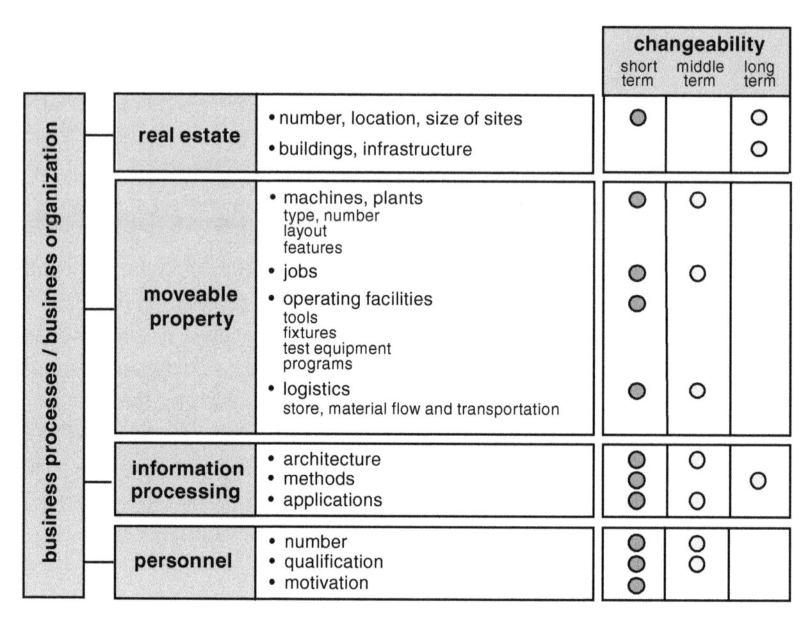

Fig. 5.7 Changeability of company structures (Westkämper). © IFA 10.053_B

anticipation. These activities can work towards changing the system as well as the environment."

Management, people, technology and organization are singled out as the basic starting points for designing the changeability. Stemming from there, Westkämper provides an extensive description of the Stuttgart approach to a changeable company in [West09].

Wirth defines a further developed form of changeable factories as *flexible temporary factories* which only serve a specific market with a specific product for a limited time [Wir00]. The knowledge that the length of the lifecycles of the products, processes, factory buildings, and area-use continually drift farther and farther away from another is decisive for this approach (see Fig. 1.4).

In addition to the previously known discussion about the product and production processes, the type of building (universal, low cost or modular mobile buildings) and the role of the factory grounds are focused on within the frame of the city/town planning. Along with that, Wirth sees a change in the roles and function of factory planning. In addition to the traditional "core planning" of resources, personnel and areas he includes also the local lifecycle of a temporary factory with preparations, ramp-up, dismantling and relocation as well as its external network and logistics.

As a further development of this approach, Schenk and Wirth suggest a factory based on a network of competences integrated in a heterarchical (in comparison to hierarchical) network organization [Sche04, p. 364 f]. It consists of the smallest value-adding units capable of surviving and of changing, i.e., so-called 'competence cells'.

The Institute of Production Systems and Logistics (IFA) also began to research the subject of changeable factories early on and has made concrete contributions in the form of talks, papers and factories which have been built [Wie00, Wie01, Her03]. The system of changeability Hernández developed together with Wiendahl at IFA stems from system theory and provides the foundation for this book with regards to the changeability of a factory [Her03]. In doing so, particular emphasis is placed on considering the architectural requirements early. From there, recommendations for incorporating process and spatial perspectives in the initial stages of the factory planning were derived, and further developed by Nyhuis and Reichardt into an approach referred to as 'synergetic factory planning' [Nyh04, Rei07].

The IFA approach to changeability starts by defining a factory as a system which, in this context, possesses the basic properties mentioned in Fig. 5.8 [Ulr95].

The notion of *wholeness and parts* emphasizes that the quality of a factory is not the sum of the qualities of its parts, but rather the *interaction* of its parts as a whole. The *degree of interlinkage* describes the density of relationships. The individual elements are not simply linear, but rather are linked together in intermeshed control loops and to some extent backwards coupled. The *openness* of the factory results from the strong correlation to the environment. There is no doubt that a factory has a high degree of complexity—complexity is in fact necessary for its survival as it allows the factory to take on different states quickly [WEM12]. Complexity is based on the number of elements and the possible relationships between them and their surroundings. The *dynamic* of the 'factory' system describes the behavior while processes are being conducted and results from the change in the system elements. The ability to then regulate the system is described with *control*. This is accomplished to some extent automatically, however predominantly via employees. The *ability to develop* can be interpreted as the ability to learn and react to impulses by adjusting or changing. Finally, the purpose and task orientation is the driver to suffice the environment's expectations and demands e.g., from the market, politics, local surroundings etc.

A system always strives towards a state of equilibrium with its environment, which in the case of environmental changes necessitates adjustments. If it does not possess this as a change enabling quality, it loses its balance becoming unstable, even to the point of its destruction. System theory recognizes two types of changes which are identified as structural coupling and transformation (see Fig. 5.9).

With *structural coupling* only the relations between the system elements change. It can thus be interpreted as a flexible reaction which proceeds with the aid of defined control mechanisms, such as redirecting an order to an alternative machine. *Transformation* on the other

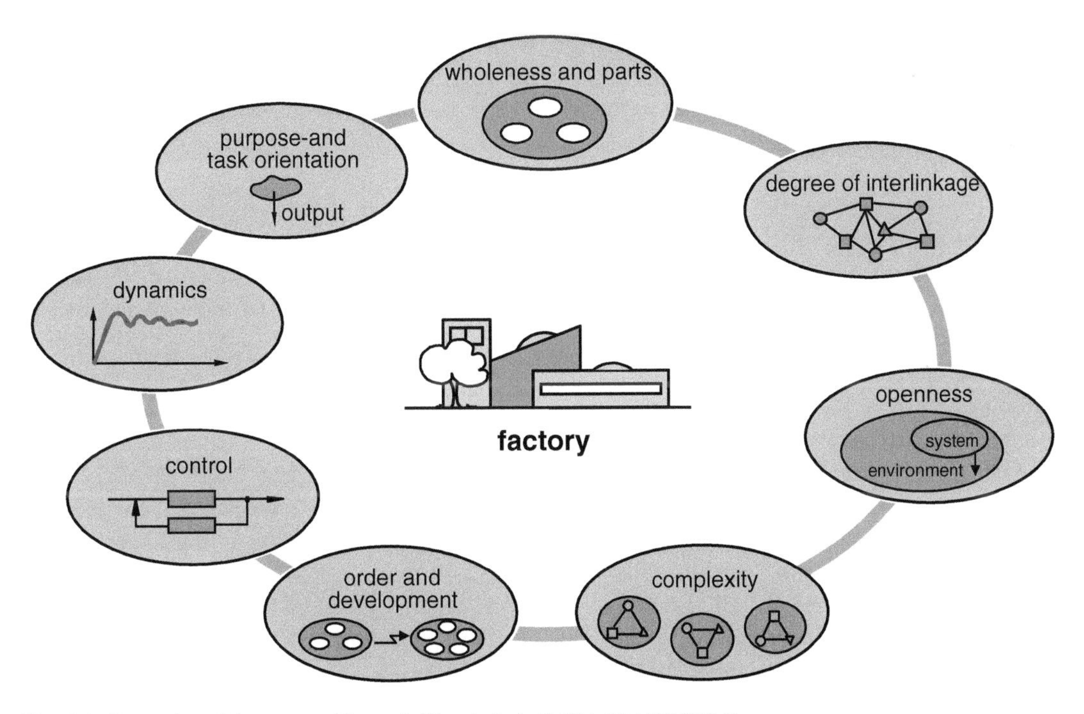

Fig. 5.8 Properties of the system 'factory' (Hernández). © IFA 10.137BSW_B

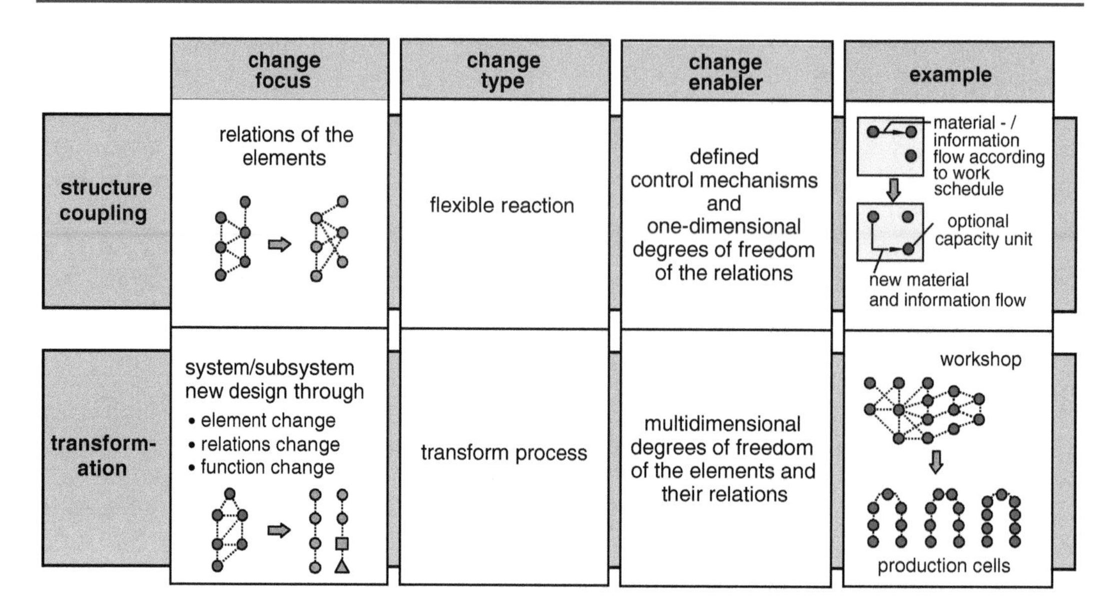

Fig. 5.9 Types of factory changes from the perspective of system theory (Hernández). © IFA 10.146_Wd_B

hand, changes not only the relation of the elements, but also their qualities and functions up until the point that new structures and systems are created. An example here is the transformation of a workshop production into a number of manufacturing cells.

The changeability of the 'factory' system thus allows the transformation of a system and is supported by three of the eight characteristics of a system mentioned in Fig. 5.8. They are presented once more in Fig. 5.10 accompanied by the relevant system properties [Her03].

In order to be able to realize a transformation, the system has to possess specific characteristics, which in the following will be referred to as *change enablers*. These are inherent characteristics which can be activated in a specific time period and create a desired change. The change enablers identified in Fig. 5.11 can be derived from the three system properties we described above as being relevant to change [Her03].

Mobility as well as *expandability and reducibility* can be allocated to the system's dynamics. They characterize the objects' ability to change with regards to the location and extension. *Modularity* as well as the *function and utilization neutrality* are linked to *complexity* and describe the ability to take on different system states. Finally, the change enablers *linking ability* and *disintegration/integration ability* are derived from the *degree of connectivity*.

In practically applying these concepts it became evident that they could be simplified further and reduced to the five enablers depicted in Fig. 5.12 along with their corresponding definitions.

When it comes to practically implementing changeability, in addition to these considerations about the system from a technological perspective, it is important to consider the actors in the enterprise who decide about the degree of changeability and how it will be concretely realized.

- From the perspective of *management* the question of interest is how quickly an entire enterprise should react to risks and opportunities, whereby aspects such as market and product strategy, financing, cooperation, organization and site are in the foreground.
- The *business economics* is concerned with the opportunities and risks as well as the cost-benefit relation of changeability. Is it worth investing e.g., in increasing the changeability of a production through a flexible manufacturing

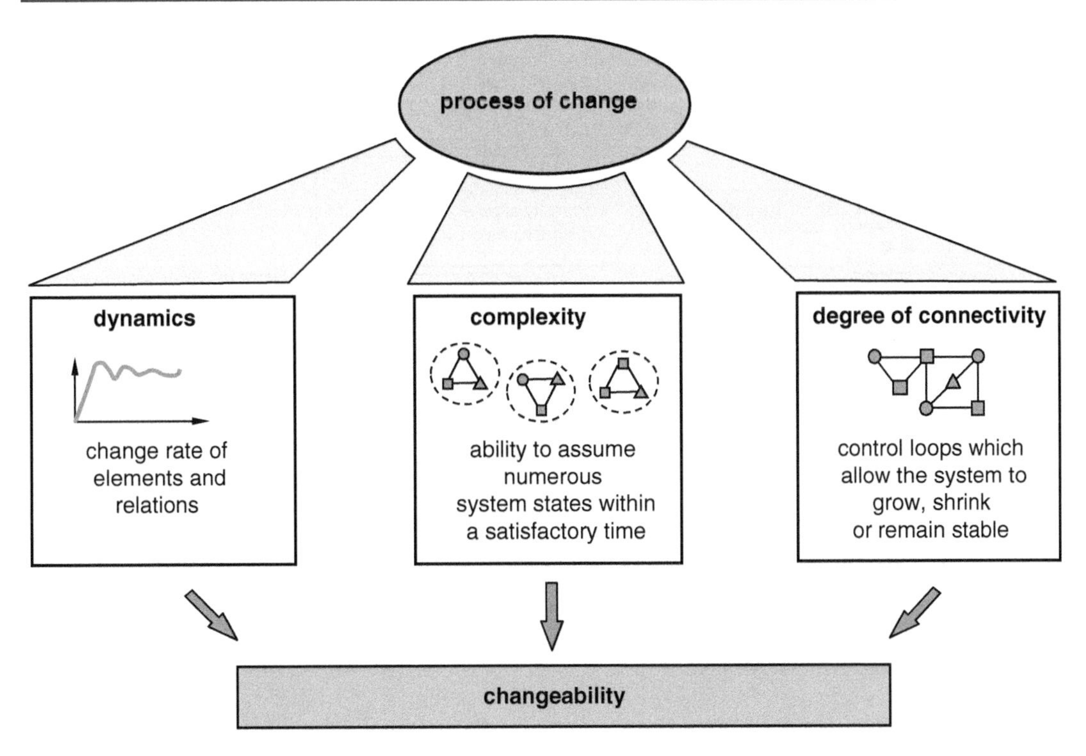

Fig. 5.10 Deriving changeability from the properties of a system (Hernandez). © IFA 10.149A_B

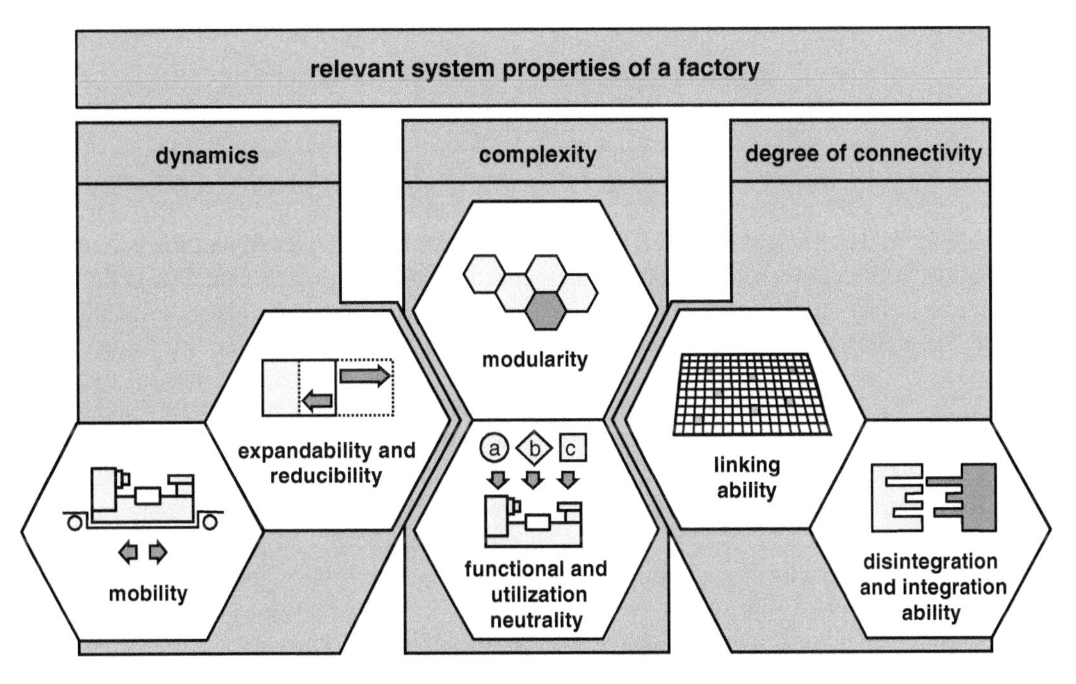

Fig. 5.11 Deriving a factory's change enablers. © IFA 10.211D_Wd_B

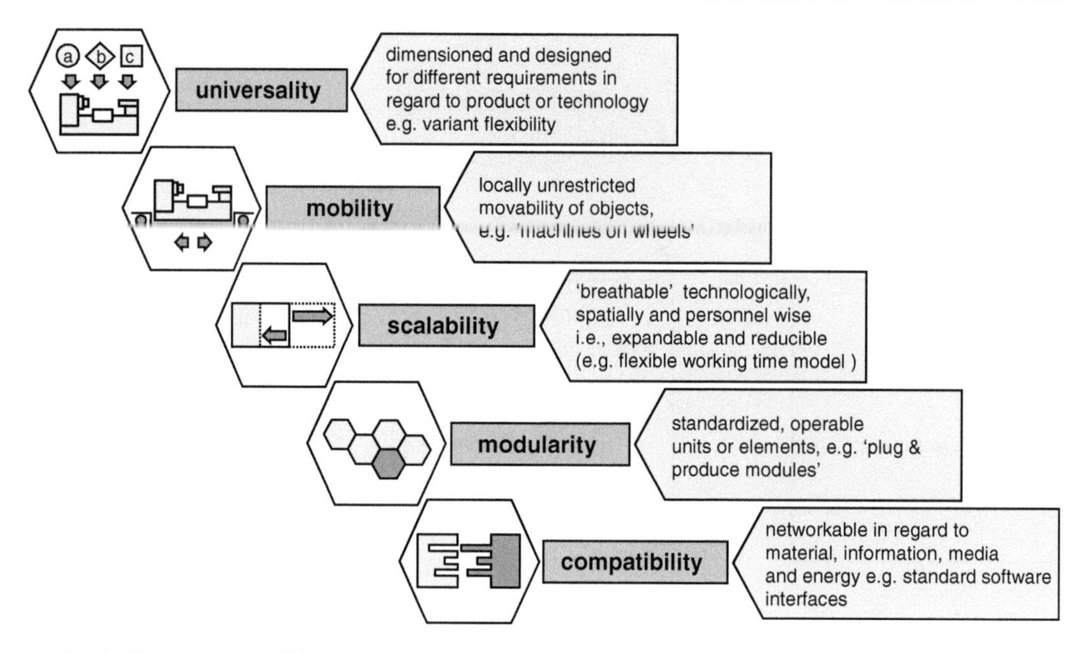

Fig. 5.12 Change enablers. © IFA 15.053_B

system which only pays itself off after the second or third product change?

- A third perspective concerns the *technical realization* of the changeability of the individual factory elements, beginning with the manufacturing and assembly facilities, as well as the logistic systems and their control up to and including the buildings and their facilities.
- Finally, from the *industrial engineering* perspective there is the question of which conditions need to be met on the level of the employees with regards to their motivation, qualification and remuneration in order to ensure the production is smoothly adjusted.

5.5 Aspects of Designing Changeability

For the individual enterprises the question now is how to define and concretely design the flexibility, reconfigurability and transformability that is demanded by all sides. In order to do so, it seems practical to first select a generic term for the different types of adjustability and to later put it into concrete terms for the various classes and orders of a factory's objects. In the following, based on numerous discussions on the international level, the term 'changeability' has been selected (see [Wie07]).

The next step is to identify the aspects which need to be designed in addition to those in a traditional factory planning in order to attain the desired changeability; an overview of this approach is provided in Fig. 5.13.

We will start by clarifying the external and internal *change drivers* (see Fig. 1.8), which present themselves in the demand volatility and the variety of goods and services forced by the market. One of the frequent change drivers is a new business strategy, triggered by a change in ownership or management. The enterprise can react by re-designing the market offering or the production performance. In both cases the changeability has to be adapted via the outlined *change enablers*. In regard to the market performance these change enablers include e.g., constructing modular products or services, introducing a platform concept, or using programming to build variants at a later time. Generally speaking, in the case of the production

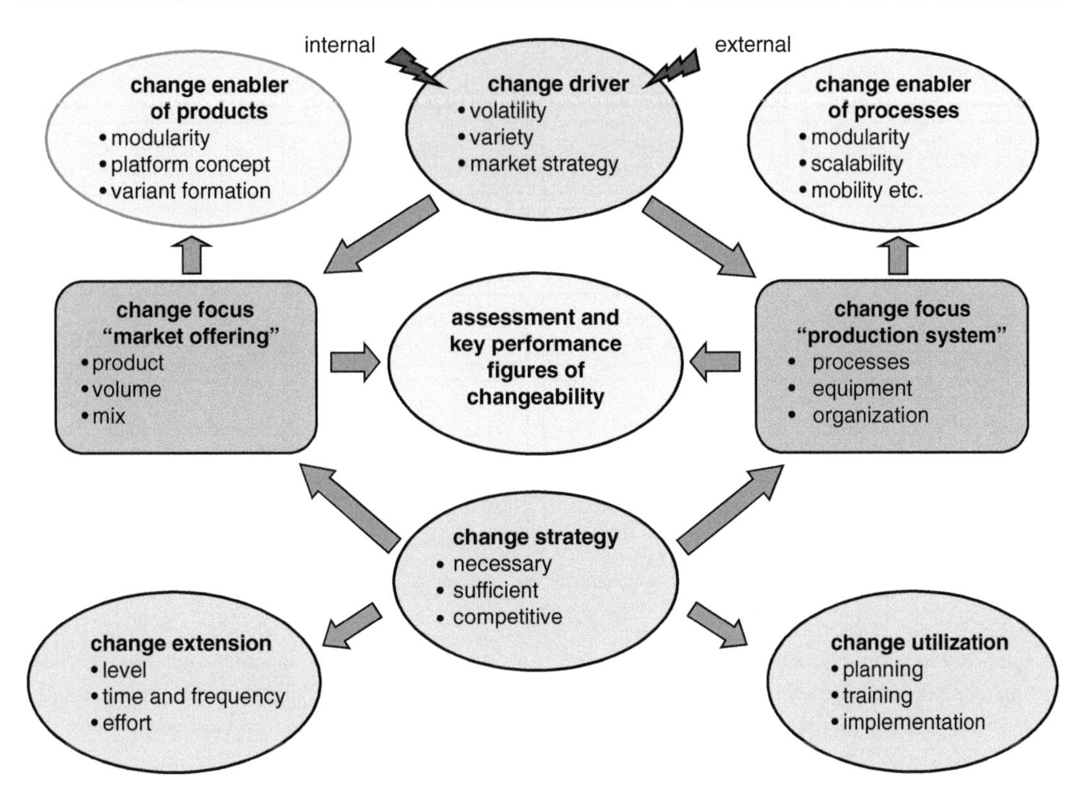

Fig. 5.13 Factors that impact the changeability of market and production performance. © IFA 14.790_B

performance the manufacturing processes, production facilities and possibly the organization should be designed to be changeable. In addition to modularity, scalability and mobility are useful change enablers here.

The *degree* to which the changeability should be increased is dependent on the strategy selected, which—as already mentioned—ranges from 'immediately required' to 'temporarily sufficient' up to a 'strategic orientation'. The degree of changeability that should be pursued can only be determined from here and is characterized by the level of change and the acceptable duration of the change as well as the costs deemed permissible e.g., extra charges for having technological building services that are easy to modify.

The improved changeability remains worthless if it cannot be quickly activated when there is an impulse for change. It is thus necessary to develop a concept for utilizing the changeability i.e., in the form of a plan of action, trainings for required personnel as well as ensuring that the technical means for implementing it are available. This approach can be created in analogy to a concept for fast setup processes.

Finally, it is preferable to be able to economically evaluate planned or existing changeability and to be able to prove it as far as possible with *key performance figures*.

5.6 Morphology of Changeability

A morphology matrix for the changeability of a production enterprise can be developed from the diversity of influential factors and their characteristics (see Fig. 5.14). Theoretically, in such a matrix each shaping of a factor can be combined with each of the others; changeability can therefore appear in a vast array of forms. In order to apply these practically, it is useful to break them down into different types. Before doing so however, we will first briefly introduce the factors and how they express themselves.

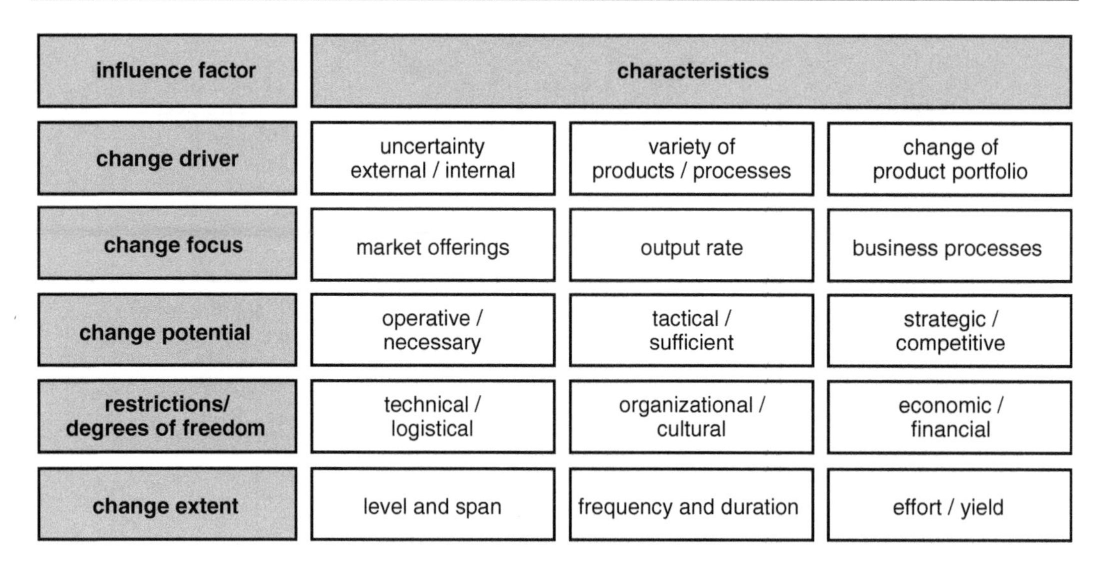

influence factor	characteristics		
change driver	uncertainty external / internal	variety of products / processes	change of product portfolio
change focus	market offerings	output rate	business processes
change potential	operative / necessary	tactical / sufficient	strategic / competitive
restrictions/ degrees of freedom	technical / logistical	organizational / cultural	economic / financial
change extent	level and span	frequency and duration	effort / yield

Fig. 5.14 Morphology of the changeability of manufacturing companies. © IFA 9903SW_B

First, *change drivers* are not only influenced by the markets uncertainty and variety of products in the sense of risks, but also contain opportunities with the availability of new manufacturing methods (in particular: laser technology, information and communication technology as well as micro-, Nano- and RFID technology). New forms of co-operations already supported by the internet are also used in development, supplier, production and logistic networks.

The second influential factor, the *change focus*, comprises three objects and is depicted in Fig. 5.15 along with the abovementioned change drivers. In addition to the product mix (comprising of functionally superior products with significant customer benefit), the performance required from the market perspective also consists of an ability to adjust delivery volumes with demand fluctuations while reducing delivery times, increasing delivery reliability and at the same time decreasing production costs. The production performance (as the generic term for the function of the production to fulfill orders) is considered here on the level of six enabling elements, which can serve as the focus of the changeability. These basically entail the manufacturing technology and the related production logistics, the hierarchical and process organization including employees, as well as the production buildings and the land they are built on. The interactions of these elements with the market offering need to be concretely identified within the scope of factory planning.

The two expressions of changeability mentioned above are initially oriented on improving the market offering (external view) or the production performance (internal view). However, there are also interactions between these two. A new product requires new production performances. Inversely though, an initially new production technology that is product-neutral e.g., the introduction of an electron-beam welder can offer new possibilities for designing products.

The third change focus thus refers to business processes (see Fig. 2.7). In addition to the main processes (market opening, product development, order obtainment, order fulfillment and service) the supportive processes (human resources, information and communication technology, accounting, general services and quality management) deserve equal consideration with regard to a company's changeability. In view of the growing significance of services as a field of operation of its own, particular attention should be dedicated to its increased changeability.

Usually, the primary change focus is the market offering. Based on an analysis of the business processes the demands on the

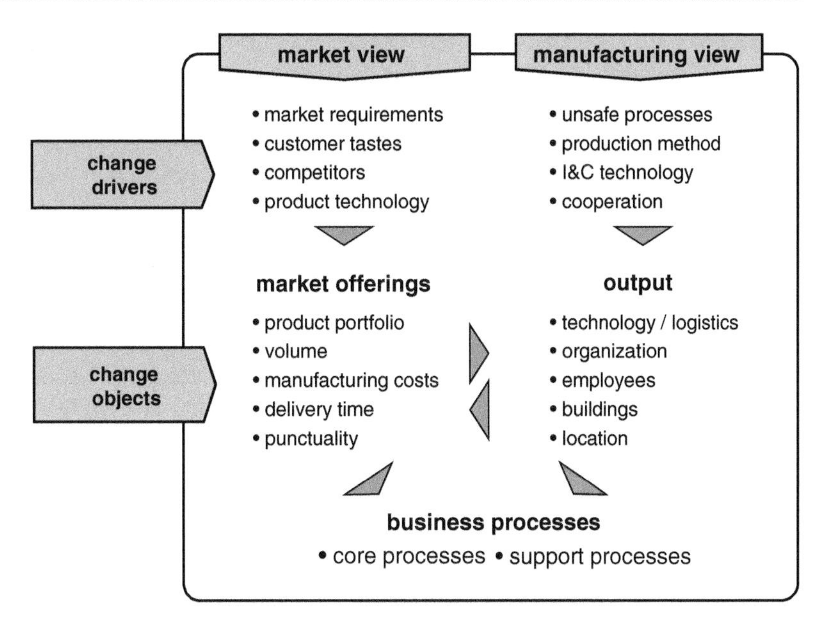

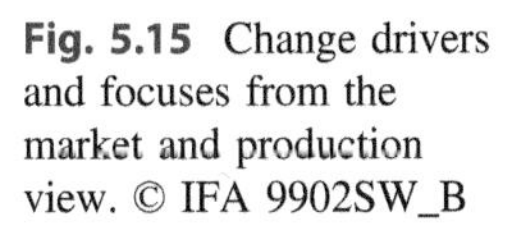
Fig. 5.15 Change drivers and focuses from the market and production view. © IFA 9902SW_B

production process and their changeability are determined from it. However, in other cases the progressive deterioration of the technical, logistical and economic production performance can also provide the reason for a fundamental change.

After deciding upon the focus of the change, the question arises of how much changeability should be 'built-into' the market offering or production performance along with how great the *change potential* needs to be. This is ultimately dependent on the chosen strategy regarding the desired change potential. There are three types of change potential 'necessary', 'sufficient' and 'competitive' which can also be characterized as operative, tactic and strategic.

The sort of changes allocated to the *operative change potential* concern the usual market fluctuations and disruptions that are unavoidable even in a relatively stable environment. These are reacted to spontaneously within the frame of practiced routines and do not, for example, require products or production systems to be structurally changed. From the product side, examples include designing variants or modular systems that can be tailored to each customer. From the production side this might mean the change-over of a machine or assembly station including changing the control program, tool and fixture in order to obtain the necessary change potential.

In comparison, the *tactical change potential* is concerned with the consistent ability to deliver a defined product spectrum in the medium term with sufficient certainty in regards to the quality, costs and logistic objectives 'delivery time' and 'delivery reliability'. This includes for example, measures for introducing manufacturing methods requiring no setups, but also allows manufacturing, assembly and logistic structures to be quickly changed for instance by introducing manufacturing segments, reducing the manufacturing depth or having components supplied just-in-time.

Lastly, the *strategic change potential* is aimed at being able to introduce new product variants, products and processes very quickly. In doing so, the firm should gain competitive advantages in regard to the price or delivery times, which surprise both the customer and competitors. The strategy here is to productively generate turbulence instead of only managing it reactively.

As already discussed an enterprise does not have an unlimited *degree of freedom;* describing it as precisely as possible therefore serves to expose the actual or supposed limitations concerning the changeability. First, we have to differentiate between the *technical and logistical degrees of freedom*, which we can also refer to as the hardware degrees of freedom. Here, we are concerned with which types of materials, manufacturing processes, assembly techniques, handling/transportation and storage processes can be managed at all including the planning, controlling and testing processes. In comparison, the *organizational-cultural degrees of freedom* are more "soft" in nature. They affect the possibility of changing the structural and procedural organization without considerable resistance of the employees and attaining the necessary qualifications, learning abilities, and readiness for change. The latter is obviously a question of the corporate and in particular management culture. Finally, the *economic degrees of freedom* are often decisive for a desirable changeability. These can be demands on the economic efficiency of an investment, such as a specified group-wide payback period or also a financial limitation in the form of a given investment sum for converting a production or for building a new factory.

The last basic influential factor of changeability according to Fig. 5.14 is defining the *extent of change*. Here the *level* and *span* of the changeability that will be pursued for the product or production has to be clarified. On the product side, it can range from single pieces and their material, form, size and precision up to the product mix, whereas on the production side it can extend from individual workstations up to the location in a production network.

A *time related change* characteristic is the frequency of the possible changes, which from the product side is coupled with the rate of order change-overs, product modifications, introduction of new products or changes in the product portfolio. In comparison, extreme cases in the production sometime include setup changes a number of times per day, capacity changes a number of times per week, structural changes a number of times per month or site changes every few years. Closely related to the frequency of the changes is of course their duration. Generally it can be determined that operative changes from one order to the next should lay within the range of minutes where possible, whereas structural changes with tactical character are required within the span of weeks to months. Even strategic changes in the products themselves or an entire production must be possible within a span of a year, in order to be able to cover the conversion costs still with premium prices for originator products. A type of changeability that goes beyond these is related to the enterprise as a whole, which is searching in a global market for future areas of operation in which the product portfolio is established via a sales and product network.

Finally, the permissible effort for a change (measured on the internal and external personnel capacity as well as the related earnings) are highly significant for efficiently designing changeable technical, organizational or personnel elements of the market offer or product output.

5.7 Classes of Changeability for Production Performance

As already indicated, when applied in industry it is not very practical to define only one of the changeability aspects for an entire production enterprise. Rather, changeability serves as a generic or umbrella term for different classes of changeability corresponding to the different levels of a production, which can be allocated to corresponding levels of the market offer.

These levels of the production performance or market offer can each be characterized from the perspective of factory planning with six terms that follow the traditional hierarchy of a factory and its products. These in turn can be allocated to different types of changeability. Figure 5.16 provides an overview (see also [HEM09]).

The lowest level corresponds to the individual *workstation*, which usually consists of one *machine* and an *operator*. Here a defined operation is executed on a workpiece with the aid of specific manufacturing methods e.g., a turning operation, a surface treatment, etc. This leads to a

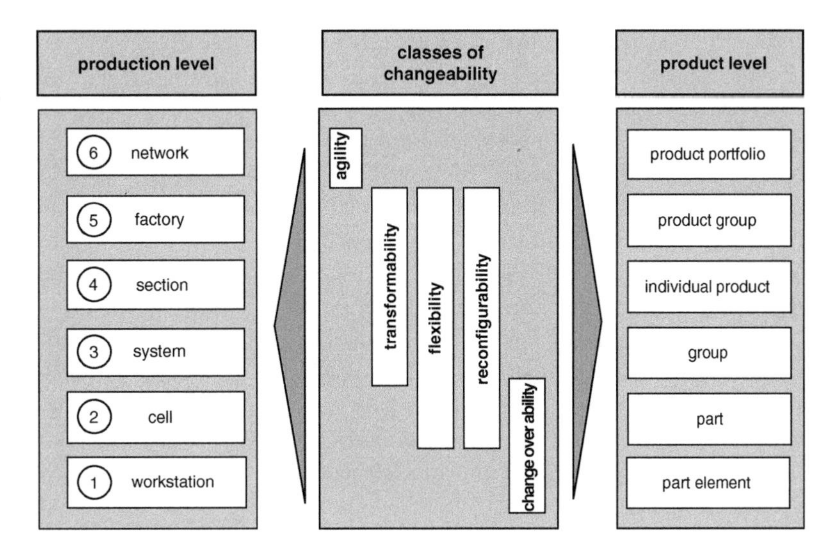

Fig. 5.16 Corresponding levels of production, changeability and products. © IFA 14.791_B

'part element', such as a drill-hole, gearing, or a surface area. Accordingly, a number of parts are joined into a sub-component on an assembly station. In order to alter the process, so-called *change-over ability* is required; on automated stations this is attained by changing the control program.

The next level comprises a *manufacturing cell*, which can execute a sequence of operations in order to produce a ready-to-use workpiece and variants of it. Usually such cells are numerically controlled and automatically change tools. Similarly, a more or less automated process forms a functional component in an assembly cell. Such cells not only have to possess change-over ability but also *flexibility* with regards to new parts or components.

A *system* generally consists of a number of stations or cells and represents a manufacturing or assembly system depending on the operations it conducts. It can be equipped with or without an interim buffer and can arise in different configurations e.g., circle, line, network etc. These systems serve to produce a group of different parts or components, which however own a certain similarity. Since not all of the variants of parts or components are known when the system is being installed, it has to also be possible to change it structurally by inserting or removing components as well as spatially re-arranging these components. Thus in addition to being flexible, they also have to be *reconfigurable*. If these systems in addition own the change enablers defined in Fig. 5.12 they are *transformable*.

Combining a number of such manufacturing or assembly systems together, creates a *section*, whose manufacturing and assembly units are supplemented by logistic systems such as storage, transportation and handling systems. Their task is to produce different components, which are in fact complete products that have been tested and are able to be used. The sections have to be flexible as well as *reconfigurable* when there is a product change. If the sections in addition own the change enablers, they are transformable.

The *factory* level joins a number of such production sections together, each of which yield a defined market offer. In order to do so, in addition to the manufacturing, assembly and logistics, it needs certain infrastructural facilities for supplying materials, energy, media and information as well as for disposal. Here, in addition to the subsystems being reconfigurable, the planning and control as well as the infrastructural systems and employees have to be able to adapt to new tasks. Are the change enablers available for all sections, the factory is said to be transformable, otherwise flexible. The presentation clarifies that a certain factory level may well possess different classes of changeability, depending on the extent to which change enablers are available.

Finally, a factory is generally part of a *production network*. Such networks consist of a number of factories on different locations and are often closely linked with suppliers of product components and sub-products. Changes on this level are usually driven by strategies e.g., entering into a new market, changing the product portfolio by introducing or removing products from those offered, or merging with a newly acquired firm. This requires *agility* and is first and foremost a responsibility of management.

The types of changeability thus described are defined as follows:

- *Change-over ability* describes the operative ability of a single machine or workstation to be able to quickly execute defined operations on a known workpiece or part family at any desired point in time with minimal cost. The change-over is reactive and can occur manually or automatically.
- *Flexibility* refers to the operative ability of a manufacturing or assembly system to be able to reactively adjust itself to a predefined number of workpiece types or components by inserting or removing individual functional elements quickly and with minimal costs in regards to hard/software. The adjustment is to some extent manual but also includes automated functions.
- *Reconfigurability* refers to the tactic ability of an entire production or logistic section to be able to mostly reactively adjust itself to a new—but similar—family of components including the corresponding in-house manufacturing and purchased parts. This adjustment is accomplished by changing manufacturing methods, material flows, and logistic functions over the mid-term with an average amount of effort in regards to hard/software. The adjustment is mostly done manual and generally requires pre planning as well as a ramp-up and optimizing phase.
- *Transformability* refers to the tactic ability of an entire factory, section or system to reactively or proactively adjust itself to a—usually similar—product family and/or to change the production's capacity. This requires structural interventions not only in the production and logistic systems, building structures and their equipment, but also in the structural and procedural organization as well as personnel. The adjustment requires a longer planning period, but can then usually be put in place relatively quickly. It is usually implemented in sub-projects with strict project management and includes both a ramp-up and optimizing phase. On the levels below it, transformability requires flexible, reconfigurable systems that can be changed-over.
- *Agility* refers to the strategic ability of an entire enterprise to mostly proactively open up new markets, develop the necessary market offering and production performance and possibly to do this across a number of sites. It requires considerable abilities in the areas of management, financing and organization.

If we now try to differentiate entire production enterprises with regards to their changeability, in addition to the described types of changeability on the different levels and their objects we also have to consider their ability to network.

Figure 5.17 depicts a portfolio developed from this for strategically positioning a production enterprise with respect to their abilities to adjust. The portfolio is described by these two characteristics (changeability and networking ability) and the degree to which these are expressed (low, medium, high and very high).

Changeability is expressed in correspondence with the concepts mentioned in Fig. 5.16 i.e., change-over ability, reconfigurability, flexibility, transformability and agility. In comparison, the

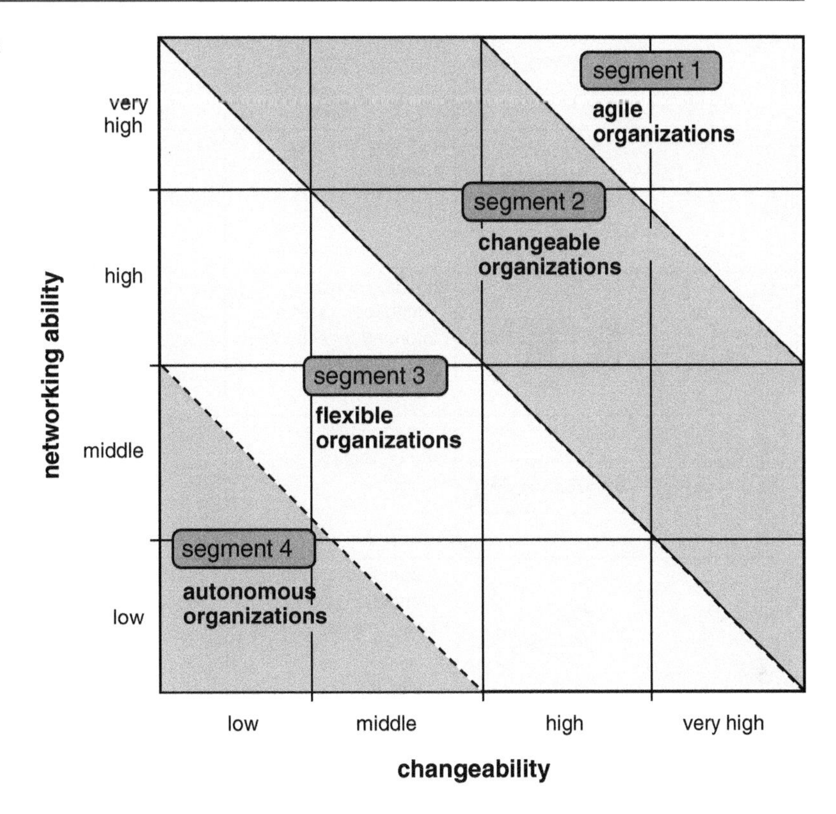

Fig. 5.17 Characterization of production enterprises from the perspective of changeability and networking ability. © IFA 9900SW_B

stages of *networking ability* refer to the intensity of the co-operations with suppliers, development partners, production partners and customers. The lower degree of networking ability corresponds to the networking ability of traditional relationships between suppliers and production enterprises to counter peak capacity demands. A middle degree of networking describes when smaller groups of articles or components are passed onto suppliers, who are already participating in the technical development. With high level networking, basic components or sub-systems are already developed and delivered by a collaborating partner. The production enterprise also has a number of sites, and the work related to products or their components is divided among these.

In situations where there is a high degree of networking, the local production enterprise becomes the integrator for specific market offers by coordinating payments and possibly services for a specific market, which are organized geographically or according to customer groups. Co-operations include development partners (for sub-systems), production partners (for part and component families) as well as logistic partners (for supplying parts, distributing goods and interim storage).

In a field such as this, four segments can be defined whose descriptions center around the types of change i.e., agile organization, transformable organization, flexible organization and autonomous organization. Segments 1, 2 and 3 are self-explanatory. Segment 4 is comprised of autonomous organizations, which only maintain a weak external network with suppliers, whereas internally only the workstations and manufacturing/assembly systems can be changed over or reconfigured.

5.8 Evaluating Changeability

In order to make the concept of changeability in practice manageable, the factory objects first have to be systemized. It is thus recommendable

factory levels \ factory fields		technology	organization	space
site		• building services - centers	• hierarchical structure	• property • site development plan • outdoor areas
factory		• building services - distribution facilities • information technology	• production concept • logistics concept • structure	• layout • building form • building structure • shell • appearance
section sub-section		• storage facilities • transportation • facilities	• work organization	• development
workstation		• production technology • production facilities • other facilities	• quality management • concept	• workplace design

Fig. 5.18 Systemization of factory objects. © IFA 13.440_B

to organize the objects affected by the change according to the factory level of detail on the one hand, and according to the type of changeability, on the other hand. On the left hand-side of Fig. 5.18 the levels of detail for a factory are depicted. Research [Nyh04, Wie05, Rei07] and practical experience in numerous factory projects has shown that in comparison to Fig. 5.16, such a detailed level classification is not necessary.

The network level is thus replaced with 'site' (since here, only the external relationships are of interest) and the cell, system and section levels are summarized into 'section/sub-section'. The types of changeability refer to the technology, organization and spatial arrangement of factory objects. In the matrix formed by these, 26 factory objects can now be assigned to the first order. Each of these objects is then further broken down, resulting in a total of 116 factory objects on the second order (for descriptions of these see Appendix A1). Further informations can be found e.g. in [Step09]

Furthermore, it needs to be kept in mind that the significance of each factory item is different on each of the factory levels. This is clarified in Fig. 5.19, which in comparison to Fig. 5.18 has swapped the columns and rows. To prevent objects from being considered a number of times within the planning frame, it is practical to assign them to a specific level as marked in the figure.

The changeability of a production is evaluated according to the control loop depicted in Fig. 5.20. The control loop was developed within the frame of a research project and tested practically by the participating industrial enterprises [Nyh10]. An extensive description is found in [Kle13], however, we will briefly explain the process here.

The starting point is a running factory which is being impacted by a change driver. The driver requires modifications that facilitate attaining the target output. Initially it is assumed that the available changeover and reconfiguration possibilities fail to suffice the required change. If both the existing flexibility and transformability are insufficient, the changeability has to be adjusted i.e., the flexibility corridor either has to be shifted or expanded. The following steps are then practical [Nyh13, p. 30 ff.] and are conducted by a team of internal and/or external experts:

factory fields		factory objects	site	factory	section/ sub section	work-station
techno-logy		building services - centers	●	●	◑	◔
		building services - distribution facilities	◔	●	●	●
		information technology	◔	●	●	●
		storage facilities	◑	◑	●	◑
		transportation facilities	◑	◑	●	●
		production technology	○	○	◑	●
		production facilities	○	○	◑	●
		other facilities	○	○	◑	●
organi-zation		hierarchical structure	●	●	●	●
		production concept	○	●	●	◑
		logistics concept	◔	●	●	◑
		structure	○	●	●	◑
		labor organization	◑	◑	●	●
		quality assurance concept	◔	◑	◑	●
space		property	●	○	○	○
		site development plan	●	○	○	○
		outdoor areas	●	○	○	○
		layout	○	●	●	●
		building form	◑	●	○	○
		building structure	○	●	◔	◔
		shell	○	●	◔	◔
		appearance	◑	●	◑	◔
		development	○	◑	●	◑
		workplace design	○	◔	◑	●

○ no importance ◔ minor importance ◑ medium importance ● high importance [●] assignment to the factory level

Fig. 5.19 Allocation of factory objects to factory levels and their respective significance. © IFA 13.441A_Wd_B

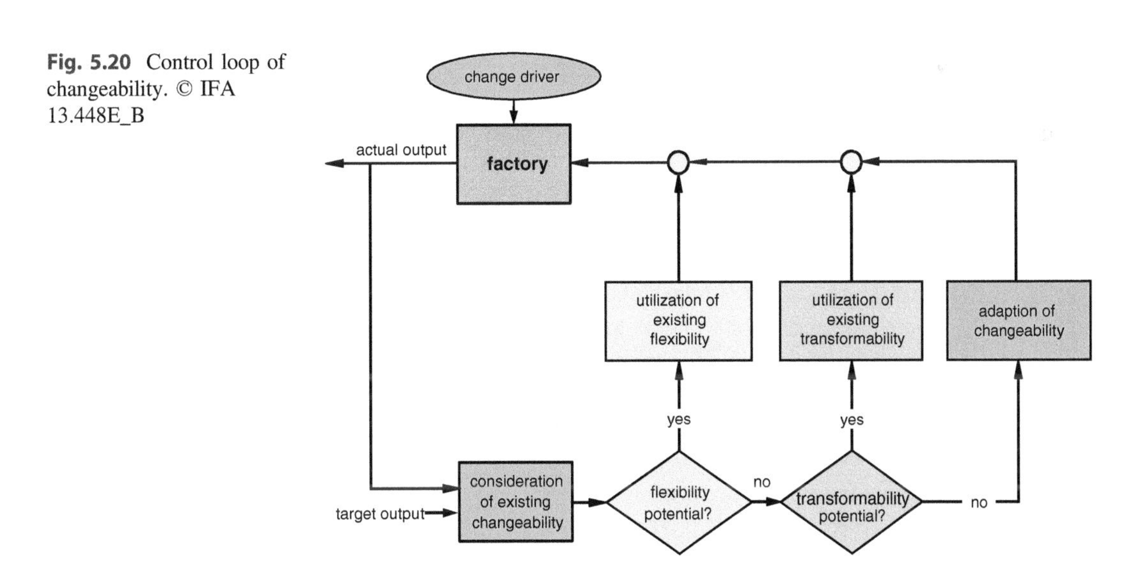

Fig. 5.20 Control loop of changeability. © IFA 13.448E_B

- The process begins with delineating the field to be investigated e.g., a factory, a division or a product group.
- The change drivers, which can be broken down into seven groups (legislators, customers, market, suppliers, competitors, business

Fig. 5.21 Excerpt from a change driver catalogue (example). © IFA 13.448E_B

cluster	driver	source of information	dimension of change	affected areas: technology	affected areas: organization	affected areas: space
enterprise	increase production volume	marketing and sales	☒ quantity	☒	☐	☒
			☐ variants	☐	☐	☐
			☒ costs	☒	☐	☒
			☐ process quality	☐	☐	☐
			☒ time	☒	☒	☒
	introduction of a new technology	technology market	☐ quantity	☐	☐	☐
			☐ variants	☐	☐	☐
			☒ costs	☒	☐	☐
			☒ process quality	☒	☐	☐
			☒ time	☐	☒	☐

and network, technology and employees) then have to be more precisely analyzed. A list of the change drivers and their definitions can be found in Appendix A2. Further steps are based on the identified change drivers. An expert panel identifies the factory elements affected in order to gain a preliminary estimate of the degree of change required for each element. Figure 5.21 depicts this principle strongly simplified, based on the example of the change drivers 'increasing the production volume' and 'introducing a new technology'.

- Next the factory areas that are affected by each of the drivers need to be described. Four different perspectives are possible here. By means of traditional business process analysis, the process view describes the production process. The spatial view visualizes the spatial relations in the layout of the facilities. The organizational view describes the hierarchical structure of the company's organization including its employees, their principle responsibilities (planning, control and operative) and the communication between them. Finally, the logistics view describes all of the logistics tasks under the headings procurement, production and distribution as well as the underlying model (see Sect. 6.2.3).
- The actual changeability corridor now needs to be evaluated from these four perspectives against the background of the identified drivers. This reveals whether the changeability is sufficient or if it needs to be adjusted. A questionnaire helps find a differentiated answer on a detailed level. In the example shown in Fig. 5.22, the driver is the development of a welding transformer which should reduce the previous variety. The object being considered is the assembly system with its sub-elements. It turns out that the only existing possibility for

adaption questionnaire "assembly system"			
term	assembly system for transformer		
location	assembly II		
observed driver development	development of unit transformer to enable the assembly system and product to be prepared for a faster change		
factory element (2nd level)	can the driver be dealt with?		what adjustments are already available to cope with the drivers development?
	yes	no	
material supply		X	---
tools		X	installation of a replaceable tool
manipulator		X	---

Fig. 5.22 Excerpt from adaption catalogue (example). © IFA 17.595_B

adaption is for the tools, whereas new solutions have to be developed for the tool handling and manipulator.

- The last step serves the decision as to whether adaption measures need to be implemented and if so what those measures should be. The basic procedure for this is shown in Fig. 5.23. Criterion 1 asks whether there is already an existing solution to adapt the element to the driver. The answer to this is based on the adaption questionnaire developed in the previous step. Criterion 2 requires solutions which are first evaluated based on the available activation period, while Criterion 3 questions the costs in comparison to the existing budget. Frequently a quick solution is more expensive than a cost effective variant. In consideration of the prioritized strategy (operative, tactical, strategic) the team then decides which solution should be planned in detail.

Systematically analyzing the factory components with regards to their changeability already makes it possible to derive approaches for a changeable factory, which have a large potential for success (Fig. 5.24).

A carefully considered market strategy which orients the entire organization on the customer's benefit is always the point to start. This leads to product structures which meet their demands as is for example described in the Global Variant approach to production (see Sect. 4.12). The technologies and methods that are implemented have to be aimed at manufacturing the exact lot size that the customer has ordered, while logistic strategies for the supply and order processing have to follow the flow principle. Moreover, buildings should be designed adaptively. Finally, it is imperative that employees be involved in designing and operating the factory. In the following chapters, each of these aspects will be discussed on the various levels of the factory and further explained.

In this chapter, we have been able to see that the concept of changeability can be made

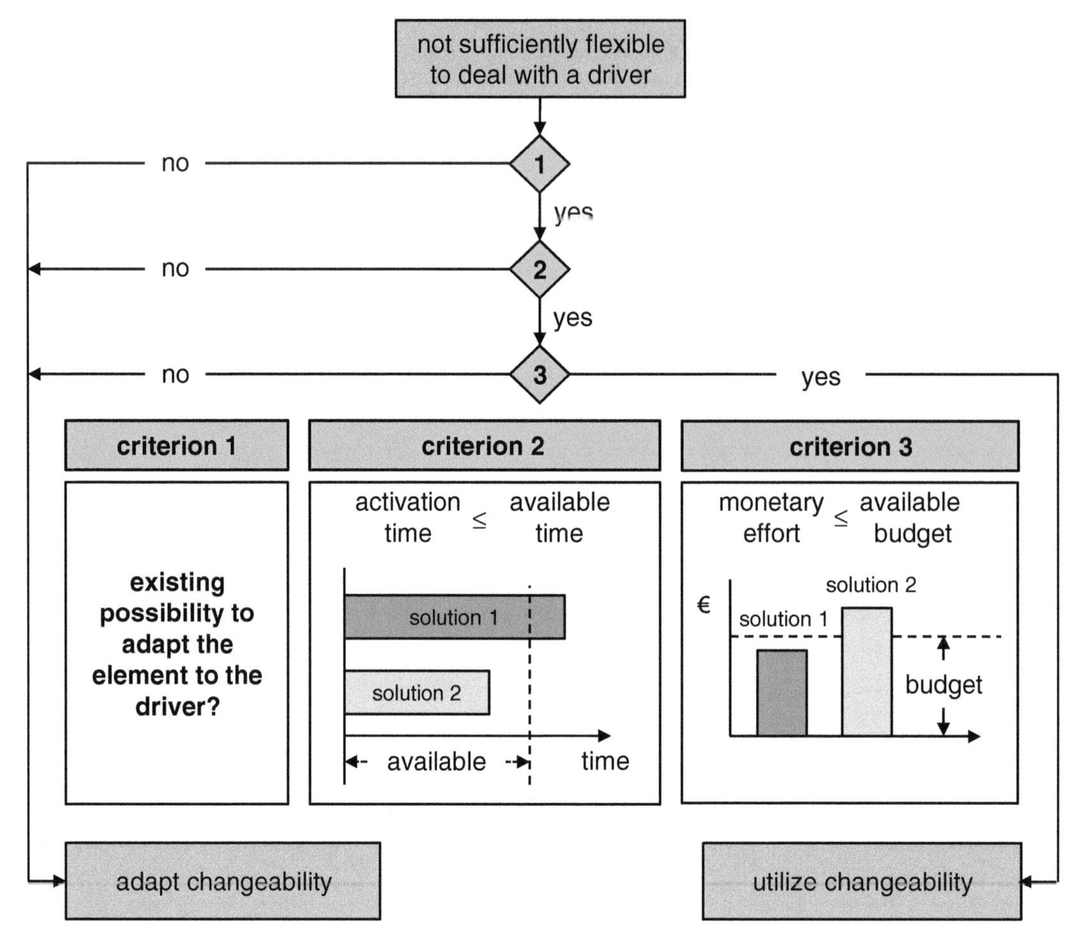

Fig. 5.23 Criteria for utilization and adaption of changeability (per Klemke). © IFA 14.792E_B

tangible. When changeability is understood as a strategic factor for success, the interactions depicted in Fig. 5.25 should be kept in mind [Heg07].

A change is only successful, when the change process is (a) understood as a strategic approach which always keeps a mind on the balance between the target and actual changeability, and (b) is oriented on the speed demanded by the market but never forgets to keep an eye on the costs. It is thus not enough just to attain the necessary changeability in the sense of a process capability; rather when a change is necessary, the changeability must also be utilized within the required time. This in turn requires employees to be competent in managing a change.

5.9 Vision of the Changeable Factory

Based on our discussion up to here we are able to develop a vision of a changeable factory based on the model of a sustainable production. In correspondence to Figs. 2.9 and 5.26 differentiates this vision according to an external view (production as a strategic mean) and an internal view (factory as a physical enabler).

Divergent from the conventional factory, which is characterized by change-resistance and internal optimization, the future production has to be oriented on market strategies and the products derived from them. This requires teams who, based on clearly communicated goals,

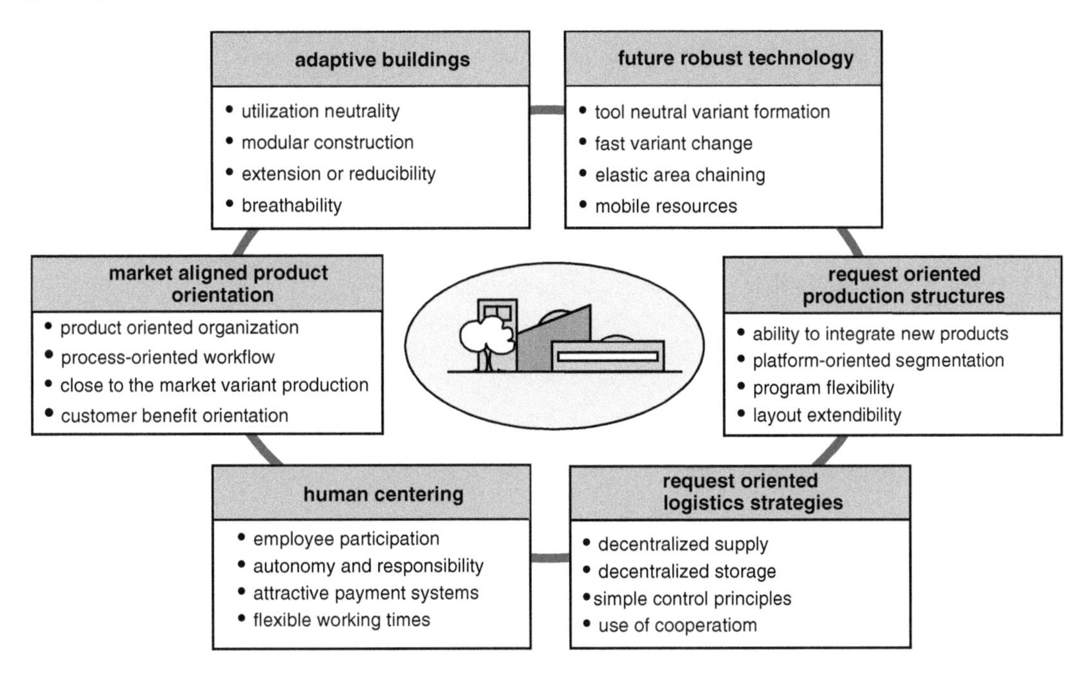

Fig. 5.24 Components and features of changeability from the factory planners view

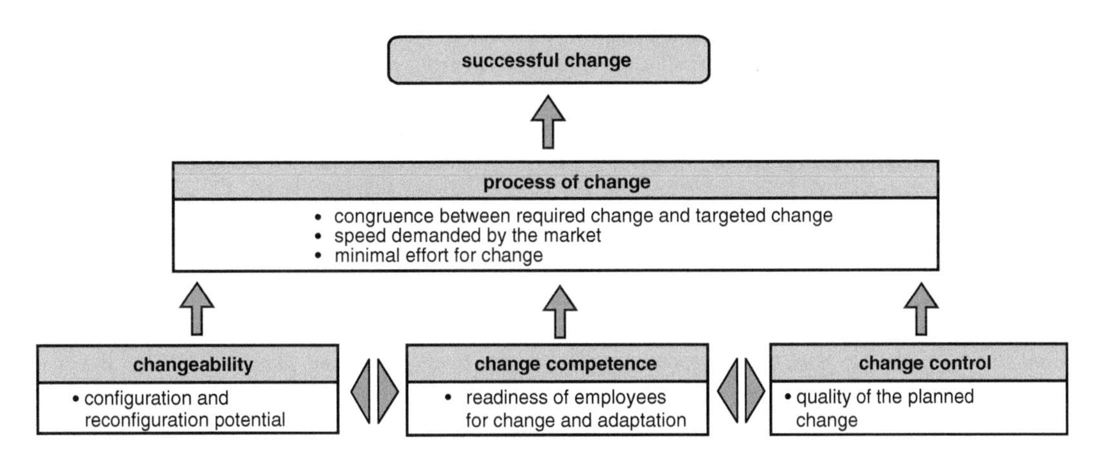

Fig. 5.25 Factors for successful change. © IFA 10.150_Wd_B

independently plan and operate business processes. In doing so, they are oriented on the technical and operational limits of the shop floor, but also on physical and logistical restrictions.

The basic principle in realizing the factory is ensuring that the resources and organization is appropriately changeable and mobile across all of the structural levels from the factory site to the buildings, to the manufacturing and assembly systems down to the individual workstations. This requires 'usage-neutral' buildings which survive generations of products and processes yet echo a design that mirrors the enterprise's self-image and its market offerings.

Finally, a clear ability to network externally with respect to logistics, organizational aspects and communications technology has to be ensured in order to effectively co-operate with suppliers, development partners and customers. The notion of sustainability comprises a long-

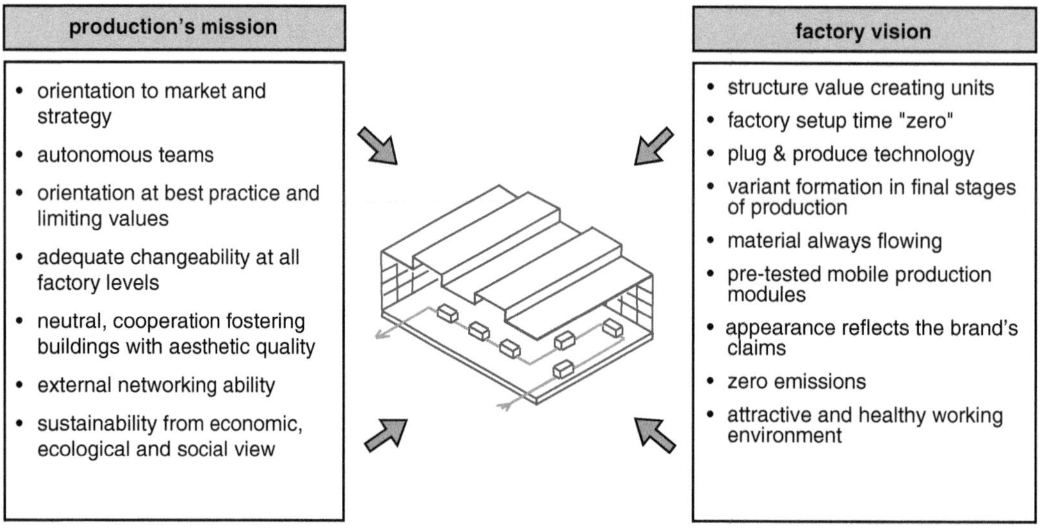

production's mission

- orientation to market and strategy
- autonomous teams
- orientation at best practice and limiting values
- adequate changeability at all factory levels
- neutral, cooperation fostering buildings with aesthetic quality
- external networking ability
- sustainability from economic, ecological and social view

factory vision

- structure value creating units
- factory setup time "zero"
- plug & produce technology
- variant formation in final stages of production
- material always flowing
- pre-tested mobile production modules
- appearance reflects the brand's claims
- zero emissions
- attractive and healthy working environment

Fig. 5.26 Vision of the changeable factory. © IFA G9536BSW_Wd_B

term economic success, which however, takes into consideration the employee's social concerns and acts environmentally responsibly.

As a result, a vision for factories arises, which —organized according to value-adding units for different market requirements—can be converted quickly and economically.

A modern theater can be seen as a metaphor here, whereby its stage technology allows scenes to be changed noiselessly in shortest time with the curtains open. In a factory, this change-over ability requires production modules, which are reconfigurable within minutes or hours, because thanks to the ease with which they can be moved and locally controlled they can communicate with a higher level control.

Due to the necessity of managing variants the traditional separation between pre-manufacturing and assembly has to be called into questioned. Variants are thus formed in so-called 'production end stages' during the latest possible step of the final assembly by integrating variant defining manufacturing operations in the assembly process. Motivated by a logistic perspective, a further vision is that of a steady flow of materials without any stops through the value-adding stages. This ensures the lowest inventories, shortest throughput times and subsequently, the greatest responsiveness. A maxim for this vision could be "produce in one day, what the customer ordered by the end of the day before—no more, no less". Ultimately, the changeable factory can go as far as pre-tested, mobile factory modules, which can be moved on the factory site, but also to other sites. Whereby, a zero emission factory is the benchmark for a healthy and attractive work environment.

Before we extensively discuss the necessary planning process under the heading of synergetic factory planning in Chap. 15, we need to consider what objects the planning has to look for in order to create a real factory. Thus in the next chapters we will describe this corresponding to the levels depicted in Fig. 5.18 (workstation, subsection/section, factory and site). In doing so, we will describe each level from both functional and spatial design perspectives.

5.10 Summary

In this chapter changeability is defined as an umbrella term for five classes of adaptability, which are applicable to the different levels of a factory: change-over ability on the workstation and cell level, flexibility on the cell, system, section and factory level, transformability on the factory, section and system level and agility on the network level. In order to enable the factory objects on the different levels to change, changeability enablers are necessary which are: universality, mobility, scalability, modularity and compatibility. To install the appropriate changeability in a real case, a balance has to be found between the desirable and the affordable. These considerations in turn give rise to practical hints about how to design the different aspects of a factory beginning with product design, technology, building, logistics and organization.

Bibliography

[Abe06] Abele, E., Versch, A., Wörn, A.: Reconfigurable manufacturing systems (RMS) for machining of case and similar parts in machine building. In: Dashchenko, A.I. (eds) Reconfigurable Manufacturing Systems and Transformable Factories, 1st edn, pp. 327–339. Springer, Berlin (2006)

[Agg87] Aggteleky, B.: Fabrikplanung und Werksentwicklung. Band 2: Betriebsanalyse und Feasibility-Studie (Factory planning and plant development. vol. 2: Operation Analysis and Feasibility Study). Hanser, Munich Vienna (1987)

[Eve83] Eversheim, W., Kettner, P., Merz, K.-P.: Ein Baukastensystem für die Montage konzipieren (To design A modular system for the assembly). Industrie Anzeiger **105**(92), 27–30 (1983)

[Heg07] Heger, Ch. L.: Bewertung der Wandlungsfähigkeit von Fabrikobjekten (Review of the changeability of factory objects). Ph.D. Thesis, Leibniz University Hannover 2006. Verlag PZH Produktionstechnisches Zentrum GmbH, Garbsen (2007)

[HEM09] El Maraghy, H., Wiendahl, H.-P.: Changeability—an introduction. In: ElMaraghy, H. (ed.) Changeable and Reconfigurable Manufacturing Systems. Springer, Berlin (2009)

[Her06] Heragu, S.S.: Facilities Design, 2nd edn. iUniverse, Lincoln (2006)

[Her03] Hernández Morales, R.: Systematik der Wandlungsfähigkeit in der Fabrikplanung (Systematics of changeability in factory planning). Ph.D. Thesis, University Hannover 2002. Fortschrittberichte VDI, Series 16, No. 149, Düsseldorf (2003)

[Kal89] Kaluza, B.: Erzeugniswechsel als unternehmenspolitische Aufgabe. Integrative Lösungen aus betriebswissenschaftlicher und ingenieurwissenschaftlicher Sicht (Product change as a company policy task. Integrated solutions from a business and engineering point of view). Berlin (1989)

[Kal95] Kaluza, B.: Flexibilität der Industrieunternehmen. Diskussionsbeiträge des Fachbereichs Wirtschaftswissenschaften der Gerhard-Mercator-Universität Gesamthochschule Duisburg (Flexibility of industrial companies. Discussion Papers of the Department of Economics of the Gerhard-Mercator University Duisburg, No. 208, Duisburg (1995)

[Kal05] Kaluza, B., Blecker, Th.: Flexibilität—State of the Art und Entwicklungstrends (Flexibilty—State of the Art and development trends). In: Kaluza, B., Blecker, Th. (eds.) Erfolgsfaktor Flexibilität. Strategien und Konzepte für wandlungsfähige Unternehmen (Flexibility as a Success Factor. Strategies and Concepts for Changeable Enterprises). Erich Schmidt Verlag, Berlin (2005)

[Ket84] Kettner, H., Schmidt, J., Greim, H.-R.: Leitfaden der systematischen Fabrikplanung (Guideline of Systematic Factory Planning). Hanser, Munich Vienna (1984)

[Kle13] Klemke, T.: Planung der systemischen Wandlungsfähigkeit von Fabri-ken (Planning of systemic changeability of factories). Ph.D. Thesis, Leibniz Universität Hannover. Publ. PZH Verlag, Garbsen (2013)

[Kor01] Koren, Y.: Reconfigurable manufacturing systems. In: Proceedings of the CIRP 1st International Conference on Agile, Reconfigurable Manufacturing. 20/21 May 2001. University of Ann Arbor Michigan, USA

[Lot06] Lotter, B., Wiendahl, H.-P. (eds.): Montage in der industriellen Produktion - Ein Handbuch für die Praxis (Assembly in industrial production—A Practical Guide), 2nd edn. Springer, Berlin (2012)

[Mut89] Muther, R., Hales, L.: Planning of Industrial Facilities. R. Muther and L. Hales Management Ind. Res. Publ, Kansas City (1979)

[Nyh04] Nyhuis, P., Elscher, A., Kolakowski, M.: Prozessmodell der Synergetischen Fabrikplanung - Ganzheitliche Integration von Prozess- und Raumsicht (Process model of the synergetic factory planning. Holistic process and spatial view). wt Werkstattstechnik online 94 (2004) Issue 4, pp. 95–99

[Nyh08] Nyhuis, P., Reinhart, G., Abele, E. (eds.): Wandlungsfähige Produktionssysteme. Heute die Industrie von morgen gestalten (Changeable production systems. Today, Shaping Tomorrow" Industry). Verlag Produktionstechnisches Zentrum GmbH. Garbsen (2008)

[Nyh10] Nyhuis, P., Klemke, T., Wagner, C.: Wandlungsfähigkeit - ein systemi-scher Ansatz (Changeabilty - a systemic approach). In:

Nyhuis, P. (Hrsg.) Wandlungsfähige Produktionssysteme. Schriftreihe der Hochschulgruppe für die Arbeits- und Betriebsorganisation (Changeable Production Systems. Series of the scientific group for the working and operational organization e.V. (HAB)), pp. 3–21. GITO-Verlag, Berlin (2010)

[Nyh13] Nyhuis, P., Deuse, J., Rehwald, J.: Wandlungsfähige Produktion. Heute für morgen gestalten (Changeable Production. Today designed for tomorrow). PZH Verlag, Garbsen (2013)

[Rak00] Narain, R., Yadav, R.C., Sarkis, J., Cordeiro, J. J.: The strategic implications of flexibility in manufacturing systems. Int. J. Agil. Manag. Syst. **2/3**, 202–213 (2000)

[Rei07] Reichardt, J., Pfeifer, I.: Phasenmodell der Synergetischen Fabrikplanung. Stand der Forschung und Praxisbeispiele (Phase model of the synergetic factory planning. State of research and practical examples). wt Werkstattstechnik online **97**(4), pp. 218–225 (1997)

[Rein97] Reinhart, G.: Innovative Prozesse und Systeme – Der Weg zu Flexibilität und Wandlungsfähigkeit (Innovative processes and systems—the way to flexibility and adaptability). In: Milberg, J., Reinhart, G. (eds.): Mit Schwung zum Aufschwung (With verve on recovery). Münchener Koll. '97. Landsberg/Lech (1997)

[Rein00] Reinhart, G.: Im Denken und Handeln wandeln (Change in thinking and acting). In: Reinhart, G. (Hrsg.) Proceedings of Münchener Kolloquium 2000. Munich Vienna (2000)

[Rein08] Reinhart, G., Kerbs, P., Schellmann, H.: Flexibilität und Wandlungsfähigkeit - das richtige Maß finden (Flexibility and adaptability—finding the right balance). In: Hoffmann, H., Reinhart, R., Zäh, M.F. (eds.) Münchener kolloquium. Innovationen für die Produktion (Innovations for production). Proceedings of Production Congress 9, pp. 45–55 (2008)

[Sche04] Schenk, M., et al.: Factory Planning Manual. Situation-Driven Production Facility Planning. Springer, Berlin (2010)

[Step09] Stephens, M.P., Meyers, F.E.: Manufacturing Facilities. Design and Material Handling, 4th edn. Prentice Hall, New Jersey (2009)

[Tom10] Tompkins, J.A., et al.: Facilities Planning, 4th edn. Wiley, Hoboken (2012)

[Ton98] De Toni, A., Tonchia, S.: Manufacturing flexibility: a literature review. Int. J. Prod. Res. **36**(36), 1587–1617 (1998)

[Ulr95] Ulrich, H., Probst. G.J.B.: Anleitung zum ganzheitlichen Denken und Handeln: Ein Brevier für Führungskräfte (Instructions for holistic thinking and action: A Breviary for executives), 3rd edn. Haupt, Bern Stuttgart (1995)

[WEM12] ElMaraghy, W., et al: Complexity in Engineering Design and Manufacturing, vol. 2, issue 61, pp. 793– 814. International Academy for Production Engineering, CIRP Annals—Manufacturing Technology (2012)

[West99] Westkämper, E.: Die Wandlungsfähigkeit von Unternehmen (Changeablity of enterprises). wt Werkstattstechnik **89**(4), 131–139 (1999)

[West00] Westkämper, E., Zahn, E., Balve, P., Tilebein, M.: Ansätze zur Wandlungsfähigkeit von Produktionsunternehmen (Approaches to the changeability of manufacturing enterprises). wt Werkstattstechnik **90**(1½), 22–26 (2000)

[West09] *Westkämper, Engelbert; Zahn, Erich (Hrsg.):* Wandlungsfähige Produktionsunternehmen. Das Stuttgarter Unternehmensmodell (*Changeable Manufacturing Companies The Stuttgart Enterprise Model*). Springer, Berlin (2009)

[Wie81] Wiendahl, H.-P., Mende, R.: Produkt- und Produktionsflexibilität – Wettbewerbsfaktoren für die Zukunft (Product and production flexibility - competitive factors for the future). wt Zeitschr. f. industrielle Fertigung **71**, 295–296 (1981)

[Wie00] Wiendahl, H.-P., Hernandez, R.: Wandlungsfähigkeit – neues Zielfeld in der Fabrikplanung (Changeablity—a new target field in factory planning). Industrie-Management **16** (5), 37–41 (2000)

[Wie01] Wiendahl, H.-P., Reichhardt, J., Hernandez, R.: Kooperative Fabrikplanung – Wandlungsfähigkeit durch zielorientierte Integration von Prozeß- und Bauplanung (Cooperative factory planning—changeability through targeted integration of process and construction planning). wt Werkstattstechnik **91**(4), 186–191 (2001)

[Wie05] Wiendahl, H.-P., Nofen, D., Klußmann, J.H., Breitenbach, F. (Hrsg.): Planung modularer Fabriken. Vorgehen und Beispiele aus der Praxis (Planning modular factories. Approach and practical examples). Hanser, Munich Vienna (2005)

[Wie07] Wiendahl, H.-P., et al.: Changeable manufacturing—classification, design and operation. Annal. CIRP **56**(2), 783–809 (2007)

[Wir00] Wirth, S. (eds.): Flexible, temporäre Fabriken - Arbeitsschritte auf dem Weg zu wandlungsfähigen Fabrikstrukturen (Flexible temporary factory - steps on the way to transformable factory structures). Wiss. Berichte FZKA-PFT 203 Forschungszentrum Karlsruhe, ISSN 0948–142 (2000)

6 Functional Design of Workplaces

A workplace can be viewed from different perspectives. First, there are the functions to fulfill at the workplace from a technological point. This will be met by equipment in interaction with human labor. This chapter provides an overview of the technology and equipment for the production of piece products, to the extent which is necessary for the factory planner. Chapter 7 then considers the workplace design from the perspective of work organization, while Chap. 8 examines the spatial integration of ergonomic and architectural aspects of workplaces. These three chapters form the foundation of the factory design.

6.1 Design Aspects

From a factory planning view, workstations are the smallest units in designing processes. In combining people and equipment, workstations execute a task aimed at increasing the value of a single part, component or assembly with the least possible effort. Figure 6.1 outlines the design aspects pertinent to a workstation.

The starting materials in the form of raw materials, semi-finished products, pre-processed parts, prefabricated parts, sets of parts, or partially assembled components enter the workstation as *input*. In order to execute a task, energy (e.g., electrical power, steam, fuel gas) and media for the process (e.g., water, protective gas, lubricant) are generally required. Necessary information is available as drawings, work plans, control programs and work instructions. The actual process is conducted at the *workstation* with the aid of process equipment (machine tools, assembly devices, annealing furnaces etc.) as well as tools and possibly clamps. Depending on the degree of automation, humans are integrated more or less intensively. The workstation requires a specific *space* for the equipment, worker and materials supplied to the workstation and completed on it. The equipment, people and surface areas are thus the basic constituent planning parameters for a workstation.

The results of the work—generally referred to here as the product—is found at the exit of the workstation. However, unwanted results also accumulate in the form of material waste (e.g., chips and remnants, tinder, auxiliary material) which has to be disposed of appropriately. Emissions such as noise, vibrations, heat, gases, dust, vapors are equally undesired and need to be managed especially with regards to health issues. Finally, the workstation also delivers information e.g. about the results with respect to selected characteristics of quality, the process duration and the quantity.

A workstation usually forms a part of a process chain and is characterized by its integration into the material flow, information flow, communication flow, and work organization of the

H.-P. Wiendahl et al., *Handbook Factory Planning and Design*,
DOI 10.1007/978-3-662-46391-8_6, © Springer-Verlag Berlin Heidelberg 2015

next higher level. Finally, the environmental conditions of the workstation from a process view (e.g., climate control, cleanroom) and a human view (ventilation, lighting, coloring) are also design elements. The latter are aspects of the building design and are discussed extensively in Chap. 10.

In designing workstations, a technical/economic perspective stands in opposition to a human/organizational perspective. The aim then is to merge the two into a long term successful synthesis. Figure 6.2 depicts these two poles under the headings of "economic principle (added value)" and "humane work design" from which four goals are developed [Mar94].

Technical/economic objectives demand that technical and human performances be correctly allocated according to their function. This means optimally combining system elements, coordinating work requirements with human abilities and economic deployment of humans. *Health and safety* encompasses aspects that prevent work related illness and injuries, reduce demands that are too high or too low and create a sense of well-being at work. *Social orientation* addresses the importance of ensuring social norms, promoting inter-personal relationships as well as participating in designing the workstation. Due to the extensively discussed environmental changes the latter proves to be increasingly essential for quickly and effectively planning as well as ensuring employees acceptance. Finally, *personal development*, aims at designing the workstation so that workers can gain confidence, prove and realize themselves while developing their skills, competencies and gaining autonomy.

Based on these preliminary considerations, the concrete aspects of designing a workstation from a process view within the frame of planning a factory correspond to those depicted in Fig. 6.3. It has to be kept in mind that the planning tasks can't be fulfilled by a single person, but rather

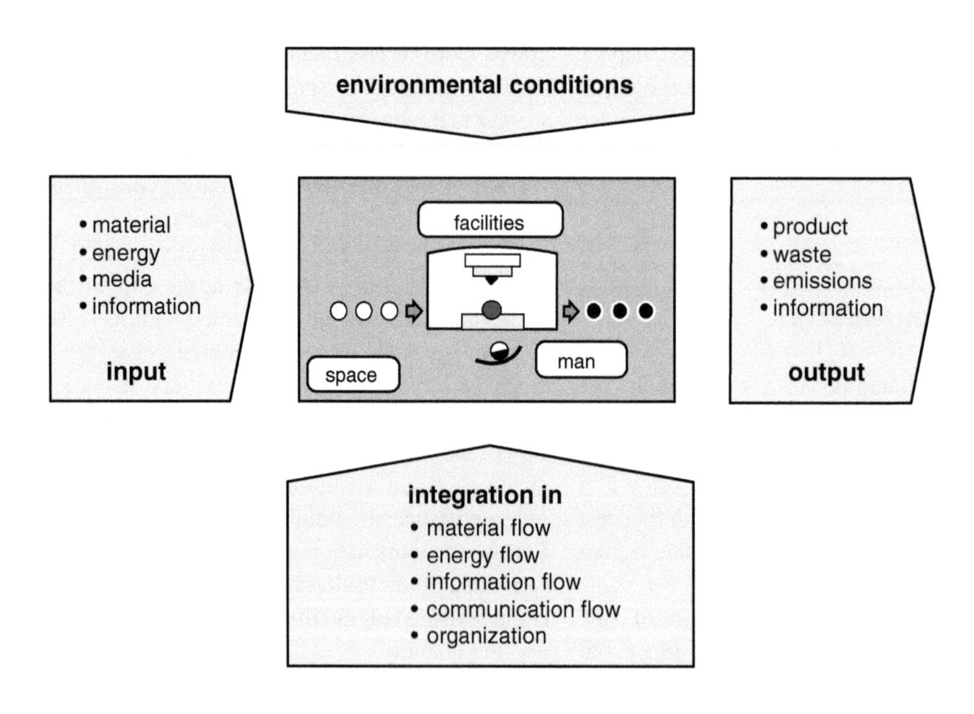

Fig. 6.1 Aspects of a workstation's design

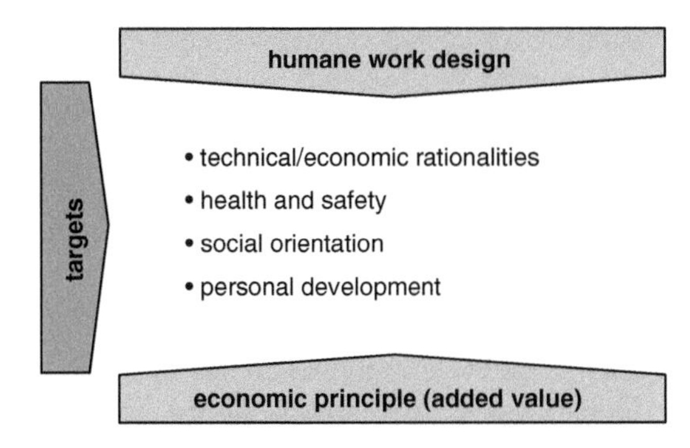

Fig. 6.2 Constraints and objectives of work design [Mar94]

always require the collaboration of an expert planning team. In Chap. 15 we will take a closer look at the composition of such a team and how it proceeds.

The initial situation is provided by the task of manufacturing or assembling a part or group of components that is to be produced in the future. When *planning the technology,* manufacturing or assembly functions need to be specified and the local logistics for supplying and disposing of materials, including required information, needs to be defined.

Strategic considerations play a significant role here with regards to the future of key technologies based on the so-called '*technology differentiation*'. These are assessed according to the competency and attractiveness of the technology [Zeh97]. Technological competence refers to both the production's technical ability with regards to a specific technology as well as the resources available for fulfilling a manufacturing task. The potential for further developing the technological competence is described by the attractiveness of the technology. It takes into consideration the availability of technologies, the interdependency between technologies and possibly substitutive technologies [Nyh09b].

The technological competence of a production process and the attractiveness of its technology define the position of the process in the 'technology portfolio'. The portfolio makes it possible to identify the strategic relevance and development potential of production processes and to divide them into key competency, differentiation and standard processes. Key competency processes are invested in, standard processes are disinvested in and differentiation processes are further divided into either key or standard processes. Knowledge about these strategic considerations is significant for a factory planner in that they can lead to new requirements for the design of the processes and buildings. Embedding this approach in the strategic site planning will be further discussed in Sect. 14.6.

The technology defined thus far should then be implemented through the *facility planning* in the form of equipment (e.g., tool machines, assembly stations, storage/transportation systems). In

Fig. 6.3 Aspects of designing a workstation from the process view

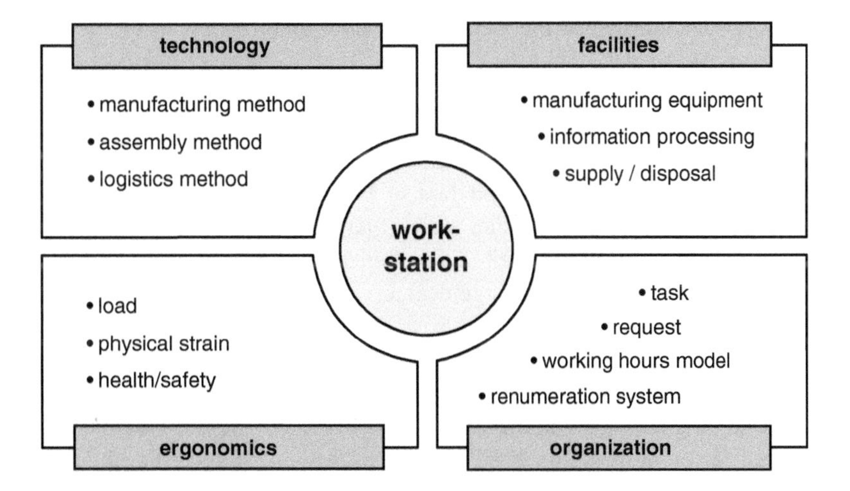

addition, the supply of information to equipment and workers also needs to be designed and finally, equipment should be implemented for supplying and disposing materials (storage, transportation and handling devices) as well as disposing of waste (chip conveyors, coolant filtration units). Planning technology and facilities is only necessary when new products are being introduced. Usually when planning workstations the majority of the equipment already implemented is adopted. Work plans which contain the work sequence, required equipment, necessary setup time and cycle time for each work-piece are generally available for products that already exist. They form the basis for later planning the capacities and material flow.

The aspects of design that we refer to as *organization* and *ergonomics* considers the humans who execute the work processes. The solutions found here are subject to an especially large number of regulations and in particular the agreement of the works council; involving them early on is thus of utmost importance. *Organization* is aimed at setting the tasks to be conducted at each of the workstations. These can range from pure monitoring and monotonous activities up to demanding bundles of tasks e.g., as found in manufacturing cells and segments. As a result of the work task, demands are placed on the worker, who in turn requires specific abilities. In this way, the basis for personnel planning and recruiting or for qualification measures is derived. Among the responsibilities involved in designing the organization is also the job of developing a work time model. This task is particularly important due to the need to be able to quickly respond to customer needs. Finally, the remuneration method has to be determined. Remuneration is progressively aimed less at buying employees capacity to work and more at paying them for the results of their work, whereby workers gain autonomy to a large degree.

The fourth aspect of designing a workstation deals with *ergonomics* (Greek: *ergon* = work, *nomos* = natural law). Ergonomics is concerned with keeping the stress resulting from the strain placed on employees by their work at a level which is non-detrimental, practicable and reasonable for the individual over the long term, while promoting satisfaction and being socially acceptable [Mar94]. In doing so, a balance between expecting too much from a worker and expecting too little has to be found. The first and foremost goal of ergonomically designing a workplace is maintaining the health and physical well-being of the employee.

In the following, we will turn our attention to examining these four aspects of design more closely.

6.2 Production Technology

Here, we will use the term 'production technology' to summarize all of the processes that serve in manufacturing geometrically defined workpieces, in assembling parts, components and end products as well as transporting or storing parts, components and products. From the perspective of factory planning, we have chosen this classification of production technology, because these three sub-areas later allow us to clearly allocate resources and equipment to them and in a further step also production and/or logistic sections. This result is in contradiction to DIN 8580 (manufacturing processes) [DIN03a], which does not recognize the 'assembly' concept in the main group "joining" but describes the key function of assembly. We will introduce all of the processes in the following only to the extent that the principle behind how they work is made clear for factory planners but without explaining the physical or chemical basis. While planning a factory, processes are selected by process experts.

6.2.1 Manufacturing Processes

In DIN 8580 manufacturing processes are categorized according to six main groups which take on the features mentioned in Fig. 6.4 [DIN03a].

When manufacturing a workpiece, drawings documenting the materials, geometry, dimensions, tolerances and surface roughness of the

creating the form	**changing the form**				**changing material properties**
creating physical cohesion	creating physical cohesion	reducing physical cohesion	increasing physical cohesion		
main group 1 primary shaping	**main group 2** forming	**main group 3** separating	**main group 4** joining	**main group 5** coating	**main group 6** changing material property

Fig. 6.4 Features of the main groups of manufacturing processes [DIN03a]

part guide the production which occurs (a) either by creating the form out of a formless material, (b) by changing the form of an initial material or (c) by changing the properties of the material.

The group of *primary shaping processes* (main group 1) is comprised mainly of casting methods which brings metal, ceramic, or plastics from a fluid, pulp or paste-like state into a geometric defined form. The castings which are thus created from metal and ceramic generally undergo post-processing, while plastics can usually be used immediately due to both the precision of their form and dimensions as well as the quality of their surfaces.

Forming processes (main group 2) can be differentiated from either forging methods or sheet metal forming [Doe97]. With forging a solid or pasty body is changed into another geometric form, whereby, the mass and its material coherence remains the same. This can result in semi-finished products such as sheets, tubes and bars or workpieces such as screws, crankshafts or gears which have great strength. Steel can be forged at room temperatures (cold forging), between 600 and 900 °C (warm forging) and with temperatures ranging from 1000–1200 °C (hot forging) [Alt05]. Sheet metal forming shapes a sheet into a three dimensional part with approximately the same wall thickness [Alt12].

Among the *separation processes* (main group 3), shear cutting dominates the splitting processes for processing sheets and producing raw pieces from rod materials for forging. Further methods include cutting tools with geometrically defined cutting edges e.g., in lathes, drills, planes, reamer and saws. Here, layers of a material are mechanically separated from a raw material by cutting them away in the form of chips with the aid of a tool in order to change the form and/or the surface of a workpiece [DIN03b].

By implementing continually new cutting materials, the cutting speed has constantly increased and reaches up to 2000 m/min with an unalloyed metal and ceramic cutting tool tips. Cutting with tools that do not have a specific geometric form (grinding, honing, lapping) requires tools that use abrasive particles bound on a wheel or band to remove material, whereby grinding with geometric-specific abrasive disks is the most significant. Ablation also continues to be an important separation method and can be described as thermal, chemical or electro-chemical ablation. With thermal ablation energy is generated by sparks or energy rich beams (laser or electron beams). In comparison, chemical ablation uses chemical reactions between the workpiece material and an active medium (acid, alkaline) to specifically etch away material. With thermal ablation methods, processes using laser beams have taken on a dominant role because the beam is programmable and wear-free. Laser beams can be used in practically all manufacturing methods, though they are more predominant in separation, joining and surface treatment processes.

The diversity of the *joining processes* (main group 4) serves to permanently or temporarily join workpieces to components, assembly units and products. Challenges on joining processes are result of more complex component forms, increasing functional needs, growing safety requirements and the ability to easily disassemble products for the purpose of recycling. Other, basically competing joining processes include welding, soldering, adhesive bonding, riveting, clinching, seaming and using threaded fasteners. Special joining processes, which can be used with large batches and short cycle times, have been developed in the assembly of precision engineering and electronic devices. With joining processes particular attention is paid to the process reliability, which is ensured by measuring basic process parameters during processing. This can occur for example by monitoring the current when welding or measuring the torque curve during the threading process.

Numerous criteria play a role when selecting a joining process. These criteria range from functional (materials, form, strength, corrosion) to process engineering (pre/post processing, production means, possibility of automation and flexibility) up to cost, personnel and environmental aspects (investments, operating costs, environmental compatibility, ergonomics, employee requirements).

The main group 5 referred to as *coating* encompass all of the manufacturing processes used to apply anti-wear and anti-corrosion layers on components. Since wear and corrosion resistant materials are generally very expensive, locally applied protection layers can lead to enormous cost savings, whereby the cost advantages of increasing the life of the component have to be compared to the cost of coatings [Stef96]. The different coating processes are based either on electro-chemical effects (cathodic electrocoating, currentless plating, anodic oxidation) which are used to coat base metals, plastics and ceramics with layers from a couple of micrometers thickness up to 100 μm. Organic coating systems apply a fluid or paste-like polymer to a part and firmly add on it an adhesive film by means of chemical or physical changes. In comparison, powder coatings work with solvent-free powders, which, beside thick polymer layers, also allow layers of polyethylene, nylon and fluoropolymers to be created. The workpieces that are to be coated have to be heated up to a point that the powder melts on the surface and forms a contiguous film whose thickness can grow into the millimeter range. This process is thus limited to metallic and ceramic materials.

Very resistant, but brittle layers can be gained with *enameling*. With this process, purified oxidized minerals and fluoride are applied in layers to a usually metallic base material and baked at temperatures of 550–900 °C into an enamel coating. Enameling is primarily used for components used in chemical industry, food processing engineering and in household appliances, which have to endure acids and alkalines as well as temperatures between −50 and 450 °C.

A method which is particularly effective in protecting metallic parts from corrosion is *hot-dip galvanizing*. The pre-treated components are immersed in a liquid metal dip comprised of aluminum, tin or zinc and with repeated submersions acquire a coating between 20–80 μm thick.

Vapor phase coatings have gained particular importance for coating nitride and carbide items e.g. tools and joints with thin, hard, resistant layers which have both anti-corrosive and anti-frictional properties. With physical vapor deposition (PVD), coating materials are first vaporized or atomized before being deposited on objects in film-like layers between a few nanometers and tens of micrometers thick. These methods are used in light-weight coatings for optical, magnetic and micro-electronic components, but also in applying decorative layers.

Chemical vapor deposition (CVD) involves coating objects with a gaseous metallic compound at temperatures between 600 and 1000 °C. This gas is released into a chamber where it passes over the object. As it contacts the object, it reacts with the substrate depositing a solid phase layer between 0.1 and 20 μm on the base material.

Another important coating method is *thermal spraying*, which can be broken down further into three basic techniques whose names (flame, electric arc and plasma spraying) indicate the energy source with which the sprayed materials are fused to molten particles. These reach the substrate with a high kinetic energy forming a more or less porous layer between 50 μm and a few millimeters thick. Since these processes can occur at relatively low temperatures (between 100 and 250 °C) a broad range of coating materials and base materials can be combined. Generally the sprayed layers have to be further processed in order to improve their porousness, adhesion, strength and hardness, whereby machining processes serve to develop a defined geometry and surface roughness.

We will conclude our discussion of coating techniques with one final method: overlay welding. In comparison to spraying methods, *overlay welding* heats the substrate to the point that it fuses with the generally high-alloy welding filler. Similar to joint welding methods, the energy source is either combustible gas, electric arcs, resistance heating or laser beams. Highly wear-resistant and non-corroding coatings with strong adherence to the parts are thus created and are applied to high-stress components used in machines, chemical plants and power plants. This method is also especially well-suited for repairing worn or damaged components.

The last main group 6 up of manufacturing processes we will consider is *changing material properties.* It refers mainly to heat treatments for metallic materials. For the usually ferrous materials here, these can generally be broken down into thermal, thermo-chemical and thermo-mechanical processes [MaM96]. The aim of heat treatments is generally to improve the material properties with regards to their formability, machinability, hardness, strength etc. Properties are altered as desired in heat treatment plants, whereby either the entire volume of the items (penetrating) can be changed or just the areas in the proximity of the surface (non-penetrating).

With non-ferrous materials thermal and increasingly thermo-mechanical methods are being used. With *thermal methods* (annealing, quenching, isothermal transformation and ageing), the parts undergo a controlled time-temperature transformation in which components are heated up to a specific temperature in an oven, maintained at this temperature for a period of time and then cooled, resulting in structural changes that completely penetrate the object. Alterations that are to be limited to areas close to the surface are usually generated through immersion, induction, or flame treatments, but can also be attained using laser or electron beams. With *thermo-chemical methods* solid, fluid or gaseous substances are diffused into materials, creating the desired properties and subsequently quenching or tempering the object. *Thermo-mechanical treatments* employ form changing processes that yield a specific time-temperature transformation.

From the factory planning perspective, the methods are primarily interesting with regards to their changeability on the five factory levels that were discussed in Chap. 5. Figure 6.5 depicts an evaluation of the six main groups of manufacturing processes with three levels of changeability (low, medium and high).

This very rough classification only provides us with an initial basis and should serve to identify supposed or—depending on the state of the technology—currently existing obstacles regarding changeability and to find ways for overcoming them. Approaches here can include:

- Replace shaping form tools with programmable tools.
- When shaping tools (e.g., forging dies or deep-drawing dies) are not avoidable, implement rapid tooling methods and modular tools.
- Avoid environmentally unfriendly raw or auxiliary materials e.g., through dry machining without lubricant.
- Make it possible to coat or change material properties of individual workpieces and/or workpiece zones.

Following this brief overview of manufacturing processes, we would now like to turn our attentions to assembly methods. As already

manufacturing method	level: station	section	factory	site	obstacles
primary shaping	⊖	○	○	○	molds, input materials, plant size
forming	○	⊖	○	○	forming die, emissions
separating	⊖	⊖	⊖	○	tools, emissions
joining	●	⊖	⊖	●	cleanliness
coating	●	○	○	○	emissions, aggressive fluids
changing material properties	●	○	○	○	emissions, plant size

changeability: ○ low ⊖ medium ● high

Fig. 6.5 Changeability of manufacturing processes

mentioned, this term is not standardized and unlike manufacturing processes there is no in-depth classification of assembly methods with corresponding standards available. However, due to the role they play in factory planning, it is essential to have an understanding of the basic sub-processes.

6.2.2 Assembly Methods

Assembly comprises all of the operations for joining individual geometrically defined parts and components as well as possibly software (in the form of operating or application programs) into functional products. Amorphous operating and ancillary materials such as grease, glue etc., are also frequently required. Moreover, parts can either be joined so that they can be taken apart using non-destructive methods (e.g., fastening screws) or joined so that destructive means need to be used (e.g., riveting). The assembly process is generally determined by the structure of the products to be mounted as outlined in Fig. 6.6.

Products can be structured according to functional or assembly oriented criteria. Decisions about how many assembly stages are required and on which component level variants should be defined are determined by the overall organization of the assembly sections. Generally, the aim is to design the assembly so that variants are formed as late as possible in the assembly sequence, in order to avoid unnecessary interim storage of semi-finished components and products. Figure 6.7 clarifies this approach. In the example on the left, variants are formed very early in the assembly sequence. This means a large number of variants on every product level which then have to be planned and temporarily stored. In comparison, the optimized structure on the right of Fig. 6.7 results in clearly fewer interim variants.

A simple, basic scheme (depicted in Fig. 6.8) can be helpful in considering the operations required for assembling a product from individual parts and components. The traditional Black-Box system selected here assumes that each station represents an element in a production flow. A transport system from the preceding workstation conveys a partially assembled component on a workpiece carrier, thus allowing the assembly-object to be fixed and identified. Items are then joined with the object on the workpiece carrier and usually the object is tested for certain quality features before subsequently being transported to the next station.

An independent, transverse part flow ensures that the parts supplied by internal logistics are

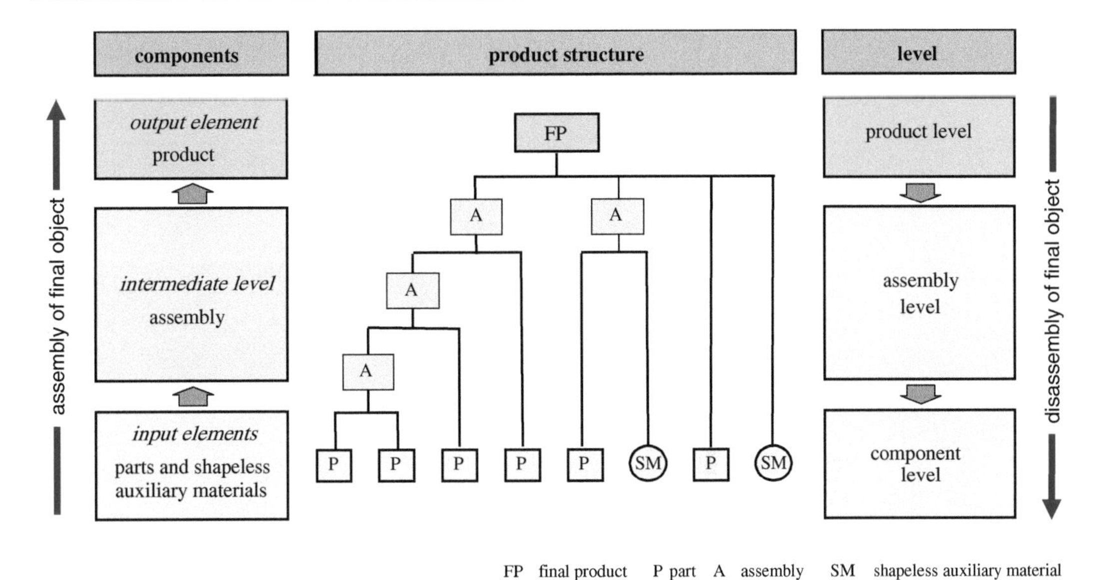

Fig. 6.6 Components and structure of an assembly product (Spur)

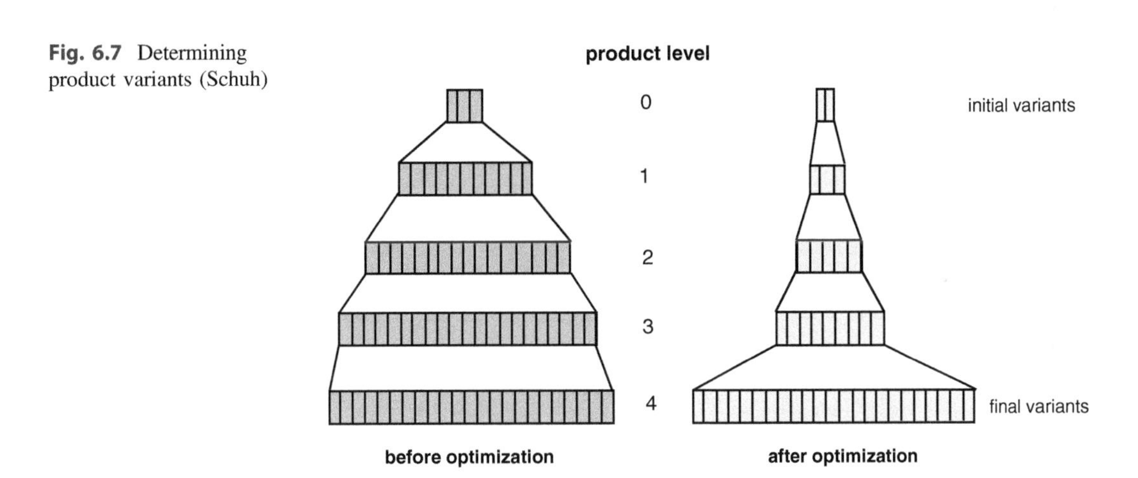

Fig. 6.7 Determining product variants (Schuh)

first stored locally in a defined quantity. This function—referred to as buffering—follows the separation of parts out of the buffer and the subsequent spatial orientation before being placed in a predefined position on the workpiece carrier. In order to bridge small disruptions and differences in cycle times, interim buffers are occasionally maintained. Parts are also typically delivered in a pre-ordered state. We will delve into the functions and forms of delivering parts from the supplier up to the site of consumption in the factory from a technical perspective in the section on logistic processes (Sect. 6.2.3) and from a strategic perspective in our discussion of procurement models in Sect. 9.5.1.

VDI Guideline 2860 defines the assembly and its sub-functions according to Fig. 6.9 whereby, in addition to the already mentioned joining, handling and controlling (in the sense of testing quality), adjusting and special operations are listed [VDI2860]. Adjusting is understood as geometrically fine-positioning parts functionally in relation to other existing parts in a component, whereas with special operations, components are

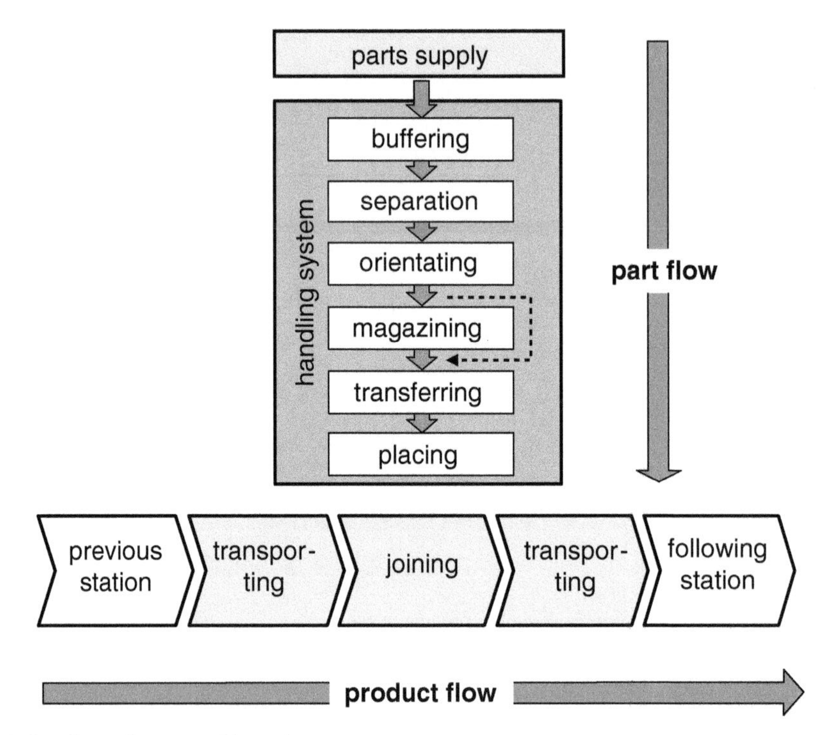

Fig. 6.8 Sub-functions of an assembly station

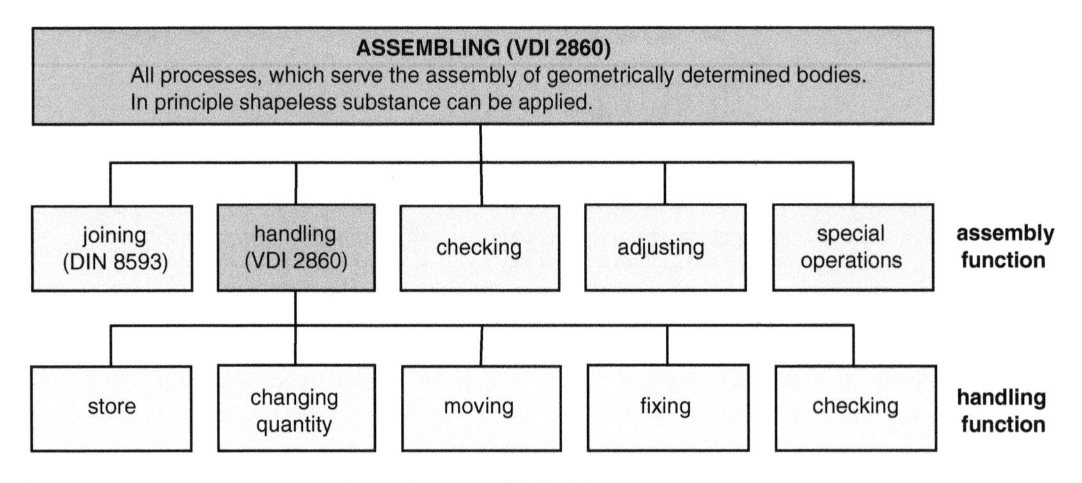

Fig. 6.9 Sub-functions of an assembly station (acc. [VDI2860])

for example, cleaned, printed or marked. From the perspective of factory planning, the stations necessary for these operations do not differ from a joining station.

Accordingly, the key functions of the assembly that need to be considered more closely are joining and handling. The 4th main group of manufacturing processes "joining" already briefly characterized in Sect. 6.2.1 is further subdivided by DIN 8593 into 9 sub-groups as depicted in Fig. 6.10.

Connected joints involve putting parts together by laying them on or in one another, nesting them together, hinging them as well as latch

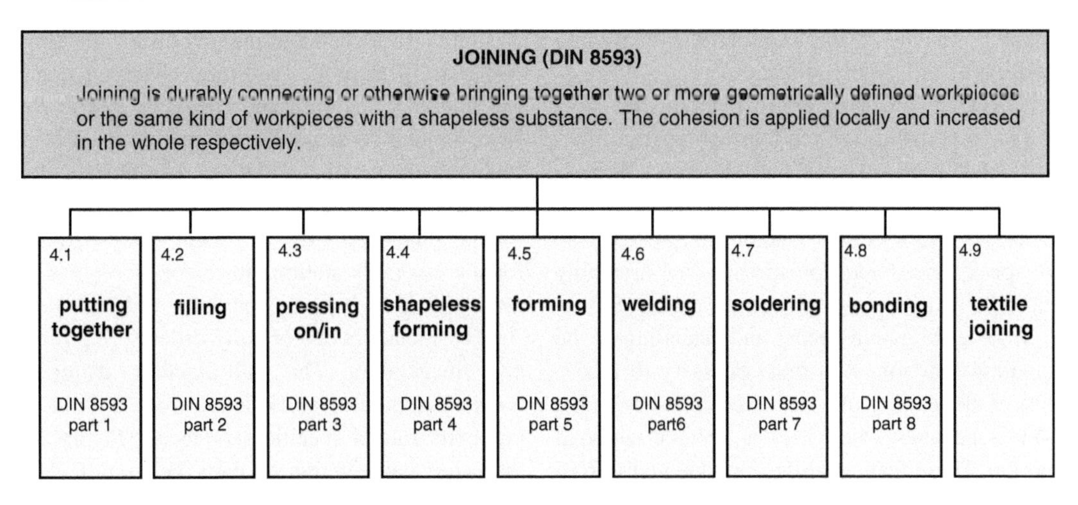

Fig. 6.10 Classification of the main manufacturing process joining [DIN03c]

them. The joined state is maintained by gravity, friction and interlocking. Some methods use the elastic distortion of the parts involved or ancillary parts to secure the connection. Connecting joints can be separated non-destructively without causing damage.

Filler joints involve pressing gaseous, vaporous, fluid, pulp-like or pasty materials into hollow or porous objects either by pouring, soaking or saturating the items. These types of joints are usually easily dissolved e.g., by applying heat.

Press joints connect parts through forced-closure, whereby the parts are generally only deformed elastically. Sub-groups of these processes include screwing, clamping, cramping, nailing, joining via press fittings, hammering and spreading. Similar to the other joining processes mentioned above, these too can generally be separated non-destructively, sometimes with the aid of special tools.

Formed joints generally involve connecting parts together by forming specific areas of parts—the way the parts then interlock secures the joint. Sub-groups here include forming wire and tape-like bodies as well as riveting processes. These joints can only be removed by damaging or destroying the joint parts.

Welded Joints connect parts together by using heat and/or force (fusion or pressure welding) with or without welding fillers. Generally, the aim of welding joints is to ensure the joint is as strong as the base material; the joint can only be destructively dissolved.

With *soldered joints* DIN 8593 differentiates between soft, hard and high temperature joints. The melting point of the filler metal that flows into the joint is lower than that of the workpiece materials. Consequently, the joint can be dissolved (with some limitations) using a desoldering tool.

Adhesive bonds use non-metallic bonding agents which harden physically or chemically joining the workpieces through adhesion and cohesion. The bond can be dissolved within limits.

Textile joints insert textile fibrous materials from the production of threads, filaments and materials up to the joints of semi-finished and finished products. These processes have not yet been standardized further and since they are rarely applied in manufacturing—except for modern car and air plane bodies made of carbon fibers—the mechanical and electrical products we are considering here, we will not consider them further.

According to VDI Guideline 2860, *handling* which we refer to here as the second key function of the assembly, is also to be interpreted (in addition to conveying and storage) as a sub-function of "creating a material flow". As such, handling is the final link in the material flow from the company's boundary up to the joining position on the workpiece carrier or joining position of the assembly station.

Together with handling, conveying and storage (warehousing) are the internal logistic functions that have external counterparts. This is discussed more in-depth in the section on logistic processes (Sect. 6.2.3). Handling represents, so to speak, the micro-logistics of an assembly station directly before the joining processes.

In order to gain a deeper understanding of the handling function, it is first necessary to take a look at the features of a workpiece that determine how it behaves when it is at rest or when in motion. These features differ fundamentally from the workpiece properties that the design engineer determines in view of its function. Figure 6.11 classifies the workpiece features on the one hand, with regards to the geometry, characteristic form elements and physical features and on the other hand, with regards to the basic properties when at rest or in motion.

From the geometric workpiece data the shape is dominant, since it determines the handling of the workpieces. 'Tangley parts', which can hook on to one another in the feeding process (e.g., snap rings, coil springs) and frequently disrupt part feed systems, are considered difficult to handle. Two factors are key with regards to the characteristic form elements: The first is whether the elements are in the inner or outer contours of the workpiece, while the second concerns their physical properties i.e., parts that are less stiff (e.g., rubber gaskets) and parts with sensitive surfaces are difficult to handle. These workpiece features each have a different impact on each of the behaviors listed in Fig. 6.11, e.g., according to how they are positioned on the workpiece carrier as well as on behaviors when in motion, say on a conveyor belt.

This topic is therefore significant for those planning assembly systems and factories because handling systems frequently represent a source of considerable disruption which impacts very much the degree of utilization, accessibility and required space.

Figure 6.12 defines and organizes handling into 5 sub-functions: buffering, changing quantity, moving, securing and checking. These can be combined so that self-contained sequences of functions are created which can then be executed by a single device.

The *buffer* provides the local stock of parts for the assembly station either not ordered (e.g., as a pile in a bunker), partially ordered (e.g., stacked metal sheets) or fully ordered (e.g., in a part magazine). The sub-function *changing quantity* separates workpieces from a bulk or put parts into a specific sequence. *Moving*, on the other hand, refers to parts being turned or shifted so that they are placed in a desired spatial position, whereas *securing* serves to clamp parts in a certain position or release them. This takes place while parts are in motion or, as is the usual case, in the joining position in order to withstand the acceleration and/or joining force. The *checking,* which occurs subsequently, does not verify the quality of the joining operation's results, but rather verifies the characteristics relevant for handling e.g., identity, position or number. The checking device then signals that these characteristics have been verified and in doing so triggers the following operation such as picking, joining or transporting the workpiece or workpiece carrier further. This further transportation is in turn part of the material flow chain to internal suppliers (e.g., discrete part manufacturing) and beyond that to external suppliers.

Before we introduce the involved logistic processes in the next section, we would like to turn our attention to the changeability of the assembly processes as summarized in Fig. 6.13.

It can generally be assumed that there is a high degree of changeability with the joining processes implemented in the assembly. With the handling processes however, changeability is, by definition, primarily determined by the form elements of a part that serve to differentiate the position and action of force for changes into the desired position.

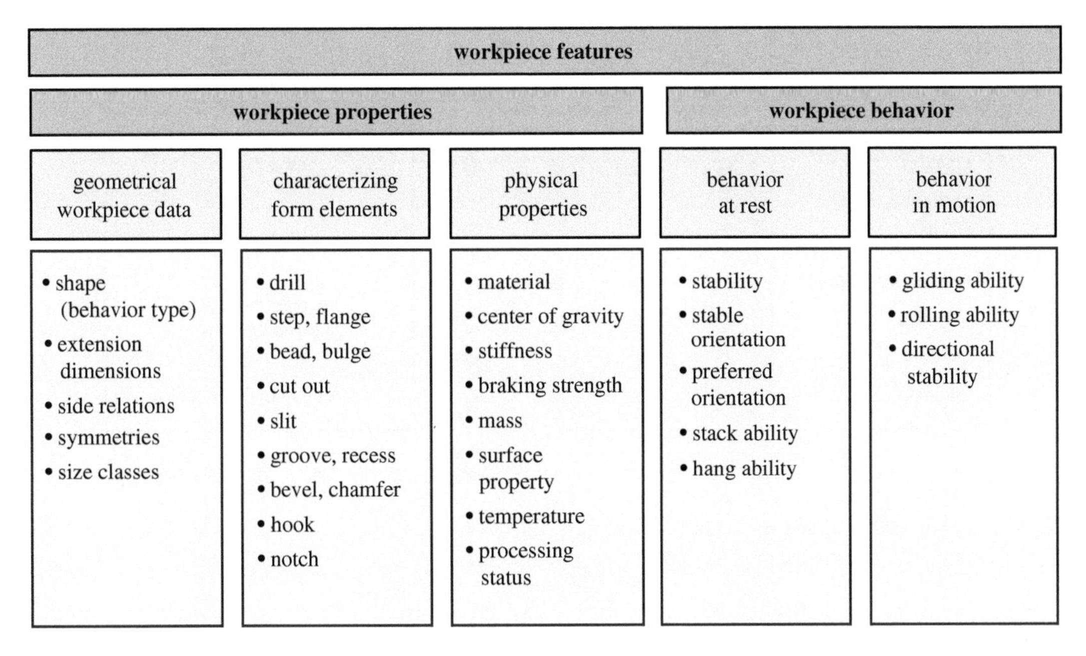

Fig. 6.11 Workpiece features relevant for handling (FhG IPA)

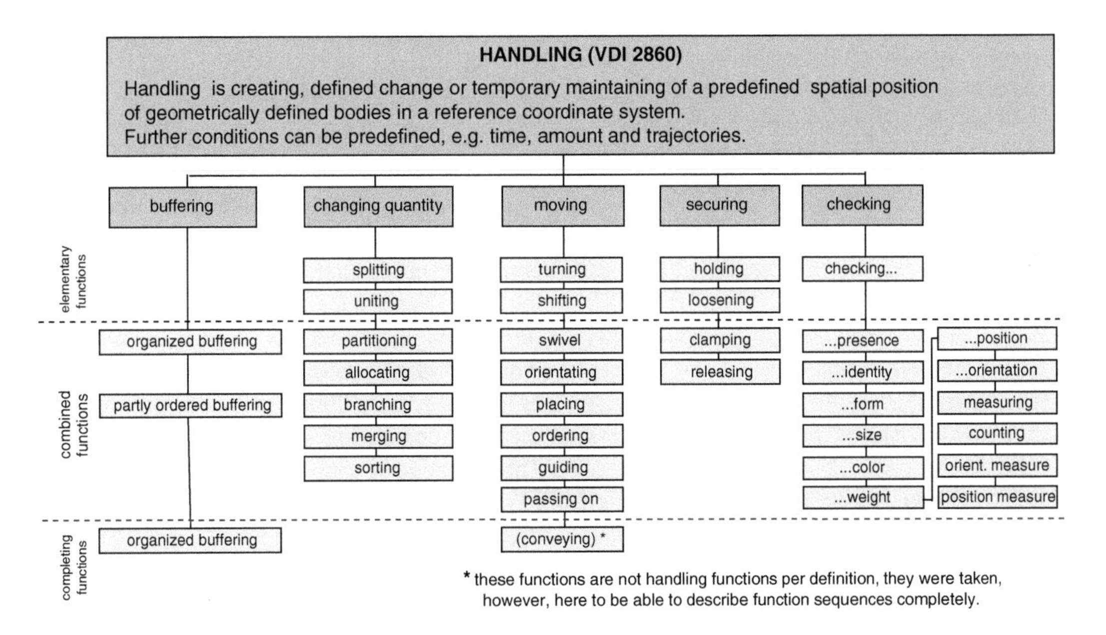

Fig. 6.12 Handling sub-functions [VDI2860]

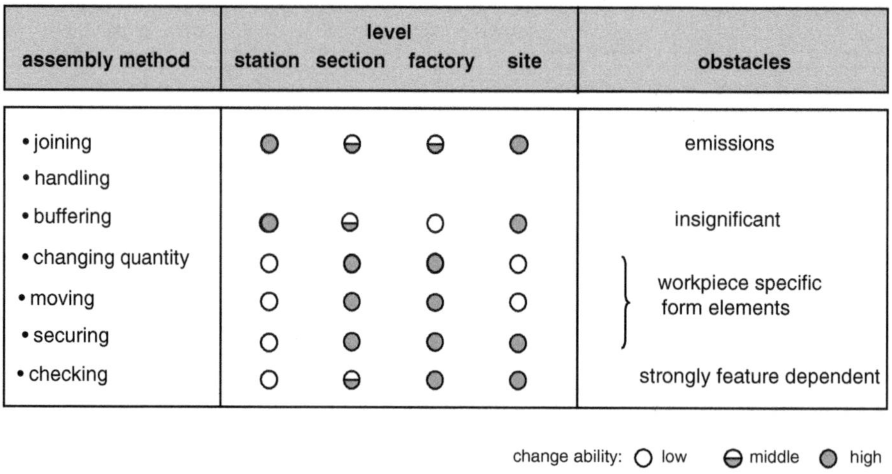

assembly method	level: station	section	factory	site	obstacles
• joining	●	⊖	⊖	●	emissions
• handling					
• buffering	●	⊖	○	●	insignificant
• changing quantity	○	●	●	○	workpiece specific form elements
• moving	○	●	●	○	workpiece specific form elements
• securing	○	●	●	●	workpiece specific form elements
• checking	○	⊖	●	●	strongly feature dependent

change ability: ○ low ⊖ middle ● high

Fig. 6.13 Changeability of assembly processes

A number of methods that can be used to overcome change-resistance in assembly operations on a workstation level can also be identified here:

- Change the features of a workpiece that are relevant for handling, without impairing the function of the workpiece. Recommendations on how to design products suitable for assembly can be found in e.g., Boothroyd and Dewhurst [Boot83], Lotter [Lot13], as well as Redford and Chal [Red94].
- Replace mechanical forces with non-contacting forces such as electrical or magnetic.
- Check workpieces using image processing instead of mechanical sensors.

In the next section, we will turn our attentions to logistic processes that operate on a workstation level, i.e., directly within the manufacturing and/or assembly workstation proximity.

6.2.3 Logistic Processes

The term 'logistics' presumably stems from the military field and encompasses all of the tasks that serve to support the armed forces. Inspired by this, business logistics—divided into industrial, trade and service logistics—developed beginning in the 1970s and rapidly in the 1980s (see e.g. [Mur10]). Generally, logistics is concerned with changing objects spatially over time, thus making them available in the correct quantity, combination and quality at the right time and place. In doing so, it has to be ensured that the costs are kept to a minimum and that the delivery service is oriented on the customer.

The area we are interested in here is industrial logistics, which can be broken down into procurement, production, distribution, and disposal logistics. Disposal logistics will not be considered further here because it has no great impact in the industrial goods production; the remaining core processes can be further divided into sub-processes as depicted in Fig. 6.14 [Gud10, AIK02]. These functions can be seen as part of supply chains. A generic model named SCOR for this has been developed and is regularly updated by the Supply Chain Council (http://supply-chain.org/).

With *procurement processes,* external transport processes deliver ordered goods to the incoming store where they are unpacked and transported to an intermediate store. In general, the subsequent manufacturing, assembly and distribution require not one, but a number of articles at a specific time, usually in varying quantities, referred to as 'orders' or 'commissions'.

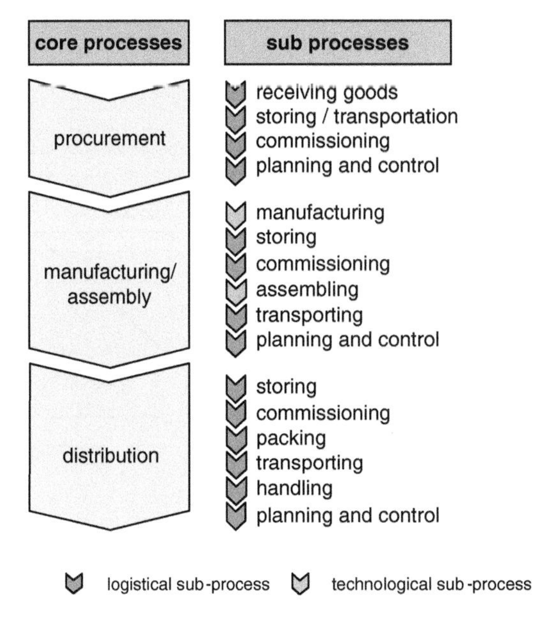

Fig. 6.14 Core and sub-processes of industrial production

The corresponding process, of collecting items for the orders is often referred to as 'order-picking' or commissioning. All of the sub-processes of procurement have to be planned, controlled and monitored.

The subsequent *manufacturing and assembly* with their technological sub-processes we have already discussed, also includes storage and transportation sub-processes. In this case raw and semi-finished products are moved and stored. If the assembly has local storage facilities, order picking processes will also apply. Depending on the production location, the manufacturing and assembly should also be more or less precisely planned and controlled.

The *distribution* is responsible for supplying the produced or purchased article to the customer, which can be a distribution center, a retail store or an end customer. Besides storing, picking and transporting this also entails protecting the goods through packaging, arranging the goods into transport units and when necessary the handling of transport units e.g., transshipping. Distribution processes also have to be planned, controlled and monitored.

From the view of factory planning the sub-processes can be reduced to the reference processes: production (manufacturing, assembly), transportation and storage. In order to depict these, Kuhn's Process Chain Elements [Kuh95] have proven to be useful as well as Nyhuis and Wiendahl's Logistic Operating Curves [Nyh09a]. The first can be used to visualize the logical relations between the elementary functions and the so-called 'process chain plans', whereby each of these apply to an article or article group with the same sequence of functions. Logistic Operating Curves, on the other hand, describe the functional interaction between the logistic objectives (WIP Work in Process), throughput time, output rate and schedule compliance) on a workstation or in a manufacturing sector.

Figure 6.15 depicts in the lower part a sample process chain plan for producing microelectronic chips. The article is an electronic circuit, created on a silicon wafer, separated and integrated into a casing. The consumer—a laptop producer—calls up the components from a store as required. Every step in this process chain fulfills one of the three basic logistical functions and requires resources, personnel, space and control information in order to fulfill its function. A process chain element can visualize a company as a whole or it can be divided up hierarchically over the factory levels down to an individual work operation.

In our current discussion of planning a factory on the "workstation" level the Logistic Operating Curves for the production, transportation and storage (as indicated in Fig. 6.15 upper part) play an important role in dimensioning workstations and their buffers, therefore we take a closer look at these now.

We first have to identify what the logistic objectives are for these three reference processes as is summarized in Fig. 6.16.

The objectives can be organized according to an internal view and an external view. Schedule compliance and throughput times are perceived by the customers and are thus attributes of the logistic performance. In comparison, the output rate (and thus the related utilization), WIP and

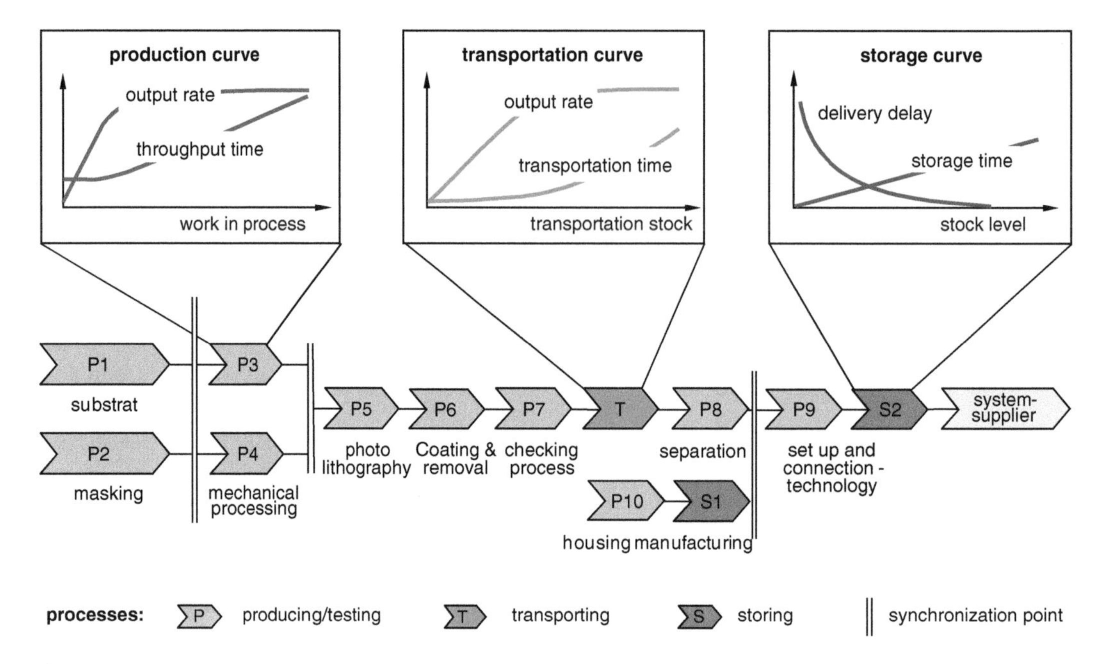

Fig. 6.15 Elements and logistic operating curves for production processes in a process chain plan

	objectives	production reference process		
		producing	**transporting**	**supply**
customer view	**schedule compliance**	highly punctual	highly punctual	minimal delivery delay
customer view	**throughput time**	short throughput time	short transportation time	short storage time
enterprise view	**output rate**	high utilization	high utilization	high utilization
enterprise view	**WIP**	low WIP	low stock in transport	low stock level
enterprise view	**costs**	low costs per yielded unit	low costs per transportation process	low storekeeping costs

Fig. 6.16 Logistic objectives for the production's reference processes

costs are internal objectives and should be maximized or minimized respectively.

The question of course is how can the three processes and their objectives be modeled, dimensioned and designed from a logistic perspective? The Funnel Model and Throughput Diagram have proven to be useful for the 'production' reference process (see Fig. 6.17, [Nyh09]).

The workstation appears as a funnel, whereby the balls represent the waiting orders (WIP) and the variable opening symbolizes the set capacity.

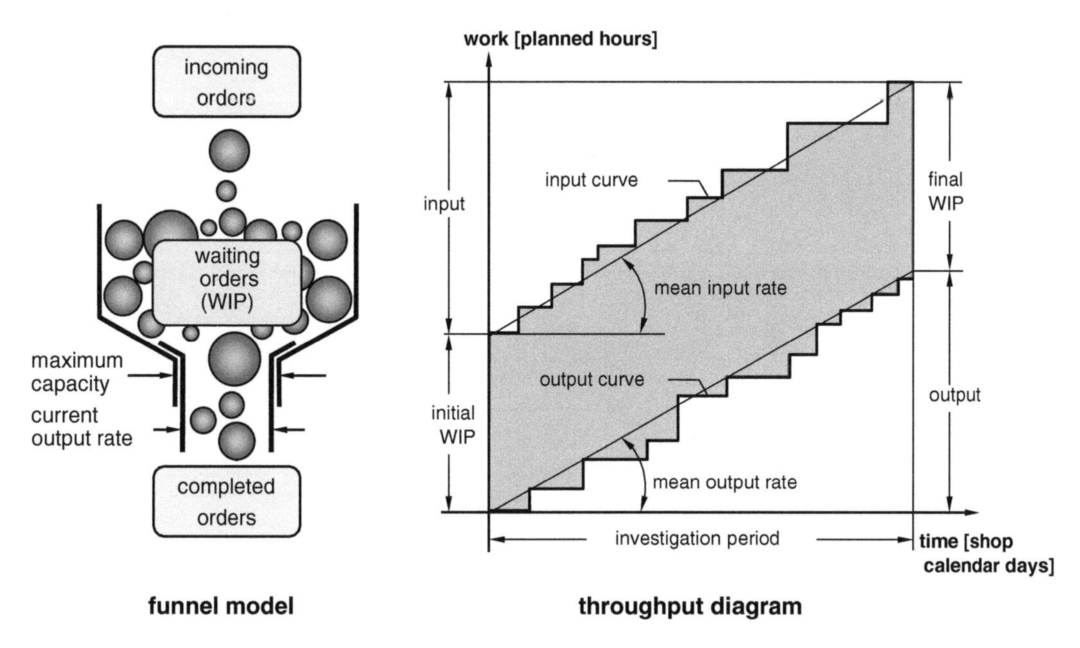

Fig. 6.17 Funnel model and throughput diagram of a workstation

Moreover, the volume of the balls is a measure of the work content in planned hours. If we observe this system over a longer investigation period, we could plot the input and output events in a so-called 'Throughput Diagram'. In which case, the lower curve would encompass the progression of the cumulated output and the upper curve the progression of the cumulated input. Usually, there would be an initial WIP level at the start of the measuring and a final WIP level at the end of the investigation period. The slope of the output curve corresponds to the mean output rate, measured in planned hours per workday, whereas the slope of the input curve corresponds to the mean input rate of the workstation during the same measurement period.

If the input is stopped at any point in time, the available WIP will be sufficient for a period equal to the ratio of the output rate and WIP. Accordingly, this value is referred to as the range, whereby the relationship in which the "range equals the WIP divided by the output rate" denotes the Funnel Formula. In comparison, the mean throughput time results from the mean value of the throughput times of the individual finished orders [Nyh09a].

With that, we have managed to explain two of the objectives identified in Fig. 6.16 for the process "producing" which leaves us still with the utilization and schedule compliance (or lateness). All of these parameters can be visualized in a Throughput Diagram as shown in Fig. 6.18.

In the center of the figure is the so-called 'logistic target cross' with the external perceived performance objectives 'throughput time' and 'lateness' and the internally perceived objectives 'utilization' and 'WIP'. Each of these objectives is allocated a Throughput Diagram which illustrates how it is represented. The WIP level appears here as the blue area between the input and output curve. The throughput of each order is depicted by a rectangle, whose length corresponds to the throughput time and whose height relates to the work content. The lateness also appears here as a rectangle, nonetheless, its length is determined by the time difference between the planned and actually attained output date. The difference can be positive (too late), negative (too early) or zero (punctual). Finally, the utilization appears as the ratio between the actual output and the planned output. In order to develop a Throughput Diagram, only the planned

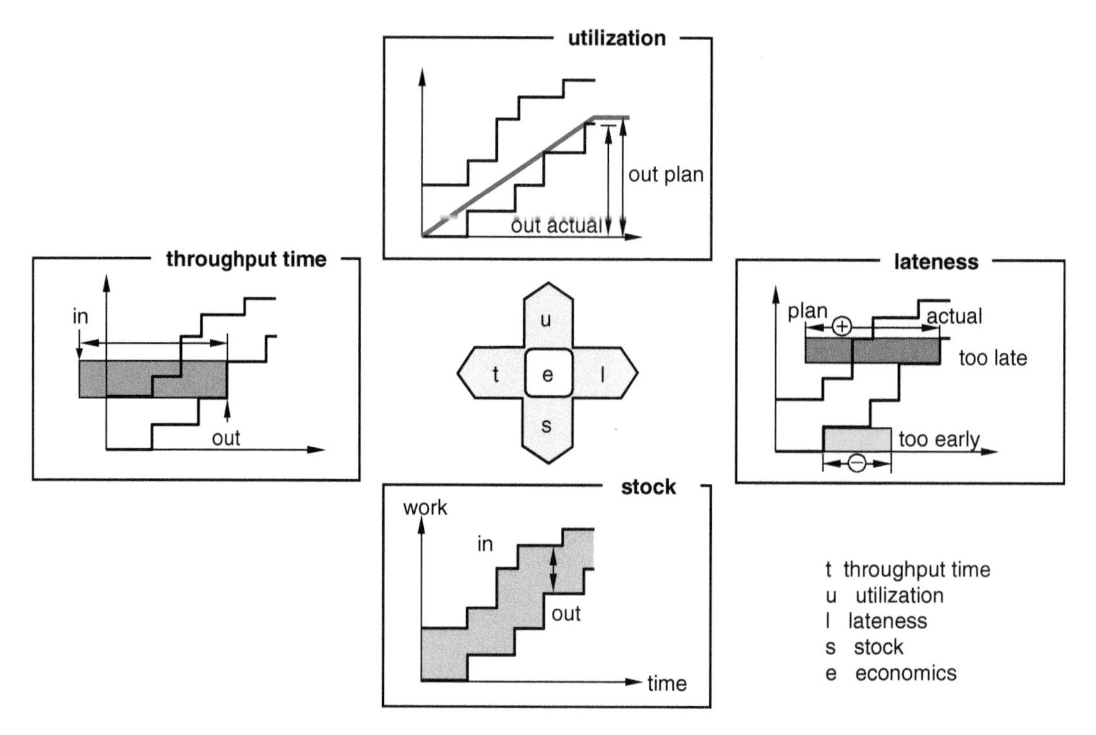

Fig. 6.18 Logistic objectives in the throughput diagram

and actual input and output dates are required along with the planned work content for each order.

The next question that arises is how these objectives interact on a workstation. Figure 6.19 provides the answer in the form of an exemplary set of Logistic Operating Curves, known as the Production Operating Curves (see Fig. 6.15). According to them, the output rate (also known as the yield) of a workstation first grows proportionally with increasing WIP and then slowly levels out until it has reached the limit of the capacity. The range and with that the throughput time increase corresponding to the Funnel Formula with the WIP. The inserted points identify the operating state of the workstations at that point in time. Accordingly with 70 h of WIP the station's operation is fully utilizing the capacity of 16 h per shop calendar day (i.e., two shifts daily), whereby an order throughput time of approximately 3.8 shop calendar days is attained. The operating curves can either be determined point-by-point using simulation experiments or calculated using approximation equations [Nyh09a].

Obviously, the logistic behavior of the system is more favorable the steeper the Output Rate Operating Curve progresses. It is then possible to attain a high output rate and utilization with low WIP levels and thus with short throughput times.

The technology and factory planner plays a decisive role in determining the progression of the curves, whereas it is the responsibility of the production control to set the operating point on the operating curve according to the selected targets. It is thus vital to know the parameters that determine the operating curves in order to be able to specifically influence them.

Figure 6.20 depicts a structured outline of these parameters differentiating between operating curves for an ideal process and those adjusted to a real situation [Nyh09a]. The ideal Output Rate Curve assumes that the input and output processes on the workstation are aligned

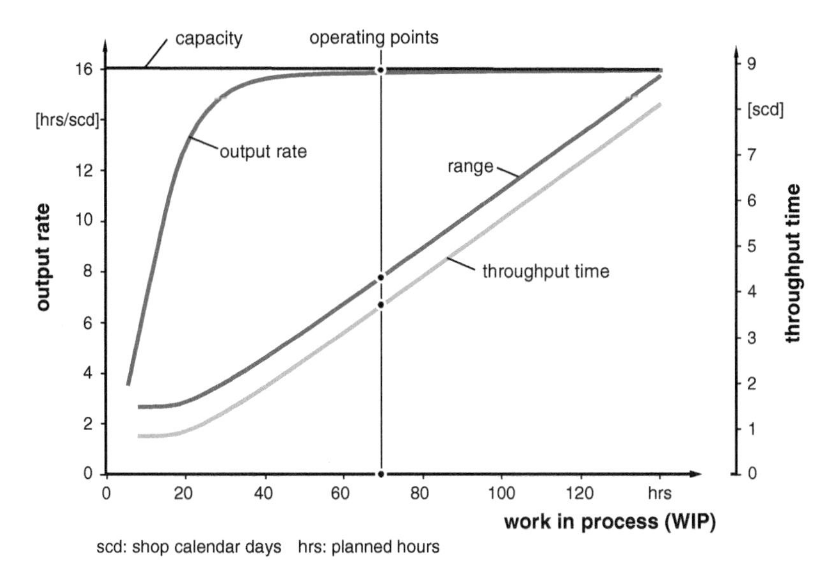

Fig. 6.19 Production operating curves for a workstation (example)

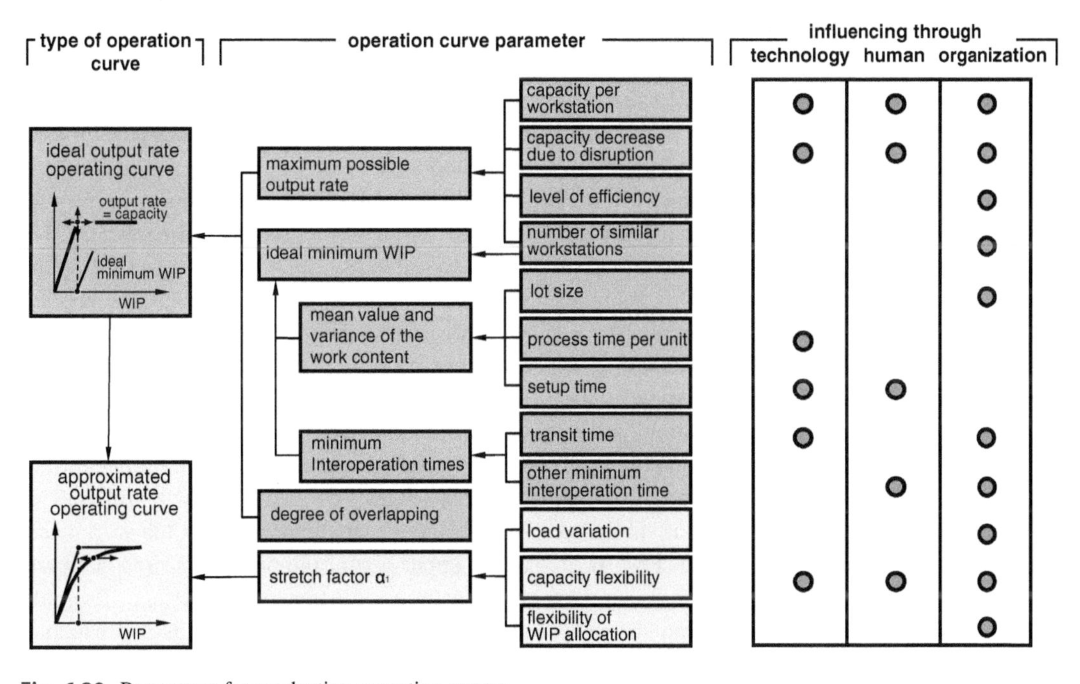

Fig. 6.20 Parameters for production operating curves

with one another so that immediately before an order is completed, the next order arrives. Thus neither the workstation nor an order has to wait. In this case, the operating curve can be calculated exactly. It is determined (a) by the workstation's maximum possible output rate, (b) the so-called 'ideal minimum WIP' and (c) the chronological degree of overlapping for the individual parts of a lot. The ideal minimum WIP refers to the WIP bound by the processing.

On the one hand, it depends only on the individual processing time, the order lot size and the workstation's setup time and on the other hand, on the minimum inter-operation time that passes up until the order is delivered to the next workstation.

On the shop floor though, these ideal conditions do not exist: the input is irregular, resulting in a more or less strongly fluctuating input. Frequently, the workstation's flexibility is not sufficient to follow these fluctuations and when there are a number of similar workstations, there is a more or less greater flexibility in allocating the work in front of the workstations.

A glance back at the right side of Fig. 6.20 clarifies which possibilities there are for favorably influencing the design of a workstation from a logistic perspective when planning a factory. Based on the three traditional approaches (technology, human and organization), the factory planner sets the output rate through the capacity structure i.e. the number of workstations and how they are connected. Skillfully selecting the processes and resources ensures minimal setup times, smaller lot sizes and with that lower values for the work content's mean and variance. Finally, a well thought-out layout with short transportation routes, as well as quality checks close to processes ensures that the tied-up WIP is minimal during transportation and testing operations.

The second reference process mentioned in Fig. 6.15 affects the transportation, or more precisely an individual means of transportation e.g., a fork-lift truck or an electric overhead conveyor. These units can also be described with Logistic Operating Curves whose progression is similar to the Production Operating Curves, Fig. 6.21 [Wie00].

In comparison to the Production Logistic Operating Curves there are two Transportation Operating Curves: The Output Rate Curve only considers those trips where a load is carried, whereas the Total Output Rate Curve also includes the idle travel time. The main parameter for the Transportation Operating Curves is the transport work tied into the system, not as the quantity or volume of the transported goods, but rather in the form of transport hours. With increasing WIP the output rate initially grows proportionally. In the transition zone of the operating curve the waiting orders increasingly compete for the means of transportation, so that queues arise and the utilization then only grows disproportionally slow until the full utilization point is met. The same is true for the progression of the Throughput Operating Curve (not shown here), which corresponds to the sum of the mean operation time and the mean idle travel time. A detailed discussion on deriving the Transport Operating Curves is found in [Egl01].

Similar to the Production Logistic Operating Curves the factory planner first determines the transportation capacity based on the necessary movement of materials resulting from the layout. The next task consists of minimizing the idle travel time e.g., by establishing transport circuit tracks. Once again, the aim here is to use the system's capacity as completely as possible while maintaining the smallest store of goods tied into the transport system as possible. By doing so the goal of shortening transport times is also supported.

The last logistical elementary process we will consider is the storage of goods or articles. This process is always required when two sequential processes are not synchronized with one another with regards to their output and input behavior and quantity-wise. Synchronized processes can be found for example in a large multi-stage press with a forced cycle, where, for example, a sheet metal is formed into a finished automobile roof in 5–7 tool stages. With every down-stroke all of the tools work simultaneously on the workpiece, as the upper part of the press then rises, the parts are synchronously transported further. The only stock that is tied into the system is that which is absolutely required for executing the process and for further transporting.

Of course, the goal is always to avoid inventories of all kinds in a factory, because they tie-up capital, necessitate inventory management and occupy space. Thus, in many companies the terms 'inventory' or 'stock' have been struck from the vocabulary and instead dynamic interim

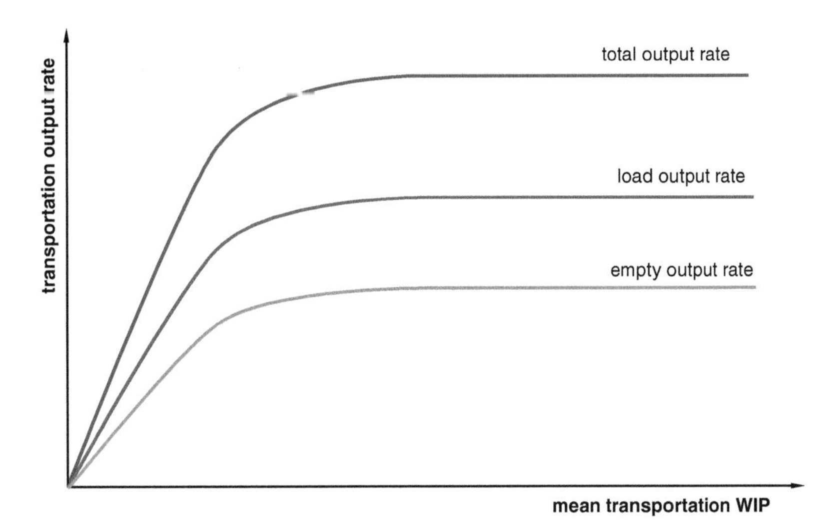

Fig. 6.21 Transportation operating curves

buffers are spoken of for example. At times, visions or guidelines for an inventory-free factory evolve from there. Guidelines such as these are of little use for planning a factory as they are unrealistic. Instead, the logistic laws of a store should be applied to each situation and limits should be sought (as discussed in Sect. 6.2.3). The Service Level Operating Curves, developed by Lutz [Lut02] provide a logical approach for this. Figure 6.22 depicts the basic progression.

A single article or a logistically similar group of articles is first considered with its input and output in or from a store. Similar to the Production Logistic Operating Curves, an ideal operating state can also be defined for the Service Level Operating Curve and from there an ideal Service Level Operating Curve can be calculated, determined only by the input lot size. In reality however, there are quantity and schedule deviations both in the input as well as the output and fluctuations in the demand which make safety stock necessary. These then lead to the actual Service Level Operating Curves. Nyhuis and Wiendahl [Nyh09a] and Lutz [Lut02] both describe equations that can be used to calculate the operating curves.

The stock is almost exclusively determined by how the input and output processes behave. It is

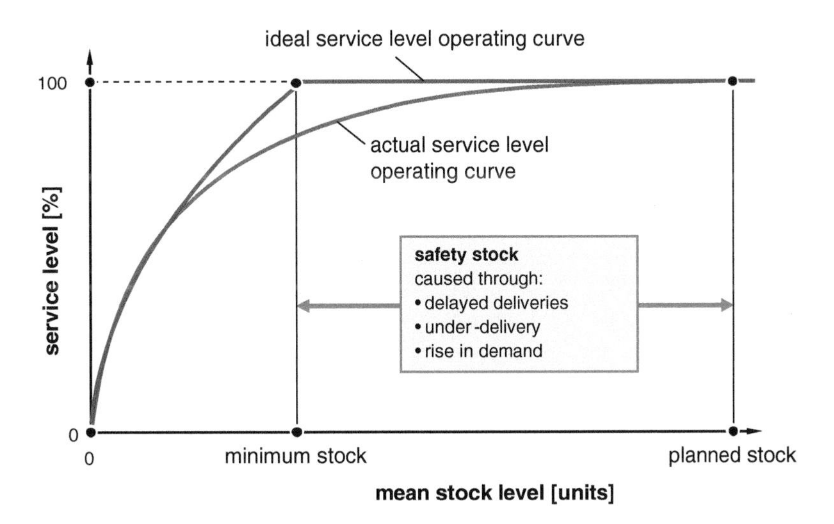

Fig. 6.22 Ideal and actual service level operating curves for an inventory article

at first surprising that accordingly there is no possibility to influence it with technology or the workforce. The reason for this is that storage, as a method, is a process that changes the material spatially and temporally but does not change its state or feature. Technology and people only have influence when the storage process is executed using logistic equipment and resources. Thus for factory planning, the basic possibilities for intervening are concerned with harmonizing the input and output processes. This includes adjusting the lot sizes for the input and output (i.e., synchronized production), homogenizing the rate of consumption, minimizing delivery time and quantity deviation and finally, shortening replenishment times.

On a workstation level, these approaches are mainly predetermined by higher level structures and strategies. These are developed when planning the factory structure and have to be derived from the factory's general logistic concept, e.g. applying the push- or pull principle. This will be explained in detail in Sect. 9.5.2. Basically, the factory planner, together with the logistician, can only influence the workstation in regards to this target with a well-organized material supply and disposal. With an assembly station, the feeding technology fulfills this task and similar devices are to be found on automated machine tools.

Generally the question of changeability does not arise in regards to the logistic transportation and storage processes, but rather first poses itself in interplay with the preceding and subsequent processes and beyond that, when the technological features of the required resources are known. Both the aspects of influencing through technology and human as well as the aspect of changeability are addressed on the next higher level in conjunction with the logistic methods described there.

Before discussing these though, we will turn our attentions in the next section to the factor identified in Fig. 6.3 as the second aspect of designing a workstation i.e. technology. Once again our aim here is to uncover the basic features of the facilities used in manufacturing, assembly and logistics that are relevant for a factory planner. These are then supplemented with the information technology necessary for operating workstations as well as the supply and removal technology. After which we will also examine how the factory planner can influence these and the characteristics of changeability that can be found in this area.

6.3 Facilities

Technical plants, devices and equipment serve to implement the manufacturing, assembly and logistic processes described in the preceding section. For our purposes, we will generally refer to these in the following as 'facilities' although the term 'resources' is also commonly used from a management perspective. In addition to the facilities, resources also encompasses the workforce (human resources), financial means (financial resources) and raw materials. Figure 6.23 outlines the facilities from the factory planning perspective according to their use in manufacturing, assembly and logistics.

The figure differentiates the workstation and section level according to the structural levels of a factory (see Fig. 5.18). Facilities are special in that they are understood as physical units which are inter-connected in view of the material, information and energy flows and are thus not able to function as separate parts. An exception here is very large systems that have a continual manufacturing and/or assembly process for large series and mass production. Examples are an integrated pre-manufacturing and assembly plant for washing machines, a plant for completely manufacturing combustion engines, a painting plant for automotive car bodies, a paper machine or an integrated print-and-fold plant for a newspaper.

From the perspective of factory planning it is practical to first separate facilities into sub-systems as depicted in the first column of Fig. 6.23. From the process view, the facility's workspace is the center of focus. It is within the workspace that the product is changed. The workspace makes it possible for the facility and equipment to conduct the process and can thus be referred to

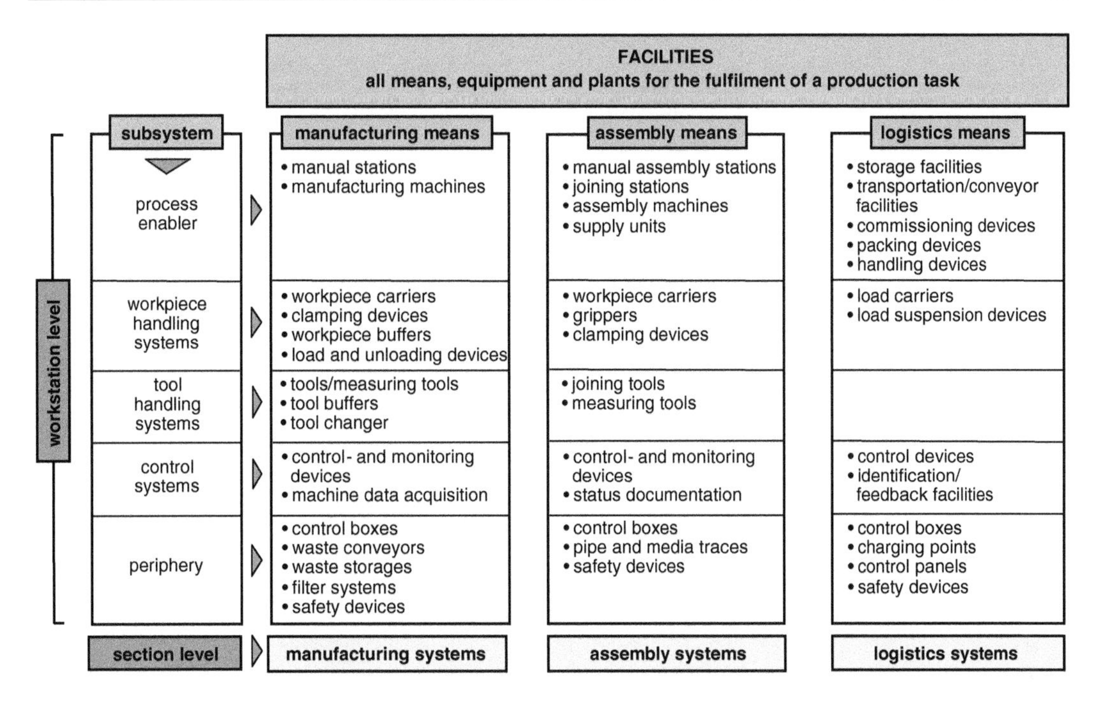

Fig. 6.23 Facilities from the perspective of factory planning

as a 'process-enabler'. From a technical perspective, facilities and equipment are concerned with manual workplaces or more or less automated workstations and machinery. In order to be implemented, many of the processes require facilities and equipment for handling, securing and moving workpieces within the workspace, as well as for loading and unloading them. Similarly, tools and measurement devices are usually required. It is also necessary to be able to automate the control, monitoring and feedback of the testing operations—usually requiring a considerable periphery for removing waste materials, ensuring industrial health and safety standards and last but not least accommodating the housing for the energy and information control module.

6.3.1 Manufacturing Facilities

Manufacturing facilities allow the manufacturing processes outlined in Sect. 6.2.1; Fig. 6.4 to be executed. *Manual workplaces* consist of a workbench which holds the workpiece that is to be processed, possibly clamps (e.g., a vice) and tools for executing the work operation. Manual workplaces are found in the industrial production rather seldom e.g., for bending tubes, deburring, welding etc. By far and wide, the largest proportion of facilities falls under the category of *manufacturing machinery*. The various forms which machinery can take are outlined in Fig. 6.24 and organized according to the manufacturing processes outlined above [Spu96]. We will not discuss these in greater detail though since their properties (geometric dimension, weight, media supply and emission) are all similar from a factory planning perspective.

For the factory planner however, it is more important to divide the manufacturing machinery into single machines and systems of machines, which can then be further divided according to the degree of productivity and flexibility (see Fig. 6.25).

With regards to flexibility, individual machines range from single purpose machines, which can only manufacture one specific part, to convertible single purpose machines to

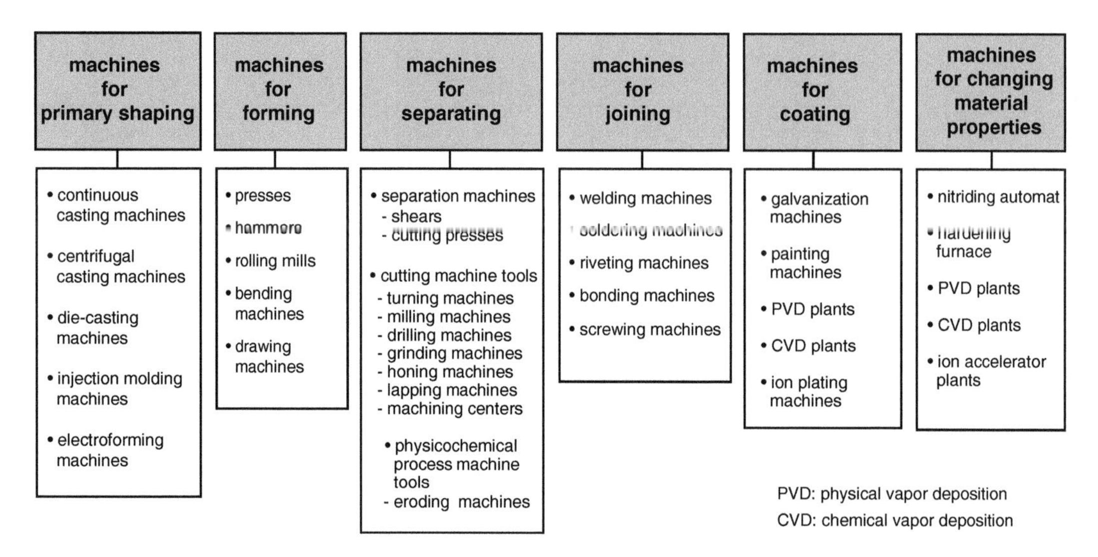

Fig. 6.24 Classification of manufacturing machinery (Spur)

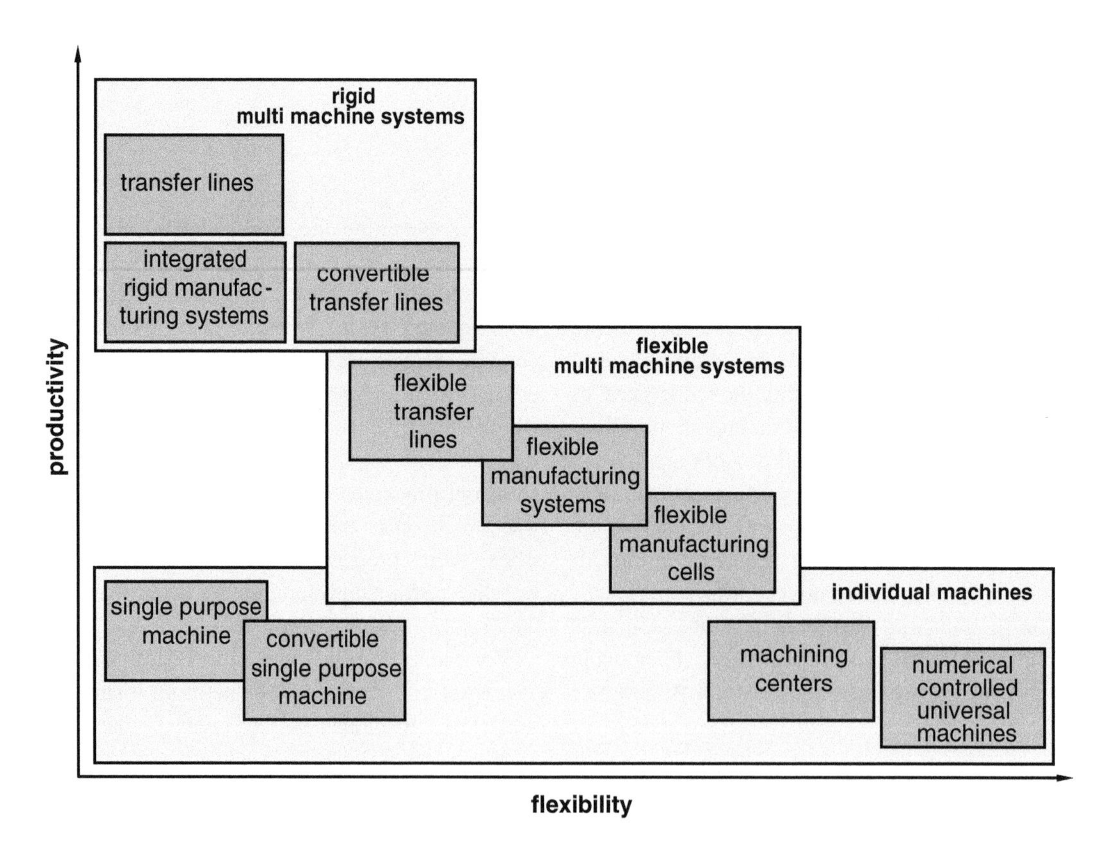

Fig. 6.25 Productivity and flexibility of manufacturing equipment (acc. Weck)

machining centers with limited part spectrums up to numerically controlled universal machines that are only limited in regards to the dimensions of the workpiece. Flexible multi-machine systems (flexible transfer lines, flexible manufacturing systems, flexible manufacturing cells) link a number of single machines into an automated workpiece flow in which the sequence of operations is more or less flexible.

The group of machine systems that has the highest productivity and lowest flexibility are takted transfer lines that can be setup to varying degrees for different workpieces. However the setups are limited here to certain dimensions and features of a very limited part group such as motor blocks for example. From the perspective of factory planning, the forms of manufacturing machines that we have briefly indicated here can be reduced to a comparably few characteristics. An example here is the numerically controlled *universal machine*. Whereas, its general structure is depicted in Fig. 6.26 [Toe95], Weck and Brecher provide an extensive introduction [Wec05]. A comprehensive overview is given by Miller and Miller [Mil04].

The machine's frame determines the spatial structure of the machine with its moving and fixed parts. Guides allow the moving parts to be shifted or turned and generally determine the precision of the manufactured workpiece. Drives, on the other hand, provide the mechanical energy for generating the main and secondary feed motion. Finally, control influences the motors and actuators (power control) as well as the control and monitoring of the movements of workpieces and tools (information control). The latter is accomplished using Numerical Control. The control unit has connections to a local data network (Local Area Network or simply LAN) that allows the electronic exchange of data with a higher level system.

From the perspective of factory planning the frame of the machine determines the required floor area, the required height the space and the weight (load) which the floor has to support as well as whether or not the machinery has to be anchored in a floor-bed. In comparison, the drives and processes determine the power and media supply such as compressed air or cooling water.

Tool handling systems consist of tools and measurement devices that the tool preparation department commissions and supplies order-by-order. With larger quantity tools both integrated tool magazines as well as separate magazines, possibly with special loading and unloading

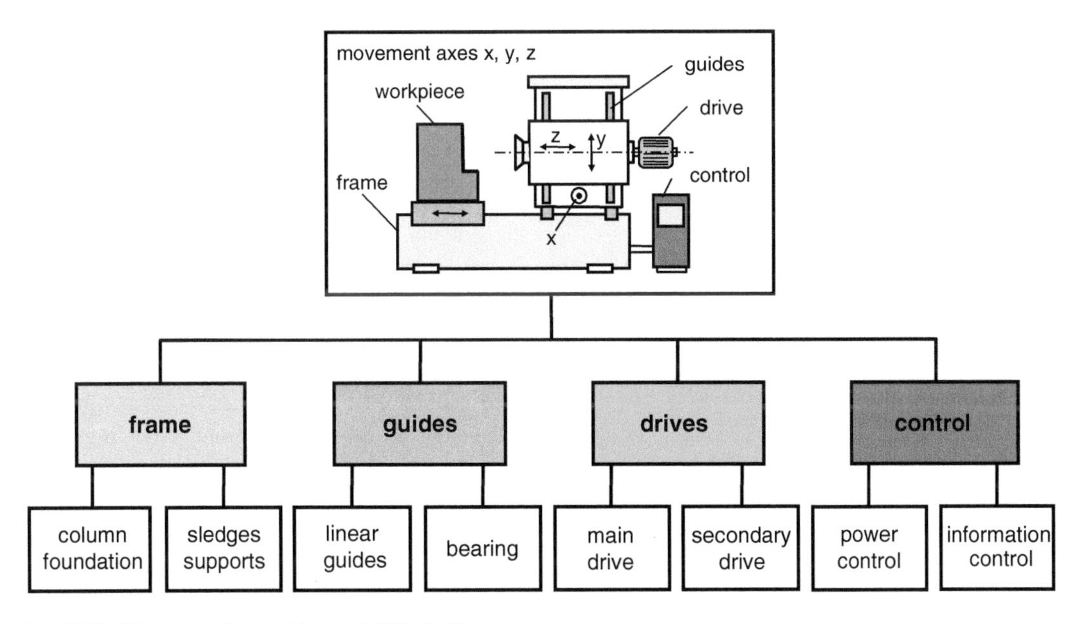

Fig. 6.26 Elements of a machine tool (Tönshoff)

devices, can be implemented. From a factory planning perspective tools systems are floor space consuming, ancillary systems that have to be designed ergonomically and considered in the spatial planning and organization.

Following the scheme of Fig. 6.23 *workpiece handling systems* serve to move and store workpieces in front of and within the machine workspaces. Workpiece palettes (also known as workpiece carriers) that frequently carry special fixtures ensure that the workpieces are correctly positioned spatially and secured in a set position even when they are subject at times to considerable cutting forces. Workpieces can be changed manually. However there are also separate workpiece magazines common that use automated loading and unloading devices. Depending on the size of the workpieces, the subsystems for handling them require considerable floor space. They create the interface with the internal material flow and are decisively designed by the factory planning and logistics.

Control Systems are typically components of machine tools and appear to the user as control or operating panels. A control process can result in data about quality, but organizational feedback such as the quantity of goods and time of completion can also be entered via the operating panel thus closing the logistical control loop. From the perspective of the factory planner, control systems only require space and operational resources for control cabinets and connective lines.

In addition to the workpiece and tool handling systems we have already mentioned, the *periphery* of the manufacturing machinery is for the most part determined by the removal of waste materials and their interim storage as well as health and safety guidelines that result from environmental and safety regulations. On the one hand, these require significant spatial areas and on the other hand, they have to be integrated into the factory disposal system without disrupting the production flow. Since cooling lubricants lead to considerable disposal costs when mixed with chips and sanding dust they deserve particular attention and should therefore either be avoided (dry cutting), reduced (using systems which require minimal amounts of cooling lubricants) or substituted [Wei99].

When selecting manufacturing machinery, it is important to consider the required technology (determined by the processing task) as well as their flexibility and productivity; an approximate classification for this is already provided in Fig. 6.25. The flexibility and productivity of a manufacturing facility is dependent not only on them being organized in one or more units, but also strongly on the degree of automation of the units [Spa97]. The automation refers to the program sequence of the tools and workpiece motion as well as the changing of the tools and workpieces. Based on a recommendation from Spath, Fig. 6.27 depicts the resulting stages of automation for the individual machines [Spa97] and with that further subdivides them.

Starting with a workspace in which the actual manufacturing process occurs, a manufacturing machine is created by adding a power drive and a local control. If this machine is able to execute different processing operations in one setup on the same workpiece by means of a local tool magazine with various tools (e.g., drill, cutter, thread cutter) and an automated tool changer as well as an integrated measuring device, then it is referred to as a *machining center*.

Figure 6.28 depicts a machining center for manufacturing small rotationally symmetric parts. The raw parts are inserted into a circulating workpiece magazine. A vertically and horizontally movable turning spindle grasps and clamps a raw part by means of a chuck. The part is shaped via a chip removal process whereby numerical control is used to move the rotating part along a still-standing cutting tool which is fixed in a drum turret. The falling chips are removed by a conveyor belt into a bin. The drum turret contains all of the tools for completely processing the part; drilling and threading operations are also possible. By turning the turret head to a defined position the individual tools are implemented. After the processing is completed, the spindle sets the completed part down on the workpiece magazine, which in turn moves a step forward and transports the next raw part to the

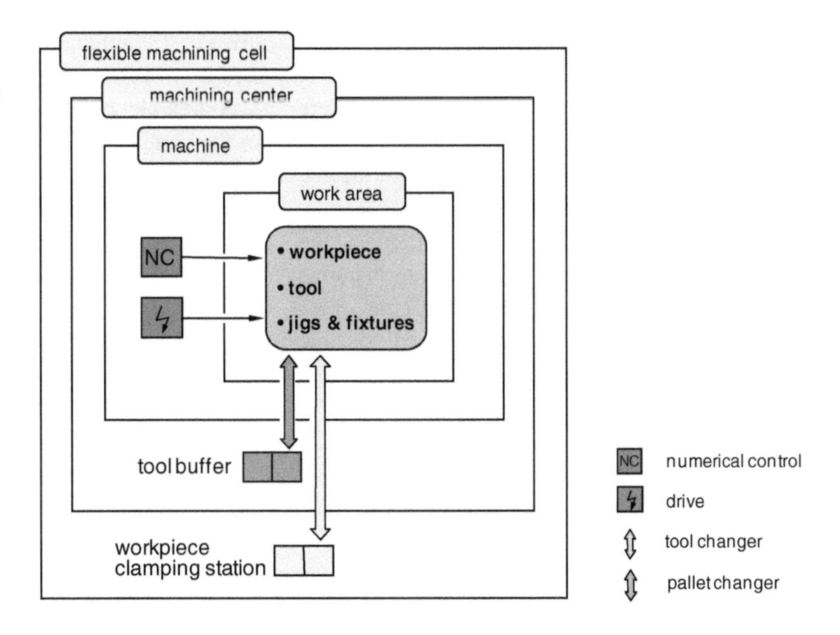

Fig. 6.27 Stages of automation for individual machines (based on Spath)

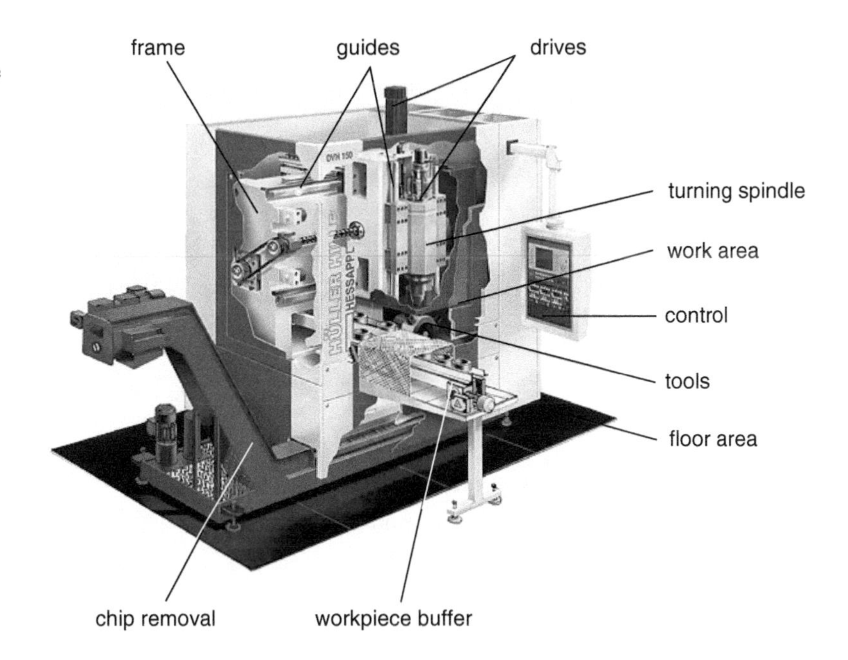

Fig. 6.28 Structure of a tool machine based on the example of an automatic Lathe (courtesy Hüller Hille Hessap)

start position. Also visible in the figure is the floor area required by the machine which is a fundamental piece of information for the factory planner. With machining centers for prismatic parts the workpiece clamping occurs outside of the machine on a so-called workpiece pallet, which is then transported in and out of the machine by means of a pallet changer.

A *flexible manufacturing cell* (FMC) is then created when tool changing and workpiece clamping stations are added outside of the workspace. The typical machines tools and systems are flexible within the frame of their defined part spectrums. However, extensively changing the workpiece spectrum, which requires the integration or removal of processing units, is generally not possible. As we discussed in Chap. 5, reconfigurability can offer a solution in cases where a quick changeover is required. Figure 6.29 introduces a concept for a reconfigurable machine tool developed within the frame of a research project.

Starting with a platform that comprises the framework and axis of motion the machine allows modules and sub-modules to be combined. Different configurations are then possible depending on what workpieces require—three such examples are included at the bottom of Fig. 6.29.

Similarly a *flexible manufacturing system* (FMS) is created by linking a number of workstations via a central workpiece and tool magazine as well as corresponding equipment for changes as far as the material and information flow goes. Figure 6.30 provides a scheme of how such a system is structured.

A flexible manufacturing system is capable for the completely automated manufacturing of a workpiece spectrum—e.g., levers, gears, shafts etc.—in a wide range of quantities and an arbitrary sequence. The individual stations not only conduct processing tasks but also take on ancillary functions such as measuring and washing parts. The system is controlled via a master computer that exchanges order data with the PPC system and controls the cell computers of the sub-systems, whereby the latter communicate with the local machine controls. From the perspective of factory planning, the FMS represents a closed unit that should be tied into the material, information, energy and personnel flow of the next higher planning level.

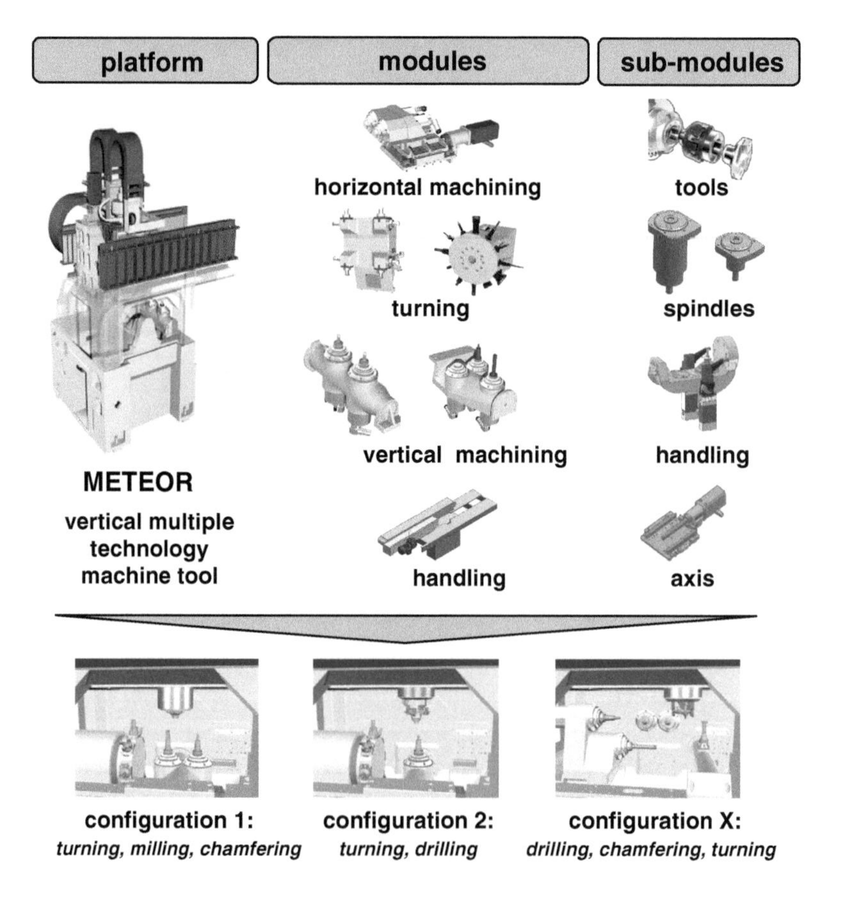

Fig. 6.29 Reconfigurable machine tool (Abele)

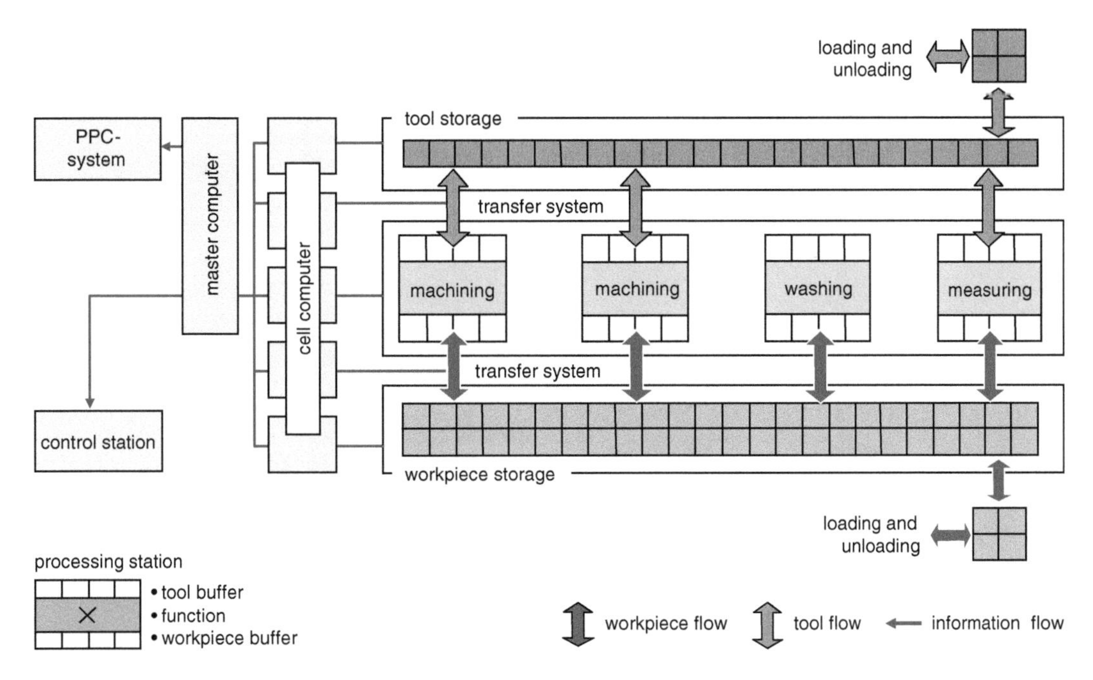

Fig. 6.30 Principle of a flexible manufacturing system

Generally speaking, machine tools do not pose any special requirements for the factory planning. The exception of course is machinery that moves vertically when working due to the process involved and that handles larger workpieces; in this situation special foundations cannot be avoided. This applies in particular to metal-forming machines which are characterized by strong vibrations and sound emissions. Figure 6.31 depicts a hydraulic double column press equipped with a single cylinder drive that yields a maximum force of 2500 kN and is located in the upper cross-head. A main electrical motor supplemented with a secondary motor provides the drive with an installed engine performance between 15 and 55 kW. A layout such as this allows a slower initialization and adjustment mode by operating only one unit, but can also attain higher ram speeds by operating both engines simultaneously in rapid-transverse for the flow and return flow of the press. At 7 m × 5 m × 6 m (height × width × depth) and 40 tons the press represents an almost immovable fixed point in the factory.

In regards to a factory's changeability, the features of manufacturing machines that impact their reconfigurability are of primary significance. Based on their sub-systems (Fig. 6.32 left) these features are summarized in the middle of Fig. 6.32 and allocated to one of three categories: machine, material/tool flow or information flow.

With increasing feature values the changeability on the individual levels is influenced differently. The larger and heavier a machine is, and the more difficult it is to take apart (as far as installation and de-installation is concerned), the more change resistant a machine is. If a machine requires its own foundation and special safety barriers/equipment, it becomes a fixed point in the layout that practically reduces the reconfigurability to zero on all five levels.

Based on the obstacles listed in the right of Fig. 6.32, we can derive the following recommendations for a high degree of reconfigurability:

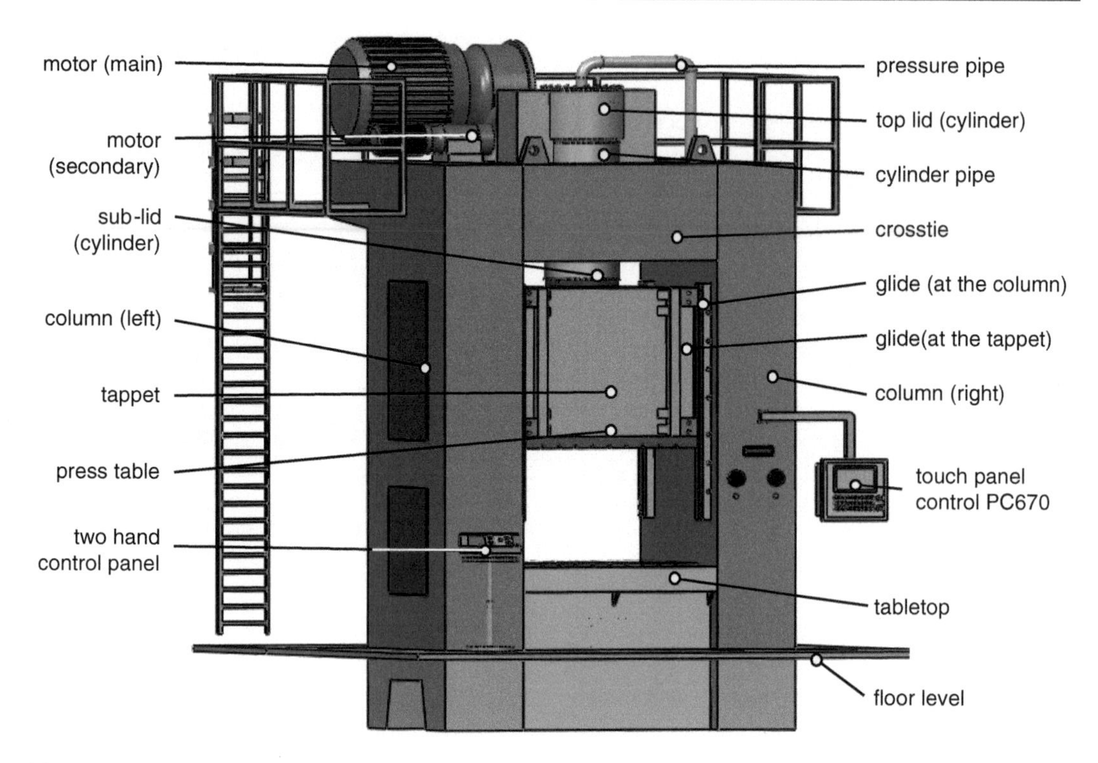

Fig. 6.31 Hydraulic double column press (courtesy Dunkes GmbH)

	manufacturing machines	level: station	section	factory	site	obstacles
machine	outer measurements	low	low	low	low	loading profile truck, train, container
	weight/foundation	low	low	low	low	floor load capacity
	disassembly from installation view	low	high	high	high	mono block construction
	reconfigurability from functional view	high	high	low	low	missing function modules
	energy-/ media demand	low	low	low	low	pipe / trasse routings
	cooling lubricant demand	low	middle	middle	middle	central storage
	noise/vibration emission	low	low	middle	middle	damping housing, floor insulation
	work and safety protection	low	middle	middle	middle	space requirements
material / tool flow	automation workpiece change	low	middle	middle	middle	workpiece carrier, gripper, clamping device
	local work buffer	low	middle	middle	middle	variant number and workpiece volume
	automation tool change	low	middle	middle	middle	tool clamping systems
	local tool stock	low	high	high	high	tool number and wear
information flow	connection to PPC system	high	high	high	high	data transmission
	information provision	high	high	high	high	data transmission

changeability with growing feature value: ○ low ⊖ middle ● high

Fig. 6.32 Reconfigurability of manufacturing machines

- The machine frame should be constructed as self-supporting in order to avoid the need of special foundations.
- The machine should be constructed from modules that are functional and pretested.
- The weight and dimensions of the modules should be adjusted to typical maximum load and profiles for streets and rail ways so that heavy and/or special transports can be avoided.
- The energy and media supply should be routed in modular form with plug-in connections.
- Machinery should be designed to include modules for connecting media and control units, filter and cleaning equipment as well as modular chip conveyors so that the machinery can be adapted to changes in the layout without laborious installation work.
- Machinery should allow modules which can be quickly and flexibly installed for changing and/or storing workpieces and tools.
- The division between supplying technical (numerical control) and logistical (PPC system) information to manufacturing machinery should be eliminated with the aid of universal operator information systems.

It is clear that these measures primarily influence the flexibility and changeability, whereas the change-over ability and reconfigurability are determined by the machine concept and the structure of its sub-systems.

6.3.2 Assembly Systems

From the perspective of the factory planning and operation, the following properties distinguish assembly from manufacturing:

- In order to produce a workpiece a manufacturing machine requires an initial material that usually consists of a single piece of a semi-finished product made from the same or similar material in a few geometric variants. In comparison, in an assembly system, depending on the product structure, a large number of completely different parts have to be joined to some degree in a number of variants and checked to ensure that they are properly mounted.
- With manufacturing processes, the preciseness of the part—which is one of the important characteristics of quality—comes first and foremost. Thus with the significant forces applied mostly in the processes, the rigidity and dynamic behavior of the tool machines play a critical role. Assembly processes on the other hand, use comparably less force so that the focus is on the precision of positioning the joint parts and the reliability of the joining processes.
- The work contents for a manufacturing process in a serial production (which usually has a wide range of variants) typically ranges between 0.5 and 20 h depending on the complexity of the parts and lot size. Manufacturing lines for high volume productions with cycle times in the minute range are an exception. In comparison, with assembly processes in a serial production rich in variety, cycle times typically range between seconds and minutes due to how quickly parts can be picked and joined; lower limits are 2 to 3 s.
- The mean continuous (interruption-free) runtime of manufacturing processes is within an hourly range and the mean duration of repair between 10 and 20 min. In comparison, automated assembly processes generate more frequent disruptions due to the many parts to be handled and shorter cycle times. These periods of interference can occur minutes after one another and typically last for a few minutes. This is of course the reason why manufacturing machines can often be operated during

night shifts without supervision, while assembly stations and even linked systems require on-site personnel who continually monitor the processes and are prepared to intervene as needed.

- As a result of the above mentioned reasons manufacturing machines are thus in principle better suited for automation, whereas assembly systems are rarely fully automated.

As outlined in Fig. 6.33, the assembly systems can be classified according to the output rate and number of parts to be mounted or number of assembly operations, whereby in this field a more or less flexible manual assembly is distinguished from an automated assembly [Lot13].

Manual assembly workstations are centered on personnel including the space they need to move within as well as their physical and mental load. Figure 6.34 depicts the basic specifications of a workplace that allows work to be conducted while sitting or standing [Mar94, Lot13]. It should be emphasized from a logistics viewpoint that the products are assembled lot-wise. When there is a product or variant change, the contents of the picking bins are partially or completely exchanged.

Starting with personnel, the resources required for conducting the work include a work desk, a place to sit and/or stand, picking bins and a joining tool. The results of the work are either deposited in a transportable container, preferably in an organized manner, or on a conveyor belt. The spatial arrangement of the work equipment takes into consideration the measurements for the height of the work desk, viewing distance, gripping range and the space available for unobstructed movement of the operator.

The operations executed by the worker can be broken down into five basic movements: reach, grasp, move, position, release [Bok12]. The aim is to develop a sequence requiring the least amount of time while avoiding difficult movements by simplifying them. Further improvements can be made by simultaneously executing similar or different movements with both hands and eliminating activities that do not add value.

According to Lotter [Lot13] a measure for the effectiveness of an assembly workplace is defined as the sum of all primary assembly operations divided by the sum of all assembly operations (=sum of primary assembly operations + sum of secondary assembly operations)

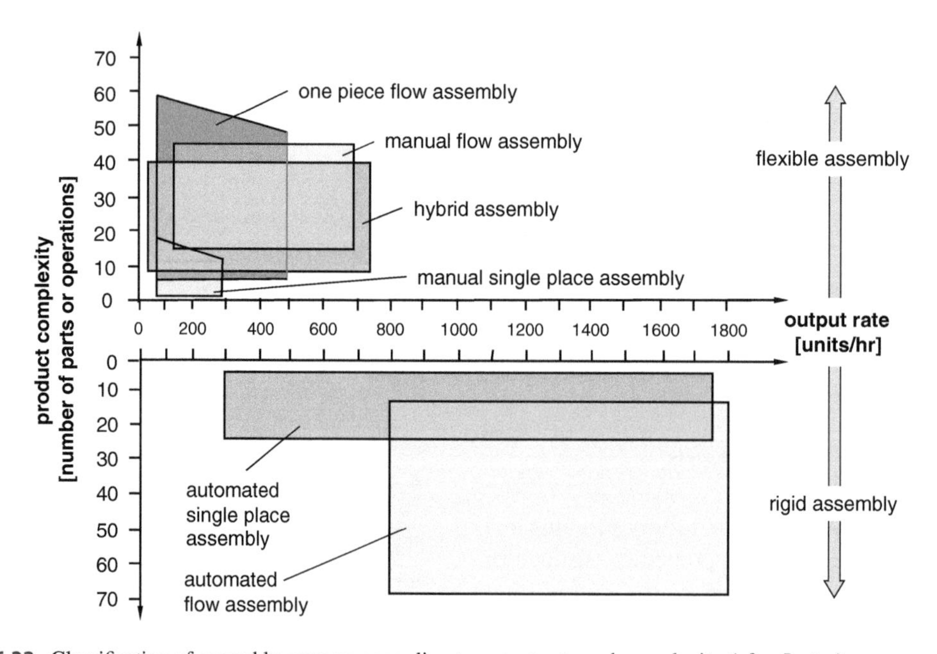

Fig. 6.33 Classification of assembly systems according to output rate and complexity (after Lotter)

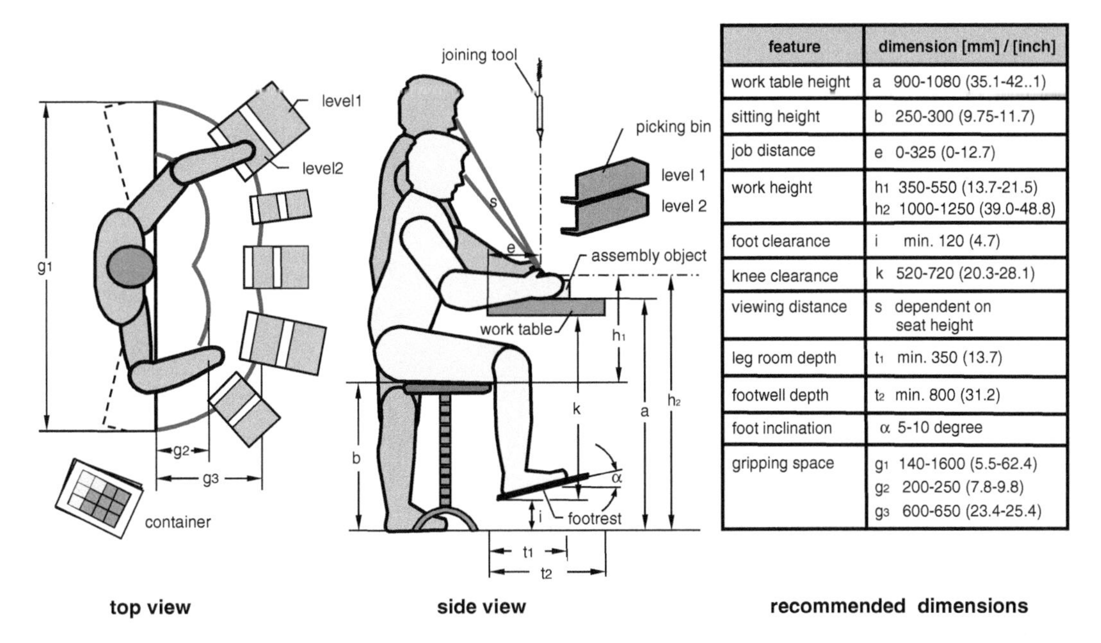

feature	dimension [mm] / [inch]
work table height	a 900-1080 (35.1-42..1)
sitting height	b 250-300 (9.75-11.7)
job distance	e 0-325 (0-12.7)
work height	h_1 350-550 (13.7-21.5) h_2 1000-1250 (39.0-48.8)
foot clearance	i min. 120 (4.7)
knee clearance	k 520-720 (20.3-28.1)
viewing distance	s dependent on seat height
leg room depth	t_1 min. 350 (13.7)
footwell depth	t_2 min. 800 (31.2)
foot inclination	α 5-10 degree
gripping space	g_1 140-1600 (5.5-62.4) g_2 200-250 (7.8-9.8) g_3 600-650 (23.4-25.4)

Fig. 6.34 Assembly workplace for manual assembly (Bosch Rexroth)

expressed in percent [Lot13]. Primary assembly operations include all operations that increase the value of a product during its assembly, whereas secondary assembly operations represent the necessary, unavoidable operations that do not create any value.

If with the measures mentioned above it is not possible to increase the output rate by the personnel because of too short cycle times, a partial mechanization and automation is appropriate. This leads to batch-wise and partially automated assembly. Figure 6.35 depicts an assembly workplace where base part T1 is inserted from a picking bin onto a turntable with fixtures for 18 workpieces. Thereby the turn table moves always to the optimal joining position Lotter [Lot13]. Following that turntable 2 is turned so far that part t2 is in the optimal position for grasping. This part is then joined to part t1 eighteen times. The operation repeats itself until all 6 parts are mounted 18 times and deposited in order in a container. The short grasping distances (reducing secondary efforts) and simultaneous use of an automated joining press are worth noting here. Assembly solutions such as this are thus also known as hybrid assembly systems.

If instead of processing products batch-wise, they can be worked on separately from one another piece-by-piece. This is referred to as One-Piece-Flow Assembly. An example for this type of solution is depicted as a single workplace in Fig. 6.36 [Lot13].

Here the worker picks the base part and puts it on a workpiece carrier that can be shifted using an 'assembly sledge' on a ball roller table. The worker then moves the sledge manually into the optimal position for grasping and joining the next part. In this case there are a total of 11 parts (P1 to P11) to be joined before the finished product can be orderly deposited. This system is beneficial because it minimizes the worker's movements and the external material supply which turns out to be advantageous when there is a variant change.

With strongly fluctuating demand it is also possible to implement a flexible workforce. A possible solution for this is depicted in Fig. 6.37 in the form of a U-shaped system in which product A and B are assembled. Depending on the demanded yield, one to three workers can for example work in area A and two workers in area B who execute a smaller or larger number of assembly operations. The commissioning for the

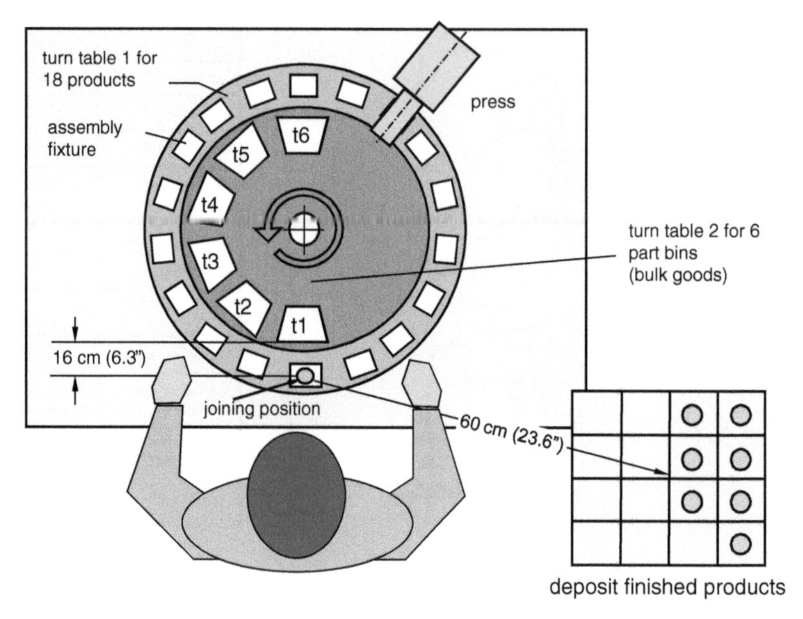

Fig. 6.35 Assembly workplace for batch-wise assembly (Lotter)

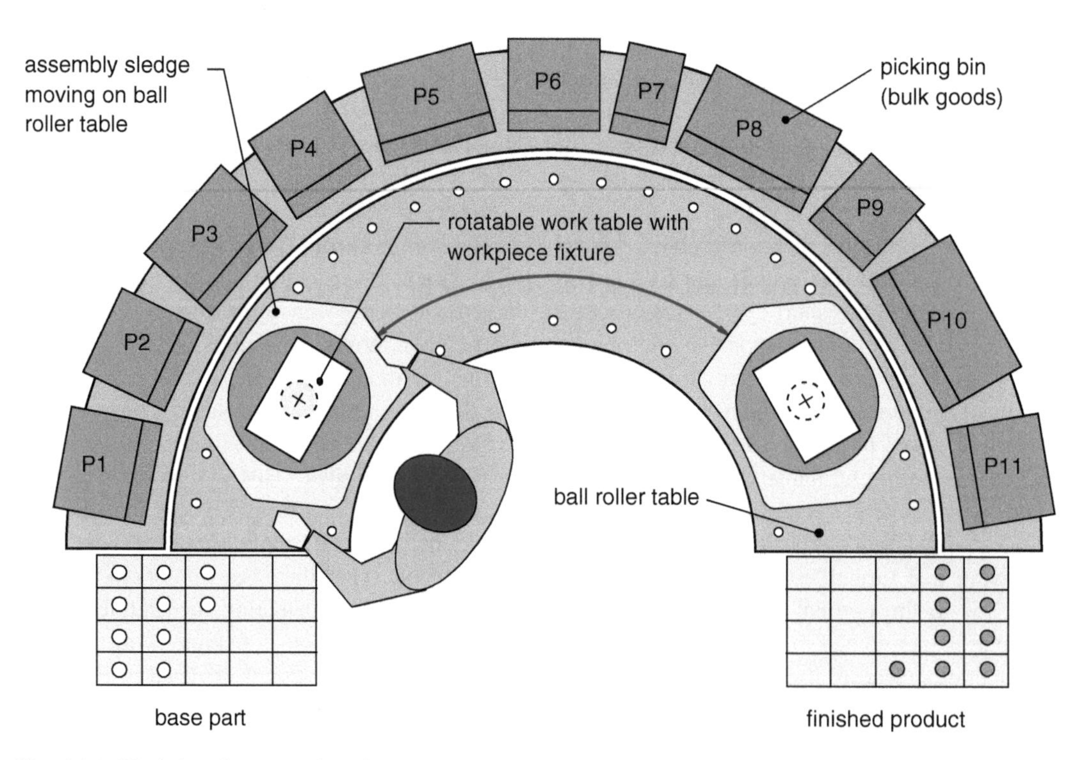

Fig. 6.36 Workplace for a one-piece-flow assembly (LP Montagetechnik)

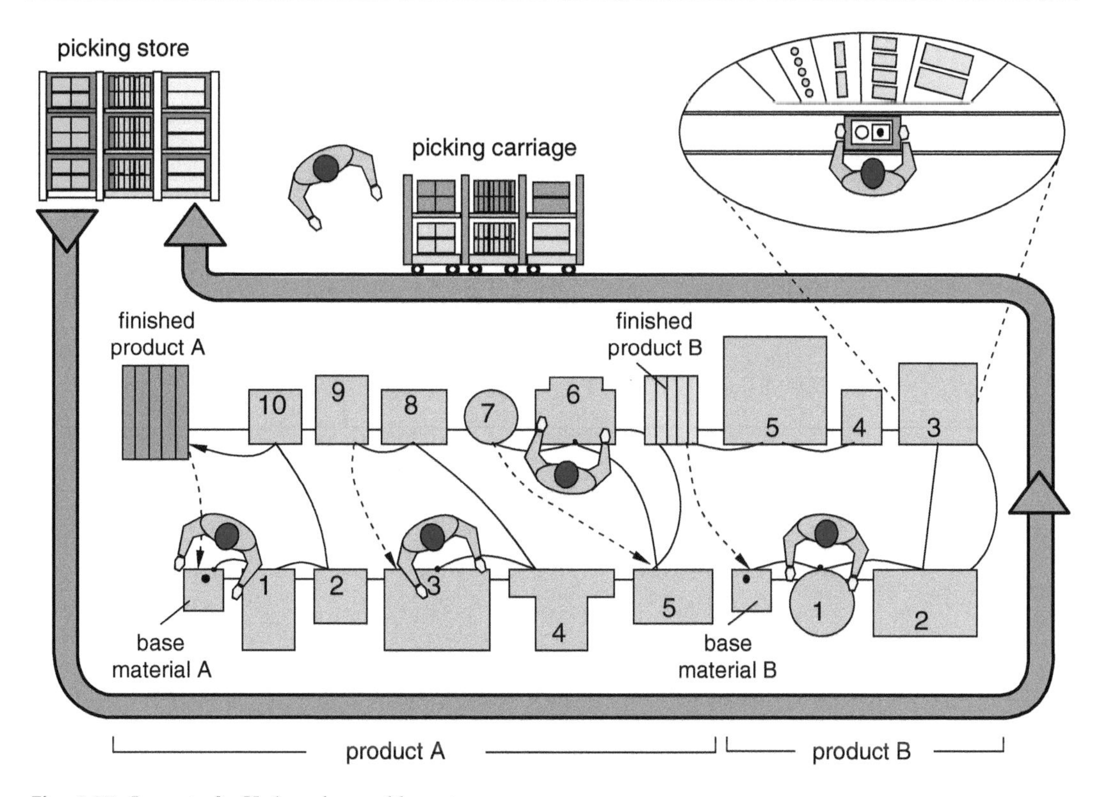

Fig. 6.37 Layout of a U-shaped assembly system

parts from an internal supermarket that are to be built-up is completed by a special worker, who supplies the parts per variant at each of the individual stations.

If more assembly stations are required and automated stations are not economically efficient a manual flow assembly comes into play. In such cases the workstations can be linked by arranging transportation means in different ways (see Fig. 6.38).

In accordance with Fig. 6.33 the output limits of a manual assembly system are thus reached. If the variants are not too extensive and the lot sizes are sufficient for operating continuously for a number of hours, automated assembly stations can be implemented. Joining operations than occur synchronously in short intermittent intervals. Figure 6.39 depicts a multi-station machine using an index rotary table. This particular example has two pick-and-place stations for parts A and B and a joining station. The advantage of this concept is the yield which can be up to 1800 pieces per hour. However the lack of variant flexibility and susceptibility to disruptions in the feeding system are disadvantages. When a workstation is disrupted, the entire machine stops. Finally, the number of workstations is also limited by the size of the rotary table; the upper limit tends to be 18 stations.

If even more parts need to be assembled and linked automated stations without buffers are to be avoided, it is necessary to connect stations using a transfer system; Fig. 6.40 depicts the various principles for linking and organizing assembly systems.

Line systems can be implemented with or without buffers and open or closed. Rectangular arrangements (again with or without buffers) simplify the return of empty workpiece carriers; however, since their inner space cannot be practically accessed, materials have to be supplied either behind or beside station operators.

An example of a modular assembly system is depicted in Fig. 6.41. Here, the stations can be

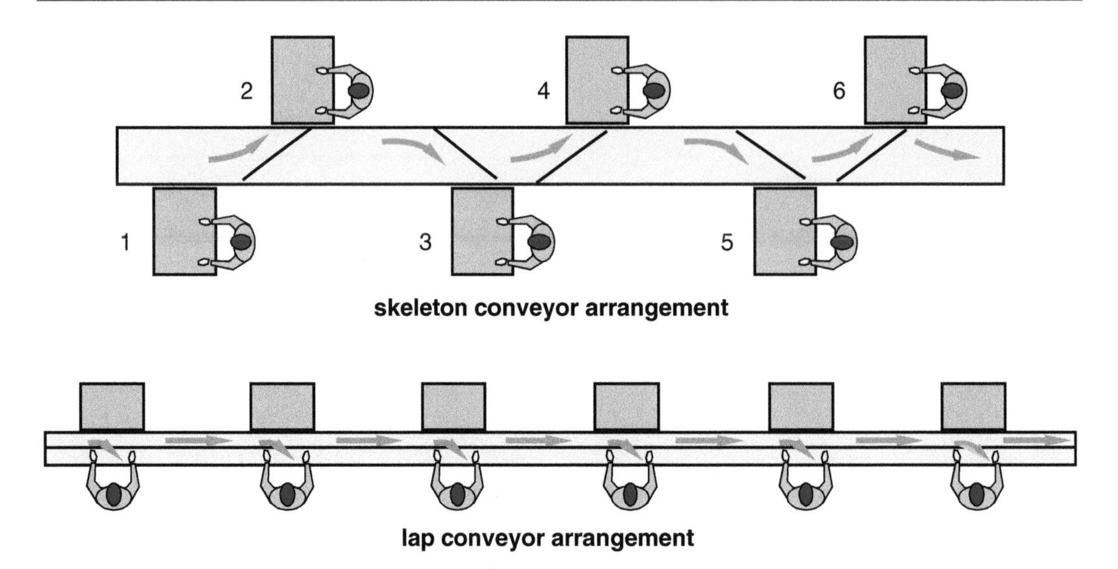

Fig. 6.38 Manual flow assembly

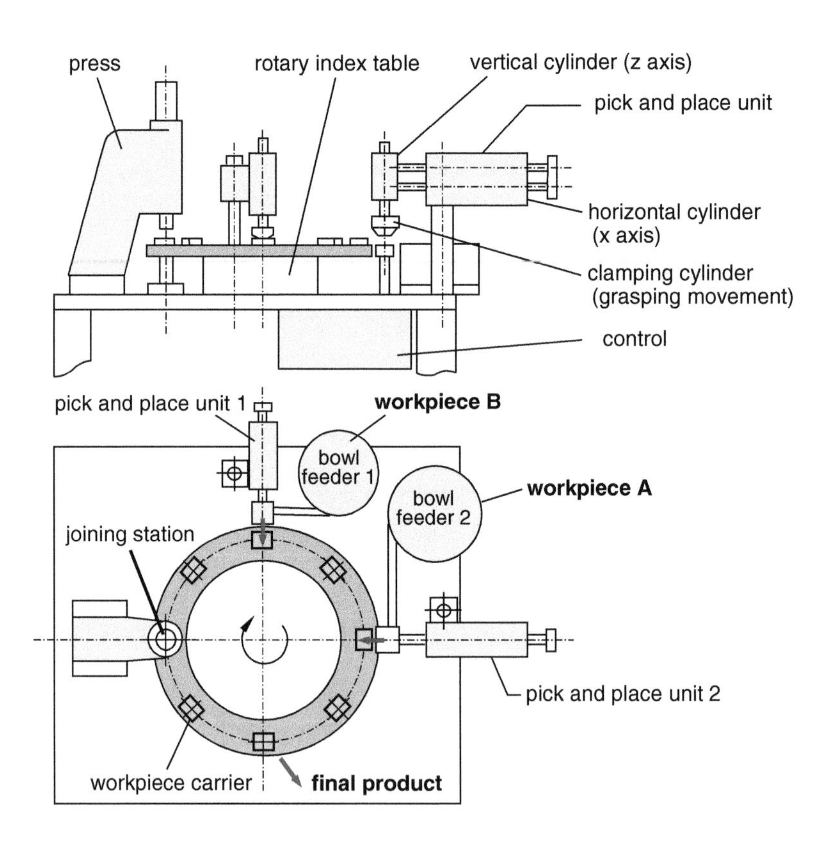

Fig. 6.39 Multi-station assembly machine using a rotary table (Lotter)

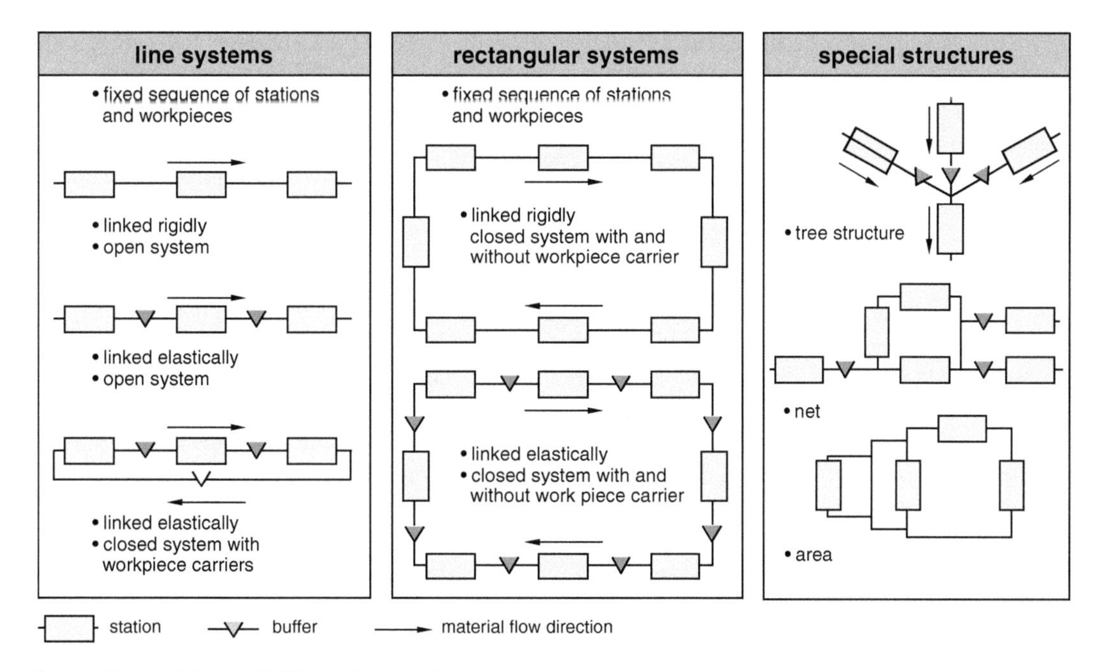

Fig. 6.40 Principles for linking and organizing assembly systems

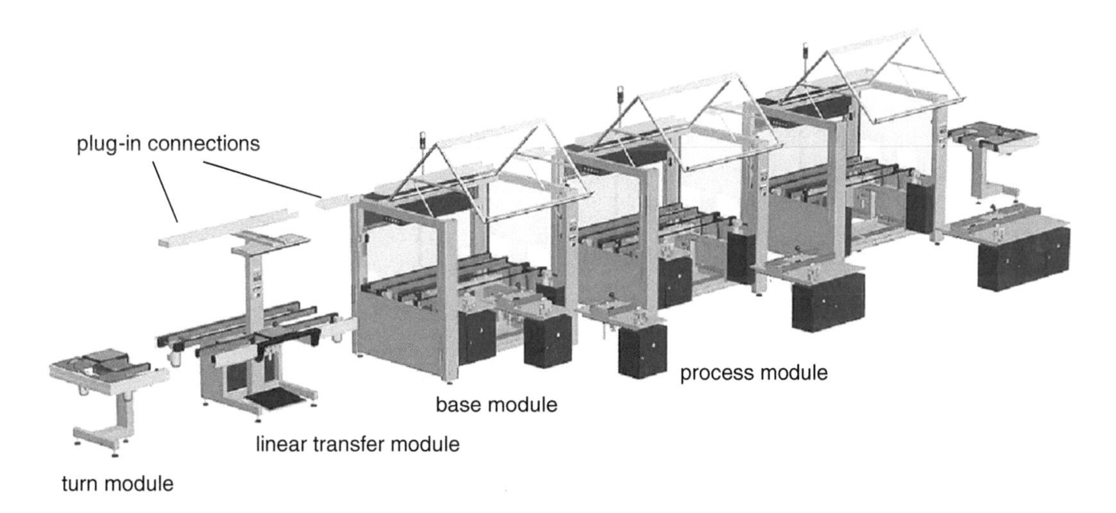

Fig. 6.41 Modular system for linear transfer assembly systems (courtesy of team technik)

equipped with different systems for joining and measuring which can then be linked using a modular transfer system. The system also allows manual workstations to be integrated.

As already mentioned, in contrast to the *part feed* in manufacturing, the part feed in the assembly is almost as important as the joining process itself. In a manual assembly, parts are usually fed by station workers who pick them out of a bulk of parts, supported by bins with an easy access to the parts.

If the aim is to automate these manual operations, a magazine, positioning system and a handling device are required as already indicated in Fig. 6.8. A storage bin separates the parts (e.g. via vibration, conveyors, sliders etc.) which then fall into a separation and ordering device. At a defined transfer point a handling device then picks a part and deposits it in the joining position.

From the factory planning perspective, feeding systems are considered space consuming and failure sensitive. It is thus critical that these systems can be constantly monitored and quickly accessed when there are incidents. One possibility of circumventing this is to build parts directly on-site and to enter them immediately into the joining position. However, disruptions in the manufacturing process then directly impact the assembly. Another possibility is to sort the parts outside of the assembly system into magazines and have the handling device remove them from the magazine.

In addition to the technical solutions we have described above, *pick and place* units and *industrial robots* represent an important component of assembly technology. The first have up to 3 degrees of freedom in space, which generally only allow linear movements. In comparison, robots are characterized by at least three independently programmable linear or rotational axes that can be combined in different ways. Figure 6.42 depicts typical kinematic configurations, primarily implemented in assemblies [Lot13].

Horizontal and vertical articulated robots are predominantly found in assembly technology, whereas portal robots serve more to load and unload tool machines. Increasing speed and precision along with sinking prices are making robots together with quick gripper changing systems and the implementation of image processing into universal devices with abilities close to those of humans. This means that the vision of random bin picking has meanwhile become reality. The occasionally extensive forces and quick sweeping movements of robots however, necessitate strict health and safety requirements as well as rules concerning accessibility.

Finally if we consider assembly stations and machinery from the perspective of their changeability. Similarly to manufacturing facilities and equipment, we can classify them according to the stations and machines as well as the material and information flows (see Fig. 6.43).

The dimensions for assembly stations and machines are comparable with manufacturing machines. Nevertheless, assembly stations are, without exception, relatively lightweight because the drives are much less powered and the frame basically only has to carry weight forces while the joining force remains within the workspace of the joining station. Due to the large dimensions of the systems, the ground floor plan and building grid can become obstacles when linking

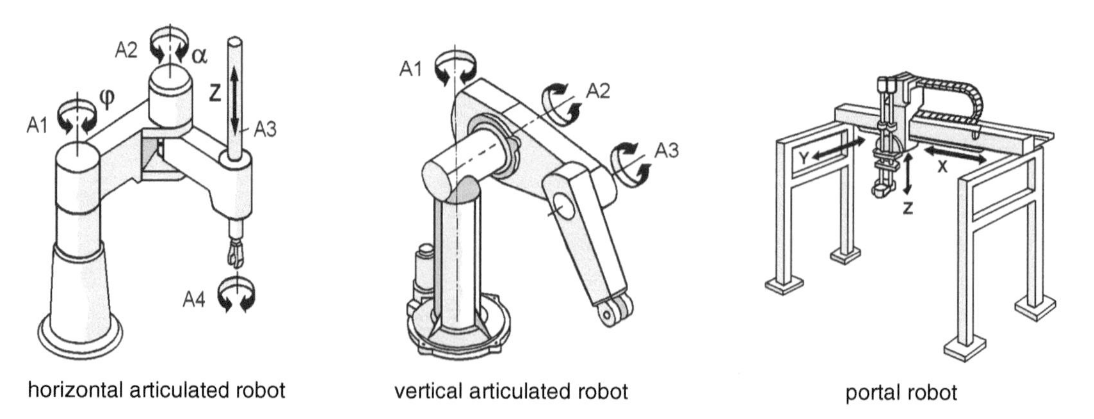

Fig. 6.42 Types of assembly robots (Hesse)

	assembly facilities	level: station	section	factory	site	obstacles
station/machine	outer measurements	low	low	low	low	loading profile, building grid
	weight	low	low	low	low	floor load capacity
	decomposability from installation view	low	high	high	high	mono block construction
	reconfigurability from function view	high	high	high	high	missing function modules
	energy and media supply	high	middle	low	low	special media
	noise emission	low	low	middle	middle	noisy joining processes
	work and safety protection	low	middle	middle	middle	space requirements
material flow	automation of workpiece change	low	low	middle	middle	takt time
	station linkage	low	low	low	low	too low buffers
	number of workpiece carriers	low	low	low	low	small lots
information flow	connection to PPC system	high	high	high	high	connection PLC - PPC
	information supply	low	low	middle	middle	number of variants
	condition recording / quality assurance	low	low	middle	middle	number of features

PLC Programmable Logic Controller
PPC Production Planning and Control

changeability with growing feature value: ○ low ⊖ middle ● high

Fig. 6.43 Changeability of assembly stations and machines

a large number of stations. Generally, a functional oriented modular construction supports changeability on all levels and is to be strongly encouraged in every case. Media requirements, noise protection and health and safety do not play a large role in the assembly, the only aspect that tends to be time consuming in a change-over is the energy and media feeds at the joining point due to the many hoses, cables and pipes. Modular solutions such as plug-in and snap-on fasteners should thus also be strived for here.

With increasing automation of the workpiece exchange, its cycle time can become more important than the joining cycle time itself with regards to determining bottlenecks. However, from the view of changeability, the more important factor is how the joining stations and the workpiece exchange system are structurally linked with one another.

The more stations that are linked together and the less buffer space there is between them, the more difficult it is to make changes on all of the levels. Finally, the information flow is generally not as extensive as it is with manufacturing stations and systems, because joining processes require much less control data than a manufacturing process. Instead, due to the shorter cycle times, the supply and retrieval frequencies are usually considerably higher and can hinder changes on the station and system levels, especially when there are a large number of variants with quick job changes.

Finally, assembly stations and systems tend to be characterized by continual quality checks either during or immediately following the joining process. These checks have to occur reliably and, due to the required process certainty, can prove to be an obstacle in reconfiguring or changing over systems when variants or products are modified.

As with the manufacturing facilities and equipment, we can derive requirements for a high degree of changeability from these obstacles:

- The joining stations, assembly machines and systems should be organized in transportable, functionally independent and pre-tested modules that can be quickly exchanged.

- Feeding systems should be easily reconfigured to variants of feeder parts by avoiding mechanical sorting elements.
- Assemblies should be located in buildings with a wide span width in order to facilitate easy expansions.
- Energy and media connections should be ensured via a modular supply system, organized as a grid e.g., via raised floors or an overhead supply system.
- Workers should be kept informed via monitors at their workplaces instead of paper documents.

These requirements directly impact changeability on the stations, systems and sections levels. In comparison with manufacturing facilities, the factory structure and production site can be quickly changed when these requirements are met.

6.3.3 Logistical Resources

The logistical resources already outlined in Fig. 6.23 serve to fulfill the core and sub logistic processes listed in Fig. 6.14. These are functions with which piece-goods are stored or changed according to the criteria quantity, time and location without modifying their functional properties. Similar to feeding technology in the assembly, the geometry, dimensions, weight and sensitivity of the packaged goods are decisive in selecting logistical resources. In our discussion here, we will focus on in-house logistics that link the previously described manufacturing and assembly functions. We will then discuss designing external logistics later in Chap. 9.

Before taking a closer look at the logistical resources, it is helpful to consider the elementary processes that the logistic sub-processes are based upon since this allows us to more clearly understand the importance and characteristics of the classification features (see Fig. 6.44).

The elementary processes involved in *storing* include receiving and identifying goods, determining the storage location, storing, commissioning, removal and dispatching them to the agreed upon transfer point e.g., a loading dock [Hom07].

Conveying consists of loading and unloading the conveyor as well as the loaded and idle runs of the conveyor.

Packaging serves to protect goods whereby both the packing materials and goods have to be supplied for the packaging process and subsequently used to form cargo units. Within a factory, goods are usually only packaged when they are meant to be dispatched. Generally packaging should be avoided and instead special returnable carriers are to be used which not only ensure the organization of the goods, but also protect the cargo during transport.

Commissioning (or order picking) consists of supplying a sorted quantity of an article from a store, the removal of the requested amount from the article bin by the commissioner, delivering these in a picking bin, collecting other articles in the picking bin into a commission and returning the empty article bins.

Trans-ship processes from one container to another are rarely required in a factory and are usually reserved for external transport hubs e.g., container terminals or rail dispatching stations.

Sorting also tends to play a smaller role in factory logistics. It serves to divide a heterogeneous flow of goods—as is, for example, typical of package distribution centers or luggage sorting systems in airports—to various target points. The elementary processes are naturally similar to storing or transporting; the units to be handled are usually just larger and protected by packaging against transportation and weather factors.

All of these functions require *planning, monitoring and control processes* which consist of planning the quantities and capacities of the logistic units and logistical resources, followed by processing the orders including generating the accompanying documents before finally

sub process	elementary process
1 storing	• receiving goods • identifying • determining storage location • storing • commissioning • removing • dispatching
2 conveying	• loading • unloading • load transporting • idle transporting
3 packing	• providing packing means • providing package goods • packing • building loading units
4 commissio-ning	• provisioning amount of articles • moving commissioning person • removing demanded amount of goods • delivering amount of orders

sub process	elementary process
5 handling	• loading • unloading • sorting • storing • removing
6 sorting	• carrying to • preparing • identifying • distributing • carrying off
7 planning	• volume and capacity planning • order handling • information provisioning (ahead and accompanying) • generating feed back data • controlling

Fig. 6.44 Sub and elementary logistic processes (based on Fleischmann, Gudehus and ten Hompel)

releasing and tracking the orders. Information that so to speak 'rushes ahead' of the orders (mostly electronically) plays an important role, however more and more it is also taken for granted that the accompanying information (i.e., in the sense of identifying where the order is at any point in time) is also always available. Finally, feedback data e.g., about goods removed from stores or completed transport operations also needs to be generated so that it can be entered into statistical evaluations and order tracking systems.

The facilities necessary for these logistics sub and elementary processes usually fulfill their functions together with more or less automated devices and logistics personnel. We will first turn our attentions to the storage facilities.

Figure 6.45 identifies components of a piece-good store. The *load carrier* (also known as a storage aid or charge carrier) can include pallets, boxes, girders or containers. It serves to form a cargo or transport and can be carried by conveyors. Depending on protection requirements or requirements for the automated loading and unloading of individual piece-goods into/from the charge carrier, goods are fixed using holding strips or intermediate layers with suitable pits. From the perspective of the factory planning, the diversity of the charge carrier should be minimized as much as possible.

Load auxiliary devices serve to handle the cargo units during the elementary processes outlined in Fig. 6.44 and are either firmly connected to the storage conveyor devices or can be exchanged with it. Rigid and adjustable forks which lift the charge carriers are wide spread. Side grippers or squeeze clamps require that the side walls of the charge carrier are correspondingly sturdy. The remaining load handling devices pull, push, lift or roll the charge carriers.

Storage conveyor devices accept charge carriers in the store and transport them from a drop-off point to a storage place (storing), from a storage place to a supply point (retrieval) or between storage places (sorting). With non-automated stores, manual vehicles in which the driver remains on ground-level assist here e.g., fork-lifts, high-lift trucks as well as reach trucks.

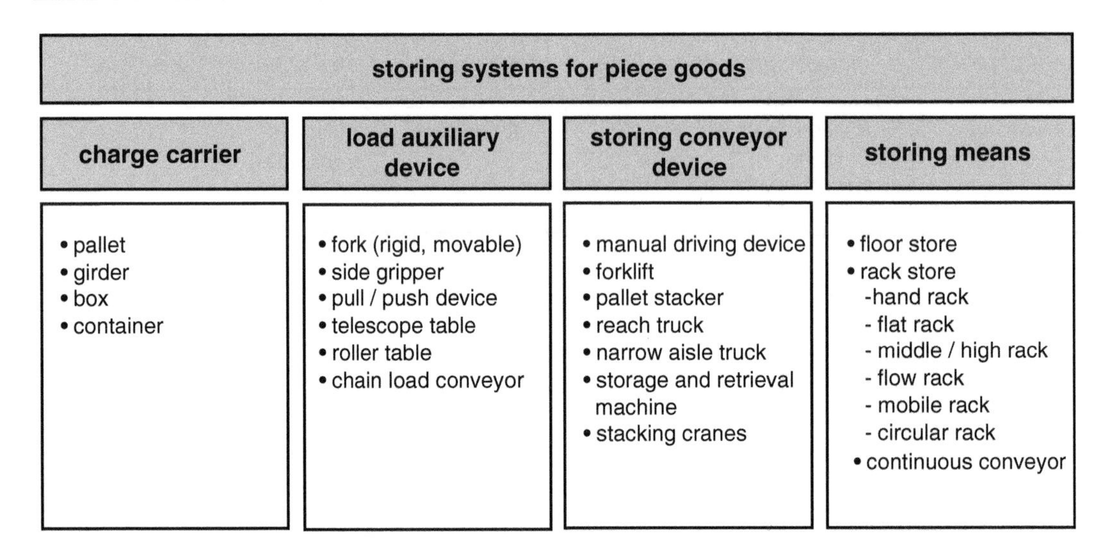

Fig. 6.45 Components of a piece-good store

With rack trucks operators drive with the device to the individual storage places.

The *storage devices* fulfill the core function 'storing' (see Fig. 6.46). The charge carriers can be stored in blocks or rows on the floor and are then referred to as block stores (Fig. 6.46 top left). Due to the need for space and accessibility storage racks—usually hand or medium to high racks—are predominantly implemented; Fig. 6.46 top right depicts a rack with stacked pallets.

When the pallets can move in the rack they are referred to as drive-through or drive-in racks (Fig. 6.46 bottom left), whereas with movable racks the individual racks move as a whole with the pallets (Fig. 6.46 bottom right).

Within a factory, rack stores mainly tend to be found in the inbound and outbound stores as well as in the dispatch area of shipping companies. Figure 6.47 depicts a visualization of possible forms and their dimensions which, depending on the overall height, can be implemented with different storage means [Hom07].

Due to the quick access, compact construction, protection from dust and its mobility the lift system featured in Fig. 6.48 is frequently used for internal intermediary storage of B-parts, tools and consumable materials. It allows articles to be accessed from a number of levels for commissioning purposes.

Continuous conveyors take on a special role, functioning both as a storage and transportation means. The storage ability of a continuous conveyor is determined by the number of charge carriers that it can accept. Stores, which allow the movement of the charge carriers and or stacks are also referred to as dynamic stores, those that do not are referred to as static stores. Extensive depictions of storage systems including components, technologies and dimensioning can be found in, ten Hompel et al. [Hom07], Furmans and Arnold [Fur08] and McGuire [McG10].

The next aspect we will consider is *means of transportation*. We use 'transportation' to refer to transports outside of the factory building, while internal transports are discussed in terms of 'conveying'. For factory planners then, the internal conveyor systems for piece goods are of primary concern; Fig. 6.49 differentiates between continous and static systems.

Sliding, roll or belt conveyors are used for comparably short distances e.g., in linking machines or assembly systems. Suspension and drag chain conveyors bridge larger distances, for example for connecting the material flow systems between production sections or halls. The latter are arranged overhead just below the hall ceiling in order to keep the hall floor free for machines and personnel. By far the majority of conveyor

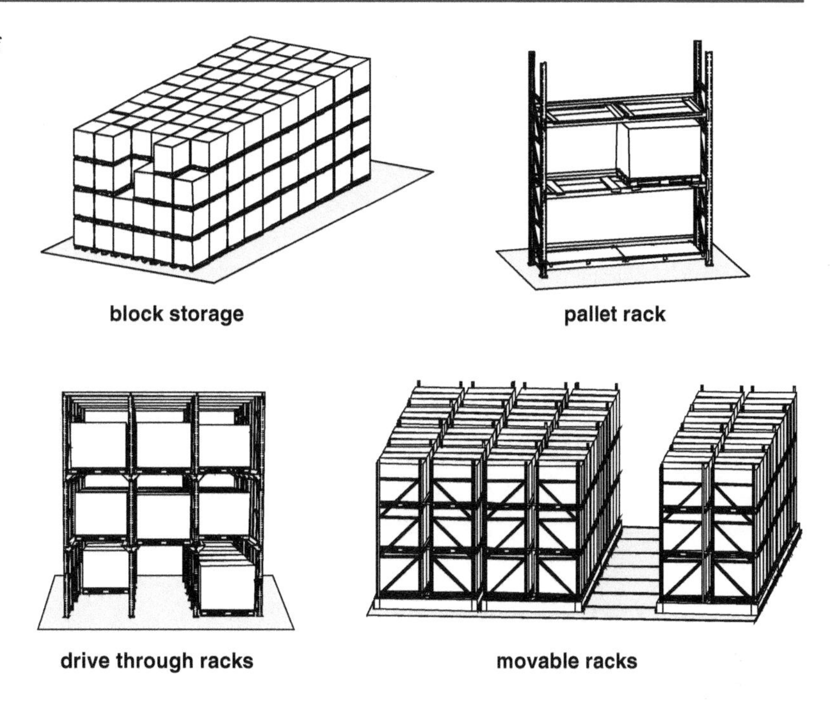

Fig. 6.46 Typical types of piece-goods storage (Schulze)

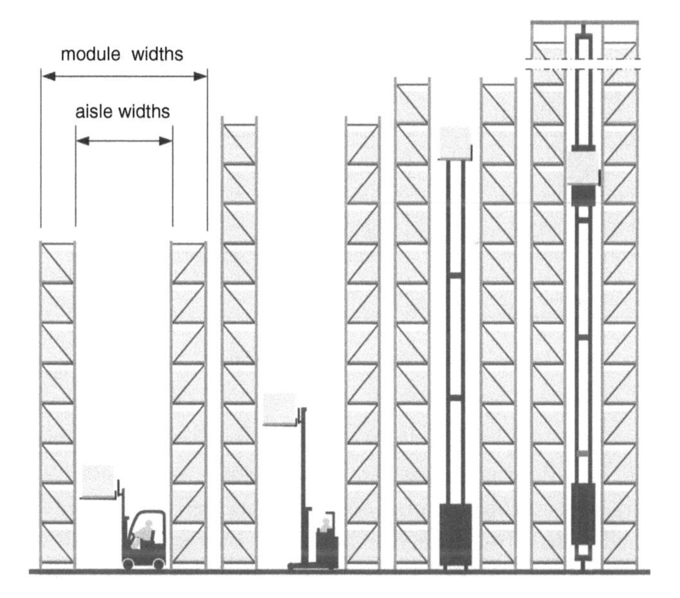

	forklift	pallet stacker	narrow aisle truck	storage and retrieval truck
lift height	8.5 m 27.9 ft.	12.0 m 39.4 ft.	13.0 m 42.7 ft.	55.0 m 180.4 ft.
aisle widths	> 3.0 m > 9.8 ft.	2.7 - 2.8 m 8.9 - 9.2 ft.	1.5 - 1.7 m 4.9 - 5.6 ft.	0.75 - 1.5 m 2.5 - 4.9 ft.
module widths	> 5.4 m > 17.7 ft.	5.1 - 5.2 m 16.7 - 17.1 ft.	3.9 - 4.1 m 12.8 - 13.5 ft.	3.15 - 3.9 m 10.3 - 12.8 ft.

Fig. 6.47 Typical rack systems (ten Hompel)

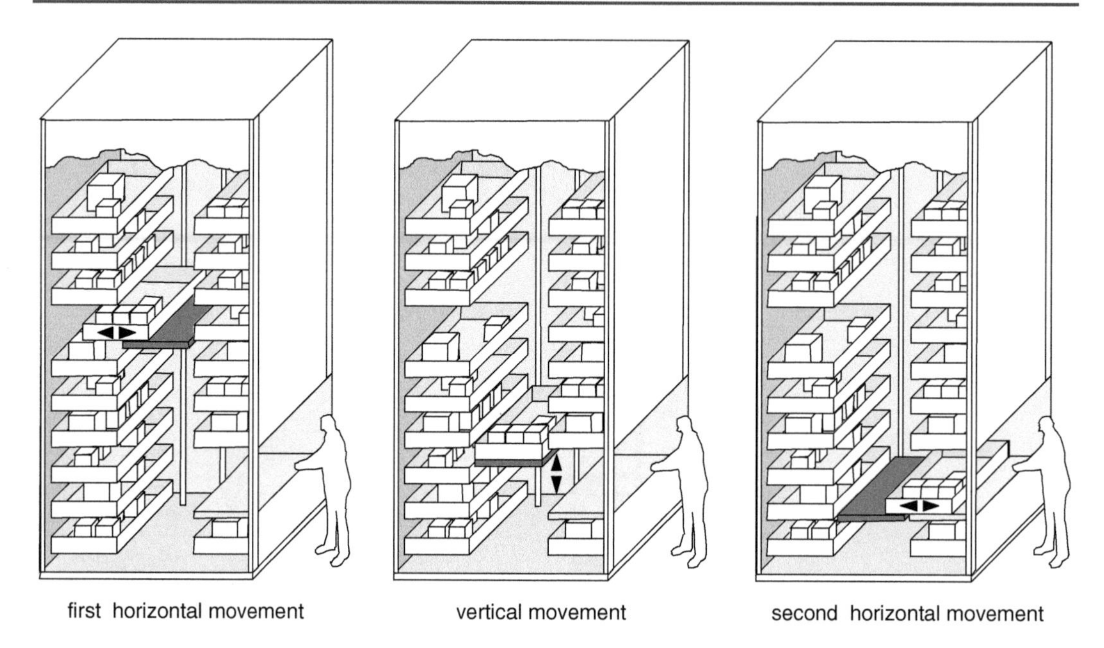

Fig. 6.48 Lift system (Kardex)

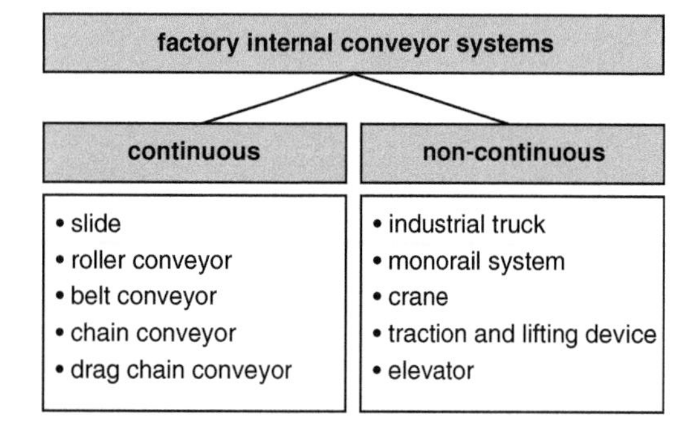

Fig. 6.49 Typical factory piece-good conveyors

systems implemented are non-continuous, whereby floor conveyors are broken down into fork-lifts, towing tractors and wagons as well as hand-pulled (or pushed) carts and overhead conveyors. Figure 6.50 depicts two typical forklifts and their respective performance ranges [Hom07].

Overhead conveyors include monorails which transport their load along an overhead track. Cranes, trains and lifting devices as well as elevators are summed up as lifting tools and primarily serve to vertically transport individual loads.

Based on the elementary logistic processes outlined in Fig. 6.44, it is clear that with the exception of large mail-order companies with large throughput quantities, *packaging processes* require comparably simple devices. In terms of logistics, orders are combined into transportable and storable packages, tied to cargo units and as such secured so that they stay together e.g., by wrapping them in plastic foils. Facilities and equipment required here include storage for packing materials, conveying systems for incoming packages as well as wrapping and strapping systems.

The *commissioning* sub-process basically consists of a combination of storing, transporting and handling processes, whereby the commissioners themselves become transported objects when they move to the goods, instead of the goods coming to them. As an example, Fig. 6.51 depicts a system with which the articles remain at their storage place (static commissioning) while the commissioner travels vertically or horizontally with the storage and retrieval systems (two dimensional movement), picks the required quantity of articles and combines them in the order box (manual removal) before depositing the order at a defined location (central deposit).

Due to the comparably few types of parts, commissioning does not play a significant role in

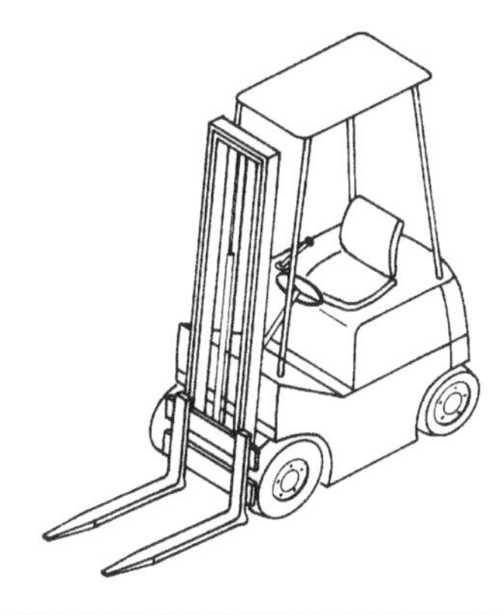

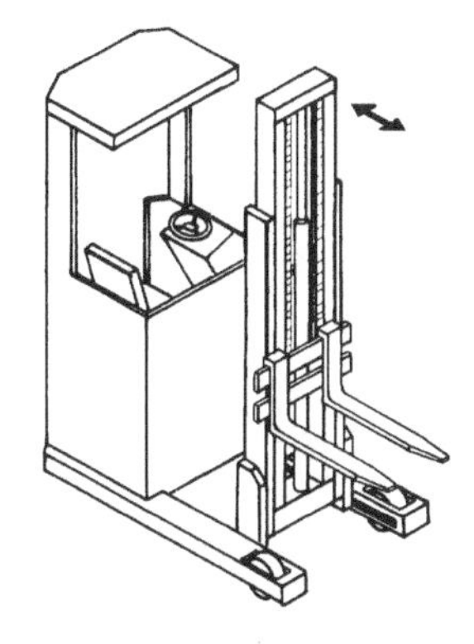

variables	fork lift truck	reach truck
speed	9 - 35 km/h / 5.5 - 21.7 mph	7-14 km/h / 4.3 - 8.7 mph
lifting speed	0.23 - 0.65 m/s / 0.75 - 1.45 ft./s	0.15 - 0.5 m/s / 0.49 - 1.64 ft./s
drive power	4-120 kW / 5.4 -160.9 HP	5- 20 kW / 6.7 - 26.8 HP
lifting height	approx. 9 m / 29.5 ft.	ca. 12 kW / 16.1 ft.
load capacity	1-16 t / 1.1 - 17.6 Short tons	1 - 2.5 t / 1.1 - 2.75 Short tons
stack height	8 and 5 pallets	12 and 7 pallets

Fig. 6.50 Two typical forklifts (acc. ten Hompel)

the part fabrication area of a factory. Assembly areas however are increasingly implementing commissioning processes for articles required for assembling components and end products. This is in part due to the growing number of variants and the related danger of mix-up them with one another. Moreover, this allows the assembly worker to concentrate more on the actual tasks of joining and checking the joints. Finally the need for space and the related complexity particularly with voluminous articles is growing so quickly and strongly that it makes sense to separate the commissioning operations from the actual assembly task. Figure 6.52 depicts two types of commissioning systems for assembly factories. In the case left the goods are transported from the store to the commissioner, whereas in the case right the commissioners travel up and down the racks as if they were in a super market and collect the articles for the order in a cart.

What is common among all of the logistical resources is that in comparison to manufacturing resources, a large surface area is required that is easily under estimated and needs to be carefully planned. This often results from safety regulations that demand appropriate distances to moving equipment and health and safety standards. Moreover, it is here in particular that valuable potential for shortening the throughput times of the orders is hidden.

We will not discuss the *handling* and *sorting* sub-processes further because, as already mentioned, they seldom arise in the factory itself. The last logistic sub-process mentioned in Fig. 6.44, *planning and control,* requires the usual equipment for processing data such as computers, monitors, a server and a data network as well as score boards and control panels. These are either parts of the respective sub-systems or are found in the offices of the corresponding employees.

The characteristics of the logistical resources that are relevant for changeability are similar to the assembly resources in view of the *equipment* that is implemented (see Fig. 6.53). Factors such

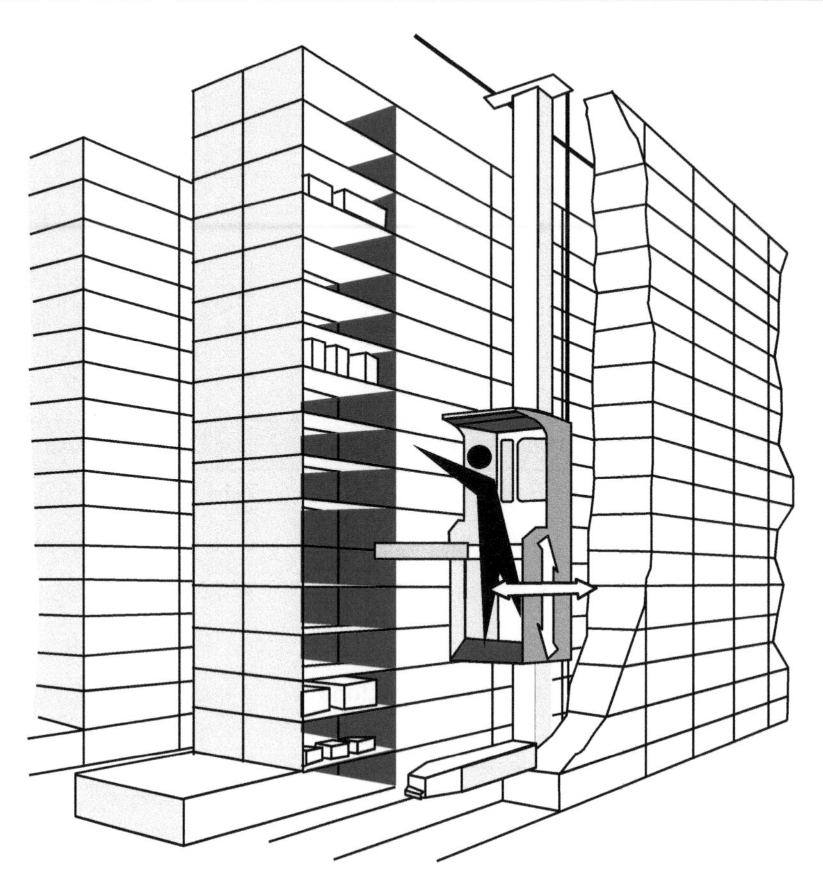

Fig. 6.51 Example of a commissioning system (Gudehus)

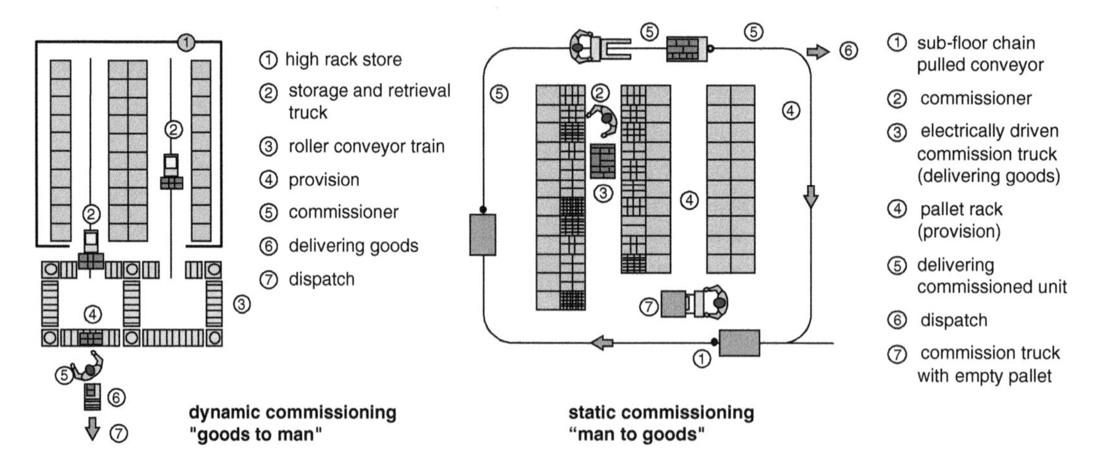

Fig. 6.52 Examples of possible commissioning systems

	change relevant features	station	sub section	section	factory	site	obstacles
logistics devices	outer measurements	○	○	○	○	○	support grid of building
	weight	○	○	○	○	○	floor load capacity
	ease of disassembly	○	●	●	●	●	mono block construction
	reconfigurability	●	●	●	●	●	mechanical interfaces
	occupational health	○	○	⊖	⊖	⊖	space requirements
material flow	charge carrier	○	○	○	○	○	execution variants
	delivery points	○	○	○	○	○	lacking standardization
	degree of linkage	○	○	○	○	○	too low buffers
information flow	connection to PPC system	●	●	●	●	●	compatibility
	order information	○	○	⊖	⊖	⊖	spatial distance
	stock information	○	○	⊖	⊖	⊖	irregular access

PPC Production Planning and Control

changeability with growing feature value: ○ low ⊖ middle ● high

Fig. 6.53 Changeability of logistic resources

as the weight and ability to dismantle and reconfigure them can be compared directly. Media requirements and noise protection do not play any significant role, whereas special attention should be paid to health and safety regulations due to the heaviness of the cargo and the overhead load. The variety of charge carriers are a key obstacle for the material flow on all levels.

Furthermore, the *material flow* is impeded by the number of transfer points from one logistical resource to another e.g., from a conveyor to a store. However, sub-systems that are too rigidly connected hinder the changeability on all levels. Finally, the information flow can also obstruct the changeability when the interfaces between the logistic and technical sub-systems are not compatible and the order information is too far from the personnel responsible for fulfilling them. One particular problem is ensuring reliable information about the amount of stock that is physically in the system i.e., work-in-process (WIP) at a given point in time. One of the basic hindrances here is unregulated access to stores, especially during night shifts.

Once again, we are able to derive a number of requirements for the changeability of the logistical resources from the obstacles we have discussed here:

- Modular, consistent charge carriers have to be created and implemented across the entire logistic chain.
- Logistic equipment and facilities also need to have modular designs.
- Buildings have to have large support grids.
- Information systems have to provide an actual overview of the physical stock levels and flow of the logistic processes.

These considerations conclude our discussion about operating facilities from a functional and technological perspective. In accordance with Fig. 6.3 the next area we will be considering is the organizational and ergonomic environment on the level of the individual workplaces. Due to the breadth of the topic, the organizational part of designing workplaces will be discussed subsequently in Chap. 7, while the ergonomic perspective will be considered in Chap. 8 due to its close relationship to spatial planning.

6.4 Summary

Workplaces are the fundamental building units of each factory. Their constituent components are equipment and people both requiring space. Within the equipment processes take place, resulting in a conversion of the starting material into usable parts and assemblies. The three main groups of processes bundled under the term production technology are: (1) part manufacturing in the sense of shaping or changing properties of parts, (2) assembly as the joining of parts into assemblies and products, and (3) logistics processes for spatial and quantitative changes of parts, components and final products. The chapter describes first the individual processes of the three groups and then explains the equipment used to perform each process in the form of machinery and equipment. For the three equipment groups with a focus on their main factory design features as well as the obstacles to change them are discussed.

Bibliography

[Alt05] Altan, T., et al.: Cold and Hot Forging: Fundamentals and Applications. ASM International, Almere (2005)

[Alt12] Altan, T., Tekkaya, E.: Sheet Metal Forming: Fundamentals. ASM International, Almere (2012)

[AIK02] Arnold, D., et al.: Handbuch logistik, 3rd edn. Handbook Logistics. Springer, Berlin (2002)

[Boot83] Boothroyd, G., Dewhurst, P.: Design for Assembly. University of Amherst, Amherst (1983)

[Bok12] Bokranz, R., Landau, K.: Handbuch für Industrial Engineering—Produktivitätsmanagement mit MTM, 2nd edn. Handbook Industrial Engineering, Productivity Management with MTM Deutsche MTM-Vereinigung (2012)

[DIN03a] DIN German Institute for Standardization (ed.): Fertigungsverfahren; Begriffe, Einteilung (Manufacturing Processes—Terms and Definitions, Division), Köln (2003)

[DIN03b] DIN German Institute for Standardization (ed.): Fertigungsverfahren Spanen; Einordnung, Unterteilung, Begriffe (Manufacturing Processes Chip Removal—Part 0: General; Classification, Subdivision, Terms and Definitions), Köln (2003)

[DIN03c] DIN German Institute for Standardization (ed.): Fertigungsverfahren Fügen (Manufacturing Processes Joining—Part 0: General; Classification, Subdivision, Terms and Definitions), Köln (2003)

[Doe07] Doege, E., Behrens, B.-A.: Handbuch Umformtechnik: Grundlagen, Technologien, Maschinen (Handbook Forming: Fundamentals, Technologies, Machines). Springer, Berlin (2007)

[Egl01] Egli, J.: Transportkennlinien: Ein Ansatz zur Analyse von Materialflusssystemen (Transport Characteristic Curves: An Approach to the Analysis of Material Flow Systems) Ph. D. thesis. Univ. Dortmund 2001. Verlag Praxiswissen, Dortmund (2001)

[Fur08] Furmans, K., Arnold, D.: (Coordinators): Innerbetriebliche Logistik (Inbound Logistics). In: Arnold, D., et al. (Hrsg.) Handbuch Logistik (Handbook Logsitics), 3rd edn, S. 614–726. Springer, Berlin (2008)

[Gud10] Gudehus, T.: Logistik: Grundlagen—Strategien—Anwendungen. 4. Aufl. (Logistics—Basics, Strategies, Applications). 4th ed. Springer, Berlin (2010)

[Hom07] ten Hompel, M., Schmidt, T., Nagel, L.: Materialflusssysteme. Förder- und Lagertechnik (Material Handling Systems. Conveying and Warehousing Equipment), 3rd edn. Springer, Berlin (2007)

[Kuh95] Kuhn, A.: Prozesskarten in der Logistik: Entwicklungstrends und Umsetzungsstrategien (Process Charts in Logistics. Trends and Implementation Strategies). Praxiswissen Dortmund (1995)

[Lot13] Lotter, B., Wiendahl, H.-P. (eds.): Montage in der industriellen Produktion. Ein Handbuch für die Praxis (Assembly in Industrial Production. A Practical Guide), 2nd edn. Springer, Berlin (2013)

[Lut02] Lutz, S: Kennliniengestütztes Lagermanagement (Characteristic Curve Aided Warehouse Management) Ph.D. thesis, Univ. Hannover (2002)

[MaM96] Macherauch, E., Müller, H.: Wärmebehandlung (Heat Treatment). In: Eversheim, W., Schuh, G (eds.) Betriebshütte, Produktion u. Management, 7th edn. Springer, Berlin (1996)

[Mar94] Martin, H.: Grundlagen der menschengerechten Arbeitsgestaltung. Handbuch für die betriebliche Praxis (Basics of Humane Job Design. Manual for Operational Practice), Köln (1994)

[McG10] McGuire, P.M.: Conveyors: Application, Selection, and Integration. CRC Press, Taylor and Francis Group, Boca Raton (2010)

[Mil04] Miller, R., Miller, M.R.: Audel Machine Shop Tools and Operations. Wiley Publishing Inc., Indianapolis (2004)

[Mur10] Murphy Jr, R., Wood, P.R.: Contemporary Logistics, 10th edn. Author: Prentice Hall, Boston (2010)

[Nyh09a] Nyhuis, P., Wiendahl, H.-P.: Fundamentals of Production Logistics. Theory, Tools and Applications. Springer, Berlin (2009)

[Nyh09b] Nyhuis, P., et al. (eds.): Globales Varianten Produktionssystem (Global Variant Production System). PZH GmbH, Garbsen (2009)

[Red94] Redford, A., Chal, J.: Design for Assembly. McGraw Hill, New York (1994)

[Spa97] Spath, D.: Fertigungsmittel (Manufacturing Facilities). In: Eversheim, W., Schuh, G. (eds.) Betriebshütte, Produktion u. Management, 7th edn. Springer, Berlin (1996)

[Spu96] Spur, G.: Produktionstechnologie (Production Technology). In: Eversheim, W., Schuh, G. (eds.) Betriebshütte, Produktion u. Management, 7th edn. Springer, Berlin (1996)

[Stef96] Steffens.: Beschichten (Coating). In: Eversheim, W., Schuh, G. (eds.) Betriebshütte, Produktion u. Management, 7th edn. Springer, Berlin (1996)

[Toe95] Tönshoff, H.-K.: Werkzeugmaschinen. Grundlagen (Machine Tools. Fundamentals). Springer, Berlin (1995)

[VDI2860] VDI-Richtlinie 2860.: Montage- und Handhabungstechnik, Handhabungsfunktionen, Handhabungseinrichtungen; Begriffe, Definitionen, Symbole (Assembly and Handling; Handling Functions, Handling Units; Terminology, Definitions and Symbols) (1990)

[Wec05] Weck, M., Brecher, C.: Werkzeugmaschinen—Maschinenarten und Anwendungsbereiche (Machine Tools—Machine Types and Application Fields), 6th edn. Springer, Berlin (2005)

[Wie00] Wiendahl, H.-P., et al.: Transportprozesse mit logistischen Kennlinien gestalten und bewerten (Design and Evaluation of Transport Processes with Logistic Characteristic Curves), vol. 5(4), pp. 16–21. PPS Management (2000)

[Wei99] Weinert, K.: Trockenbearbeitung und Minimalmengenkühlschmierung: Einsatz in der spanenden Fertigungstechnik (Dry Machining and Minimum Quantity Cooling Lubrication: Application in the Cutting Technology). Springer, Berlin (1999)

[Zeh97] Zehnder, T.: Kompetenzbasierte Technologieplanung (Competence Oriented Technology Planning) Hochschule St. Gallen, Gabler Verlag, Wiesbaden (1997)

7 Designing Workplaces from a Work Organizational Perspective by Detlef Gerst

Besides the technical equipment motivation, organizational integration, ergonomics and design of the working environment of the workplaces, employees play a significant role in the economic success of a factory. This chapter, therefore, addresses first the necessary development of the employee skills, and then the division of labor and responsibility within the organization. The design of the wage system and the working schedule conclude the chapter. The equally important issues of the ergonomic design of a workplace and its spatial environment are treated in detail in Chap. 8.

7.1 Human Resources as a Concept

Just as the term 'human capital' was voted the taboo word of the year in 2004 by German linguistics, the term 'human resources' is also debated. Critics are concerned that the term encourages a mindset in which production personnel are only considered from a monetary perspective and degraded to an object of production planning. However, in professional publications the concept of 'human resources' actually serves the opposite purpose, highlighting performance related abilities specific to humans and developing them through appropriate measures [Mat10]. From the perspective of developing human resources, when designing the workstation, possibilities are sought for increasing production efficiency by taking into consideration the motivation of employees for personal growth. This approach is based on the assumption that the workforce's competency, motivation and time based flexibility play a decisive role in the enterprise's competitiveness.

For example, when we consider employees a resource it decreases our focus on *personnel capacities*. From the perspective of planning and designing work, personnel capacities are related to factors such as hiring personnel, creating a long-term bond between personnel and the company as well as planning the work-hours schedule. As a resource though, employees are also important in a qualitative respect. Four aspects that distinguish employees from facilities should be emphasized here:

- Human work is characterized by a specific *flexibility* that technology can at best partially emulate.
- A further aspect is *creativity*. Since humans are able to deviate from programmed procedures and routines, they can develop creative and at the same time appropriate solutions for technological and organizational problems.
- In addition, employees *carry specific knowledge*. Whereas machines are capable of storing

This chapter was first published in German as Gerst, D: „Humanressourcen“ in: Arnold et al. (Ed.): Handbuch Logistik, 3rd ed., pp. 343–361. Springer Berlin Heidelberg 2008. The authors would like to thank Dr. Gerst and Springer-Verlag for the permission to reprint it.

H.-P. Wiendahl et al., *Handbook Factory Planning and Design*,
DOI 10.1007/978-3-662-46391-8_7, © Springer-Verlag Berlin Heidelberg 2015

and processing almost unlimited information, humans have extensive experience and context based knowledge that also allows them to orient themselves in new situations. Moreover, employees have free wills which influence their job performance considerably.

- As a result, *motivation* is a key topic in the development of human resources. Thus, by emphasizing flexibility, creativity, specific human competencies and motivation, the concept of human resources orients itself on an alternative model to humans as 'flexible machines'.

7.2 Human Resources and Production Performance

The correlation between human resources and production performance first becomes evident when the concept of production is no longer reduced to methods and algorithms that transform materials into parts and products as well as regulates the inventory and the material/product flows. Even when the ultimate goal in designing production processes is to reduce the impact of humans on the quality of the production and logistic performances, it should not be forgotten that humans are able to improvise and react to unanticipated technological and organizational disruptions, where the bureaucratic control fails [Wel91]. Employees thus take on a central function for the economic efficiency of an industrial production. With this in mind, developed human resources play a decisive role in the production performance, which in turn is a condition for a strong logistic performance and low logistic costs.

The extent to which human work contributes to the production performance is dependent on the competency and motivation of the workforce, including incentive systems. One of the particular problems in developing competencies and designing incentive schemes is how the human workforce influences the logistic performance. The reason for this is that whereas technological knowledge is either confirmed or refuted by direct experience, this does not apply to the same degree for logistical relationships. In addition to the frequent lack of understanding about the complex and model-based logistic correlations, incentive programs implemented in the production contribute significantly to the behavior of the people involved in the logistic control. Incentive systems can make the difference as to whether production personnel only orient on the productivity and utilization in addition to the quality, or if they also orient on the targets to minimize the work-in-process (WIP), shorten delivery times and increase the delivery reliability. Consequently, the competency, availability of personnel and orientation of production staff on targets are key input factors when planning the production [Wie05].

7.3 Competency and Human Resources Development

Nowadays research on vocational training and continuing education is no longer primarily concerned with qualifications, but rather with competencies. Whereas the term *qualification* identifies knowledge as a formal expression of recognized vocational or professional abilities of employees [Int04: 5], the concept of *competency* is more comprehensive. It identifies the expertise of an individual as an expression of the knowledge and abilities that he or she has command over in a specific context [Int04: 5]. Learning processes are understood today as self-organized. Accordingly, learning does not consist of instilling and accumulating pre-structured course content. Instead, humans learn by connecting new course content with existing knowledge, producing relations to known contexts and linking new knowledge to practical problems. One of the practical consequences of this concept is that learning processes should be designed to reveal the real-world sense and context of the course content. In addition to anchoring this knowledge in concrete questions, it is also critical for the success of the learning process that those learning are always encouraged to question and consider the course content from different perspectives. Current research in vocational training is based on the idea that this occurs above all in the work process. The demand to design work processes and tasks so that they promote learning is thus derived from here.

7.3.1 Professional Competence

In order to determine existing competencies and deficits in competencies or to depict requirements in the form of a competency profile, it is necessary to identify and distinguish the individual aspects of professional competence. A common typology that has proven itself in the practice contains four areas of competency: *technical*, *methodological*, *individual or self*, and *social and communication*. Figure 7.1 elucidates these terms and sets them in relation to a framework that will be further explained. Unlike the concept of qualifications or skills, research on competency is centered on individual and self-competence. Bergmann even equates competency to expertise:

"Competency refers to the motivation and aptitude of a person to independently further develop knowledge and ability in an area to a level that can be characterized as expertise" [Ber00: 21].

Current research also distinguishes between explicit and implicit knowledge and emphasizes that professional competence is based on both. The term *explicit knowledge* refers to conscious, logically organized knowledge that can be communicated. *Implicit knowledge* stems from experience; it allows tasks to be executed with certainty, however it is not consciously present and cannot be verbally communicated. Research assumes that explicit knowledge only accounts for 20 % of a person's professional competence [Sta99: 02]. Professional competence is therefore at least based on explicit and implicit knowledge in the four areas mentioned above. Nevertheless, according to current opinion this knowledge only constitutes the *ability to act* i.e., the cognitive prerequisites for being able to successfully execute specific tasks.

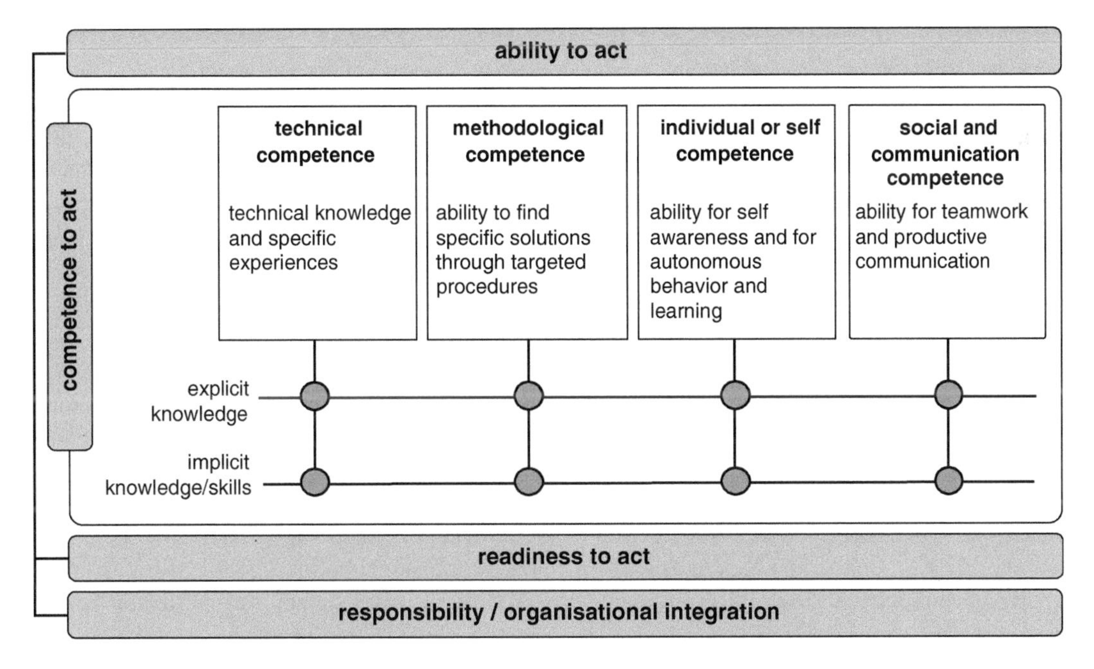

Fig. 7.1 Dimensions of Professional Competence. © IFA 11.627_Wd_B

Within the frame of researching innovation processes, the model of professional competence was extended to include two other aspects: *motivation* or the *readiness to act* and the *organizational integration* or *responsibility* of the employee [Sta02]. This means that according to the model, if someone is declared as officially responsible for a specific task their professional competence is increased.

The four previously mentioned dimensions of professional competence, along with the distinction between implicit and explicit knowledge, the readiness to act and the organizational integration, can be combined into a model, which can be used to explain a large number of unmet business goals related to employee performance. For example, expectations regarding employee participation in improvement processes are frequently disappointed because employees are insufficiently integrated organizationally. Moreover, based on the model, starting points for developing specific competencies can be determined. It is clear that operational measures for improving employee competences have to address different areas of competency.

In order to control processes for developing competences, *competency profiles*—a tool for more clearly representing employee's occupational and methodological competences—are implemented in the operational practice. The reason for this specialization is that assessing social and self-competence poses a difficult methodical problem for the practitioner. However, if only functional and methodological competences are analyzed or depicted in target competency profiles, it can result in systematically underestimating the significance of self-competence as well as social and communicative competence.

In order to clarify this, we will use the example of a competency profile that was developed during a research project with industrial partners [Nyh13] at Sartorius AG's Göttingen site [Sal13]. One of the special features of the profile presented in Fig. 7.2 is its comprehensiveness i.e., not just limited to technical and methodological competences. Usually competency profiles are developed in view of the personnel's specific tasks or functions. At Sartorius AG, the profile, shown here, serves as a representation of the target competences for the shop floor supervisor. The mid-term goal was to qualify employees at the lowest managerial level in order to increase the workforce's changeability. This target competency profile was thus developed with an eye towards changeability. The competences required for changeability are allocated to one of the four abovementioned areas of competency (technical, methodological, social and self) within the profile. By arranging and visualizing them like this, an enterprise can easily check whether or not it has neglected essential dimensions of competence within the frame of an analysis or qualification plan.

7.3.2 Strategies for Developing Competence

Learning has long been equated with formal continuing education as in formalized schooling or training programs. These traditional forms of learning are practical for certain qualification goals however when applied alone they are related to disadvantages as well. Traditional continuing education is also considered to be relatively expensive, chronically delayed and not well enough anchored in the practical problems and experiences of the participants. Nowadays, *work based learning* is considered more valuable [Bai04, Deh01, Ger04, Rae08, Son00]. According to this approach, learning should be as clearly related to work tasks as possible and should optimize in particular social and self-competence. In order to improve all of the sub-competences mentioned in Fig. 7.1 diverse settings for learning are required. In this context three forms of learning with corresponding settings can be distinguished (see Fig. 7.3).

Formalized learning is systematically and didactically guided. It is primarily found in courses, instruction and training sessions and is particularly well-suited for professional and methodological competencies as well as for communicating explicit knowledge. *Partly formalized learning* takes place in a learning environment, integrated into the workplace, within

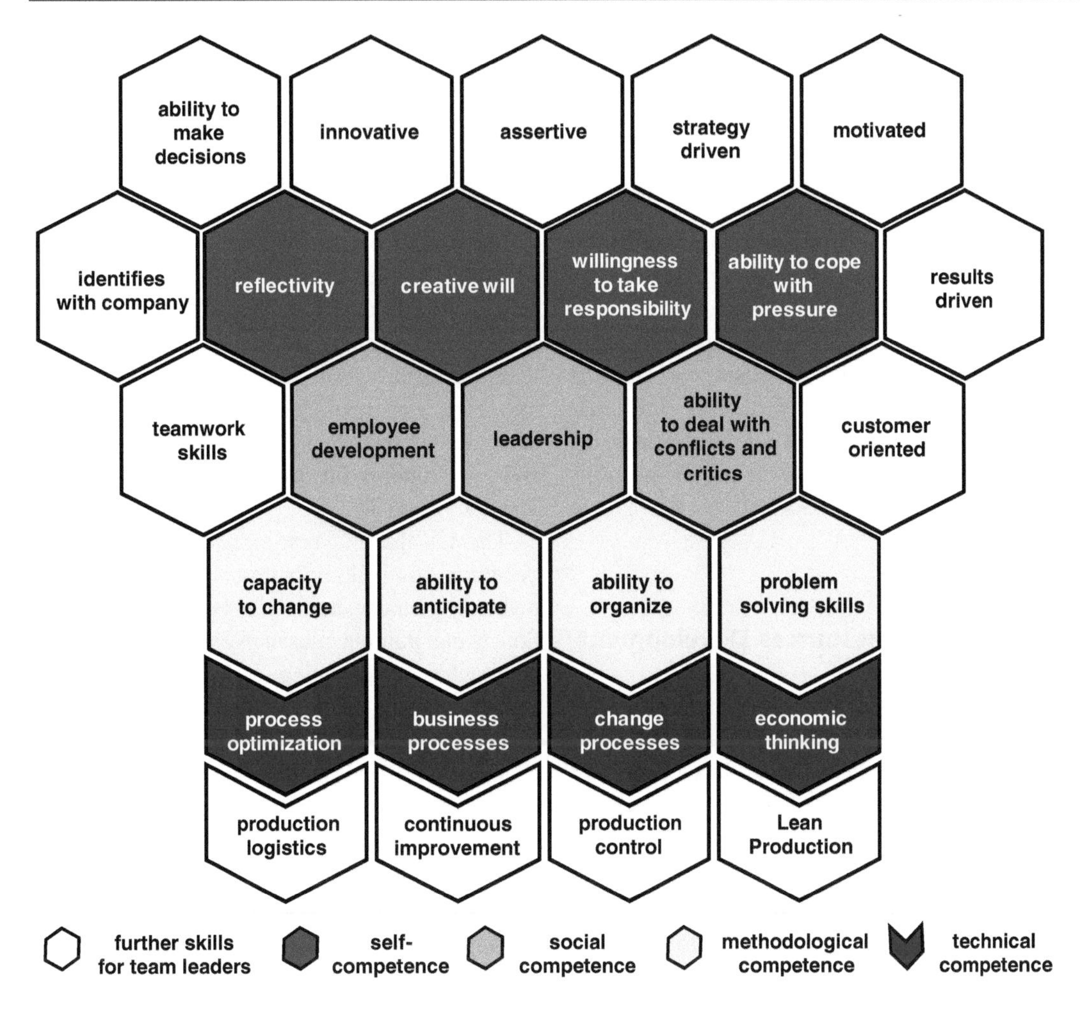

Fig. 7.2 Changeability competency profile for a shop floor supervisor (per [Sal13]). © IFA 17.596_B

Fig. 7.3 Learning forms. © IFA 14.798_Wd_B

	formalized learning	partly formalized learning	informal learning
definition	systematically didactically guided learning	partly structured learning (work integrated learning environment)	unstructured, experience guided learning
examples	• instructions • courses • training	learning • at work • in a learning place • in a learning island	learning • at the workplace

which learning processes are consciously supported but not didactically pre-structured in detail. An example here is a so called learning island or training station in which the employee independently masters complex assignments using educational materials. Both explicit and implicit knowledge in all four dimensions of professional competence can be communicated in this manner. This approach is advantageous in that it is closely linked to experience and well suited to improving social and self-competence. The one disadvantage of partially formalized learning is the limited opportunity to explore theoretical aspects, which in turn can leave knowledge gaps that can only be closed through formalized learning. The third form is *informal learning*. Learning here is unstructured and based on experience. It usually takes place at the workplace by performing work activities or in conversing with colleagues. Informal learning is experience oriented and increases the informal knowledge; moreover it is usually not consciously perceived as learning by participants. However, it is indispensable for improving skills and for passing on informal knowledge to colleagues.

7.3.3 Human Resources Development

The concepts of qualification, behavior optimization and career management distinguish three central areas of human resources development. In professional publications these are also known respectively as *knowledge oriented* [Son01a], *behavior oriented* [Son01b] and *career related* [Sch00]. The primary goal is to align personnel with the requirements of the production both in the medium and long term. Methods for developing human resources can impact individuals either directly or indirectly. A *direct impact,* for example, occurs when schooling measures are aimed at increasing work related knowledge, whereas an *indirect impact* on the individual can be conveyed through the design of the workstation. An example for the latter is when group work is introduced i.e., when it is designed so that tasks have to be mastered by a team. In this way team members stimulate one another's learning processes.

In the area of *knowledge oriented approaches*, human resources development is aimed at optimally aligning personal competences with the requirements of work activities. In order to do this, the knowledge and competences that employees need to execute specific activities have to be compiled. It is possible to orient this reactively, however, a prospective orientation is considered more suitable. Instruments that are available for developing human resources include requirement analysis, scenario building and staff appraisals. The staff appraisals deliver information about the current abilities as well as an employee's potential for development. It entails an analysis phase followed by development plans that contain concrete measures, a timeline as well as a final evaluation. Current knowledge oriented approaches consider individual learning styles and have complex learning objectives based on actual job related situations. For industrial and technology sectors, educational objective systems, learning islands as well as training offices and firms are recommended [Son01a].

There is also a diverse array of instruments for behavior oriented measures, whereby common methods include attempts at behavior modification using training measures as well as various approaches to coaching and consulting. With concrete goals, change management can be implemented as an effective measure for making changes over the medium-term. *Organization development* in comparison, is aimed at ensuring changes are accepted and at sustainably improving the organization culture. More recent behavior modification tactics distinguish themselves from others with their holistic approach. This is mainly a result of studies that question the long-term effect of training measures such as group development or outdoor-training sessions. Aspects that have been criticized include the lack of transfer potential, the selective character of the measures as well as the resulting alibi function if the work system is not simultaneously improved. Approaches that are considered more suitable combine training measures as well as measures that are concerned with designing the workstation. The reason given for this is that the structure of the work facilitates the long-term development of the employee's competence.

The aim of *career-related staff development* [Sch00] is to design a career system that aligns the requirements of the organization with the professional goals of individuals. Careers can be planned in two directions. *Vertical careers* imply a hierarchical climb but also a hierarchical descent including the special form of the pseudo-promotion. *Horizontal careers* maintain employees in their fields of work without any climbing or descending. The fact that horizontal careers have gained significance in recent years is related to the flattening of company hierarchies.

In designing the career system, human resources development takes into consideration the *motives of employees*. Their readiness to climb vertically is generally stronger than to change horizontally. The pre-eminence of the vertical career orientation is justified by the wish for autonomy, for gaining a position of power, for self-development, for prestige and last but certainly not least for higher earnings and income. Other motives are decisive for horizontal changes: the wish for more interesting and less stressful work or the desire to experience successes that the current position does not offer enough of. Changes in a horizontal direction are stopped due to interest in a stabile career development.

In view of these problems, the goal of human resources development is to make career paths transparent and to provide employees possibilities for personal and professional growth. Here it is advantageous to designs careers according to employment phases and for example, differentiate between an integration phase, the early, mid and late career years as well as leaving the company. Further aspects of human resource development can be found in [Mon12].

7.4 Work Structuring

The structuring of work is particularly important for the organizational perspective of factory planning. When we talk about structuring work we are referring to the division of work and allocation of responsibilities within and between different functional sections in a company. *People-independent strategies* define activities and tasks that are summarized in a job description without any concrete relation to a person. *People dependent strategies*, in comparison, link the organization of work systems specifically with the needs of individual employees.

Whereas the consequence of the work structure on motivation, competency, health, and job satisfaction are in the meantime almost unanimously assessed, the impact on the economic efficiency is debated. Research on so-called *High Performance Work Organization* (HPWO) first sees the economic viability of specific participative and team oriented structures when combined with performance oriented remuneration systems, flexible work time rules and training measures [App00, Com06].

The basic approaches to structuring work include job enlargement, job rotation, job enrichment and partially autonomous group work (see also Sect. 4.2—Group Work). Figure 7.4 depicts these approaches in relation to the attainable goals. Changing workplaces is not included in the table because its effect cannot generally be assessed. Depending on which tasks are integrated, it can be practiced either in the sense of a job enlargement or a job enrichment; usually though a change in workplace is restricted to enlarging a job.

With *job enlargement*, the scope of an already existing job is expanded with similar activities. This can be achieved for example by extending the assembly cycle. This generally does not increase qualification requirements but does allow employees a change in workload and a greater diversity of work within a modest range. In comparison, with *job enrichment* a position is usually supplemented with activities requiring more thought and qualifications. Job enrichment pursues the concept of *complete work tasks* [Hac98] which, in addition to the pure execution of a task, also comprises the planning, preparation and control of it. An example here is extending the job of a machine operator to include maintenance, quality testing and order control. Job enrichment is usually accompanied by a higher qualification and to some degree also greater earnings.

Partially autonomous teamwork is based on a group of employees who plan, prepare and control their work activities within a known range [Ant94]. It thus combines both job enlargement and job enrichment strategies and in doing so focuses on transferring tasks into the area of

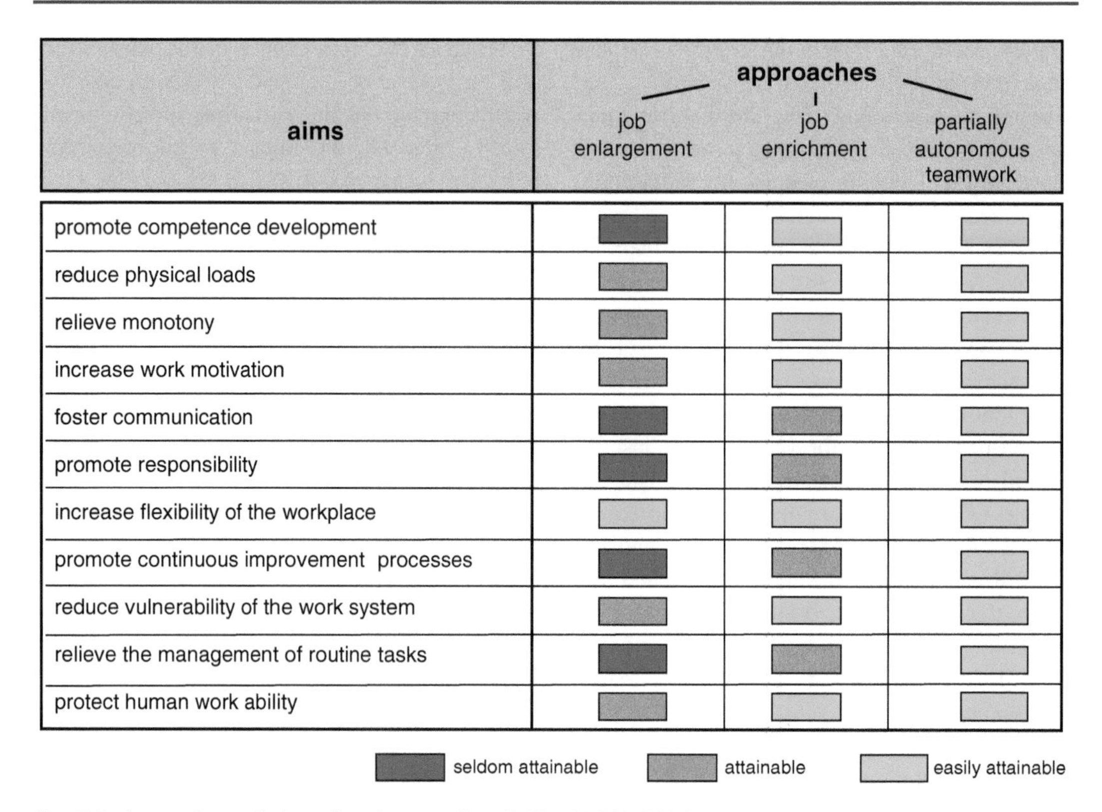

Fig. 7.4 Approaches and aims of work structuring. © IFA 14.799_Wd_B

personnel management. The more these changes enrich the jobs, the more suitable group work is for reducing loads and maintaining or expanding the employees' competences. The current increase in partially autonomous group work can be explained by the growing complexity of production processes. Its main performance benefit is the flexibility and speed with which appropriate action strategies can be developed in view of the diverse production requirements and environmental influences [Ger06]. Teamwork can improve product quality, shorten throughput times, reduce sequence-dependent waiting times and decrease standstill times [Uli01: 260]. In order to increase the work schedule flexibility and goal orientation of workgroups, it is advantageous to combine group work with a bonus scheme and flexible working hours.

In addition to partially autonomous teamwork, many companies orient themselves on the concept of "supervised teamwork". This concept stems from the Toyota Production System and entails a team leader (in Japanese—*hancho*), assigned by the enterprise. The team leader is responsible for a problem-free production as well as measures for continually improving the production process [Nom95, Shi04]. If all of the indirect work functions of the group are also transferred to the *hancho* (which occurs fairly commonly) this endangers the positive influences that the teamwork design can have on the motivation and development of competences by employees as well as the work and safety regulations [Ger99].

Work structuring concepts that are people-dependent include differential and dynamic work design. With *differential work design* tasks are structured according to the individual interests and competences, whereas with *dynamic work design* tasks are continually individually updated with the aim of addressing changing competences and interests.

Sociotechnical system design provides work sciences with a well-researched approach to structuring work, from which numerous guidelines can be derived. This method is based on the assumption that enterprises consist of technological and social sub-systems and that it is in the interest of economic efficiency to align these two systems with one another [Mum06, Pas82]. This approach draws conclusions about the structure of tasks and in doing so concentrates on teamwork, which it perceives as a more efficient work form. The key assumptions of the sociotechnical system approach are [Fri99, Ric58, Uli01]:

- The group is interested in efficiently organizing and fulfilling tasks.
- A group is efficient when it can complete tasks as a whole.
- Related tasks within a group require team members to have satisfying social relationships with one another.
- A group being responsible for a defined territory positively impacts the social relationships within the team.

In addition, the London based Tavistock Institute, which was largely responsible for developing the sociotechnical system, also formulated principles for designing work with reference to the individual [according to Fri99]. Individuals should:

- be challenged on a professional level,
- learn at their workplace,
- be able to make decisions on their own,
- receive recognition and respect,
- see their work as practical and
- consider their work as contributing to a desirable future.

In addition to the original differentiation between technological and social systems, later versions of the sociotechnical system approach recommended distinguishing between three types of systems: *people*, *organization* and *technology* [Fri99]. The aim of sociotechnical system design is to align the interfaces of these three systems. For a long time, the sociotechnical system approach was considered the basis for adjusting the social system to the technological. However, in the meantime, there has been a paradigm shift in research, according to which, the technological system should already be adjusted to the concerns of the social system during the production planning phase [Zin97]. This latter approach is rarely practiced today, thus deficits in the social system resulting from the technical design and work organization (e.g., lack of motivation, work-to-rule, high rate of absenteeism and illness) can barely be reduced.

7.5 Motivation

In addition to being dependent on their competency and the design of their workplaces, the performance of employees is contingent on their motivation. Motivation cannot be directly observed, instead only its outcomes can be seen in the form of actions and results. Generally, *motivation* as a concept refers to the energy involved in actions, the direction in which this energy is guided as well the person's perseverance in pursuing a goal [Kir05: 321, Rob12].

The reason for why a person acts or refrains from acting can be traced back to either intrinsic or extrinsic motivation. With *intrinsic motivation* it is executing the action itself that drives the person. This applies when a job is considered independent, professionally challenging and as a basis for personal growth. With *extrinsic motivation* the reason for an action is the related reward or lack of reward/punishment. Whereas a Taylorist view of man assumes a primarily extrinsically motivated employee i.e., by payment and threatened punishment, intrinsic motivation plays a growing role in modern work organization.

In motivation research, two groups of theories can be distinguished. *Content theories* aim at a substantive definition of human motivations and interpret human behavior through the need to eliminate a specific deficiency. *Process theories* explain actions against the background of complex, multi-staged decision processes. Guidelines for organizing work and managing personnel can be derived from both approaches.

The most well-known *content theory* stems from Abraham Maslow. It differentiates the 5

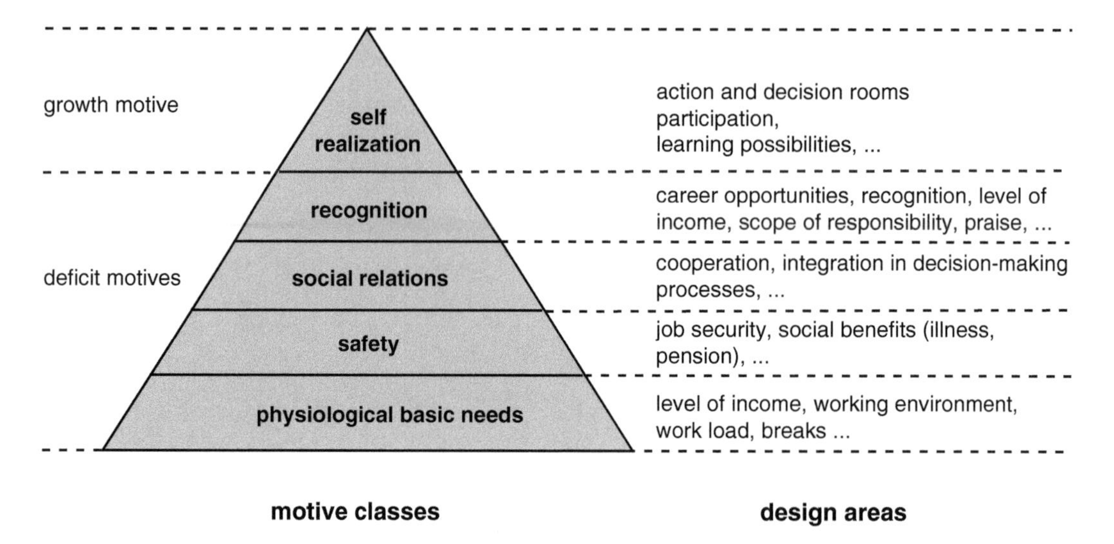

Fig. 7.5 Maslow's hierarchy of needs as applied to designing a work system [per Sch00, Spa04]. © IFA 14.800_B

classes of motives depicted in Fig. 7.5 middle. This "hierarchy of needs" is based on the assumption that only one class of motives currently determines the actions of a person, whereby the higher classes can first be activated once the needs on the lower stages have been satisfied. Unlike the four lower classes of motives, it is no longer possible to find satisfaction on the highest level, which is why Maslow speaks of a growth motive in comparison to the four deficit motives. If we follow Maslow's model, the 5 levels of needs each correspond to a specific area of organizing work (Fig. 7.5 right). Basic physiological needs can for example be satisfied in the design of the reward system, safety needs in job security, needs for social relationships in collaborative forms of work, needs for recognition in career possibilities and the need for self-realization in opportunities to learn.

As one of the first content theories of motivation, the hierarchy of needs indicates the variety of human motives. Nevertheless, the demarcation of needs and the assumption of a hierarchical order have proven to be problematic. Later theoretical approaches reduce the number of 'need classes' and eliminate the hierarchical order. However, they share the emphasis on *performance and growth motives*. By focusing on these, content theories question the assumptions made by Taylorism and indicate ways to implement workers more productively. As a result:

- Employees can be motivated in a variety of ways, not only—as Taylorism suggests—by financial compensation for efforts made and strain endured.
- The personal need for growth represents a valuable potential for a company. Room to maneuver and make decisions should thus not be limited more than absolutely necessary. Only in this way can the employees' voluntary willingness to cooperate be gained.
- It is the responsibility of management to create a work environment in which the satisfaction of employees' needs are linked together with the business goals and in which personal motivations for performance can be increased.

Process theories in motivation research require substantive nameable motives; however they explain human actions first and foremost as the result of decision processes that integrate various stages of the work process. According to these approaches, motivation originates in the work process and the mental anticipation of it.

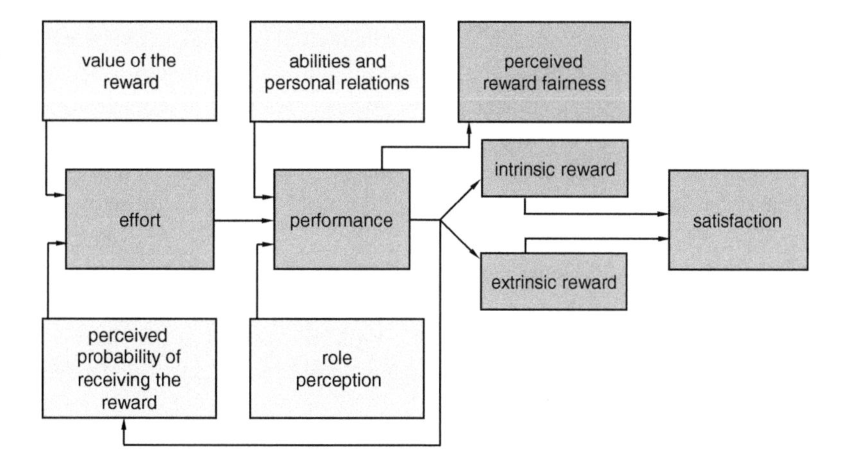

Fig. 7.6 Motivation theory according to Porter and Lawler. © IFA 14.801_B

The majority of process theories work with the *value* which specific processes and their results have for an employee. Moreover, the theories argue that an employee's expectation, as to whether or not this value is attainable, matters.

One of the best know process theories, which explains not only the work satisfaction but also the motivation to work, stems from Porter and Lawler (Fig. 7.6) [Port68]. According to the theory, employees will only make an effort when they anticipate a reward that is of value to them. The extent to which the effort leads to a performance is dependent on the one hand, on the individual abilities and on the other hand, how they perceive their role, i.e., how the employee defines the success of the action. Effort and performance lead to intrinsic and extrinsic rewards which are also judged from the perspective of fairness. The perceived rewards ultimately determine the degree of work satisfaction. Porter and Lawler's theory includes decision processes that accompany an employee from the start of a work task up until its conclusion.

Practical consequences for personnel management can be derived from process theories such as Porter and Lawler's.

- With participation from employees, management should develop a clear goal orientation and clarify operational targets. Targets which have been agreed upon together and recorded 'on paper' serve to improve the goal orientation.
- Management should establish conditions under which their employees can attain the desired results. This requires not only eliminating technological and organizational obstacles but also creating measures for developing competences.
- Management should gear their behavior to the employees' different value orientations and competences. For example a participative and performance oriented management style is suitable with competent and decisive employees, whereas a more directive style is appropriate with less competent employees and teams that have conflict-ridden and less constructive forms of dealing.

7.6 Designing Remuneration Systems

The criteria which payment rates should be oriented on are decisive to remuneration systems. Two goals are pursued here: developing a payment which is perceived as appropriate and controlling employee performance. The question of which remuneration system is best suited to the fairness criteria is dependent on normative decisions and cultural backgrounds [Wäc97, For04]. Depending on the fairness criteria selected, the remuneration is an expression of:

- the relationship between the offer and demand on the job market,
- acquired qualifications and professional certifications,
- acquired social privileges, e.g., the length of time someone has belonged to an organization or their length of service,
- social needs such as the responsibility for spouses and children,
- the general difficulty of the work task,
- the specific performance of the employee.

Today's remuneration systems take all of these into consideration although employee performance and offer/demand on the job market dominate. Those who possess rare qualifications which are in heavy demand on the job market have a relatively good earning opportunity. Moreover, an individual's performance is also greatly valued in many remuneration systems. Here, the enterprise is not only concerned with appropriate payment, but rather also in steering the employees' behavior in a desired direction. In this sense remuneration is a *performance incentive* i.e., a reward for a specific effort, see also [Ber08: 447].

Besides monetary incentives there are also non-monetary incentives that can exercise a strong influence on employees' actions. An overview of these according to Thommen and Achleitner [Tho03: 692] is provided in Fig. 7.7. Monetary incentives should therefore always be designed in conjunction with all other possible incentives.

Remuneration systems are frequently presented in the form of a *pay pillar*, in which the requirement-dependent pay forms the base, followed by performance-dependent standard remuneration components and those that exceed the agreed upon scale. The base pay is derived from a job evaluation (also referred to as job assessment).

In a *job evaluation* the requirements for a job are assessed in relation to other jobs using a uniform reference. The aim here is a pay differentiation based on the job requirements. In addition to determining the base pay, job evaluations can be implemented with an eye towards developing personnel and/or optimizing work processes.

In Germany, there is a 3-stage method available for evaluating jobs, which was developed by REFA (an association for work design, industrial organization and company development http://www.refa.de). In collaboration with researchers and tariff partners REFA develops methods and

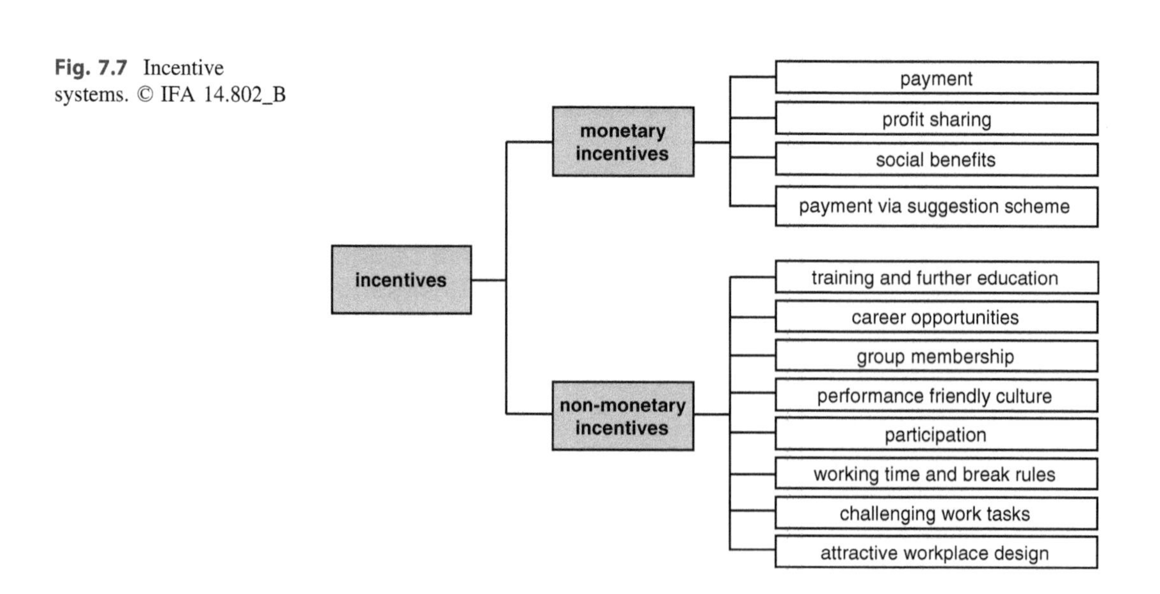

Fig. 7.7 Incentive systems. © IFA 14.802_B

training modules for work design and industrial organization. The first step in evaluating a job is compiling and describing the job activities, workplaces and organizational relationships. Step two is comprised of analyzing the types of requirements, while step three entails an assessment of the requirements and an overall evaluation of the work activities.

Job evaluation approaches can be categorized as either summary or analytical. While *summary methods* start with the activities and either compare them with one another or with an example catalogue and thus attain pay or salary groups, whereas *analytical methods* assess the individual types of requirements separately and subsequently calculate a total work value that can be allocated to a pay or salary group. Both approaches to job evaluation entail quantification processes which can in turn be categorized as either ranking or grading. Whereas with *ranking* processes, work activities or individual job requirements are classified based on the degree of difficulty, *grading* processes involve precisely defined levels either for pay or salary groups (summarizing) or for the characteristics of the individual requirements (analytical). All of the methods depicted in Fig. 7.8 quantitatively evaluate job activities [Doe97].

The *ranking method* is based on a series of pair-wise comparisons through which all of the activities in a company are ranked. In contrast, the *pay/salary group method* is oriented on a catalogue that characterizes work activities in a gradated form. An example here would be the category "difficult skilled work requiring special abilities and years of experience"; so-called 'directive examples' are provided as an aid.

Analytical job evaluations utilize requirement catalogues, which, in Germany, are established by collective agreement. The Genfer Schema, formulated in 1950 at an international conference on job evaluation, provides orientation help here. It differentiates between mental and physical requirements as well as responsibilities and work conditions. Analytical methods include the rank-row method and the step method. The *rank-row method*—as its name indicates—is based on rank-rows for each type of requirement and which, depending on the difficulty of the work, contain corresponding numeric values. Exemplary activities serve to orient users. The *step method* or *step value method* is based on evaluation tables that also provide numeric values for each type of requirement depending on the difficulty. The ranking however is oriented on qualitative terms such as "very high, high, average, low and very low" or on comprehensive descriptions of the respective level of the requirement stage. Here too, examples are used as a means of orientation. As with the rank-row method, a score is determined for each requirement characteristic. The total score allows the job to be allocated to a pay/salary group. The "ease of use for the evaluator and the clarity for the employee" [Tho03] distinguishes the step method from others.

Payment schemes can be divided into two main types: pure payment or combined payment (see Fig. 7.9). Whereas *pure payment forms* are oriented either on working hours, work difficulty or performance, *combined payment forms* pool together two or more of these characteristics. Moreover, forms of payment differ from one another as to whether they react directly to the performance or if the payment remains a constant. Whereas hourly pay does not react to the performance, both the piece rate pay method and bonus pay approach do.

Fig. 7.8 Job Evaluation methods. © IFA 14.803_B

quantification methods	qualitative analysis methods for requirement profiles	
	summary approach	**analytical approach**
ranking	ranking method	rank-row method
grading	pay-/ salary group method	step method

Fig. 7.9 Payment forms. © IFA 14.804_B

performance sensitivity	pure payment forms	combined payment forms
not performance sensitive	• hourly wage • salary	• standard pay • polyvalence pay
performance sensitive	• piece rate pay	• bonus pay • target based pay • hourly pay with productivity bonus • salary with profit sharing

Hourly pay is based on a set payment for a specific time unit. It is one of the pure payment forms and does not react directly to the performance. Since the payment does not fluctuate with the performance, this means the only expectation is that the employee is present. Hourly pay is "advantageous with work,

- that requires a high quality standard,
- that needs to be conducted carefully and conscientiously,
- where is there is a high risk of accidents,
- whose performance cannot be measured or is difficult to measure (quantitatively) as is the case for example with tasks requiring creativity,
- where there is the danger that people or machines will be put under too much pressure or strain" [Tho03: 716].

The disadvantage of hourly pay is the lack of financial compensation for performance. However, hourly pay can become more incentive when combined with payment bonuses. These are ensured based on an evaluation of the individual's actions i.e., they causally reward personal contributions to the company's performance.

Standard pay, which is especially common in the German automobile industry, is a borderline case between a payment that reacts to performance and one that does not. With this type of performance related payment, employees have to attain a certain target output for a specific time period. Although deviations from the target do not impact the payment, failure to meet the target does result in a root cause analysis and implementation of corrective measures.

Among the payment forms that react to employee performance is piece rate pay. *Piece rate pay* (or payment-by-results) rewards the output related element that can be influenced by an employee (or in the case of group piece rate pay—by a group). One of the performance indicators here is the output rate which describes their performance in relation to a normal performance. Piece rate contains a financial performance incentive, however, is also related to a number of disadvantages. First and foremost, there is the danger that people and machinery are put under too much pressure and neglect the quality. This payment method thus should not be implemented in situations where there is the risk of accidents or when there is a high demand for quality.

Bonus pay is generally a more flexible payment method than piece rate pay which only reacts to an output rate related element. *Bonus pay* supplements a base pay (which is dependent on the job requirements) with an additional changeable reward. Bonuses can be oriented for example on the output rate, quality, productivity, reduction of material and time consumption as well as the utilization of production resources. Bonus pay can generally also be used to support logistic goals; nevertheless, in the production area it is usually aimed at productivity and rarely at logistic targets. Consequently, from a logistics perspective, it can lead to a faulty control and thus have an undesirable impact.

So-called *polyvalence pay* is comprised of both a requirement-oriented base pay and a bonus for abilities/skills. The main aim of this payment system is to encourage employees to

performance indicator ▷	**time at work**	**qualitatively differentiable effort**	**economically usable result**	**successfully utilized result / market success**
criterion ▷	working time	• accuracy of work • flexibility • work deployment • team behavior	• productivity • cost-cutting • delivery times • quality	turnover, profit, ...
payment form ▷	hourly wage	performance orientated payment		success and result oriented payment

Fig. 7.10 Relationship between performance and payment. © IFA 14.805_B

develop additional qualifications and their individual competences. In practice, this raises the question of how to weight abilities and how to evaluate competence [Uli01]. One possibility is to express the competence through the amount of time required to learn it. This indicator accumulates the time required to learn a task at a workstation up to the point that a worker is able to conduct the work "independently and within a typical time limit" [Pla04].

Target based pay was formulated for the German industry within the new remuneration framework for the collective bargaining area of the metal and electric industry (ERA). Starting with a base pay it is supplemented by bonuses oriented on agreed upon targets.

Another possibility for a combined payment form is to supplement a salary with profit sharing. Usually corporations then orient themselves on the turnover or earnings before taxes. This type of payment system is aimed at increasing the employees' identification with the company as well as limiting the enterprises risk as far as remuneration is concerned. Research has shown though that one of the disadvantages of result oriented payment systems is that performance is devalued when it does not lead to success on the market [Bah01].

Remuneration is the reward for the results produced by an employee. The concept of human job performance can however be defined differently and moreover has been subject to a historical transformation. At the high point of Taylorism performance was equated with the speed with which a worker produced a fault-free product. Managerial staff was responsible for different tasks than production workers, resulting in different salary structures.

Currently production workers are increasingly recognized as co-collaborators and play a role in planning, controlling and optimizing processes. Consequently, the activities of production workers and managerial staff are becoming more similar. This posed new requirements for the design of payment systems which in turn led bargaining partners in Germany to negotiate uniform collective wage agreements for workers and managerial staff first in 1988 in the chemical industry area and later in the electric and metal industry (ERA).

Payment systems correspond with how an enterprise views the content of the job performance and with that the function of the workers. Four different performance indicators can be distinguished here (Fig. 7.10). Companies can remunerate employees for making their time available, for making a certain effort, for targeting an economically valuable product and thus for success on the market. Figure 7.9 outlines the criteria for the performance indicators which are reflected in the corresponding forms of payment.

Modern enterprises increasingly orient their payment systems on economically valuable results. As a consequence, the function of the employee is shifted. Instead of rewarding them for a specific effort or for the time they expended in the company, the skillful actions an employee contributes to increase productivity and save costs or other directly valuable results are viewed as the desired performance. In professional publications this development is referred to as the change from a causal to a functional performance conception [Ben97].

7.7 Planning Working Times

Planning working times is a critical calculation parameter in designing the resource capacities. Thereby, the focus is on the duration, location and distribution of working hours within a defined period e.g., a week or month. This includes rules for vacation as well as planning rest breaks. The aims in planning working times include adjusting the personnel capacities to the production planning, creating sufficient possibilities for employees to rest and recover, maintaining employees' health and their continued ability to perform as well as taking into consideration individual interests. Which of these goals are emphasized can strongly vary between the different working time models and within their individual layout [Whi03].

When planning working times, *contractual freedom* applies between the employer and employee; nevertheless, legal and collective bargaining rules have to be maintained. The most important legal foundation in Germany is the Working Hours Act (ArbzG). In the US the according laws can be found in the Regulatory Library of the U.S. Department of Labor (www.dol.gov).

Historically considered, the Working Hours Act is a protective act for employees. Whereas at the start of industrialization in Germany, an average workday was typically 15 h and even children (after the age of 6) were forced for up to 12 h of heavy physical work, according to the German law current work hours are generally 8 h per day. This can be extended to 10 h; however the average number of hours per work day over a 6 month period cannot exceed 8 h per day. Additional rules for planning the working times are also found in collective bargaining agreements and individual work contracts.

In Germany working times are primarily regulated legally and through collective bargaining agreements. In other countries legal guidelines play a smaller role than individual guidelines when planning work hours. For example in the USA the Fair Labor Standards Act (FLSA) provides a federal labor law. Although, it does not set a limit for the maximum work hours, each hour above a 40 h work week provides for a minimal 50 % surcharge in addition to the base wage. This acts as an indirect influence in limiting the work hours.

Models that can be implemented for planning working times can generally be distinguished from one another according to their degree of flexibility (Fig. 7.11). Nevertheless, there are different definitions of flexible and rigid work hours. The reference point for characterizing work hour models is usually the differently defined normal work hours. Usually with the term *normal working hours,* a regular and strictly regulated workday between 7 a.m. and 7 p.m. is understood. Currently, the majority of people employed, work under the conditions thus defined, however this model is being replaced more and more by other variants.

Furthermore, not every deviation from these so-called normal working times leads to flexible work hours e.g., *reduced regular hours* within the context of part-time work. *Shift work* is also a deviation from normal work hours; however, since a shift rhythm can be planned, it does not yet fall into the category of flexible work hours. A further group of work hour models is based on an *irregular distribution of work hours*, such as seasonal work or working á la carte so to speak. Here too, it does not necessarily deal with flexible work hours. Seasonal work can follow a strict yearly rhythm and with work á la carte, the work can be distributed on set days or a fraction of days in the week. According to Nachreiner and Grzech-Šukalo [Nac97], from the view point of occupational science a practical concept of *flexible working times* has to emphasize the "room to negotiate and plan the duration, location and distribution" of the work hours. Otherwise, so many different models of working times would fall under the concept of flexible working times that it would be impossible to make any generalized statements about this group of

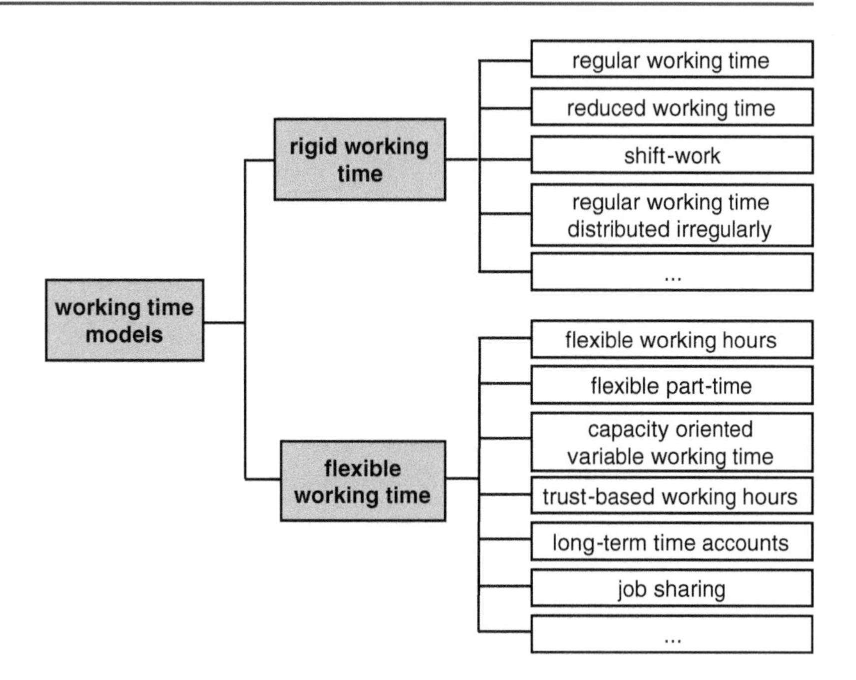

Fig. 7.11 Systematic of working times models. © IFA 14.806_B

models. As a result of this suggested definition, *rigid working times* are characterized by periodic repetition of blocks of work and time-off with the same duration and location within a period.

The evaluator's perspective and the question of who is in the position to schedule the work hours are critical in the evaluation of flexible working time models. With capacity oriented variable work hours (in which workers are 'on-call'), only the employer is able to plan working times—for this reason this model has not been agreed to by unions. Other forms of flexible working times such as flexitime, flexible part-time, trust-based working times (honor system) can allow employees to have a certain degree of autonomy in scheduling their work time and at the same time address the employers need for flexibility. Thus with job-sharing it is possible for two employees to share a position and independently divide their work time. In addition to unpaid leaves (sabbaticals) other forms of flexible working hours include variations of so-called "lifelong work accounts", which for example allow a smooth transition into retirement.

In today's economy, night work and shifts are required for a variety of reasons. It can, for example, be necessary to utilize a particularly expensive technology as close to twenty-fours-a-day as possible in order to justify the involved investments costs. Further economic reasons for night and shift-work include the speed at which technologies are changing, which in turn shortens the amortization time for manufacturing resources: Moreover, shifts may be required when technologies requiring continual operation are implemented e.g., in the steel and chemical industry or when a company is concerned with supplying a population with energy or medical services that cannot be limited to normal working hours.

According to German labor law though, night and shift work need to be planned according to sound scientific knowledge about humane working conditions. Humane working conditions in this context means that not only the well-being and health of employees have to be taken into consideration (ergonomics) but also that an appropriate level of participation on a social level is facilitated [Kna97].

The "physiological performance curve" [Gra61, Sch93] provides a basis for maintaining the health and well-being of employees. The curve (Fig. 7.12), which describes the progression of the physical and mental willingness to

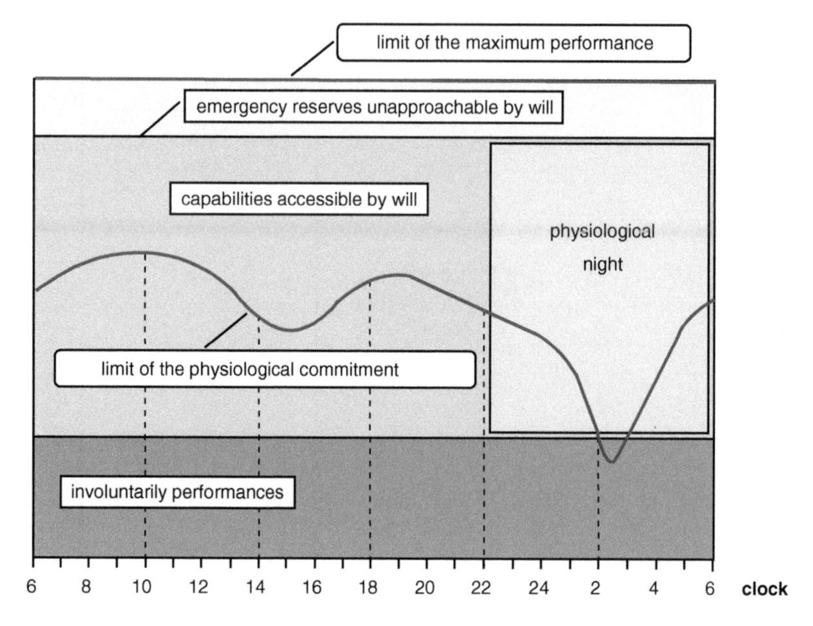

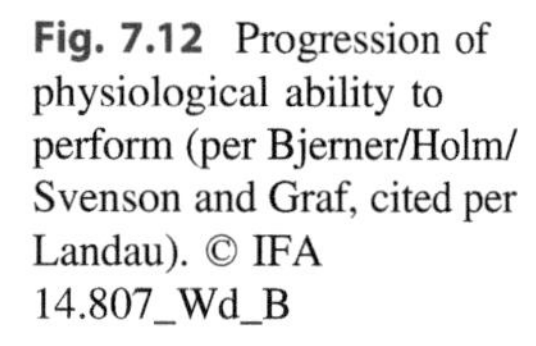
Fig. 7.12 Progression of physiological ability to perform (per Bjerner/Holm/Svenson and Graf, cited per Landau). © IFA 14.807_Wd_B

work, is for the most part genetically set [Land01, p. 40]. During the physiological night time the curve lies within a range accessible to free will. The course of the curve can vary for an individual and between individuals but offsetting the curve by a shift or a half day is biologically impossible: The body adjusts to the shift work however early and (in particular) late shifts push employees into a permanent struggle with their inner clock.

This permanent struggle should be minimized with scientifically proven measures—otherwise, there is the threat of neglecting employees' well-being and of serious health risks. During phases of low capabilities, the risk for mishandling items and for accidents increases. As research has proven, early shifts are related to sleep disruption and fatigue, while late shifts and weekend work interferes with employees' social lives [Kna97]. These problems also arise with night shifts. Moreover, night shifts have also been proven to be associated with further disruptions in employees' well-being e.g., diminished appetite, gastrointestinal complaints and gastrointestinal or cardiovascular diseases [Kna97].

A study in which 9000 shift workers were examined shows that when planning shifts the following recommendations should be observed [Kna97]:

- The number of back to back night shifts should be kept as low as possible. The maximum number is considered four, although less than three consecutive night shifts are considered ideal. This applies equally to early and late shifts.
- Priority should be given to blocked free weekends over individual free days during the week.
- Forward rotating shift systems (early, late, night shifts) are better than backwards rotating shift systems (night, late, early shifts).
- With greater work strain the length of shifts should be shortened.
- Early shifts should first start at 7 a.m. instead of 6 a.m.
- Shift plans should be predictable and should not be changed in the last minute by employers.

Publications in the area of work sciences also make the following recommendations:

- A nightshift should be followed by a pause that is as long as possible and at the very least more than 24 h.
- Performance requirements should be decreased during nightshifts; this includes refraining from work incentives.
- Additional occupational health measures should be taken for night workers. In Germany the rights to work related medical screenings is regulated.

One possibility to observe as many of these recommendations as possible is to transition from a 3-shift system to a 4 or even 5-shift system.

One of the reasons flexible working hours are increasingly prevalent is that in order to satisfy the desire for shorter delivery times and higher delivery reliability, plant operating hours are being adjusted to fluctuating demands. New manufacturing and logistic concepts such as lean production or just-in-sequence enforce this trend. The relative growth in services which are oriented time-wise on the customers, also contributes to this. This can be observed not only in retail but also in industry related services. In industry work is frequently oriented on projects which also require flexibility with regards to time. Moreover, flexible working hours are becoming more socially acceptable and, last but not least, employees are interested in having greater flexibility with regards to the availability of their time.

There are a wide variety of possibilities for flexibly designing working hours. Two long known approaches include overtime and on-call service whereas relatively new models include so-called ‘flexitime’, trust based systems, long term work hour accounts and annual or working life models. Overtime remains the most common method for making working hours more flexible, however, it is slowly being replaced by more flexible and, from the employer’s side, more economically efficient solutions. In the following we will introduce flexitime as a frequently practiced model of designing working times and the honor system as a frequently debated model.

Flexitime models are based on agreements that set a flexible time frame within which the employee can distribute their working times (see Fig. 7.13 [a.o. acc. Luc98, Mar94]). Usually a

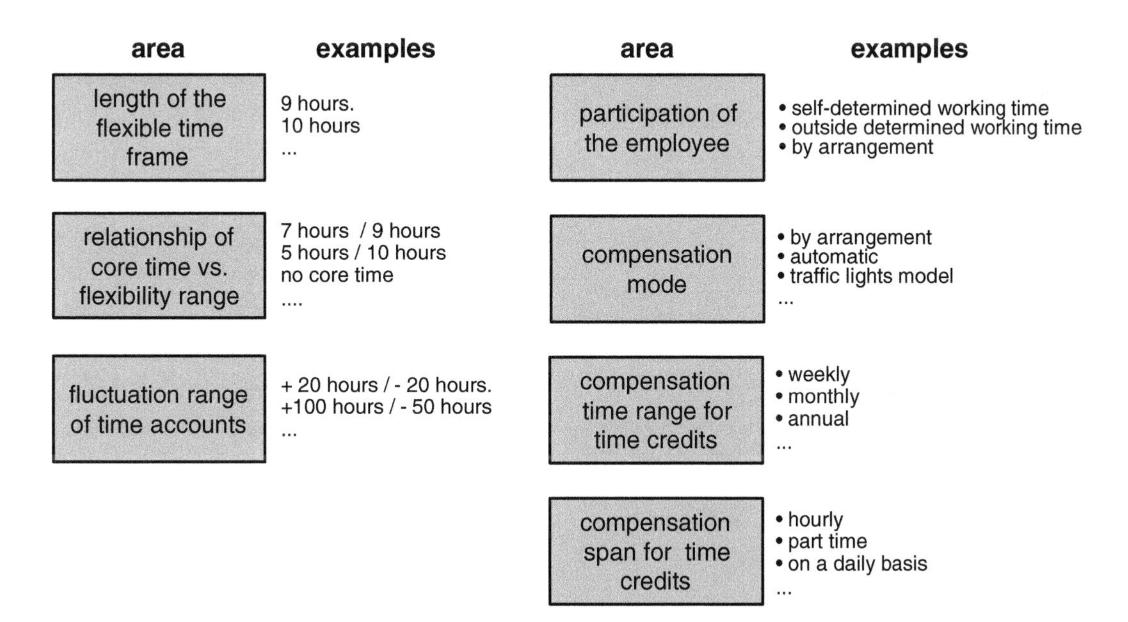

Fig. 7.13 Design areas for flexible working times. © IFA 14.808_B

core time is set during which employees have to be present along with a fluctuation range for the time accounts. The aim of designing flexitime models is to optimally plan working hours while providing employees with a high degree of autonomy. Broad fluctuation ranges and longer compensation periods provide the greatest flexibility. The consequences of a flexitime system are quite complex and case sensitive though. For example, when an account has a wide range of fluctuation there can be an urgent need for action when the fluctuation limit is reached.

From the perspective of employees, flexible working times are attractive when they can choose or at least be part of the decision about where and for how long they work. Greater time autonomy can also be attained when the worker can flexibly select the range and compensation period for balancing positive time accounts.

The mode of compensation is also an area that needs to be planned. This can consist of an agreement between the employer and employee, however, this form is no longer sufficient when the limit of the fluctuation range is attained. In cases such as this, so-called 'traffic light models' are implemented, according to which employees are required to undertake actions in order to balance their accounts. One possibility is automated balancing which occurs as soon as the limit is met. Moreover, when designing flexible working times, it is necessary to regulate the time period within which the account has to be zeroed out. Typically these periods are longer e.g., 6 months to a year.

The *trust based* model for working times is characterized by the lack of a time clock; instead, employees are responsible for regulating their working hours themselves. The decisive factor then in evaluating the work performance is the result for which the employee is accountable. Since trust-based working hours are highly unbureaucratic and flexible it offers a great opportunity for both the employer and employee. Nevertheless, it is a highly debated model particularly among unions. Critics fear working times will be extended since the protective mechanism of regulated working times is lost and the responsibility for results could wrongly lead to "self-exploitation" and a loss of solidarity among workers. As was shown in a study though the "unabated increase in performance" is rather a rare exception [Böh04].

Increasingly previously used flexible time systems are continued to be practised even after a trust based system has been implemented. In addition, rather than an unabated performance increase, advantages are realized more frequently for both the company side and employee side. When introducing a trust based system that allows both flexibility around work times from the employer side and autonomy for employees in scheduling their work hours, one of the decisive factors is a compensation system that prevents employees from being over strained. The following measures have proven to be efficient [Böh04]:

- make it possible to individually document working hours,
- implement virtual 'traffic light' accounts with mandatory graduated actions,
- establish a committee for clarifying contentious questions and complaints (clearing places),
- make paid extra work possible,
- include optional models that allow a return to time clocking,
- increase efforts for developing teams with the aim of promoting solidarity among members.

With these comments, we will conclude our discussion of the organizational work aspects of planning that are fundamental for factory planning. Beyond these, the design and layout of the immediate spatial work environment is vital for a reliable and healthy work performance. In the next chapter, we will consider these from the perspective of spatial planning.

7.8 Influence of Demographic Change

According to the 12th coordinated population projection by the Federal Statistics Office of Germany, the labor force in Germany will continue to age up until 2024. Whereas, the share of the working population aged between 50 and 65 years was 31 % in 2008, it will climb to 40 % by 2024. During the same period, the percentage of workers between 20 and 30 years old will shrink by 2 % and the share of those between 30 and 50 by 7 %. This structure will remain more or less constant up until 2060. At the same time, however the size of the working population will dwindle. This shrinkage will continue until at least 2060 at which point approximately 36 million people between the ages of 20 and 65 will live in Germany. This corresponds to a decrease of 27 % in comparison to 2009 [Ege09].

The aging and decreasing working population impacts human resources and from a business perspective entails the following risks:

- uncertain supply of skilled workers,
- retirement of those with knowledge,
- increased costs due to illnesses and restricted performance abilities,
- increased recruiting and personnel costs,
- tensions and conflicts between generations.

From the perspective of individual employees, new risks arise due to the gradual increase of the legal retirement age to 67 and the simultaneous termination of transitional steps towards retirement (e.g., partial retirement after 62) as well as more difficult conditions for entering into retirement for those with health restricted performance abilities. For older employees this means: If they are no longer able to cope with growing psychological pressure and physical demands at work, there is a threat of unemployment, early retirement and significant concessions in retirement levels.

From a business perspective the urgent question is: Do the performance abilities of employees, consequent to the biological aging process, change to such an extent that special measures customized according to the various phases of life are required in designing and organizing work?

Scientific studies have reached the conclusion that on average there are age related changes in the population in terms of sensory, physical and mental performance, however that these only influence professional performance minimally. As we age, our sight, fine motor skills, physical strength and reaction speed when taking in information and processing signals decrease. Nevertheless, these changes do not affect most work tasks; only our abilities for high performance are negatively impacted. Activities that are affected include heavy physical work, for example in the forging industry, in competitive sports or where extreme demands are placed on abilities to process information e.g., air traffic controller.

Changes in performance clearly vary from person to person. Research has shown that there is strong variance in the performance of older employees. This is the result of a complex interplay of genetic predisposition, personal behaviors around health as well as pressure and training effects through vocational work. Furthermore, the human performance range does not change overall nor does it change in the same direction, but rather it varies greatly in different dimensions. Older workers' capabilities are thus no longer determined by the deficit model, which one-sidedly emphasizes emerging performance deficits. Rather in the scientific debate, a competency model, which takes into account both increasing competences and possible decreases, has prevailed (see Fig. 7.14). Older workers are frequently rich in experience and knowledge, are skilled in working with others, can integrate new knowledge quickly into known contexts and can thus deal well with unexpected events and stresses. Based on the competency model, older workers therefore appear to be highly efficient. It is thus crucial to design work environments and tasks so that older workers can contribute their

increasing ⇧	decreasing ⇩
• practice and working experience • conversation ability and judgment • people skills • reliability and awareness of responsibility • quality awareness • company loyalty • coping with stress and chaos • learning ability for structured contexts, associable with familiar situations	• muscle strength and physical performance • sensory abilities • reaction speed • willingness to take risks • retention • learning ability for abstract contexts • currency of education

Fig. 7.14 Age related changes in performance according to the competency model (per [Kie08]). © IFA 7.597_B

particular strengths. Many of the deficits that come with aging, such as sight restrictions or fading maximum strength can be compensated with creative measures.

The problem for enterprises is thus not a possible general decline in performance with age but rather the difficulty that there is a strong burden of disease in the age span of 45–65 years in the working population. This is largely due to the cumulated effects of stress and strain over the course of living and working. Chronic diseases impact the performance of employees and are the cause of sick leaves. Accordingly, companies have to take an interest in designing and organizing work so that it positively impacts the aging process of their employees. In view of the increasing average age of the work force, this is the only way factories can maintain capable human resources.

Designing work adapted to processes of aging aims to allow employees to age within the company without losing competence, health or motivation. This addresses employees in all age groups. Age-adapted work design in comparison takes into account specific age groups with particular needs for protection. This includes younger employees, who are for example relieved from having to lift heavy loads or older employees requiring a broader range of protective measures. Many of the measures entailed in an age-adapted work design, especially for older employees are difficult to justify from a work sciences perspective. Negotiating them in collective and plant agreements often reflects corporate ethical considerations and recognition of lifelong achievement. Aging-adapted work design on the other hand, is based to a large extent directly on the requirements outlined in the occupational health and safety acts, which dictate in many industrial countries that health risks be preventatively eliminated at the source of their creation.

Aging-adapted work design is nevertheless more comprehensive than the legal requirements for hazard-free work. It also contains trainings for maintaining employability as well as measures for transferring knowledge from generation to generation.

Before an enterprise can plan measures for age and aging-adapted work design, a thorough analysis needs to be made including:

- An age structure analysis in order to determine the current age distribution as well as the one expected in 5–10 years. This should be broken down according to production area and level of qualifications.
- A personal risk analysis to detect losses of specific know-how in the future as well as recruiting or cost risks.
- Risk assessments on the potential impact of strain resulting from work tasks as well as the physical and social environment.
- Training needs analyses to determine current and future qualifications bottlenecks and potentials.

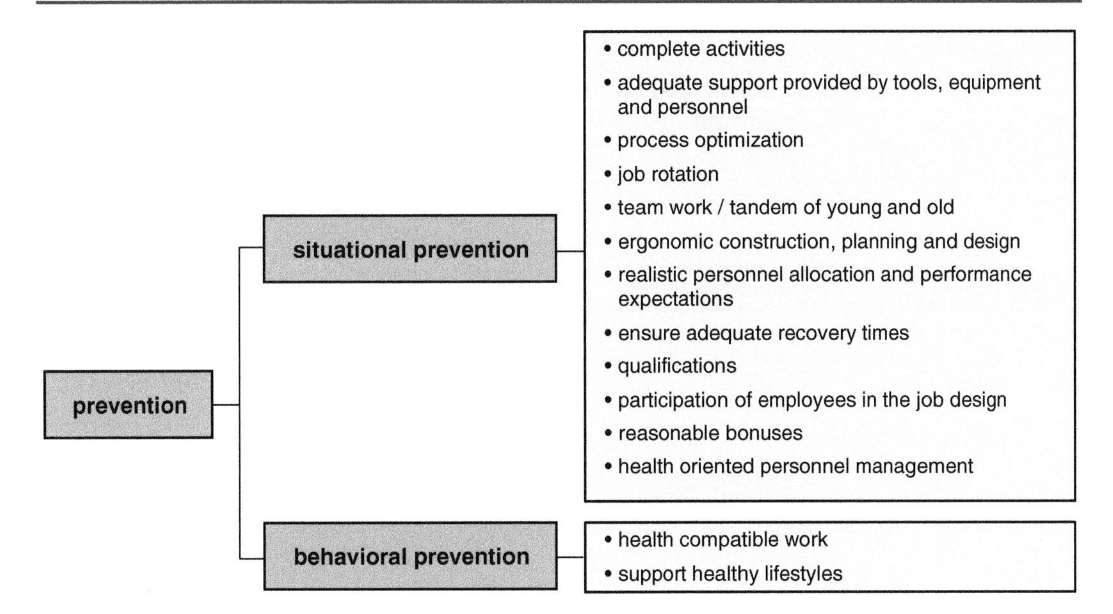

Fig. 7.15 Preventative measures for retaining performance and employability. © IFA 7.598_B

Following this comprehensive analysis both preventative and compensatory measures are planned.

The preventative measures mentioned in Fig. 7.15 are aimed at retaining as well as increasing the performance and employability of workers over the long term. A subsection of these measures is concerned with situational prevention. Situational prevention is a legal obligation for enterprises, dictated by the occupational health and safety act and workplace protection law. The goal here is to eliminate health hazards posed by how work environments and work tasks are designed. A second subsection of these measures is concerned with behavioral prevention. It is aimed at making work health compatible and supporting a healthy lifestyle for employees; measures in this area are voluntary for an enterprise. Studies have shown that behavioral prevention (e.g., back training, physiotherapy consultations and stress management courses) can be very effective in terms of employees' health. It is crucial that companies succeed in sustainable behavioral changes and that the situational prevention does not provide reasons to neglect behavioral prevention. Ideally, situational and behavioral prevention are interlinked.

In addition to preventative measures, enterprises also take into consideration compensatory measures. Compensation is necessary when older workers' performance and operational capabilities are already limited or are expected to be in the future. Compensatory measures can be broken down into three subsections (see Fig. 7.16). Employees can change work areas e.g., from the foundry to the assembly. The plant could reduce the performance requirements e.g., by increasing breaks or exempting workers from shift work. Lastly, exit options for retirement come into consideration. Within the sphere of an enterprise's influence are partial retirement and lifetime work accounts. Further options for flexibly exiting into retirement are political in nature and reserved for legislature.

This concludes our discussion of the work design and organization factors pertinent for factory planners. Beyond these, the immediate spatial work environment is critical to a reliable and healthy work performance; the following chapter addresses this in view of spatial planning.

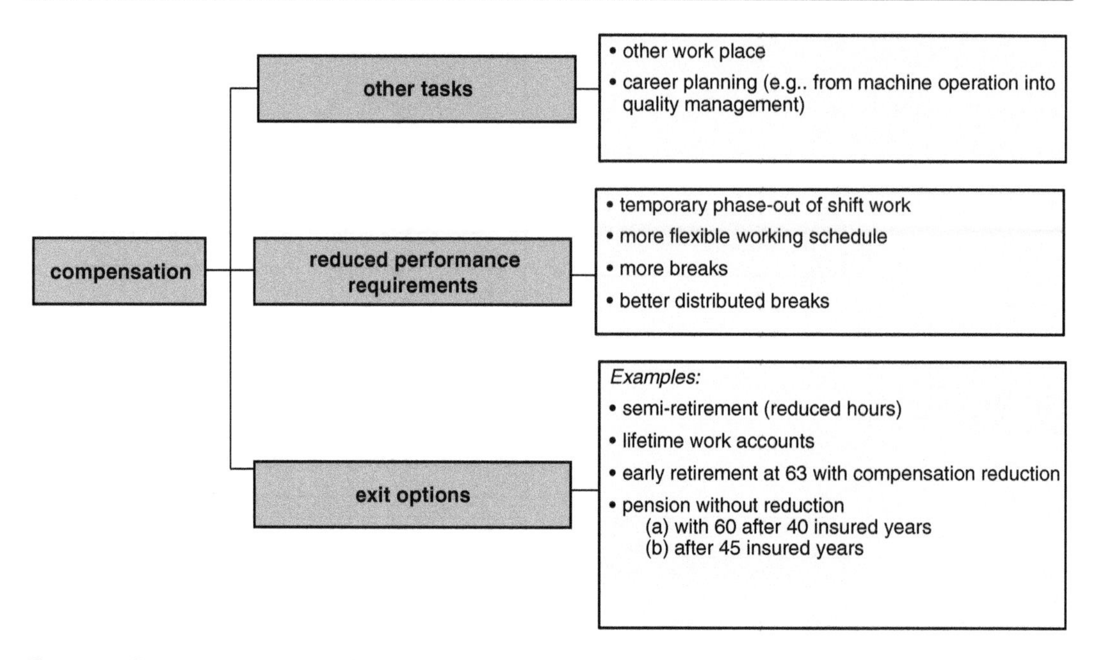

Fig. 7.16 Compensatory measures for older personnel. © IFA 7.599_B

7.9 Summary

For the economic success of a company the employees play a central role. For this they first must possess the necessary capacity to act, which consists of the technical expertise, methodological competence, the individual skills and social skills. It is the responsibility of the work structure design to incorporate the employees in the company organization which leads with increasing task complexity to group work. The permanent commitment of employees in the company requires their motivation which is depending on whether the tangible and intangible rewards are a personal value for them. One of the main motivating factors is the working hours and remuneration design; for which a vast variety of forms exist.

Bibliography

[Ant94] Antoni, C.H.: Gruppenarbeit - mehr als ein Konzept. Darstellung und Vergleich unterschiedlicher Formen der Gruppenarbeit (Group work - more than a concept. Presentation and comparison of different forms of group work). In: Antoni, C.H. (Hrsg.) Gruppenarbeit in Unternehmen. Konzepte, Erfahrungen, Perspektiven (Group work in companies. Concepts, experiences, perspectives), S. 19–48. Beltz, Psychologie Verlags Union, Weinheim (1994)

[App00] Appelbaum, E. u.a.: Manufacturing Advantage. Why high-performance work systems pay off. Cornell University Press, Ithaca (2000)

[Bah01] Bahnmüller, R.: Stabilität und Wandel der Entlohnungsformen. Entgeltsysteme und Entgeltpolitik in der Metallindustrie, in der Textil- und Bekleidungsindustrie und im Bankgewerbe (Stability and changes in forms of payment. Pay systems and pay policies in the metal industry, the textile and clothing industry and the banking industry). Rainer Hampp Verlag, München (2001)

[Bai04] Bailey, Th.R., et al.: Working Knowledge: Work-Based Learning and Education Reform. Routledge Farmer, London (2004)

[Ben97] Bender, G.: Lohnarbeit zwischen Autonomie und Zwang. Neue Entlohnungsformen als Element veränderter Leistungspolitik (Wage labor between autonomy and coercion. New forms of remuneration as an element of change in performance policy). Campus Verlag, Frankfurt am Main (1997)

[Ber00] Bergmann, B.: Arbeitsimmanente Kompetenzentwicklung (Intrinsic work skills development). In: Bergmann, B. u.a. (Hrsg.): Kompetenzentwicklung und Berufsarbeit

(Skills Development and Professional Work), pp. 11–39. Waxmann Verlag, Münster (2000)

[Ber08] Berger, L., Berger, D.: The Compensation Handbook, 5th edn. McGraw-Hill, New York (2008)

[Böh04] Böhm, S., Herrmann, C., Trinczek, R.: Herausforderung Vertrauensarbeitszeit. Zur Kultur und Praxis eines neuen Arbeitszeitmodells (Challenge working time. Culture and practice of a new working time model). Edition Sigma, Berlin (2004)

[Com06] Combs, J., et al.: How much do high-performance work practices matter? A meta-analysis of their effects on organizational performance. Pers. Psychol. **59**, 501–528 (2006)

[Deh01] Dehnbostel, P.: Perspektiven für das Lernen in der Arbeit. In: Arbeitsgemeinschaft Qualifikations-Entwicklungsmanagement Berlin (Hrsg.): Kompetenzentwicklung 2001: Tätigsein - Lernen – (Innovation Prospects for learning at work. In: Association for qualification development management Berlin (ed.) Competence Development 2001: Being Active–Learning—Innovation), pp. 53–94. Waxmann, Münster (2001)

[Doe97] Doerken, W.: Arbeitsbewertung (work assessment). In: Luczak, H., Volpert, W. (Hrsg.) Handbuch Arbeitswissenschaft (Handbook Ergonomics), S. 994–998. Schäffer-Poeschel, Stuttgart (1997)

[Ege09] Egler, R.: Bevölkerungsentwicklung in Deutschland (Population Development in Germany). Statistisches Bundesamt (2009)

[Fri99] Frieling, E., Sonntag, K.: Lehrbuch Arbeitspsychologie (Textbook Work Psychology), 2nd edn. Verlag Hans Huber, Bern (1999)

[For04] Forth, J., Millward, N.: High involvement management and pay in Britain. Ind. Relat. **43**(1), 98–119 (2004)

[Ger99] Gerst, D.: Gestaltungskonzepte für die manuelle Montage. Selbstorganisierte versus standardisierte Gruppenarbeit? In: Angewandte Arbeitswissenschaft (Design concepts for manual assembly. Self-organized group work versus standardized group work? In: Applied Ergonomics), No. 162, pp. 37–54 (1999)

[Ger04] Gerst, D.: Arbeitsorganisation und Qualifizierung. In: Wiendahl, H.-P., Gerst, D., Keunecke, L. (Hrsg.) Variantenbeherrschung in der Montage. Konzept und Praxis der flexiblen Produktionsendstufe (Work organization and qualification. In: Wiendahl, H.-P., Gerst, D., Keunecke, L. (eds.) Variant Control in the Assembly. Concept and Practice of the Flexible Production Stage), S. 95–118. Springer, Berlin (2004)

[Ger06] Gerst, D.: Von der direkten Kontrolle zur indirekten Steuerung. Eine empirische Untersuchung der Arbeitsfolgen teilautonomer Gruppenarbeit (From the direct control to indirect control. An empirical study of the effects of semi-autonomous group work). Rainer Hampp Verlag, München (2006)

[Gra61] Graf, O.: Arbeitsablauf und Arbeitsrhythmus. In: Lehmann, G. (Hrsg.) Handbuch der gesamten Arbeitsmedizin. Bd. 1: Arbeitsphysiologie (Workflow and work-rate. In: Lehmann, G. (ed) Handbook of the Total Occupational Medicine, vol. 1, Applied Physiology), pp. 789–824. Urban und Schwarzenberg, Berlin (1961)

[Hac98] Hacker, W.: Allgemeine Arbeitspsychologie: Psychische Regulation von Arbeitstätigkeiten (General Industrial Psychology: Mental regulation of work activities). Huber, Bern (1998)

[Int04] Internationales Arbeitsamt: 92. Tagung (2004), Bericht IV (2B) (International Labour Office: 92nd Session (2004), Report IV (2B). Genf (2004)

[Kie08] Kiepsch., H.-J. et al.: Mensch und Arbeitsplatz (Human and Work Place). BG-Information 523, Köln (2008)

[Kir05] Kirchler, E., Walenta, C.: Motivation. In: Kirchler, E. (ed.) Arbeits- und Organisationspsychologie (Industrial and Organizational Psychology), pp. 319–408. UTB, Wien (2005)

[Kna97] Knauth, P.: Nacht- und Schichtarbeit. In: Luczak, H., Volpert, W. (Hrsg.) Handbuch Arbeitswissenschaft (Night and shift work. In: Luczak, H., Volpert, W. (eds.) Industrial Engineering), pp. 938–946, Schäffer-Poeschel, Stuttgart (1997)

[Land01] Landau, K., Wimmer, R., Luczak, H., Mainzer, J., Peters, H., Winter, G.: Anforderungen an Montagearbeitsplätze (Requirements for assembly workstations). In: Luczak, K. (Hrsg.) Ergonomie und Organisation in der Montage (Ergonomics and Organization in Assembly), S. 40. Hanser, Wien (2001)

[Luc98] Luczak, H.: Arbeitswissenschaft. Springer-Lehrbuch (Work Science. Springer text book), 2nd edn. Springer, Berlin (1998)

[Mar94] Martin, H.: Grundlagen der menschengerechten Arbeitsgestaltung. Handbuch für die betriebliche Praxis (Basics of Humane Job Design. Manual for operational practice). Bund Verlag, Köln (1994)

[Mat10] Mathis, R.L., Jackson, J.H.: Human Resource Management, 13th edn. South-Western Cengage Learning, Mason (2010)

[Mon12] Mondy, R.W.: Human Resource Management, 12th edn. Prentice Hall, New Jersey (2012)

[Mum06] Mumford, E.: The story of socio-technical design: reflections on its successes, failures and potential. Inf. Syst. J. **16**(4), 317–342 (2006)

[Nac97] Nachreiner, F., Grzech-Šukalo, H.: Flexible Formen der Arbeit (Flexible forms of work). In: Luczak, H., Volpert, W. (Hrsg.) Handbuch Arbeitswissenschaft (Handbook Work Science), pp. 952–957. Schäffer-Poeschel, Stuttgart (1997)

[Nom95] Nomura, M., Jürgens, U.: Binnenstrukturen des japanischen Produktionserfolges: Arbeitsbeziehungen und Leistungsregulierung in zwei japanischen Automobilunternehmen (Internal Structures of the Japanese Manufacturing Success: Labor Relations and Regulatory Performance in Two Japanese Automotive Companies). Edition Sigma, Berlin (1995)

[Nyh13] Nyhuis, P., Deuse, J., Rehwald, J. (eds.) Wandlungsfähige Produktion. Heute für Morgen gestalten (Changeable Production designed Today for Tomorow, PZH Verlag, Hannover (2013)

[Pas82] Pasmore, W., et al.: Sociotechnical systems: a north American reflection on empirical studies of the seventies. Hum. Relat. **35**(12), 1179–1204 (1982)

[Pla04] Plaut, W.-D., Sperling, H.-J.: Qualifikationsgerechte Entlohnung. Das Konzept der Lernzeit zur Grundlohneinstufung (Reward according to qualification. The concept of the learning time for the basic wage classification). In: Gergs, H.-J. (Hrsg.) Qualifizierung für Beschäftigte in der Produktion (Qualification for Production Employees), Eschborn (2004)

[Port68] Porter, L.W., Lawler, E.E.: Managerial Attitudes and Performance. Ill, Homewood (1968)

[Rae08] Raelin, J.A.: Work-Based Learning: Bridging Knowledge and Action in the Workplace. Jossey Bass, San Francisco (2008) (New and rev. edn)

[Ric58] Rice, A.: Productivity and Social Organisation. The Ahmedabad experiment. Routledge, London (1958)

[Rob12] Robins, S.R.: Organizational Behavior. Concepts—Controversies Applications, 15th edn. Prentice Hall, Englewood Cliffs (2012)

[Sal13] Salazar, Y., Regber, H., Große-Heitmeyer, V., Goßmann, D.: Personal – Der Mensch als Wandlungsbefähiger (Staff – Man as changeability enabler). In: Nyhuis, P., Deuse, J., Rehwald, J. (eds.) Wandlungsfähige Produktion. Heute für Morgan gestalten, pp. 152–185. PZH Verlag, Hannover (2013)

[Sch00] Schanz, G.: Personalwirtschaftslehre (Human Resource Management), 3rd edn. Vahlen, München (2000)

[Sch93] Schmidtke, H. (ed.): Ergonomie (Ergonomics). Hanser, München (1993)

[Shi04] Shimizu, K.: Reorienting Kaizen Activities at Toyota: Kaizen, Production Efficiency, and Humanization of Work. Okayama Econ. Rev. **36**(3), 255–278 (2004)

[Son01a] Sonntag, K., Schaper, N.: Wissensorientierte Verfahren der Personalentwicklung (Knowledge oriented process of human resources development). In: Schuler, H. (ed.) Lehrbuch der Personalpsychologie (Textbook Personnel Psychology), pp. 242–263. Hogrefe, Göttingen (2001)

[Son00] Sonntag, K., et al.: Leitfaden zur Implementation arbeitsintegrierter Lernumgebungen. Materialien zur Beruflichen Bildung (Guide to Implement Work Integrated Learning Environments. Materials for Vocational Training). Bertelsmann, Bielefeld (2000)

[Son01b] Sonntag, K., Stegmaier, R.: Verhaltensorientierte Verfahren der Personalentwicklung. In: Schuler, H. (Hrsg.) Lehrbuch der Personalpsychologie (Behaviour-Based Method of Personal Development. In: Schuler, H. (ed.) Textbook of Human Psychology), pp. 266–287. Hogrefe, Göttingen (2001)

[Spa04] Spath, D.: Der Mensch im Arbeitssystem. Manuskript zur Vorlesung Arbeitswissenschaft 1 (The Human in the Work System. Manuscript for the lecture work Science 1), Stuttgart (2004)

[Sta99] Staudt, E., Kriegesmann, B.: Weiterbildung: Ein Mythos zerbricht. Der Widerspruch zwischen überzogenen Erwartungen und Mißerfolgen der Weiterbildung. In: Arbeitsgemeinschaft Qualifikations-Entwicklungs-Management Berlin (Hrsg.) Kompetenzentwicklung '99. Aspekte einer neuen Lernkultur. Argumente, Erfahrungen, Konsequenzen. (Training: A myth shattered. The contradiction between excessive expectations and failures of training. In: Association for qualification Development Management Berlin (ed.) Competence Development '99. Aspects of a new culture of learning. Arguments, experiences, consequences, pp. 17–59. Waxmann, Münster (1999)

[Sta02] Staudt, E., Kriegesmann, B.: Zusammenhang von Kompetenz, Kompetenzentwicklung und Innovation. Objekt, Maßnahmen und Bewertungsansätze - Ein Überblick (Connection of competence, competence development and innovation. Object, action and evaluation methods - an overview. In: Staudt, E., et al. (Hrsg.) Kompetenzentwicklung und Innovation. Die Rolle der Kompetenz bei Organisations-, Unternehmens-, und Regionalentwicklung (In: Staudt, E., et al. (eds.) Competence Development and Innovation. The Role of Expertise in Organizational, Business, and Regional Development), pp. 15–70. Waxmann, Munster (2002)

[Tho03] Thommen, J.-P., Achleitner, A.-K.: Allgemeine Betriebswirtschaftslehre. Umfassende Einführung aus managementorientierter Sicht (General Business Administration. Comprehensive Introduction from Management-Oriented Perspective). Gabler, Wiesbaden (2003)

[Uli01] Ulich, E.: Arbeitspsychologie (Work Psychology), 5th edn. Schäffer-Pöschel, Stuttgart (2001)

[Wäc97] Wächter, H.: Grundlagen und Bestimmungsfaktoren des Arbeitsentgelts. In:

Luczak, H.; Volpert, W. (Hrsg.) Handbuch Arbeitswissenschaft (Foundations and determinants of earnings. In: Luczak, H., Volpert, W. (eds.) Industrial Engineering), pp. 986–989. Schäffer-Poeschel, Stuttgart (1997)

[Wel91] Weltz, F.: Der Traum von der absoluten Ordnung und die doppelte Wirklichkeit der Unternehmen. In: Hildebrandt, E. (Hrsg.) Betriebliche Sozialverfassung unter Veränderungsdruck (The dream of the absolute order and the dual reality of the business. In: Hildebrandt, E. (ed.) Occupational Social Constitution Under Pressure to Change), pp. 85–97. Edition Sigma, Berlin (1991)

[Whi03] White, M., et al.: High performance management practices, working hours and work-life balance. Br. J. Ind. Relat. **41**, 175–196 (2003)

[Wie05] Wiendahl, H.-H., Wiendahl, H.-P., Von Cieminski, G.: Stumbling blocks of PPC: towards the holistic configuration of PPC systems. Prod. Plan. Control **16**(7), 32–35 (2005)

[Zin97] Zink, K.J.: Soziotechnische Ansätze. In: Luczak, H., Volpert, W. (Hrsg.) Handbuch Arbeitswissenschaft (Socio-technical approaches. In: Luczak, H., Volpert, W. (eds.) Industrial Engineering), pp. 74–77. Schäffer-Poeschel, Stuttgart (1997)

8 Spatial Workplace Design

Workplaces, work related equipments, materials and the work flow should all be designed to ensure safe and healthy work within an aesthetically stimulating environment.

From a spatial perspective this means providing a changeable yet orderly structure within the immediate visual field of a work area. Workplaces have to be ergonomically (from Greek argon = work; nomos = law, rule) designed and provide optimal working conditions. The space and connecting pathways for each of the workplaces should be laid out keeping in mind accessibility requirements for the physically challenged workers, also.

Depending on the requirements of the processes being conducted, interiors have to be designed to include detailed solutions for the floor, walls and roof. Moreover, they need to be hygienic, easy to clean, dust free and/or where necessary sterile. An overall color concept should unite all of the components of a workplace into an integrated color scheme. Standardized safety colors, media indicators, psychological effects of color as well as aesthetic requirements should be combined into a convincing harmony of colors. Special precautionary measures against injury, fire and damage protect employees from occupational hazards. Thus within the framework of occupational health and safety, structural and technological measures should be planned so as to prevent all possible hazards (including fires and explosions) from operational facilities, processes, hazardous materials, supply, disposal systems and media. Figure 8.1 depicts the fields and elements of a workplace that result from these requirements that can be designed. In the following sections, we will take a closer look at these and their significance for changeability.

8.1 Ergonomics

The goal of designing an ergonomic workplace is to optimize working conditions with regards to the manufacturing tasks and environmental conditions to human characteristics and abilities. In doing so, it should to be ensured that the required output satisfies the quality requirements while at the same time keeping the costs for the work system to a minimum. Furthermore, it must be assured that the workload and demands on the human are bearable over a long-term without endangering their health and safety. When designing workplaces proven knowledge gained from the field (and summarized here in the upper part of Fig. 8.2) needs to be considered, while environmental conditions need to be designed through the spatial planning keeping in mind the factors mentioned in the lower part of Fig. 8.2 [REf91].

Anthropometric (Greek anthropos = human) design is concerned with organizing the workpieces, tools and control panels in accordance with body measurements. Here, the ranges of movements, reach distances and visual fields resulting from the skeletal and muscular framework of the human body (including where

H.-P. Wiendahl et al., *Handbook Factory Planning and Design*,
DOI 10.1007/978-3-662-46391-8_8, © Springer-Verlag Berlin Heidelberg 2015

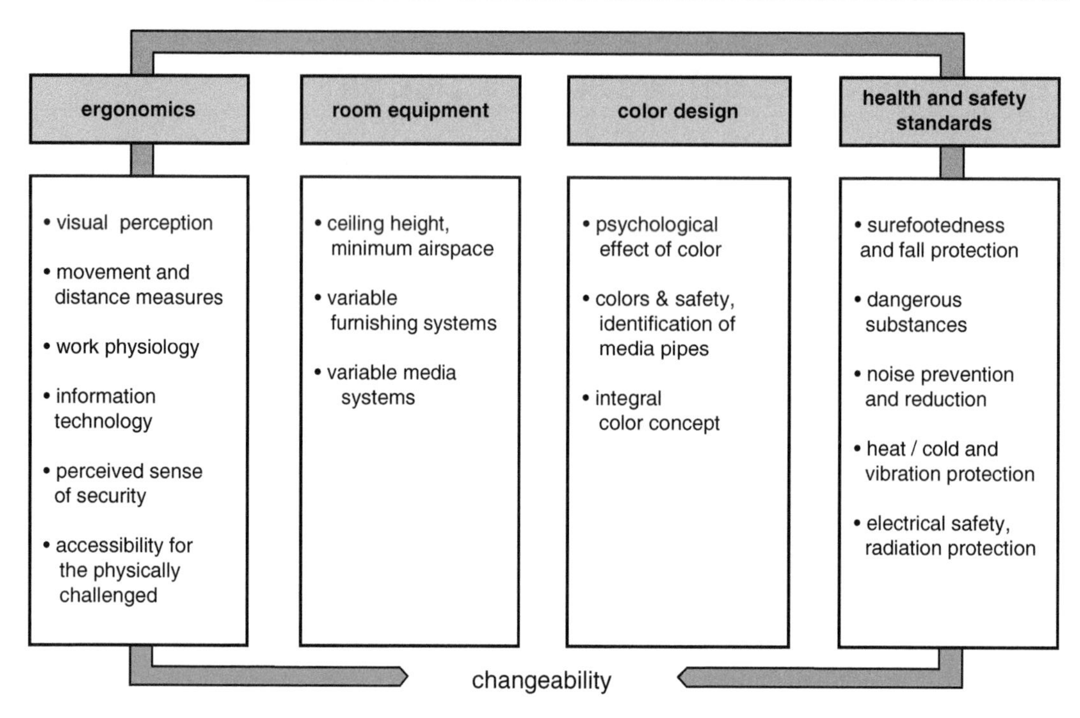

Fig. 8.1 Design fields and elements of a workplace

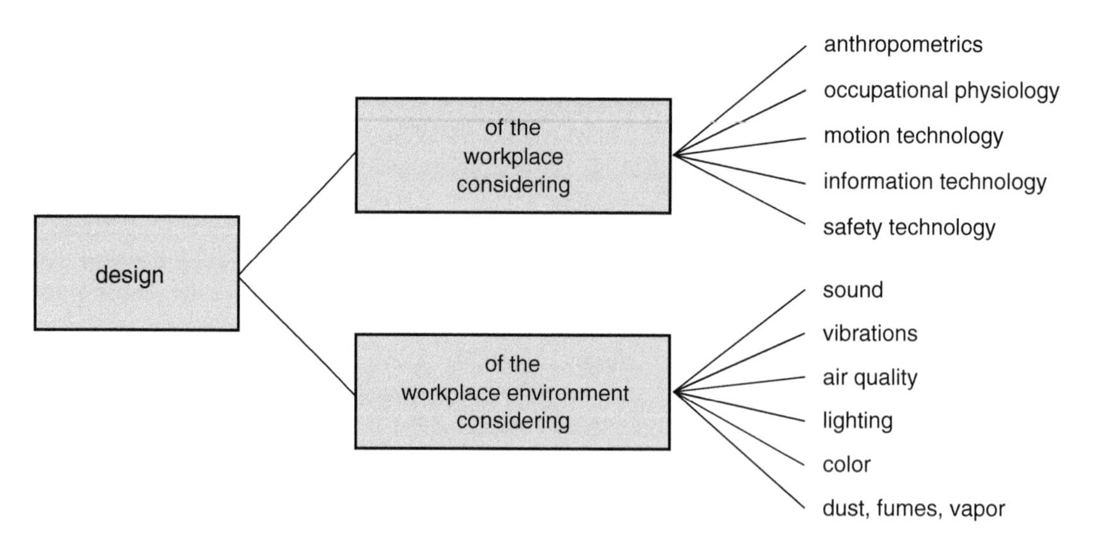

Fig. 8.2 Workplace design factors (acc. to REFA)

applicable protective clothing) need to be taken into account. All of the operations should be executed effortlessly and in harmony with the physical possibilities. Mobile assembly lines, workbenches or office work should promote natural visual and muscular movements. An overview of this can be found in [Rüs06, Lan06, Sal12, Til15].

Figure 8.3 depicts two different aspects, whereby the left side focuses on the maximum and optimal field of vision. These characteristics are particularly important with work that involves

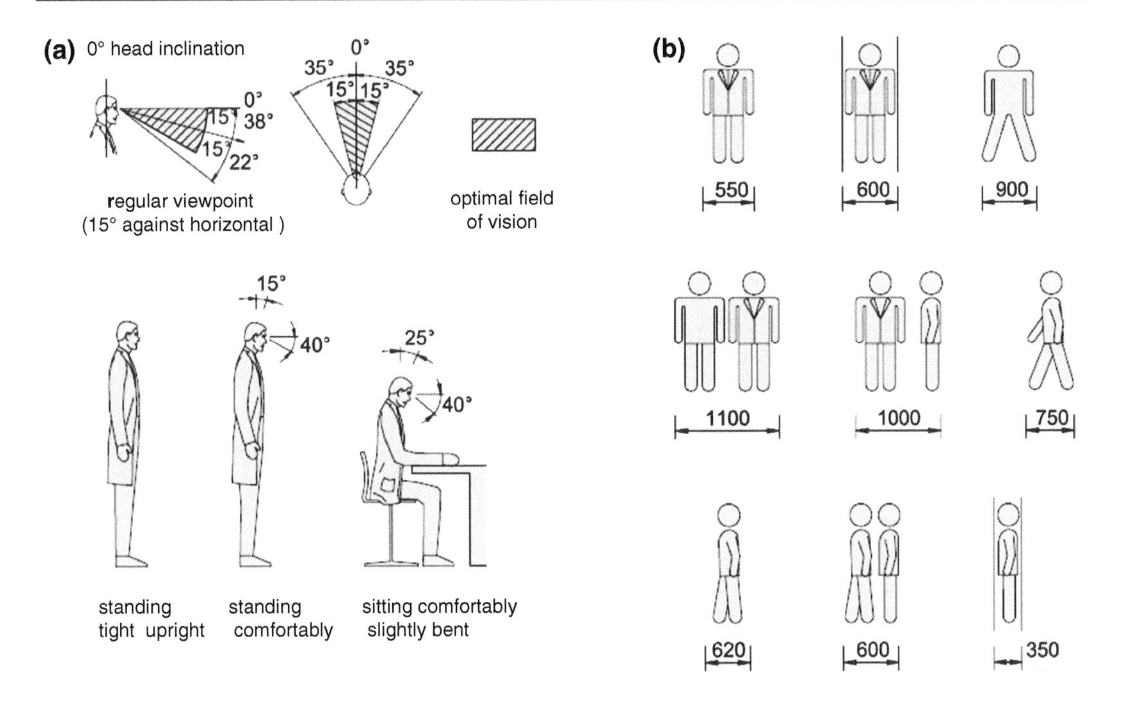

Fig. 8.3 Ergonomic aspects of a workplace. **a** optimal field of vision, **b** distance and movement measures (in mm)

short cycles and a constant posture e.g., with flow assemblies. Next to this, Fig. 8.3 right illustrates a number of dimensions related to movement and distances for humans. These need to be taken into consideration while deciding the dimensions and areas for standing and moving at workplaces as well as traffic routes. Accessibility in the work environment should be a matter of course—this means both accessible workplace's as well as the possibility to make adjustments where needed.

Figure 8.4 depicts an ergonomically designed workplace which allows work to be completed in both standing and sitting positions. The shell surrounding the worker indicates his workspace. Different body sizes are accommodated by adjusting seat heights and footrests. The recommended dimensions for the space of movement and the assembly table can be taken from Fig. 6.34.

Occupational physiological design focuses primarily on stress and strain caused by muscular work. In doing so, particular attention is paid to avoiding unilateral muscle strain and bent or stooped postures. The strain has to be kept under the permissible limits. This limit refers to the level of human performance that can be sustained daily over the long-term without significant work fatigue and damage to the worker's health.

Movement technique design pursues three basic principles: (a) simplification of movements, (b) consolidation of movements and (c) partial mechanization and automation. The simplification of movements is based on five elements of movement including joining, grabbing, executing, reaching and retrieving. The aim here is to identify a sequence of movements which require the least amount of time and in view of diligence and accuracy avoids unnecessary ones. This is achieved by minimizing the distance of the movements and organizing materials accordingly. Movements can be consolidated by conducting similar or different movements at the same time with both hands. Further improvements can also be attained by eliminating non-productive i.e. non-value-adding activities. Partial mechanization and automation can be

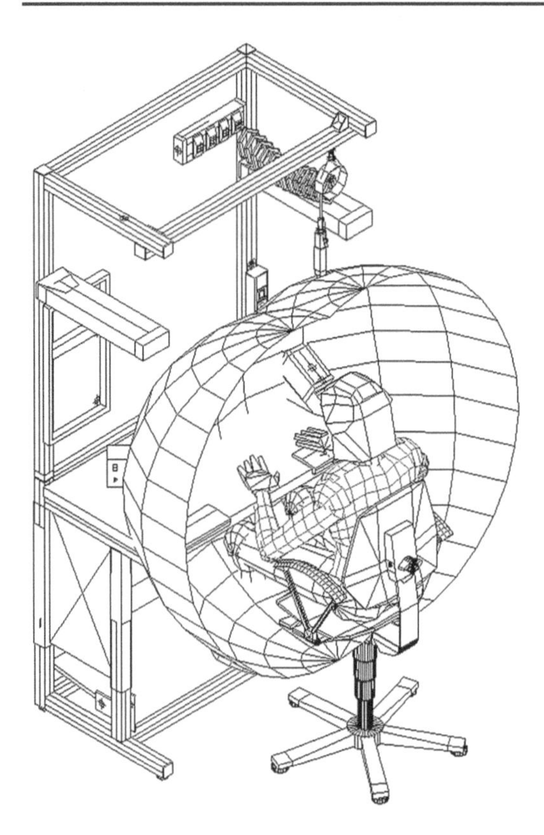

Fig. 8.4 Standing-sitting workplace (Bosch Rexroth)

encouraged after motions have been consolidated as much as possible, since the investment costs are disproportionate to the additional savings in time. An example was shown in Fig. 6.35.

Information technology design is concerned with the flow of information between humans, materials, work objects and the work environment, predominantly through visual and acoustical signals. Here, reliability and clarity are decisive guidelines for organizing and designing process status monitors, machine control panels and display masks.

Safety technology design serves to prevent accidents and occupational illnesses. DIN 31000 distinguishes between three types of safety technology: direct, indirect and warnings [DIN79]. Through its design, *direct safety technology* inherently avoids danger; other types should only be implemented when direct safety measures are not possible. *Indirect safety technology* aims at integrating protective measures in hazardous areas where there is a danger of injury. In this case safety distances need to be complied with in order to ensure that the hazardous areas are inaccessible. Since hazardous areas can never be completely eliminated, safety symbols or warning systems need to be implemented to identify them and if needed suitable safety equipment and means of protection have to be provided. An overview of safety technology measures implemented in designing workplaces can be found in [Leh05, Rüs07, MCol07, Col01].

In order to facilitate designing workplaces from an ergonomic and economic standpoint various computer-aided methods were developed especially for manual assemblies. An integrated work-place model comprising of humans to scale their probable movement patterns, furnishings, equipment for supplying parts and work-pieces can help in identifying optimal settings and depicting them three-dimensionally with algorithmic support. As an example, Fig. 8.5 depicts a visualization of an assembly workplace generated with the aid of a simulation program and integrated into the process flow of a large plant.

8.2 Room Interiors

Each workplace should be planned to include a sufficient ceiling height and a minimum of air volume for all permanent employees. In Germany, in accordance with §23 of the German Workplace Regulations, depending on the floor space of the work area, a ceiling height range between a minimum of 2.5 m (8.2 ft) up to or beyond 3.25 m (10.7 ft) (see Fig. 8.6). Ceiling heights for flexible workplaces with variable floor spaces should be developed with an eye to the future requirements of the space. The same applies to the minimum air volume per employee. According to guidelines in Germany, flexible, transformable workplaces should provide at least 15 m^3 (530 ft^3) of air volume per person.

According to the ASHRAE Standards [US07] in the United States of America a minimum of 5 cubic feet of external air per minute (CFM) and per person (141.5 l/min/person) should be

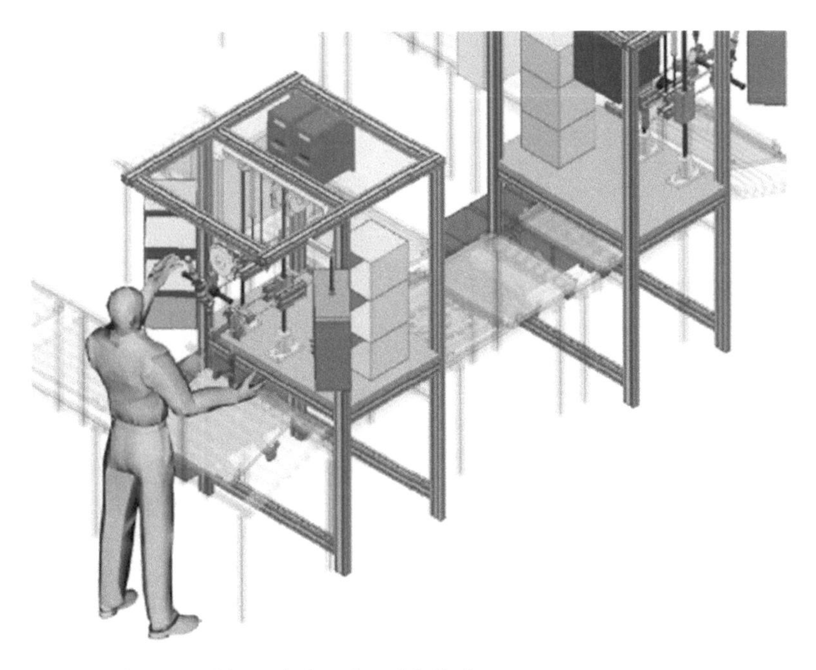

Fig. 8.5 3D-simulation of an assembly workplace (acc. Modine)

(a)

base area for work space [m²]	head room [m]
8 to 50	≥ 2.50
> 50 to 100	≥ 2.75 1)
> 100 to 2000	≥ 3.00
> 2000	≥ 3.25

1) for offices as well as workrooms with predominantly easy or sitting activity reduction by 0.25 m is possible

(b)

activity type	minimum air volume per person [m²]	examples
predominant sitting activity	12	office work assembly of smaller parts
predominant non sitting activity	15	mechanical metal processing, transportation of light loads
predominant non sitting activity	18	manual transportation of heavy loads, forging of large work pieces

Fig. 8.6 Minimum ceiling heights and air volume for workplaces (acc. §23 German Workplace Regulations). **a** Minimum headroom for workrooms, **b** Minimum air volume for employees permanently present

ensured. This value should be increased to 15 CFM (424.5 l/min) for reception areas and 20 CFM (566 l/min) for offices with moderate working conditions. The standard value for designated smoking areas with local mechanical ventilations is 60 CFM (1.698 l/min).

Within the context of planning a media routing system, the design at the workplace should allow variable air feeds and outlets for varying workplace arrangements. In the semi-conductor industry, special media needs to be allocated for clean-room technology. With regards to furniture, modular and variable storage and desk systems are preferable. Furniture should be easy to move, with durable, easy to clean surfaces and include integrated installation space for electrical wires, data and voice cabling.

8.3 Color Design

"People need color to live. Color is as necessary an element as water and fire" (Fernand Léger). By generating a stimulus in the eye, colors directly influence the cerebral cortex, impacting the people working in various ways both physiologically and psychologically. The color design should thus integrate all of the components such as materials and resources, buildings, media and furnishings into a spatially coherent master plan. Positive effects on employees include increased worker satisfaction and performance, improved working atmosphere and safety. It also encourages order and hygiene. A comprehensive color concept combines the psychological impact of colors and easy identification of media guides into an aesthetic and functionally harmonized color scheme.

8.3.1 Psychological Impact of Color

In English, a variety of expressions for emotions have become common place. We can temporarily "see red", "be blue" or "be green with envy". Moreover, we can become "as white as a sheet" when we are ill and "get a bit of color" when we start to feel better. Additional color phenomena are summarized in [Gek07]. Since humans tend to differ emotionally they prefer certain colors for clothing or furnishings which serve to either emphasize a personal note or to increase our well-being through their lively or calming tone (depending on what the affected person needs). This too can be applied in the work place. The basic colors have been ascribed characteristic properties by color researchers:

Red The color of fire and blood, it expresses life and energy. Red is inseparably bound to passion, heat, anger and war. It is considered a stimulating color.

Blue The color of the deep sea, intuitively we connect it with infiniteness and the vastness of the sky. Blue speaks to the intellect, while red impacts the emotional sphere. It is the symbol of truth, corresponding to a calm consideration and never to a rushed decision.

Yellow The brightest of the basic colors, the color of the sun and an expression of brilliance, radiating and liveliness.

Orange A mixture of red and yellow, it thus combines the strength of red with the luminous, bright sheen of yellow.

Green A mixture of blue and yellow; as the color of nature it stands for serenity, resurrection, peace. Symbolically, it is the color of hope.

Purple A mixture of red and blue, to which concepts such as pomp, splendor, and royal grandeur are linked; similar to green though it has a calming and soothing influence.

Spaces and behaviors are related to one another. The color scheme of a workspace should be planned depending upon the type of operations. With monotonous work, stimulating color elements are recommended (columns, doors); however, such colors should not be applied to large surfaces (walls, ceilings). If the workspace is large, it can be spatially subdivided with special color elements. When the work executed in a space requires a great deal of concentration, the color scheme should be conservative in order to avoid unnecessary distraction. In this case, walls, ceilings and other structural elements should be painted with colors that are as light as possible and/or mildly toned.

In order to create color contrasts, the color schemes for large surfaces (walls, furniture, etc.)

should be distinguished from small surfaces (eye catchers for switches, grips, lever, etc.). For larger surfaces colors that have a similar degree of reflection should be selected. Moreover, with larger surfaces bright colors should not be used since they unilaterally strain the retina, which in turn generates after-images. Orienting and detecting workpieces is facilitated by creating a color contrast between the workpiece and the immediate surroundings (workbench or machine). Here too, differences in brightness should be avoided.

The architectural "appearance"—its sensory, visual impression—is primarily determined by the color of the materials and the structure. The choice of construction materials is thus directly related to the appearance of the completed building. From the outside, either the actual color of the materials (exposed concrete, metal, wood, natural stone, synthetic materials) or a colored coating determines the character of the building. Internally, the choice of materials plays a significant role with regards to the levels of comfort. In addition to the color of the materials, the material properties and surface texture (visual and haptic perception) are key criteria for the "indoor climate". Thus for example, glossy enameled walls have a different character than those with a matt finish.

8.3.2 Safety Colors and Identification of Media Lines

Nowadays, most countries regulate specific colors for designating defined dangers. Pre-assigning colors to specific information promotes the development of automatic protective responses. Similar to the coding of traffic lights, DIN 4818 in Germany regulates mandatory colors from the RAL color system for specific dangers.[1] The color RAL 1004 (golden yellow) thus signals caution and indicates possible dangers with conveyor belts, traffic routes and stairs. Fire-fighting equipment and systems should be identified with RAL 3001 (signal red). Similar to street traffic signals, red stands for forbidden, stop, and danger. According to DIN 4844, RAL 5010 (gentian blue) dictates additional safety regulations for e.g., preventing noise, while RAL 6001 (emerald green) signals safety and first aid. Emerald green is used in pictograms for escape paths, emergency exit doors as well as rooms and devices for first aid.

In addition to the color of fire-fighting equipment and systems, the identification of pipelines is defined by the German DIN 2403. Based on these work regulations, many enterprises have developed their own coding for process media and building services. It is also practical to mark the direction of the pipe flow using for example, arrowheads.

In the US the "Safety color code for marking physical hazards" has been published by the Occupational Safety and Health Administration under standard 1910.144.

8.3.3 Holistic Color Schemes

Industrial premises as a whole including everything from general work areas right down to individual workplaces should present an organized visual structure. Color schemes can be used to clarify and emphasize spatial forms, the arrangement of operating systems and lighting. Helpful organization principles include spatial axis, reference planes, groups or geometric patterns. Processing sequences can be clarified through lines or arrows. In an extension of the actual industrial premises, [Ben07] also expressly includes the environment of the neighboring industrial buildings in a comprehensive color scheme.

In addition to the general spatial effect that surfaces have as boundaries of building components, colors also play a significant role in a functional and aesthetic sense when it comes to machines and interior furnishings. The psychological impacts of color, safety colors, media indicators and prioritized corporate colors should

[1] First established in 1927, RAL is a German committee that regulates delivery conditions and quality assurance. Among other things it determines these color codes.

be combined into a functional and aesthetic whole. With the aim of consciously developing a recognizable corporate identity, it often proves practical to differentiate between accented "primary colors" and more subtle "secondary colors". Moreover, within the frame of a synergetic factory planning, it is advisable to comprehensively simulate the colors for the processes, building, media and furnishings using a 3D model (see Chap. 15).

Traditional studies use color collages to develop color harmonies. Color studies using a 3D model however are more advantageous in that one is able to assess the spatial effect by selecting any point of view; "primary colors" and "secondary colors" can be realistically distributed and color variations can be comparatively evaluated. Moreover, areas which are particularly important, such as the furnishings of the foyer, can also be photo-realistically simulated by defining material and light characteristics.

8.4 Occupational Health and Safety Standards

8.4.1 Overview

When it comes to planning and operating a factory there are a large number of laws, rules and regulations that play a critical role. Both the factory planner and architect need to be familiar with these; the architect in particular needs to be more aware of the details. Two issues are especially important because they directly impact their activities. First, legislators worldwide have developed occupational health and safety standards for workers along with detailed guidelines and regulations for finishing, furnishing, fixtures, fittings and operations of industrial premises. Second, statuary co-determination (as prevalent in countries like Germany) ensures wide reaching participation of individual workers and the workers council in social, personal and economic issues. Increasingly, however, environmental legal aspects have also been influencing company operations.

8.4.2 Workplace Regulations

Figure 8.7 [Ave76] provides an overview of the basic laws and regulations concerning workplaces in Germany. We can see that in addition to the industrial code there are also issues that overlap with the civil code, commercial code and the work constitution act. Moreover, it is evident that the industrial safety code, the safety of machinery law, accident prevention policies, building regulations and specific collectively agreed upon work regulations are correlated and inter-dependent. In international projects, depending on the countries involved, specific local factors or a mix with existing traditions and expectations in the originating country are to be considered.

In addition to the Workplace Guidelines (ASR) [ASR06], the German Workplace Regulation (ArbStättV) [Arb04] is the most recent and comprehensive set of measures protecting the interests of the workers in Germany. These regulations have successfully combined a number of individual guidelines regulating workplaces, updating and supplementing them with the latest knowledge about work safety, protection, occupational medicine and hygiene based up the latest data from occupational sciences. Protection for non-smokers is the newest addition. The regulations apply to all workplaces regardless of whether they are industrial, commercial or service oriented [OSP05].

According to §3 of the German Workplace Regulations, the employer has to furnish and operate the workplace (including the circulation routes, warehouses, machinery, side rooms, social areas, washrooms and first aid areas that belong to it) according to these regulations. Moreover, employers are obliged to satisfy all applicable regulations that protect workers and prevent accidents (including technical safety, occupational medicine and hygienic standards) as well as to apply proven knowledge gained from research in occupational sciences.

The regulations for workers' protection and injury prevention relevant to our discussion are mostly object related. In addition to the equipment safety laws, key regulations address issues

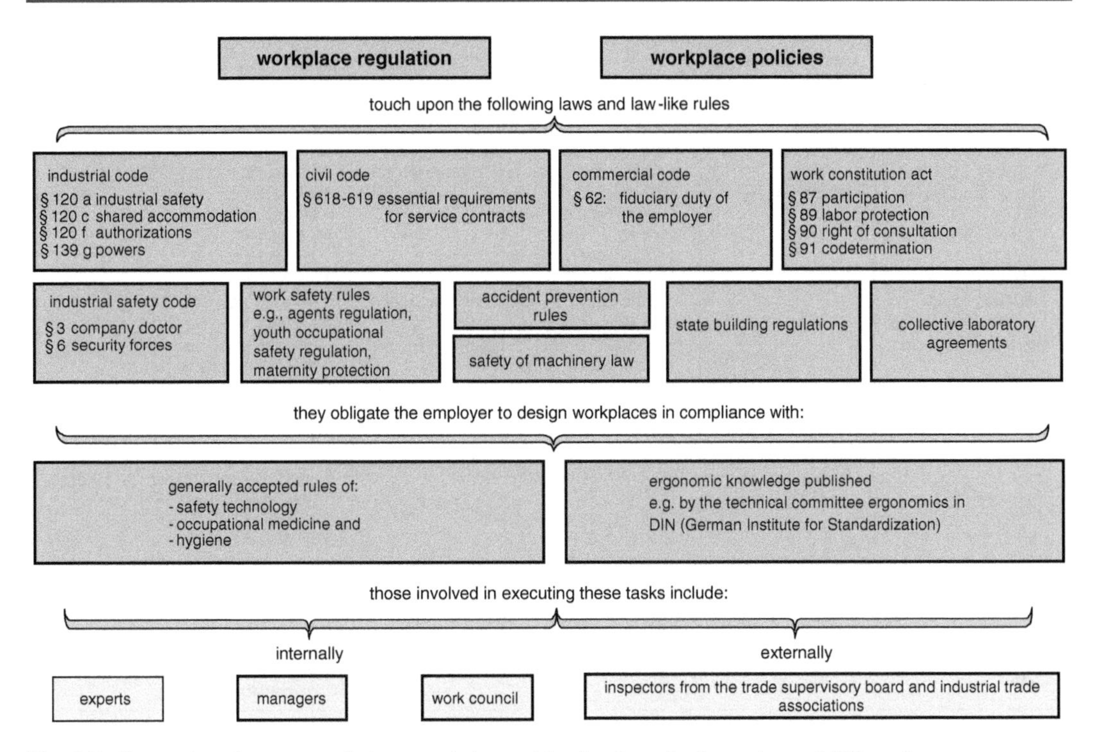

Fig. 8.7 Connections between workplace regulations with other laws (to Avenarius and Pfützner)

like systems requiring monitoring, compressed air, hazardous substances and radiation protection. Personnel related regulations include the Youth Employment Act, Maternity Protection Act and accident prevention guidelines of the industrial trade associations.

Technical safety, occupational medicine and hygienic regulations include relevant standards such as the VDI (Association of German Engineers) and VDE (Association for Electrical, Electronic and Information Technologies) guidelines as well as rules that are generally recognized by professionals and that have proven themselves in practice. These are an important part of 30 workplace guidelines (ASR) published in Germany by the Ministry of Labor and Social Affairs [ASR06]. It is anticipated that a majority of these will be adopted across Europe soon. Similar rules and regulations exist in British (see [UK08, UK92]) as well as the American environments too (see [US07, US08]).

Figure 8.8 provides an overview of the German Workplace Guidelines. As we can see they mainly apply to lighting, room climate and social areas. These guidelines are not legally binding; however these and the related recommendations can only be subverted when measures that are equally effective are met. These guidelines are supplemented by state building codes, which in some cases overlap [OSP05]. The federal states are responsible for monitoring compliance with the workplace regulations and the remaining legal guidelines protecting workers. The states have in turn made the trade supervisory board and the industrial trade associations responsible for monitoring them.

European Legislation gained increased significance when, on June 12, 1989, the European Council established Directive 89/391 EWG (Occupational Safety) concerning the introduction of measures for the safety and health of employees during working hours. The directive equally obligates employees and employers to pursue preventative safety and health measures. The corresponding measures are to be implemented via an occupational health and safety management [Koe01, Kubi05]. An overview of the US regulations is given in [Buc13]. The US

exposure	room climate	developments	social areas
ASR 7/1 - visual contact to the outside ASR 7/3 - artificial lights ASR7/4 - safety / emergency lights ASR 41/3 - artificial lights for jobs and public traffic ways	ASR 5 - ventilation ASR 6/1,3 - room temperatures	ASR 8/1 - floors ASR 8/4 - walls allowing for light transmittance ASR 8/5 - not to be entered roofs ASR 10/1 - doors and gates ASR 10/5 - glass doors, doors with glass insert ASR 10/6 - protection against digging, falling out of doors and gates ASR 11/1-5 - automatic doors and gates ASR 12/1-3 - protection against fall and objects falling down ASR 13/1,2 - fire-fighting equipment ASR 17/1,2 - transport routes ASR 18/1-3 - escalator and travelators ASR 20 - crampon / non-slip walks and ladders	ASR 15/1 - seating ASR 2971-4 - break-out rooms ASR 31 - lounge room ASR 34/1-5 - changing rooms ASR 35/1-4 - wash-rooms ASR 35/5 - washing facilities outside of required wash-rooms ASR 37/1 - toilet compartments ASR 38/2 - first aid room ASR 39/1,3 - means and equipment for first aid ASR 45/1-6 - day accommodations on construction sites ASR 47/1-3,5 - wash-rooms for construction sites ASR 48/1,2 - toilets and toilet compartments on construction sites
GermanWorkplace Guidelines (ASR) are not the same as European Workplace Guidelines 89/654/EWG			

Fig. 8.8 Overview of german workplace guidelines (acc. Lehder)

occupational safety and health standards for general industry promulgated by the Occupational Safety and Health Administration (OSHA) are summarized in [OSHA11].

Specific operation facilities and processes as well as the processing of hazardous materials require appropriate media for supplying and disposal. Particular attention has to be paid to preventing fires and explosions. Identifying and evaluating dangers are vital to protecting employees from accidents and work related health risks.

Since workers' protection is also a focus within the frame of changeability, legal requirements for ensuring slip resistance and protection against falls, handling hazardous substances, noise protection and reduction, protection from heat/cold/vibrations/radiation as well as electrical safety also have to be discussed. However, before doing so we will briefly consider the workers' co-determination in designing workplaces.

8.4.3 Participation

As previously mentioned, in Germany the inclusion of employees and the workers council is required by law when designing workplaces. The current regulations provide for two distinct types of rights for employees and the workers council: the right of participation and that of co-determination. With *participation* employees have the right to be informed, to be heard or to advice—however, the decision is ultimately made by the employer. *Co-determination*, on the other hand, means the right to participate in the decision making process. There are three different types of co-determination:

- With the right to initiate, employees or the works council can demand or force certain measures.
- The right to object allows the workers council to speak up against measures that the employer can independently decide upon.

- With the right of assent, the workers council's agreement is required for certain employer's measures.

In Fig. 8.9 we can see which of the various workers council's and employee rights for those participating cover which issues [Dlu08].

Thus the employee has the right to be informed, to be heard and to initiate measures in relation to their workplace, work routine and their person. The workers council has the right to participate in this as well as other considerably extended areas. In addition to the general issues that impact the council itself, their involvement applies mostly to social and personnel related issues for employees. In correspondence to the legislators' intention to protect employees from possible negative impact of management decisions, the strongest rights to participate are here. As the lists for the two groups show, the workers council is ensured extensive rights, outlined in detail in the law, with regards to information, consultation and approval for all issues concerning the design and occupation of the workplace [Fit06]. The installation of a so called economics committee enables the workers council to be informed about the company's financial situation and to participate in decision making processes leading to major operational changes having financial implications.

Ultimately, the dominating principle of Germany's Workers Constitution Act that needs to be emphasized is a trust-based co-operation between the employer and the workers council for the benefit of the employee and the company. If, in certain cases, no agreement can be reached between the employer and the workers council or employee, the issues goes to arbitration. Only after this has proven to be ineffective does the case proceed to the labor court.

With that, we will now return to our discussion on work safety as mentioned above and consider a few aspects in depth.

8.4.4 Tread Certainty and Protection Against Fall

Falls resulting from slipping, misplaced footsteps or stumbling, need to be prevented by appropriately designing tread surfaces. Accidents mostly occur in the absence of non-slip surfaces or uneven floors. Depending on the requirements a

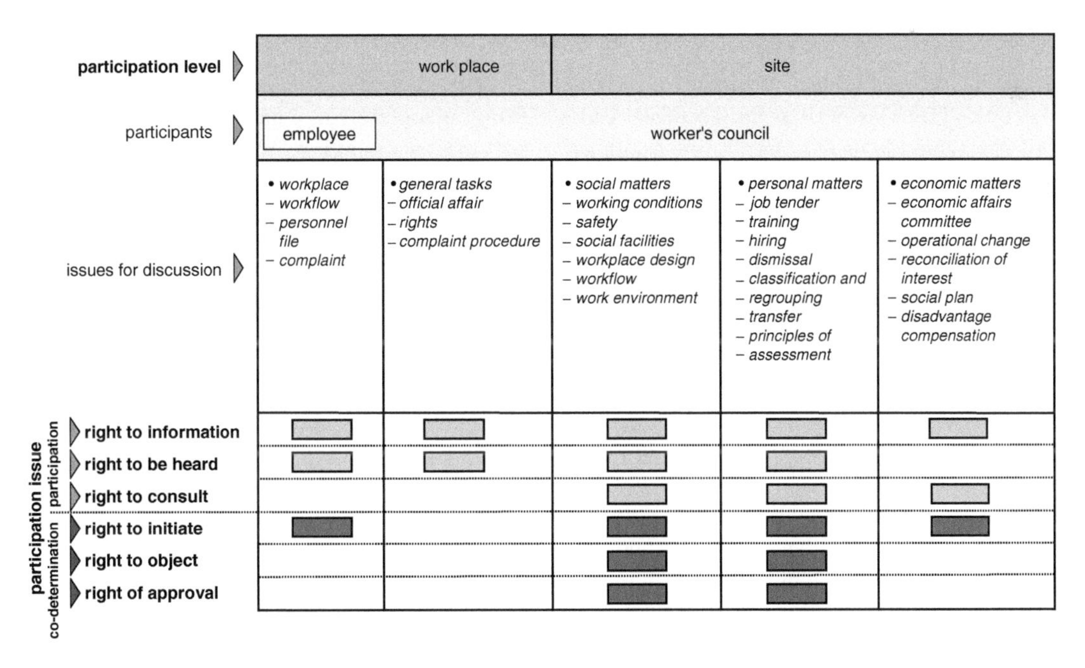

Fig. 8.9 Principles of co-participation (Dlugos)

slightly rough, rough or grooved surface may be specified. According to DIN 51130 as well as industrial trade associations' guidelines, slip resistant properties of tread surfaces are classified according to 'R groups'. If considerable traffic and/or soiling are to be anticipated in workspaces or open areas, grooved floors would be a better solution. Along with tread surfaces, riser inclinations or declinations of more than 25 % as well as raised edges greater than 6 mm (0.23 in) are to be avoided. Moreover, there is a danger of falling when there is a change in level greater than 1.0 m (3.3 ft). In accordance with workplace regulations and guidelines, guardrails or equivalents are required to prevent falls in dangerous areas such as these.

8.4.5 Protection from Hazardous Substances

Airborne hazardous substances such as gases, vapors, mists, smoke or dusts can enter the human body through respiratory organs, skin and/or the gastro-intestinal tract. In Germany measures for airborne hazards are defined in the Chemical Act (ChemG) as well as the Hazardous Substance Act (GefStoffV). Further specific rules are set-out in the Technical Rules for Hazardous Substances (TRGS), Technical Rules for Dangerous Work Materials (TRgA) as well as the Technical Rules for Flammable Fluids (TRbF). For the health and safety of employees, permissible concentration levels for various industrial toxins in the air at workplaces (MAK—values) are legally regulated. In the US the limits for Chemical Hazards and Toxic Substances are listed by the Occupational Safety and Health Administration and can be found under https://www.osha.gov/SLTC/hazardoustoxicsubstances/.

In particular, constructional safety measures are aimed at immediate collection of the hazardous materials at the point of origin, storage in specially designed storage facilities and discharging at specific points. In Germany, in accordance with the Technical Rules for Hazardous Substances, storage facilities for hazardous materials must include special fire protection systems, safety clearances and designated access for the fire department, escape routes and lighting.

Moreover, it also needs to be kept in mind that there is an increase in hazardous substances in the actual construction due to growing use of chemical based construction materials (e.g., bitumen, floor adhesives and coating materials). In the sense of the transformability, implementing eco-friendly materials is highly recommended. Many building owners nowadays emphasize on the usage of eco-friendly materials when setting the objectives for the building structure. Refraining from solvents or PVC in building related components is sensible from an economic standpoint since this protects the health of employees over the long-term. Furthermore, in terms of structural changeability, it is much easier to convert a building structure that is, for example, not contaminated by asbestos.

8.4.6 Noise Protection and Reduction

Noise is as an undesirable form of audible sound at most industrial sites. The overall loudness of noise that is found to be disruptive is a combination of sounds from various equipment as well as other sources. The equipment sounds arise from manufacturing processes, conveyor or hoisting technology as well as supply and disposal technology. Other sounds include all kinds of possible disruptions including background noise in the room or street noise from outside.

The sound induced oscillations spread like waves from various media (air-borne, structure-borne and fluid-borne). Frequently machines and processing operations primarily generate structural sounds (i.e. vibrations) and emit these as air-borne sounds; nonetheless, air-borne sounds also arise directly from flow processes e.g., exhaust systems, jets or fans. Figure 8.10 depicts sources and their noise levels in various scenarios; Industrial workplaces stand out with regards to their intensity [UK05].

Noise is measured in decibels (dB). An 'A-weighting' written as 'dB(A)', is used to

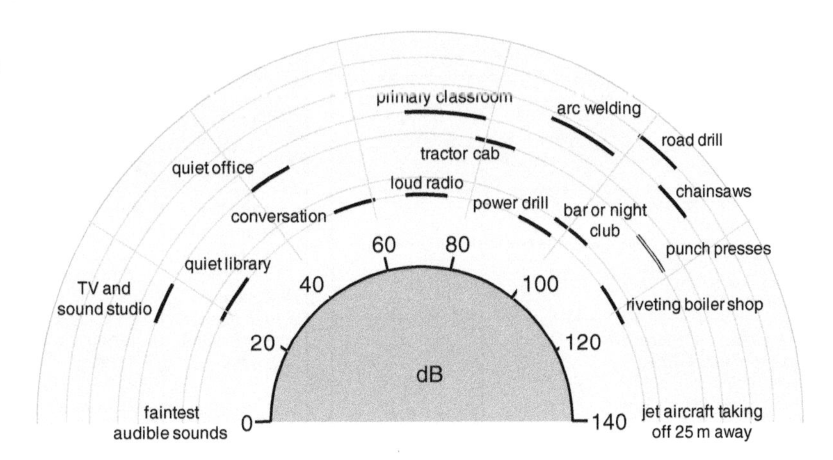

Fig. 8.10 Examples of typical noise levels (*Source* Health and Safety Executive UK 2005)

measure average noise levels. Noises that are repetitive or that continue for extended periods are irritating and harmful to one's health; noises that are frequent and impulsive are generally more dangerous than those which are less frequent and continuous. Hardness of hearing, safety risks and decreased job performance with greater chances of making errors or even vegetative disorders can result from damaging noise emissions.

Measures to reduce sounds include implementing low-noise machines, reducing sounds at the sources and decreasing air-borne sound transmissions. According to [Fas03], intermediate layers of rubber, cork or synthetic materials decrease the transmission of structure-borne sounds in fixed components. A further structural measure for dampening air-borne sounds involves soundproofing ceilings, wall surfaces as well as interior surfaces by lining them with soft, thin porous materials which absorb sounds. Soundproof cabins or sound barriers primarily serve to immediately reduce noise at the workplace. Closed cabins with their own ventilation system can reduce the noise level by up to 30 dB. Based on [Sch96], Fig. 8.11 outlines the limiting values for noise protection and technical possibilities for reducing noise. From the perspective of changeability, the structural system for reducing noise system should generally be flexible, movable and easy to convert.

In the US permissible levels are defined under Standard Number 1910.95 of OSHA the Occupational Safety and Health Administration (https://www.osha.gov). Figure 8.12 depicts the permissible noise exposure and duration determined by the US administration [US95]. In accordance with Footnote 1 of this regulation it is important to note that: "When the daily noise exposure is composed of two or more periods of noise exposure of different levels, their combined effect should be considered, rather than the individual effect of each."

In the UK "The Control of Noise at Work Regulations 2005" define the exposure limit values for noise, obliges employers to assess the risk of exposure to noise and lists actions to be taken to prevent health damages [UK05]. Similar regulations can be found in all industrial nations.

8.4.7 Protection from Thermal Radiation and Vibrations

Some production or storage processes result in considerable cold or heat stress; such stresses cannot be allowed to endanger the health of the employees. Dangers arise from extreme environmental conditions, surfaces that radiate heat or cold as well as from direct physical contact with surfaces, fluids or gases with extreme temperatures. Figure 8.13 depicts the correlation between heat stress and the sensation of pain (according to [Poe85] p. 10). Special structural measures include shielding from dangerous radiations by reflecting or absorbing it using mobile partitions, protective shields, chain screens, wire netting, reflective coatings and protective glass. Ventilation and air conditioning

	noise level [dB(A)]	
	values to be achieved according to DIN and ISO 1999	limiting value acc. to §15 WPR
predominantly mental activities	30 -45	55
break-out, stand-by, lounge areas and restrooms	-	55
simple or primary automated office activities or comparable activities	45 -55	70
other activities	75 -80	85 90 (in exceptions) 1)

1) at levels > 90 dB (A), in accordance with accident prevention regulations, among other things
- loud areas (i.e. > 85 dB(A) are to be indicated
- noise level reduction programs are to be drawn up and carried out

noise reduction at the source (reduction of emission)	noise reduction during transmission (increase of the damping at the entrance and transit)	noise reduction at the place of origin or work place (reduction of noise emission or noise exposition)
constructive measures for noise reduction selection and use of quieter machines and processes	reduction of sound propagation - arrangement of noise sources - capsules - sound absorbing ceiling and wall covering - acoustical barriers - partition walls reduction of noise transmission - vibration isolation - floating creeds - separating joints of construction elements	cabins acoustical barriers

WPR German Workplace Regulations (Arbeitsstättenverordnung)

Fig. 8.11 Limiting values and technical possibilities for noise protection and reduction

Fig. 8.12 Permissible noise exposure (US Dpt. Of Labor 1910.95)

duration per day, [hours]	sound level [dB(A)]
8	90
6	92
4	95
3	97
2	100
1 1/2	102
1	105
1/2	110
1/4 or less	115

measures serve to support greater comfort. In the US the relevant information can be found in the OSHA Technical Manual (OTM) Chap. 4, heat stress (https://www.osha.gov/).

Factory equipment, transportation devices and tools generate mechanical vibrations when in use. These are transmitted through contact points into human body parts such as the hands and

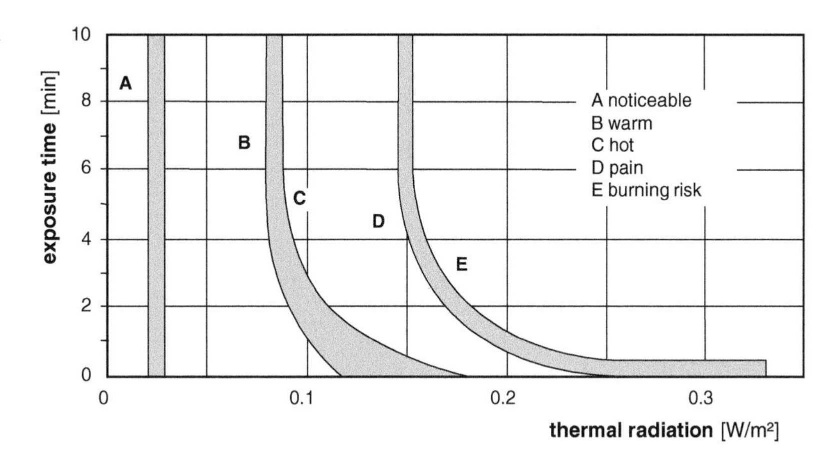

Fig. 8.13 Human reaction to increasing thermal radiation

arms or into the body via floors or seats. With activities requiring greater concentration or fine motor skills, noticeable oscillations can cause strain and decrease performance. They can also lead to damages to the cardiovascular, nervous and/or muscular-skeletal systems.

Primary vibration protection reduces or eliminates the cause of the vibrations e.g., by changing processes or implementing different equipment. On the other hand, secondary vibration protection tends to calibrate the oscillating system such that the vibrations of the system touched by humans are minimal. Structural measures for abating the transmission of vibrations include reducing the natural frequency of machines by setting them on springs or insulators made from steel, rubber or cork. This requires flexible connections for media and transport systems. Furthermore—especially in multi-storied buildings—one should be worried about harmonics resulting due to stimulation of the natural frequency of the structural slabs and systems in contact with a vibrating machine. Figure 8.14 depicts the relation between the frequency of vibrations, the natural frequency and damages to buildings [Leh05].

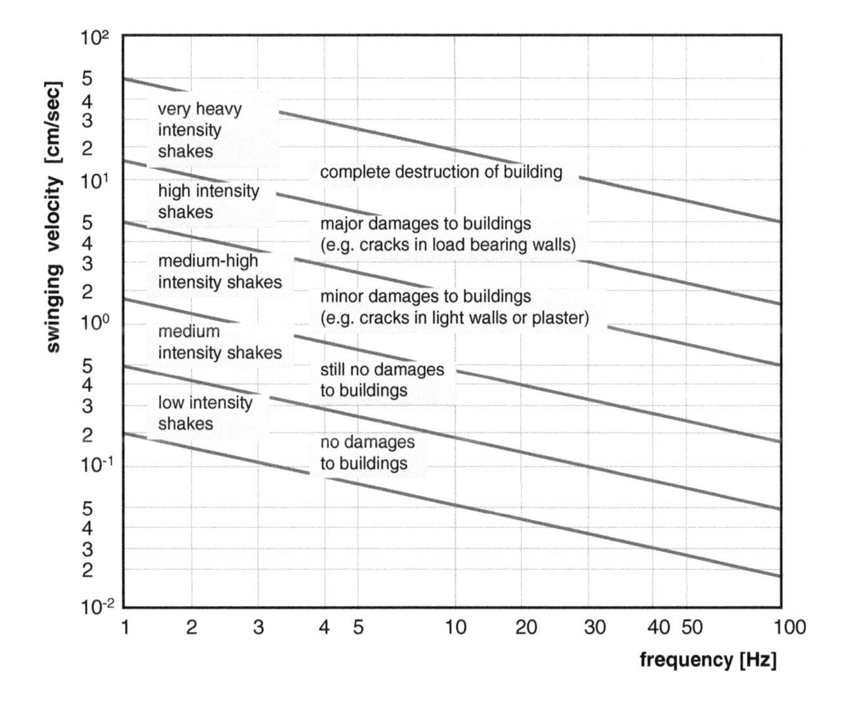

Fig. 8.14 Damaging effects of vibrations on buildings (acc. Lehder)

8.4.8 Electrical Safety and Protection from Radiation

Electrical equipment has to be reliable and operated without disruptions. Transformers and rectifiers need to be in closed electrical service rooms. Switchboards have to be protected from touching live components as well as from being penetrated by foreign objects and especially water. Health risks arise when electricity flows through the human body when it comes into contact with parts that are carrying voltage. Along with all measures that provide protection through an automated shutdown, equipotential bonding needs to be implemented in the building. An equipotential bonding bar joins the switchboard with various metallic building structures, conductive parts from technical systems as well as metallic pipes.

Recent research indicates that electro-smog at the workplace may pose health risks. Electrosmog refers to the undesired electromagnetic radiation resulting from electrical, magnetic and electro-magnetic fields. When selecting electrical devices it is thus critical to prioritize components with low radiation and electrical load such as flat screen monitors. Radiation arises both as electromagnetic radiation and as corpuscular radiation. The most important source of radiation is the sun. Dangerous effects mainly arise from electromagnetic radiation with a wavelength under 10^{-8} m such as x-rays or gamma rays as well as radioactive corpuscular rays. Usually the intensity decreases with the square of the distance to the radiation source. Appropriate measures should be planned in consideration of the specific characteristics each type of radiation present.

Structural protective measures against beta radiation include thin metal sheets, whereas reflective surfaces are effective against infrared radiation and metal shields against radio waves and alternating currents. Thicker materials such as iron shields are implemented as protection against x-rays and gamma-rays.

Devices and systems that project a more dangerous and higher intensity radiation should preferably be located far away from areas that employees frequently use. Fixed shields made of concrete or brick, flexible walls made from lead bricks or mobile shields made of iron or textile materials provide effective protection.

This concludes our discussion of design considerations on a workplace level from a spatial perspective. These flow together with our comments on functional workplace design (Chap. 6) and organizational workplace design (Chap. 7) into designing sections or divisions. We will consider this next from a functional perspective (Chap. 9) and then from a spatial perspective (Chap. 10).

8.5 Summary

The spatial design of a workstation needs to be done on the basis of the function to be performed (see Chap. 6) and integrated with specific organizational requirements (see Chap. 7). An ergonomic design approach leads to humane dimensions, comfortable work and environmental conditions. In particular emphasis needs to be given for labor protection. One needs to avoid any risks to employees due to falls, hazardous substances, noise, heat and cold, vibration, 'live' electrical components and radiation. It is to mention that the design of work places governed by law in all industrialized countries, are in most cases a matter of co-determination.

Bibliography

[Arb04] Arbeitsstättenverordnung (Workplace regulations): ArbStättV. Bundesgesetzblatt Jg. Teil I Nr. **44**, S. 2149–2189 (2004)

[ASR06] Arbeitstättenrichtlinien. Vorschriften und Empfehlungen zur Gestaltung von Arbeitsstätten (Workplace guide lines. Regulations and Recommendations for the Design of Workplaces). Verlagsgesellschaft Weinmann, Filderstadt (2006)

[Ave76] Avenarius, A., Pfützner, R.: Arbeitsplätze richtig gestalten nach der Arbeitsstättenverordnung (How to Design Workplaces Properly According to the Workplace Regulation). München (1976)

[Ben07] Benad, M., Opitz, J.: Architekturfarben—Lehre der Farbgestaltung nach Friedrich Ernst v. Garnier (Architectural Colors—Theory of Color Design After Friedrich Ernst von Garnier). Siegl, Anton, München (2007)

[Buc13] Buckley, J.F., Roddy, N.L.: State by State Guide to Workplace Safety Regulation, 2013 edn. Wolters Kluver, Alphen aan den Rijn (2013)

[Col01] Collins, R., Schneid, Th.D.: Physical Hazards of the Workplace (Occupational Safety and Health Guide Series). Lewis Publishers, Boca Raton (2001)

[DIN79] Allgemeine Leitsätze für das sicherheits gerechte Gestalten technischer Erzeugnisse (General rules for the Safety-Conscious Design of Technical Products). Beuth, Berlin (1979)

[Dlu08] Dlugos, G.: Mitbestimmung (Participation). In: Grochla, E. (ed.) Handwörterbuch der Organisation, 2nd edn. Poeschel, Stuttgart (1980)

[Fas03] Fasold, W., Veres, E.: Schallschutz und Raumakustik in der Praxis—Planungsbeispiele und konstruktive Lösungen (Sound Insulation and Room Acoustics in Practice—Design Examples and Constructive Solutions), 2nd edn. Verl. Bauwesen, Berlin (2003)

[Fit06] Fitting, K., et al. (Hrsg.) Betriebsver fassungsgesetz (BetrVG) Handkommentar (Works Constitution Act, Handbook of commentaries), 23rd edn. München (2006)

[Gek07] Gekeler, H.: Handbuch der Farbe - Systematik, Ästhetik, Praxis (Handbook of Color. Systematic, Esthetic, Practice), 6th edn. Verl. Dumont Buchverlag Köln (2007)

[Koe01] Koether, R., Kurz, B., Seidel, U.A., Weber, F.: Betriebsstättenplanung und Ergonomie. Kap. 10.3: Arbeitsschutzmanagement S. 335 ff., (Worksplace Planning and Ergonomics, Sect. 10.3: workplace protection management), München (2001)

[Kubi05] Kubitscheck, S., Kirchner, J-H.: Kleines Handbuch der praktischen Arbeitsgestaltung (Small Manual of Practical Work Design), München (2005)

[Lan06] Lange, W., Windel, A.: Kleine Ergonomische Datensammlung (Small Data Collection of Ergonomics), 11th edn. Bundesanstalt für Arbeitsschutz und Arbeitsmedizin (2006)

[Leh05] Lehder, G., Skiba, R.: Taschenbuch Arbeitssicherheit (Pocket book workplace safety), 11th edn. Schmidt (Erich), Berlin (2005)

[MCol07] MacCollum, D.: Construction Safety Engineering Principles—Designing and Managing Safer Job Sites. McGraw-Hill Construction Series, New York (2007)

[OSHA11] OSHA Standards for General Industry as of 01/2011. Washington, DC (2011)

[OSP05] Opfermann, R., Streit, W., Pernack, E.F.: Arbeitsstätten (Workplaces), 7 ed. Hüthig Jehle Rehm, Heidelberg (2005)

[Poe85] Poeschel, E., Köhling, A.: Asbestersatzstoffkatalog. Band 2: Arbeitsschutz (Asbestos substitute catalog, vol. 2. OSH). Hauptverband der gewerblichen Berufsgenossenschaften. Sankt Augustin (1985)

[REf91] REFA (Hrsg.) Methodenlehre der Planung und Steuerung (Methodology of planning and control), 6 volumes, München (1991)

[Rüs06] Rüschenschmidt, H.: Ergonomie im Arbeitsschutz—menschengerechte Gestaltung der Arbeit (Ergonomics in OSH—Human-Centered Design of Work), 2nd rev. edn. Verl. Technik und Information Bochum (2006)

[Rüs07] Rüschenschmidt, H., Reidt, U., Rentel, A.: Gesundheitsschutz am Arbeitsplatz - mit Ergonomie gestalten (Health at Work—With Ergonomic Design). Technik & Information Bochum (2007)

[Sal12] Salvendy, G.: Handbook of Human Factors and Ergonomics, 4th edn. Wiley, Hoboken (2012)

[Sch96] Schirmer, W.: Technischer Lärmschutz (Technical Noise Protection). VDI-Verlag. Düsseldorf (1996)

[Til15] Tillmann, B., et al.: Human Factors and Ergonomics Design Handbook, 3rd edn. McGraw-Hill, New York (2015)

[UK05] Noise at work. Guidance for employers on the Control of Noise at Work Regulations 2005. Published by Health and Safety Executive, UK (2005). Health & Safety Offences Act 2008. Legislation Government UK. http://www.legislation.gov.uk

[UK92] The Workplace (Health, Safety & Welfare) Regulations, no. 3004, 1992. Legislation Government UK. http://www.legislation.gov.uk

[US95] Occupational Noise Exposure (1910/1995). US Dept. of Labor, Occupational Safety and Health Administration. http://www.osha.gov/

[US07] ANSI/ASHRAE Standard 62.1-2007, Ventilation for Acceptable Indoor Air Quality. American Society of Heating, Refrigerating and Air-Conditioning Engineers, (2007). http://www.ashrae.org/

[US08] ASTM E2350-07 Standard Guide for Integration of Ergonomics/Human Factors into New Occupational Systems. ASTM International (2008). http://www.astm.org/Standards/E2350.htm

9 Functional Design of Work Areas

As shown in Fig. 5.16, a work area combines several manufacturing and assembly areas organizationally together, linked by storage, transportation and handling systems. The purpose is to serve the production of a salable product. For the functional design of a work area, the order type (customer or stock production), the type of procurement, the organization of production and assembly and the type of production planning and control must be determined. This is described in more detail below and forms the basis for the spatial design, which is discussed in Chap. 10.

9.1 Overview of Design Aspects

From the perspective of planning work areas, there are various options for reacting to the factory's internal and external influences and to design, plan and control production processes with a focus on their function. Figure 9.1 depicts the design aspects that are relevant here and are focused on the purchase, make and deliver processes.

Clearly, considerations need to be centered on the production (make) process and need to resolutely focus on both the enterprise's competitive strategy and the customer demands. In view of delivery time requirements, replenishment times and economic deliberations, it needs to be determined which part of the product will only be produced once a concrete customer order has been placed (made-to-order) and which part will be made-to-stock on a forecast basis. This point in the production is referred to as the customer order decoupling point. Subsequent to this decision, key requirements can be identified for the way in which orders are handled. Keeping in mind the product structure and technological restrictions, suitable manufacturing and assembly principles also need to be selected and manufacturing segments have to be formed. Following that, the planning and control of production needs to be configured so that the created potential of the product structure can be utilized sustainably and as well as possible.

Production, however, cannot be designed separately from purchasing and delivery processes. Frequently, the customers directly influence the design of production when, for example, they demand that supplies be delivered every hour on the hour. Similarly, purchasing requirements also need to be considered when, for example, there is an agreement with the supplier that certain articles can be ordered with a 2-day lead-time, while others are stocked in-house.

The selection of the purchasing and supply models is decisive not only with regards to how orders are triggered, but also in regards to where and how articles are stored as well as how items are supplied and who is responsible for operating the stores. These aspects, which we will expand upon here, need to be integrated into the process

H.-P. Wiendahl et al., *Handbook Factory Planning and Design*,
DOI 10.1007/978-3-662-46391-8_9, © Springer-Verlag Berlin Heidelberg 2015

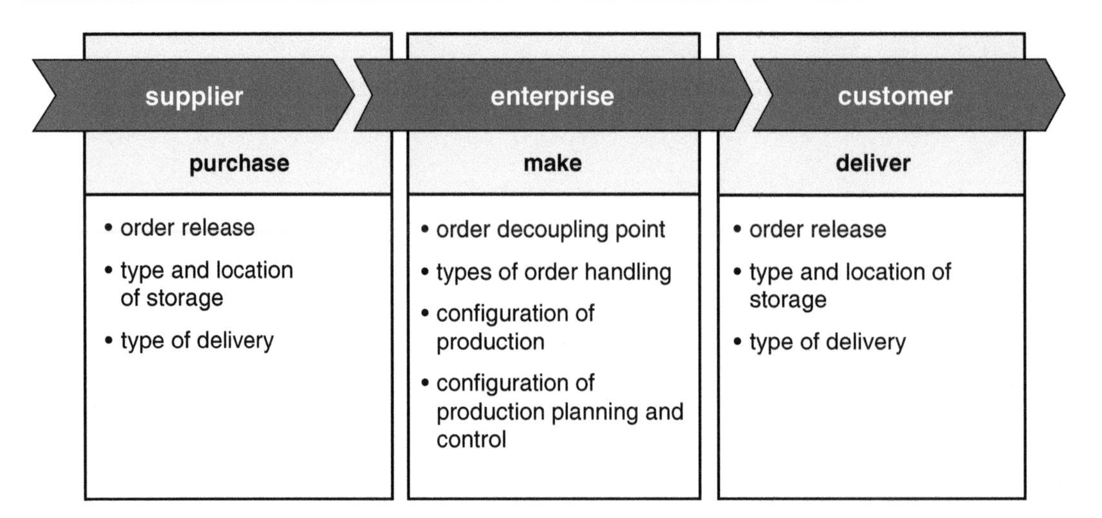

Fig. 9.1 Design aspects of a factory from process view

of designing work areas and have to be considered by the factory planner together with the planning for production and logistics. In industry, due to the different products and sales markets, this frequently results in a variety of purchasing, production and supply structures that can change over time. The changeability of the factory is thus not only determined by production but also by the selected purchasing and supply models.

9.2 Customer Order Decoupling Point

The term 'order decoupling point' was coined by Hoekstra and Romme from the Philips Company and is defined as follows: "It separates the part of the organization oriented toward customer orders from the part of the organization based on planning" ([Hoe92], p. 6). The key criterion in setting the customer order decoupling point (also called order penetration point) is the difference between the required delivery time and the replenishment time of the product from material procurement to delivery.

Especially with complex products the customer order decoupling point is generally not set for the product as a whole, but rather is attained through a differentiated allocation of components and individual parts before or after the decoupling point. Parts are then assigned based on an analysis of the product structure, the steps of that are depicted in Fig. 9.2. For a given product or representative of each product class (formed according to criteria such as customer specifications, product complexity, dynamic of demand, proportion of sales etc.) the product structure—described in the multi-level bill of materials (Fig. 9.2 left top)—is transferred into a so-called 'order schedule' (Fig. 9.2 top right). This depicts all of the relevant components with their replenishment times over the time axis for parts purchased and the planned throughput times for components assembled/manufactured in-house. This information is supplemented with the time TMT required to manage the order (from receiving the order up until the order is set in the PPC system) as well as the time to transport the final product up until the point where it is handed-over to the customer.

The total replenishment time for the product is a result of the order schedule. If the wanted delivery time (also known as the delivery time window) is then plotted, we can identify which components need to be procured, manufactured and/or assembled independent of an order (i.e., order-anonymously). The start time of these components' corresponding time elements is before the delivery time window.

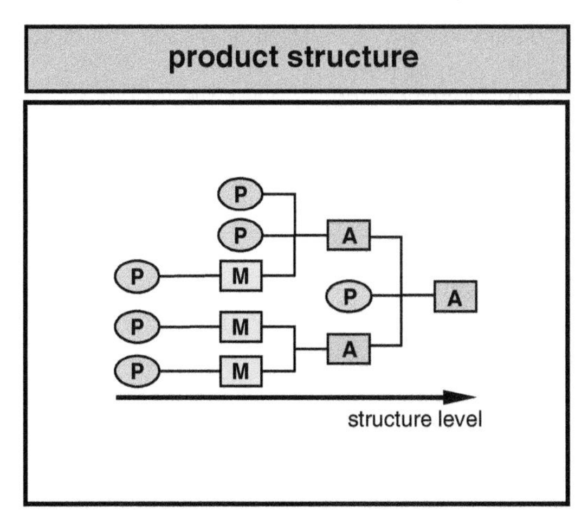

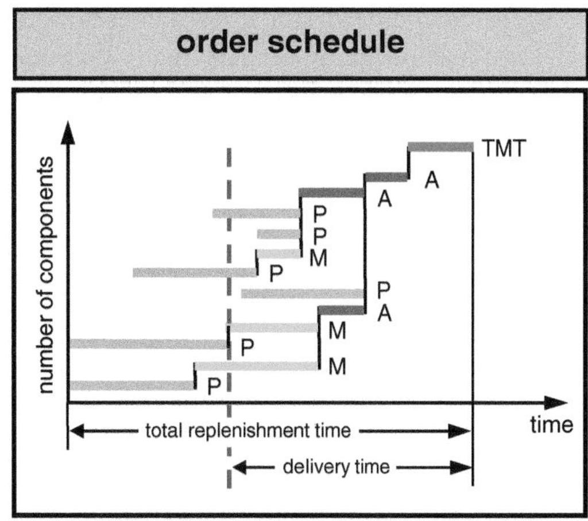

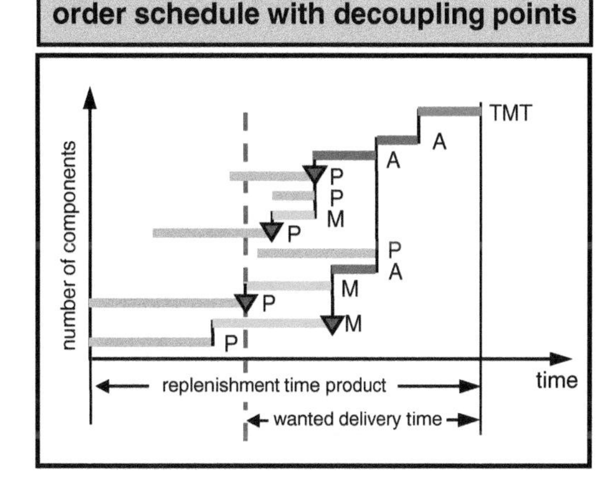

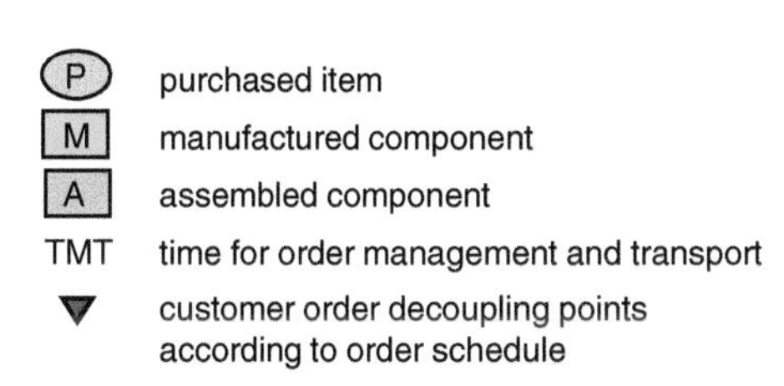

Fig. 9.2 Determining the customer order decoupling point

Components whose start time is within the delivery time window can be purchased or made either customer specifically or customer-anonymously. In view of the necessary certainty of supply and economic efficiency, it has to be verified which of these alternatives is more practical. Basic criteria here include reusability, dynamics of demand and the value of the components.

Once these decisions are made the customer decoupling points can be entered into the order schedule (red triangles in Fig. 9.2 bottom). In those cases where there is still a possibility to choose, this point should first be fixed once corresponding investigations have been conducted for all of the related product groups. Based on these results, the types of order handling and process models (see Sect. 9.5) that are to be implemented across all of the products are then determined.

9.3 Approaches to Handling Orders

Figure 9.3 depicts the four main processes purchasing, production (sub-divided into manufacturing and assembly) and delivery. Also, depending on the position of the customer order decoupling point, there are four basic approaches of handling orders: make-to-stock, assemble-to-order, make-to-order (each from standard products) and manufacturing products with a portion of customer specific engineering.

With a *make-to-stock* approach, a saleable product is produced and stored based on a sales forecast even when there are no existing orders. Cameras, household appliances and printers are good examples of such products. The advantage of this type of order handling (shorter delivery times) has to be weighed against the greater costs of the capital tied-up in warehoused goods.

With an increasing number of variants and a higher product value, this type of order handling is not economically justifiable; an alternative then is *assemble-to-order*. When orders are handled using this principle, they first enter the assembly once an order from a customer has been received. In doing so the assembly draws upon prefinished standard components. A famous example of this is the internet-based ordering of laptops from Dell. During the ordering process the customer decide what setup he would like by using a product configurator to make decisions regarding the hard drive, RAM, processor, software, etc. With standard components stored in a warehouse, custom-tailored products could be assembled and delivered within a short time.

However, it is not always possible to pre-finish the components for all possible customer wishes, even when it concerns standard parts. Among the possible reasons for this could be costs or storage restrictions. In this case the manufacturing of the components will be triggered after the customer's order has been received. With this type of order handling, also known as *make-to-order*, it is assumed that the work plans already exist and that no customer-specific adjustments are required. The initial material is usually procured and hold on stock based on sales forecasts. A typical example of make-to-order is the manufacturing of a high-value, variant-rich component in the automobile manufacturing industry such as interior trim or seats.

The fourth type, *engineer-to-order*, comes into play when a customer's specifications for a product requires a design process for at least one component of the product to be delivered. In cases where items are manufactured in-house, individual work plans and bills of materials are required. This form of order handling is typical of plant engineering.

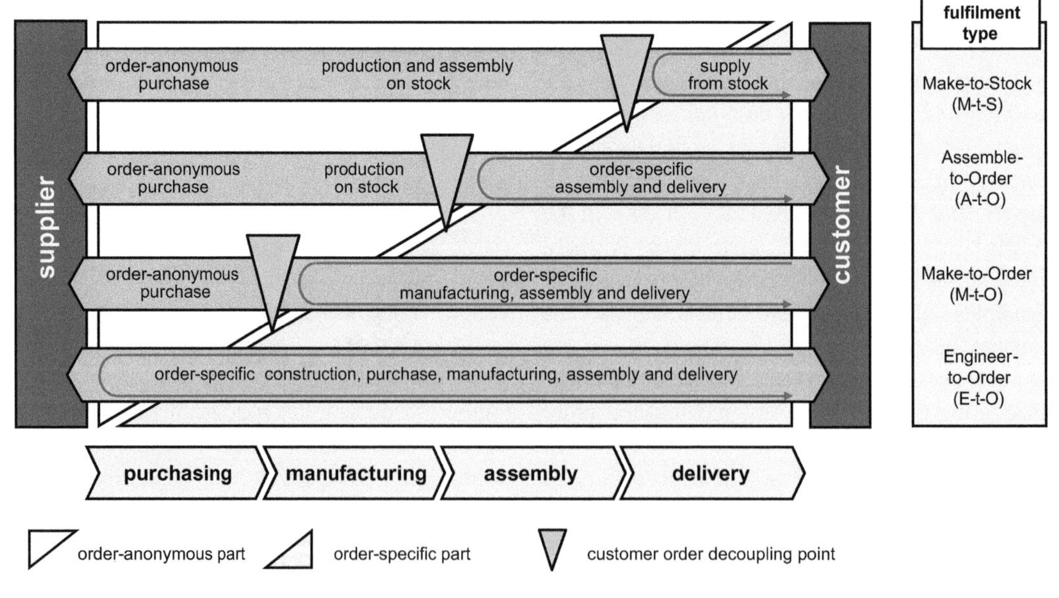

Fig. 9.3 Approaches to handling orders (modified to Hoekstra and Romme)

The logistic objectives that are prioritized for pursuance also differ depending on the type of order handling. Generally the logistic targets before the customer order decoupling point are focused on reducing the logistic costs. Thus the objectives are to utilize available resources and minimize WIP and stock. After the decoupling point the focus is on logistic performance, i.e., delivery time and delivery reliability. In dimensioning the decoupling point (which in reality is a store for the order anonymous parts) it is thus important that the supply for the subsequent process is ensured while at the same time keeping WIP and stock at a minimum.

9.4 Order Types

Parallel to the types of handling, we can also derive different types of orders as well as the demands that these place on planning processes. Figure 9.4 depicts the general types of orders, differentiated based on how the orders are triggered.

With *customer neutral process chains*, orders for procurement, manufacturing and assembly are formed from the production program and the manufacturing/procurement programs derived from it. The orders serve to maintain the stock in the decoupling point at a defined level thus ensuring the supply for the subsequent process. The production program presents the primary requirements (required saleable products) including the necessary quantities and due dates. The production program is developed based on historical data and market indicators, frame work agreements and customer queries.

With a customer neutral process chain it is also possible to procure or to produce items based on stock replenishment orders (pull principle). In this scenario, orders are triggered as a direct result of materials being withdrawn from the decoupling store or when the stock falls below a set order point. In this case, the manufacturing and procurement programs generally

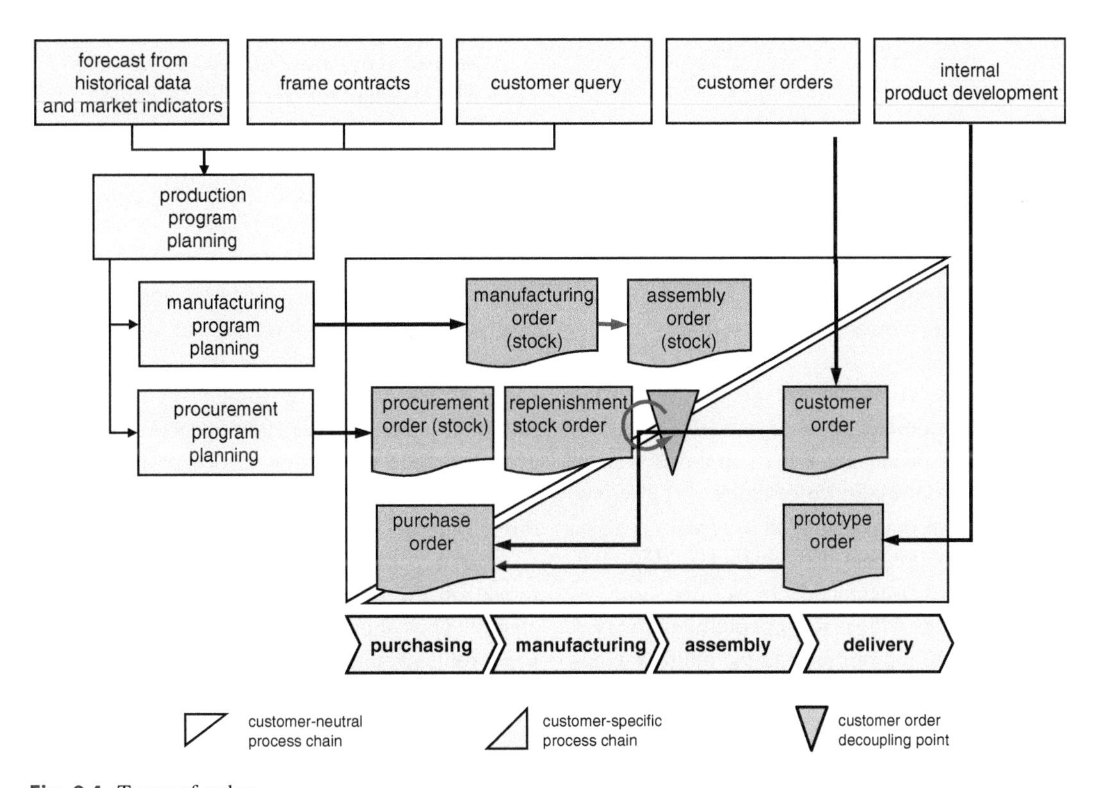

Fig. 9.4 Types of orders

serve to dimension the stock (maximum stock, stock order point) in the decoupling store.

With *customer specific process chains* concrete customer orders or internal prototype orders are processed and/or procurement orders are triggered. Article numbers, quantities and due dates are provided by internal or external customers and order handling is restricted to scheduling the individual processing steps and implementing them according to schedule.

9.5 Process Models

Further designing production requires processes to be broken down in greater detail than just customer neutral or customer specific chains. The individual processes have to be designed as efficient, even cross-company business processes focused on optimally fulfilling both the customers' demands and the business' goals. To facilitate this, we will now take a closer look at the main procurement, production and supply processes introduced in Fig. 9.1.

9.5.1 Procurement Models

The *procurement process* includes all activities that ensure the efficient supply of an enterprise with the required manufacturing and assembly materials, commodities and external services. It represents the connection between the supplier and production.

In the course of the last decade a growing number of procurement forms have developed all of which aim at ensuring the supply as well as possible with a minimum of stock and low process costs. An overview of the current six basic types of procurement is depicted in Fig. 9.5. These are distinguished from one another according to the trigger for procurement, the type and location of the storage and the point at which ownership is transferred.

With traditional *reserve stock procurement* the purchaser executes all of the procurement activities, i.e., planning and ordering, receipt and incoming goods inspection, storage and delivery to the place of consumption. Material stores are consciously maintained for the purpose of ensuring that subsequent processes are supplied.

With all of the other procurement models, stores are maintained by the supplier or service provider (in the following referred to together as the supplier) or are not maintained at all. Accordingly, these models require that different forms of storage areas are maintained in the immediate vicinity of the place of consumption and thus should be taken into consideration when planning the factory.

A *consignment store* is a warehouse (including a return area) maintained by the supplier in which the supplier's articles are hold with a contractually agreed upon minimum stock. The stock is constantly at the disposal of the purchaser; however, the supplier remains the owner up until the time when the goods are withdrawn from the store. The consignment concept is generally implemented for articles with a high value. Due to the special ownership relationship, the goods frequently have to be secured, e.g., in a lockable storage area.

Standard part management is suitable for procuring standard articles with low values. With this scenario, the supplier regularly fills the material buffer found in the immediate vicinity of the workplace up to a stock level contractually agreed upon with the purchaser.

With the *reserve stock model*, procurement occurs based on the procurement program, whereas with the consignment and standard part models it occurs based on the withdrawal of materials. With the remaining three concepts (contract stock, single item procurement and synchronized production processes), procurement occurs as the result of a concrete customer order.

Contract stock involves a warehouse maintained by the supplier located close to the purchaser. This allows frequent delivery of goods, synchronized with demand and 'on-call'. These calls generally occur based on customer orders. It is also possible that calls are triggered by a withdrawal of goods from an additionally required buffer.

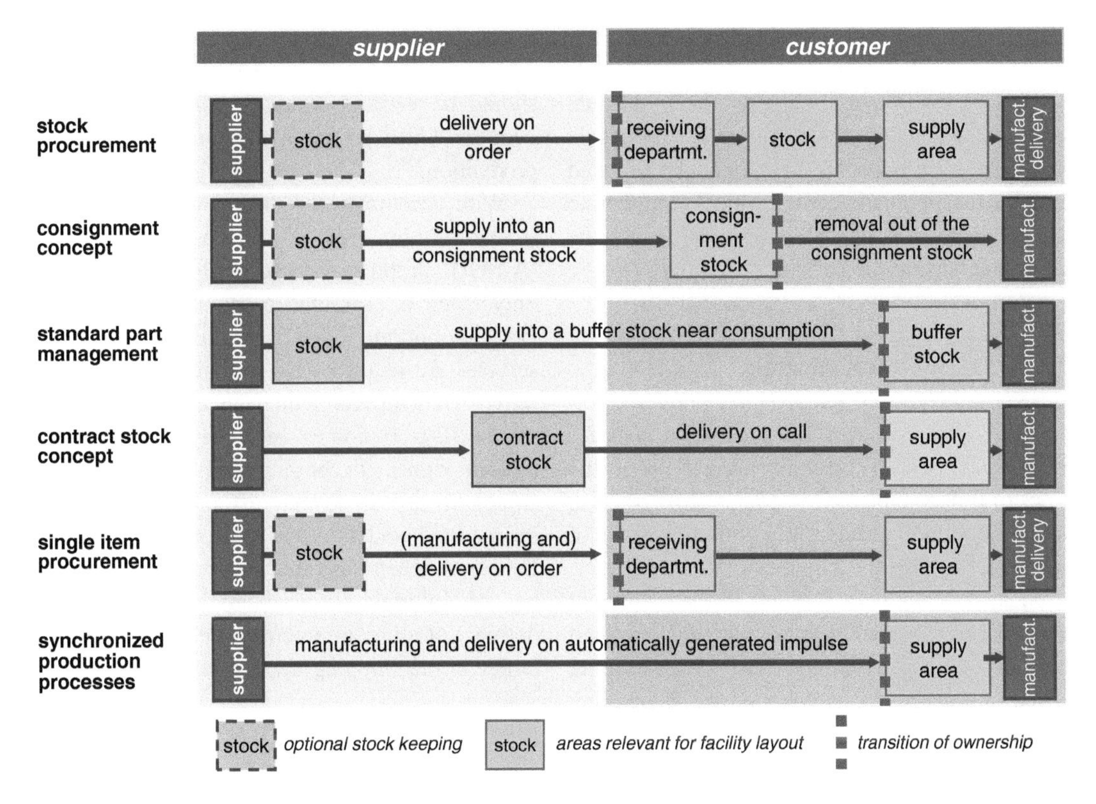

Fig. 9.5 Overview of typical procurement models

Concrete customer orders are also the trigger for the *single item procurement* and *synchronized production process* models. The material is delivered directly to the place of consumption without any interim storage. As the name implies, the synchronized production process concept is characterized by the supplier and the customer synchronizing their manufacturing cycles, whereby the supplier's production site is also located close to the purchaser (sometimes also on the purchaser's premises).

With the last three mentioned procurement models, the purchaser does not have any storage processes. Nevertheless, provision areas that allow a timely decoupling of the supplier and consumer need to be included in the factory layout.

In order to select the best suited procurement model in concrete situations and to allocate suppliers and material numbers the following criteria should be particularly kept in mind:

- the relationship between the procured good and a customer order,
- the significance of the article,
- the certainty of the supply from the procurement side, and
- the consistency of demand for the procured good.

If a procured good is not used more than once, only single item procurement comes into question. The same applies to procurement goods that have a high value (A-parts) and are not stored due to very sporadic demand both for the purchaser and the supplier. For all other A-parts and for most B-parts the consignment concept, contract stock and synchronized production processes should be tested with regards to their possible applicability. All three of these models are primarily aimed at reducing inventory costs while at the same time working towards reducing process costs. Process costs refer to all of the costs that arise from managing orders.

In contrast, standard part management is only focused on reducing process costs and is thus best suited for C-parts. Finally, the traditional *reserve stock concept* is primarily applicable when the price advantages that can be targeted through the purchase demonstrably compensate for the increased logistic costs and can at the same time ensure the certainty of supply.

9.5.2 Production Models

For the production area, the process models for manufacturing and assembly result from the types of orders explained in Sect. 9.4. Figure 9.6 depicts the basic models along with their flows of information.

Make-to-stock (M-t-S) refers to the traditional scenario in which products are manufactured and assembled customer-anonymously before being stored in finished goods warehouses. Depending on the control principle, two process models can be distinguished. A push model is characterized by a process in which production orders (jobs) are formed based on a manufacturing and/or assembly program (incl. quantities and due dates) in view of existing inventories and pushed through the corresponding production areas after their release (see Fig. 9.6 top left). In comparison, a pull principle functions based on a consumption control, i.e., the order is triggered after a product is withdrawn from a store (and the stock falls below a minimum WIP); thus, it is generally simpler (Fig. 9.6 top right).

The decision concerning which of the two possible make-to-stock production models should be implemented is basically dependent on the consistency of demand, the multiple uses of the affected article and the number of variants. The pull principle is suitable when there is a highly consistent demand, a large number of uses and few variants. Despite the greater effort required for planning and control processes the push principle is then generally recommended for the other cases. However, the production program must be able to provide a good forecast of the demand trend. If the demand is rather more sporadic (generally related to a low number of demands and a greater number of variants) and can only be forecasted with difficulty, it should be verified whether or not the required delivery times permit a make-to-order (M-t-O) production.

With customer order related productions (make-to-order 'M-t-O' and assemble-to-order 'A-t-O'), products are scheduled and released only after a customer order is received (see Fig. 9.6 bottom left). If additional engineering services are required, e.g., in order to adjust the design of a product, then an engineer-to-order model (E-t-O) comes into play (see Fig. 9.6 bottom right). In both cases, the product is delivered directly to the customer after it is has been tested and released without being stored. However, when customer orders include a number of different requirements and a complete delivery of all requirements is desired, an interim buffer in the shipping area is required.

9.5.3 Delivery Models

The delivery process includes all of the information and value streams from the transmission, processing and control of a customer order (order handling) from the time the order is received as well as the flow of goods from the production site to the storage site up until the agreed upon delivery point at the customer's site.

Every procurement process on the buyer's side has to correspond with a matching delivery process on the supplier's side. Consequently, in accordance with the six previously described procurement models (see Sect. 9.5.1) there are also six delivery models referred to with predominantly similar names: customer neutral store, consignment, contract stock, standard part management, single item delivery, and synchronized production [Frü06].

The key characteristic that first distinguishes them is the way in which the orders are handled and materials transferred to the customer (see Fig. 9.5). These characteristics however are of secondary significance in factory planning, instead it is much more important whether finished products have to be stored or not and, if

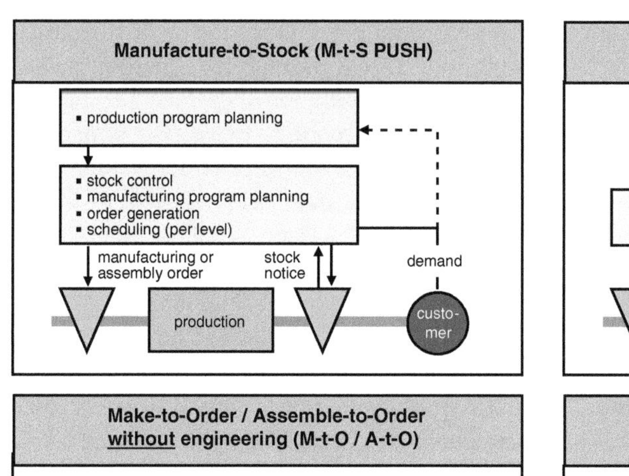

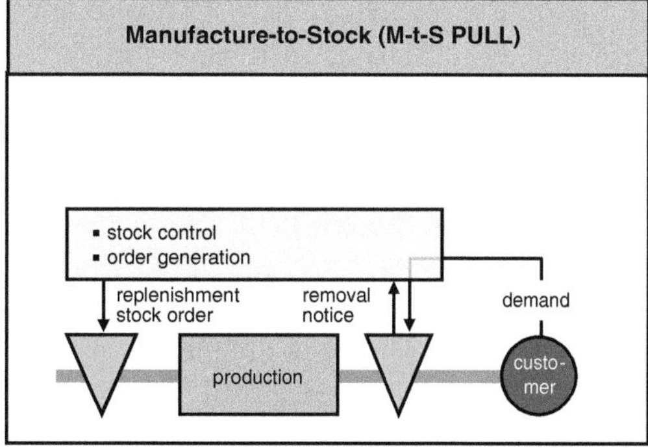

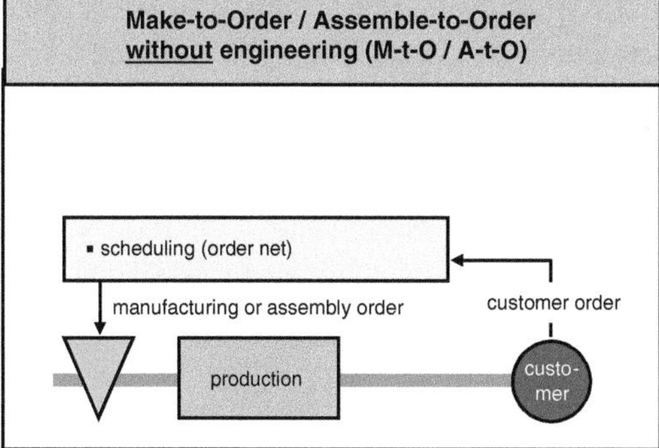

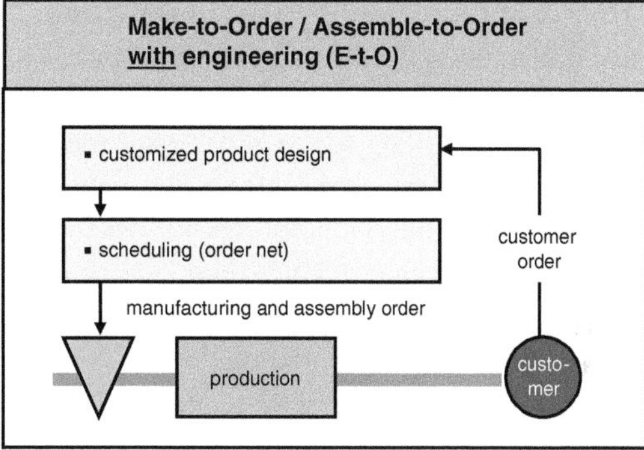

Fig. 9.6 Flow of information with different production models (simplified representation)

so, on which site. A finished goods store is required on the supplier's own premises when orders are processed from a customer neutral store as well as with standard part management. With consignment as well as single item delivery, maintaining an in-house store can be practical as long as there are different purchasers for the affected product. However, if the products are customer-specific, it should be verified whether an M-t-O process is suitable for their production and thus whether they are able to refrain from maintaining their own store of the product. With contract stock, a separate store in the close proximity to the customer needs to be maintained. Synchronized production has the greatest impact on in-house production. With this model, in which neither the purchaser nor the supplier maintains a store, the production of the supplier has to be completely aligned with the requirements of the purchasers both in regards to capacities and in controlling the range of variants. Due to very short delivery times, it is usually necessary for the supplier to develop a production site close to the customer's. From the perspective of the supplier, the conditions for a delivery model of this type are almost exclusively found in the automobile supply industry—and then only for valuable components. The reason for this is that the investment costs for such an arrangement can only be justified when there is a contractual connection and when there is a large sales volume.

9.6 Manufacturing and Assembly Principles

In industrial practice there is an almost unconceivable variety of manufacturing principles upon which the basic system components (workpieces, people and equipment) are arranged. Every real manufacturing system can however be distinguished by its motion structure, its spatial structure, and its temporal and organizational structure.

A practical classification of organizational types can be achieved when we consider the spatial structure related to various manufacturing principles. Figure 9.7 lists the basic manufacturing principles found in industrial practice according to their classification criteria, the typical terms used to refer to them and their spatial structure as well as a couple of examples. Because manufacturing according to the job shop principle, the flow principle and the group/segment principle are the most common organizational forms we will take a closer look at them first.

Manufacturing according to the *job shop principle* (or functional principle) arranges the workplaces according to the processing method. The work item (the workpieces or components) are transported individually or lot-wise from workplace to workplace. Once there, they have to wait until the items already queued are processed. The job shop principle has the great advantage of being flexibly adjustable to different products and to their different operating sequences. Furthermore, resources can be utilized well. The long throughput times and high WIP in the production process though are disadvantageous.

In contrast to the job shop principle, production organized according to the *flow principle* is built based on the sequence of operations on the products and is thus also referred to as the product principle. Here the throughput of the parts is very quick because a workpiece is transported immediately to the next workstation after it has been processed and does not have to wait for the completion of others. Flow oriented production in which the individual workstations are connected via a buffer zone is referred to as loosely or elastically chained. However, also rigidly chained flow production can be found that does not have any interim buffers for temporarily receiving workpieces. With such rigid chains, small disruptions on individual workstations lead to a standstill of the entire system.

One of the key disadvantages of the flow principle is that, due to its focus on a specific product, the system can only be laboriously converted when there are technical changes.

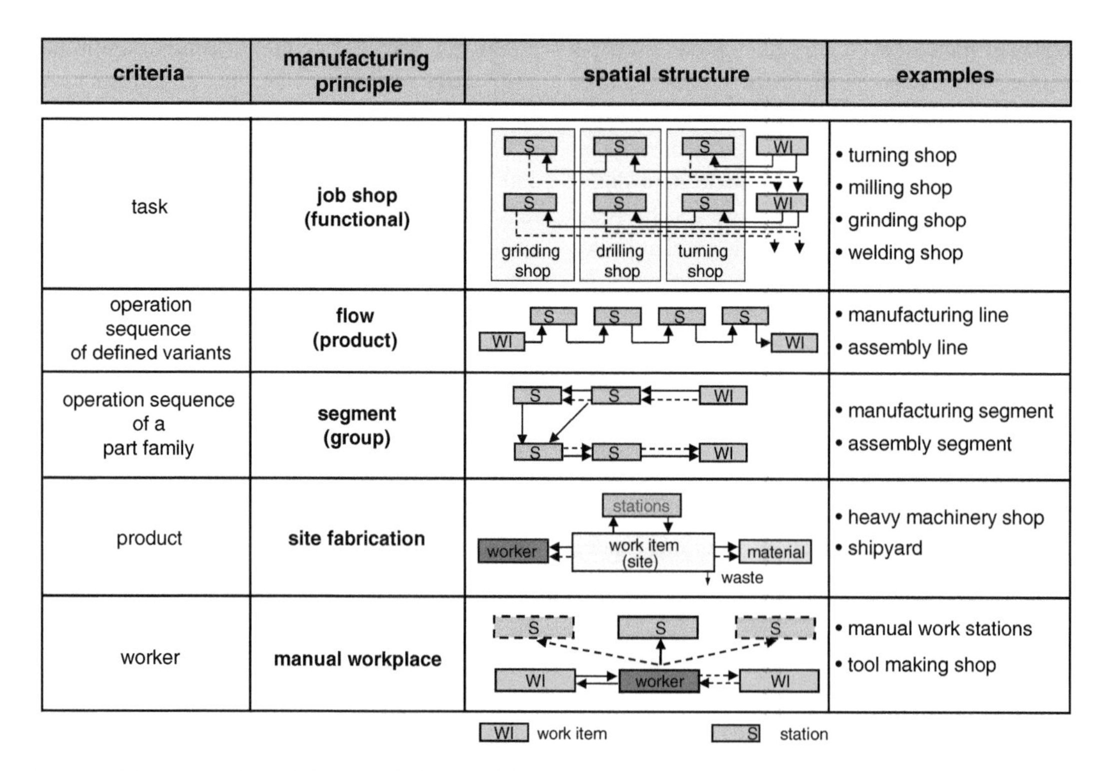

Fig. 9.7 Structure of industrial manufacturing principles

Furthermore, producing parts then becomes more expensive when the facilities cannot be efficiently utilized due to a lack of demand for the intended workpiece or product.

Manufacturing segments (or cells) are spatially or organizationally consolidated arrangements of all the equipment required producing a group of similar workpieces or products as completely as possible. Waiting times between the individual work steps are avoided by arranging machinery based on the operations sequence and through overlapped manufacturing (one-piece-flow). Overlapped manufacturing means that instead of forwarding parts in lots, parts are immediately forwarded to the next station (one-by-one) after they have been processed.

In a manufacturing segment, the actual manufacturing operations are transferred along with the organizational, planning and control functions to a group of employees, who are then responsible to a large extent for operating the segment. These functions are comprised of material requisition, finite scheduling, order sequencing and work planning including the control program for the numerically controlled tools and measuring machines. Eliminating the strict division between planning and executing activities combined with spatially concentrating the machinery and resources results in shorter throughput times for products with significantly increased room to maneuver for the employees involved.

In addition to these three key organization types, *site fabrication* plays a role in the manufacturing of large and heavy workpieces. These situations arise when building systems or large machinery such as pressure housings for hydroelectric turbines or large generator shafts. In these cases, the workpieces are organized on a base plate and the tool machines are placed where the work is to be done. The extreme case of site fabrication is when workpieces can only be assembled and completed on-site because they are no longer transportable.

One of the less common types of organization in the industry is the *manual workplace principle*, which is generally implemented when operations are predominantly completed by hand and without extensive machine work. Such workplaces are for example found in tool and fixture shops.

Similar to part manufacturing, organizational principles can also be defined for assembly. In Fig. 9.8 and analogous to manufacturing principles, the relative movement of the assembly object

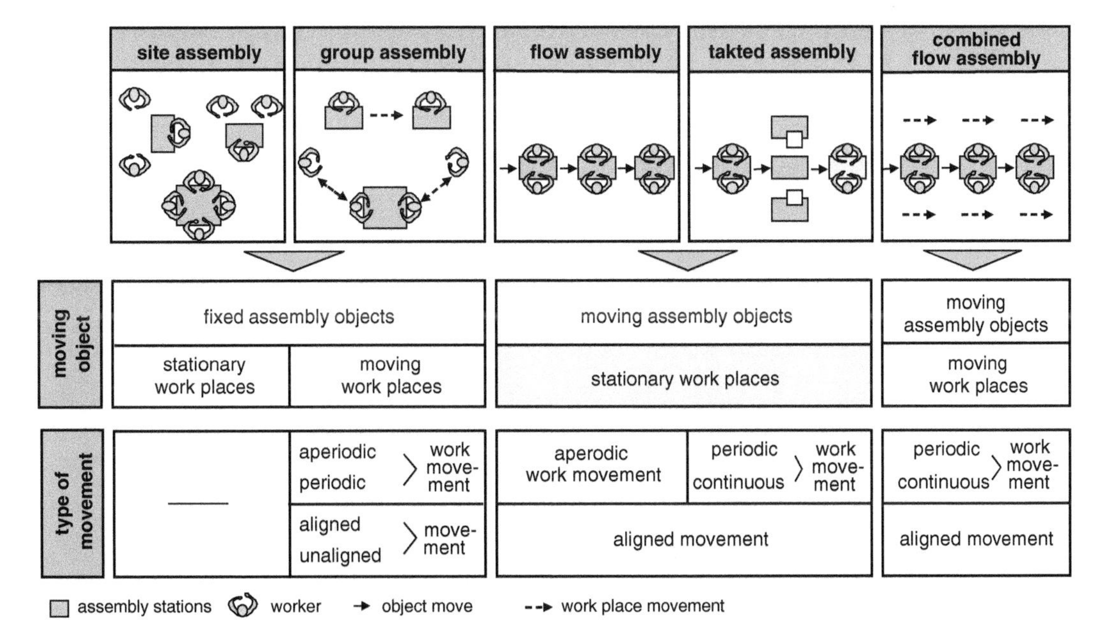

Fig. 9.8 Organizational forms of assemblies (acc. Eversheim)

to the assembly workplaces serves as the organizational criterion. *Site assembly* corresponds to site fabrication, whereas *group assembly* works on a fixed object with periodic or aperiodic movement of the workplaces and with an aligned or unaligned movement of the objects. As sub-categories of *flow assemblies*, clocked or combined flow assemblies are characterized by moving assembly objects in one direction and are differentiated from one another by how the humans involved are tied into the clock cycle or flow rate with continually moving workpiece carriers.

9.7 Production Segments

Depending on the range of the production program the requirements of a production are generally more or less heterogeneous. This is expressed in customers differing demands for specific delivery models and performance in regards to various logistic characteristics. It is also visible though from the product side, for example, in the product structure and range of varieties. It is thus frequently not possible to meet all of the requirements with only one manufacturing principle and one assembly principle.

When the number of pieces permit it, it seems practical to create production segments as product oriented decentralized organization units [Wil98]. Production segments are characterized by a specific competitive strategy focused on reducing costs, shortening throughput times and/or improving quality. A higher degree of autonomy is strived for by integrating planning and indirect functions. Moreover, production segments stand out in that a number of stages in the logistic chain are integrated into them. Thus a production segment can consist of a number of manufacturing or assembly cells.

As an example, Fig. 9.9 depicts three production segments of a company that produces water pumps. In this case, segments are formed according to the rate of pieces. One of the

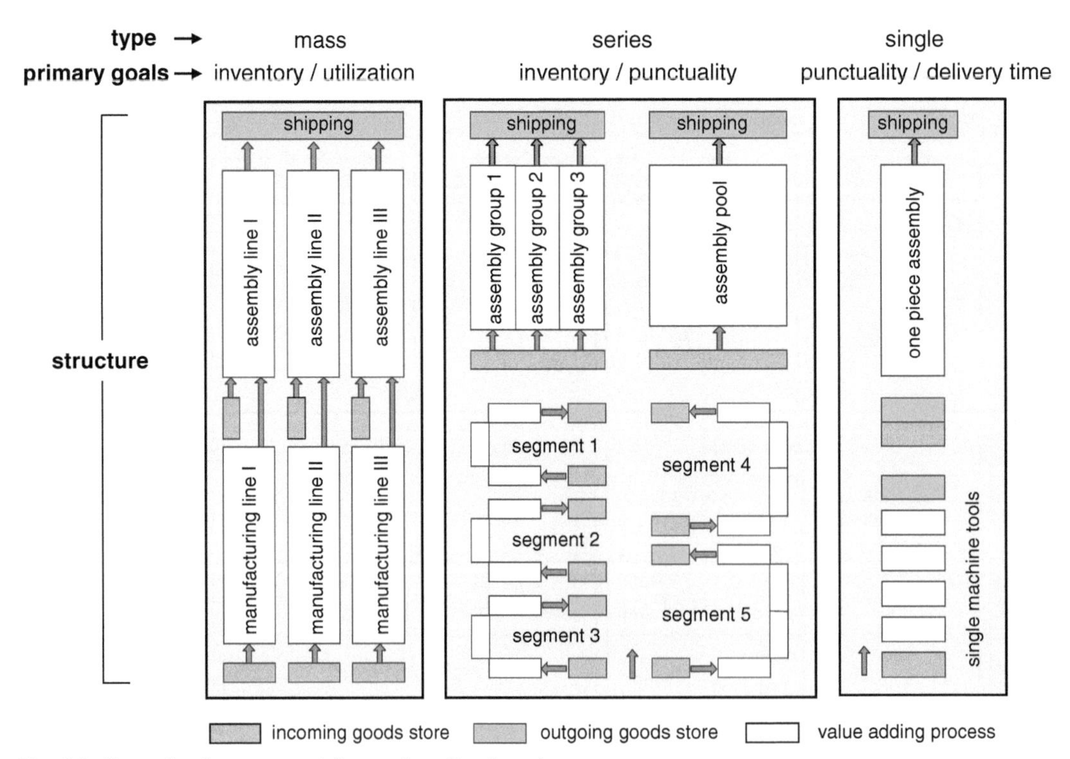

Fig. 9.9 Example of a segmented factory (acc. Brankamp)

segments is mass production, which for the most part has constant sales. The primary goal of this segment is for production to be as efficient as possible whereby the segment's success is measured by the utilization rate and WIP level. A second segment covers the requirements of serial production with a broad product variance. The integrated manufacturing cells allow one-piece flow and thus low WIP and short throughput times. Depending on how the customers are tied in (see supply models Sect. 9.5.3), the primary goal can also be due date reliability. The third manufacturing segment is focused on the requirements of single part production. Here, custom-tailored products are produced with a broad range of variants and with short throughput times. Due to the different operating sequences, the machinery is organized according to the function or job shop principle.

Due to the diverse requirements and restrictions of the three segments, different planning and control methods are also required. We will now turn our attention to the basics of production planning and control as well as the possible approaches to selecting and configuring the methods.

9.8 Production Planning and Control

The control loop, as depicted in Fig. 9.10 is well suited as a model for planning the production. The target agreement process and/or strategic positioning along with customer demands provide target values. Based on these a production planning and control system (PPC) uses various methods to generate a production plan that is then further broken down into in-house production plans, procurement plans and supply plans. These procurement, production and delivery processes are then executed. Once these processes have been completed, the actual values attained through an operating data collection system are converted into key figures and graphics with the aid of a logistics monitoring system. These then are compared with the planned, i.e., target values, and any deviations found are analyzed and measures for improving the target attainment are recommended.

The key tasks of production planning and control include planning the production program, planning the production requirements, and planning and control of external procurements and

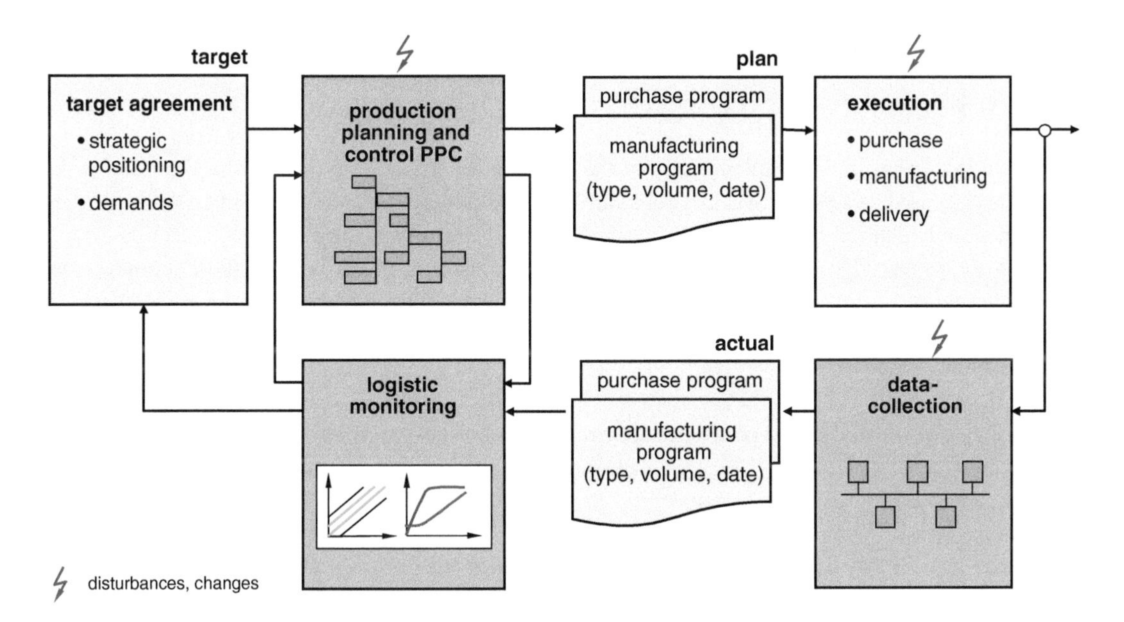

Fig. 9.10 Production planning and control loop

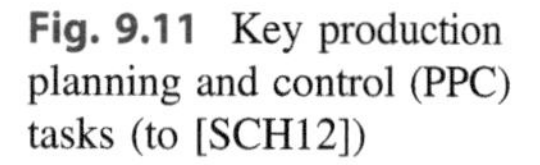

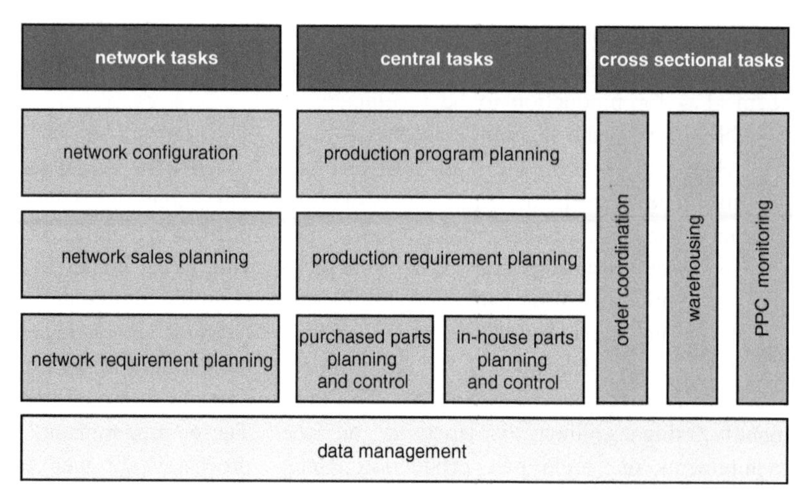

Fig. 9.11 Key production planning and control (PPC) tasks (to [SCH12])

in-house manufactured items (see Fig. 9.11). *Production program planning* determines which products should be produced in which quantities during the next planning periods. In doing so the underlying sales plan containing both forecasted sales and customer orders is tested for its feasibility in close collaboration with sales, production and purchasing. Usually the primary requirements of saleable products or product groups are listed in table form for a planning horizon of one or more years.

Production requirements planning determine the required materials and resources from the production program. In view of existing or planned inventories, secondary requirements for parts and components are then determined, manufacturing orders are scheduled, loads of the manufacturing and assembly workstations are determined and required due dates for procured materials are derived. The resulting procurement, manufacturing and assembly programs are then further planned by the external and internal PPC systems, and subsequently released for execution.

Nowadays, in order to avoid plans that appear to be more accurate than they are, production planning and control tasks are transferred as much as possible over to autonomous manufacturing sections; the keywords here being decentralization and segmentation. Employees are then not only responsible for executing the program but also—to a certain degree—for selecting the methods and measures used. The depth and complexity of planning is thus reduced while at the same time increasing the quality of task fulfillment.

In order to ensure the efficiency of the value adding chain, three cross-cutting tasks are required in addition to the four main PPC tasks. *Order coordination* aligns the processes, procedures and schedule across the various areas of the enterprise and beyond. *Warehousing* is responsible for managing inventories and for ensuring that both the manufacturing and assembly areas are supplied as well as the customers. *PPC monitoring* measures the target attainment from the customer's side as well as from the companies side. The key tasks and cross-cutting tasks are reliant upon master data and order status data being carefully managed.

Nowadays, in addition to the key and cross-cutting tasks, other network tasks are allocated to PPC [Sch12]. These combine all of the planning tasks that can be found in a production network. The core of these tasks is concerned with strategically designing the network configuration, planning the overall sales and planning the network requirements.

The MRP II concept (MRP stands for material resource planning) represents a very common planning approach for fulfilling the key PPC tasks. It was developed from the MRP (material requirement planning) approach originally established in the USA during the 1970s. Simultaneous planning is not pursued because it is assumed that the diversity of parameters and variables, their interdependencies and the uncertainty of the underlying planning data leads to a complexity that cannot be controlled even with today's computing abilities. Instead the entire PPC is divided into sub-problems or modules (see Fig. 9.12).

MRP II allows a solution to be found for the above mentioned planning tasks. This is accomplished in a gradual (level by level) coordination process, whereby the variables determined in each level provide the input variables for the next levels. The hierarchy of the planning levels is oriented on the chronological horizon of the sub-problems. Based on the strategic business plan, the material and capacity requirements are continually refined up to the loading of each machine for individual operations. Backwards coupling between each of the levels ensures the feasibility of the sub-plan on each of the next levels. Where there are deviations, actions either from the order side (due date shifts) or the resource side (capacity adjustments) is required.

MRP II represents the basis for many manufacturing planning software systems often called manufacturing execution systems MES. Nevertheless, our discussions regarding the customer order decoupling point, the types of orders and how they are handled as well the process models (see Sects. 9.2–9.4) show that the specific design of the PPC, the planning object and the planning depth have to be focused on the selected structures and processes. All of the PPC tasks can be impacted by this.

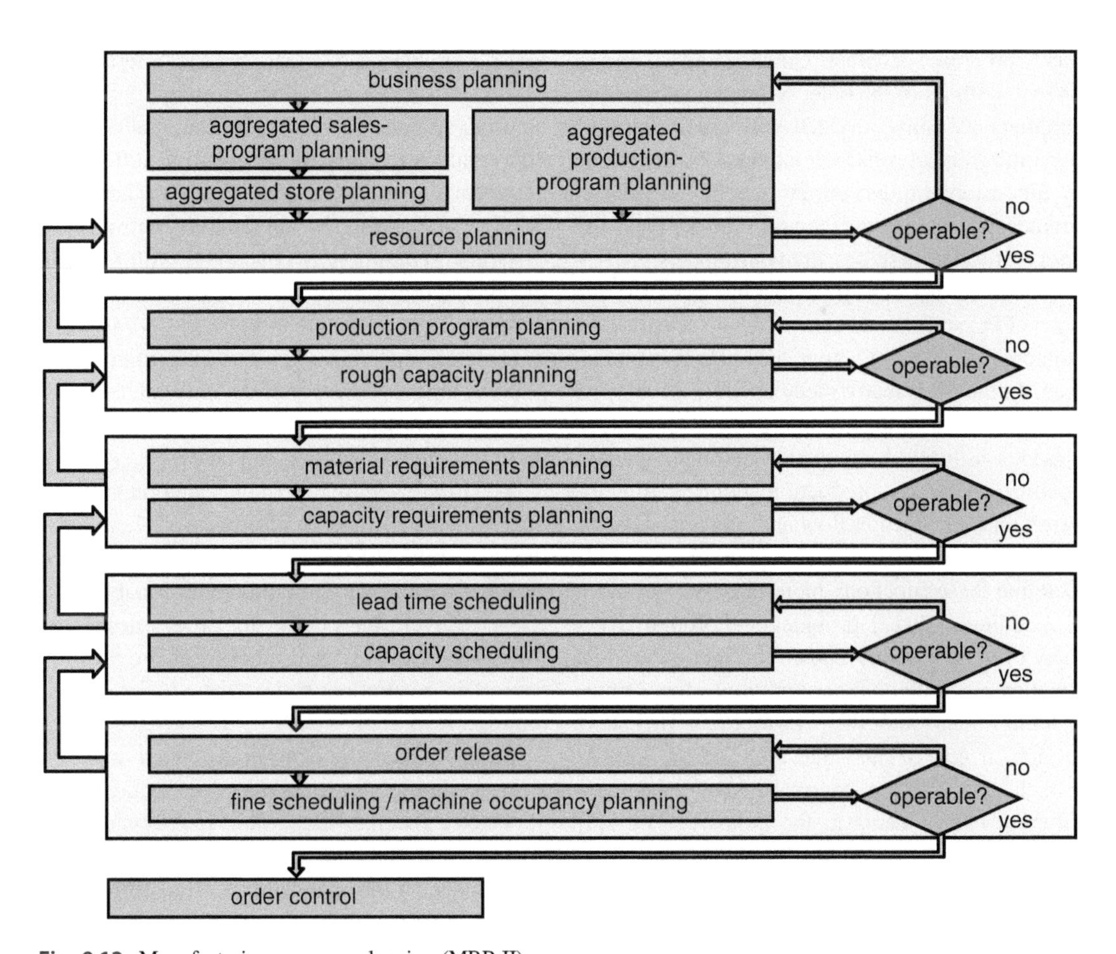

Fig. 9.12 Manufacturing resource planning (MRP II)

How detailed the production program planning must be is strongly dependent on the extent to which orders are produced neutrally with respect to customers. For production areas before the customer decoupling point, the production program serves to generate customer neutral manufacturing and procurement orders. In comparison, after the customer decoupling point concrete customer orders trigger manufacturing and/or procurement operations. In this case, the production program primarily serves to estimate the required capacity and materials, and is then usually only created on the level of the product groups.

The choice of process model strongly influences the PPC functions. The implemented procurement model is thus decisive for planning and controlling external procurements. Only when the traditional reserve stock model or single item procurement is implemented does the procurement program generate concrete procurement orders. With all of the other procurement models, suppliers are only provided with minimum and maximum stock levels and/or materials are called-up for customer orders currently being processed. Framework agreements then provide the basis for cooperation in these procurement models.

These factors strongly influence how the in-house PPC is designed as well as the selection of suitable methods and how they are parameterized. In the past, many methods were developed for controlling production processes and were aimed at fulfilling individual tasks under specific conditions. Since production control directly determines production flow and has a number of consequences for the factory planning, we will examine these functions more closely.

Lödding [Löd13] provides a thorough overview of how currently known production models function including the conditions for implementing them and their restrictions. In short though, it can be said that none of the known models can claim to comprehensively meet the different requirements in the industry. The aim therefore should be to test the control methods that come into question in regards to conditions, requirements and abilities of the production area and to the possibilities of implementing them adjusted to one of the respective tasks. In the next section we will consider a procedure for doing this.

9.9 Selecting and Configuring a Production Control Method

It is the responsibility of production control to implement the targets determined during the planning stages as well as possible, even when there are unavoidable disruptions. Based on this fundamental understanding, Lödding [Löd13] developed a universal model of production control (or manufacturing control as he refers to it) that defines four tasks for the production control. Figure 9.13 left depicts these using a funnel model for a production, while Fig. 9.13 right depicts the basic actuating and control variables as well as the target variables influenced by the interactions. The tasks set the actuating variables, while the control variables that determine the quality of the goal attainment each result from the differences between two actuating variables. We now will briefly clarify these four production control tasks.

Order generation is concerned with creating the production orders (jobs) based on the customer orders, production program and/or withdrawal of materials. It thus determines the planned input and output of the individual orders. Moreover, in scheduling the orders it also indirectly sets the processing sequence.

The *order release* determines the point in time after which an order can be processed. Generally, the availability of materials is also then verified and, where necessary, the delivery of materials is triggered. Relevant criteria for order release are, for example, the planned input dates provided by the order generation and/or the availability of required resources.

Order sequencing determines which orders in a queue should be processed next. Each order is allocated a specific value that is based on defined criteria and that determines its priority in comparison to the other orders remaining in the queue. There are a number of known sequencing rules, each of which support different primary

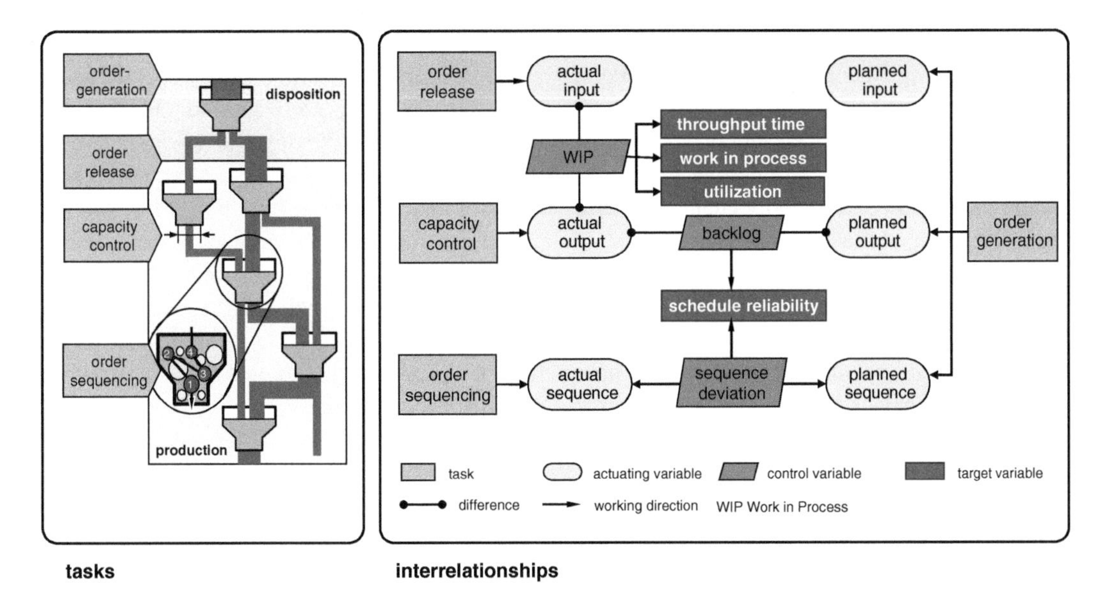

Fig. 9.13 Model of production control—tasks and interrelationships

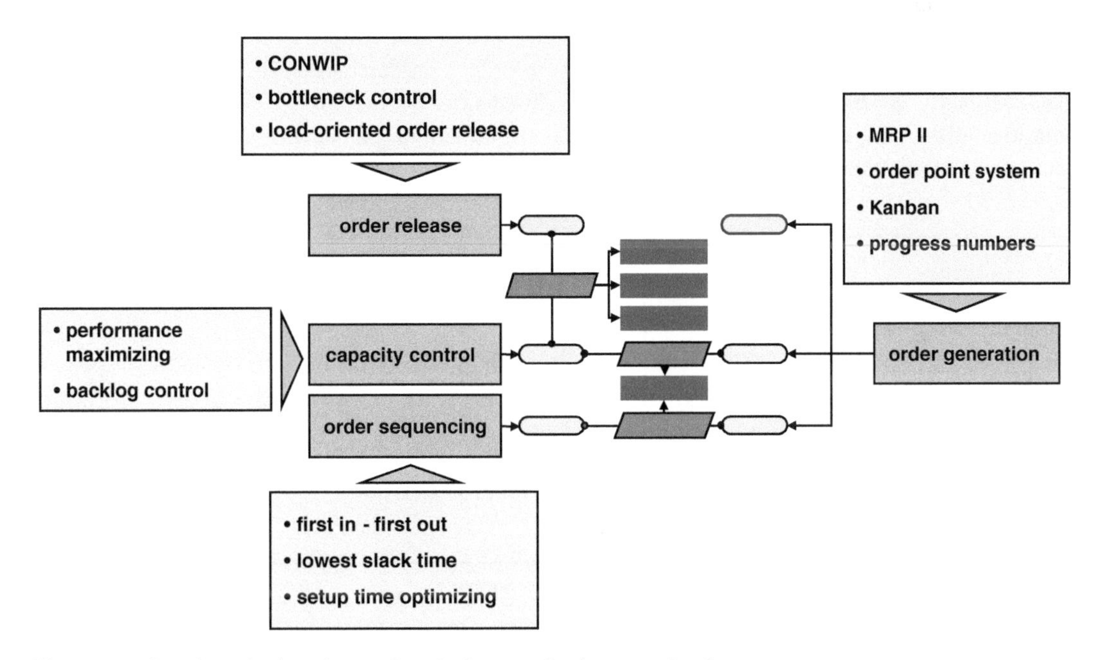

Fig. 9.14 Allocation of selected control methods to production control tasks

targets such as short throughput times or saving setup times.

Finally, *capacity control* is responsible for determining the working hours for each of the workstations and allocating employees to individual systems or groups. It thus influences the actual output of the production.

Examples of selected methods that can be drawn upon to fulfill the tasks are depicted in Fig. 9.14. Due to the abundance of known methods, we are only able to provide a brief presentation of these here. For further information, we recommend [Löd13] and [Sch11] in particular.

Of the methods for generating orders, MRP II is commonly implemented as is the Order Point system and Kanbans (see Sect. 4.7 for an explanation of Kanbans), whereas progress numbers (or cumulative production figures) is often found in networked supply chains and used across a number of manufacturing levels.

For the order release, methods such as Load Oriented Order Release (LOOR) [Wie95] or CONWIP [Hop96] are commonly implemented. Both are aimed at controlling the throughput times with the aid of WIP regulation. Another approach is bottleneck control, e.g., Optimized Production Technology (OPT) [Gol04].

Methods for controlling capacities are frequently aimed at maximally utilizing workplaces, whereas others support due date compliance by regulating backlogs.

Finally, sequencing rules should support shorter throughput times, greater due date compliance or minimum setup times. Nevertheless, it has been demonstrated that the impact of priority rules is heavily dependent on the WIP and that with low WIP levels, all of the sequencing rules come down to the 'first in-first out' rule.

What is important to note at this point is that not only does each method have a specific range of application, but each method also requires specific conditions in order to be implemented and has specific restrictions. Moreover, some methods also address several of the four tasks.

When configuring the production control for a specific application case it is important to keep in mind the strategic goals of the enterprise, the customer requirements, the restrictions resulting from the product structure and the abilities of the manufacturing system.

To begin, the customer order decoupling point (Fig. 9.15 top left) and individual control systems for production (Fig. 9.15 top right) are set and the objectives for each of the control systems are quantified (Fig. 9.15 bottom). In each control system different primary goals can be pursued. As already discussed, in make-to-stock productions areas the predominating goal is an economically efficient production, whereas in make-to-order production segments it is logistic efficiency.

Furthermore, it is important that the targets for the control systems are aligned with one another

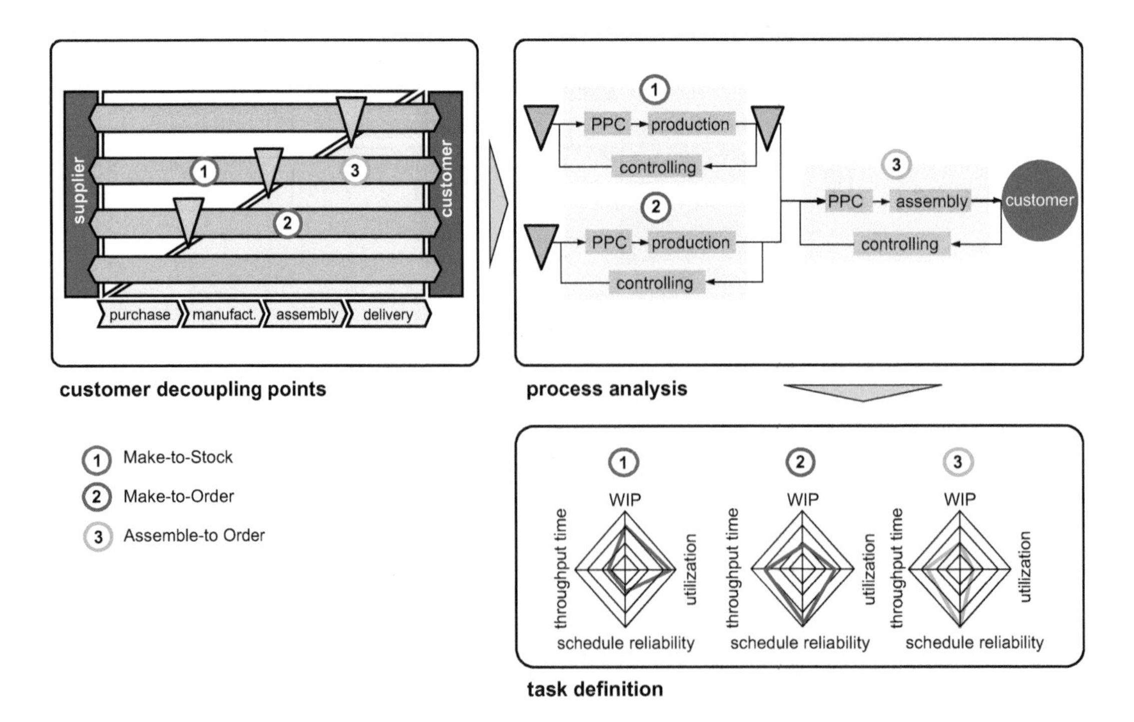

Fig. 9.15 Configuration of Production Control—process analysis and target definition

	criteria	characteristics				
product	product complexity	low-part product	multi-part product with simple structure	multi-part complex product		
product	product value	very low	low	middle	high	very high
product	sales volume per year	very low	low	middle	high	very high
product	variant number	very low	low	middle	high	very high
product	demand fluctuation	very low	low	middle	high	very high
product	storability	very low	low	middle	high	very high
product	load flexibility	very low	low	middle	high	very high
product	work plan quality	very low (not available)	middle	high	very high	
production	manufacturing principle	workbench principle	construction site principle	job shop principle	segment principle	flow principle
production	way of manufacturing	single production	individual and small series production	series production	mass production	
production	part flow	batch production	lot-wise transportation	overlapped manufacturing	one-piece flow manufacturing	
production	material flow complexity	very low	low	middle	high	very high
production	production bottlenecks	many (changing)	several (constant)	some	one	no-one
production	fluctuation of capacity demand	very low	low	middle	high	very high
production	capacity flexibility	very low	low	middle	high	very high
production	supply reliability by predecessors	very low	low	middle	high	very high
production	data availability	very low (only per order)	middle	high	very high (per work place)	

Fig. 9.16 Characteristics of control relevant features

as long as they collaborate in fulfilling customer orders. Thus for example, if the target for the make-to-order part of a value-adding chain is formulated as high due date reliability with short throughput times, then the supply certainty for this part of the process chain has to be ensured either with a suitably sized decoupling store or the ability to quickly react, thus requiring short throughput times in the preceding systems. If however, they are parallel manufacturing segments (see, e.g., Fig. 9.9) that produce various products for different customer groups, then the primary goals for the segments are independent.

The individual paths should also be analyzed for control-relevant features that result from the products completed there as well as the production itself (Fig. 9.16). *Products* should be differentiated with regards to complexity (few parts or multi-part complex products), value, and a variety of demand criteria (sales quantities, fluctuations, number of variants). Restrictions regarding storage (e.g., short shelf-life) can also be relevant for decisions. Finally, load flexibility expresses whether or not purchases (internal or external customer) can accept due date shifts to a certain degree. Generally this is possible when productions are make-to-stock. In exceptional cases though, due date shifts are also negotiable with the customer in make-to-order production.

The production itself influences production control through the selected manufacturing principle, the way of manufacturing as well as the type of part flow. The complexity of the material flow indicates whether there is generally a linear or a strongly networked material flow with a number of back loops for all of the products to be processed. The assessment of the bottleneck situation is thus also generally included. With a very linear material flow there is a tendency towards few but constant bottlenecks,

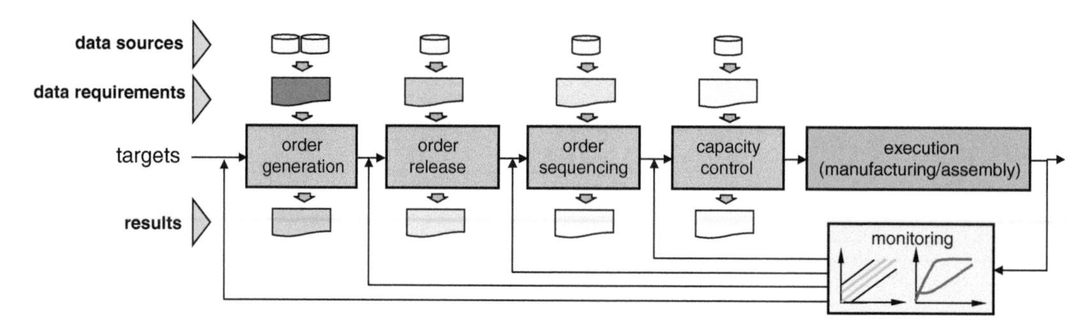

Fig. 9.17 Configuring Production Control

whereas with complex material flows there are frequently shifting bottleneck situations. Moreover, fluctuations in the capacity demand should be analyzed. Although they arise from demand fluctuations for the product, when a number of products are manufactured, it is possible that due to compensatory effects, the capacity requirements on the workstation level will be comparably constant despite strong fluctuations on the single product level. The more heterogeneous the product spectrum is the greater the probability of this. Finally the reliability of the supply from the preceding process must be considered. The above-mentioned requirements and restrictions have to be weighed against capacity flexibility as well as data availability.

In the next step, based on this information the production control can be configured (Fig. 9.17). Starting with the quantified objectives and the characteristics of the control-relevant features, it should be assessed for each production control task which of the methods that could in principle be implemented, should be implemented. In view of the expected results (required as input for the subsequent task) and the methods' requirements, the data requirements and sources need to be identified. Supplementary to this, a monitoring concept needs to be developed with which target attainment can be tested and which can be used to help derive measures for improving processes when necessary.

Next, using the practical example of a system supplier for a commercial vehicle manufacturer, we will discuss the procedure for configuring the production control.

During a workshop focused on defining targets, the significance of each logistic objective is first determined and then compared to the current performance profile (Fig. 9.18 top). This reveals that there is considerable need for action especially in regards to the delivery reliability. In the past, the supplier circumvented this problem by manufacturing-to-stock and supplying the customer from there. Weaknesses in the due date reliability were thus compensated for with large stores of finished goods. The project team thus agrees that the second main goal of configuring the production control is to reduce the finished goods stores and thus the tied-up capital costs; where possible, stores of finished goods should be completely eliminated.

The company manufactures customized products in single and small-sized series with a large number of variants (Fig. 9.18 bottom). Due to the customer's requirements regarding due date compliance, the load flexibility is very limited. The production is organized according to the job shop principle and is characterized by a very complex material flow. Previously, the enterprise attempted to meet the logistic objectives—in particular preventing schedule deviations—by using flexible working-hours, a group of cross-trained employees and outsourcing. The results however have not met the requirements, thus the production control is to be re-configured.

Initial discussions consider implementing a Kanban control. Due to the large number of variants and the low number of pieces per variant though, it is clear that such a method would not reduce the finished goods stores as required.

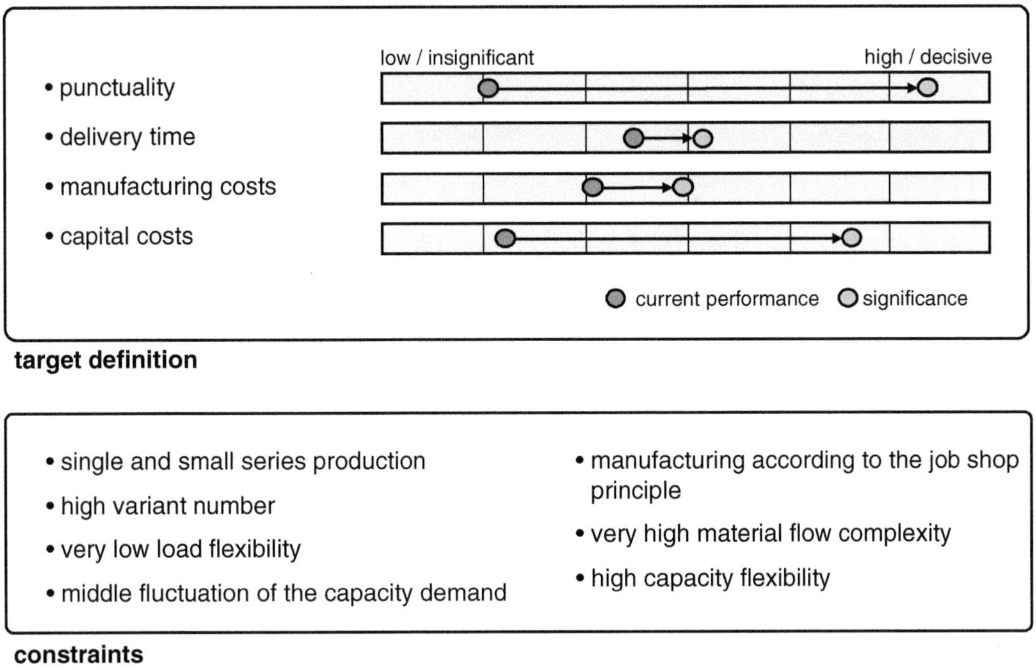

low / insignificant — high / decisive

- punctuality
- delivery time
- manufacturing costs
- capital costs

current performance — significance

target definition

- single and small series production
- high variant number
- very low load flexibility
- middle fluctuation of the capacity demand
- manufacturing according to the job shop principle
- very high material flow complexity
- high capacity flexibility

constraints

Fig. 9.18 Configuration of the production control—initial situation for case study

Generally speaking, pull-controls such as Kanban inherently produce a store (Kanban buffer), thus stores are a necessary condition for this method.

In addition the team checks to see if a WIP regulating method (in this case the Load Oriented Order Release) could be implemented. With this method an order is released when the WIP on a workstation falls below a given value. The aim of this method is to not only maintain a constant WIP, but also to stabilize throughput times and thus increase the planning certainty and in particular the due date certainty. In order to implement this method though, it has to be possible to shift the loads time-wise on the workstations. This is only possible if the customer's requirements allow load flexibility at least to some extent or when the production is decoupled from the customer via a stock buffer i.e., store. Since in this case, neither of these are an option, a WIP regulating method was also ruled out as an option.

The configuration that the enterprise chooses in the end is depicted in Fig. 9.19. Orders are generated based on the weekly customer demand. In the first step, backwards scheduling is used to schedule the orders down to the level of the individual operations. The information gained by this is used, for example, to determine the capacity requirements with three different planning horizons (3 weeks, 6 weeks and 3 months) and is suitable for suggesting measures for aligning the capacities (mainly planning special shifts and out-sourcing).

Orders are now released strictly according to the calculated start date; other criteria such as ensuring utilization of individual workstations by pulling orders are no longer permitted. Sequencing the production orders also occurs based solely on the remaining slack and therefore of due date criteria.

For controlling capacities in the short-term range, an approach for regulating backlogs is ultimately chosen. When differences between the planned output and the actual output arise, measures that can be quickly implemented are used to align the capacities e.g., flexible work-hours and relief workers.

Since operation related feedback data are required for sequencing and for controlling

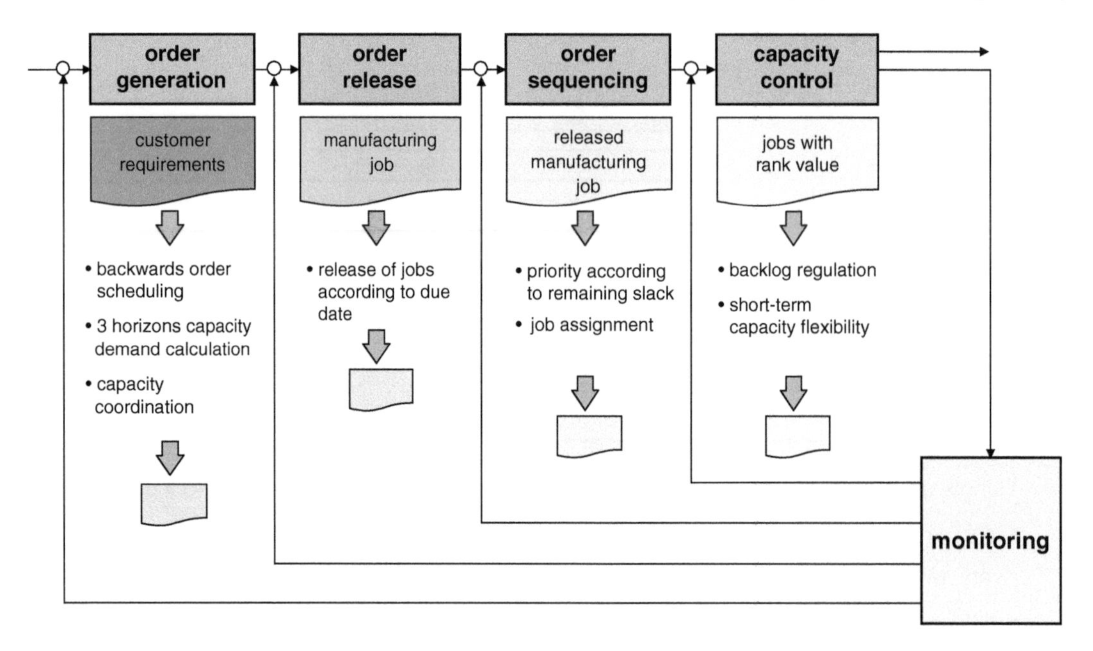

Fig. 9.19 Configuration of the production control—end situation for case study

capacities, a system for monitoring logistics that can supply this data is also introduced. Moreover, the system serves to continually review important planning parameters such as the throughput times and capacities of the workstations as well as to enable sustainable changes in production control.

We will let this case study be the end of our discussion on designing factories at a work area level from the perspective of production systems and logistics. In the next chapter we will consider the corresponding spatial design.

9.10 Summary

For the spatial layout of a work area, an exact knowledge of the functions to be fulfilled therein is essential. These are highly dependent on the so called Customer Decoupling Point and the resulting order types. Before the customer decoupling point orders are made to stock, while after the decoupling point orders are customized. The sub-processes necessary for this are procurement, manufacturing—usually divided into parts manufacturing and assembly—as well as the delivery of finished goods to a warehouse or to the customer. For procurement different models are used, which differ according to the procurement initiation, the location of goods and the transfer of ownership. The delivery models comply with the procurement models where the production company is considered as supplier.

Furthermore, the structure of production and assembly including the spatial arrangement of resources (machines, equipment) and utilities must be determined (parts, assemblies, and finished products) as well as the number of required employees. The main manufacturing principles are the job shop or functional principle (machines with the same function are grouped in workshops), the flow or product principle (workstations for a product are located in the sequence of operations), the segment or group principle (work stations for the production of a group of similar parts are combined and autonomously controlled) and the site fabrication principle (a product is manufactured in a place and machines may be brought to this place). For the assembly, there are analogous principles.

Finally, organizational planning and management of contracts must be determined. For this purpose, a Production Planning and Control system first plans the internal and external items with respect to the amount and date. The main tasks of the following shop control are order generation (batch size), order release, order sequencing and capacity control. These are configured according to the selected order type with different methods.

Bibliography

[Frü06] Frühwald, C., Wolter, C.: Prozessgestaltung (Process Design). In: Hagen, N. u.a. (Eds): Prozessmanagement in der Wertschöpfungskette (Process Mangement in the Value Chain). Haupt Verlag, Bern, Stuttgart, Wien 2006

[Gol04] Goldratt, E.M., Cox, J.: The goal. A Process of Ongoing Improvement. 3rd edn. North River Press, Great Barrington (2004)

[Hoe92] Hoekstra, S.J., Romme, J.M.: Integral Logistic Structures: Developing Customer-Oriented Goods Flow. Mc Graw-Hill, Berskshire (1992)

[Hop96] Hopp, W.J., Spearman, M.L.: Factory Physics. Chicago, Irwin (1996)

[Löd13] Lödding, H.: Handbook of Manufacturing Control: Fundamentals, Description, Configuration. Springer, Berlin (2013)

[Sch11] Schönsleben, P.: Integral Logistics Management. Operations and Supply Chain Management in Comprehensive Value-Added Networks, 6th edn. CRC Press, Boca Raton (2011)

[Sch12] Schuh, G., Stich, V. (Eds.): Produktionsplanung und -steuerung (Production Planning and -Control). Vol. 1: Grundlagen der PPS (Fundamentals of PPC). Springer Berlin (2012)

[Wie95] Wiendahl, H.-P.: Load-Oriented Manufacturing Control. Springer, Berlin (1995)

[Wil98] Wildemann, H.: Die modulare Fabrik. Kundennahe Produktion durch Fertigungssegmentierung (The Modular Factory. Customer-Oriented Production by Manufacturing Segmentation), 5th edn. TCW, Munich (1998)

10 Spatial Workspace Design

The work environment is a part of who we are. As a personally perceived part of a production, workshop or office, the workspace should be a natural and pleasing extension of our personality. Frankly, it is unthinkable to reduce the quality of our lives down to our time outside of work and to differentiate between our work and our spare time. Up until now, the legal requirements for minimum lighting and ventilation, and/or maximum noise levels have set the standards for whether human needs have been met in factories or not. Fulfilling these requirements was considered to be the essence of 'humanizing' the workplace. However, it is not easy to quantify subjective parameters such as the changes in daylight and a harmonious environment since they are just as important to our physical well-being as the easily measureable parameters.

When planning for workspaces, it is thus important to investigate the various possibilities for promoting the physical and mental well-being of employees, their willingness to work and their ability to perform their jobs. Figure 10.1 depicts the most important elements with regards to the design fields: communication, lighting, comfort, relaxation and fire protection. In the following sections, we will clarify and consider their roles with regards to changeability.

10.1 Communication

While the influence of communication on productivity is undeniable, creativity and innovation are prerequisites for a progressive and changeable organization. Buildings play an important role in revealing hitherto unknown possibilities of communication. People are increasingly spending a greater proportion of their time at work participating in project teams, decision making committees and workgroups as well as discussion rounds. An old Japanese expression roughly translated as "the gold lies in the minds of the employees" suggests that it would be wise to focus once again on the unique talents of human minds. In Taylorist mass production theories, thinking, decision-making and acting was separated. The mental and physical aspects cannot be separated any more and needs to be reunited. The traditional 'blue collar'/'white collar' division has to be dissolved and replaced with solidarity as the new ideal. Material and communication flows, need to be re-integrated. Communication has become a decisive production factor. Whereas mistakes in the physical material flow become evident sooner or later, those in the mental communication flow usually remain hidden.

H.-P. Wiendahl et al., *Handbook Factory Planning and Design*,
DOI 10.1007/978-3-662-46391-8_10, © Springer-Verlag Berlin Heidelberg 2015

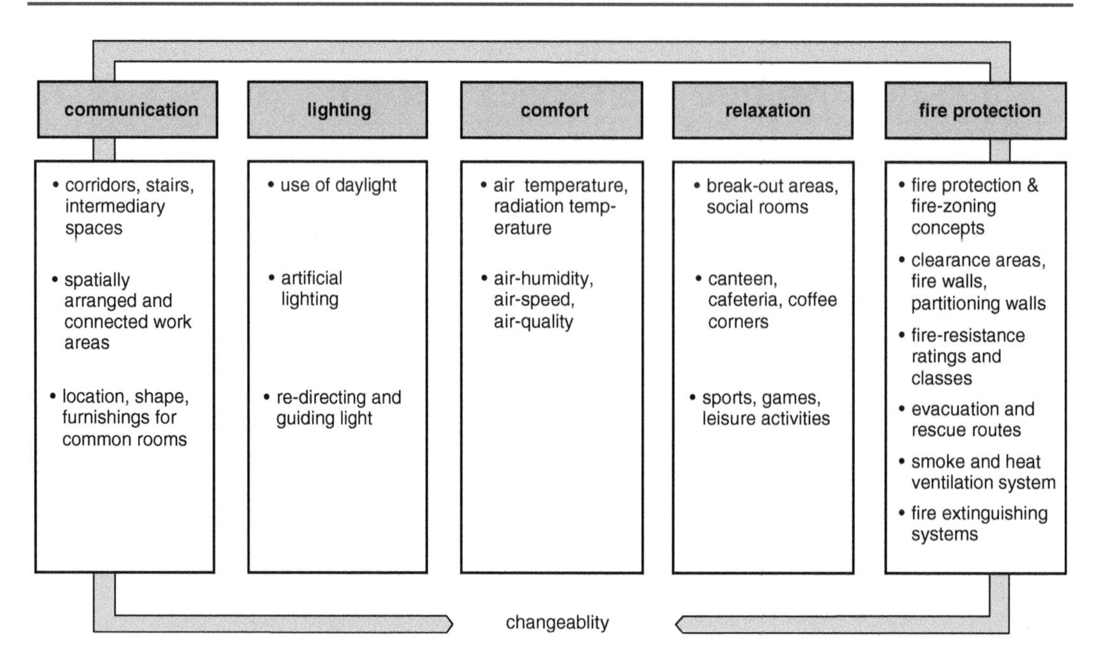

Fig. 10.1 Overview of design fields and elements of a workplace. © Reichardt 15.180_JR_B

For the building concept this means providing building structures that stimulate the creativity of the users. Outside the formal work-spaces, seminar halls and meeting rooms, communication occurs more by chance informally in the hallways, connecting corridors, etc. [Rei01] has shown that appropriately designed circulation areas, workspaces and common rooms increase the level of communication, especially in larger industrial plants.

Figure 10.2 illustrates the impact of the spatial design depending upon the type of communication, which can be oriented either internally or externally. Finding the 'right' mix between internal and external orientated communication for a project should be an important point of consideration when formulating the concept and objectives. Consequently, the architectural design could offer combinations of larger (common) group rooms with smaller (individual) rooms for structured 'think tanks'.

Communication takes place on purpose in common rooms and rather spontaneously in connecting areas. Thus, the need to provide stimulating impulses for communication is therefore a matter of designing 'intelligent' spaces.

10.1.1 Corridors, Stairwells, Intermediary Spaces

Circulation areas are often conceived unilaterally for functional necessities. In the truest sense of the words these "evacuation routes" are mostly narrow, claustrophobic and dimly lit. They 'force' people to rush through them without stopping for a moment. There is no impetus to pause and spontaneously exchange a thought or an idea with a colleague.

Architectural elements such as daylight, an appealing view and others offer possibilities to linger along these "evacuation routes" (Fig. 10.3, left). Building structures which do not naturally lit stairwells and corridors should be a thing of the past. Varying sunlight and ever-changing shadows induce a calming effect on humans, while attractive views provide a variety of stimuli as employees gather visual impressions from their surrounding environment. The natural curiosity of humans is satisfied. Moreover, the feeling of being a part of a whole facilitates spontaneous contacts i.e., by signaling one another with the wave of a hand or by calling-out.

group allocation ▷	several groups in one room	entire group in one room	group distributed among rooms	
persons per room	13 - 30	3 - 12	2 - 4	1
spatial zoning ▷ / group type ▽				
strongly internally oriented	disturbance due to too much information	suitable for high degree of spontaneous communication	too long corridors	too long corridors
weakly internally oriented	not necessary	not necessary	adequate for situations with a minimum of communication	too greatly isolated (functional and social)
externally oriented	preferable for contacts to neighbouring groups external orientation not ensured	external orientation not ensured	external orientation restricted	suitable when there is a large number of customers/ visitors and confidential talks

Fig. 10.2 Communication oriented interior design. © Reichardt 15.161_JR_B

1 corridor area
2 stairwells between floors
3 stairway halls
- daylight
- attractive visual link
- opportunities to linger

1 hall level
2 gallery level
3 service sector
4 office area
- variable room shapes
- variable/transparent room boundaries
- building structure favors communication spaces

1 hall level
2 gallery level
3 service sector
4 office area
- bright, airy, attractive
- changeable views
- leisure flair
- green areas, plants

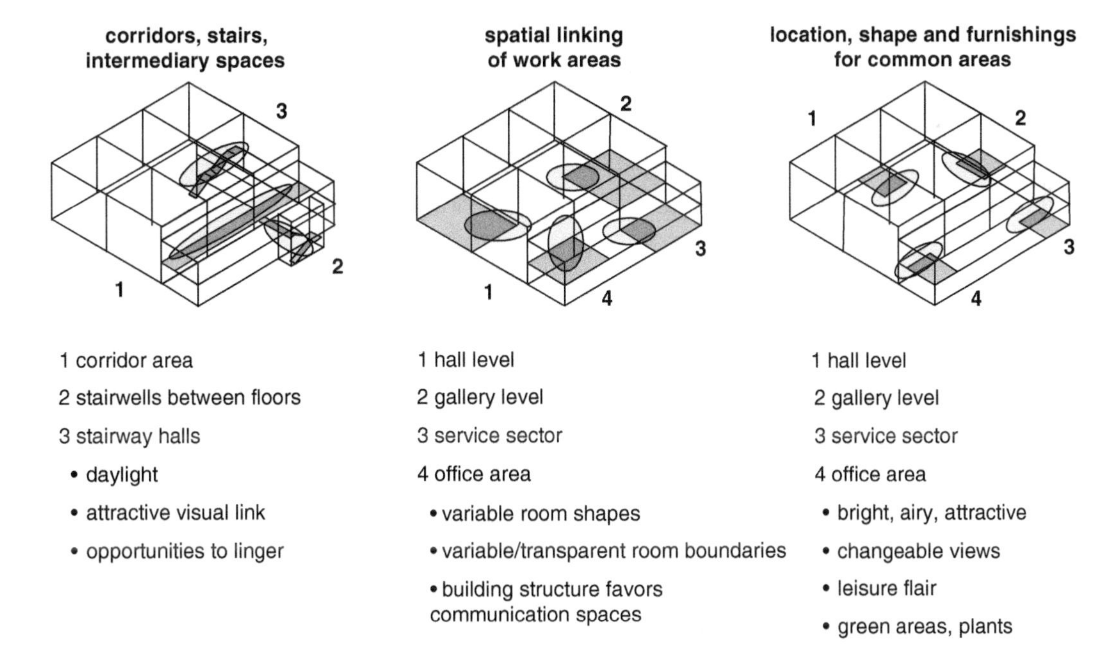

Fig. 10.3 Structural features promoting communication. © Reichardt 15.162_JR_B

The standard dimensions need not have to be used for corridors, nor do staircase landings always have to be the smallest possible space that permits only maneuvering. "Casual" encounters can be planned by creating larger spaces that encourage temporary lingering. Depending on needs, coffee corners or copying stations can be allocated to these spaces. Studies by Ebadi and Utterback [Eba84] and Bismarck and Held [Bis98] show that 80 % of all innovative ideas originate through direct personal contact and that informal communication promotes collaboration at the workplace. Architecture which facilitates communication can be highly rewarding.

10.1.2 Arranging and Linking Workspaces

Changeable spaces, which allow mutual exchange and interaction, should be provided for the different sections of a factory from the manufacturing, assembly and prep-work areas to the PPC and quality management right up to research and development. Unfortunately many factories are still designed as production halls with independent administration buildings connected via bridges. Narrow, isolated structures are frequently docked onto the production halls across firewalls without a view into the production. Sterile architectural solutions, strongly limit exchanging and intertwining thoughts, decisions and actions at meeting points, thus restricting a critical aspect of communication.

Building structures having mutable spaces with expandable and transparent room boundaries as well as a range of multi-purpose communication spaces are definitely advantageous (Fig. 10.3, middle). In office buildings, the relation between the shape of the building and the depth of the rooms is obviously dependent on the working style. With regards to communication, traditional office typologies having a corridor in the middle and cellular offices on both sides have been found to be disadvantages when compared to newer working styles such as combi-offices, group rooms, or combinations derived from these. With cellular offices, the depth of the building normally adds up to 12–13 m (39–42 ft) including a minimum of 2 m (6.7 ft) width for corridors and 5 m(16 ft) depth for small offices on either side. On the other hand, combi-offices require a building depth of approximately 15–18 m (49–59 ft) since corridors are substituted with multi-functional spaces for common use or group activities.

Whereas up until now, designing office spaces was considered to be a distinct and separate activity when compared to planning the factory premises, the notion is increasingly undergoing a change with both places—for manual labor as well as for mental work—being now designed by the factory planner. Without going into too much detail on this, we will outline a few approaches here. For a comprehensive overview of the topic the readers may wish to refer [Spat03]. Further references can be, also, found in [Mar00].

Figure 10.4 organizes currently known office concepts based on the parameters of 'time present' and 'mode of operation', while Fig. 10.5 depicts an overview of layouts for these concepts.

Traditional *cellular offices*, in which employees are present continuously and operate independently, offer a personal space to retreat in order to avoid disruptions. However, they are in

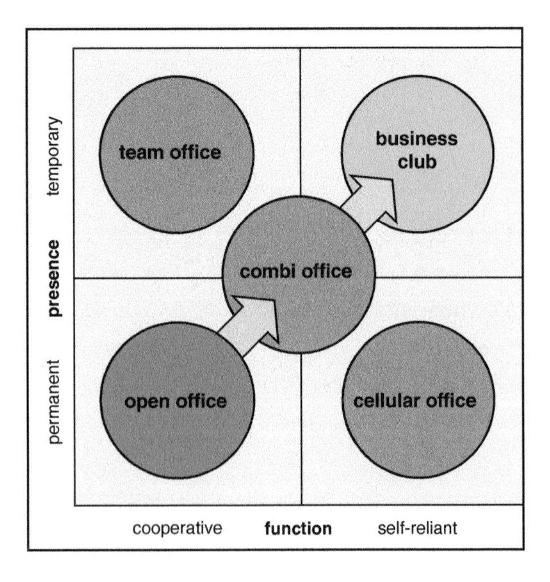

Fig. 10.4 Office concepts depending on workers' presence and mode of operation (Congena GmbH acc. to Wikipedia). © IFA 15.270

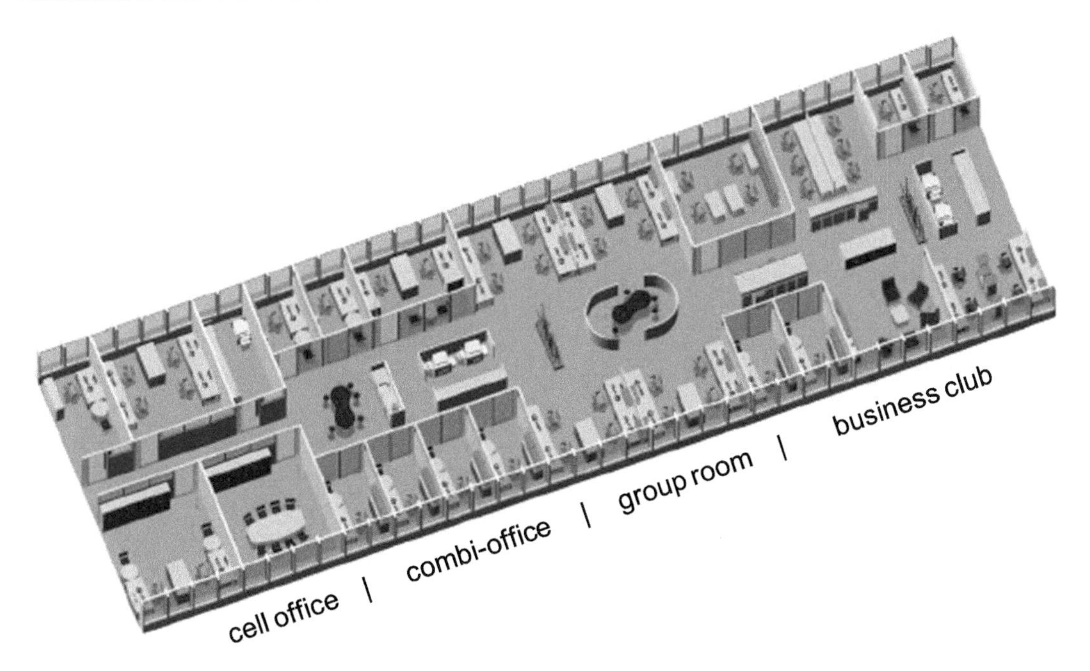

Fig. 10.5 Overview of office concepts (Congena GmbH acc. to Wikipedia). © IFA 15.271

principle resistant to change and adverse to communication. The opposite of this design are *open offices*, which are aimed at strengthening collaborative work. Unfortunately, open offices are often not preferred because telephone conversations, discussions and people moving around generally make it difficult to concentrate on work. A number of organizations have made attempts to counter these disadvantages by dividing this space into smaller areas and agreeing upon a code of conduct. *Team offices* provide spaces for temporary project teams.

It has since been proven though that a combination of office forms is practical. Such offices are referred to as *combi-offices*. In this case, individual offices are established in a large area, partitioned by transparent and possibly mobile dividers with bigger work areas arranged in front of them. Employees can then concentrate and work alone or spontaneously work as part of a group. Finally, the *business club* idea is increasingly gaining followers. Similar to the model of business centers found in airports, train stations and hotels, temporary workplaces are established in enterprises for individuals or workgroups who wish to withdraw or search for new solutions with internal or external teams. Approaches such as this seem to be a natural choice for marketing, research, product development and engineering divisions.

While designing office spaces, the question of how long and how often employees are present is increasingly becoming important. For example, when sales personnel are only present in the office approximately 50 % of the time and in all probability they are busy in meetings half of that time, it seems logical not to allocate any permanent workplace for them. Instead they might have portable containers with their personnel files, which they can pick-up from a depot on days that they are in the office. They can then seek out a free desk, log themselves into the company network with their laptop and are ready to work within a few minutes.

If additional spaces for research and development, labs or workshops are required, it can easily mean another 20 m or more is added to the building depth. In case parts of the production can be 'stacked' above one another i.e., on different floors, intelligently structured spaces with generous floor plans can be conceived as "industrial lofts" without disruptive walls and supports. Variable spaces and room depths provide the best options for changing work or communication forms. The room closures for the different sections could be reversible and interchangeable thus allowing spaces to be enlarged or reduced with minimal costs and effort as well as facilitating exchanges between sections and allowing them to be interconnected. These mobile 'transformation elements' should be transparent for the most part, allowing people to see in and through the different areas. This transparency between the individual groups helps develop and maintain a sense of community i.e., of a team realizing a common vision.

By clever reinterpretation and extension of the function of circulation areas, organizations can profit from a communication idea that is deep rooted in the building floor plans as well as cross-sections and is the guiding principle behind how the various structures, spaces and functions are interconnected. Based on Scandinavian examples like the innovative Swedish car assembly plant of Saab in Malmö, a number of variants for a communication backbone have been developed. In a reversal of traditional models, the middle is no longer dedicated to the material flow and logistics, but rather is occupied by communication and recreational areas for employees. Skoda's car plant in Mladá Boleslav near Prague is a good example of this concept.

10.1.3 Location, Shape and Furnishings of Common Areas

Various possible forms of formal and controlled communication could be distributed across key strategic positions in the building. It seems logical to place seminar and training rooms at the junction or along the production areas. Venetian blinds or slatted window coverings make it possible to change the views inside or outside in a variety of ways. Discussion or meeting areas can also be located here—or in case of frequent visitors—positioned close to the vestibule. Musty break-out and/or change-rooms in the basement should be a thing of the past, replaced by bright, airy, attractive possibilities (Fig. 10.3, right). A cafeteria or canteen located on an upper level, such as a roof terrace with leisurely flair, beckons workers to visit them outside of the main meal times and thus exchange information with one another.

The two-storied production and engineering building depicted in Fig. 10.6 is an example of a holistic solution wherein communication and changeability are the main drivers. High quality audio devices are produced here [Fi09]. The offices for all of the divisions involved in developing technology, preparing the work and producing the devices are accommodated in the intermediate galleries along three sides of each level. The large column grid, measuring 16.8 × 8.4 m 8 (18.3 × 9.2 yd), ensures a high degree of changeability. A break-out area on the ground floor, glass office fronts and a centrally located meeting room allow for options for quick communication. In addition, certain areas are encapsulated using light-weight construction systems to avoid emissions or for special climatic requirements. The galleries can be extended outwards thus providing the possibility for "internal" growth without having to intervene in the building structure itself.

10.2 Lighting

In the 1970s, windowless artificially lit factories were propagated as a replacement for conventional north lighting in production halls. Since then however, there has been a concerted effort to return to the environmental/psychological advantages of good lighting for workplaces; the goal now often being to light factories using daylight. The reasons for this are increasing environmental knowledge and the overall awareness of the employees who expect "natural" working conditions.

Natural lighting provides a significant economic and ecological benefit by saving the energy that would have been otherwise wasted in lighting up rooms during daytime. Moreover, lighting and especially daylight is one of the most important factors for humane work.

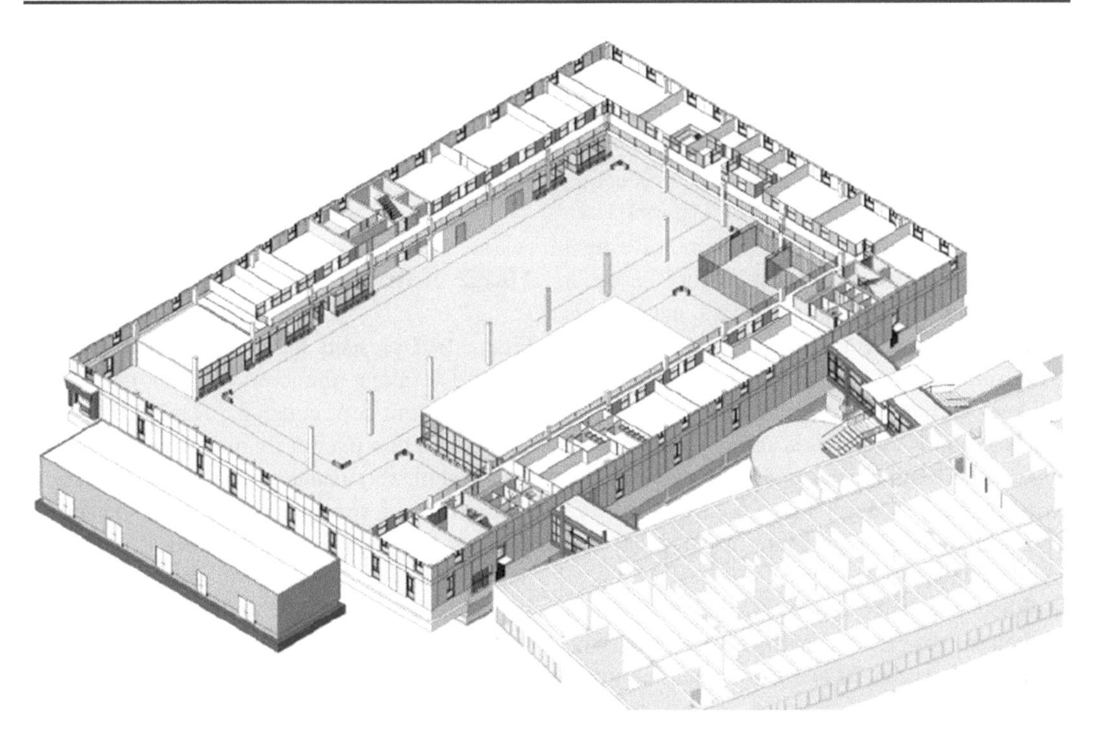

Fig. 10.6 Communication friendly factory building with a high degree of changeability (Reichardt). © Reichardt

Humans absorb 80–90 % of their information visually. Light influences our levels of motivation and our sense of well-being. Varying light intensity and atmosphere impacts how organisms are stimulated. It's proven beyond doubt that the weather conditions physically influences the moods of humans. Furthermore, the daily 24 h rhythm determines our sleeping and waking behavior and is indispensable as a regulator for our internal bodily functions and thus our overall well-being (see also Fig. 7.12).

It is also proven that changes of the sunlight reduce the error rate in production. For times when it is dark and in order to supplement the natural light, there needs to be a comprehensive plan encompassing all possible conditions for artificial lighting. The results need to be merged together with the natural lighting sources into a comprehensive lighting design. The distribution of light in rooms, especially in case of a deep building, can be optimized with systems that redirect light. Based on their structural configurations as well as their significance for the changeability of a factory, we will now turn our attention to a more detailed discussion of light related design elements i.e., daylight, natural lighting, artificial lighting, and redirecting light.

10.2.1 Daylight

Ever-changing components of daylight (e.g., intensity, direction and color spectrum) transmit considerably more information than the static ones found for example in artificially lit spaces. Daylight not only makes it possible to see things better, it also facilitates visual perception processes through its changes, increasing the amount of information being received and decreasing mental stress. The heightened levels of awareness increase our levels of attention and consequently prevent mistakes. However, using intelligently a free resource such as daylight is by no means common practice. The lighting designer Christian Bartenbach feels that—especially in industrial construction—there is a lack of understanding about daylight. "Unfortunately the measures for using daylight are often completely lacking. This is, in fact, incomprehensible, since mistakes especially in this area have an immediate financial impact, moreover, investments in resources for lighting, seeing and perceiving measured against the investments in buildings and production resources are infinitesimally small" [Bar98].

Sunlight is a combination of short wave light as well as long wave thermal radiation; as such it can enter in through glass surfaces and heat up interior spaces. The thermal properties of large glass surfaces thus have to be kept in mind. Whereas earlier research on luminous efficacy and heat ingress was based on assumptions and manual calculations made across the building cross-sections, today there is a wealth of advanced tools available such as 3D light and energy simulations.

When we talk about suitably lighting a workplace we are primarily concerned with ensuring consistent illumination and luminous efficacy. Illuminating a space evenly prevents shadows and glare. The consistency of the light distribution measured on the work level is dependent on the distance from the opening in the roof, whereas the size and type of windows determines the quality and consistency of the luminous efficacy. Furthermore, the relation of the available interior daylight level to the external light level when there is a cloudy sky is determined by means of the daylight factor DF.

10.2.2 Natural Lighting

Figure 10.7 presents the progression of daylight factor DF along the cross section of various possible room configurations. Daylight factor is defined as DF = interior illumination (lx)/external illumination (lx) under a cloudy sky condition. For each of the spaces in the diagram, the sum of daylight openings is assumed to be 1/6th of the floor space. This parameter is referred to as the 'window factor' WF and defined as WF = window area/floor space.

The typical progression of daylight for a side window with the assumed WF is shown in the upper left portion of the figure. It is clear that due to the positioning of the window the mean attainable DF value is not optimal in this case. The exponential drop in the intensity level of

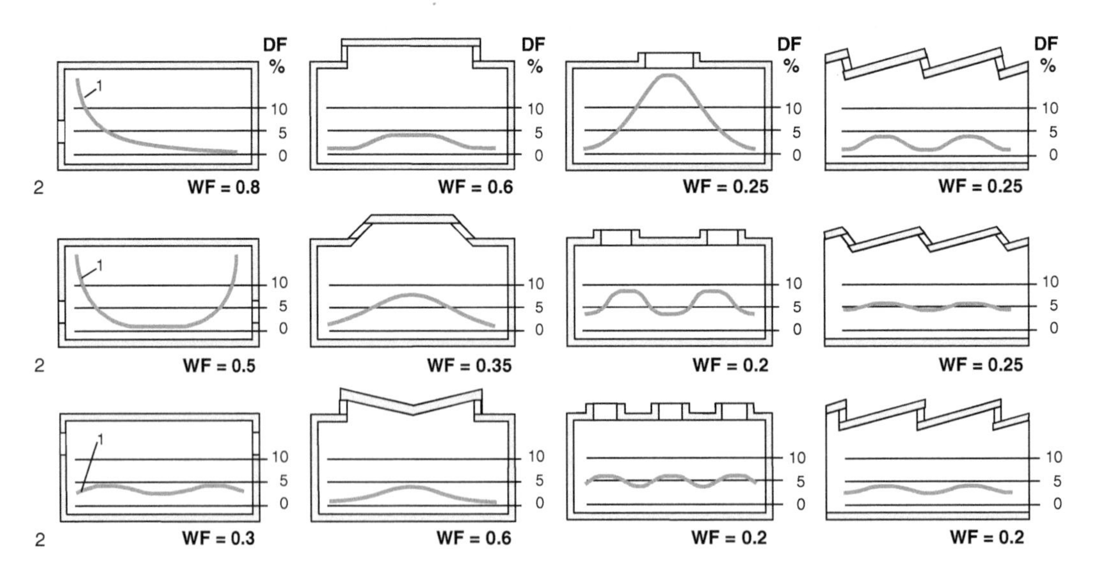

Fig. 10.7 Impact of room profiles on the daylight factor. © Reichardt 15.164_JR_B

daylight entering into the room through windows means that the same is barely useable. In short, it is almost impossible to use deeper rooms whose only natural light source is side windows without supplementary artificial lighting.

Generally speaking, workplaces whose daylight factor is 2 % or less should not be considered without supplementary artificial lighting. A rough rule of thumb is that points in a room from which a piece of the sky cannot be seen would probably require supplementary artificial lighting. An influx of light from above provides the best supply of daylight for the middle of a room. The entry surface should be designed such that the work area is uniformly illuminated.

A number of rooftop lighting systems and configurations have been developed for industrial buildings based on the criteria mentioned here. As variants of so-called shed (or pitch) roof forms, they are oriented in the northern hemisphere for glare-free northern light.

Shed roofs are most common for large-scale buildings. By installing a number of small ridge-like superstructures one behind another, the height of the roof can be controlled. The slope of the two sides for each of these super-structures is usually different; one side is steeper so that the building requires less support for the roof structure. The other side is usually made out of glass.

Figure 10.8 presents' conceptual schemes of the light distributions within halls for roof forms commonly found in industrial buildings. Thereafter, Fig. 10.9 lists typical advantages and disadvantages of the different overhead lighting arrangements.

Since incoming sunlight also brings heat into the space, on sunny days the heat generated can be uncomfortable. Thus, in addition to possible glares and reflections at workplaces, measures have to be taken to ensure that the room and equipment temperatures do not exceed permissible levels due to the incoming sunlight. To some extent, this can be achieved by using shading devices which impede solar radiation. Unfortunately, the intensity of the light also reduces proportionally. In short, when shading systems are effective more artificial light is required on nice sunny days.

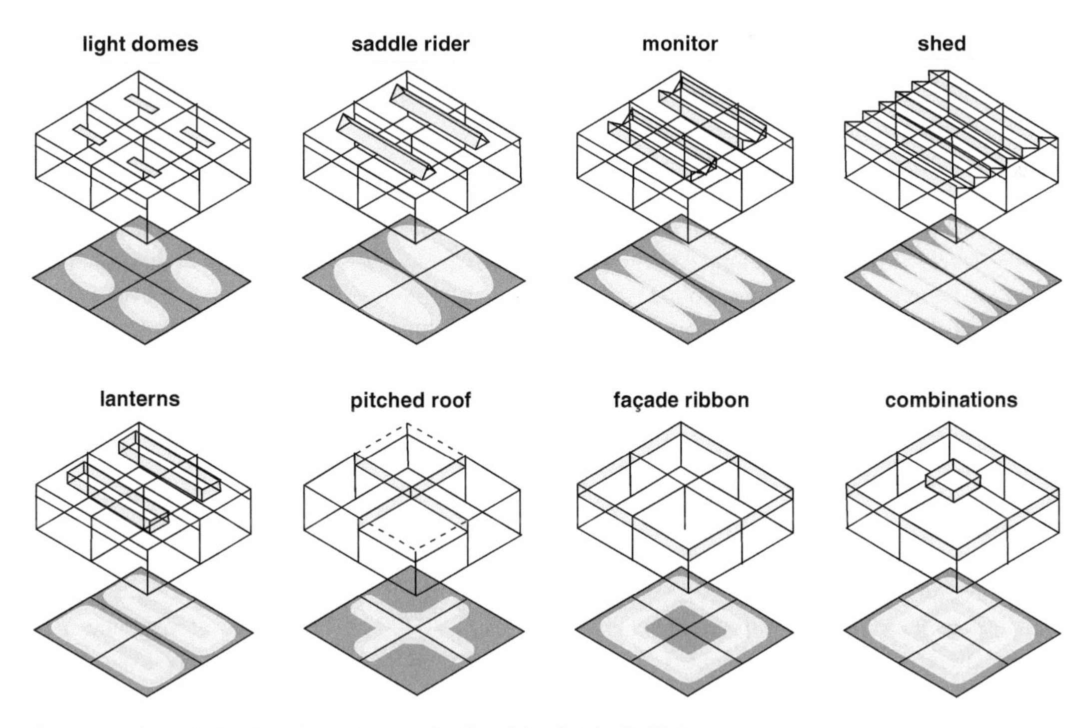

Fig. 10.8 Light distributions for various roofs. © Reichardt 15.165_JR_B

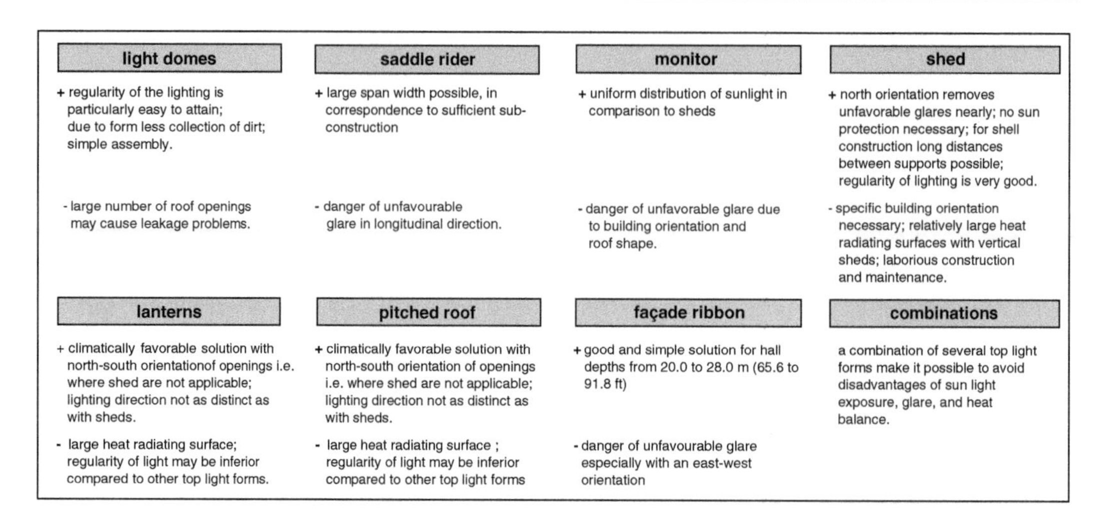

Fig. 10.9 Advantages and disadvantages of different overhead lighting arrangements. © Reichardt 15.166_JR_B

Intelligently designed industrial sheds, having north-lighting systems (in the northern hemisphere), ensure uniform illumination without glare. Also, they do not require shading devices and/or cast shadows.

According to [Bra05], a multitude of factors including processes, logistics and users over the long-term have to be considered in order to develop the lighting design for an industrial project. A comprehensive overview is given in [IES11].

The degree of changeability regarding daylight utilization is directly dependent on choosing the most advantageous forms of roof configuration and room section to maximize the natural lighting levels inside the building. When combining a number of skylight forms, three-dimensional lighting simulation software can be used to analyze different configurations to establish the expected distribution of light and avoid disadvantageous exposures. Typical results of a daylight simulation exercise for a large hall of 126 m × 40 m (131 × 43.7 yd) are given in Fig. 10.10. The important parameters for modeling the light include the geographical location, the cardinal direction of the hall, the geometry of the hall as well as the color and reflectance of all interior surfaces.

10.2.3 Artificial Lighting

According to medical science, humans require approximately 75 % of their total energy reserves for an undisrupted seeing process. Artificial lighting, as a supplement to natural light, thus plays an important role at the workplace. It makes the working environment more humane and efficient. The basic characteristics that determine the quality of lighting include the level of lighting, the uniformity of the lighting intensity, the absence of reflections and glares, the direction and color of the light and its efficiency. Illuminance, measured in lux (lx), is a criterion used to measure the quality of the apparent brightness. One lux is defined as the luminous flux or luminous power (measured in lumen) per m^2. The luminous flux of a candle is approximately 10 lm.

Based on research in physiological-optics, work-physiology and psychology the following recommendations are made:

- 200 lx as the minimum illuminance for a continuously occupied workplace,
- 500–2000 lx as the optimal range for workplaces located in buildings and,
- 2000–4000 lx as the range for very fine work requiring concentration over longer periods.

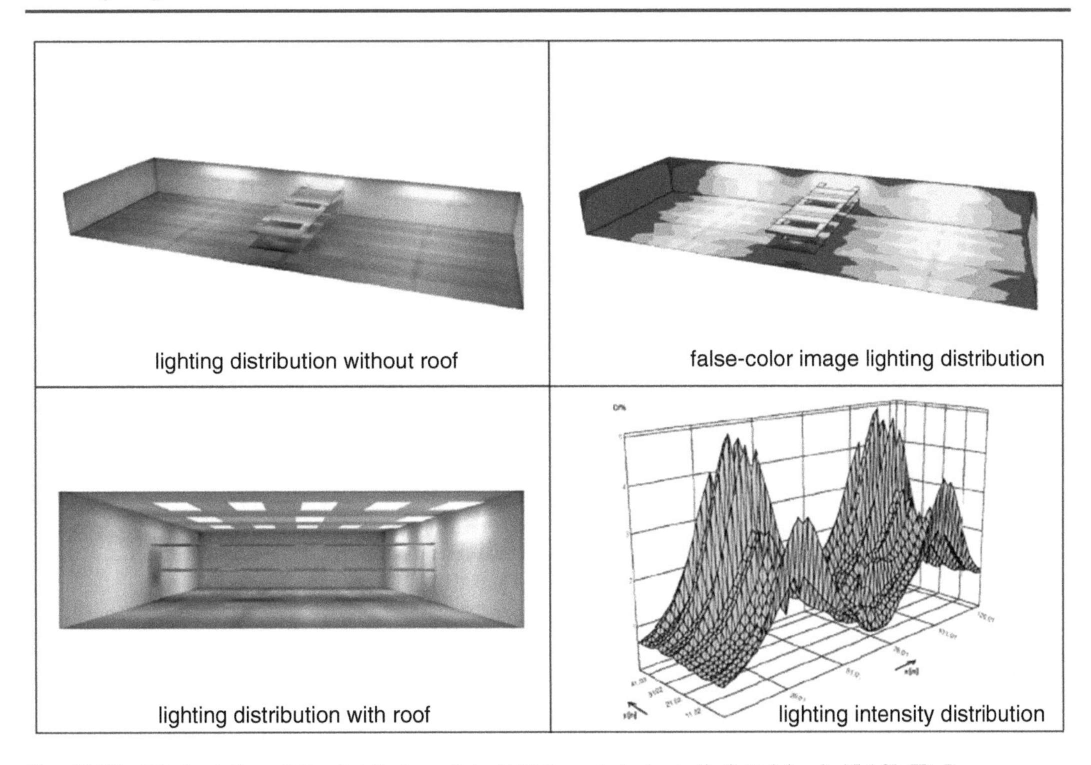

Fig. 10.10 3D simulation of the distribution of daylight in an industry hall. © Reichardt 15.167_JR_B

DIN 5035 Part 2 lists the minimum values for the nominal illuminance for 176 different activities. The values for most workplaces are between 200 and 1000 lx. Figure 10.11 depicts a table with the specified illuminance, light temperature, color rendering index and anti-glare factor for frequent industrial activities according to DIN 5035.

Surfaces that are exposed to a relatively strong illumination density cause distractions (glare, reflections) within the visual field thus impairing the worker's well-being. When users express that the lighting is too bright it usually means that the system is not designed well enough to limit glares and reflections. For more information readers are requested to refer to DIN 5035. Figure 10.12 depicts the minimum illumination density in foot candles for typical operations acc. to the OSHA Standard 1926.56, where one foot candle is equal to one lumen per square foot or approximately 10.764 lx.

Reflected glare is caused by reflecting strong illuminance off of shiny surfaces. This can be reduced by properly arranging lights and using matted surfaces on equipment and/or tinted walls and ceilings.

The direction of the light and the resultant shadow are important to recognize spatial objects, whereas unnatural lighting can cause 3-dimensional forms to be misinterpreted. Silhouette effects created when objects or persons are viewed in front of bright window surfaces can, for example, be decreased with a sufficiently high percentage of top lighting. In industrial set-ups the recommended ratio for top to side lighting is 1:3.

According to [Deh01], variable light and the light color visually and thus mentally impact the moods and well-being of humans through more than just the perceived brightness. The color of the light also referred to as the color temperature is determined by the spectral power distribution and measures the impression of color the light source provides. It is defined as the temperature that a black body has to be heated-up to in order to radiate a light that (with the same brightness and under fixed observational conditions) emulates the described color as closely as possible. Color temperature is measured in Kelvin (K).

Electrical lamps are divided into three color temperatures (or qualities) based on their

activity	lighting required acc. to type of work [lux]	color quality[1]	color rendering index[2]	anti-glare factor[3]
closets and other secondary rooms	50	ww, nw	3	-
storage rooms	100	ww, nw	3.4	-
locker rooms, washrooms, toilets	100	ww, nw	2	2
office work with easy visual tasks	300	ww, nw	2	1
office work with normal visual tasks, data processing	500	ww, nw	2	1
open offices	750	ww, nw	2	1
technical drawing	750	ww, nw	2	
laboratories	250	ww, dw	2	1
forging, rough assembly	200	ww, nw	3.4	2
welding and locksmith work	300	ww, nw	3	2
setup of machine tools, assembly	500	ww, nw	3	1
marking-out, fine assembly, inspection	750	nw, dw	1.2	1
fine assembly, precision work, inspection areas, measuring and test rooms	1000	nw, dw	1.2	1
assembly of fine parts, electric components, jewellery	1500	nw, dw	1.2	1

1) dw - daylight; nw - neutral white; ww - warm white 2) indices: 1, 2, 3,4 3) factor: 1, 2, 3

Fig. 10.11 Nominal lighting levels for industrial activities (to DIN 5035). © Reichardt 15.168_JR_B

TABLE D-3 - MINIMUM ILLUMINATION INTENSITIES IN FOOT-CANDLES

General. Construction areas, ramps, runways, corridors, offices, shops, and storage areas shall be lighted to not less than the minimum illumination intensities listed in Table D-3 while any work is in progress:

Foot-Candles	Area of Operation
5	General construction area lighting.
3	General construction areas, concrete placement, excavation and waste areas, access ways, active storage areas, loading platforms, refueling, and field maintenance areas.
5	Indoors: warehouses, corridors, hallways, and exit ways.
5	Tunnels, shafts, and general underground work areas: (Exception: minimum of 10 foot-candles is required at tunnel and shaft heading during drilling, mucking, and scaling. Bureau of Mines approved cap lights shall be acceptable for use in the tunnel heading)
10	General construction plant and shops (e.g., batch plants, screening plants, mechanical and electrical equipment rooms, carpenter shops, rigging lofts and active store rooms, mess halls, and indoor toilets and workrooms.)
30	First aid stations, infirmaries, and offices.

Other areas. For areas or operations not covered above, refer to the American National Standard A11.1-1965, R1970, Practice for Industrial Lighting, for recommended values of illumination

Fig. 10.12 US OSHA standard 1926.56—illumination intensities

impression of color: ww—warm white (up to 3300 K), nw—neutral white (3300–5000 K), dl—daylight (5000 K and above). Color rendering, on the other hand, relates to the color an object appears under a given light source and is characterized by a color rendering index Ra (1 = incandescent lamp to 4 = sodium vapor lamp), whereby incandescent lamps are the least color-distorting. True-Lite lamps are a relatively new development. Unlike traditional lights they emit infrared radiations and thus correspond more closely to the color rendering of natural sunlight. It is advisable to study various lighting systems and combinations under parameters which are in fact comparable. From an economic perspective, it is important to consider the electro-mechanical structure of the luminaire, for ease of assembly and regular maintenance.

For flexible, changeable workplaces the electrical power supply should be dimensioned with an eye to the future. Good lighting enhances performance or willingness to perform, prevents early fatigue, supports retention, logical thought and promotes safety and speed while reducing the frequency of errors and accidents [Rüs05]. Figure 10.13 depicts the impact of increased illuminance on work performance and fatigue as well as on the reduction of accidents. 3D lighting simulation software could not only help to simulate lighting conditions during the day but also during the night with reference to the proposed lighting scheme.

10.2.4 Redirecting Light

The quantity of side lighting is influenced by the room height, depth as well as the neighboring buildings that restrict light entering from the side. Typically, rooms with a depth greater than 7 m (22 ft) can no longer be naturally lit; one of the reasons why historically multi-story buildings seldom had a room depth of more than 15 m (49 ft)—even with higher ceilings. This thus raises a pertinent question as to how glare-free sunlight can be directed or guided into the depths of the room. Systems for redirecting light prove to be particularly helpful because they can transport sunlight up to 20 m (65 ft) deep into a room.

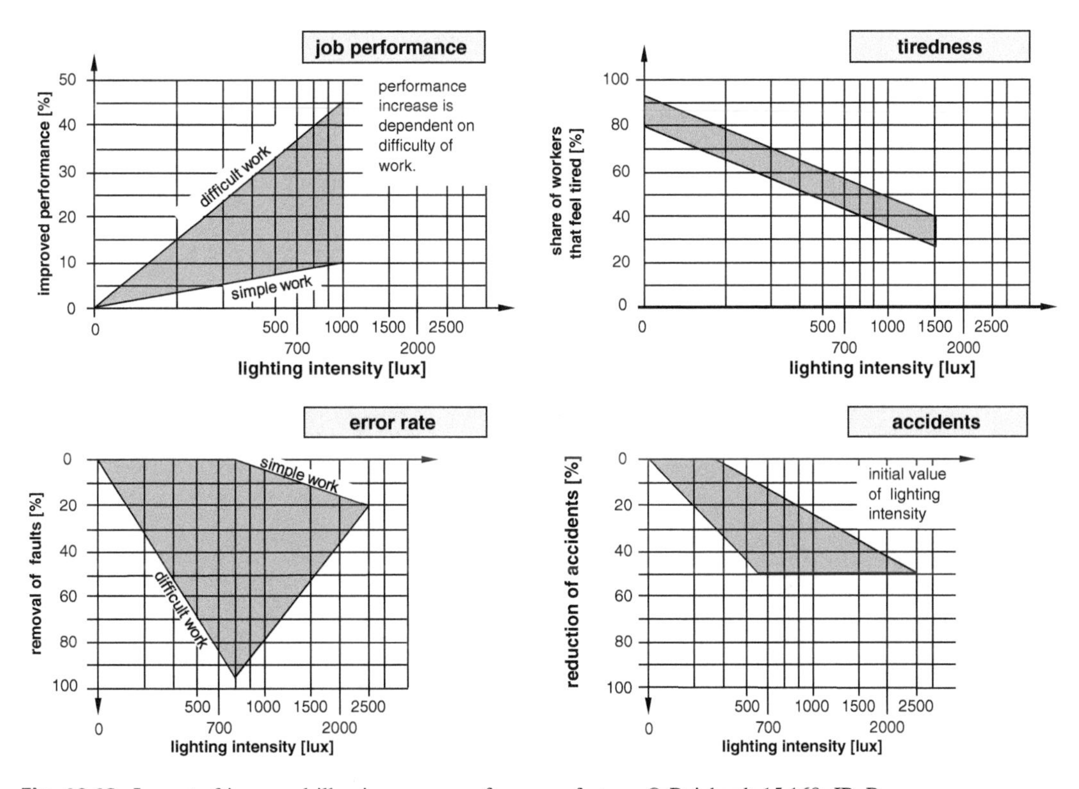

Fig. 10.13 Impact of increased illuminance on performance factors. © Reichardt 15.169_JR_B

The idea of redirecting light is not new; in fact a patent for redirecting sunlight into the interior of building using mirrors was taken way back in 1900. For many years now, the lighting expert Christian Bartenbach has been researching the possibilities of uniformly lighting greater room depths by re-directing daylight through roofing systems and facade elements or using glazing with integrated elements. LED (light-emitting diode) systems may be particularly well suited for this. Holographic sheets, developed by Müller [Mül01], make it possible to control the direction of light: Using relatively small strips on the façade the full depth of the room can be illuminated while the rest of the facade offers an unobstructed view. The diagrams in Fig. 10.14 illustrate possible light directing mechanisms for an office space integrating currently available techniques for controlling light. Using reflective mirrors, light shelves or holographic sheets the light can be redirected into the depths of the room and considerably increase the daylight factor useable for office work.

10.3 Comfort

For humans, the indoor climate has a comforting effect when a body's regulatory mechanisms maintain the required internal temperature undetected from the person. The sensation of comfort differs considerably and varies according to the type and duration of work as well as age, gender, health and clothing. Unfortunately, there are no standard or nominal values for the thermal well-being. Individual thermal comfort is a result of interactions and interconnected components comprising of the physical indoor climate (room temperature, radiant temperature, humidity, air flow and air purity). Further comfort aspects such

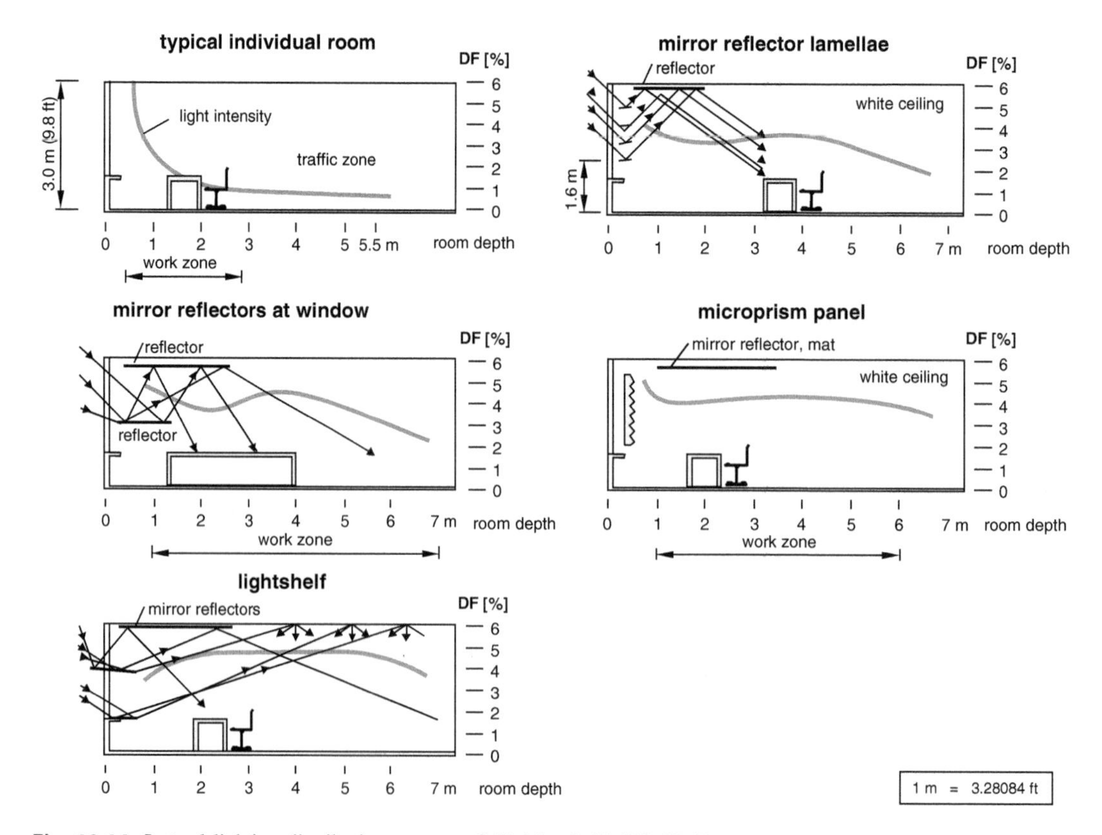

Fig. 10.14 Lateral lighting distribution systems. © Reichardt 15.170_JR_B

as the color scheme and noise thresholds are discussed in Sects. 8.3.1 and 8.4.6.

The question of an individual's level of comfort has to take into consideration not only the heat produced due to the activities in which the individual is involved but their ability to expel excess heat both dry heat (convective and radiant) as well as damp heat (evaporation). Heat emitted on a cold surface will be perceived as cooling just like a draft, while the radiation of highly tempered heating surfaces or heavily heated shading systems present an annoying source of heat. Thus in areas where heavy work is being executed, it is important to pay particular attention to preventing drafts near to gates.

Due to existing production methods or the external climatic conditions it is often barely possible to maintain a 'tolerable' internal environment. 'Tolerable' here means that there are no anticipated health risks related to the internal conditions. When the body is under stress or when external temperatures are above comfort levels, the body's internal regulator increases the production of sweat and thus the expulsion of heat via evaporation. According to [Opf00], when air quality decreases due to the presence of pollutants, heat and/or excessive moisture, it is important to renew the air in order to maintain a certain minimum level of comfort condition within the various workshops.

The temperature zone surrounding a workplace is characterized by the air temperature and the radiation temperature. Depending on the activity, the air temperature should range between 18 and 24 °C (64.4–75.2 °F). The lower value applies to work that is slightly flexible; the upper for a sedentary state. The mean radiation temperature can be 3–4 °C (5.4–7.2 °F) lower than the room temperature. The arithmetic means of the air temperature and radiation temperature corresponds roughly to the perceived temperature. Thus an air temperature of 22 °C (71.6 °F) and a radiation temperature of 18 °C (64.4 °F) are perceived as a uniform ambient temperature of 20 °C (68 °F). In the summer months a temperature of 26 °C (78.8 °F) can still be referred to as comfortable for light work.

With appropriate thermal insulation, the temperatures of the various room enclosures (walls, ceilings, floors) should develop so that they deviate as little as possible from the mean values. Figure 10.15 depicts the comfort zones of

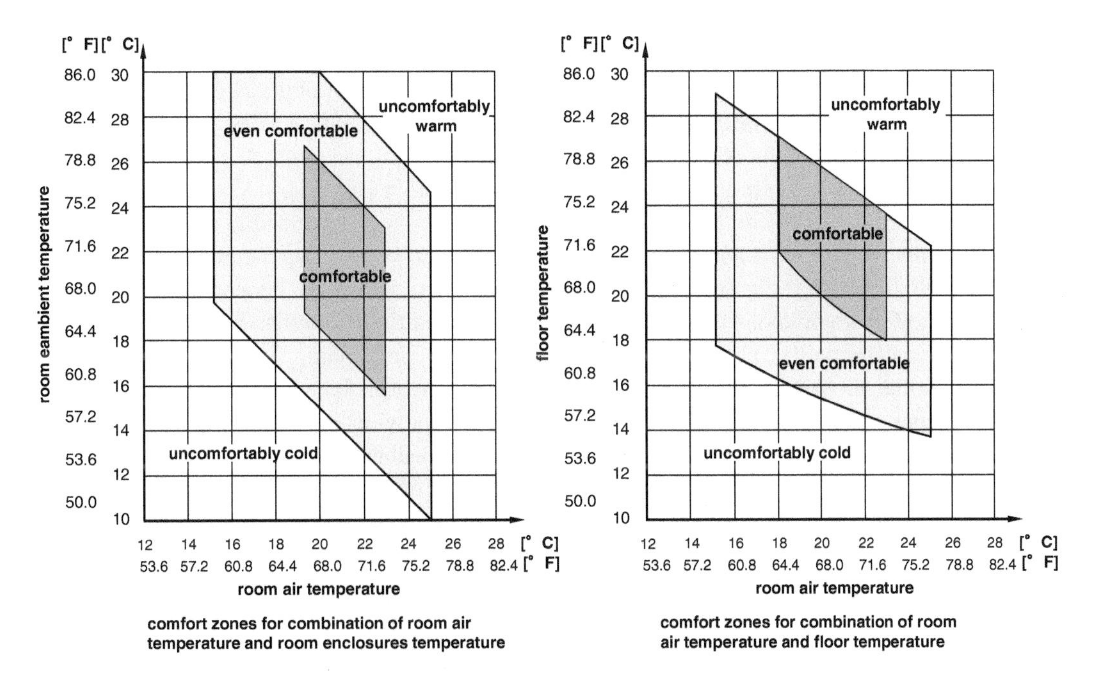

Fig. 10.15 Comfort zones for room air temperatures, ambient and floor temperature. (acc. Frank), © Reichardt 15.171_JR_B

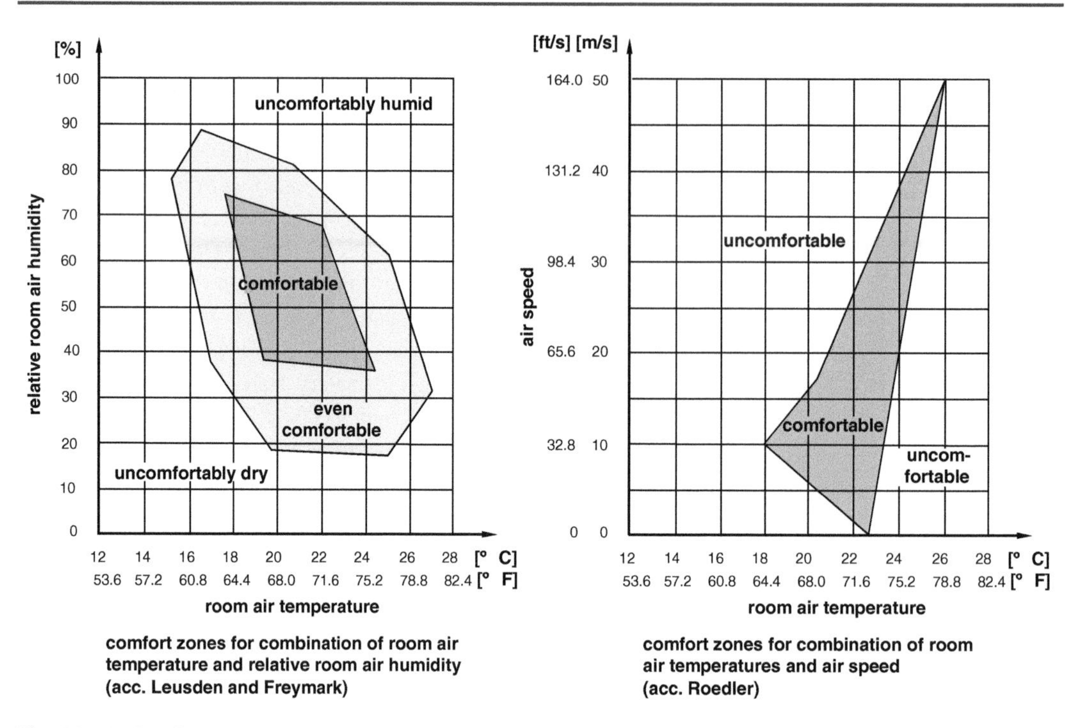

Fig. 10.16 Comfort zones for room air temperature, humidity and air velocity. © Reichardt 15.172_JR_B

humans for the ambient room temperature, as function of temperatures for the surfaces enclosing the space walls (left), ceilings and as function of the floor temperature (right). Together with the heating and cooling systems, the construction materials for the building have to be oriented towards the required comfort values.

Figure 10.16 depicts the comfort zones of the ambient room temperature as function of humidity (left) and air speed (right). Accordingly, humidity should generally be within the range of 35–65 %. When air temperatures are higher, lower air humidity should be aimed for in order to increase the percentage of heat emissions through evaporation, otherwise the air will be noticeably muggy.

The permissible air speed is dependent on the indoor air temperature. If a temperature of 20 °C (68 °F) is attributed to an air speed of 0.15 m/s (0.5 ft/s) then 22 °C (71.6 °F) would relate to 0.20 m/s (0.65 ft/s). If a job involves physical activity the air speed can be increased. Air pollutants including dust, gases, vapors and odors that are uncomfortable are considered as a measure of the air purity.

The amount of fresh air that needs to be fed into the interior is dependent on the number of people in attendance as well as the nature of pollutant. If gases and vapors arise in specific areas, the hourly air exchange cannot be generalized; instead it has to be determined as a function of the accruing gas and vapor concentration. Approximate values can be found in [Leh98, Ski00]. A comprehensive overview for US is given in [Goe11].

It is often recommended to ensure a higher air exchange rate by placing a number of smaller air-supply units in various strategic locations, thus allowing partial operations when compared to a central air-conditioning system. The necessity of having transformable work areas as well as being able to individually control the surrounding requires sensible planning for small volume work areas having as much natural ventilation as possible.

In terms of sustainable changeability, flexibility in controlling the room climate (temperature, humidity, etc.) should be ideally defined based on a holistic consideration of the process and spatial views. The building services and structure can then be adjusted to new requirements in regards to temperature and humidity without any significant changes.

10.4 Relaxation

'Breakout' or relaxation areas should contribute to the employees need for recreation after a period of concentrated work. Clever integration of the recreational area into the building structure along with an attractive ambience is imperative for a healthy working atmosphere for all concerned. Increasingly stressful work processes may entail negative psychological consequences in terms of either being over- or under burdened. Relaxation phases that offset this mental stress promote a sense of community, social competencies and team-building. Recreational areas thus offer a large potential for employees to identify with "their" company.

These periods of recovery, should be architecturally supported with appropriately designed break-out areas, social spaces, canteens and cafeterias as well as sporting or gaming facilities, libraries and other recreational areas. Keeping in mind both the existing as well as future requirements, it would be imprudent to limit required recreational areas to the absolute legal minimum. In this context, changeability also means the spatial possibility to adjust to the desires and requirements of the user and to convey a touch of "leisurely flair".

10.4.1 Break-Out Areas and Social Rooms

According to the prescribed rules outlined in the German Workplace Guidelines, a break-out area needs to be provided when there are 10 or more employees. Moreover, due to specific recovery requirements or special types of activities separate break-out rooms may be required (German Workplace Guidelines—§29). In the absence of such areas or when break-out areas are too far away, employees often consume their snacks or lunches while standing next to the machinery and/or work-place. Attractive break-out areas or zones should thus offer the possibility to relax especially in factories that have manufacturing processes which directly influence the level of comfort. There is often the possibility for an 'oasis'—a visually attractive, 'green' area right in the middle of the production.

In Germany, the term 'common areas' summarizes the basic requirements according to the workplace guidelines for locker-rooms, wash areas and lavatories. These "secondary rooms" are often situated in a basement devoid of any connection to the outside world; allocating it to a friendlier location with natural light and possibly a view of the outdoors substantially improves the atmosphere.

10.4.2 Canteen, Cafeteria, Coffee Corners

Areas for having a meal or a snack or for preparing a drink should be located in visually attractive locations in the building, preferably having some relation to the outdoors. In most countries, a sunlit terrace offers a pleasant place to eat along with relaxed conversations promoting social competencies and team-building skills. The choice of materials along with lighting, furnishings and color scheme should support the overall character of the place and pursue a holistic design concept. Coffee corners should be ideally located in places where different paths intersect in a building—upgraded hallways near stairwells, gallery areas overlooking the production, landscaped areas or foyers.

10.4.3 Sport, Recreation and Spare Time

Concerned organizations support the recovery phase of their employees with visits to a sauna, fitness studio or to tennis courts. Group activities outside work hours promote team-building and can also contribute to reducing personal conflicts between employees. It seems logical to reserve areas like terraces or outside facilities for such activities, in the long term. Measures such as roof landscaping or furnishing a recreational trail with sporting equipment for example, can then be made gradually.

10.5 Fire Protection

Fire protection includes all of the features and measures that serve to protect people and material assets from fire. The general relationship between the building usage, fire protection, firefighting and restoration following a fire are presented in Fig. 10.17 together with the building components that play a role. The properties of the building elements can influence requirements, impose restrictions on the building's usage, prevent and/or retard fire, insulate noise and influence comfort levels. Moreover, the properties also determine fire precautions, related risks and the potential for later restoration.

Since industrial buildings have a diverse array of uses, this topic is quite complex in view of the stringent building regulations [ICC12, Max13]. A well designed building structure effectively contributes towards passive fire protection as the same is integrated in its design. For example, a more transparent factory makes it possible to detect sources of fires early-on. Further, its network of emergency evacuation routes would be easy to comprehend both horizontally and vertically across the plant. Areas which concentrate building facilities and are therefore susceptible to fire should be positioned close to one another.

A comprehensive fire protection concept tries to holistically combine spatial, process, logistical and delivery requirements in order to arrive at structural and technical measures that would be required to protect both people and material assets. Calculation of the fire load helps to determine the minimum distances between buildings and the maximum allowable size of fire compartments within. These calculations help in configuration of the firewalls, the partitioning walls as well as the specifications regarding the classes of fire resistance or building materials.

building elements ▽	utilization phase	fire prevention	fire fighting measures	fire damage restoration
building materials	comfort, heat and humidity protection, sound insulation.	prevention and spreading of fire and smoke	risk reduction in fire–fighting	reusability
distance between buildings, parking	sufficient daylight	prevention of fire spreading to neighboring building	movement, parking for fire brigade vehicles and equipment	noise and odour protection
ceilings, walls roofs	structure of the building	prevention of fire spread	fire-fighting zones, reducing danger of collapse	sufficient stability
doors	delimitation of fire zones, safety and security	prevention of fire-and smoke spread	strategic fire-fighting zones	
stairs, corridors and stairwells	exit ways for able / physically challenged people	rescue paths	attack-and rescue paths	circulation routes
media supply lines	drinking and industrial water	water supply for sprinkler systems, hydrants/riser lines	water supply for stationary and mobile fire-fighting	corrosion protection
windows	daylight, ventilation and visual connection	escape routes, smoke outlets	attack-and rescue paths, smoke outlets	

Fig. 10.17 Influence of building elements on use and fire protection. © Reichardt 15.173_JR_B

For those interested in an overview of requirements of building permits within Europe are requested to refer to [Max13]. For the US, [Cot03] is a reliable reference. Moreover, as mentioned above, the location, distance, layout of evacuation and rescue routes need to be considered. Finally, active fire protection measures such as smoke and heat detectors and firefighting equipment need to be provided. Next, we will clarify structural aspects of these elements and examine changeability factors.

10.5.1 Fire Protection Concept and Fire Compartment

According to [Max13] when developing a fire protection concept for industrial buildings, the latest versions of the relevant regional building codes (municipal, provincial, state etc.), guidelines concerning supervision of construction processes, building regulations as well as model guidelines for construction fire protection in industrial buildings all need to be considered (see e.g., for Germany http://www.bauordnungen.de or the US: NEPA 5000: Building Construction and Safety Code 2012). The regional building codes contain a plethora of material specifications with regards to fire protection, especially risk scenarios in residential or similar types of buildings. Therefore, experts such as fire protection engineers in Germany generally inspect larger buildings based on the model industrial building code (Muster-Industriebaurichtlinie M IndBauRL, März 2000). In other countries the relevant code is applied. Once plans have been completed, basic aspects need to be evaluated as to its compliance with reference to the related building codes.

The basic terms of object description, the legal specifications, a risk evaluation from the perspective of fire protection engineering and the presentation of the fire protection concept based on empirical calculations and planning documents are explained hereunder.

The object description outlines the construction of the building and additional structural characteristics as well as the planned use. The subject of changeability should be addressed at this point in order to avoid future problems with the building, systems and operations (e.g., in underestimating the effort for fire protection engineering) when the factory undergoes its first modifications.

The fire protection risk assessment summarizes both the specific fire load for the material flow at the site according to safety categories and the fire load of the building construction itself. According to DIN 18230, two important results of calculating the fire load include the largest size of a fire sub-compartment and its required fire resistance rating. The fire resistance rating indicates how many minutes a building component can fulfill its function when exposed to fire (DIN 4102, Part 2). Fire resistance classes, also referred to as fire grading period or fire protection class, differentiate between fire resistant (at least F 30), fire retardant (at least F 60) and best possible flame retardant (at least F 120), whereby F 30, F 60 and F 120 each refer to the fire resistance rating provided in minutes.

Figure 10.18 depicts the permitted size of fire compartments according to German fire regulations and DIN 18230. The term 'fire compartments' refers to the allowable connected floor spaces, separated by firewalls or complex partitioning walls. Moreover, the fire resistance rating of load-carrying and reinforcing building components is provided depending on the safety classification of the usage and the number of floors in the building. With regards to the changeability of the building in general, requirements for clearances, firewalls (F 90) and separation walls (F 120, F 180) are crucial.

Complex partitioning walls are often required for storing goods that have a high fire load (e.g., paper) and significantly limit options for changeability due to the in situ construction in reinforced concrete.

10.5.2 Clearances, Firewalls and Complex Separation Walls

Figure 10.19 depicts an overview of the clearance requirements between buildings as well as access roads, deployment areas and free

fire resistance class *) / safety class **)	number of building stories: ground floor		2-storied			3-storied		4-storied	5-storied
	fire resistance of the bearing and reinforcing components								
	no requirements	F30	F30	F60	F90	F60	F90	F90	F90
K1	1800 1)	3000	800 2) 3)	1600 2)	2400	1200 2) 3)	1800	1500	1200
K2	2700 1)	4500	1200 2) 3)	2400 2)	3600	1800 2)	2700	2300	1800
K3	3200 1)	5400	1400 2) 3)	2900 2)	4300	2100 2)	3200	2700	2200
K3.1	3600 1)	6000	1600 2)	3200 2)	4800	2400 2)	3600	3000	2400
K3.2	4200 1)	7000	1800 2)	3600 2)	5500	2800 2)	4100	3500	2800
K3.3	4500 1)	7500	2000 2)	4000 2)	6000	3000 2)	4500	3800	3000
K4	10000	10000	8500	8500	8500	6500	6500	5000	4000

*) fire resistance class: a section with F30 can resist fire for 30 min, F60 for 60 min and F90 for 90 min
**) safety class: K1: no special measure, K2: fire alarm system, fire alarm and fire brigade, K 3: fire zones or fire fighting sections with automatic fire alarm system, K4: sprinkler
1) width of the industrial building ≤ 40 m (131.2 ft.) and heat exhaust area (according to DIN 18230-1) ≤ 5%
2) heat exhaust area (according to DIN 18230-1) ≤ 5%
3) permitted size of 1600 m² for low buildings

1 m² = 10.7639 sq ft

Fig. 10.18 Permissible fire compartment sizes (acc. to German fire regulations). © Reichardt 15.174_JR_B

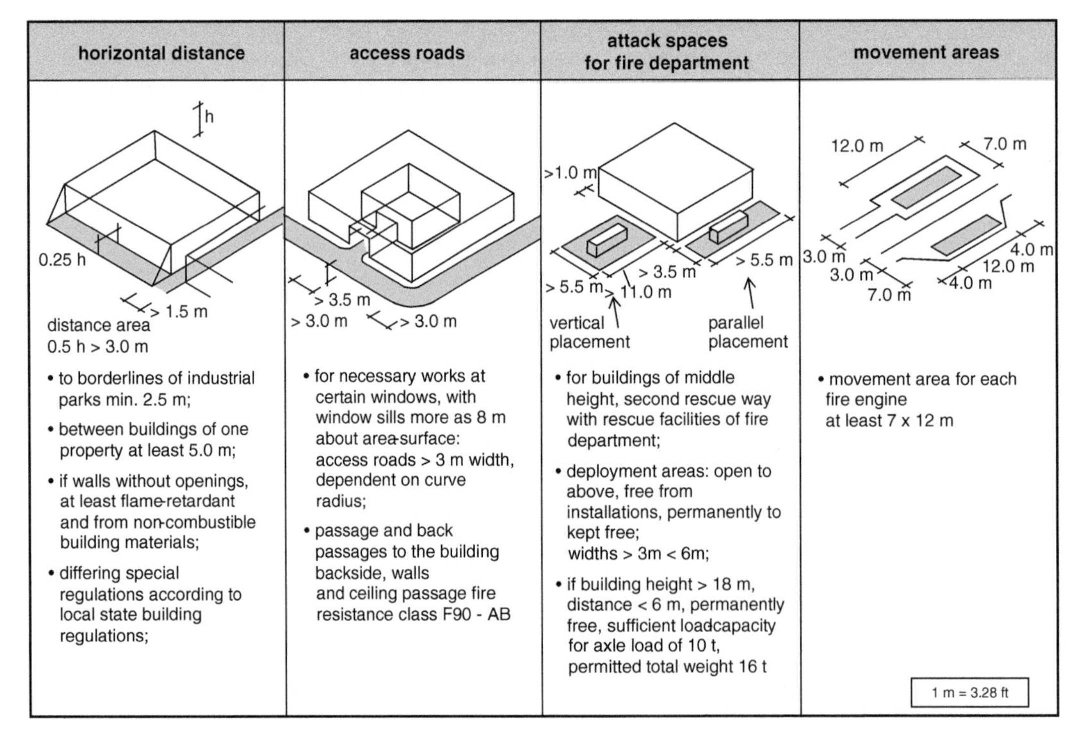

Fig. 10.19 Fire protection requirements for surrounding building areas. © Reichardt 15.175_JR_B

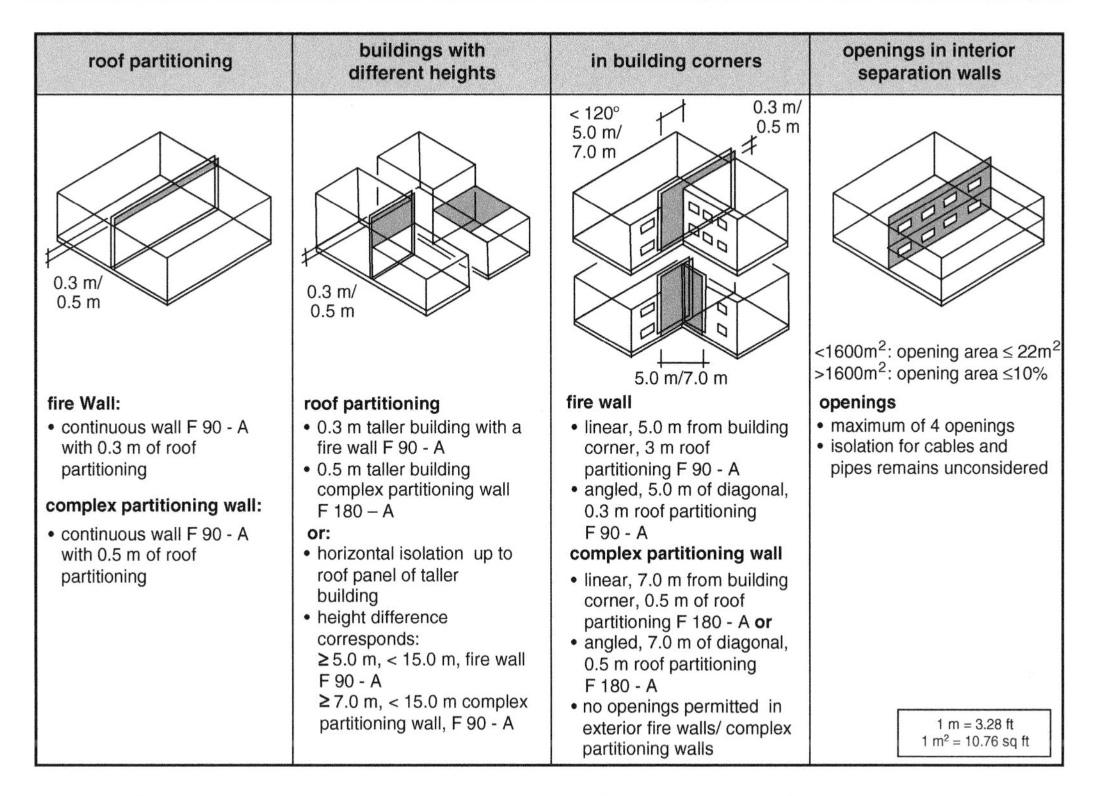

Fig. 10.20 Types of firewalls and complex partitioning walls. © Reichardt 15.176_JR_B

movement zones for the fire department. It can be seen that, in addition to the regional building code specifications, the clearances between buildings are determined by the firefighting zones and rescue paths. It is thus advisable to tackle fire protection issues early-on in order to ensure acceptance of the building concepts by the relevant authorities at a later date.

Fire compartments limit fire related damages and permit safe evacuation routes in areas that are less susceptible to fire by limiting spaces using fire resistant enclosures (e.g., floors, walls and doors). According to state building codes in Germany, fire compartments are limited to 40 m × 40 m = 160 m^2 (131.2 ft × 131.2 ft = 17.222 ft^2). Implementing industrial building regulations and in particular special measures such as sprinkler systems permit larger connected production areas.

Fire compartments are limited by firewalls. In case of an accident, they are meant to restrict fire from developing beyond a site. A minimum of one-and-a-half-hour fire resistance (F 90-A) would ensure structural safety in a fire accident. In areas that are at greater risk due to storage of inflammable goods, partitioning walls are mandatory. Also, in such scenarios higher requirements with regards to the fire resistant ratings and structural detailing need to be adhered to. Figure 10.20 illustrates different ways of preventing fires from spreading with roof partitioning, walls in building corners and openings in interior partitioning walls. So called Complex Partitioning Walls have to fulfil certain requirements like fire resistance of 180 min, resistance of three times an impact of 4000 Nm and ensuring a complete room closure (VdS 2234).

10.5.3 Fire Resistance Rating Classes

Use of fire resistant materials for construction of the supporting structures, shell, media and building services as well as furnishings would help in countering the development and spread of fires. Ideally, the building should remain stable at least until all life saving measures have been conducted. Depending on the local fire department's requirements, the building regulations and the fire protection concept, a fire resistance rating of F 0–F 90 is normally required for the supporting structures while F 90–F 180 for walling material for the fire compartments and materials for the shell and furnishings. Chapter 7 (Fire and Smoke Protection Features) of the International Code Council provides the reader with a good overview [ICC12]. Specific details and regulations can be found e.g., in the European standard DIN EN 13501-1 and for Germany in DIN 4102.

The design and detailing of building services also have to undergo similar considerations. Installation ducts, cables, lines and pipes should predominantly using fire-resistant materials. All media and utility lines which pass through fire-resistant walls also have to be sealed in a manner that protects against fire spread. Moreover, all openings such as doors, gates and flaps in partitioning components have to be authorized by a building inspector.

10.5.4 Evacuation and Rescue Routes

Vertical and horizontal evacuation and rescue routes ensure a quick safe exit from a building when there is a fire. Entrances to rescue routes and deployment areas for firefighters need to be clearly marked and permanently kept free. Other architectural means for ensuring a quick orientation with easy detection of the evacuation routes include use of daylight, transparency and providing points that allow overviews or views in/out.

In Germany, the state building regulations, workplace regulations and guidelines, assume that a person from any given point within the building should be able to reach the outdoors or other secure areas within 35 m (114 ft) through an exit or a series of emergency exits. Depending on the ceiling clearance heights and the fire protection engineering of the structure, the model industrial building guidelines permit greater rescue distances.

Figure 10.21 illustrates requirements for rescue routes depending on the location, the fire

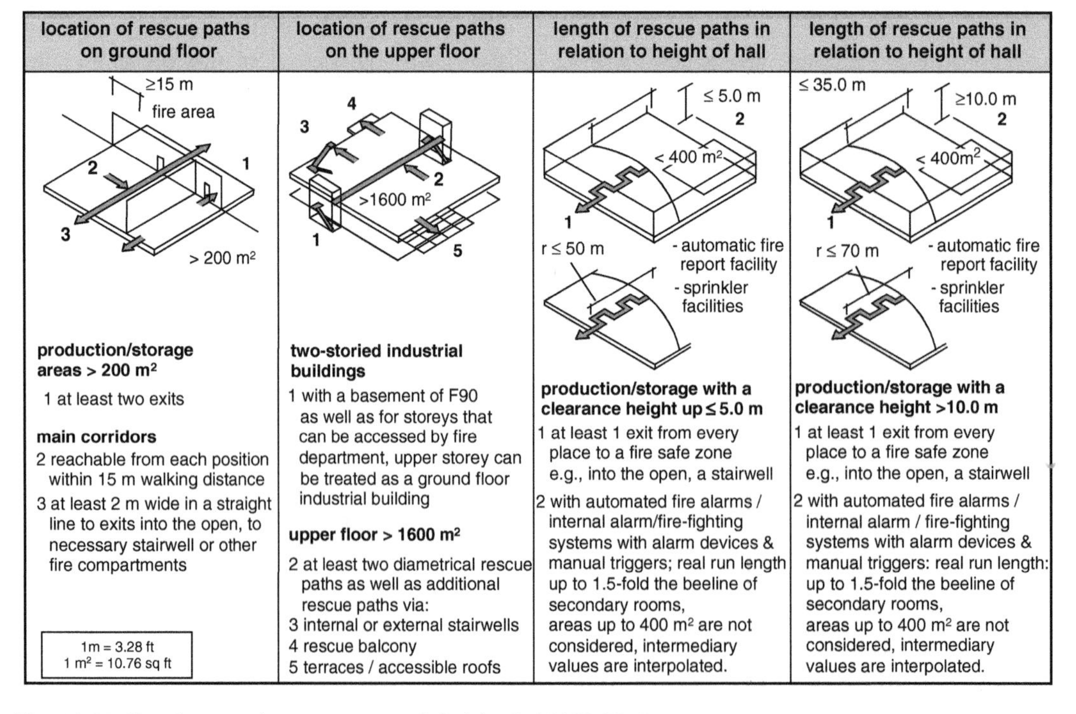

Fig. 10.21 Requirements for rescue routes. © Reichardt 15.178_JR_B

resistance rating of the building and the clear ceiling height. Accordingly, due to the advantages of a high ceiling regarding smoke exhaust, an actual rescue route of up to 105 m (344 ft) is allowable for halls having ceiling heights of 10 m (33 ft) or more (see Sect. 10.5.5). It is mandatory to connect all the floors in a building in one continual stairwell, designed and built such that —along with the entrances and exits to the open air—the same can also be safely used as a rescue route. According to German fire regulations, the internal walls should have a fire resistance rating of at least F 30. In accordance with DIN 4844 all the evacuation and rescue routes are to be clearly identified. Moreover, workplace regulations and guidelines dictate that back-lit pictograms and safety lighting are also required.

10.5.5 Smoke and Heat Ducts, Fire Extinguishing Equipment

Determining the fire load giving due consideration to the space, processes and logistics provides information necessary for installing smoke ventilation systems on the side walls or hall roofs to facilitate removal of life threatening smoke and gases. The thermal efficiency of the smoke outlets only comes into play when there are suitable air intake openings. A 3D flue gas simulation is recommended for assessing the thermal process for specific cases. For a production hall or storage area with an automatic fire-extinguishing system and an area of 1600 m^2 (17,260 sq ft), the general industrial regulations in Germany assume that smoke can be removed sufficiently through smoke vents having an effective cross sectional area of 0.5 % in relation to the floor surface. Moreover, the smoke removal has to meet the protection objectives and equipment specifications outlined in DIN 18232.

The ratio of the air supply's geometric opening surface to the geometric input surface of all the smoke duct surfaces in the largest smoke compartment has to be 1.5–1. Figure 10.22 illustrates the requirements for the smoke and heat ducts according to German fire regulations for industrial buildings having surface areas <200 m^2 (2152 sq ft) and areas >1600 m^2

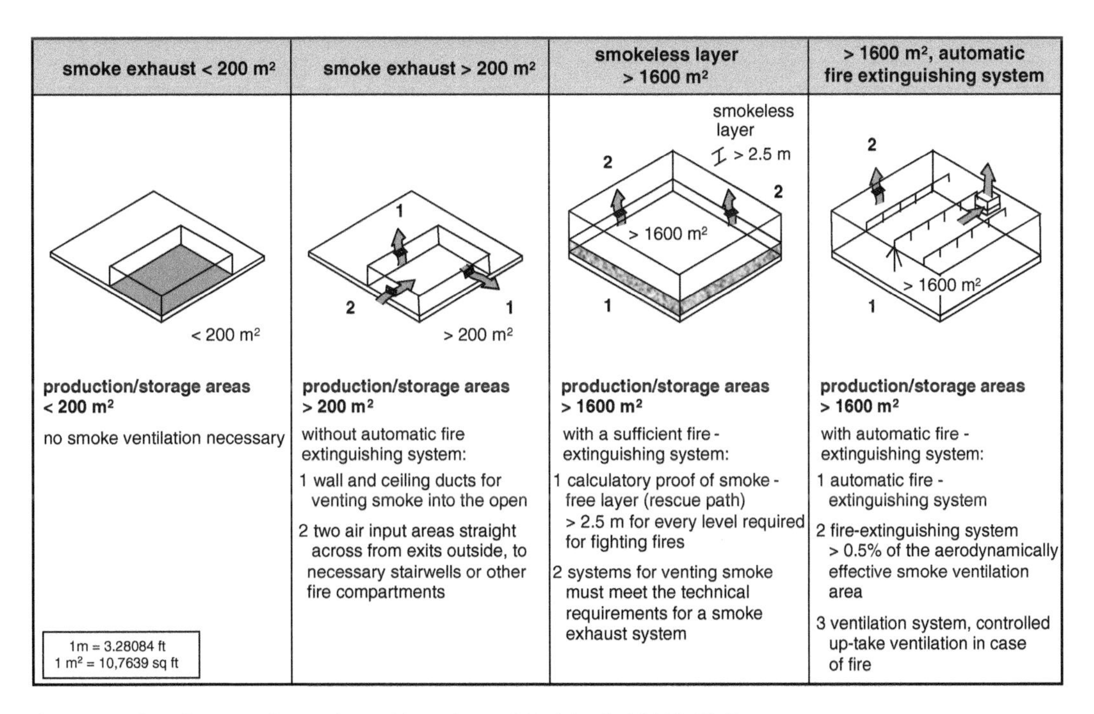

Fig. 10.22 Requirements for smoke and heat ducts. © Reichardt 15.179_JR_B

(17,200 sq ft). At every level, i.e., including galleries, one needs to verify a 2.5 m (8.2 ft) high smoke-free layer (breathable air without toxic smoke) from where people can be safely rescued.

In terms of fire protection, additional hall clearance is therefore a long term investment keeping in mind future changeability requirements. Active fire suppression measures include an automatic sprinkler system controlled via fire detection and alarm system, as well as the installation of external and wall hydrants. A mandatory sprinkler system should be planned so as to comprehensively include all the fire compartments. Design and installation of a sprinkler system requires expert knowledge and a well-coordinated effort including the Insurance Company responsible for the plant, experts in fire suppression systems and the local authorities in charge of fire-safety permissions.

With this we conclude our discussion on spatial design on the factory section level. In consideration of the sectional design from a functional view discussed in Chap. 9, this information flows into the design of the factory, which we will examine next in Chap. 11.

10.6 Summary

The spatial design of work spaces where people are permanently or temporarily located, results primarily from the type of cooperation and the required levels of communication. Secondly, it is important to promote mental and physical well-being. Finally, under all circumstances, health risks need to be minimized. In terms of communication, planning and implementing tasks of production should be close to each other or "under one roof", while offices could be planned in various forms—the office cell via combined offices and group rooms up to business clubs—as they are known from the service centers of airports. Natural light adds a substantial "feel good factor" to the design and the ambient environment. The level of comfort is further enhanced through well-designed break-out areas, cafeterias, sports, and leisure and recreation facilities. Of special significance is finally the fire protection, which is regulated in detail in all countries. It includes dividing the building into fire zones with fire-resistant walls, ceilings and other materials. It also includes detailed requirements for emergency exits, escape routes as well as smoke and heat extraction in case of fire.

Bibliography

[Bar98] Bartenbach, C., Corthesy, R.: Bartenbach Lichtlabor. Bauen mit Tageslicht Bauen mit Kunstlicht (Light Laboratory Bartenbach. Building with Natural Light. Building with Artificial Light). Vieweg+Teubner Wiesbaden (1998)

[Bra06] Brandi Licht, U. (ed.): Lighting Design: Principles, Implementation, Case Studies. Birkhäuser, Basel (2006)

[Bis98] Bismarck, W.-B., Held, M.: Ergebnisbericht der Befragung zur Anwendung innovativer Kommunikationstechnologien (Results of the Survey Report on the Application of Innovative Communication Technologies). Universität Mannheim (1998)

[Bra05] Brandi, U.: Detail – Tageslicht – Kunstlicht: Grundlagen, Ausführung, Beispiele. 1. Aufl. (Detail – Daylight – Artificial Light: Fundamentals, Design, Examples). 1 Ed Intern. Institute of Architecture Doc. Munich Institut f. Intern. Architektur-Dok. München (2005)

[Cot03] Cote, A.E. (ed.): Fire Protection Handbook, vol. 2. National Fire Protection Association, Quincy, Massachusetts (2003)

[Deh01] Dehoff, P.: Die sinnvolle Veränderung des Lichts am Arbeitsplatz (The sensible change of light in a workplace). Architekten-Magazin **4**, 36–39 (2001)

[Eba84] Ebadi, Y.M., Utterback, J.M.: The effects of communication on technological innovation. In: Management Science 05/84, Massachusetts Institute of Technology, Cambridge (1984)

[Fi09] Fischer, A., Schmidt, A.: Wettbewerbsfähig in die Zukunft. Audiospezialist Sennheiser baut neues Produktions- und Technologiezentrum (Competitive into the Future. Audio Specialist Sennheiser Builds New Manufacturing and Technology Center). wt Werkstattstechnik online 99 (200) H. 4, S.199–204

[Goe11] Goetsch, D.L.: Occupational Safety and Health for Technologists, Engineers and Mangers, 7th edn. Prentice Hall, New Jersey (2011)

[IES11] IESNA Lighting Handbook.: Reference and Application, 10th edn. Illuminating Engineering Society of North America, New York (2011)

[ICC12] International Code Council (ed.): International Building Code, Chapter 7 Fire and Smoke protection Features. Ann Arbor (2012)

[Leh98] Lehder, G., Uhlig, D.: Betriebsstättenplanung (Plant Planning). Weinmann Filderstadt (1998)

[Max13] Max, U., Schneider, U.: Baulicher Brandschutz im Industriebau: Kommentar zu DIN 18230 und Industriebaurichtlinie (Structural Fire Protection in Industrial Buildings: Comment to DIN 18230 and Industrial Construction Guideline), 4 edn. Beuth, Berlin (2013)

[Mar00] Marmot, A., Eley, J.: Office Space Planning: Designing for Tomorrow's Workplace. McGraw-Hill, New York (2000)

[Mül01] Müller, H.F.O.: Die verschiedenen Systeme der Lichtlenkung (The Various systems of light control). Architekten-Magazin **3,** 34–41 (2001)

[Opf00] Opfermann, R., Streit, W.: Arbeitsstätten (Work Places). Forkel-Verlag, Heidelberg (2000)

[Rei01] Reichardt, J.: Kommunikationsorientierte Fabrikstrukturen (Communication-Oriented Factory Structures). In: Proceedings of Fabrik 2005+. 3rd Deutsche Fachkonferenz, Fabrikplanung Stuttgart (2001)

[Rüs05] Rüschenschmidt, H., Reidt, U., Rentel, A.: Licht Gesundheit Arbeitsschutz (Light Health Safety), 6th edn. Technik und Information Bochum, Verl (2005)

[Ski00] Skiba, R.: Taschenbuch Arbeitssicherheit (Handbook Occupational Safety). Erich Schmidt Verlag, Bielefeld (2000)

[Spat03] Spath, D., Kern, P. (Hrsg.): Zukunftsoffensive 21 - Mehr Leistung in innovativen Arbeitswelten (Future Offensive 21 - More Power in Innovative Work Environments). Egmont, Köln (2003)

11 Building Design

The architectural and structural design of a building consists of four components which impact its form: the load-bearing structure, shell, media routings (pipelines, wiring etc.), and interior finishings. The "performance" of a building —that is its ability to serve both current and future purposes—is determined by the characteristics of the selected technical and structural solutions together with these four components. For those interested in examples of industrial buildings, a wide range of designs are presented in [Ada04].

In Fig. 11.1, we have illustrated our approach to a comprehensive discussion about the variations of structural characteristics from the perspective of the processes and the enveloping space. It is important to differentiate between unchangeable, difficult to change and easy to change characteristics in comparison to the anticipated changes in requirements because these decisions determine the future changeability of the building.

The load-bearing *structure* or structural framework is the most permanent component of a building and thus the most difficult to change. Generally, it is designed to last the full duration of the building's use. The structure consists of the surface and column-like components, reinforcements and foundations required to ensure the stability of the building. The components used are either constructed on-site or modular and are made from steel, reinforced concrete, wood, light alloys or combinations of these materials. The selected structure strongly influences the long term functionality of the building as well as the interior and exterior architectural design.

The *shell* separates a protected interior space —as an independent climatic area—from an external space. It consists of stationary, closed or transparent elements for the façade and roof as well as moveable parts such as gates, doors, windows or vents. Aspects such as natural lighting, the afforded views and communication in particular, determine the long term quality and changeability of the building shell.

We use the term *building services* to refer to all of the equipment necessary to ensure the production processes, users' comfort and building security including the technical equipment centers, pipelines, wire routings, connections, etc. It covers all measures which guarantee the spatial comfort of users and provide the necessary technical media for the production facilities.

The technical facilities associated with the building are treated in literature under the term TBE technical building equipment, and include facilities for sewage, water, gas and fire extinguishing systems, heat supply, ventilation, electrical systems and building control systems. In particular, aspects such as modularity, upgradeability and accessibility (i.e., for maintenance purposes) determine the degree of the building equipment changeability.

With *interior finishings* we are referring to all of the stairways, building cores and special built-in units like elevators or wet rooms as well as static optional components (walls, windows etc.).

H.-P. Wiendahl et al., *Handbook Factory Planning and Design*,
DOI 10.1007/978-3-662-46391-8_11, © Springer-Verlag Berlin Heidelberg 2015

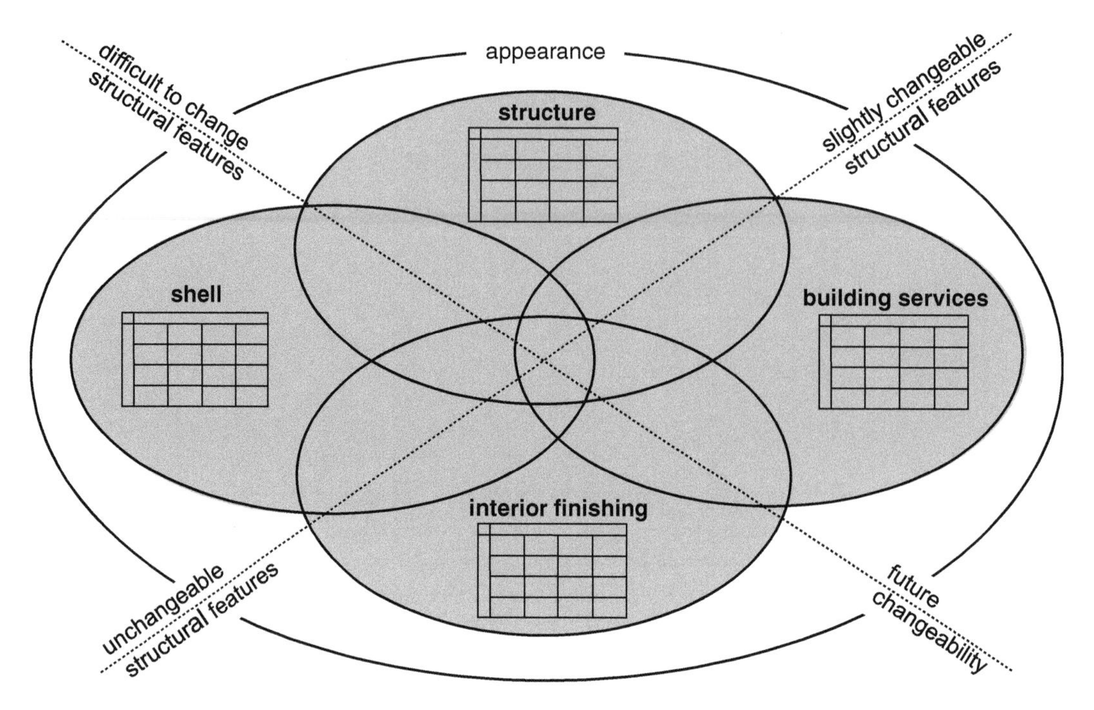

Fig. 11.1 Design fields of a building. © Reichardt 14.782_JR_B

Generally speaking, permanent interior finishings should be kept to a minimum so as not to limit or impede the changeability of processes.

Finally, the appearance of industrial and commercial architecture is created by the overall atmosphere, structural order and clarity as well as the balance between unity and diversity. The resulting feeling of harmony is attained through internal cohesiveness of the elements in relation to their entirety much like a living organism. The clearly articulated structures and architectural forms as well as the immediate understanding and legibility of each component's purpose play a decisive role in the overall construction.

A high aesthetic quality does not necessarily mean high costs nor does it require special fascia-work or unique treatments. The principle of simplicity should not be mistaken for banality, lack of imagination and primitivism found in common commercial construction. Rather, the demand for economic efficiency goes extremely well with the concept of minimalism as do a lack of fascia-work and the avoidance of embellishment.

As mentioned above, the "performance" of a building, (i.e. its ability to serve current and future purposes) is basically determined by the features of the chosen technical and structural solutions together with the building structure, shell, media and finishings. The most important objective when planning a building is an in-depth discussion through which everyone involved in the planning reaches a mutual agreement upon all of the performance related features.

An optimized layout designed with modular resources and spaces is helpful in effectively coordinating the processes and spatial planning. Choosing a common dimensional scheme simplifies the allocation of media systems to production units and facilitates modifications in the sense of changeability. Moreover later, it expedites cost-effective addition of building elements such as hall bays on the façade.

In order to exploit the building's changeability with regards to current tasks, tasks extended in the future as well as to tasks not yet known, it is important to strategically combine unchangeable, difficult to change and changeable structural

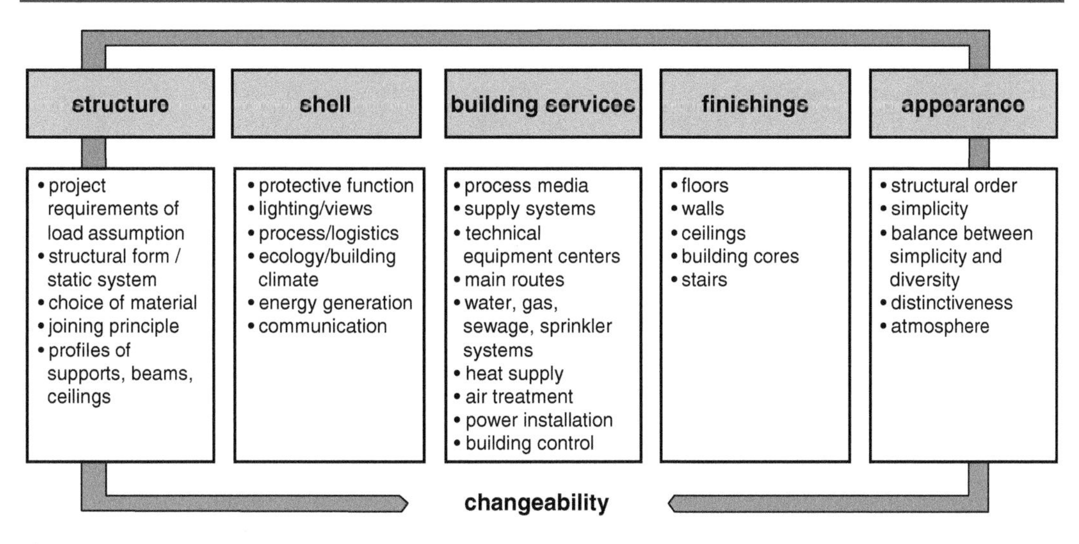

Fig. 11.2 Structural features of a building. © Reichardt 15.181_JR_B

features. The term 'unchangeable structural features' refers here to the load-bearing capacity of the foundation and base plate, whereas 'difficult to change' structural features include the load-bearing capacity of support columns and beams profiles as well as the diagonal braces of the static reinforcement which limit the expansion of a hall. Changeable structural features include moveable, closed or transparent façade elements, which, depending on the need, allow daylight inside the building either from the hall façade or roof openings.

Figure 11.2 provides an overview of the structural features of the building's design fields described above. In this chapter we will discuss each of these in detail with regards to their significance for changeability.

11.1 Load-Bearing Structure

11.1.1 Project Requirements and Load Assumption

In order to meet the manifold and in some cases contradicting requirements of a specific project, it is necessary to find a way to design the load-bearing structure oriented towards the long term production strategy of the company. Figure 11.3 depicts examples of various project requirements for the structure. The geometric and technological parameters derived from the production and logistic processes are the primary factors influencing the load-bearing structure and determine how the building services are installed. As explained extensively in Chap. 5, the changeability of size and utilization areas as well as built-in units plays a dominant role. During the utilization phase the protection and safety of people and property as well as comfort levels are equally important criteria. Ultimately though, the main priority is economics i.e., the costs and construction time.

Pursuing what we call a 'synergetic factory planning' (presented in detail in Chap. 15), fuses the project specific requirements together into a so-called 'requirement profile'. The concrete parameters are then aligned by all those participating in the planning process, especially in view of the long term changeability.

Once the basic project requirements are defined, the dimensions for the structural members should be estimated early-on; this can be accomplished using tables such as those found in [Kra07]. In order to do so, the load assumptions for the traffic loads, static loads and dynamic loads need to be determined. In most cases, countries have set standards, which help in determining the applicable load assumptions for

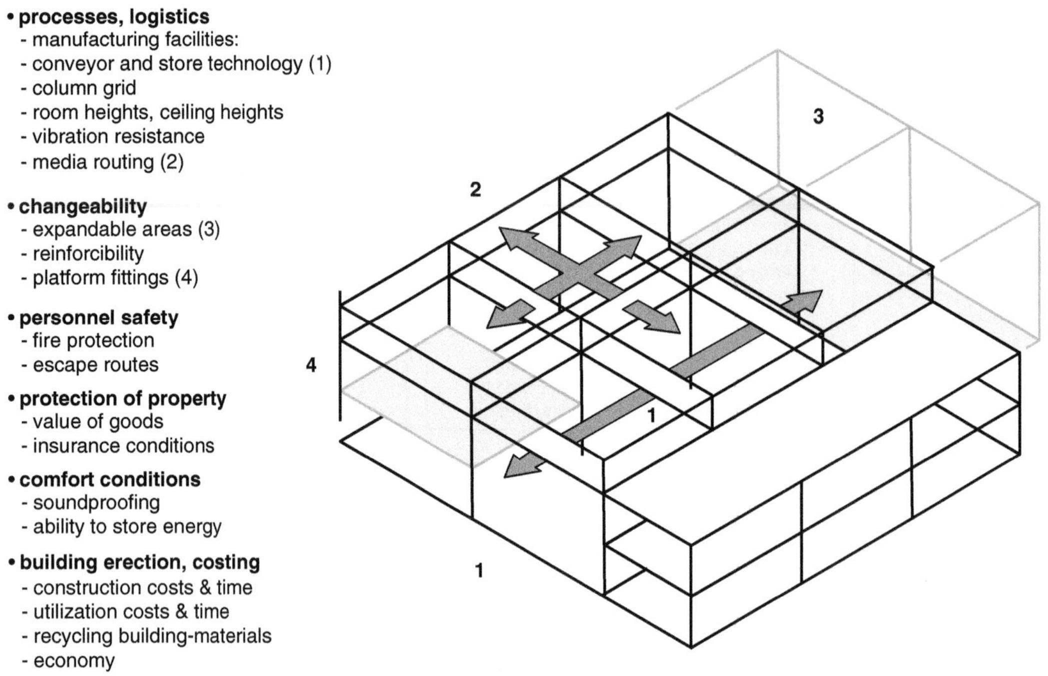

- **processes, logistics**
  - manufacturing facilities:
  - conveyor and store technology (1)
  - column grid
  - room heights, ceiling heights
  - vibration resistance
  - media routing (2)
- **changeability**
  - expandable areas (3)
  - reinforcibility
  - platform fittings (4)
- **personnel safety**
  - fire protection
  - escape routes
- **protection of property**
  - value of goods
  - insurance conditions
- **comfort conditions**
  - soundproofing
  - ability to store energy
- **building erection, costing**
  - construction costs & time
  - utilization costs & time
  - recycling building-materials
  - economy

3
2
4
1
1

Fig. 11.3 Project requirements for the load-bearing structure. © Reichardt 15.182_JR_B

various building typologies. For example in European countries, loads generally need to be established in compliance with the EN Eurocode standards, Similarly one could refer to BS 6399 "Loading for Buildings". Part 1 of the practice code refers to *dead and imposed loads*, Part 2 to *wind loads* and Part 3 to *imposed roof loads* [BS6399] for projects which need to be in compliance with the same.

In the absence of the relevant data, the planning team has to make common assumptions. For example, the location of the main technical centers for the equipment or suspensions for media needs to be addressed at an early stage of the project with care. It is therefore advisable to document all findings and assumptions related to the project in segments and to continually refine these as the project progresses (see also Chap. 16 Project Management and Chap. 17 Facilities Management).

Figure 11.4 provides an overview of important load assumptions for the main elements of the structural framework. Some of these are conditional on processes, others are due to local conditions e.g., the snow load. The effort to create foundations and the base plate is largely determined by the quality of the specific subsoil, especially by groundwater level and location as well as the existence of load bearing soil layers, which is why a qualified soil report must be submitted when the project starts. The floor plate plays a special role with regards to changeability as it decisively determines the possibilities for changing the location of equipment and machinery or re-designating floor surfaces e.g. from logistics to manufacturing or vice versa.

It is well worth taking time for a detailed discussion on the possible additional future loads. Although the initial cost of the structure increases if the building is designed for additional future loads it also provides for a corresponding higher degree of changeability. When making decisions it is critical to practically weigh both strategic and economic factors. Expanding floor areas or overhead areas have to be considered with foresight, in the same way as anticipated changes in the production e.g., a new generation of machinery and equipment along with their structural requirements.

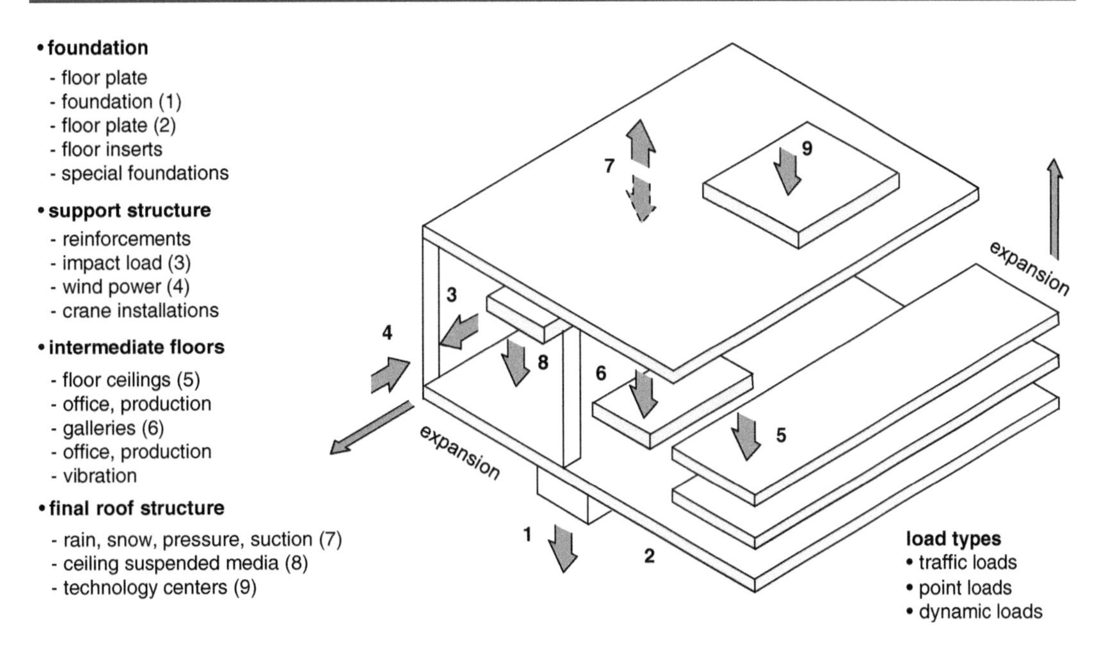

Fig. 11.4 Overview of load assumptions for a building structure. © Reichardt 15.183_JR_B

11.1.2 Building Structure Form as a Static System

The properties of the building structure are determined by the architectural usage of beams and supports systems, ceiling systems, floor plates, foundations, load-bearing walls and cores. In general, the resultant distributions of the static system's forces define the structural form of a building. In turn, the selection of the structural form and the principle behind the load transfer and bracing determine which directions the building can be extended as well as its suitability for accommodating special loading conditions. Based on [Eng07], eight families of structures can be roughly distinguished: beam structures, grids, frames, arches, cable suspensions, cable trusses and domes as well as cable nets and textile constructions.

According to [Gri97] structural forms for halls can be simplified into four groups: supports and beams, frames, arches and space frames. Figure 11.5 provides an overview of structural forms for halls which employ either combined support/truss structures or just truss structures; arch systems and space frames are seldom used in factory buildings and thus not addressed here.

Generally, every hall and multi-story building has to be braced longitudinally and transversely. Space frames or arch structures are inherently stable in transverse direction and thus only need to be braced longitudinally. Directed structures are clearly identifiable by the location of the main and secondary support beams. Here, the vertical load is directed along a single axis over the main beam into the support columns. In comparison, non-directed structures distribute the vertical load bi-axially over all the structural members into the support columns. Non-directed structures are thus usually only efficient on square support fields. Nevertheless, it is easier to extend in two directions with non-directed structures than with directed structures.

According to Fig. 11.6, it is also critical to discuss how the modularity, span width, reinforcements, load distribution and extensions can be combined for every project from both the process and spatial perspective. When doing so the horizontal and vertical extensions should be distinguished from one another. It is advisable to employ 3D modeling techniques so as to identify possible areas of conflicts between process facilities and building services early on. In general, a higher degree of changeability leads to a number

SUPPORTS AND BEAMS
construction without purlins
halls with purlins
lattice purlins and trussed rafters
suspended trusses
FRAMES
fixed truss
double articulated frame
three-girdle truss frame
guyed truss frame

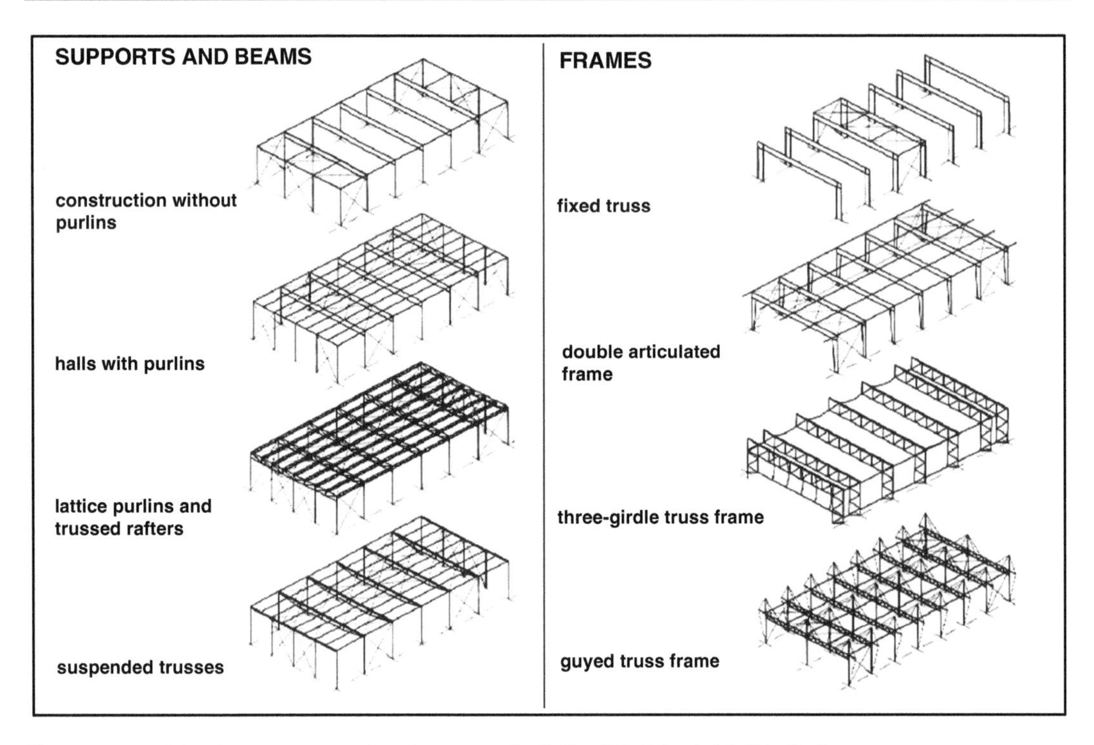

Fig. 11.5 Load-bearing structures and static systems for halls. © Reichardt 15.184_JR_B

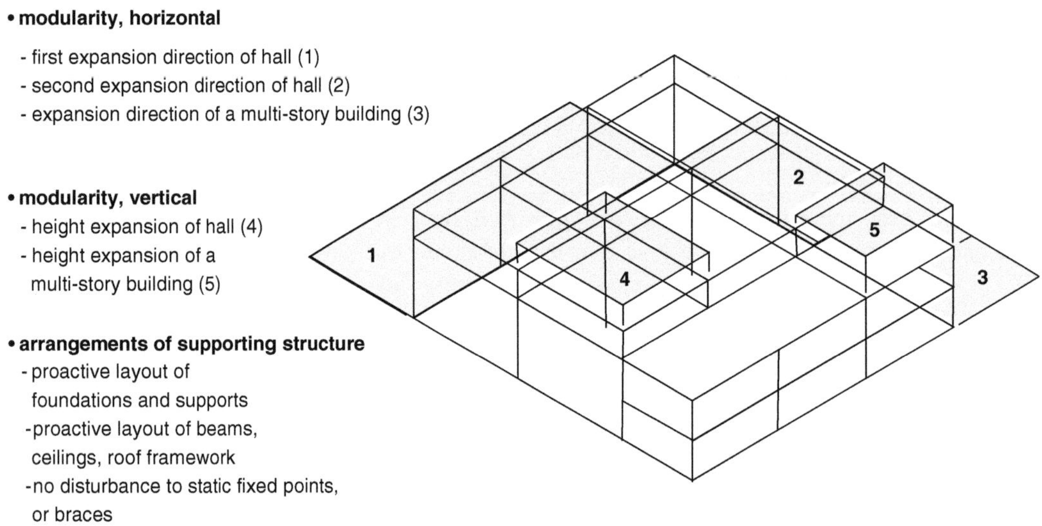

• modularity, horizontal
- first expansion direction of hall (1)
- second expansion direction of hall (2)
- expansion direction of a multi-story building (3)
• modularity, vertical
- height expansion of hall (4)
- height expansion of a multi-story building (5)
• arrangements of supporting structure
- proactive layout of foundations and supports
-proactive layout of beams, ceilings, roof framework
-no disturbance to static fixed points, or braces
1
2
3
4
5

Fig. 11.6 Possible structural framework forms. © Reichardt 15.185_JR_B

of possibilities for extensions and expansions. The building structure for large halls, for example, should permit galleries to be built-in for administrative functions closely related to the production (work planning, production control, quality assurance etc.) and should thus allow interiors to be easily modified. If necessary, these auxiliary levels could be suspended from the hall structure in order to provide for an unobstructed column free space below.

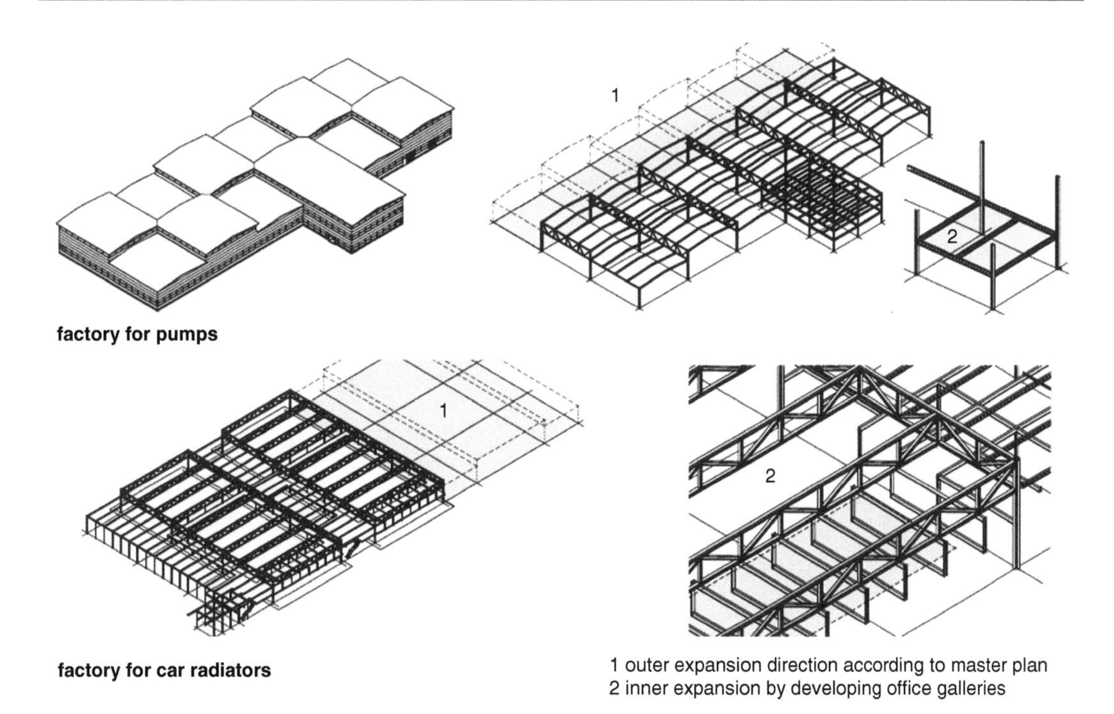

Fig. 11.7 Building structures examples. © Reichardt 15.186_JR_B

Figure 11.7 shows two projects that were realized and the options they offer for external or internal extensions. In the case of the pump factory, six additional modules of 21 × 21 m (68 × 68 ft) could be added on to the length without disrupting the production. The interior of the module set at the front is furnished with office spaces that continue into the factory and thus facilitate direct communication. Similarly, in the car radiator factory large and independent modules spanning 18 or 36 m (59 × 118 ft) can be added without disruption of operation. Instead of a separate administrative building, an integrated office gallery offers room for the management with a direct view on to the production processes (see also Fig. 11.43).

Skylights are recommended for providing naturally lit work areas in low-rise buildings and halls. These naturally lit areas either run parallel or perpendicular to the span of the load-bearing structure. If naturally lit areas are installed perpendicular to the structural span, the structural members would then lie within or outside the temperature regulated area underneath the building's shell. This type of construction can rarely be justified in the sense of energy consciousness due to the cold bridges that are thus created. It is therefore much more practical to install exposure areas in relief structures parallel to filigree beams so that the daylight can penetrate the girder area. Examples of various applications of lighting elements as well as the relationships between the light openings, room heights and room depths are discussed further in Sect. 11.2.3 on natural lighting.

In regards to the economy of hall structures with large spans, it is also important to pay attention to the impact loads resulting from the movement of the fork lift trucks on supports as well as the sum of the loads to be suspended by the roof structure. Furthermore, the use of a floor based distribution system for media should be investigated as a means of minimizing the load with broad-spanning hall roofs.

In order to find solutions for a specific project it is advisable to discuss optional plans from different perspectives. Figure 11.8 depicts a comparative analysis of a few possible modular

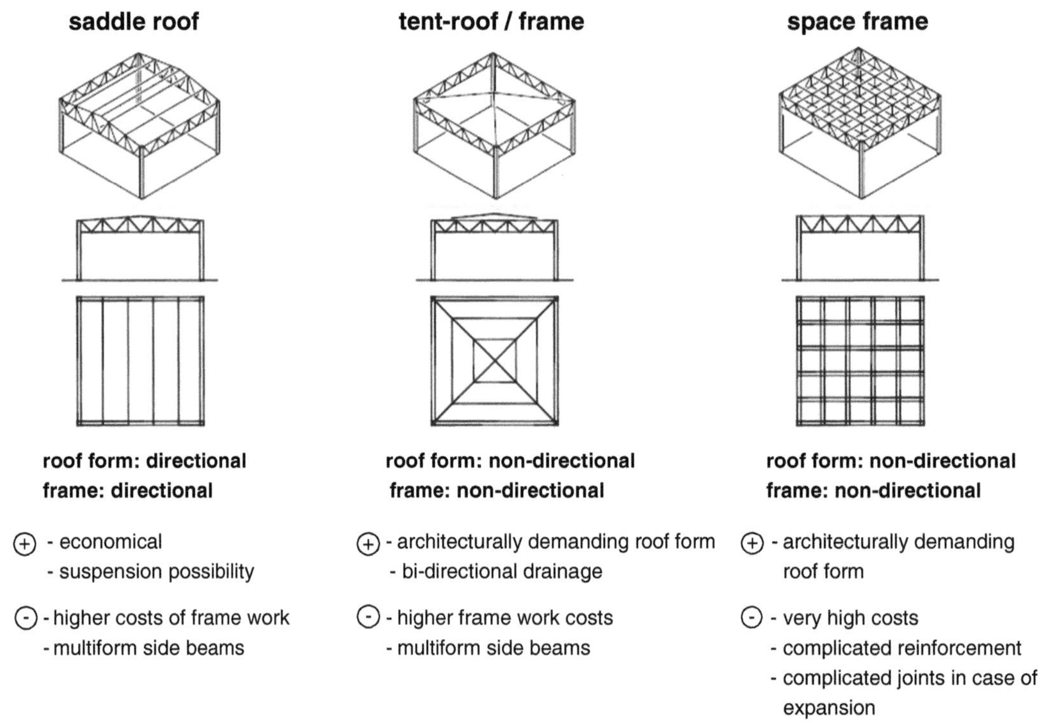

saddle roof	tent-roof / frame	space frame
roof form: directional frame: directional	roof form: non-directional frame: non-directional	roof form: non-directional frame: non-directional
⊕ - economical - suspension possibility	⊕ - architecturally demanding roof form - bi-directional drainage	⊕ - architecturally demanding roof form
⊖ - higher costs of frame work - multiform side beams	⊖ - higher frame work costs - multiform side beams	⊖ - very high costs - complicated reinforcement - complicated joints in case of expansion

Fig. 11.8 Possible building structures for a car assembly hall. © Reichardt 15.187_JR_B

structural forms and static systems based on a typical car assembly hall. A comprehensive assessment includes studying the advantages and disadvantages of the processes, architecture, structure, building services and economic efficiency.

Steel, a material that can withstand a great deal of stress, is a preferable option for changeable load-bearing structures for a number of reasons. Wide spanning structures can be developed with relatively minimal weight. It also allows for comfortable routing of services pipes through the truss-work. In order to ensure that the building structure complies with fire protection regulations composite steel constructions (i.e., load-bearing elements encased in concrete) can be used.

In Europe, the Euro codes, which were introduced as construction standards, regulate corresponding requirements for building structures e.g., DD ENV 1993 Eurocode 3: *Design of Steel Structures* [ENV1993], DD ENV 1994, Eurocode 4: *Design of Composite Steel and Concrete Structures* [ENV1994], DD ENV 1998 and Eurocode 8: *Design of Structures for Earthquake Resistance* [ENV1998]. In the United States (I) E72–E1670 regulate the entire construction process with reference to 108 subsystems, processes and quality controls [E1670].

11.1.3 Span Width

Determining the span width for halls or multi-storied buildings is one of the important, if not the most important decision from the perspective of processes and space. The aim to have as few supporting columns as possible obstructing the use of space has to be weighed against the efficiency of the building structure; the optimum targets a compromise that is tolerable from both perspectives. Polonyi examined the cost development as a function of the span, roof load and material for an approximately 300 m^2 (3,229 sqft) hall with a clearance of approximately 6.50 m (21 ft) [Pol03]. A double articulated frame, a

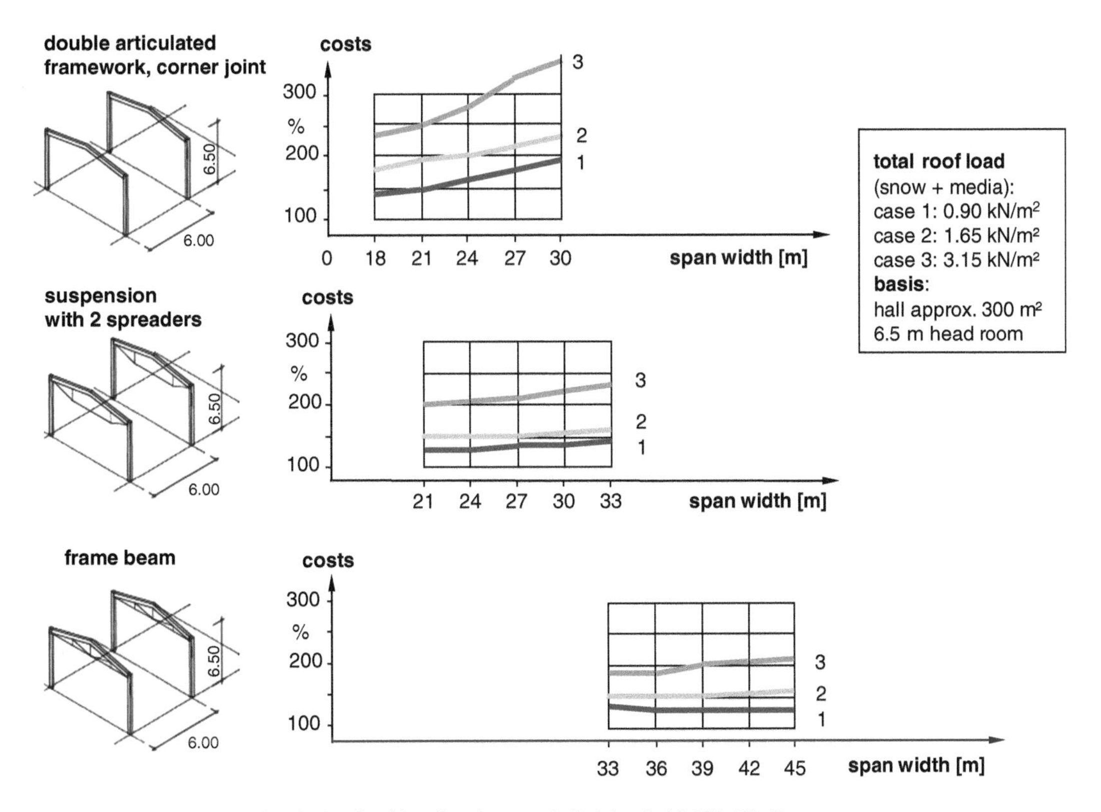

Fig. 11.9 Relative costs for timber load-bearing frames. © Reichardt 15.188_JR_B

cable-supported double-spreader structure and a frame-beam system were chosen as the basic optional static systems to be compared for both steel and timber cross-sections. According to Figs. 11.9 and 11.10, the relative costs can be derived as a function of the span width.

Accordingly, in industrial hall construction, in comparison to the standard solutions with a 20 m (65 ft) span, structures with a span of 30–50 m (98–164 ft) are also possible without additional construction cost burden provided that the roof loads (i.e. snow load, suspended loads, etc.) are minimal. In some cases, cable supported timber constructions divided in compression and tensile zones having spans ranging from 21 to 30 m (68–98 ft) may provide cost effective alternatives.

Additional studies were conducted on the afore-mentioned load-bearing frames for the car assembly hall with span widths of 15 × 15 m (49 × 49 ft), 20 × 20 m (65 × 65 ft) and 24 × 24 m (79 × 79 ft). Various alternatives of fixed supports in steel and concrete as well as roof structures made of steel, pre-stressed concrete and timber were also considered. An in-depth analysis showed that when compared to the original 15 × 15 m (49 × 49 ft) column grid, a 21 × 21 m (68 × 68 ft) column grid offered value addition over the long term with an approximately 10 % higher construction cost.

11.1.4 Selecting the Materials and Joining Principle

In industrial construction there are a multitude of materials available for the occasional large span construction with roof loads. Since steel is capable of carrying large loads without buckling, it is particularly well suited for modular construction as well as halls with large spans. Laminated timber and cable supported timber constructions are appropriate for halls having average spans while light-weight metal constructions facilitate building and dismantling

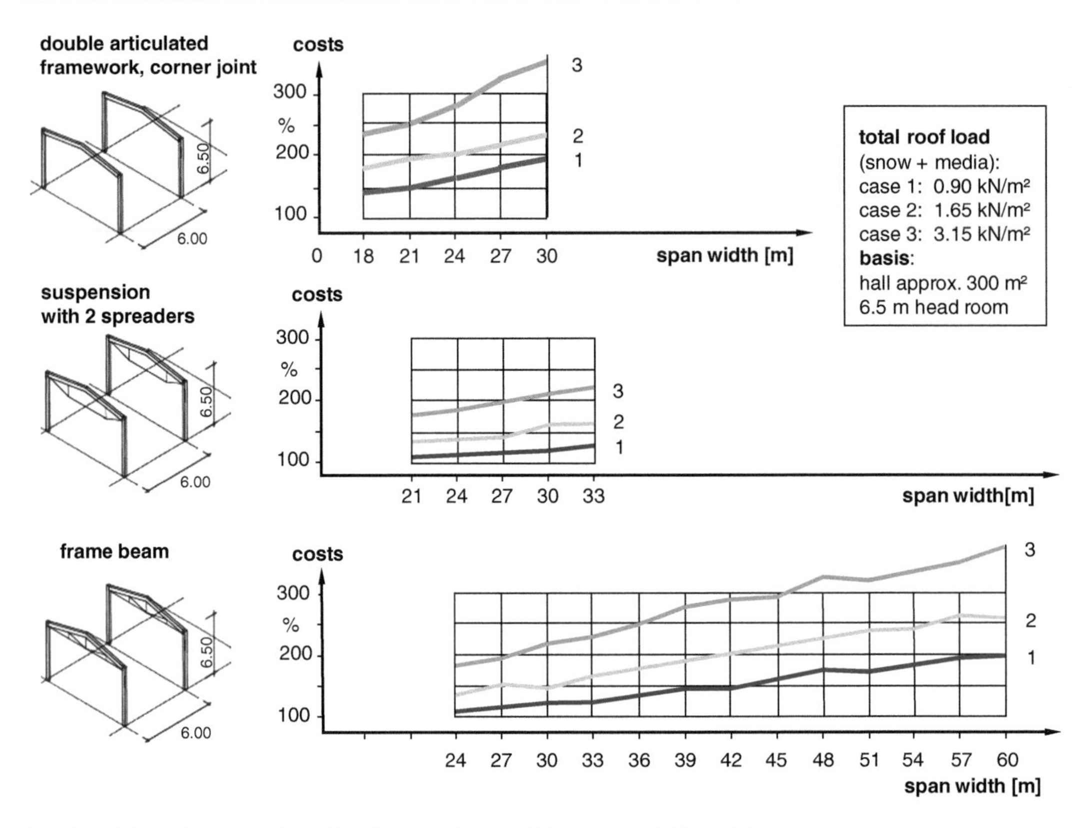

Fig. 11.10 Relative costs of steel load-bearing frames. © Reichardt 15.189_JR_B

temporary building structures quickly due to their lighter components.

When selecting materials it is also important to take into consideration fire safety issues. In addition to the fire rating of the materials, the fire load due to processes, operating facilities and logistics is also critical. By applying appropriate coatings, fire safety ratings up to F 90 (fire safety up to 90 min) can currently be attained for steel halls, while fire rating values up to F 60 are currently achievable for timber constructions with suitable profiles or intumescent coatings. Multi-storied buildings, which have to meet higher safety requirements with regards to the fire resistance of supports, beams and ceiling slabs, are built with concrete steel or composite steel. Figure 11.11 depicts the dimensions of beams and supports with various materials or combinations of materials. In comparison to reinforced concrete, steel composites allow a leaner profile for a lower static height.

Similarly, resistance to corrosion and weathering could be other criteria for selection of construction materials. Building structures made out of steel, timber or reinforced concrete needs to be protected with appropriate measures from driving rain. For example, steel is protected by hot galvanization or coatings against corrosion. In general, industrial buildings need to be constructed as quickly as possible; modular structures that are pre-fabricated are often advantageous than in situ concrete constructions. Modular components not only allow the structure to be assembled almost regardless of the weather and even in snow, they also have basic advantages over monolithic constructions. The choice of joining principle (welding, screwing or inserting) determines the geometry of the joints as well as the planning, manufacturing and assembly schedule. The ability to easily dissolve or undo construction joints makes it possible to later strengthen or retrofit support beams with greater loads, a feature

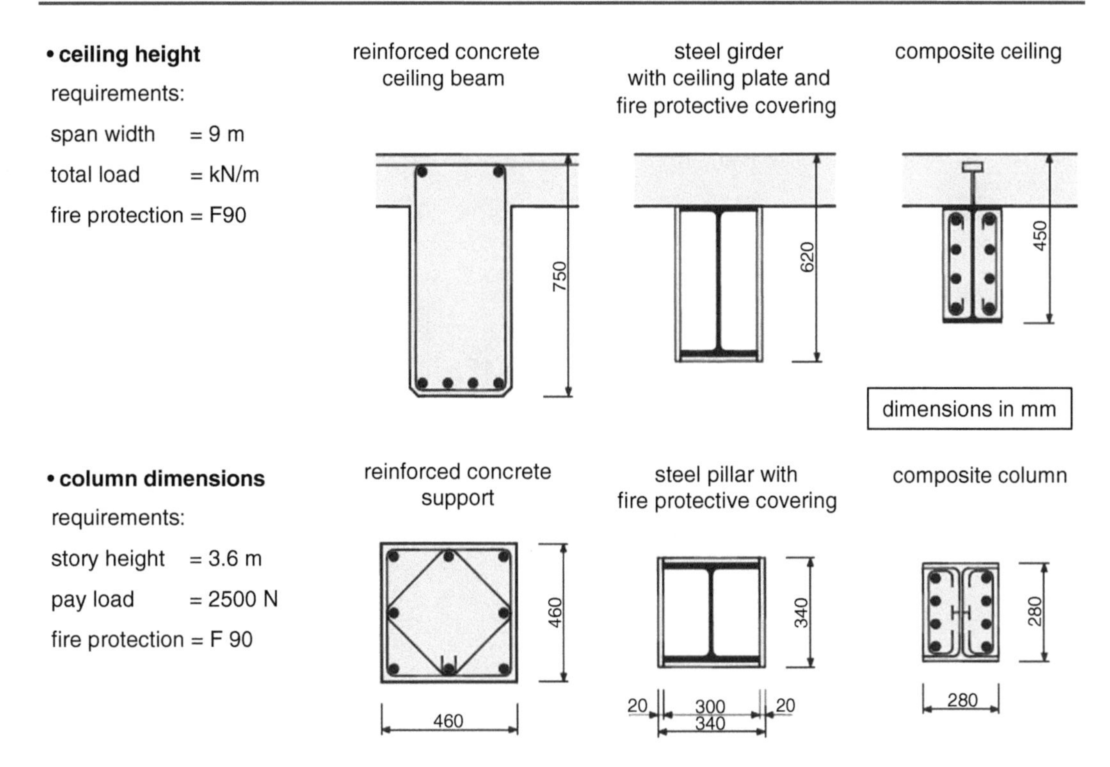

Fig. 11.11 Relevant measurements for structural members from various materials. © Reichardt 15.190_JR_B

that supports changeability. Furthermore, when the building is dismantled, materials need not be separated for recycling purposes.

Based on the degree of prefabrication and modularization, we can identify four types of systems for joining structural members: monolithic constructions (with so-called in situ concrete) are built with homogenous, non-dissolvable joints on-site. Homogenous but dissolvable joints are created by welding steel joints. With a corresponding amount of effort, such building structures can once again be separated into individual structural members. With partial prefabrication, structural members that are partially reinforced can serve as lost formwork; their final stability being provided by the concrete applied on-site. A steel skeleton bolted together represents the highest degree of modularization with complete prefabrication of all components, whereby the structural members can be fundamentally changed with additional joining.

Figure 11.12 depicts an overview of several materials and joining principles for halls and multi-storied buildings. According to [Ack88], there are a large number of materials available for halls, all of which have different properties especially with regards to fire protection classes. For multi-storied buildings these materials are limited to reinforced concrete, steel and steel concrete composites. Nowadays, as indicated in [Rei08], innovative fire-resistant timber constructions can also be considered for industrial and commercial construction.

For the changeable factory, modular constructions (see also examples in Fig. 11.42) seem to be the preferred approach to developing structural frameworks:

- Completely prefabricating a load-bearing structure with easy to dismantle joints facilitates changes wherein beams can be equally quickly and easily reinforced or replaced.
- Removable ceiling panels, allows for future vertical openings between stories e.g., for conveyor systems.
- Required internal extensions can be realized through gallery areas, which when prepared appropriately can be hung on existing structural members (see also the example in Fig. 10.6).

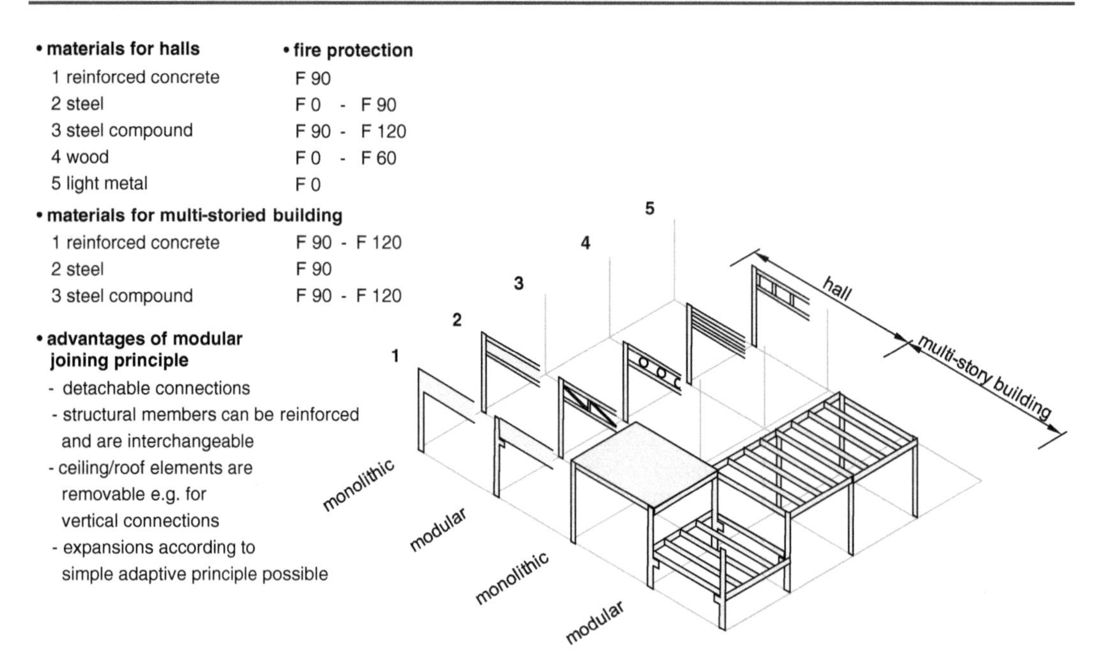

Fig. 11.12 Materials and joining principles for load-bearing structures. © Reichardt 15.191_JR_B

According to [Lac84], the design idea behind the construction can generally be implemented in steel, concrete steel, timber and light steel. Particular details such as plug-in connections allow structural members to be quickly assembled and disassembled thereby meeting the requirements for temporary or even mobile factories. However, it is advisable to weigh in the additional costs against the increase in changeability (i.e., quicker retrofits and fewer disruptions with reference to the on-going operations).

11.1.5 Profiling Support Columns, Beams and Ceiling Slabs

Depending on the structural form, span width, materials and joining principle a multitude of designs are possible for the support columns, beams, roof coverings, intermediate floors and ceiling slab components. Ideally fewer supporting columns are preferable; however, supporting columns are equally important for integrating conveyor facilities and building services. Guiderails for overhead or slewing cranes as well as other lifting equipment needs to be closely coordinated for safe anchoring into support columns. Columns can also be used to fix vertical trusses for building services such as uptakes, downspouts or air ducts. Moreover, the framework for the machinery supply and disposal systems (e.g., electricity, compressed air or water pipes) can be attached to supports. Cross-shaped supports allow media routings to be suspended within the arms of the cross.

With modular building structures the construction time can be reduced with foundations molded around the support columns. Impact sockets with a height of ca. 1.2 m (3.9 ft) anchored in the foundation plate dissipate the impact load of e.g. trucks without subjecting the support to a further static load. Lowering the upper edge of the foundation on the support shaft can make it easier to horizontally route media lines or retrofit them below the floor plate. In our car assembly example, the use of concrete and steel with various profiles for the hall's support columns was investigated in accordance with the roof load and clearance to determine the costs of the supports, beams and foundation for a module area of 20 m × 20 m (65 × 65 ft) (see Fig. 11.8). In this case, concrete supports with a molded foundation proved to be an efficient solution.

The profiling of the beams as solid web girders, castellated beams, truss girders or cable-supported trusses influences both the distribution of media and daylight. For example, filigree beams allow media to be routed within the girder area whereas in case of solid web girders the clear usable height of the hall is reduced by the space necessary for installing media lines below the beams. At times, floor-high truss girders allow maintenance bridges or technical galleries to be integrated into the roof structure. Since permeable beams support the spreading of natural light in the roof area, profiled beams can be drawn upon to optimize the lighting of work areas by redirecting the light. A flat roofing system is advantageous for systematically installing media services as well as arranging lighting. In comparison, moderately sloped surfaces and trusses profiled to facilitate roof drainage, while requiring slightly more effort to design and construct offer the added benefit of beams at equal distances, in turn facilitating suspension systems for cables, pipes etc. for media and processing or conveyor equipment. A consistent and well-coordinated plan for the media supply and disposal system is extremely important for planning, coordinating and/or making future changes (see also Sect. 16.9 Building Information Modeling).

The structural system, choice of materials and modularized components of the roof should allow skylights or other necessary roof penetrations (e.g., vents, etc.) to be retrofitted. The roof detailing should also allow for sound absorption where necessary. Surfaces that reflect light more increase the brightness of the room, whereas porous surfaces made from perforated metal or hanging sails decrease the sound level. Corrugated or profiled sheeting materials allow for a number of possibilities for hanging elements or systems in future, if necessary.

The advantages of a modular system for a storey-high roof compared to a monolithic construction lie in the possibility to integrate building service equipment, the future changeability of the structure and the speed of the building process.

Equipping modules with details such as "tracks" allows installed media to be varied with minimal effort. Moreover, with modular elements vertical connections between multiple floors can be easily added. With greater spans, media systems can be cleverly organized and routed within the static height of ceiling or roofing elements when they are implemented as ribbed panels.

Another critical factor with regards to long term changeability is the load-bearing capacity of the floor plate. The entire surface—without any exception—should be consistently able to bear the same load. Moreover the consistency in level floor construction needs to be considered. In case of vacuum de-watered floors, mistakes in flooring levels cannot be altered at a later date and needs to be planned accordingly with an eye to the future, especially keeping in view of the requirements for processing technologies.

Detailed planning for the floor plate may include: media routings, conveyor technology, process related waste removal, special foundations and fire escape routes. If built-in cable trays, guide rails or chip removal systems cannot be avoided, their cover plates should be standardized and inter-changeable. In addition, special foundation systems for specific machines limit the changeability considerably; similarly location of escape tunnels as well as their entrances should be examined with an eye to the future.

If we adhere to the principles introduced here, we can, in accordance with [Pol03], see the futility of decorating a building with unnecessary or impractical constructions.

Figure 11.13 summarizes the characteristics of changeability that are applicable when profiling the supports, beams and ceiling/floor slabs.

11.2 Shells

Encasing the building structure involves vertical façade surfaces as well as horizontal or sloped roof surfaces. These generally consist of

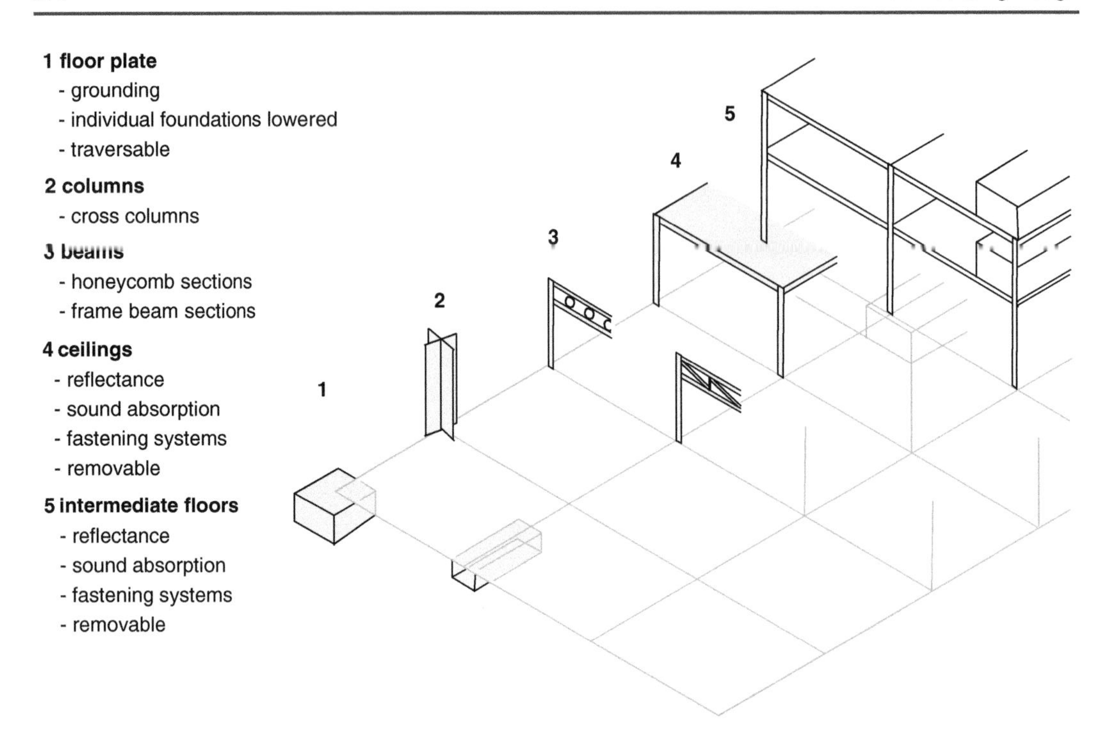

Fig. 11.13 Profiles of structural members. © Reichardt 15.192_JR_B

combinations of closed and transparent surfaces. Openable elements such as windows, doors or gates are integrated as required into the shell surfaces. A shell plays a protective role, fulfills production and logistic requirements, aids in providing light as well as outdoor views, facilitates communication; the shell could also be eco-friendly and can produce energy. A plethora of possible solutions along with types of construction and examples are presented in [Her04, Sch06].

11.2.1 Protective Functions

The shell should be designed to provide suitable climatic protection depending on the geographical location of the building. Meteorological data from the nearest weather station can provide the mean temperatures of the hottest and coldest periods along with data on wind, rain and snow. A comprehensive 3D analysis of the shell's energy losses and gains within the frame of a synergetic factory planning is advisable before defining the wall and roof structures. The analysis should also include the energy generated by the processes (e.g., machinery heating-up) and is best conducted for a number of alternative wall constructions as well as possible future scenarios regarding the cost of energy.

The orientation of the building in relation to prevalent wind directions influences the arrangement of the gates, canopies or air exhausts. If building laws and regulations require certain noise levels to be adhered to, closed surfaces and openings in the shell need to be planned accordingly. Similarly, if the building is located near a highway or an airport it may be necessary to provide sound insulation from external sounds (e.g., by implementing special noise reflecting measures on the façade).

Figure 11.14 provides an overview of the basic features which play an important role against cold, heat, rain, wind and noise. The value of the heat transfer coefficient (U-value) determines the thermal insulation, whereas the solar heat gain coefficient (SHGC or g-value) measures the amount of heat gained through

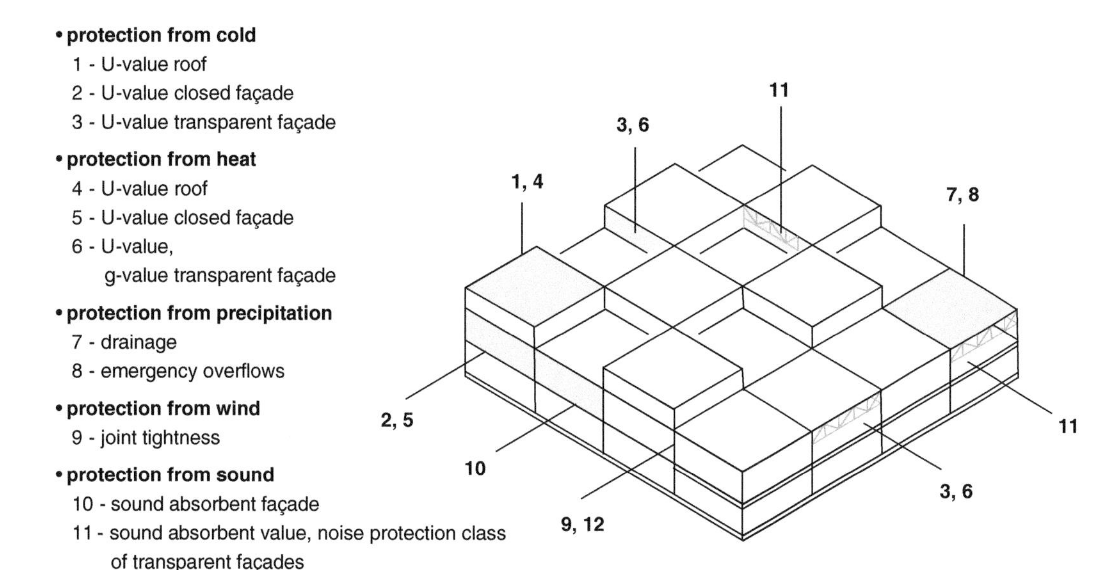

Fig. 11.14 Building shell features with protective functions. © Reichardt 15.193_JR_B

solar radiation. Depending on the geographic location, the maximum anticipated rainfall could be critical selecting the roofing system. The demand for the shell to have greater changeability generally means avoiding load-bearing outer walls since monolithic constructions are difficult to modify or extend. In comparison systems that are modular can be more quickly and economically adapted to new requirements.

11.2.2 Production and Logistics

In terms of the shell, structural requirements that result from production and logistic are related mostly to: receiving and dispatch points, escape routes, assembly openings and media systems for processes, fire protection, and building services that penetrate the shell. Examples of these requirements are presented in Fig. 11.15.

With the aim of increasing the degree of structural changeability monolithic constructions should be avoided for the roof and walls; instead it is advisable to plan them as a number of non-rigid or easily changeable zones. For the vertical façades, modularized transparent or translucent components of approximately 3.00/4.50 × 4.50 m (9.8/14.7 × 14.7 ft) are well suited. Supplementary exits as well as truck entrances into the halls may be designed based on this grid, to enable changes at a later date.

In order to provide protection from the weather when loading and unloading vehicles, it should be possible to erect canopies with their own foundations anywhere along the periphery of the hall. Interchangeable façade columns extending between the floor plate and girders permit larger openings in the façade for bringing larger machinery into the hall. Furthermore they make it easier to quickly extend the hall. Skylights, heat exhausts as well as other process related air ducts can be integrated in band-like roof structures. Modularized transparent and closed panel systems for sheds and floor to ceiling windows can always be adjusted to meet new requirements. Two façade systems that were developed based on this strategy for a plant that assembles automobile cooling systems are illustrated in Fig. 11.16.

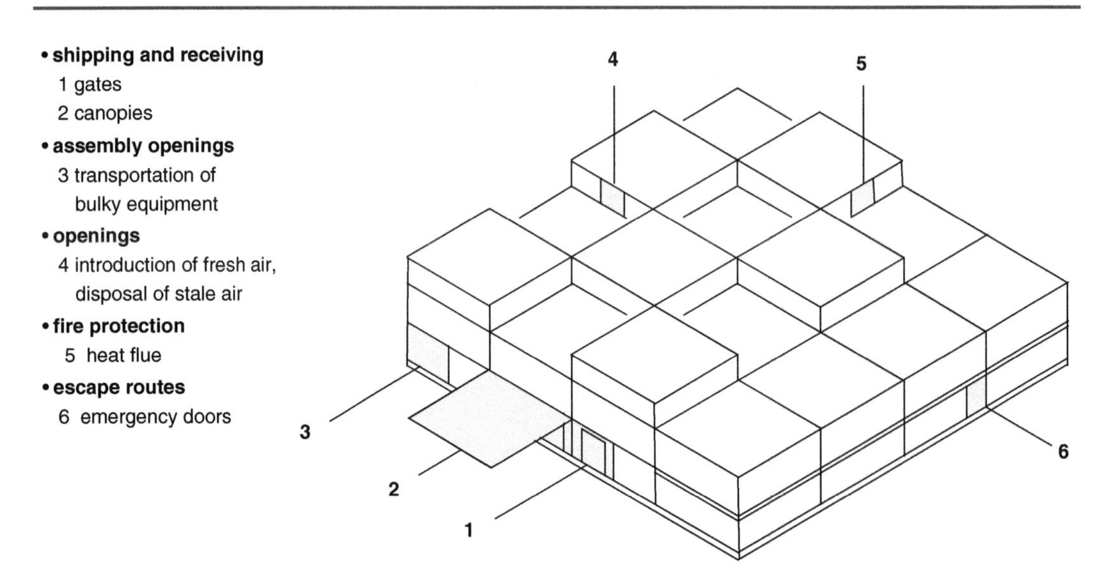

Fig. 11.15 Features of a building's shell from the perspective of production needs. © Reichardt 15.194_JR_B

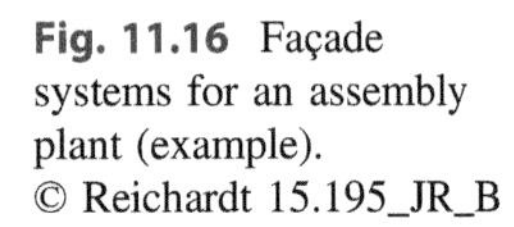

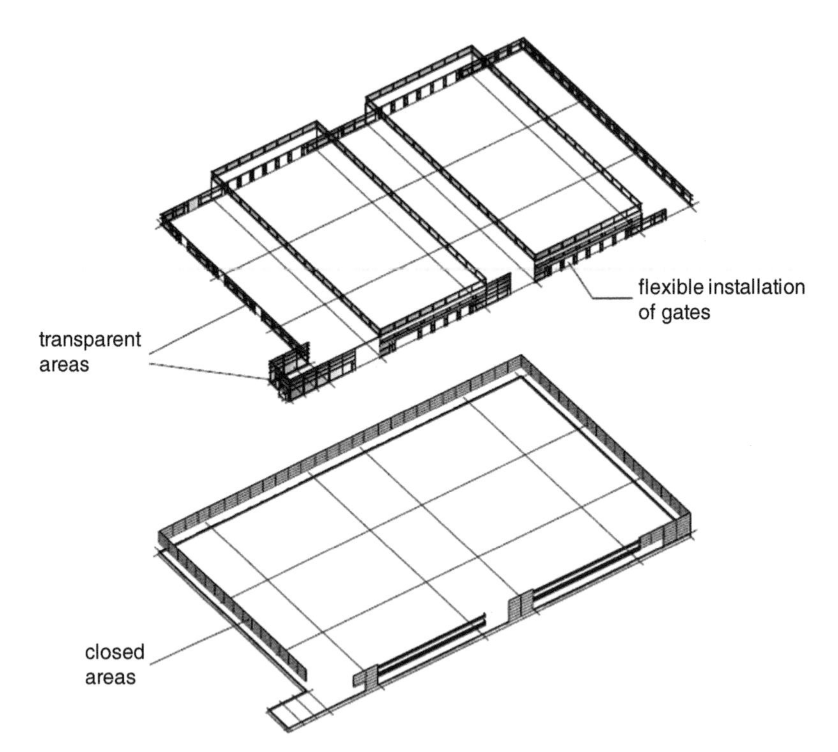

Fig. 11.16 Façade systems for an assembly plant (example). © Reichardt 15.195_JR_B

11.2.3 Lighting, Views, Communication

According to the specifications of German workplace guidelines 10 % of the ground floor area of halls up to 2000 m^2 (21,500 ft^2) in size should be reserved for transparent façades at eyelevel. For larger halls there is no requirement for a direct outdoor view due to the depth of the hall, instead the focus shifts to skylights providing natural daylight to the work area. In order to provide a comfortable and livelier atmosphere for their employees, enterprises are increasingly trying to create brighter work places. Changeable work areas are not dark areas; rather openings in the shell allow workers to be aware of the time of the day and changes in the weather outside, if any. A roof that has been properly profiled would provide work areas with glare-free and naturally diffused light for comfortable working environment. Beyond providing outside views and natural light, façades can contribute meaningfully to the communication between a building and its surroundings as well as convey a sense of identity and significance. Figure 11.17 provides an overview of the elements of a building's shell which help to determine these features.

In addition, particular attention needs to be paid to the drainage of roof extensions. An appropriately profiled roof ensures that every point of the roof has a slope of at least 2 % to remove rain in a clear direction. Large span building structures necessitate girders with structural camber. Damages to roof structures and buildings due to sudden and heavy torrential downpours overloading the rain water drainage system led to a comprehensive revision of the DIN and EN standards. This in turn has impacted the design and detailing of even the emergency overflows.

11.2.4 Ecology and Energy Production

Façade surfaces are extremely well suited for implementing measures that improve both the ecology of a building and its energy balance. Green façades and roofs can be integrated as ecologically valuable measures when calculating equalization areas. Furthermore, roof plants help in reducing the amount of rain water flowing into the public storm water drainage network and are thus advantageous when applying for necessary approvals for discharge.

Nowadays a number of tested systems for actively producing energy on façades are available. Thermal collectors for generating warm water, photo-voltaic collectors for generating power and wind turbines are plausible either as independent units or as systems integrated into façade and roof elements. With a synergetic factory planning a 3D computer model is used to optimize their efficiency, depending on the orientation of the building as well as their economic considerations.

Figure 11.18 provides an overview of features from an ecological perspective as well as with regards to energy production and illustrates how they can be integrated into the building shell. In

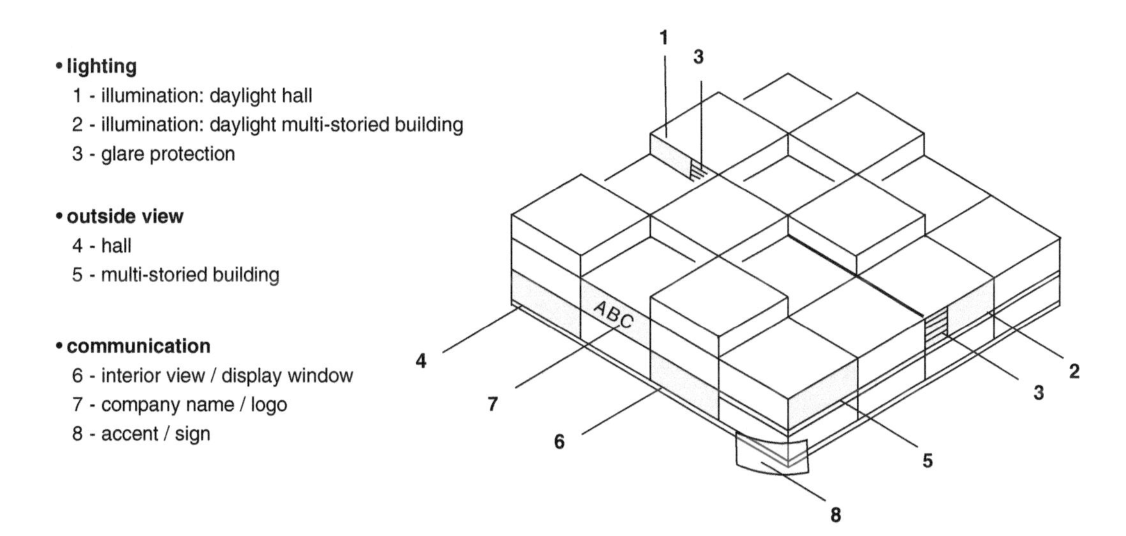

Fig. 11.17 Requirements of a building shell in terms of lighting, views and communication. © Reichardt 15.196_JR_B

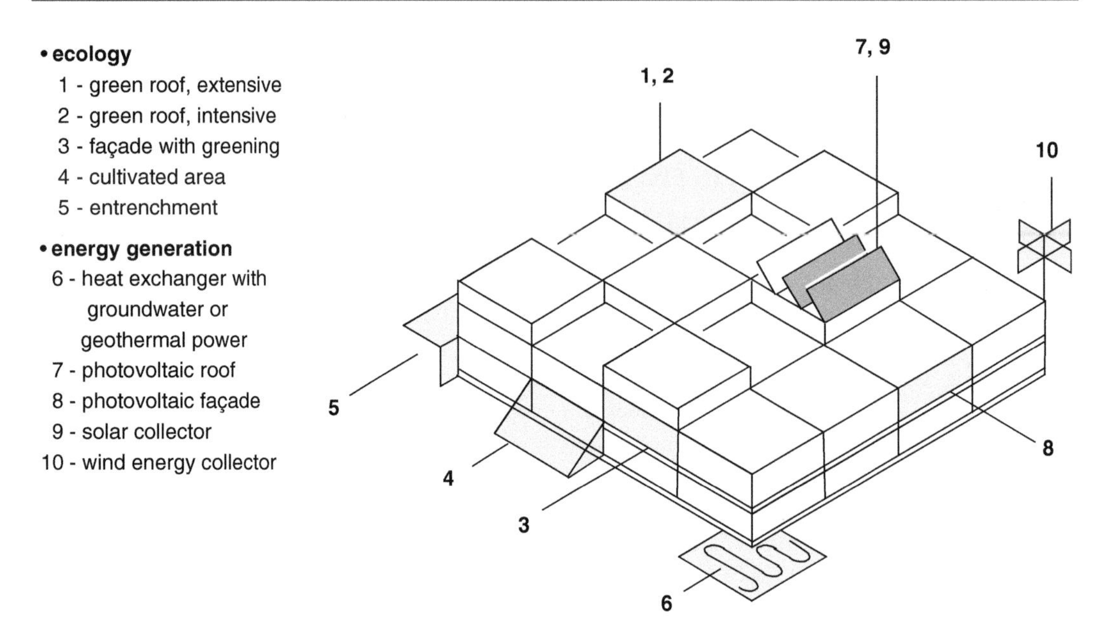

Fig. 11.18 Features of a building shell from the perspective of ecology and energy production. © Reichardt 15.197_JR_B

Germany we anticipate that, in accordance with [Hau07], the "passive building standard" (yearly required energy for heating less than 15 kWh/m^2), already in place for residential buildings, would be soon enforced for commercial and industrial buildings too. Currently a large number of façade and system manufacturers are developing integrated systems for generating environmentally friendly solar and geothermal "passive" energy in particular. Additionally, certification systems for attaining a "Green Building Standard" have also being developed for industrial and commercial buildings (see Sect. 15.7.2). Ecological and energy conscious façades thus also hold the promise of future sustainability.

11.3 Building Services

As mentioned earlier, we are using the collective term "building services" to refer to the building service equipment, the supporting structures needed to run the building services as well as the wires, cables, pipes etc. for the operation of the manufacturing facilities. In industry, "building technologies" (planned by architects) are commonly differentiated from "process technologies" (planned by technology planners). Traditionally, these were assumed to be independent of each other; however, they pose a significant risk, if their potential synergy is not harnessed. In view of the increasing complexity, all the media systems should be holistically optimized keeping in mind future savings in energy and natural resources (see Fig. 3.28). Examples of optimizing energy in industrial buildings can be found in [Hat06]. In order to facilitate integrated planning, a variety of advanced software programs are available making it possible to model all the systems in 3D, simulate energy generation and losses and provide thermodynamic flow simulations (see Fig. 11.35).

In [Dan96], Daniels describes general requirements and principles for planning building services:

- Find the optimum between investment and consumption costs.
- Minimize the use of energy and natural resources as well as pollutant emissions within the frame of a cost-benefit analysis.

- Discuss the type and provision of electricity and heating or cooling energy (e.g., the possibility of the company operating its own block-unit heating power plant).
- Carefully determine requirements and dimension them according to specifications with sufficient reserves for growing needs.
- Carefully plan supply systems for processes and buildings, where possible with short distances, flexibility and adaptability for new technologies with as little downtime as possible.
- Separate building service centers, main routes, line paths and outlets that need to be always accessible according to a superordinate system plan.
- Clearly identify common main and secondary lines and mark the direction of media flow.

Changeability plays a particularly important role in the design of the supply and disposal systems, service centers, main routings, line paths and outlets. The hierarchy of these systems in the building structure is clarified in Fig. 11.19. For the factory planner the location and development of centers and main routes are particularly significant and should facilitate both current as well as future production needs.

11.3.1 Supply and Removal Systems

The building's media structures have to be designed with regards to supply and disposal systems and within the overall framework of the building's master plan. Transfer points, quality, quantity and especially possibilities for extending them determine the distribution and collection structures. In general the supply systems include power supply, heating, ventilation and air-conditioning, pressurized air, water, coolant, lubricant, etc. while removal systems would include sewage and drainage networks, coolant and lubricants. In their overviews, Pistohl [Pis07] and Krimmling et al. [Kri08] point out basic designs for supply and removal systems.

To begin with, one needs to decide whether a centralized or decentralized layout is better. A centralized supply is definitely a practical solution when media is required in large quantities throughout the plant, considering the investment and operating costs. Also, implementing heat recovery technology for ventilation systems is often economically efficient. However, one of the disadvantages with centralized systems is that defects and malfunctions can cause complete operating disruptions. In addition, there tends to

1 external media lines

2 technical equipment centers
- penthouse hall
- penthouse multi-storied building
- basement

3 main routes
- hall
- multi-storied building

4 media paths
- hall
- multi-storied building

5 outlets
- hall
- multi-storied building

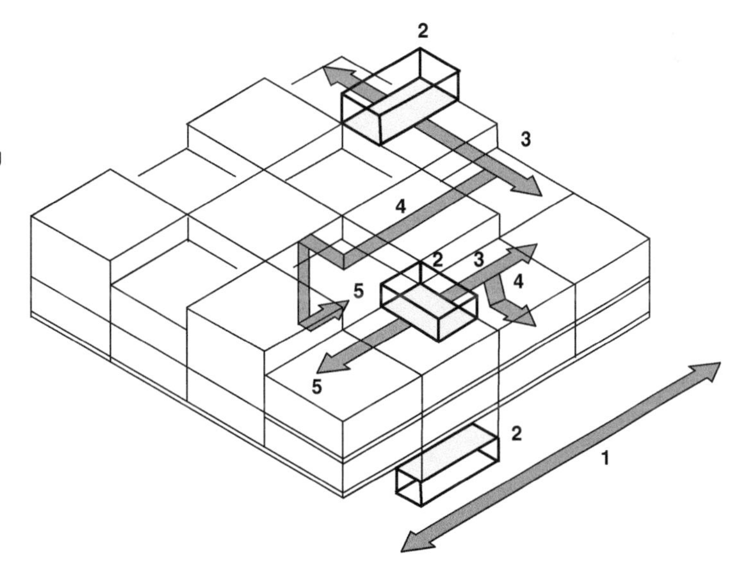

Fig. 11.19 Hierarchy of media systems. © Reichardt 15.198_JR_B

be a far more number and length of cables and pipes than with decentralized systems which in turn means a greater number of horizontal and vertical lines in the building. A further disadvantage is the loss of efficiency that goes hand in hand with the longer line paths.

Compact systems with a higher efficiency factor lead to decentralization. With regards to planning, operating and changes, the greater independency of decentralized systems means a considerable increase in flexibility and changeability especially where modular factory concepts are concerned.

Figure 11.20 depicts an example of a decentralized solution for the supply system of a motor assembly plant with four "sub-factories". Each of the approximately 5000 m^2 halls is a three-story service building with technical centers for ventilation systems and others; space for the electrical systems was allocated on the roof top. A comb-like branch flow system was proposed for the distribution network for ventilation, pressurized air and electric system. This modular concept allows the supply system to be adapted to the sub-factories' future requirements at any time without disruptions.

11.3.2 Technical Centers

In order to generate, operate and monitor media (pressurized air, steam, cool air etc.) the main machinery are located in so called technical centers. Planning them should result in a comprehensive and pragmatic concept including where the centers are located, their spatial requirements and room specifications as well possibilities for future expansions.

The location of the technical centers in a building is determined by the combination of whether they are centralized or decentralized as well as whether they are built in-house or erected separately. Ventilation and air conditioning systems are frequently located near to heating centers (boiler rooms and distribution centers) and refrigerating machines. Due to fire regulations

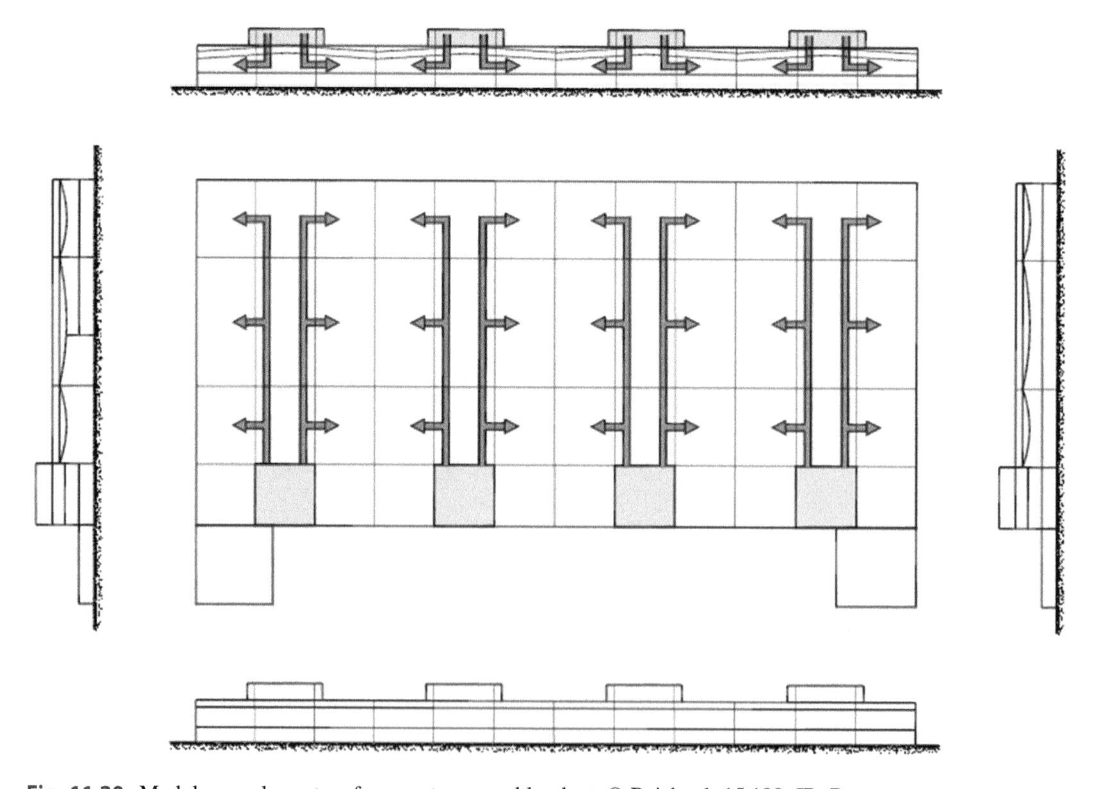

Fig. 11.20 Modular supply system for a motor assembly plant. © Reichardt 15.199_JR_B

locating the ventilation plant and heating system in the same area or close to each other may not permissible. It can however, be advantageous to link the centers close to the vertical installation shafts of the building's core. Other factors that are an integral part of the planning discussion include:

- Calculating the load assumption of the technical centers in the final stage as well as provisions for changing technical units early on.
- Required partitioning against noise, fire and vibrations.
- The location and layout of sprinkler systems and tanks needs to be discussed together with insurance agencies and public authorities. Moreover they need to be closely coordinated with the strategic extension of the plant outlined in the master building plan.

The advantage of a centralized design lie in its lower investment costs with minimal floor space requirements and simplified machinery installation in a synergetic way. In comparison, decentralized systems are far more beneficial in terms of their changeability; it is easier to convert individual production areas and their associated media systems locally with minimal disruption to the on-going production during the replacement of lines or enhancement of the networks.

Locating technical centers in the basement is advantageous since it helps to economically shield against noises and vibrations; also the heavy equipment load is directly transferred to the ground thereby enabling a more economical structural framework. Nevertheless, especially with high halls, long distances and a loss of floor space in the ground floor plan negatively impact the main routings; exchanging systems however can be facilitated via simple lifting equipment.

Locating centers on the ground floor should be avoided, if possible, since they represent fixed points in the floor plan when it comes to options for extending the building. Nowadays there are possibilities to even install electrical and transformer systems at upper floors.

The basic advantage of locating a center on a mezzanine floor lies in the generally smaller channel cross-sections, nonetheless greater measures have to be taken to avoid transmission of noise and vibration (e.g., implementing shielding, installation of shock absorbers, etc.) through the building.

Structurally independent and weather-proof technical components set on the roof floor or on the periphery of the building provides a greater possibility for changeability especially when buildings are not too tall. As a structural measure, an appropriately constructed load-bearing frame should be able to accommodate additional technical modules for potential factory alterations or extensions taking into consideration future anticipated loads. Recent developments are heading towards "ship and plug-in" modules i.e., mobile construction units consisting of completely pre-equipped technical containers that need only to be connected (plugged-in) to one another on-site forming supply and disposal systems that are ready-for-use.

With wide spanning halls, the structural framework zone is often a floor high and can be used to house technical centers as well. The lower joist of the frame can support transverse beams for equipment platforms or maintenance catwalks. Moreover, the entire roof layout is available for integrating the technical building systems in a holistic manner. It is advisable for this strategy to use a 3D computer model to closely coordinate the structural framework design with the media layout.

Compared to centers located in structures independent from the main building, housing centers within the building structure are advantageous because they can be re-arranged overtime; main routings and supply grids can be easily threaded, continuously monitored, maintained or repaired in a protective environment. However, in some cases special encasings for sound or fire protection needs to be integrated.

Figure 11.21 depicts three possible variations for housing technical centers including: a modular penthouse solution atop a multi-storied engine plant, a container type "ship and plug-in" technical center on the hall roof of a cooling system assembly plant and a gallery built into the frame work of a tire factory. The transformer systems were integrated in the penthouses for both the engine and assembly plants.

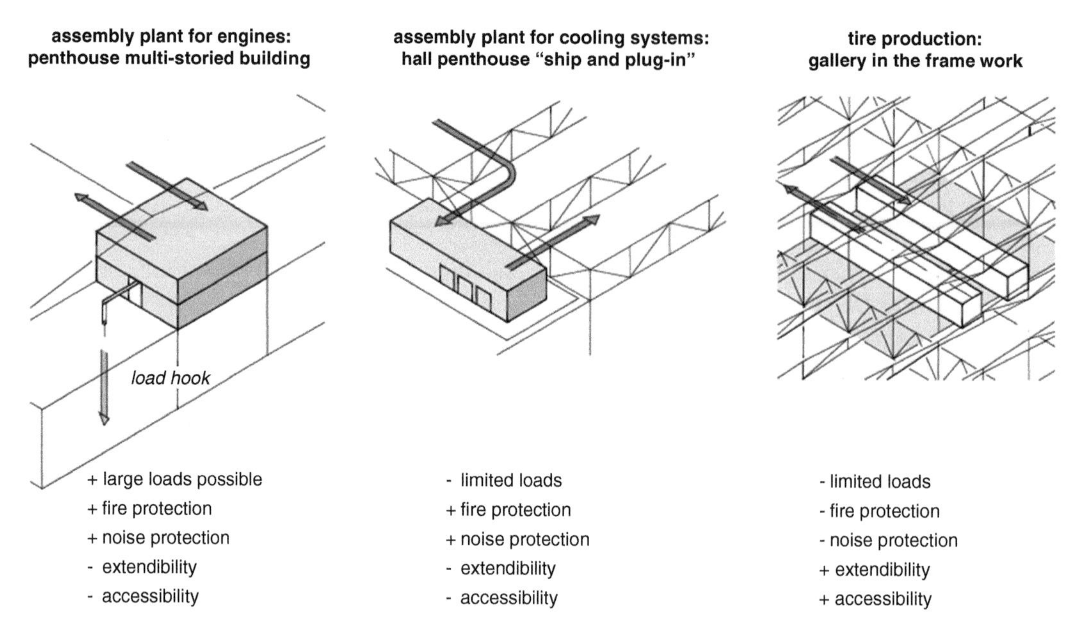

Fig. 11.21 Trouble-free locations for technical centers. © Reichardt 15.200_JR_B

Accordingly, individual loads of up to 40 kN were taken into consideration by constructing grated platforms for landing areas for equipment and providing access via lifting equipment or mobile cranes. The advantages and disadvantages of these basic solutions are stated below each option, based on the criteria of possible loads, fire protection, noise protection, extension possibilities and accessibility.

The spatial requirements for technical centers should be determined in the early planning stages, since the building design and structure are considerably influenced by them. Sound damping requirements need to be considered early on as they greatly influence spatial requirements. The necessary ceiling clearance is often underestimated especially with low floor to ceiling heights in the basement floor; restrictions result in chaotic media routings in the technical centers and limit changeability at a later date. In the course of planning a factory in the long run, it is important to think ahead about possible step-by-step solutions for easing the control and maintenance of the plant. Here too, it is critical to integrate these ideas when planning future spatial requirements.

Since the lifecycle of the technical systems vary from 5 to 15 years, the center's machinery components should ideally planned and constructed as modules which could be easily exchanged and extended. Once again, 3D modeling of the technical centers within the frame of a synergetic factory planning is advisable along with a facility management system which incorporates documentation and maintenance of the components (see also Chap. 17).

11.3.3 Main Routings

The main routing's vertical and horizontal manifolds lead from the technical centers to the distribution systems. Vertical lines are often routed in the shafts of the buildings core, while horizontal lines are routed through or below the structure of the hall roof or ceiling slabs. When selecting the anticipated positioning of the main routings, special attention has to be paid for future options of vertical or horizontal extensions. It is difficult to reroute ill-conceived media packages, often blocking meaningful growth of building facilities and literally stifling any further development.

Shafts and canals for the main routings should be planned so that they meet the requirements for structural stability, fire safety, dampness protection, thermal insulation and hygiene. Moreover, they should be designed so that they are easily accessed for maintenance and cleaning either externally or internally. Shafts having control valves and systems requiring periodic maintenance needs to be large enough for comfortable human movement.

When selecting systems and planning the details of the building structure the layout of the main routings plays a special role. The following points should be taken into account:

- With multi-storied buildings, the connection between the horizontal trays and the vertical shafts are critical points.
- The static reinforcement that should be provided by the core can be negatively impacted when the layout of building services and other media is not designed well.
- The connection openings have to be wide enough to accommodate retrofitting the main routings.
- Horizontal main routings frequently traverse the zone of building structure's static beams.

Moreover, the choice of materials and the overall form of the building structure are decisive for flexible routing of media, especially with regards to their capacities for future retrofitting. Often during the planning, all the possible points of conflicts between the building structure and routings may not be detected. During construction, this could lead to further decrease in clearances when the routings have to be laid unplanned below the load-bearing frame. With the aid of a 3D model, conflicts can be detected early on thereby ensuring the required ceiling clearance as well as preventing unpleasant joints and problems with extensions and retrofitting.

In case of factories, the layout of the electrical power supply for the production facilities as well as the IT-connection of the production and offices is often not considered adequately. International Electrotechnical Commission (IEC) documents 60374-5-51 to IEC 60364-5-54 (Electrical Installations of a Building [IEC11]) provide necessary guidelines in this regard and have been introduced as national standards in 170 countries.

11.3.4 Line Nets

Once the main routings are planned, a practical distribution network to the outlets needs to be developed. The branched media runs behave similarly to the main routings in terms of their vertical and horizontal pathways through the building. The modular system of the building structure should meet the needs of a coordinated system plan for all the proposed line paths. The density of the supply and disposal networks has to consider issues related to both the building services and processes. All of the pathways need to be easily accessed and modified without disrupting the production. Horizontal and vertical points at which the networks of cables, wires and pipes can be installed onto the building should be identified and finalized in accordance with the superordinate dimensions and based upon a standard assembly system. Moreover, implementing a factory wide color coding system and labeling the direction of the media's flow serves as a valuable aid to quickly identify the distribution networks.

Figure 11.22 depicts the 3D media routing plan developed for the fresh air, ventilation, lighting and sprinkler systems for a meeting room of an assembly plant. The distribution networks were coordinated using a system grid of 1.25 m (4 ft); the modularized ceilings with perforated boards for acoustic purposes were accordingly aligned in the plan.

11.3.5 Inlets and Outlets

Many work processes create dust, gases or vapors. Unwanted or disturbing particles are best filtered out at the point of their creation with the aid of ducts. These are the transfer points where media enter the rooms (air supply duct) or flows out of them (exhaust vents). Ducts have to be

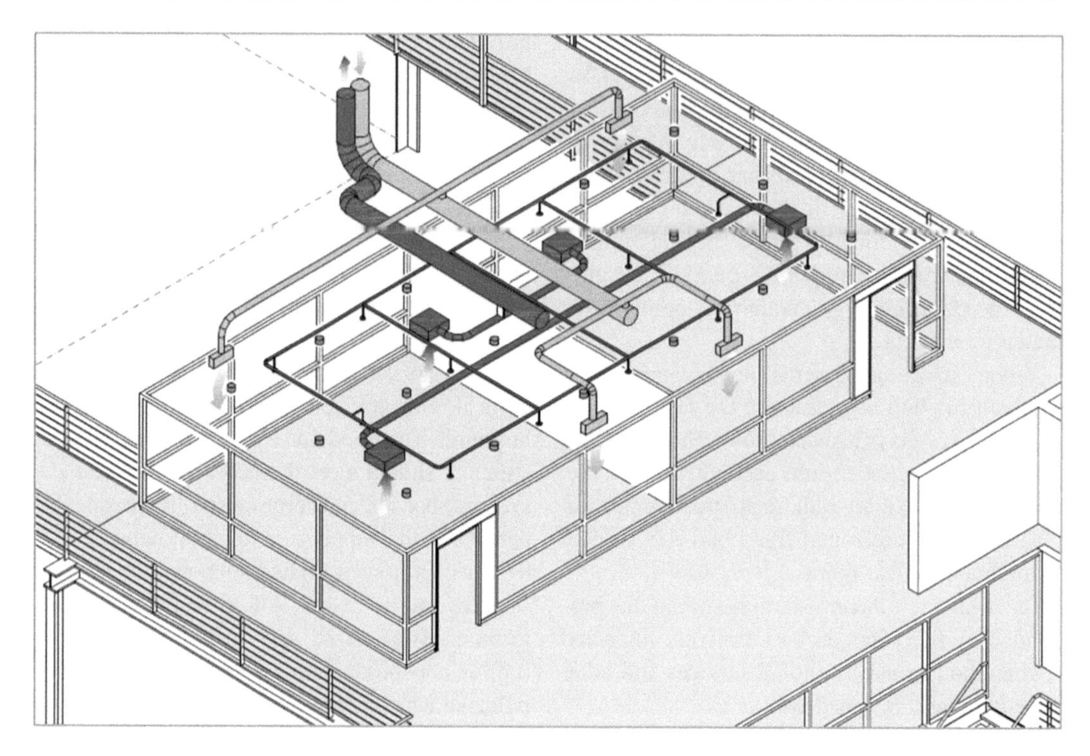

Fig. 11.22 3D modeling of media routings for a meeting room (example). © Reichardt 15.201_JR_B

extremely carefully dimensioned and executed in order to prevent disruptions or breakdowns due to drafts, contamination or soiling of workplaces.

Media intakes and outlets should be easy to locate so as to not interfere with re-arrangement of machinery or installation of new equipment with different requirements.

Once again we can see that creating a 3D model when planning the location and size of the inlets and outlets helps to prevent different media from colliding with each other. With respect to ensuring a high degree of changeability the locations of lights, data lines and air ducts should be closely examined. In addition to the required quantity of illumination covering the floor area of the hall, the distribution of light and its quality should planned such that the hall could be used for various purposes. Similarly, data lines should be distributed throughout halls and multi-story buildings so that equipment and offices could be re-arranged as freely as possible.

Figure 11.23 depicts an example of an adaptable air supply duct made out of fabric for a large bakery (see also Fig. 15.75). The hose-like system filters the air through the material's micro-fine pores distributing it without any noticeable air draft. In addition, the ducts can be easily cleaned (machine washable) and quickly adapted both with regards to its location and its length.

11.3.6 Building Services

By Gerhard Hoffmann[1]

11.3.6.1 Introduction

As discussed in the introduction to this book, the development cycles in industrial manufacturing are becoming shorter while the demands for economic efficiency increase. In addition to

[1]This section was made kindly available to the authors by Gerhard Hoffmann, Managing Shareholder of ifes GmbH Cologne. We would like to express our sincere gratitude to him.

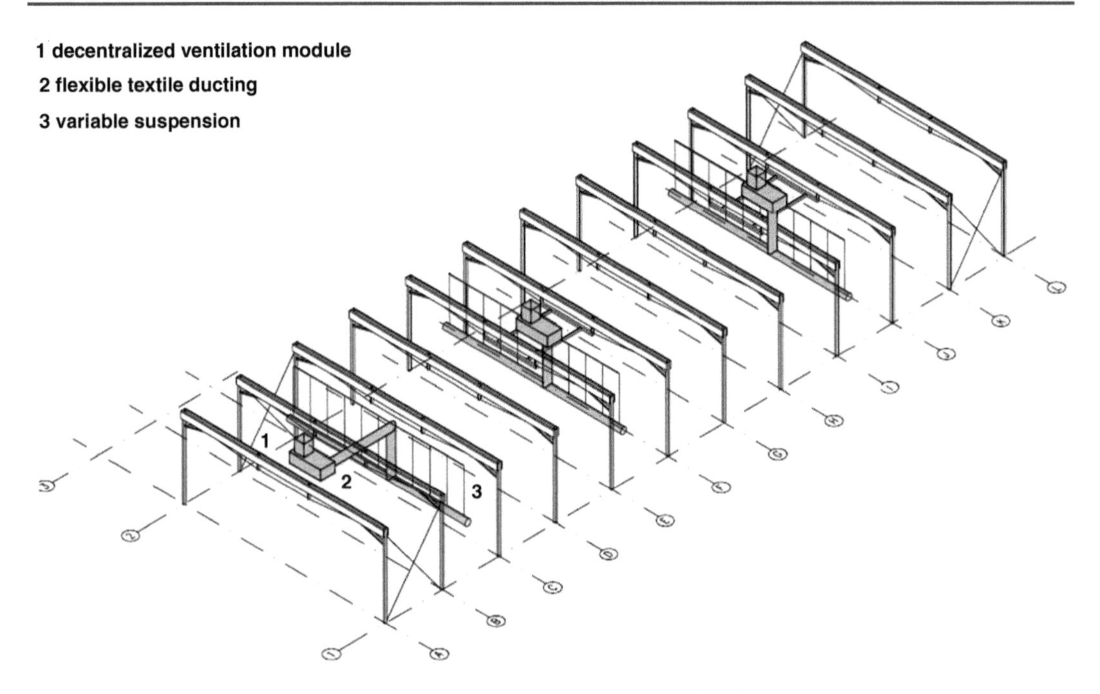

Fig. 11.23 Example of an adaptable air supply system. © Reichardt 15.202_JR_B

highly flexible production equipment, shorter product and innovations cycles require the technical systems in factory halls to be equally flexible. These are generally referred to as building services sometimes also as Facility Systems [Tom10]. When we consider the breakdown of the total costs for building a factory, the building construction, the shell and interior finishing usually represent 30–40 %, the facade approximately 12–30 % and building services the remaining 30–40 %.

DIN 276 subdivides the construction costs of a building into 7 main groups [DIN08]. Building services is further broken down under DIN 276 Cost Group 400 with the heading "Building—Technical Equipment". It encompasses the costs for all of the technical systems or system parts that are built into the building, connected to it or securely fastened to it. Whereas Sect. 16.7 will closely examine how to calculate and control costs of a building, Fig. 11.24 depicts the subgroups of Cost Group 400 relevant for our discussion. Each of these is briefly clarified below.

Media supply (cost group 410) to industrial buildings' includes water, various gases and fire extinguishing systems. The latter serves to protect both the structure and the technical equipment. Since manufacturing halls are specialized structures, the fire protection concept has to be developed by relevant fire protection experts. The expert planning team should also determine which technical fire protection systems need to be installed; for example, a sprinkler system is not always required for a building. However, systems that are able to remove heat and smoke in case of a fire are generally mandatory requirement for industrial buildings. The specifics of this are discussed in detail in Sect. 10.5.

Heating systems (cost group 420) comprise all systems and equipment for generating, distributing and utilizing heat, while air conditioning and ventilation systems are concerned with controlling temperature, humidity, and air circulation within given a space. Air conditioning systems are further sub-divided into different types (see Sect. 11.3.6.4).

Electric power plants (cost group 440) supply the factory with energy for electrical drives and processes. In view of increased flexibility, the cables for supplying electricity to systems in

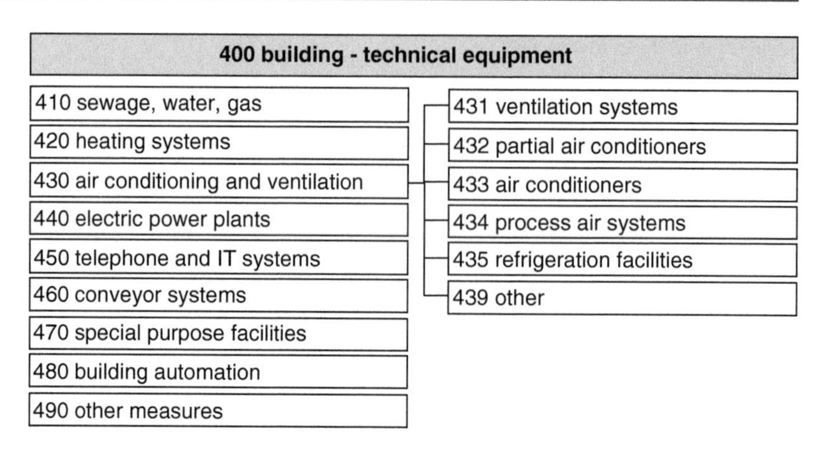

Fig. 11.24 Cost structure—technical equipment (per DIN 276). © IFA 17.600_B

industrial halls should be ideally routed above the ground whenever possible. This allows for process-related conversions to be quickly implemented. In addition to supplying processing equipments with electricity, the hall lighting is critical to the employees' performance. Standards and guidelines, like DIN EN 12464-1: Lighting of Workplaces [DIN11], describe the principles and conditions for work-appropriate lighting.

Telephone and IT systems (cost group 450) ensure internal and external communication. Due to specific security and environmental requirements, special rooms may be required for the server clusters.

Conveyor systems (cost group 460) refer here to facilities that move parts in progress or finished goods and need to be securely installed to the building structure at higher levels above the production area. Typically such systems are common in automobile plants. Special purpose facilities are rarely found.

Building automation (cost group 480) refers to all equipment which is required to monitor, control, regulate and optimize the building systems and is thus an important part of Facility Management (see Chap. 17). The aim here is to automatically perform operational sequences throughout the plant in accordance with given settings while minimizing the required energy and simplifying their operation and monitoring.

Due to the direct impact of a factory's layout and its facilities on the indoor climate, we will only take a closer look at the heating, ventilation and air conditioning systems (also referred to as HVAC).

11.3.6.2 Requirements

Building services have to meet the production requirements as well as ensure the health of personnel working in the hall. The key comfort factors relevant to the workers' health (incl. room temperature, humidity, air circulation and air purity) are clarified in Fig. 11.25. These are discussed in more detail in Sect. 10.4.

Although flexibility and changeability are desirable, the key parameters which continue to decide most of the technical systems are economic efficiency of investment and future operating costs. Most companies expect a ROI (Return of Investment) of 2 to a maximum of 5 years for capital investments. Due to growing concerns of environmental protection and sustainable—energy efficient production (i.e., in terms of a "Green Factory") it is possible to extend the ROI in individual cases to a maximum of 10 years.

The following criteria should be taken into account in order to comprehensively design the technical systems for heating, cooling, aeration, ventilation, electricity and compressed air:

- the products' demands on the hall climate and the production processes',
- demands on building services,
- demands on the flexibility and changeability of the building services health and safety requirements,

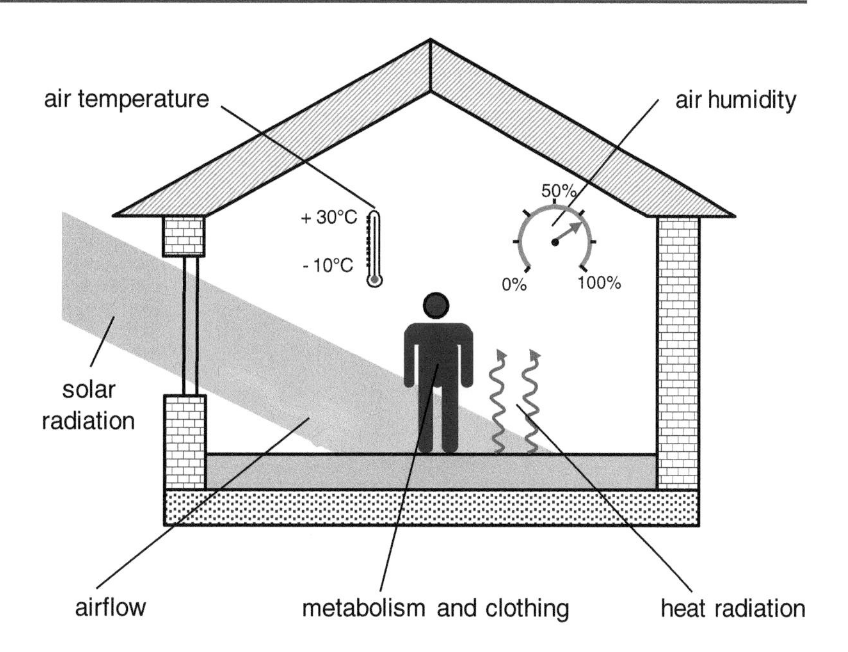

Fig. 11.25 Influences on thermal comfort. © IFA 17.601_B

- dissipation of thermal loads and contaminants for improving the quality of the workplace,
- the location and local climate.

The steps documented in Fig. 11.26 have proven to be a practical and necessary checklist for analyzing and determining the basis for planning building services. The steps refer to the building with its physical properties, the production along with the consumption of energy and media, the thermal and pollutant load created by the production process as well as the air movement with their characteristic airflow, temperature and humidity.

The concentration and thresholds of substances need to be taken into account depending on the type of production (foundry, painting, mechanical manufacturing, metal forming or assembly). Readers are requested to refer to the current regulations, standards and guidelines as explained in Sect. 8.4.

The planning and costing of building services are completed in accordance with DIN 276 Table 2 for Cost Group 400—Technical Equipment. Estimating and controlling costs are discussed in their entirety in Sect. 16.7. This division allows the individual planning and/or construction costs to be objectively compared with other plans. In accordance with HOAI [HO13], these costs are also used as a basis for calculating the planning fees and determining the lifecycle costs for a comprehensive feasibility study.

11.3.6.3 Heating Systems

Heating systems include systems required for generating the heat (e.g. a boiler), the distribution network, individual devices for heating rooms and others. Industrial heating systems could be broadly differentiated between centralized and decentralized; heat could be introduced into a building via convection or as radiant heat via ceiling mounted radiators.

Heat radiators are generally the most economical option for maintaining a steady and comfortable temperature at body height in a typical production hall as explained in Fig. 11.27 [AS13]. The left side of the diagram shows how heat is introduced radiantly. This type of heating does not warm the air, but rather heats the fixed bodies underneath it and the floors by thermal radiation. As these bodies warm up, they in turn release convectional heat (center diagram). Finally the warm bodies, in turn heat up the air in the absence of airflow (Fig. 11.27 right).

With convective systems, centralized indoor air systems are distinguished from decentralized

Fig. 11.26 Partial steps and data required for the design of building services [VDI10]. © IFA 17.602_B

data group	required data
1. building data	dimensions, building structure, building shell physics etc.
2. production data	production flow, production volume, production equipment data (dimension, simultaneity factor etc.), energy and material flow, floor plan, construction and layout of the conveyor equipment etc.
3. heat and substances release data	*thermal load*: location of release, type (convection, radiation), duration, time function (delay and damping by buffering) recording the energy consumption (meter reading etc.), recording of fluid flows and temperatures, recording of surface temperatures *substance load*: location and type of release, type of dissemination, type and characteristics of substances, substance flow, time function, characteristics of release (temperature difference, characteristics of individual movement, contaminated air flow, dissemination direction etc.) evaluation from existing work area analyses or by analyzing consumed substance amount or by measuring (e.g. acc. to DIN 33981 or EU guide lines)
4. air treatment data	construction, quality features and airflow of the existing extraction equipment, airflow and construction of supply and exhaust air systems, space for climate controls (site of central systems, air purification systems, opportunities for the installation of air ducts) etc.

air recirculation systems which are either equipped with gas burners or hot water conduits.

11.3.6.4 Air Conditioning Systems

Overview

Indoor ventilation and air conditioning systems are responsible for maintaining the target values for the air quality while maintaining thermal comfort and optimizing energy costs. One of the key ways of doing this is by planning and installing energy-saving and well-coordinated ventilators.

The basic functions of a ventilation system in a production hall are depicted in Fig. 11.28. In the left side of the hall, the conditioned air is introduced close to the work environment, while to the right of the workplace the air pollutants are extracted by the exhaust system. The halls are heated with ceiling radiators; the exhaust air is extracted from the entire hall while the leftover heat is used to pre-heat the fresh outdoor air.

Figure 11.29 provides an illustration of how, in accordance with [DIN04], the air equipment used in production facilities is classified and allocated. Compressed air systems are designed as closed units thus avoiding losses associated to large networks. Free flow ventilation systems use the natural air currents created by the different temperatures to provide a trouble-free extraction. Ventilation and air conditioning systems either force a flow into the circulating air without a fresh air feed or introduce fresh air into it. In the latter case, the fresh air is partially mixed with the recirculated air for economic reasons. Air equipment is further sub-divided according to the

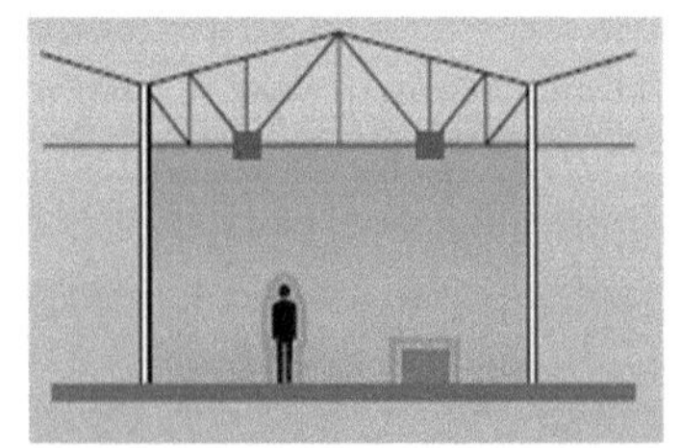

Fig. 11.27 Types of hall heating [AS13]. © IFA 17.603_B

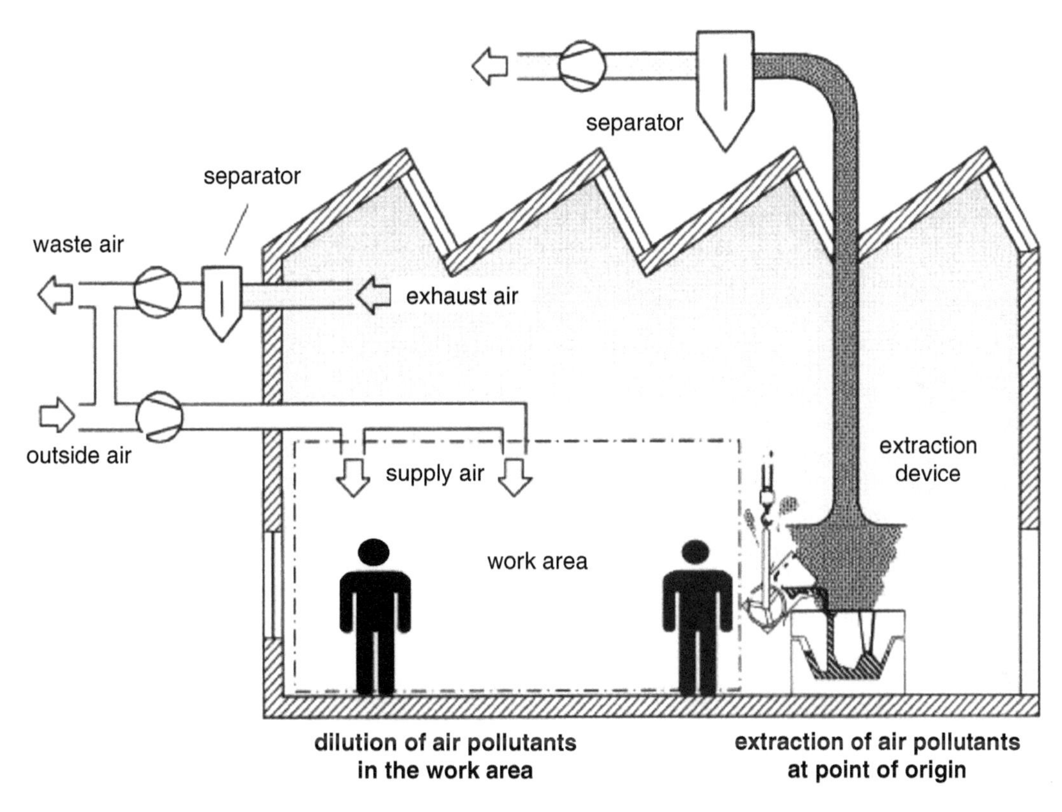

Fig. 11.28 Functions of a ventilation system in a production hall (Kolarik [Kol02])

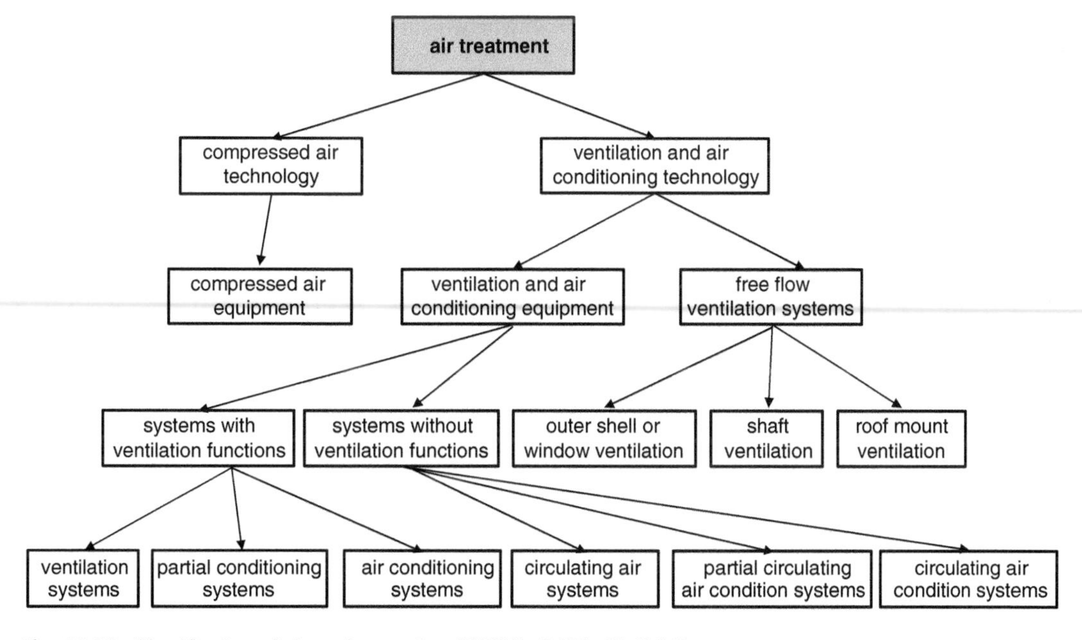

Fig. 11.29 Classification of air equipment (per DIN04). © IFA 17.605_B

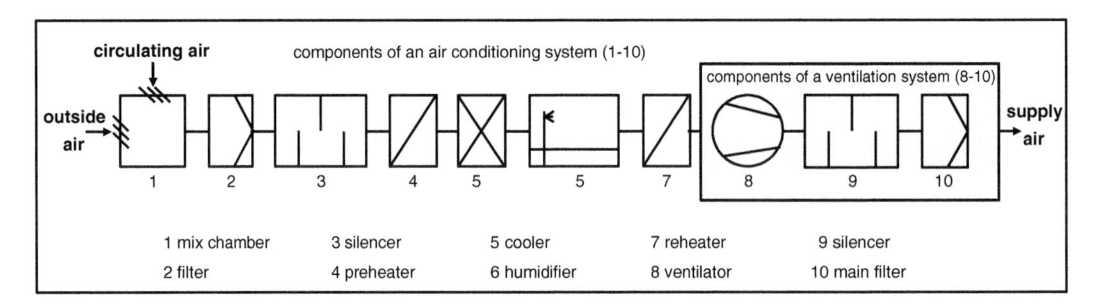

Fig. 11.30 Structure of an air conditioning system (per Recknagel [Rec13]). © IFA 17.606_B

four thermal dynamic methods of treating air: heating, cooling, active humidifying and active de-humidifying.

When a ventilation and air conditioning system conducts all four functions it is referred to as a 'full air conditioning system', whereas those which conduct only some of these functions (heating and cooling, humidifying or de-humidifying) are referred to as a 'partial air conditioning system'. The principle structure of an air conditioner is presented in Fig. 11.30. The outdoor air is first mixed with the recirculated air that is fed back into the system and filtered. It is then routed over a silencer and through a pre-heater, cooler, humidifier and re-heater. These functions are individually activated depending on the state of the input air and the state of the mixed air. With the use of blowers, the conditioned air is then routed over silencers and filtered once more before it is introduced into the room. Depending on what functions are required and desired, placeholders can be installed for individual components, thus facilitating future retrofits.

Ventilation systems can be executed as either single-channel or dual-channel systems (see Fig. 11.31). In the first case, a centralized air conditioning plant in installed from where the conditioned air is directly supplied to the various rooms. This makes it possible to implement and use different sources of energy. In terms of sustainability, a heating system based on wood chips, for example, is possible; the hall heating would then be almost CO_2 neutral.

With dual-channel systems the air is pre-conditioned using a central system and is then routed through decentralized pipes in each zone for final conditioning before releasing the same into the space under consideration. Dual-channel systems are advantageous in that they are easy to install, can be adjusted locally and are easy to shut down to avoid smoke and fire spread. However, there are also a number of disadvantages including the higher space requirement and the need for a large number of decentralized electrical terminals. Keeping in mind a modularized and changeable structure, dual-channel systems appears to be better option since it is easier to adapt to localized requirements when for example a part of a factory is not working or a zone is undergoing conversion and must be switched off.

For decentralized heating systems, which usually work with recirculating air, suppliers offer compact ventilation systems with heating and cooling functions along with integrated heat recovery. Figure 11.32 depicts an example of a decentralized ventilation system as a roof installation. It fulfills the following functions: The outside air passes through a filter and enters a heat exchanger where it is warmed by the re-circulated air that has been extracted from the hall. The supply air fan pushes the pre-heated air through a heating/cooling register and an air diffuser into the hall. The cooled recirculated air leaves the building through a filter as exhaust. The circulating air and bypass flaps control the ratio of fresh air to exhaust air. The condensers serve to generate cold when the air has to be cooled. This example of a ventilation system does not include devices for humidifying or de-humidifying the air.

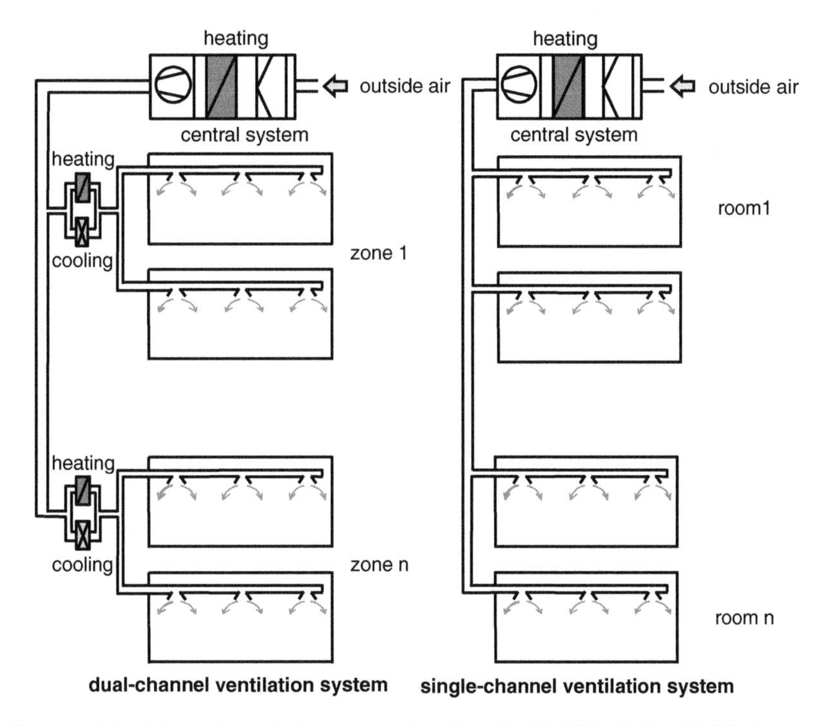

Fig. 11.31 Single and dual channel ventilation systems (per Kolarik [Kol02]). © IFA 17.607_B

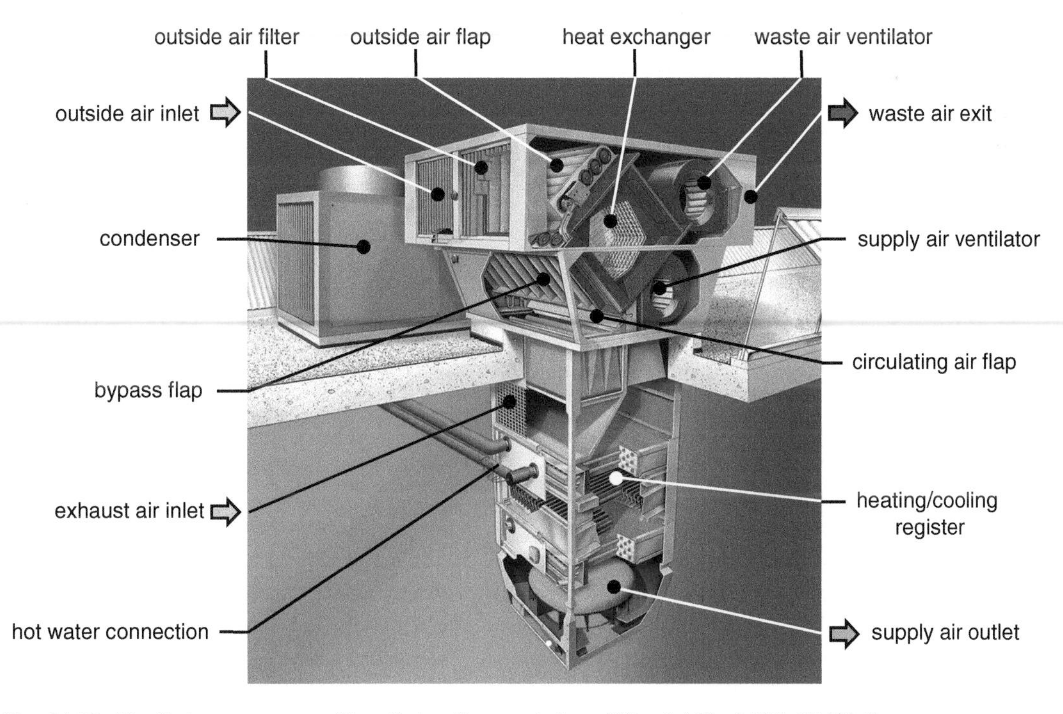

Fig. 11.32 Ventilation system—roof installation (by permission of Hoval AG). © IFA 17.608_B

The cost of HVAC systems increase with the number of functions. The following table can be used to roughly estimate the costs (as of 2013):

• Full air conditioning	22 €/m^3 intake air
• Heating and cooling	14 €/m^3 intake air
• Ventilation and aeration	8 €/m^3 intake air

Detailed investment costs can be obtained from the local manufacturers'. Obviously both initial investment costs as well as operating costs are higher for full air conditioning and air cooling systems. According to VDI 6022, both humidifiers as well as dehumidifiers require periodic and thorough maintenance. The cold water quality needs to be tested for bacterial or fungus growth every 6 months while the entire ventilation and air conditioning system needs to undergo a hygiene inspection from time to time depending on the quantum of air humidified. Based on VDI 2067, the operating costs for maintaining humidifiers and dehumidifiers is approximately 2 % of the investment costs per year while for heat recovery units it is approximately 10 %.

As already mentioned, comfort conditions depend not only on the perceived temperature and humidity, but also on the air movement in a given space. Drafts are often the cause of common colds and influenzas. Disruptive air movements are usually the result of ill-designed ventilation systems, thermal imbalances at sources of heat, cold downdrafts from poorly insulated external surfaces or doors and windows temporarily left open.

Ventilation systems thus have to be planned to ensure a comfortable air movement, which, depending on the climatic zone and the proposed system lies between 0.2 and 0.7 m/s (0.7 and 2.3 ft/s). Ventilators and air supply ducts should be selected and arranged in the hall space in accordance with these target values. The principles and details of thermal comfort in the workplace along with calculation and evaluation methods are provided in the relevant national and international guidelines and standards e.g., in [ISO08].

HVAC systems are also responsible for ensuring the air quality of a workplace. The air contaminants have to be extracted from the airflow and removed from the hall.

The flow of the contaminated air depends on the source of the substance and the manner in which the substance is dispersed. This is determined by:

- air density differences due to temperature differential (e.g., tire presses, welding plants, foundries, etc.)
- external forces such as inertial force, gravity or aerodynamic forces (exhaust blowers, cold air blowers, back flows etc.) and,
- how the air is distributed.

Types of Air Flow

In terms of a comprehensive plan, when designing an HVAC system, one of the first decisions to be taken includes the method for controlling the air distribution. There are three basic types from which to choose, depending on the needs of the production: laminar (or unidirectional), mixed and stratified.

Mixed and stratified airflows created by displacement from the bottom-up are well suited for manufacturing halls. They are described in detail in [Bac93].

- Laminar

The supply air is introduced into the hall via large diffusers. The low turbulent, unidirectional airflow displaces the contaminated air. Due to the costs involved, this type of air distribution was generally reserved for processes requiring high quality air (e.g., painting/spray booths, clean rooms for micro-electronics or pharmacies). In recent times, such systems are increasingly used for mechanical engineering environments where mechatronic or medical components are involved.

- Mixed

The fresh air supply is completely mixed with the room air in the manufacturing hall. As a result the temperature and concentration of substances is almost the same throughout the hall. This intensive mixing is attained through high impulse air jets. In order to maintain the physiological requirements, the impulses and temperature differentials need to be minimized outside of the work area. Also, the sources of contamination need to be controlled carefully.

- Stratification

The air supply is introduced at a low velocity (≤0.5 m/s (0.64 ft/s)) as well as impulse so that the thermal levels surrounding the production facilities in the manufacturing hall remain undisturbed. The hot air from the work area rises into the unused space above it in the hall. If the extracted air in the work areas is completely replaced with the fresh air supply, this results in a less contaminated zone at the workplace but a greater level of contamination at the higher level. In order for this to occur, there have to be heat sources in the work area and the majority of the contaminated flow has to be released near the heat sources into the thermal flow. The air temperature in the work area is then set higher than the temperature of the air supply. Moreover, the rate of the air supply has to be set so that it is sufficient for the thermal and collected air flows to rise up to a defined room height. When ideally designed this type of air distribution can keep the air contamination in the work area at a very low level.

Designs

- Mixed Airflow Designs

Air distribution systems which create mixed airflows throughout the entire hall are state of the art and commonly found in a variety of designs (see Fig. 11.33). When the supply air vents are placed horizontally below the hall roof, the air diffusers create zones with circular flows and return contaminated air into the work area. The supply and thermal airflows are thus mixed together. Thermal flows can be diverted into the

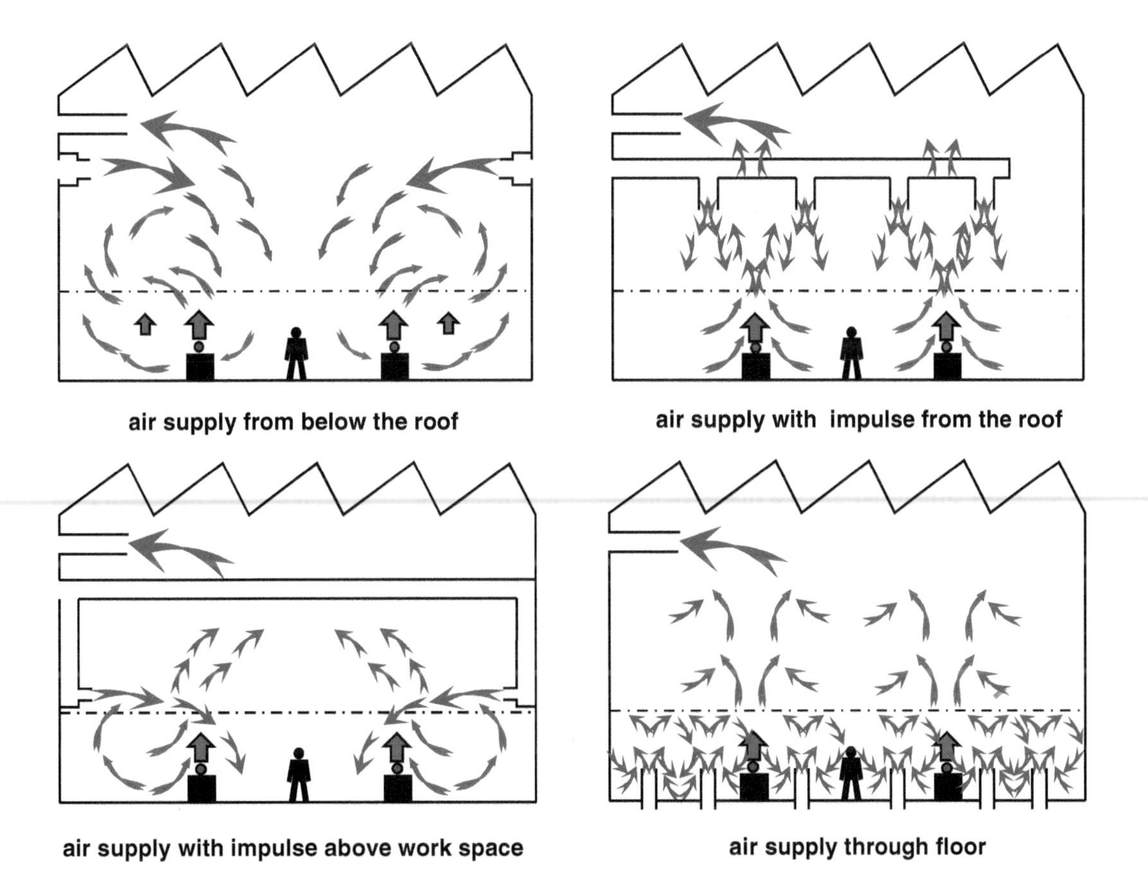

Fig. 11.33 Types of air distribution with mixed airflows (per [VDI10]). © IFA 17.609_B

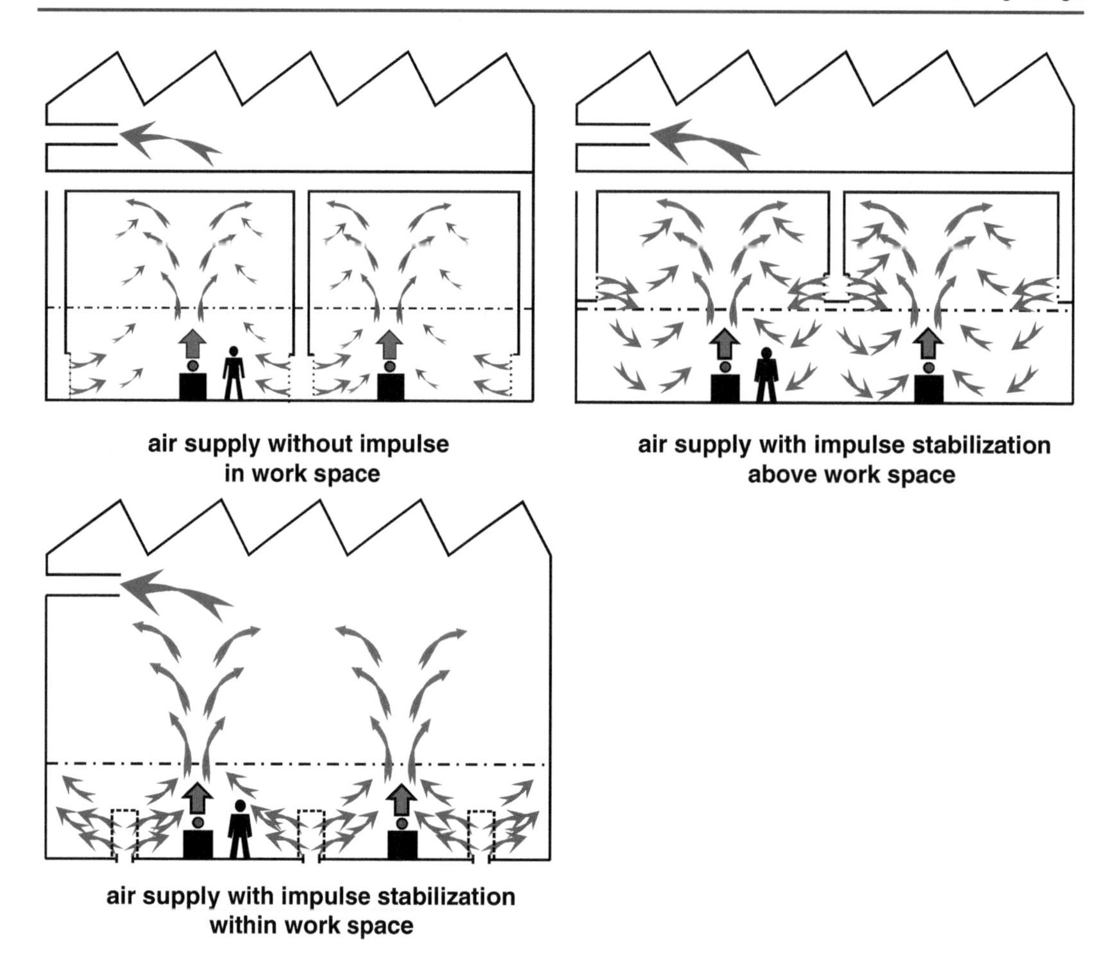

Fig. 11.34 Types of air distribution with stratified airflows (per [VDI10]). © IFA 17.610_B

work area and the air flow can be disrupted. With a large number of air supply openings in the hall roof, a mixed flow can be attained in the entire hall due to induction at the air diffusers.

- Stratified Airflow Designs

Stratified airflow designs use thermodynamics for transporting both heat as well as contaminants. The supply air has to be introduced with impulses that are low enough not to disrupt the layers or thermal currents, moreover the supply outlets needs to be installed close to the sources for the heat and/or contaminants. Figure 11.34 depicts a number of designs, which basically differ from one another in the location of the air outlets. Air impulses are stabilized through outlets with a fine mesh opening or using textile hoses that also allow the dirt to be easily washed away.

Temperature and flow profiles can be simulated and visualized nowadays with sophisticated software. As an example, Fig. 11.35 depicts the temperature and flow profile for a manufacturing hall. The requirements here included a temperature profile of ±1 K with a range of 0–6 m high and a flow rate of 0.2 m/s (0.65 ft/s). HVAC systems with requirements such as these can only be reliably and economically designed using simulations.

11.4 Interior Finishing

Once the structural frame, shell and media have been determined, the focus is directed at finishing the buildings' interior. There are a number of

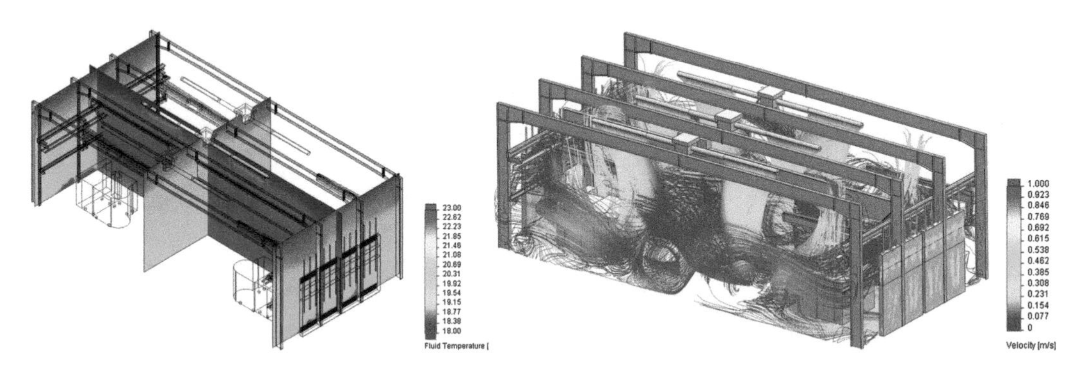

Fig. 11.35 Temperature and flow profiles for a manufacturing hall (ifes). © IFA 17.723E

systems and materials available for this. The building system for the floors, walls and ceilings should be aligned with the user's interests but should also avoid constraints that might negatively impact the buildings' changeability. The demand for a high degree of changeability for a building is similar to being able to easily change stage sets between acts in a theater or constructing exhibition spaces for a trade show. This means monolithic constructions should be avoided as much as possible in a buildings' interior finishings. Instead, modular, variable and easily convertible building systems should be favored even for the stairs and core areas.

Ideally a building's floors, walls, ceilings, core and stairs should be developed as modularized construction kits from the perspective of both engineering and aesthetics. The fundamentals of developing the interior of a construction project are summarized in [Pot07]. Lightweight constructions can also provide a high level of fire protection without monolithic structures. A 'building within a building', takes these factors into account. With due consideration to details, projects built on this approach are spatially variable and allow for changes without disrupting on-going processes.

Figure 11.36 depicts an overview of 5 design fields involved in the interior finishing along with 4 design elements for each field. In the following sections, we will discuss the structural characteristics of each of these elements in view of their changeability.

11.4.1 Floors

Figure 11.37 provides an overview of the structural features that are relevant for the changeability of floors along with their parameters. The user perceives the surfaces while the structural specifications are based upon the process and environment requirements. In terms of changeability, it is also critical that the floors are sustainably robust, easily installed and simple to modify.

The *surface* is primarily determined by the load. Construction materials for industrial floors should thus be both resistant and durable. In addition to being even and easy to maintain, they should be constructed easily, efficiently and economically. These requirements can be fulfilled by monolithic or multi-layered constructions. For centuries now, surfaces made from concrete have proven to be durable. They are found to be robust for the most part against soiling, mechanical impacts, water and frost. Concrete floors are easy to clean and keep clean while entailing extremely low maintenance costs. In addition, old concrete floors can be recycled to make new concrete. Monolithic floor plates made out of steel-fiber reinforced concrete can replace the on-site reinforcement with steel normally applied.

Multi-layered constructions are usually finished with a hard surface course or wearing coat after they are laid. The required durability and flatness of the overall structure is specified based

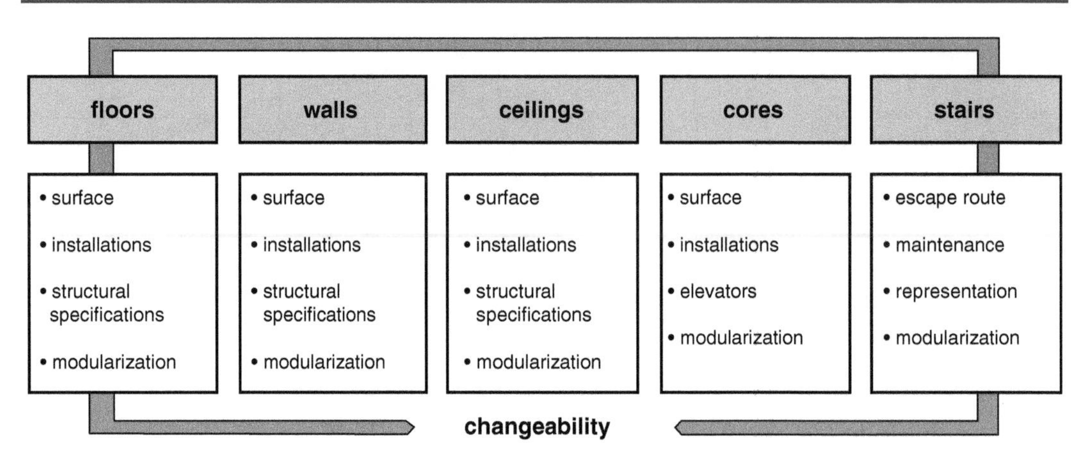

Fig. 11.36 Overview of structural elements in a building's interior finishing. © Reichardt 15.203_JR_B

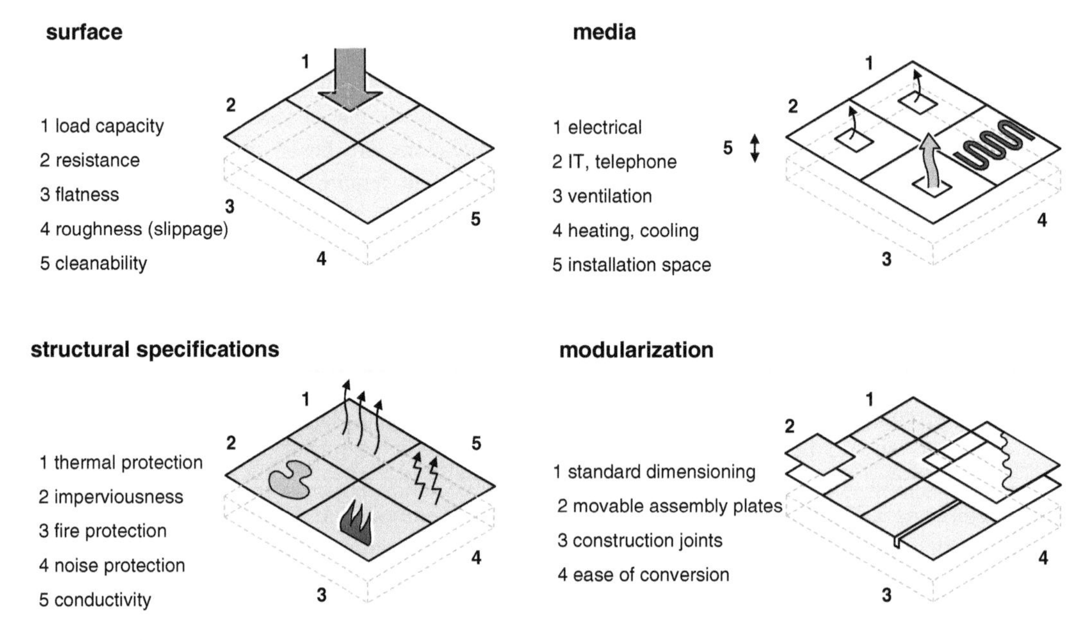

Fig. 11.37 Structural features of floors relevant for changeability. © Reichardt 15.204_JR_B

upon the needs of the processes and logistics. Since it is quite impossible to change the level and durability criteria at a later date the present requirements should be coordinated with possible future needs.

DIN 51130 provides further details regarding non-slip and easy to clean surfaces. In order to ensure a high degree of changeability in terms of process and logistics, hall floors should be kept free from media routings and ducts as much as possible. In comparison, with ceilings in multi-storied buildings it is often advisable to integrate systems like—electric, IT, ventilation, cooling and heating, together. With these types of installations it is recommended not to build them directly onto the static ceiling construction, but rather as separate layer. Systems such as double floors or floor plenums create a number of possibilities for adding media routings and ducts later.

The *structural specifications* of the floors are derived from the project's guidelines and requirements particularly with regards to protection against heat, noise and fire as well as water. In cold countries, a 3D energy simulation may be used to review and determine the floor's capacity to provide sufficient warmth in the work areas and avoid excessive heating. Better insulation values resulting in higher surface temperatures for the hall floor increase the possibilities of changing the arrangement of workplaces in the hall and support changeability. Impervious hall floors are required to protect against possible water ingress into the building from the sub-soil and surroundings. Additional coatings made of epoxy resin could be helpful under extreme circumstances. Such coatings are best applied during the construction period for perfect adhesion between the layers.

Requirements for protection against fire, noise and conductivity are usually related to floor and ceilings between the various floors of the building. Ideally, inflammable materials should be used as much as possible. Surface treatments made of out of PVC seem to be economical at first glance. However they release toxic fumes in case of a fire and require special care for disposal making them expensive in the long run and are best avoided. With proper detailing, unwanted noise generated out of footsteps on a floor can be avoided. Furthermore, with the aim of increased changeability, it is important to keep in mind that the use of computers would continue to grow in the future, floor properties should thus be planned so that they can dissipate correspondingly higher levels of electrostatic energy. Accordingly, floor systems should be installed so that they can always be easily retrofitted and simple to replace.

The degree of *modularization* significantly determines the effort required to alter floors. In order to be able to extend, dismantle or exchange systems, components should be uniformly dimensioned. If an industrial floor—characterized by its ability to withstand considerable weight, lasts long and easy to recycle—is required, the floor surfaces can be created using components like Stelcon large plates or Stelcon hexagon elements. These large plates are produced with a concrete quality of DIN C 35/45 and a standard size of 200 × 400 cm (6.6 × 13.2 ft). They are 14–16 cm (5.5–6.3 in) thick and have edges protected from damage by special mild steel angle sections. The large surface plates are used for example as finished floors in plants and storage halls, but also as anchors for port facilities and rail systems, transshipment sites in the chemical industry, access roads and gas stations. When planning the details of necessary structural joints, it is also important to consider any discontinuities that might be created (e.g., induction loops and transponder points) from the perspective of processes and logistics.

11.4.2 Walls

Figure 11.38 provides an overview of the basic structural features related to the changeability of walls. Whereas surfaces have to fulfill various purposes, the structural specifications determine the technical features. Walls that can be easily installed and changed are becoming increasingly important. And finally, here too, the aim is to find a consistent and modular system that can be harmonized with the building's grid.

Modern working methods and styles are constantly changing when compared to working in labs, workshops or operating out of offices. Spatial boundaries that can be easily adjusted depending upon the size of the workgroups and the content of their work are becoming increasingly preferable. In contrast to monolithic construction, prefabrication provides the required freedom to not only select the wall position but also select the position of the closed surfaces (for privacy), glass panels (for visual contact) and doors (for circulation). Transparent walls that connect the individual areas visually are particularly important for communication between personnel and encourage employees to identify themselves as a team.

Generally, there are two systems available for lightweight wall constructions: so-called 'drywalls' made from gypsum boards and wall panels made from sheets of wood, plastic or metal. Drywalls made from gypsum boards are fixed onto aluminum or wood stud frames, patched and painted or covered with wallpaper. In comparison,

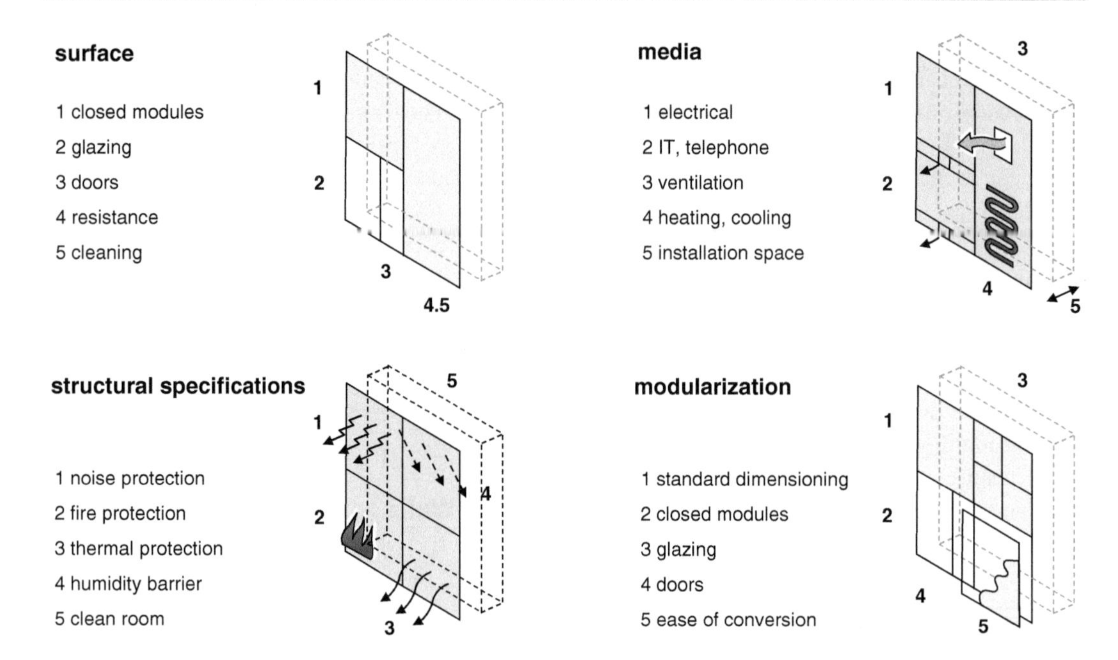

Fig. 11.38 Structural features of walls relevant to changeability. © Reichardt 15.205_JR_B

wall panels are supplied as elements with a finished surface and joint systems for connecting each other. Glass panels and doors do not disrupt wall surfaces, but are instead equally weighted elements.

Construction kits such as these can be uninstalled in a few days and re-built in another combination at another location in the factory. With the right attention to detail, systems for supplying electricity, IT as well as heating and cooling can be integrated at the skirting level or just below the window sill. In some cases, office furniture such as closets or bookshelves can be locked into the system's joints.

By providing insulation materials between the gypsum boards or wall panels, special physical requirements for noise, heat or fire protection can be met conveniently. Drywalls with suitable paneling and insulation materials can resist fire up to 3 h and satisfy F 180 fire ratings, if need be. Modularized wall panels can resist fire up to 2 h (F120). In case of precision industrial manufacturing, wall panels can be also used for enclosing clean rooms. In order to attain the required impermeability, sealants can be applied after installation. Wall panels are advantageous in that they can be easily reinstalled without the dirt and dust associated with tearing down dry walls. Thus, they support changeability especially in sensitive processing and logistic facilities.

Here too the basis for modularization is a consistent dimensional system for all closed surfaces, glass paneling and doors. Floor plans ideally should be designed in grids of 1.00, 1.20, 1.25 or 1.5 m (3.3, 3.9, 4.1 or 4.9 ft).

11.4.3 Ceilings

With regards to changeability, ceilings are subjected to almost similar characteristics as floors and walls. Figure 11.39 provides an overview of the structural features relevant for the changeability of ceilings.

Similar to wall systems, the varying spatial requirements of modern work-forms require flexible ceiling systems. So-called 'house in house construction' offers the benefit of coordinated wall and ceiling elements. Here, the precision of industrial pre-fabrication is best suited for meeting the structural specifications and allowing a carefully planned installation.

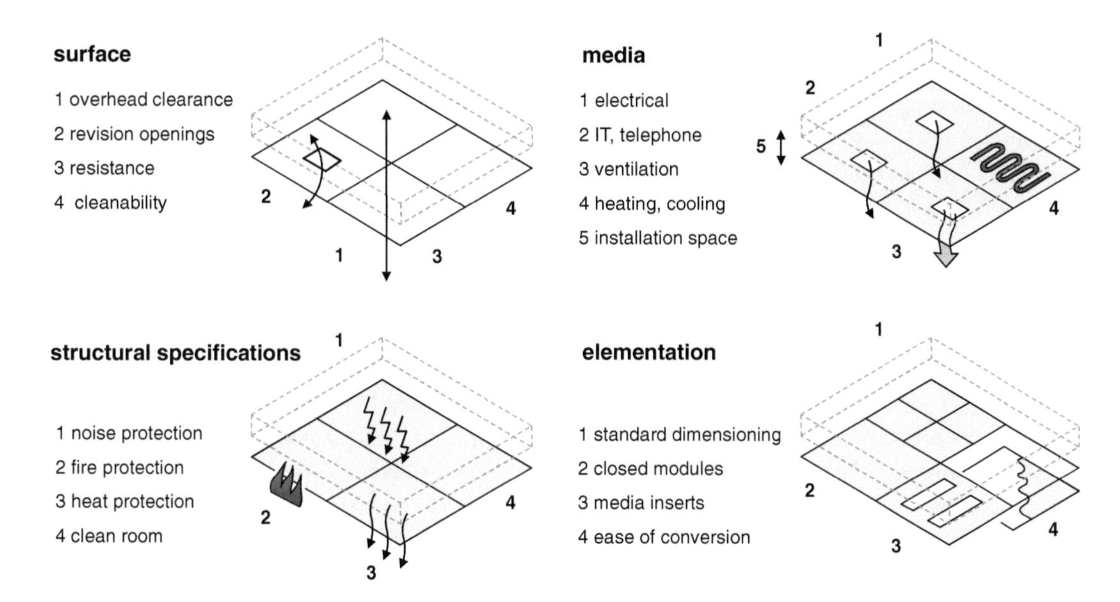

Fig. 11.39 Structural features relevant for the changeability of ceilings. © Reichardt 15.206_JR_B

One of the basic features of changeability is the overhead clearance. Rooms larger than 50 m^2 (538 ft^2) should have at least 2.75 m (9 ft) headway, whereas rooms larger than 100 m^2 (1,076 ft^2) should have at least 3.00 m (9.8 ft). When designing the floor plan the minimum clearance of the smaller rooms also needs to be verified in view of future developments. One of the design principles for modular ceilings involves mounting panels fixed below a supporting metal framework. The cavity that is thus created can be used for media routings and the ceiling panels can be made from a variety of materials.

Ideally the *surfaces* should be durable and easy to clean. System grids are commonly found in 0.50, 0.60, 0.625 or 1.00 m (1.6, 2.0, 2.05, 3.3 ft). Access areas for maintaining, or retrofitting the media routing installations that are located above ceiling panels can be created through certain removable panels. Another basic characteristic that is relevant to changeability is the actual clearance above the ceiling construction: For example, if air ducts need to be replaced with ones that have a larger cross-section in order to meet increased ventilation demands, they can only be installed provided the necessary space is available for the same.

The *routings* and media outlets for the electrical supply, EDP/telephone, ventilation, cooling and heating should also be modular and therefore adaptable without impeding each other. The physical requirements for sound proofing, fire protection, thermal insulation and clean-room specifications can be met only through careful planning of the mounting plates and the interim space below the structural slab. Perforated plates improve the spatial acoustics by increasing the sound diffusion. Insulating layers made out of non-flammable materials such as mineral (rock) wool or special fire-proof plates increase the fire resistance of ceiling structures or individual media lines. In clean-rooms, the joints (e.g., between walls) can be made impervious by applying suitable sealants.

When developing the elements for *modularization*, careful attention should be paid to maintaining standard dimensions and minimizing customization. If the layout grid is free from disruptions, peripheral areas remain available for media and elements can be interchanged in the entire floor plan. Ceiling plates and routing systems with pre-finished surfaces and clip-on/plug-in connections minimize both the time required to reconfigure them as well as the risk of contamination when doing so.

11.4.4 Cores

Cores are building areas that serve to dissipate a concentrated static load and are serve additional functions. For economic reasons it is often advisable to transfer the vertical and horizontal loads of a hall's structure over concrete or steel walls and to use the space thus created for elevators, emergency staircases or installation shafts. The general positioning and spatial layout of such core areas should be discussed together in detail along with factory planning, since once cores are planned and constructed they are more or less unchangeable. The location of the cores in the building as well as the required widths and depths of them is generally derived from the plan of the load-bearing frame.

Cores should be positioned with an eye to the future, so that possible options for changes in processes and logistics are in no way impeded. The clearance space allowances for the *installation shafts* needs to be checked to ensure that there are enough allowances for future purposes. Changes in processes or logistics often require additional media and thus the requirement of additional space inside the shafts.

A critical point for most of the shafts is the space where the verticals pipes and wires are separated and are to be routed horizontally. Finally, the openings to the shafts that are important for fire protection and should make it easy to maintain, repair, retrofit or replace routings.

One of the key characteristics in regards to changeability is the detailed design of the passenger and freight *elevators*. Since elevator technology has to be replaced every 15–25 years, it is of paramount importance that elevator clearances as well as the height and width of the elevator doors are planned with an eye to transport larger volumes in the future.

The possible *modularization* in the sense of changeability and therefore the potential to modify cores is greatly restricted due to the above-mentioned reasons, especially when constructed using monolithic (in situ) concrete. Cores can be also constructed using precast concrete elements as well as steel frameworks with infill wall panels. Implementing elements that are assembled together generally even makes it possible to change the core's width and depth. In summary, Fig. 11.40 depicts an overview of the structural characteristics relevant to the changeability of cores.

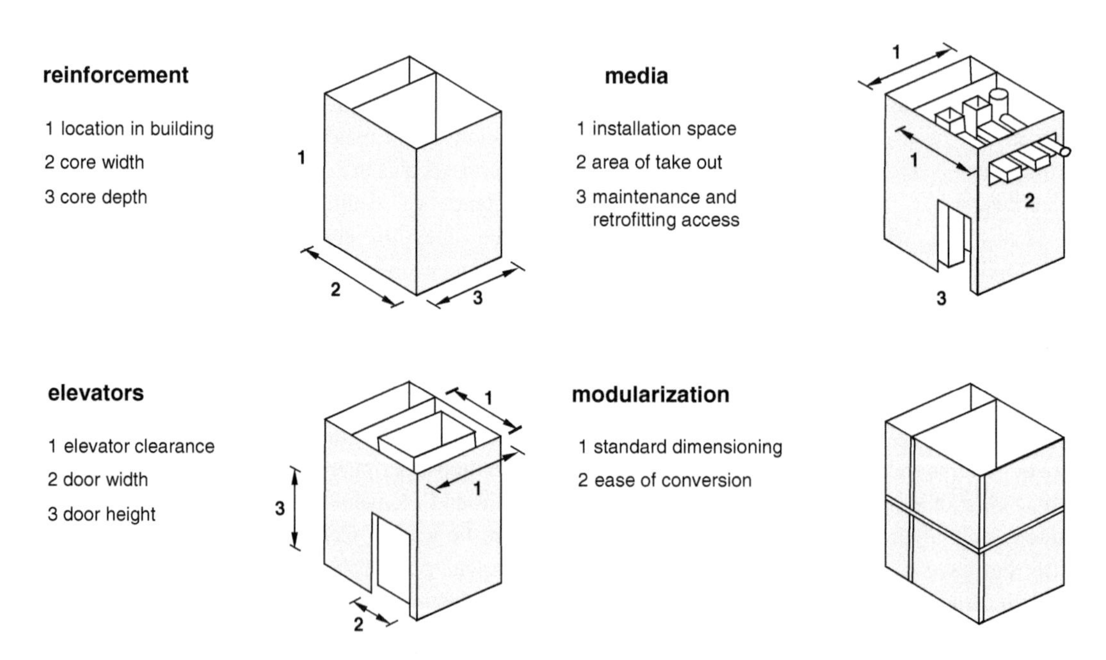

Fig. 11.40 Structural characteristics relevant for the changeability of building cores. © Reichardt 15.207_JR_B

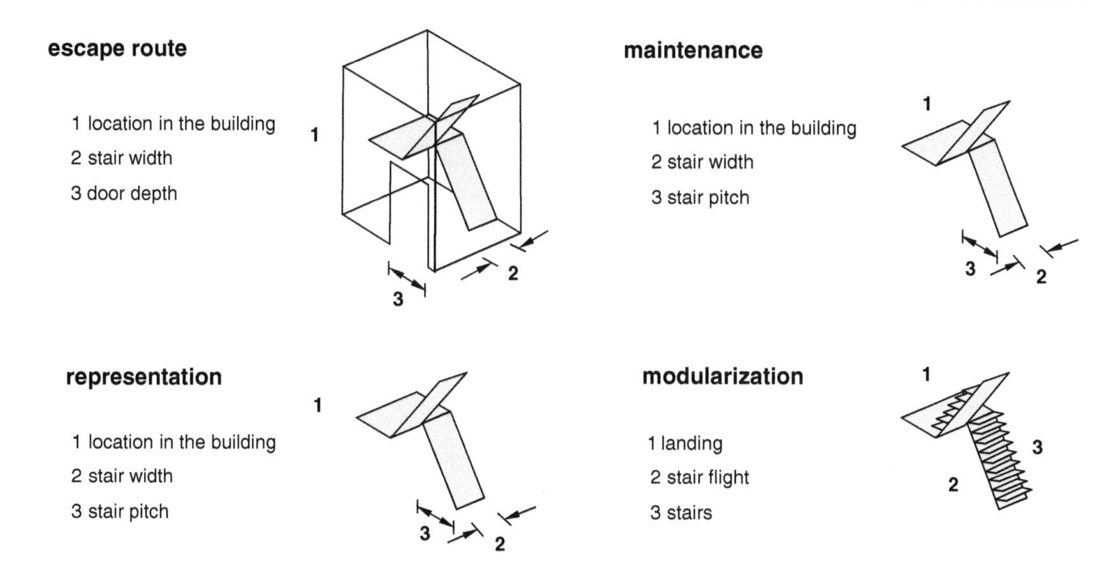

Fig. 11.41 Structural characteristics relevant for the changeability of stairs. © Reichardt 15.208_JR_B

11.4.5 Stairs

By making use of fire resistant constructions, stairs located within cores are generally planned and implemented as escape stairways. However, escape stairways can be located in other parts of the building also. According to most building regulations, the seal required for stairways that are planned as escape routes require a fire resistance of 1.5 h (F 90) and can also be realized with assembly kits made out of materials ranging from dry or completely vitrified construction. Similar to cores, the location of evacuation stairwells in buildings needs to be planned keeping the future in mind. Later modifications require not only reconstructing the stairways, but also reviewing the escape routes and obtaining re-approvals from the concerned fire department.

The required width of the stairs and doors is oriented on the number of people who would need to be evacuated in the worst case scenario, thus when planning stairwells, the area for possible office extensions should also be taken into consideration. Stairs for maintenance are not subject to the same requirements as those meant for evacuation, rather their location in buildings are more oriented towards functional needs. The necessary stair width, pitch and design details have to be built in accordance with the relevant workplace and accident prevention guidelines.

Additional stairs in foyer or galleries can be constructed as open stairways and need not be enclosed or complaint with the necessary fire regulations. The width and pitch of the stairs should also be oriented towards possible number of future users. Generally, with regards to changeability, stairways with a modularized construction should be given preference over those built in concrete and thus fixed at a certain location. As with the other finishings, the modularity allows the staircase, landings and individual stairs to be altered in a variety of materials. An overview of the structural characteristics of stairways presented here as relevant for changeability is provided in Fig. 11.41.

11.5 Examples of Changeable Buildings

Figure 11.42 depicts two factories that were actually built with focus on changeability when designing the structural frame, shell, interior finishings and media.

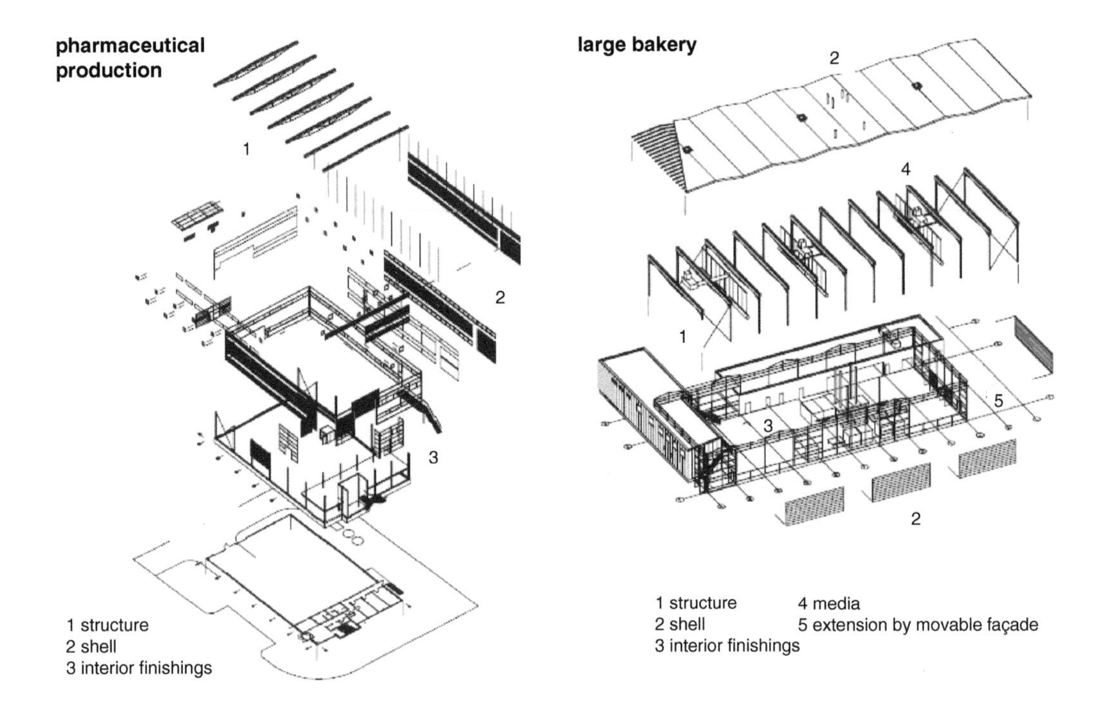

Fig. 11.42 Modular design principle as applied to a building (examples). © Reichardt 15.209_JR_B

In both cases, the building concept was implemented according to modular design principles with separate building components for the structure, shell, media and interior finishings. In the case of the large bakery, the project requirements necessitated a timber skeleton construction for the load-bearing frame, a metal and glass façade for the building shell (independent of the structure), as well as modular air circulation devices within the structural frame zone for the ventilation technology. By strictly separating the systems, the baking hall was able to be extended by one grid field (approx. 6 m × 22 m (19.7 × 72.2 ft)) 'over the weekend' without causing any disruptions.

The same applies to the pharmaceutical production in which special attention was given to ensure the building could be expanded with minimal disruption and that the interior could be easily changed.

An example of a modular and versatile assembly factory for car coolers shows Fig. 11.43, which is also available as a video animation in Appendix D3. The target was to design for trouble-free scalability of processes, systems and buildings as well as re-usability. Hall spans of 18 and 36 m (59 and 118 ft) and 8 m (26 ft) building height enable easily movable work places.

11.6 Grace and Aesthetics

Architecture reflects the forces that were at work during the building's creation. Unfortunately, this applies to most of unsightly industrial and commercial buildings across the world. It appears that most companies view their buildings as tedious necessities, which they would prefer to do without entirely. Architecture is the most fundamental and permanent expression of social culture and it is time that industries and businesses reconsider their attitude and approach to architecture. "Graceful" aesthetics arise when the design elements—that is the location, type of construction, structural framework, shell, media and interior finishings—are cleverly and creatively combined based on

Fig. 11.43 Example of a modular and changeable assembly factory (Reichardt)

synergetic planning approach. Examples of such graceful factory buildings can be found in [Uff09].

Thus according to [Mes05, Kni06], for each of the specific tasks, it is necessary to develop an innovative overall design strategy for the forms, materials and colors. Ideally this strategy should permeate the building in every detail including how they are combined both "externally" and "internally". Details such as the design of the façade's proportions or the color to be used are secondary design features. The tectonics of how the buildings' volume is developed and how the buildings are distributed are essential to the overall impact of the construction. Often when there are creative differences, 'grace' becomes a victim under pressure to keep costs down or to increase 'economic' efficiency.

Outstanding industrial architectural examples across the world however provide a strong counter-point. Henry Ford, the king of the automobile industry was supposedly a penny pincher in his days; yet his extremely functional factories, which were considered to be a model of production processes, impressed everybody with their extreme efficiency as well as their exemplary grace [Hil74]. Ford's remark that "good design pays" demonstrated his conviction and his awareness of the value of aesthetic design. By 1920s and 1930s, the Ford factories had reduced the number of people 'taking sick days' and/or quitting thereby increasing the overall profitability of operations and forging long term bonds between the company and employees.

The objective of graceful industrial architecture can be summed up with the following fundamental parameters: structural order, simplicity, balance between unity and diversity and distinctiveness. Böhme [Böh06] suggests that the emotional or 'atmospheric' quality of a building should also be taken into consideration as an additional design value.

11.6.1 Structural Order

The principle of structural order creates a feeling of harmony. Structural order is attained with a consistent relationship between the parts and the whole, one reminiscent of a living organism where there is internal coherence between the elements and their totality. The immediate sense of clearly articulated structural shapes and architectural forms along with the careful understanding and legibility of the elements' functions (e.g., the structural frame, shell and interior finishings) all play a decisive role in the overall structure. The structural order naturally unfolds with the help of its own grammar: the floor plan, sectional and vertical projections. Clarifications are not necessary to understand the building, since the relationship between the parts is easily understandable. An example of a structurally ordered type of construction is a modular plant based on hall segments.

11.6.2 Simplicity

With industrial buildings there is no need for filling, cladding or laminating. Solving a building project in this way, i.e., without "stage magic" pleasantly corresponds to the rational design principle used in medieval cities. Aesthetics need not be costly. At the same time, simplicity should not be mistaken for the banality, lack of imagination and primitivism of common commercial construction. Rather, the demand for economic efficiency goes hand in hand with minimalism and the lack of extensive and complicated fascia work.

The 'appearance' of many enterprises across the world seemed to be saying something quite different; they confuse the simple-minded with their over the top and unnecessary fascia treatment. Intelligent reduction of forms, materials and color allows for powerful aesthetic effects. Liberated from unnecessary coverings, the quality impacts the viewer directly and is more sustainable than attempts to present a deceptive package with flashy gimmickry.

11.6.3 Balance Between Unity and Diversity

The balance of visual information within the field of tension between monotony and chaos requires aesthetic comfort. Unity and diversity are mutually exclusive but require one another. They are necessary poles between which the balance has to be re-adjusted for every project. On one hand, viewers are quickly bored when they detect a total uniformity and monotony dominating in the architecture. On the other hand, an agitated mind results in a 'ball of confusion'. The excited oddity of short-lived fashions is soon forgotten and seems embarrassing when looking back after a few years. The ideal solution is a viable urban and architecturally sustainable design. Fixed building heights set over the long term or a canon of materials that can be implemented with options for design possibilities within this framework create potential for individual solutions for the future.

11.6.4 Distinctiveness

Distinctiveness stands out when compared to forgettable visual experiences. It often seems that the anonymity of insignificance is the goal for many industrial projects. How else could one explain the degree to which these projects have refrained from a notable design? In reaction to this, some enterprises attempt to gain attention with unnecessary detailing—a formal arbitrariness oriented on a surprise effect which rarely sustains in the long run or defines the character of a building or its users. A distinctive design results out a creative combination of specific agreements, special local situations and conscious selection of the components of the construction and building. This type of "value-added" architecture has a much stronger visual impact using much less capital than elaborate print and film advertising campaigns; it is the embodiment of a project's mission, accomplished through the total innovative performance of the planning team.

11.6.5 Emotional Quality, Atmosphere

Graceful buildings touch an inner chord and a positive relationship develops between the observer and the building. According to [Böh06], this emotional quality influences the deeper perception of spaces, materials, color and light far beyond the organizational and functional references. When we consciously design the type and form of the spatial environment, starting with the buildings' orientation and the internal/external views etc. we create an immediate understanding of the architectural structure and facilitate its usefulness. The way in which materials are used and joined together can convey everything from the greatest industrial precision to handcrafted artistic spontaneity. Structural parallels to the quality of the manufactured industrial products are thus directly produced or can be at least interpreted. Just as with the product design, the emotional quality of the building should be given special consideration while determining the objectives of a project. Examples and suggestions for this can be communicated and updated in goal planning workshops.

In short, the exterior of a building should reflect the company's claims and its interior the product's claims; or as [Rei05] suggests aesthetics and efficiency have to be aligned.

11.7 Summary

The architectural design of a building consists of four main components: structure, shell, building services, and grace. The performance of a building is essentially determined by the chosen technical, constructive and last but not least aesthetic solution in the interaction of these components. One has to differentiate between constant, difficult to change and easy to change structural components. In many cases, the phase of the basic evaluation, with the detailed discussion of all major building requirements, is underestimated. It is recommended that a project-related transparency and quality assurance of all structural detail solutions and a documentation into project parts, assumptions and findings takes place. A high adaptability and high sustainability ensures a construction project with a large reuse potential in second and third use. This task requires considerable coordination and planning.

Bibliography

[Ada04] Adam, J., Hausmann, K., Jüttner, F.: Entwurfsatlas Industriebau (Design Manual Industrial Construction), 1st edn. Birkhäuser, Basel (2004)

[Ack88] Ackermann, K.: Tragwerke in der konstruktiven Architektur (Structures in constructive architecture). DVA, Stuttgart (1988)

[AS13] ASUE Arbeitsgemeinschaft für sparsamen und umweltfreundlichen Energieverbrauch. Die Erdgas-Strahlenheizung (Work group for saving and environmental friendly energy consumption. The natural gas radiator). http://asue.de Kaiserslautern (2013)

[Bac92] Bach, H. et al.: Gezielte Belüftung von Arbeitsbereichen in Fabrikhallen (Targeted ventilation of work areas in factories). Forschungsbericht HLK 1-92. Stuttgart (1992)

[BS6399] British Standard Institution: BS 6399 Loadings for Buildings. Part 1: Code of practice for dead and imposed loads, 1996. Part 2: Code of practice for wind loads, 1997. Part 3: Code of practice for imposed roof loads, 1988

[Böh06] Böhme, G.: Architektur und Atmosphäre (Architecture and Atmosphere). 1. Aufl. Verl. Fink (Wilhelm), Paderborn (2006)

[Dan96] Daniels, K.: Haustechnik, ein Leitfaden für Architekten und Ingenieure (Building Services, a Guide for Architects and Engineers). Oldenbourg, München (1996)

[DIN08] DIN 276: Kosten im Hochbau (Building Costst). Beuth, Berlin (2008)

[DIN11] DIN EN 12664: Licht und Beleuchtung von Arbeitsstätten (Light and lighting - Lighting of workplaces). Beuth, Berlin (2011)

[DIN04] DIN EN 12792:2004-01: Lüftung von Gebäuden - Symbole, Terminologie und graphische Symbole. (Ventilation for buildings - Symbols, terminology and graphical symbols). Beuth, Berlin (2004)

[Eng07] Engel, H.: Tragsysteme (Structure Systems), 3rd edn. Hatje Cantz, Ostfildern (2007)

[ENV1993] British Standards Institution: DD ENV 1993-1-1: 1992 Eurocode 3. *Design of Steel Structures. General Rules and Rules for Buildings* (1992)

[ENV1994] British Standards Institution: DD ENV 1994-1-1: 1994 Eurocode 4. *Design of Composite Steel and Concrete Structures. General Rules and Rules for Buildings* (1994)

[ENV1998] British Standards Institution: DD ENV 1998-1-4: 1996 Eurocode 8. *Design*

Provisions for Earthquake Resistance of Structures. General Rules. Strengthening and Repair of Buildings (1996)

[E1670] ASTM International: Building Constructions E72 to E1670-I—Annual Book of ASTM Standards, Section 4: Construction, vol. 04 (2008)

[Gri97] Grimm, F.: Hallen aus Stahl (Halls from Steel). Stahlinformationszentrum, Düsseldorf (1997)

[Hat06] Hatz, R.: Auf Sparflamme. Energetische Optimierung von Industriebauten (On Low Flame. Energy Optimization of Industrial Buildings). In: db 8/2006, pp. 71–75

[Hau07] Hausladen, G., de Saldanha, M., Liedl, P.: ClimaSkin- Konzepte für Gebäudehüllen, die mit weniger Energie mehr leisten (ClimaSkin - Concepts for building Shells that do more with less Energy), 1st edn. Callwey, München (2007)

[Hil74] Hildebrand, G.: Designing for Industry. The Architecture of Albert Kahn. The Massachusetts Institute of Technology (1974)

[Her04] Herzog, T., Krippner, R., Lang, W.: Fassaden Atlas (Facades Atlas). Birkhäuser, Basel (2004)

[HO13] HOAI Honorarordnung für Architekten und Ingenieure (Fees for Architects and Engineers) Federal Law June (2013)

[IEC11] International Electrotechnical Commission: IEC 60364-5, 51, 52, 53, 54, Electrical Installations of Buildings Edition 3.0 (2011)

[ISO08] DIN EN ISO 13790: Energieeffizienz von Gebäuden - Berechnung des Energiebedarfs für Heizung und Kühlung (Energy performance of buildings - Calculation of energy use for space heating and cooling)

[Kni06] Knittel-Ammerschuber, S.: Erfolgsfaktor Architektur – Strategisches Bauen für Unternehmen (Success Factor Architecture - Strategic Building for Companies). Birkhäuser, Basel (2006)

[Kol02] Kolarik, F. et al.: Energieeffiziente Lüftungsanlagen in Betrieben (Energy efficient ventilation systems in enterprises). Landesgewerbeamt Baden-Württemberg, Informationszentrum Energie. Stuttgart (2002)

[Kra07] Krauss, F., Führer, W., Jürges, T.: Tabellen zur Tragwerklehre (Tables for Structural Design). Rudolf Müller, Köln (2007)

[Kri08] Krimmling, J., Preuß, A., Deutschmann, U.: Atlas Gebäudetechnik – Grundlagen, Konstruktionen, Details (Atlas Building Services - Fundamentals, Designs, Details). Rudolf Müller, Köln (2008)

[Lac84] Lachenmann, G.: Industrialisiertes Bauen (Industrialized Building). In: Ackermann, K. (ed.) Industriebau, Ausstellungskatalog, S. 118–141

[Mes05] Messedat, J.: Corporate Architecture – Entwicklung, Konzepte, Strategien (Corporate Architecture - Development, Concepts, Strategies). Av edition, Ludwigsburg (2005)

[Pis07] Pistohl, W.: Handbuch der Gebäudetechnik (Handbook Building Services), 6th edn. Neuwied, Dessau-Roßlau (2007)

[Pol03] Polónyi, S., Walochnik, W.: Architektur und Tragwerk (Architecture and Support Structure). Ernst & Sohn, Berlin (2003)

[Pot07] Pottgiesser, U., König, K.: Baukonstruktion Ausbau (Interior Building Construction). UTB, Stuttgart (2007)

[Rec13] Recknagel, et al.: Taschenbuch für Heizung + Klimatechnik (Handbook for Heating + Air-Conditioning, 76th edn) Oldenbourgh, München (2013)

[Rei05] Reichardt, J.: Ästhetik Effizienz (Aesthetics Efficiency). In: Nelte, H.M. (ed.) Ästhetik Effizienz - Industrie Gewerbe Verwaltungsbauten (Aesthetics Efficiency – Industrial, Commercial Administrative Buildings), 1st edn. Nelte, Wiesbaden (2005)

[Rei08] Reichardt, J.: Industrie- und Gewerbebau in Holz (Industrial and Commercial Buildings in Wood). Holzbau Handbuch, row 1, part 3, vol. 11 Bonn (2008)

[Sch06] Schnittich, C.: Im Detail Gebäudehüllen (Building Shells), 2nd edn. Birkhäuser, Basel (2006)

[Tom10] Tompkins, J.A.: Facilities Planning, 4th edn. Wiley, Hoboken (2010)

[Uff09] van Uffelen, C.: Factory Design. Braun, Berlin (2009)

[VDI10] VDI-Richtline 3802:Raumlufttechnische Anlagen für Fertigungsstätten. Absaugung luftfremder Stoffe an materialabtragenden Werkzeugmaschinen (VDI Guide line 3802: Air conditioning systems for factories – Capture of air pollutants at machine tools removing material) Beuth, Berlin (2010)

12 Master Building Plan

The master building plan comprehensively defines both the current and future performance of a factory's urban development concept. In creating the master plan, the characteristics of the project requirements determine not only the forms that will be built but also the safety and security criteria. Once defined, the master building plan provides the guidelines for arranging the factory's buildings as well as the zoning for traffic circulation and green landscaped areas [Stei07]. All of which are broken down into possible developmental stages. Selecting the building typology and deciding how they will be combined plays a key role in ensuring the highest possible degree of changeability. Figure 12.1 depicts the design fields and elements pertinent to the master plan. In this next chapter, we will focus on discussing each of these in detail.

12.1 Request Program

The project specifications are agreed upon by the planning team and define:

- the floor space required both currently as well as in the future,
- the modularity of the building volumes in area and height
- the principles underlying the supply and removal system and
- any other specific requirements.

In order to confirm or correct initial assumptions, the specifications of the project should be refined throughout the entire planning period and continually clarified.

12.1.1 Required Floor Space and Room List

Compiling the basic quantitative and qualitative spatial requirements as early as possible provides the basis for the spatial planning and layout. The quantification of the net floor areas needed currently as well as in the future for the various divisions is best summarized in the form of a table known as a 'space mirror' or 'room list'. Requirements regarding the quantitative spatial or qualitative design of the different sections with specifications for the spacing between columns, preferred room proportions, maximum floor loading and the ceiling height can be included in such a table. Appropriate symbols and colors may help to differentiate the specifications set together by the planning team from those initially assumed; corrections or clarifications can then be made by updating the room list as the planning progresses.

In order to define the required floor space abstract or project-specific methods can be drawn upon. Abstract determinations are based on general experiences such as standard values or indices as well as a synthetic approach e.g., generating the sections additively. Project related

DOI 10.1007/978-3-662-46391-8_12,

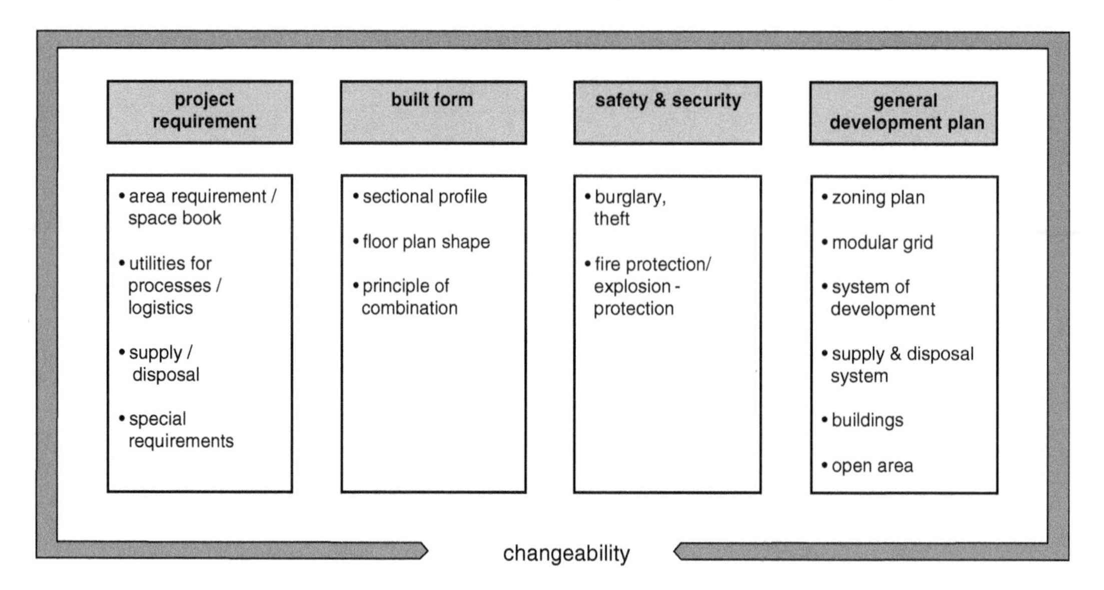

Fig. 12.1 Overview of the master building plan's design fields and elements. © Reichardt 15.210_JR_B

methods, in comparison, use an available 'current state' to derive the floor space specifications.

The floor space required for a production is determined by three factors:

- facilities (e.g., machinery and equipment),
- materials (e.g., raw materials, workpieces being processed and finished goods) and
- areas required to operate machinery and manipulate materials including both up and downstream ancillary functions.

The layout of the rooms should also be discussed early on and included in the room list or space book as one of the characteristics. An overview of important primary and secondary factors that influence the required floor space for a production is provided in Fig. 12.2. Whereas the primary influences directly determine the geometry and dimensioning of the three factors, the secondary influences result from the production operations in consideration of logistics, ergonomics and local circumstances. The procedures for determining the required floor space will be expanded upon in Sect. 15.5.2 "Detailed Dimensioned Planning".

For areas which serve functional purposes or are used as production areas, rectangular ground-plan shapes are always preferable when compared to an irregular shape because they are better suited for practical use. From the perspective of good internal communication, quality assurance, the flow of people, guiding visitors etc., the cut of administrative spaces should be selected so that length to width ratio is approximately between 1:1.5 and 1:3. Freestanding columns in functional areas should be avoided if possible and should not interfere with installing, operating or maintaining machinery and equipment. The minimum width of functional areas has to be determined from case to case and is a pre-condition for later rearrangements of facilities.

These considerations need to be discussed within the frame of the building plan and in view of a suitable degree of changeability. More often than not it leads to spacing the columns as far apart as possible (see Sect. 11.1 "Load-Bearing Structure").

The room list can also contain other requirements regarding the ceiling height as well as particular locations of divisions in the spatial structure. In setting the targeted ceiling height, it should be taken into account that mezzanine floors inserted later are the simplest form to reserve internal space. In general clear room heights are desirable; the same may not be structurally possible in all cases. In others, the utilities network installed at the bottom of the structural system may be more crucial than the

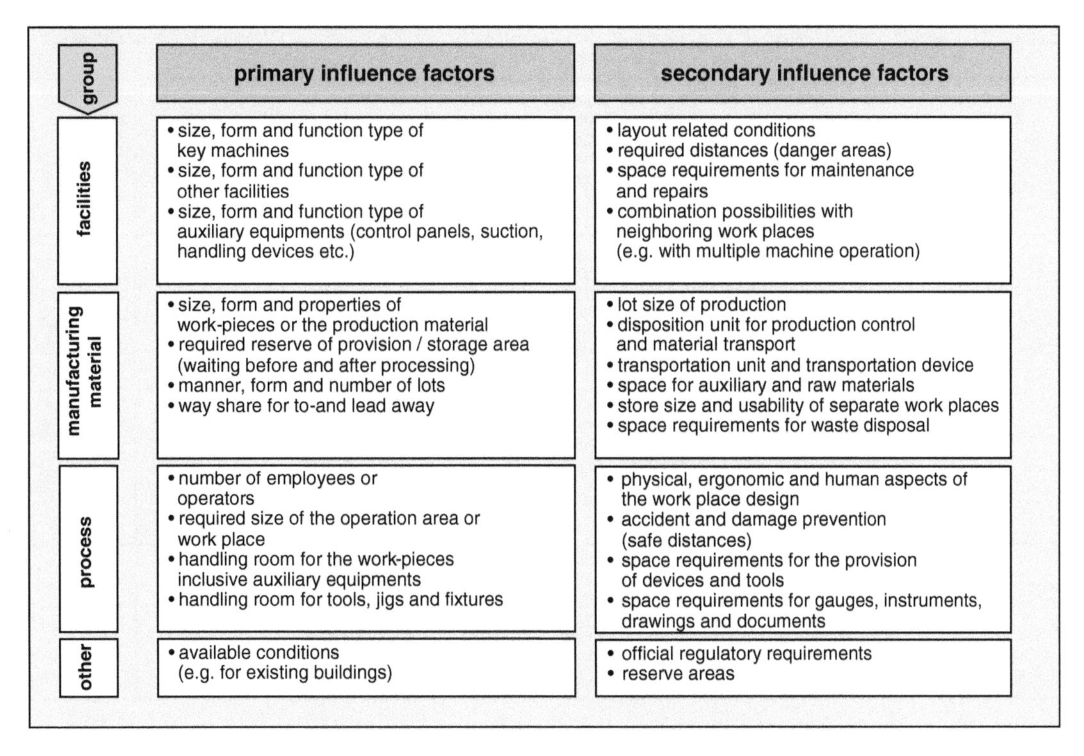

Fig. 12.2 Primary and secondary factors influencing required floor space (per Aggteleky). © Reichardt 15.211_JR_B

clearance below the structural system. Moreover, the possibilities of implementing natural lighting and using the natural outdoor environment for climate control are decisive to whether a room should be located directly adjacent to the external shell or within the building's volume.

According to DIN 277, spatial areas can be categorized as: main, secondary, circulation, or construction (e.g. where walls, columns and shafts are located). In order to prevent unpleasant surprises resulting from improperly calculated areas when estimating costs, it is advisable to agree upon a common use of language early on in a project. The contents of the room list—from the rough framework up to the fine details—can be directly linked with the parallel 3D CAD planning. In this way, redundancies can be prevented and planning statuses can be synchronized across the board with the synergetic factory planning's continually actualized evaluation. Furthermore, quantifiable trades provided with cost benchmarks can be selected and summed up into sub-budgets for on-going budget estimations. With a properly organized room list, this procedure also makes it possible to quickly and at least approximately evaluate planning alternatives.

Based on experience, it is practical to break-down the room planning according to project components and levels; the table should be organized in a way that allows the room list to be expanded and later extended into space books for each room, supplemented with interior finishings such as floor coverings or types of doors. Figure 12.3 depicts a sample room list for an assembly plant. Rooms are numbered floor-wise; geometrical data gives clear area and height, while interior qualities of floors, walls and ceilings as well as utilities systems are defined where needed.

In view of potential changeability, floor areas should be optimized for current purposes. However, plausible optional arrangements should also already be examined with regards to preventing foreseeable conflicts as well as keeping in mind potential synergies, e.g. changing a logistics area into an assembly area. During the course of the project, initial assumptions should be further clarified as they are confirmed or corrected once changes are noted.

code		name	geometry		room fitting				utilities building services		
	room		area [sqm]	height [m]	door/gate	floor	wall	ceiling	floor	wall	ceiling
	150	**hall**	7,766.30	7.35	type gate 1.1 gate 1.2 gate 1.3 gate 1.1	industrial screed monolith. reinforced concrete floor plate	industry- glazing metal- cassette/ insulation	trapezoidal sheet			LED lamp sprinkler ventilation compressed air
	151	**maintenance**	135.00	3.85	type gate 1.1 gate1.1	industrial screed monolith. reinforced concrete floor plate	industry- glazing metal- cassette/ insulation	trapezoidal sheet		8 sockets 4 double- sockets rj45 for IT	LED lamp sprinkler ventilation compressed air
	152	**workshop**	135.00	7.15	type gate 1.1 gate 1.1	industrial screed monolith. reinforced concrete floor plate	industry- glazing metal- cassette/ insulation	trapezoidal sheet		8 sockets 4 double- sockets rj45 for IT	LED lamp sprinkler ventilation compressed air
	153	**model shop**	135.00	7.15	type gate 1.1 gate 1.1	industrial screed monolith. reinforced concrete floor plate	industry- glazing metal- cassette/ insulation	trapezoidal sheet		8 sockets 4 double- sockets rj45 for IT	LED lamp sprinkler ventilation compressed air

Fig. 12.3 Room list for a factory (example). © Reichardt 15.212_JR_B

12.1.2 Process and Logistics Elements

The goal of every factory planning is to optimize the interplay between the process and logistics elements primarily from the point of view of a streamlined material flow with minimal lead time and minimum stocks. Depending on the size and weight of the required machinery and equipment a multitude of interfaces need to be considered for the spatial planning. When metal-forming processes are involved for example, critical requirements for the spatial surroundings include the soil bearing capacity, ceiling clearance, location of supports, the supply and removal of media and noise protection.

In most cases, isolation of engineering facilities, process and spatial planning leads to coordination problems resulting either in permanent functional inconveniences or structural defects. In the framework of the Synergetic Factory Planning approach, process facilities and logistics elements even in the design development phase are generated as 3D objects and inserted into the surrounding building space model. This enables, for example, early detection of points of conflict and integration of 3D structures for processes and logistics based upon a horizontal and vertical design module. Such dimensional coordination of process and space are important for further planning and the final selection of building materials for structure, utilities, media and interiors.

The use of "ID cards" for process and logistics elements has proved to be quite effective for a balanced communication between production engineering and space planning. The required information is available "at a glance" for the design team during the specific phase of factory planning. ID Cards are not only relevant for process material flow simulation but also for optimization of temperature and energy and for development of color concepts based upon the available information. Figure 12.4 shows the 3D model as well as a photo and basic geometric data of a "cleaning robot" for composite rubber elements for a new production plant. The overall dimensions are shown in the three views,

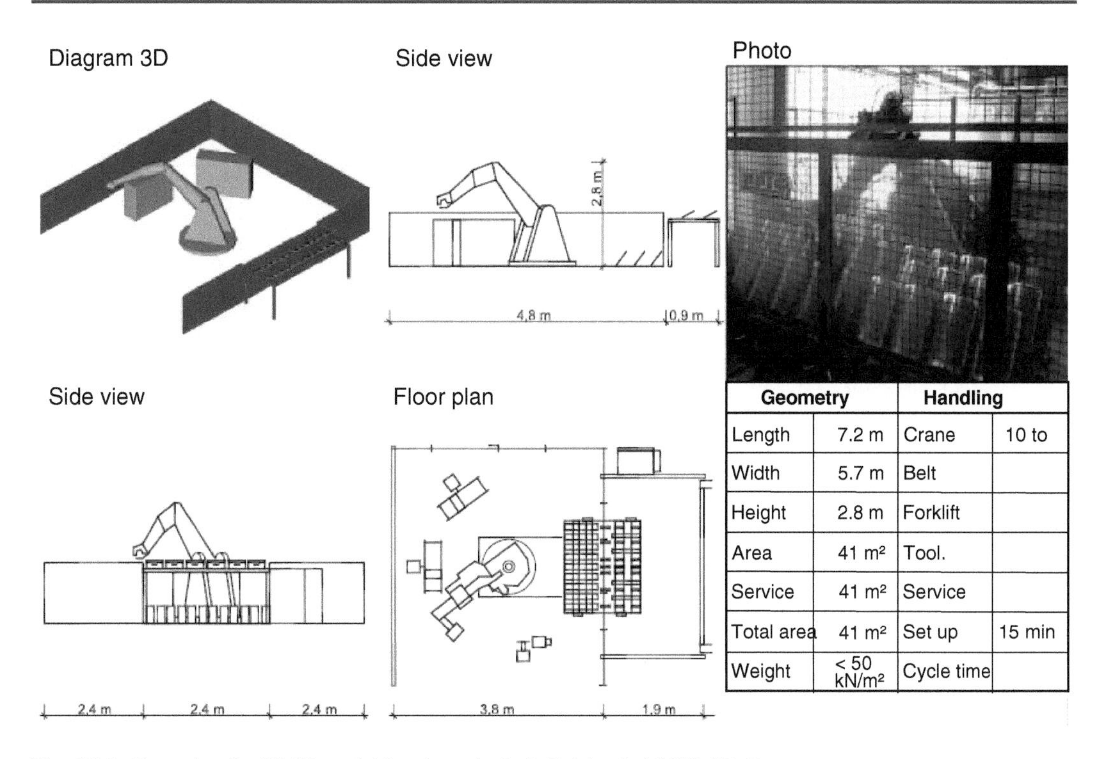

Geometry		Handling	
Length	7.2 m	Crane	10 to
Width	5.7 m	Belt	
Height	2.8 m	Forklift	
Area	41 m²	Tool.	
Service	41 m²	Service	
Total area	41 m²	Set up	15 min
Weight	< 50 kN/m²	Cycle time	

Fig. 12.4 Example of a 3D ID card (cleaning robot). © Reichardt 15.213_JR_B

complemented by footprint, working areas and weight. A special text block area may be reserved for site specific modifications or requirements along with purchase and delivery details. As and when more details are available, the contents of the ID card are refined to include informations about supply, delivery and disposal. The ID cards serve as digital process library and if maintained in a systematic manner could play a useful role in subsequent documentation during facility management and the operational phases.

The degree of spatial adaptability of the process and logistics elements increases with 'moveable' facilities in a production hall. Hence it is advisable to a start with simplest equipment layout. Figure 12.5 shows the essential features of spatial adaptability for process and logistics objects, process attachments to structural members, types of supply and disposal as well as types of conveyor system. Ideally, permanent installations to structural members such as roof slabs, foundations, etc. for waste disposal systems and conveying systems is best avoided for ease of conversion.

12.1.3 Supply and Removal

Although secondary from a planning perspective, supply and removal systems for workplaces and facilities are key subjects when it comes to spatial planning. Their often unexpected volume in addition to the difficulty of adjusting them to layout changes poses serious obstacles for changes recognized as necessary.

In order to locate, evaluate and select detailed solutions later, it is absolutely necessary to compile and document the requirements early-on in a suitable table that allows the information to be easily up-dated. Furthermore, specifications for the supply and removal system should be directly allocated to each of the divisions in the room list. Coupling the spatial and technological requirements in this way prevents different planning states and offers a solid basis for changeability scenarios since variants can be considered as a whole. In particular, the ID Cards described above can also be used in the facility planning to record the 3D plan of all the supply and removal connections as well as all the

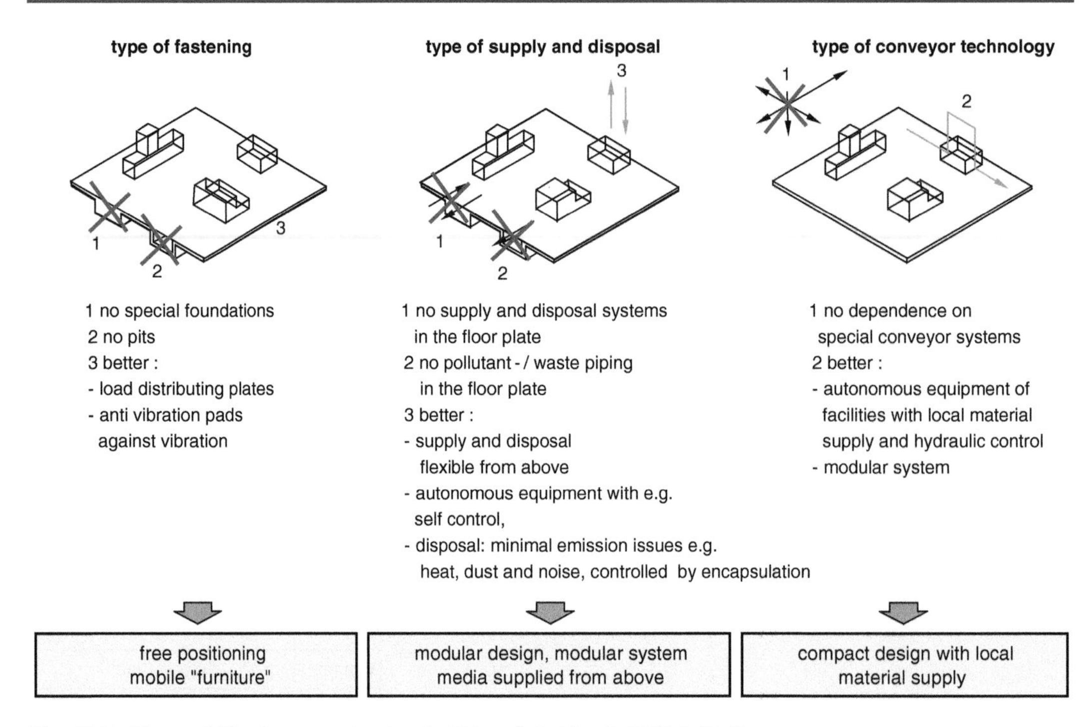

Fig. 12.5 Changeability features related to facilities. © Reichardt 15.214_JR_B

necessary text data such as required amounts of compressed air, electrical services etc.

Figure 12.6 depicts an overview of media connections based on the example of a degreaser for a rubber manufacturer. In refining the 3D construction of the element, all of the required power inputs, media connections and outputs are considered; later problems with coordinating the process technology and building services are thus avoided.

In view of the ever-growing concerns with global warming, demands to minimize energy is going to be more and more important. When analyzing technological and economic aspects, new production facilities should include an energy cost analysis of all the consumption points in the process. Within the frame of the synergetic factory planning, the data required for creating an integral energy simulation can be directly accessed from the corresponding ID cards. A detailed survey of measures towards energy minimizing in production facilities can be found in [Neu13], part V "machineries and equipments" as well as part VI "production processes".

When determining capacity data, specific usage profiles and especially the so-called 'simultaneity factor' need to be considered. The simultaneity factor takes into account the fact that even when a factory is operating at full capacity, not all of the consumption points are on simultaneously and with their nominal power. For machine building factories this can range from 0.25 to 0.4, for example. Of course, when measuring the capacity, the anticipated factory as a whole should be considered. Nonetheless, it is also important to keep in mind the continual improvement of productivity and the degree of impact implemented resources will have during their service life. Here too, the modular principle of power utilities proves to be practical.

One possible solution is that the factory draws the power it requires from a local supplier which builds and operates the necessary transformers, pressurized air and steam generators etc. on the factory grounds themself and only charges for the actual consumption.

Figure 12.7 illustrates the process and energy optimization based on the example of an industrial bakery [Rei98]. By cleverly coupling the

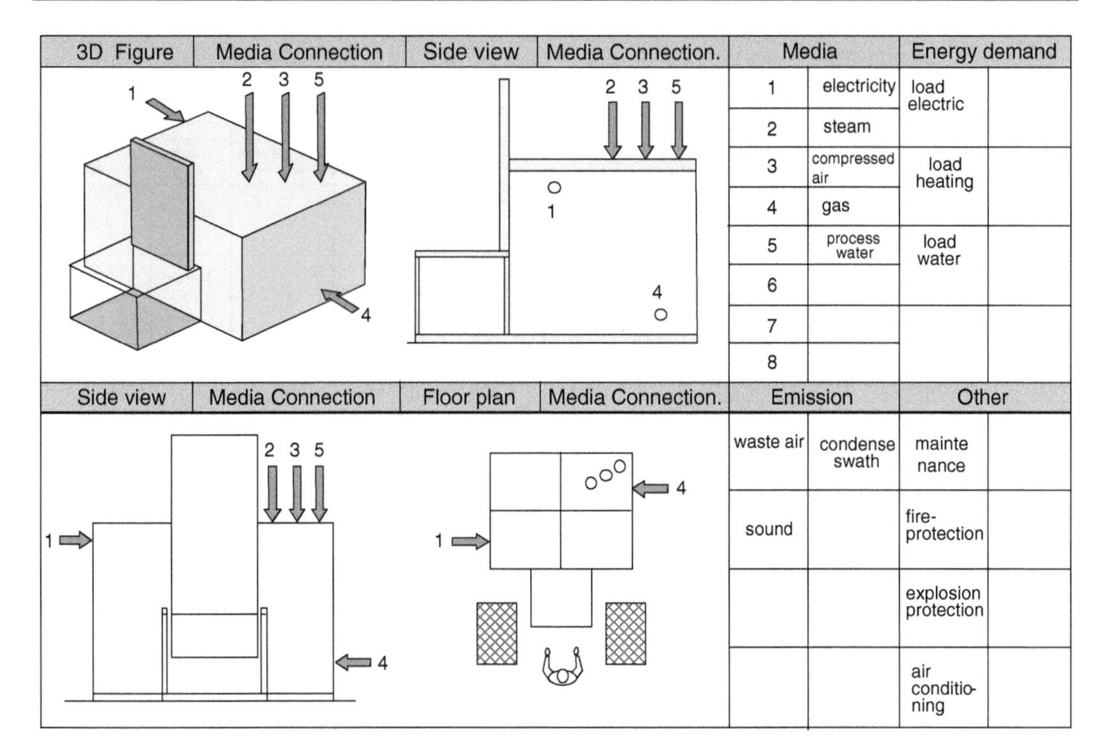

Fig. 12.6 Example of a 3D ID card (Degreaser). © Reichardt 15.215_JR_B

process engineering and building services using heat recovering technology the yearly energy required for heating was reduced by 62 % and for ventilation by 39 % in comparison to conventional solutions.

12.1.4 Special Requirements

During the course of general planning process, the planning team may be faced with a number of unique challenges due to the production requirements. In addition to specific needs for infrastructure, systems of supply and disposal (e.g. clean room technologies) there may be additional needs for manufacturing equipment, such as strength requirements for floor slabs due to heavy machinery assemblies or for roof slabs due to gantry crane systems. As pointed out before, it is recommended to collect and interlink these requirements by means of "ID-cards" of the process and logistics equipment.

12.2 Building Typology

Pevsner [Pev98] defines factory buildings as buildings of a certain size, in which products are manufactured in large quantities. Dolezalek [Dol73] and [Agg80] traditionally characterize industrial building types according to basic functionalities into part manufacturing, assembly, warehousing, management, development, social areas, exhibition and sale as well as on the shape of the specific building section.

Over the years, these building typologies following basic industrial functions have not been found to be conducive to changeability—as processes, functions and consequently building requirements, in most cases, would definitely change during the useful life of building. This in turn leads to the quest for multifunctional structures wherein the building volumes and sections are able to "breathe" and do not require major structural changes or unforeseen costs in a short

energy balance			
heat load in the hall	demand [kW]	waste heat [%]	heat load [kW]
dough preparation			
- mix, weigh	2	0.22	0.44
- knead	30	0.09	2.70
dough processing			
- roll plant	11.5	0.13	1.50
- press	1	0.28	0.28
- dough separation	2	0.22	0.44
- dough separation	5.5	0.15	0.83
boiling pastries	10	0.5	5.00
fine bakery	10	0.5	1.25
various	5	0.25	2.50
fermenting room	5	0.5	0.5
sum [kW]			19.93
- area [m²]	816		
- simultaneity	0.8		
sum energy [W/m²]			19.54

optimized annual energy requirement				
	heating		ventilation	
	load [kw]	ann. demand [kWh/a·m²]	load [kW]	ann. demand [kWh/a·m²]
initial situation	90	32	192	576
optimized situation	34	31	117	449
improvement [%]	62	4	39	22

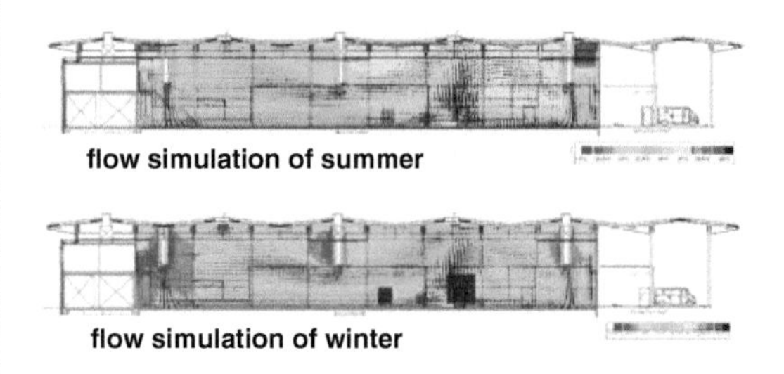

Fig. 12.7 Energy optimization of a building (example—industrial bakery). © Reichardt 15.216_JR_B

period on completion or erection of the structure. Proper selection and the choice of detailing of the building structure, in particular its section, is therefore one of the utmost important planning tasks. This also includes building performance aspects such as capacity for structural loads, clear heights, natural lighting, utilities systems and zones as well as expansion possibilities, which in the case of structural design, are irreversible. The determination of building section and site plan figure also includes aspects of energy efficiency and sustainability (see certification systems and project examples in Sect. 15.7). In addition to the material flow technology, such issues are particularly relevant for sites located in extreme climatic conditions. In line with the energy and temperature issues of production technologies, the orientation of buildings towards the sun to facilitate entry of daylight using glazed facades or to generate electricity using photovoltaic and others need to be considered.

12.2.1 Sectional Profile

Figure 12.8 shows important characteristics of building sections having inherent capabilities to either expand horizontally and/or vertically; the concepts could be adopted for low-rise, shed or multi-storey buildings, combined forms and/or specific buildings. Types of roof forms and internal heights characterize hall structures and multi-storey buildings (see also Sect. 11.2). It may be necessary to combine various forms in a given project to satisfy specific project needs. Hence, the frequently encountered building type of low production or assembly hall with offices on one side actually follow the building typology of hall structure plus a multi-storey structure. Special structures in large technical process-driven plants or systems, such as supporting scaffolds for refineries or silo constructions and steel supporting structures for high bay automatic warehouses require separate and detailed attention.

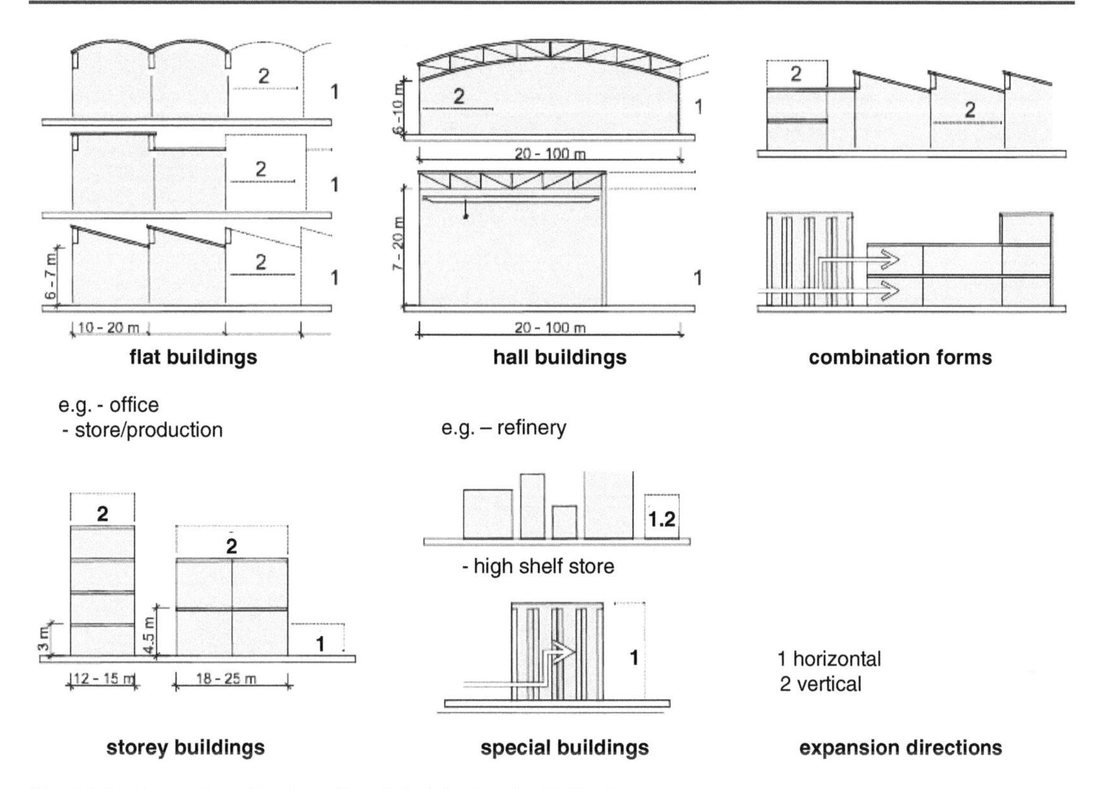

Fig. 12.8 Types of sectional profiles. © Reichardt 15.217_JR_B

Figure 12.9 provides an overview of structural features for the sectional profiles of low-rise buildings, multi-storey buildings and halls. The new multi-storey factory structure is a result of the historical industrial building. Various refinements, to accommodate day-lighting and free span, emerged in the English textile spinning mills of the 18th century; similarly refinements were also seen in the American automobile factories at the beginning of the 20th century. Currently such building types are still in use in the small part mechanics of optical and electronic industry, as well as in the food and clothing industry. Multi-storey buildings are ideal for chemical and pharmaceutical industries since they prefer to vertically transport goods making use of gravity. As a structural response to new process requirements of mass production, coherent low-rise hall buildings emerged in large numbers in the early 20th century.

The introduction of the assembly line principle in the American automobile production favored concepts of easy expandability on the ground floor level, enabling quick installation of heavy machinery with possibilities of day-lighting through the roof above.

The history of Henry Ford developing and refining hall constructions is described in [Buc02]. Up until today, hall constructions offer very good possibilities for changeability, especially when sufficient clearance is reserved for later integrating galleries and a consistent modular construction is implemented allowing multiple external extensions.

Hall constructions are called for when there are additional demands for wider spans, greater ceiling clearance, or heavy conveyor technology. A high degree of internal changeability can be targeted with a design that incorporates reserves for extensions. Whereas flat buildings are advantageous in terms of external changeability because they can be expanded in all directions, hall constructions can usually only be extended practically along one axis. Moreover, in

	storey buildings	flat buildings	halls
floor plan dependent on	escape route length to stairs max. 30 m fixed point formation with stairs, lift, sanitary rooms, installations, etc.	connection between manufacturing areas with main building or peripheral auxiliary / side rooms	rectangular, crane track, route lead in, side hall or annexe for auxiliary / side rooms
section profile dependent on	windows width/room depth span widths/inner supports	area to be lit, top light ribbons individual top lights, cranes, span/ support grid, roof pitch	span of crane, crane hook height, area to be lit, side glazing, roof surface, ventilation, roof pitch
cellar	normal	possible	very rare
supply connections	from floor or ceiling , transmission pipes at fixed points	from floor channels or ceiling	from enclosing walls and floor channels
operating costs	good heat retaining capacity	cooling off through the roof	cooling off through walls and roof
ceiling heights	3.50 m to 4.00 m	4.50 m to 5.50 m	over 5.00 m
crane payload	up to 11 to	up to 3 to	up to 100 to
vertical conveyance	elevator, lifting platform	bridge crane, column crane	crane
horizontal convey.	forklift	forklift, conveyor belt, crane	crane
flexibility of use	restricted	guaranteed for entire floor plan	possible
expansion	by annex	extension mostly in several directions	extension or side positioning
load transfer	on floor ceilings, vibration sensitive beams on cellar floor	on foundation or stiff cellar ceiling, vibration sensitive beams on ground	on foundation, rarely cellars are planned underneath
characteristic	limited floor plan, many work - places, stairs and elevators	connected area at ground level	large spans and heights, few work places, crane installations
loads	easy to medium requirements	easy to medium requirements	heavy to very heavy
net area demand	small to medium requirements	high	medium to high
building land need	low	high	medium
lighting need	normal to high	high	low to normal
lighting areas	lateral windows	overhead lighting strips, top light domes	lateral windows, roof surface
indoor climate need	normal to high	normal to high	normal
heating/ventilation	local radiators, window ventilation	heating or air conditioning	heating, roof ventilation
building form	lengthways-oriented	no specific direction	lengthways oriented

Fig. 12.9 Features of different sectional profiles (acc. Dangelmaier). © Reichardt 15.218_JR_B

comparison to flat constructions, greater expenses are to be reckoned with for the structural framework, conveyor technology and technical installations in halls.

It would be ideal to take advantage of the various building typologies and combine them to arrive at a coherent building concept, which best suits the present and future needs of changeability of the project under discussion.

12.2.2 Outline Figure

In practice, the choice of the sectional profile as the single criterion for the definition of the building structure is necessary, but may not be sufficient in most cases. Besides the sectional profile, the assignment of functional areas, their orientation and ability to change is largely determined by the outline figure. Plans in turn are dependent upon processes which could be centralized or decentralized, compact or spread out, dissected open, examples of which are given in Fig. 12.10.

With the compact, closed layouts all of the rooms or functional areas are pooled together into one overall space under a roof. Homogenous basic forms such as this often have contiguous rooms and often lead to large buildings and expansive interiors. The actual space is then often defined by room-dividing elements that are independent of the structural framework.

The specific factors which lead to such closed or compact buildings and thus to a large scale structure often include production and ancillary areas which belong together functionally and technologically. Further features include:

- cost-effective building systems with large capacities that can form larger and therefore economic units when extensive in-house technologies are required,
- decreasing transportation distances, and
- saving on the cost of land when the price of property is high.

Since large spaces are generally practical for a wide-range of requirements whether in relation to processes, logistics, or administration they offer a high degree of future changeability. From an

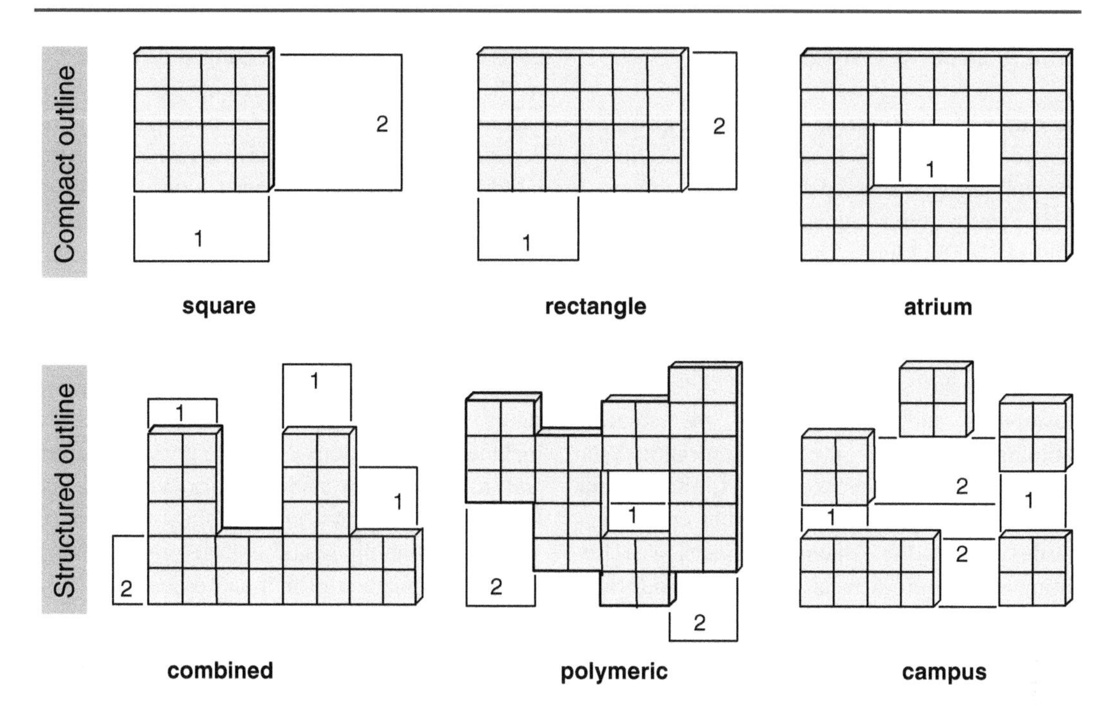

Fig. 12.10 Typology of outline figures. © Reichardt 15.219_JR_B

energy perspective they are advantageous in that the loss of heat is minimal due to a favorable A/V ratio (envelope surface A to volume V) and shorter pipes and cables for supply and removal systems. As long as constraints such as column grids that are too narrow or disadvantageously located core areas are avoided, interiors can also be modified.

In contrast, the total volume of a dissected and open outline building is divided into several equal or different sub-volumes. The subdivision may be classified by offset, stacking, terracing, gearing or coupling of the resulting building mass. Although complete separation of structures may lead to new individual building elements (see: campus), these detached buildings are regarded as part of an articulated group of structures based upon a well-thought out Master Plan.

All of the heterogeneous building layouts typically have smaller, manageable spatial and body dimensions and are characterized by a greater ratio of exterior wall area per square meter of floor space. Factors which lead to structured expansive building forms often include varying structural and spatial specification for the production such as greater loads, various heights, increased humidity or different ventilation and air-conditioning requirements etc. Further reasons for separating areas into different buildings include:

- dangerous or irritating sub-processes that e.g., generate loud noises or vibrations; emit gases, vapors, dust or smells
- areas with a greater risk of fires or explosions that thus require their own special safety regulations.

In most cases, a large envelope and building volume is generally less energy efficient and more costly. On the other hand, one might wish to increase fenestrations for more visual contact to the outside world while being spatially close to production; at the same time, decentralized buildings are extremely efficient in the form of rest rooms and canteens, changing-rooms, toilets, etc.

12.2.3 Linking Principle

In view of maximizing building capabilities towards changeability of designs, in addition to the previously discussed building typologies, a third point requires a considerable thought—the "linkage" of buildings. The entire campus—work areas and individual buildings are connected and maintained by "blood vessels", characterized by traffic routes, utilities networks and material flow routes. In the same way, once selected, the positioning and connection of the adjacent buildings determines the communication relationships among them. According to [Kar90], the chosen combination principle either encourages or discourages the "internal" expansion (retrofitting media) or external expansion (additional plant areas). Selecting the combination principle is thus of paramount importance since the multitude of essential functions that overlap at the interfaces of the building structures often means that changes are significantly more complicated to execute than in the floor plan or profiles of the individual structures.

In continuation of principles discussed so far examples for linking principles are shown in Fig. 12.11. The examples focus on long-term arrangements corresponding to expansion options of industrial building structures. The principle of coupling leads to additive design of building structures. The linear arrangement of building volumes along an axis results in a spine for people, material and utilities movement. Vacant locations on either side adjoining the axis could be filled-in with future expansions at a later date provided the total length of spine is planned well in advance. In contrast, the building volumes around a courtyard are not connected to each other but organized usually at right angle around an open space (e.g. logistics truck space). With skillful interpretation, expansions might be ring-like allowing for shorter runs to neighborhood buildings. The star shaped arrangement also provides a central open space, but the angular geometries provide lesser freedom in terms of the site plan.

It may be also possible to locate buildings in a site in a seemingly chaotic manner and link them using fractal theory. The site geometry of traffic cross allows volumes to be arranged perpendicular to the axis of movement creating four zones or sectors between the axes for potential building zones. In case of spiral and a circular structure, building volumes are arranged along imaginary lines of movement, as part of dynamic circuit geometry. The grid principle is built up with reference to a geometric coordination system allowing definitions of the current and future relationships between built and un-built areas to change.

The selection of the sectional profile, outline figure and the linking principles should be included in discussions to resolve possible current and future process and logistics requirements. Figure 12.12 shows changeability aspects of section, outline figure and linkage principle for a typical combination of hall and office wing. The arrangement of the buildings determines expansion possibilities in plan and section, but also the linking space for e.g. staff, visitors, materials and utilities media. Hereafter the characteristics of changeability of a design are clearly defined.

12.3 Property Protection

Depending on the type of plant and how much it is endangered, systems may require special measures with regards to burglary and theft as well as protection from fires and explosions. [AGI04] provides an overview of possible factors and examples.

12.3.1 Burglary, Theft

The traditional solution against burglary and theft is fencing the perimeter of the plant to provide necessary external security. Depending on where the gate is, extensive parking for employees and visitors can be located before the plant entrance if permitted by authorities. Fences often provide little protection though against trespassers as they are relatively easy to overcome for experienced offenders.

From an urban planning point of view as well as from the perspective of more changeable

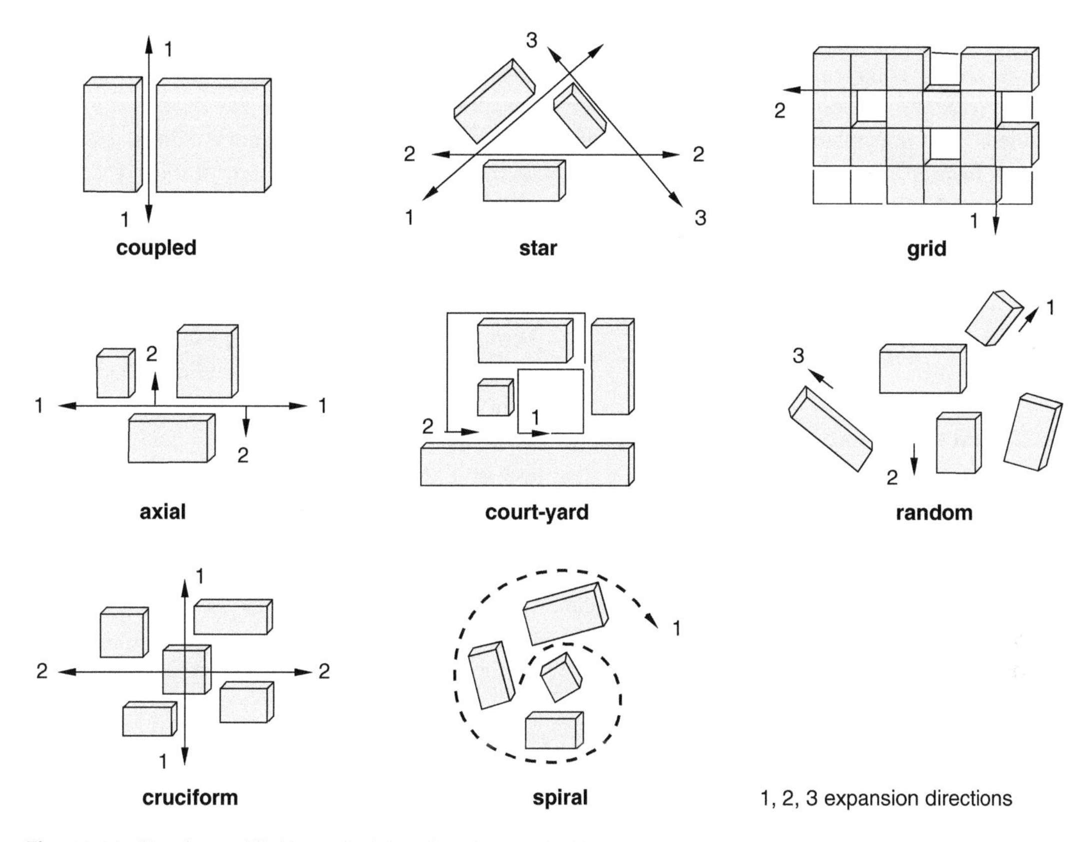

Fig. 12.11 Typology of linking principles. © Reichardt 15.220_JR_B

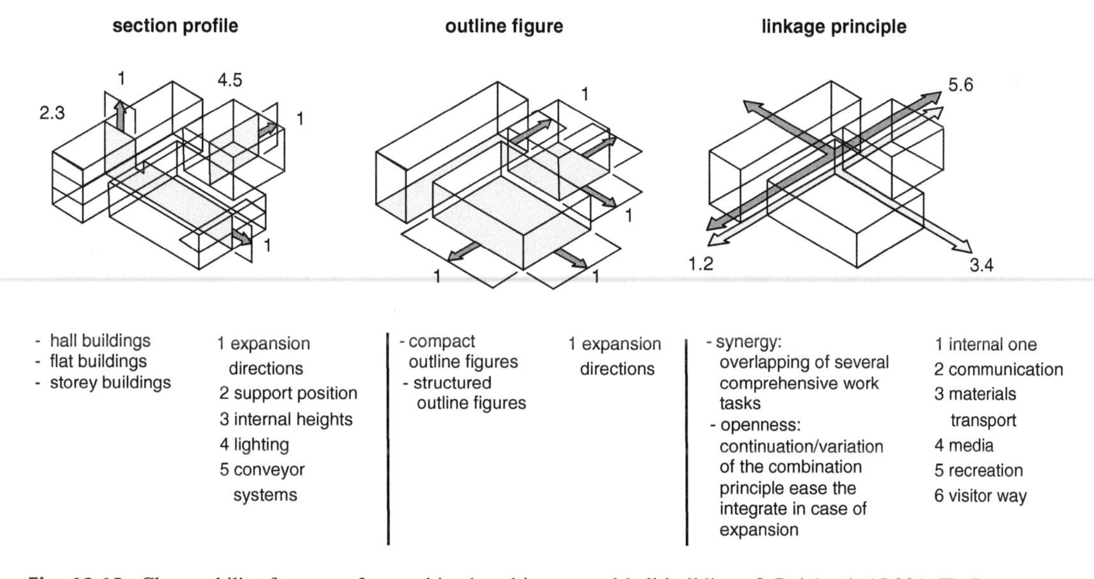

Fig. 12.12 Changeability features of a combined multi-story and hall building. © Reichardt 15.221_JR_B

grounds where extensions are concerned, fencing systems can be completely dispensed with when motion sensors and door/window sensors are installed. An integrated, coordinated security concept focused on defined, special safety areas can also gradually and easily respond to changes in uses. Moreover, modern alarm systems allow messages or alerts to be automatically forwarded to service centers i.e., plant security, the police or a private telephone number.

12.3.2 Fire and Explosion Prevention

Measures for preventing fires and explosions should be derived from a comprehensively developed fire prevention concept for the entire factory premises. The concept first regulates the safety of people by:

- providing sufficiently short escape routes,
- ensuring the fire resistance quality of the structural support systems, building shells, extension elements, rescue routes and
- extracting smoke when there is a fire.

In particular, preventing explosions means storing dangerous substances in specially designated rooms. In order to facilitate the available changeability, the developed fire prevention concept should not be obsolete as soon as initial changes are made, rather it should contain reserves for short escape routes and factors for extracting fumes. Spaces for storing explosive materials should be easily enlarged or able to be dismantled or demolished if necessary when the space is no longer required. The technical details are in-depth described in Sect. 10.5.1. Further informations can be found e.g. in [Jon14] and [Int11].

12.4 General Development (Master Plan)

12.4.1 Procedure

The general development, also called "Master Plan" shall generate a synopsis of material-flow, town planning, landscaping, energy and environmental and laws, in a way that best suits the functional utilization of the given site. The "Master Plan" brings together buildings and open spaces in present and future in clearly defined phases. A master plan looks forward to the various stages of development in a coordinated manner even though a part of the terrain is already built. The goal is to avoid blockages or bottlenecks which might impede future expansion, such as unfavorable building locations or utility systems which require expensive renovations. A comprehensive master plan optimizes not only communal regulations with regards to specifications for building dimensions, floor areas or necessary parking spaces, but also urban planning problems related to the general factory planning along with energy and ecology related factors. Examples here are:

- the functional relationships between sectors of the plant,
- the quality of the spaces,
- the orientation of the buildings according to solar criteria with regards to shading and energy balancing and
- consciously including green spaces and ventilation corridors.

For more detailed information on urban planning law regarding master plans as well as building regulation codes see [Mar11, Chi12, Kie07].

With reference to our discussion on compact and dissected outline figures, building types could be centralizing or decentralizing factors in the master plans. Figure 12.13 lists these principle factors. For example centralizing factors follow the functional principle of grouping together similar functions as close as possible, whereas decentralizing factors tend to divide the structure of a company into more or less independent operating units.

The differences between an *ideal plan* and a *real plan* should be pointed out. The ideal plan is the master building plan that would be realized when there are no local restrictions. This plan is important for developing a concept of the ideal factory and to be able to estimate the

centralizing factors	decentralizing factors
• functionally coupled operating areas • organizational and leadership identity • closely connected areas for material flow • uniform structural and spatial demands • areas related with respect to infrastructure • high land price • story building, underneath cellar • avoidance of transportation in the open field • building of larger and more economic units • building of larger buildings, more favorable investments, lower heat losses, short supply and disposal pipes	• separate organization and management areas • different areas with respect to infrastructure (e.g. fire and explosion hazard) • distance due to vibration or earthquake

Fig. 12.13 Master building plan factors (per Dolezalek, Warnecke). © Reichardt 15.222_JR_B

disadvantages that result with deviations from this ideal state. In order to then develop the real plan from the ideal plan, the features of the eligible grounds need to be verified in regards to the:

- zoning and organizational grid,
- access to public infrastructure (incl. public transportation and utilities),
- supply and removal systems,
- structures and open spaces.

Figure 12.14 shows a flowchart for developing a Master Plan [Agg80]. The necessary areas in the site are defined according to use (specific zones), the building structures and technical infrastructure are best integrated into a coherent development. The aim here is not only to satisfy the existing requirements but plan for the future extension phases, too. Usual planning horizons are five, ten or fifteen years. It is important to exactly evaluate the requirements and arrive at a precise coordination of the possible zones, buildings and infrastructure.

What is striking is the absolute necessity of establishing a cooperative planning process which involves the active participation of the authorities early-on in regards to local infrastructure, supply and removal systems and environmental protection. The zoning provides key information for the functional areas manufacturing, logistics, administration etc., whereas the planned routing for the supply and removal systems is essential for the infrastructure. Traffic plans, parking places, external green areas, streets, fencing and the building plans (cadastral maps) should all be clarified for the master plan.

12.4.2 Zoning and Organizational Grid

The basis for the zoning is the roughly dimensioned factory concept developed during the structural planning (see Sects. 15.2 and 15.3). In particular, all of the necessary options for extensions need to be thoroughly identified. The plan for the functional areas is then derived from the refined zoning. Depending on the size of the grounds and the project a grid based on square, rectangle or triangular units is recommended. In order to facilitate further development of the site, it is advantageous to use a uniform dimensional system for the construction and interior planning which then also allows a quick area control by counting the area modules. Moreover, any additions, conversions or extensions in future

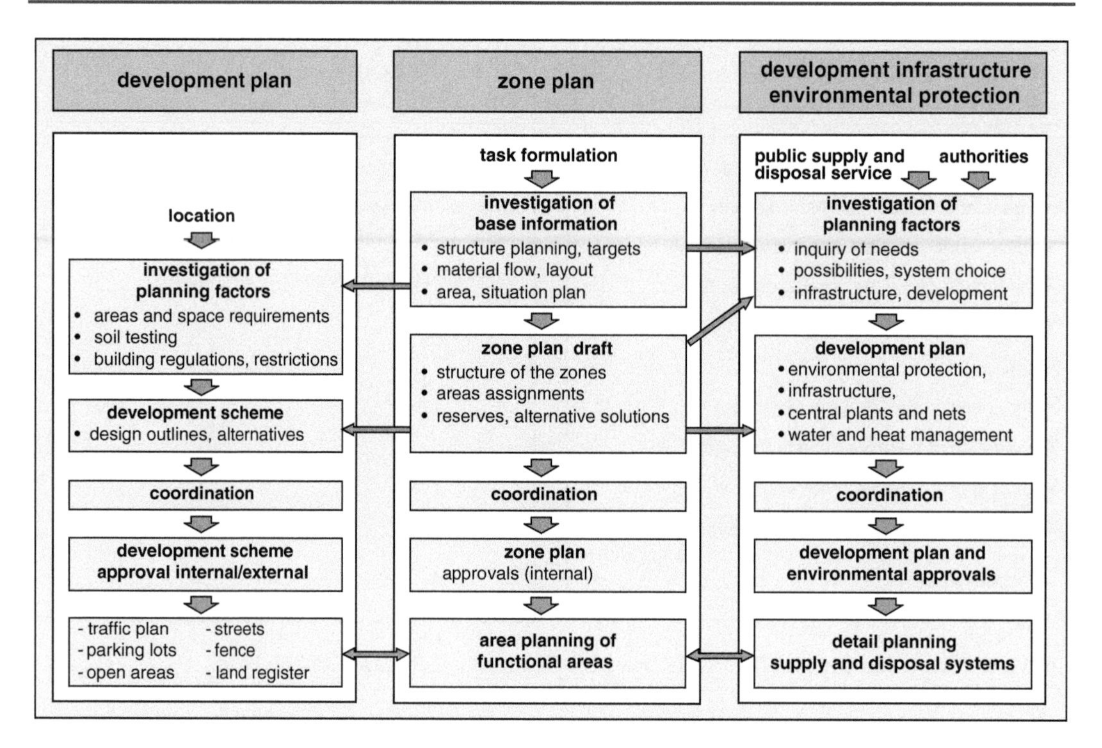

Fig. 12.14 Flowchart for developing a master building plan (acc. Aggteleky). © Reichardt 15.223_JR_B

development stages are then embedded into the arrangement of buildings and roads.

A site is generally best assessed when the modular grid points are oriented more or less parallel to the main direction of the lot's progression. Buildings can then be advantageously constructed with straight, continuous external walls. According to DIN°4171, DIN°4172 and ISO°1791 possible grid systems are standardized; grid sizes are selected depending on the objectives of the factory planning, the cut of the land and bank costs.

Figure 12.15 depicts an example of a project zoning and grid. In this example with the zoning, it is important to take into account a protected areas as well as vegetation areas that are not to be built over in order to provide a scenic nature reserve which also serves as a planned "vegetation compensation" area. Based on the material flow, it is possible to have three construction stages without disrupting an on-going production. The factory expansion takes place in accordance with the modularity of the process and logistics elements on a parent grid network for the road construction, building and space expansion.

The large scale of the zoning grid is further sub-divided into smaller grid systems for detailed planning. Typical dimensions for zoning grids are e.g. 36, 50 and 60 m. For building and construction suitable smaller grids of e.g. 18 m (possible clear span of production hall structure) or 1.20 m (possible office module) may be derived.

12.4.3 Infrastructure, Supply and Removal Systems

Infrastructure refers not only to the linking of the premises with the public traffic network, but also to controlling the extra-plant material flow on the grounds, managing the flow of employees and visitors as well as ensuring access for emergency services. In agreement with the master plan's strategic plant expansion the access grid sets the area modules for the road system. The traffic routes have to be defined together with the

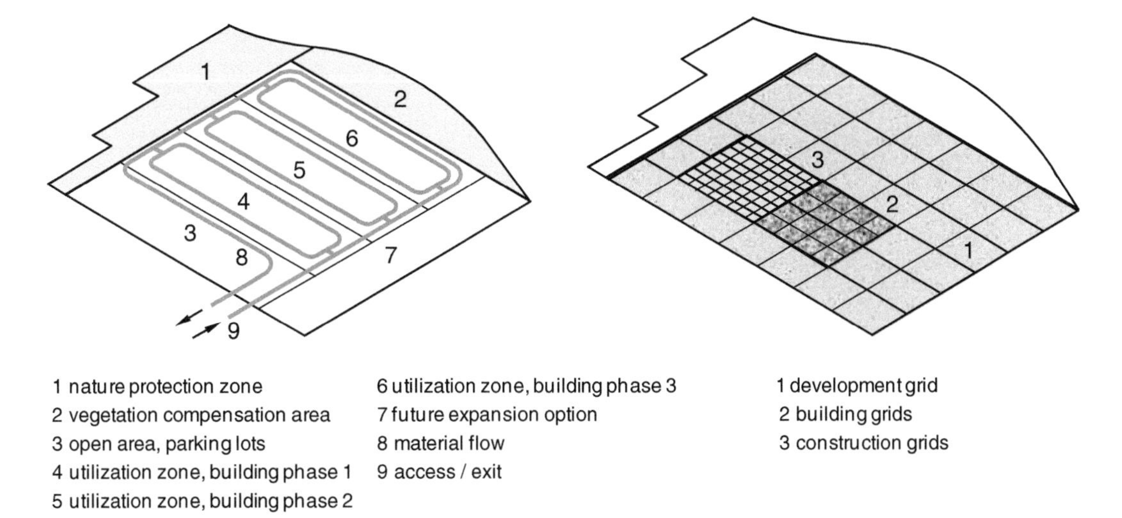

Fig. 12.15 Zoning grid of a general site development plan (example). © Reichardt 15.224_JR_B

building forms and structures. It is often practical to overlap the access systems and media routes on the property, since building over the routes for supply and removal ducts is generally against building regulations.

One of the most important tasks of the master planning is to design and position the main traffic and media routes so that they are trouble-free and suitable well into the future. Especially with modular concepts it is critical that the infrastructure is planned with farsightedness. Moreover with the supply and removal system's collection and distribution structures it has to be weighed whether a centralized or decentralized layout is advantageous. Centralized systems for the energy supply, transformers, heating or ventilation and air-conditioning systems usually mean lower investment and operating costs. In comparison, with decentralized systems there are fewer lines and thus less transmission losses. Furthermore, when there is a defect, the entire plant is not blocked and when retrofitting individual plant units, it is easier to make adjustments to the media utilities. Here too, there is a tendency towards modules which are functionally pre-assembled and tested so that they are quickly operational.

Figure 12.16 illustrates exemplary layouts of access, supply and removal systems. On the left-hand side the access routes are visible, whereas on the right-hand side variations for the media lines are vertically depicted. In view of changeability, it is important to provide possibilities for easy expansion and development well into the future with simply upgradeable additional main-lines within the cross-sections of the ducts, pipes, bridges or secondary stations.

12.4.4 Buildings and Open Spaces

In accordance with the objectives of the factory design and once the building forms and grids have been selected, modular building concepts strategically expanding the site in multiple stages are beneficial. Green areas and parking lots integrate the building into the grounds as well as into surrounding urban and natural environments. A consistent green concept is recommendable for the entire plant premises including all of the boundary areas transitioning the factory grounds into its surroundings. The concept should include seepage areas and storm water retention basins as ecologically valuable biotopes. Green areas also serve as recreational areas for employees during their breaks.

Parking areas should be shaded using trees or light-weight roofing structures, particularly in warmer climates to avoid being 'heat islands'. Impermeable parking surfaces should be

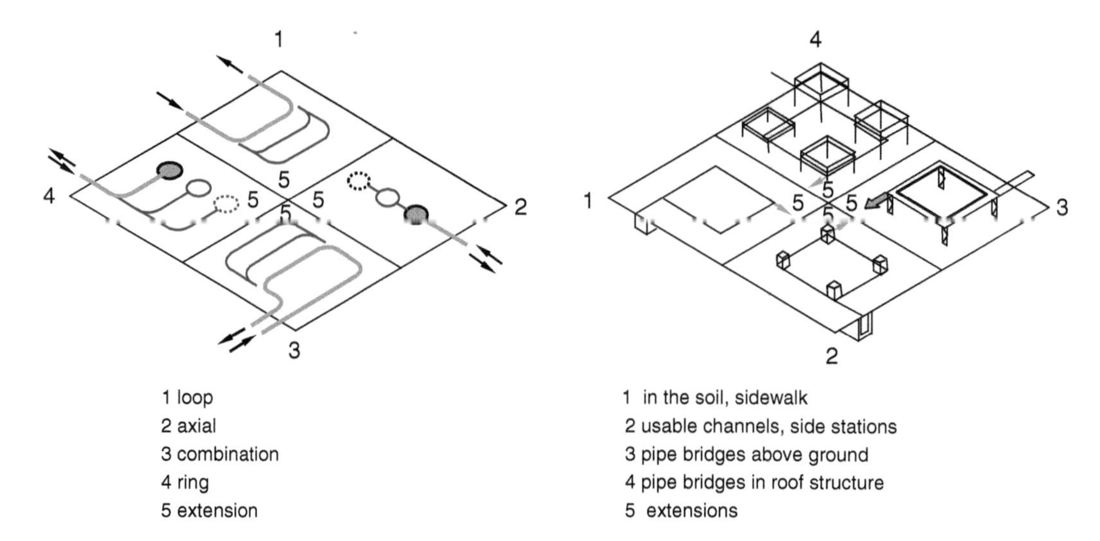

Fig. 12.16 Arrangement of traffic, supply and disposal systems in a general development plan. © Reichardt 15.225_JR_B

prevented; gravels, cobbles stones, paver blocks which allow grass to grow, are ideal for rain-water percolation. Figure 12.17 illustrates the principle of the spatial layout for the exemplary modular factory presented in Fig. 12.15.

In Fig. 12.18 an example of various options for relocating a plant from an urban setting to the outskirts of a large city are depicted. The suitability of two possible locations is analyzed based on a list of criteria. The changeable building structure evolved from the requirements of the accompanying factory planning based on a construction grid measuring 26 m × 36 m. The plant is developed in a number of construction stages. For both locations a comprehensive green concept was developed in continuation of existing vegetation structures, whereby storm water retention ponds were integrated into the landscape for the roof drainage. Since solution B offers better possibilities for developing the area as well as the option to have a number of entrances to the property it is selected for further planning.

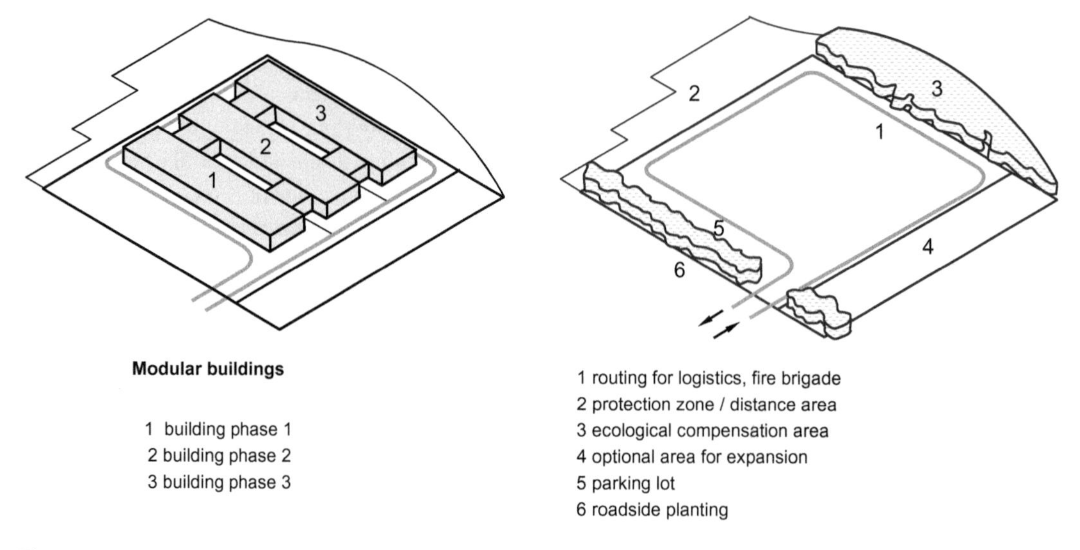

Fig. 12.17 Spatial arrangement of a general development plan (example). © Reichardt 15.226_JR_B

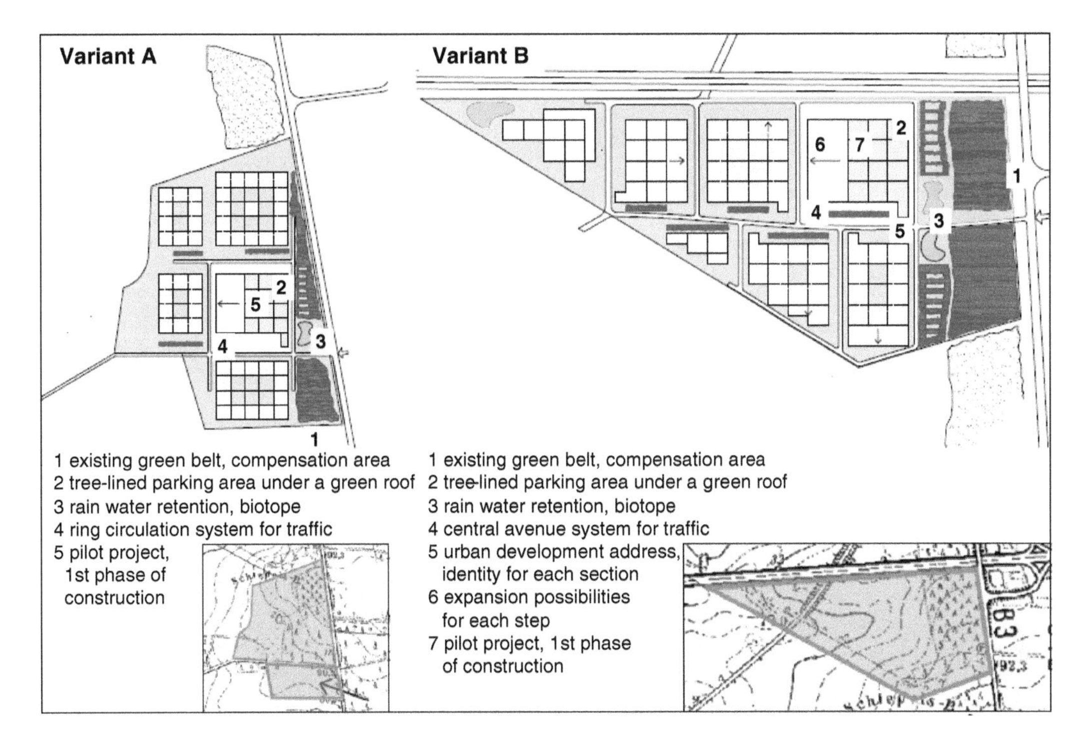

Fig. 12.18 Plant relocation options (example). © Reichardt 15.227_JR_B

With this we conclude our discussion of developing a plant site. In general, this also means that the factory planning tasks are in a strict sense complete. The next higher level of planning concerns the location and is closely linked to the production strategy of the company. We will examine it in the next two chapters, first from a spatial view and then from a strategic perspective.

12.5 Summary

Starting point of the site master plan are the specifications of the production program as well as the topographical, infrastructural, climatic and legal requirements of the location. In an in-depth discussion of variant concepts regarding future levels of adaptability aspects of energy efficiency and sustainability are to be observed. The decisions on the sectional profiles, floor plan figure and linkage principle between buildings determine the future security of the enterprise and requires a team effort with high creativity of all stakeholders.

Bibliography

[Agg80] Aggteleky, B.: Fabrikplanung – Werksentwicklung und Fabrikrationalisierung. Carl Hanser München (Factory Planning—Site Development and Factory Rationalization). Carl Hanser, Munich (1980)

[AGI04] Arbeitsgemeinschaft Industriebau e.V.: Objektschutz und Sicherheitstechnik im Industriebau (Property Protection and Safety Technology in Industry). Callwey, Munich (2004)

[BUC02] Bucci, F.: Albert Kahn—Architecture of Ford. Princeton Architectural Press, New York (2002)

[Chi12] Ching, F.D.K., Winkel, S.R.: Building Codes Illustrated: A Guide to Understanding the 2012 International Building Code. Wiley, Hoboken (2012)

[Dol73] Dolezalek, C.M.: Planung von Fabrikanlagen (Planning of Factories). Springer, Heidelberg (1973)

[Int11] International Fire Service Training Association: Fire Detection and Suppression Systems, 4th edn. Stillwater (2011)

[Jon14] Jones Jr., A.M.: Fire Protection Systems, 2nd edn. Jones & Bartlett Publishers, Sudbury (2014)

[Kar90] Karsten, G.: Fabriken für das 21. Jahrhundert (Factories for the 21st century). In: Presentation at Symposium Fabrikplanung, Jena, 30/31 Aug 1990

[Kie07] Kiepe, F., von Heyl, A.: Baugesetzbuch für Planer (Building Code for Planners), 3rd edn. Rudolf Müller, Köln (2007)

[Neu13] Neugebauer, R. (ed.): Resource Orientated Production. Hanser, Munich (2013)

[Mar11] Marshall, S. (ed.): Urban Coding and Planning. Routledge, Abingdon (2011)

[Pev98] Pevsner, N.: Funktion und Form, die Geschichte der Bauwerke des Westens (Function and Form. The History of the Buildings of the West). Rogner & Bernhard, Hamburg (1998)

[Rei98] Reichardt, J., Drüke, K.: Bäckerei mit innovativem Gesamtkonzept (Bakery with an Innovative Concept). In: industrieBAU 6, p. 34 ff (1998)

[Stei07] Steiner, F.R.: Planning and Urban Design Standards. By American Planning Association, Wiley, Hoboken (2007)

13 Site Planning from Space View

The choice of location is the most far-reaching decision in the context of a new factory. On one hand it is based upon the company's long term global supply chain policies and the corresponding networks of customers and suppliers. On the other hand, basic decisions are to be taken about the long term development of the site.

In the following discussion it is assumed that on the strategic level decisions on the role of production as such have been successfully completed even if there were a few uncertainties. It is also assumed that during the previous planning phases, the number of possible sites were restricted to few options. Both aspects will be thoroughly discussed in more detail in Chap. 14 "Strategic Site Planning".

It is now necessary to analyze these sites at the local level in terms of space planning. Figure 13.1 summarizes the relevant design fields for all elements of the local site level from space view.

13.1 Site Development

During the industrial era, textile mills were built in the immediate vicinity of water-courses in England. Prior to the availability of steam and water power, mechanical energy was essential for the transmission drives of looms and waterways were used for distribution of goods to urban markets.

In the present context, the choice of a site is dependent on a variety of interrelated global, regional and local factors. At the global level it is determined by the exchange of goods and services at corporate level. At the local level, the strategic location of the site is extremely important within the overall logistics network—availability and capacities of various transport facilities—including truck, rail, air and sea.

In order to avoid constant interference in the movement of goods, the traffic interfaces—motorways or trunk roads—should be outside problematic traffic zones. Moreover the connecting roads and bridges should be free of inner city traffic or restrictions in terms of heights, widths or weights of vehicles. Convenient access to airports is becoming increasingly important, due to competition and shortened reaction time, especially for export-oriented companies. For rail transport, it is important to have existing railway tracks in the vicinity; one may even consider the advantages of owning a railway siding in the long run. A further advantage could be the proximity to freight distribution centers or container terminals. The regional movement of goods along rivers and canals is economically viable for bulky, heavy or cheap raw materials. The effect and development of traffic and transport system of a possible location needs to be studied over a period of 15–30 years.

H.-P. Wiendahl et al., *Handbook Factory Planning and Design*,
DOI 10.1007/978-3-662-46391-8_13,

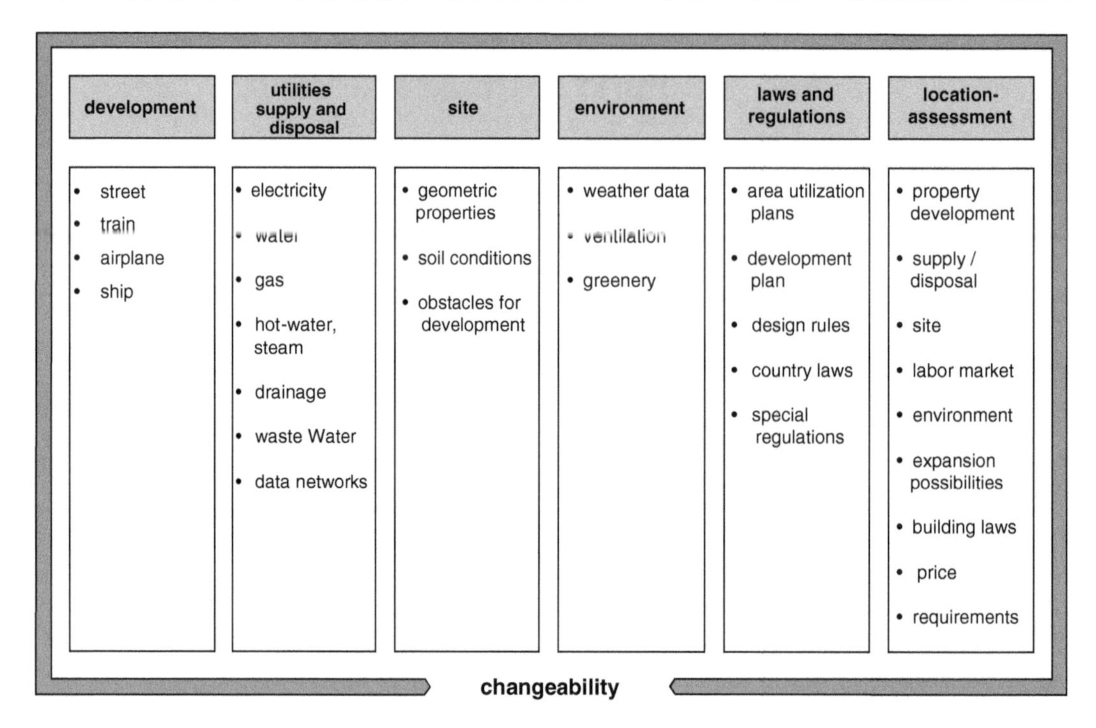

Fig. 13.1 Overview of design fields location. © Reichardt 15.233_JR_B

13.2 Utilities Supply and Disposal

A thorough inventory provides clarity on all the media existing within the premises of the property as well as the media entering and leaving the plot. During the basic data evaluation of the various utilities, the required and available quality and quantity, along with delivery reliabilities as well as current and expected future tariffs need examination in detail. Major media supplies include electricity, water, gas and telecommunications while disposals include storm water drainage and sewage systems. For existing supply and disposal systems on the property their suitability for the proposed new requirements needs to be checked. In case of revitalization of existing industrial brown-field projects, the condition of existing networks of drainage and sewers should be inspected (e.g. for leak) with the help of appropriate tests and devices before planning ahead. It is advisable to record and analyze the layout of the various existing media along with their levels using a 3D CAD model in the context of synergetic factory planning. Figure 13.2 shows excerpts of a 3D data model for the supply and disposal lines of an assembly plant. Early documentation of all existing and planned utilities and media networks of the site avoids unpleasant surprises during construction.

In order to determine the required amounts of energy and energy load profiles, an expert or a team of experts having the required knowledge of overall energy simulation of the proposed plant should attend the negotiations with local utilities authorities. Location and number of transformer stations have to be aligned with the objectives of the general development steps of the site. Long term changeability in all phases of the general development plan should be ensured through strategic expansion phases of the work, taking into account coincidence factors for the machineries. A coincidence factor is defined as the ratio of average consumption to the installed power.

These requirements should result in the required data—for example the daily average requirement of water supply or the peak consumption during the different development phases. The authorities or insurance companies at times, ask for a separate firefighting sprinkler

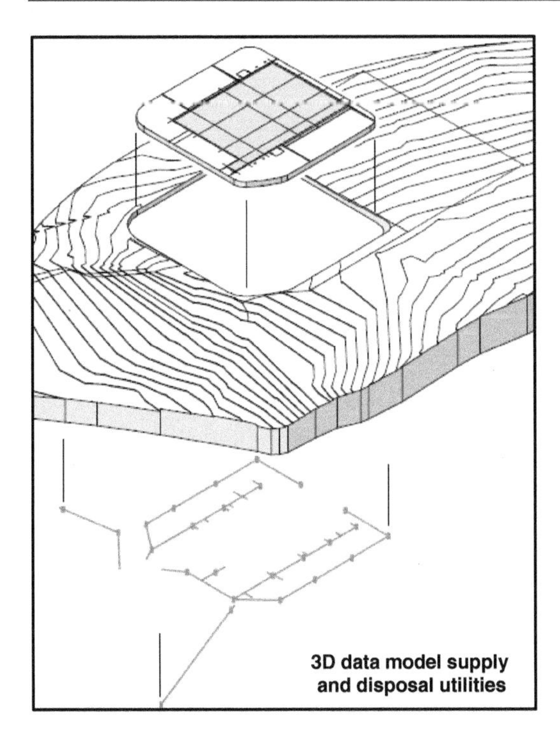

Fig. 13.2 Supply and disposal of an assembly plant. © Reichardt 15.228_JR_B

water reservoir. It is important to quickly analyze the situation and ascertain possible requirement for water pressure and other requirements for the sprinklers. Often it is sensible to examine the possibilities of trying to have own water supply through deep tube-wells thereby securing a long term source; such systems are particularly important for companies such as breweries or food manufacturers where the quality of water is of great importance. Analysis of physical and chemical properties (particularly hardness, temperature, etc.) of the water samples should be ideally a part of the agreed water supply contract.

For heating building and processes, gas is preferable to petroleum products for reasons of lower environmental impact. For smaller demands portable containers may be sufficient. Sites in the vicinity of thermal power stations for district heating can make use of hot water or steam at economical tariffs. A detailed analysis of energy demands and possible expenditure could help clarify longer term profitability when compared to owning costs.

The disposal of rain water from the roof-tops, paved roads and open spaces into the public storm water drainage system may not be possible for various reasons. It would be ideal to install rain water harvesting systems within the site to re-charge the sub-soil water tables and aquifers. A detailed geo-technical survey during the basic data evaluation phase would provide necessary clarity in terms of the existing soil conditions, location and depth of water table, etc. so that the installations for rain water harvesting could be provided at suitable locations, leaving the rest of the site free for the proposed buildings.

Liquid waste from factories are usually a mix of domestic waste, waste water from manufacturing processes and others including storm water polluted with oil and grease from the parking lots. All such liquid wastes need to be suitably treated prior to allowing the same into the public disposal system. The 3D utilities, site and building model could be a helpful tool to decide whether the waste water is able to flow out of the plot boundaries due to gravity or needs pumping due to differential levels in building, plot and the public disposal network outside. Internet and video conferencing require high-end data cabling and telecommunication networks. With the fast development of IT-technology, high speed networks are of increasing importance.

13.3 Plot

13.3.1 Geometric Properties

The size, shape and topography of the plot determine its usability. The definition of the overall space requirements at the inception of the project leads to the general arrangement of the master plan for the given site. Often questions arise as to whether it is advisable to purchase the land required only for the first phase development or for entire project including all possible extension phases. There are a number of parameters, which needs to be looked into such as current versus future land price, costs of

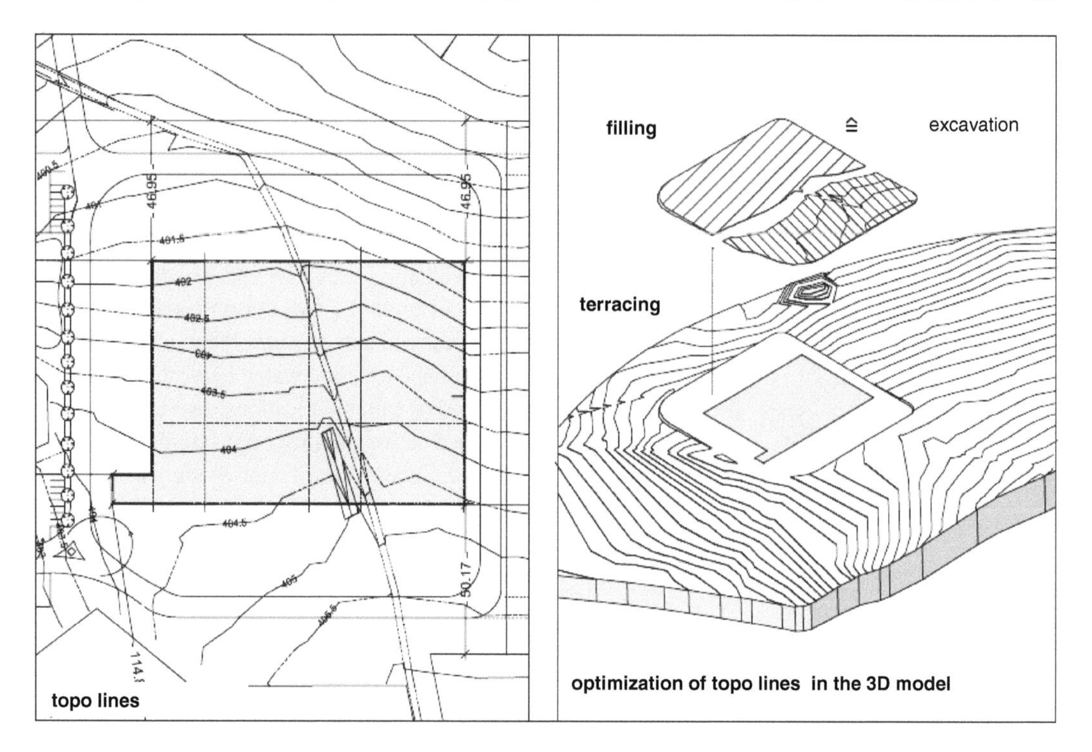

Fig. 13.3 Topography of an assembly plant location. © Reichardt 15.229_JR_B

development, capital annuities and tax, planning and sanctioning risks, maintenance costs, etc.

The ratio of length of the plot to its width, the angles at the corners and the location of entrance driveways determine the suitability of the site for the present and future processes as well as changeability or future conversion potentials. Ideally compact and square plots are suitable for a concentrated type of production while longitudinal plots are preferred for predominantly linear production processes. The location of possible driveways determines the future possibilities of development, for example regarding modular plants as "factories in the factory" with their own entrances and fire hazard escape routes.

Plots having a gentle slope up to 2–3 % are advantageous compared to more steeper slopes since the same might restrict material flow and distribution of utilities. Furthermore, it would not be advisable to level the site in terraces, since the costs and the time required for construction increases significantly. In some sloping terrains, intelligent engineering might help to connect a two-storey independent utilities block to a single storey production building at both levels without leveling of the land. In order to minimize the effects of additional costs and time it is advisable to optimize the cutting and filling of the terrain. Figure 13.3 shows an example of the optimization of excavation and filling using a 3D topography model for an assembly plant such as no external movement of soil is necessary.

13.3.2 Soil Characteristics

The various technical possibilities for building foundations are dependent on the type of soil, its physical and chemical properties, the safe load bearing capacities at various depths, depth of the ground water table and others. The water permeability of the soil determines infiltration possibilities of rain water within the boundaries of the property. A complete geo-technical analysis of the soil needs to be performed to get a basis for either planning or contracting. For a complete picture of the sub-soil strata, generally standard penetration tests are recommended over a grid of 10 m × 10 m

(32.8 × 32.8 ft) or 20 m × 20 m (65.6 × 65.6 ft) depending on the size of the plot or the scale of the constructed area. In certain cities, locations or parts of the world, the soil may contain artifacts of archeological or historical significance or even unexploded bombs. Such sites might entail detail aerial photography, studies of historical data and discussions with the local authorities. The entire process of recovery and safe disposal of the devices, artifacts, etc. could endanger the overall project and require special risk management policies for the proposed cost and time schedules.

13.3.3 Obstacles and Edifices

Existing and proposed neighborhood developments needs to be extensively studied towards possible ecological impacts of noise, smoke, fumes or special fire hazards. Public water bodies adjoining the proposed construction site may entail special earth and water retaining structures to avoid seepages or contamination. If there are special animal habitats or plantations worth protecting on the site the planning team along with the landscape experts should take into account the same while designing the master plan, the buildings and the remaining open spaces. In old mining areas, the site needs to be checked for mine shafts, to avoid un-equal foundation settlement due to inconsistent soil strata or sudden voids. Similarly existing structures either inside the plot or just adjoining the boundary wall on the other side needs to be studied, analyzed and incorporated in the master plan layout where necessary. Further existing buildings are to be examined for suitability of conversion; in most cases, demolition is the only way to guarantee a holistic—functional and aesthetic appearance of the entire factory project.

13.4 Laws and Regulations

Generally, the national, state and local planning authorities define the areas suitable for industrial development in either comprehensive developmental plans or in the relevant land-use plans along with their norms and policies from time to time. The plans, norms and policies also stipulate the extent of use of the land, zoning regulations, the footprint to surface area ratio, the total area which could be built in all the floors, maximum possible heights of all the buildings etc. Depending on the designation of the area, such as business park or industrial area, the set-back distances from the boundary wall, building heights, permissible sound level and maximum permissible limits of pollution emissions are also formulated. In many cases, questions are asked about the general compatibility of industrial and commercial areas in an urban environment. References [Kar88, Rei94a, Rei94b, Sch92] plead for positive integration of industrial and commercial areas in the urban fabric for town planning as well as environmental reasons. Existing rules and regulations of the site should not a deterrent for future changeability. Therefore, during the site evaluation process (see Sect. 13.5) all possible scenarios for future site developments must be considered. Special approval processes for pollution levels from specific governmental authorities often take longer than expected and may be decisive for the final choice of a site.

In most cases, town planning and architectural objectives shape the existing rules and regulations and design manuals of a region or locality. Sometimes, there are specific requirements with reference to roofs heights, facade materials and color or even roof-top landscaping. Figure 13.4 gives some of the legal and aesthetic guidelines for an urban development project consisting of a 400,000 m^2 (988,000 ac) business park with 1500 working places. According to [Rei97] the project was based on a public private partnership model. The comprehensive urban planning was engineered using 3D synergetic factory planning methods; the resulting 3D town planning model was continuously updated during the 7-year execution phase. The legal stipulations regulated the building volumes and landscaping through the development plan, while the aesthetic aspects of the project were clarified in advance through 3D animations. This holistic planning approach

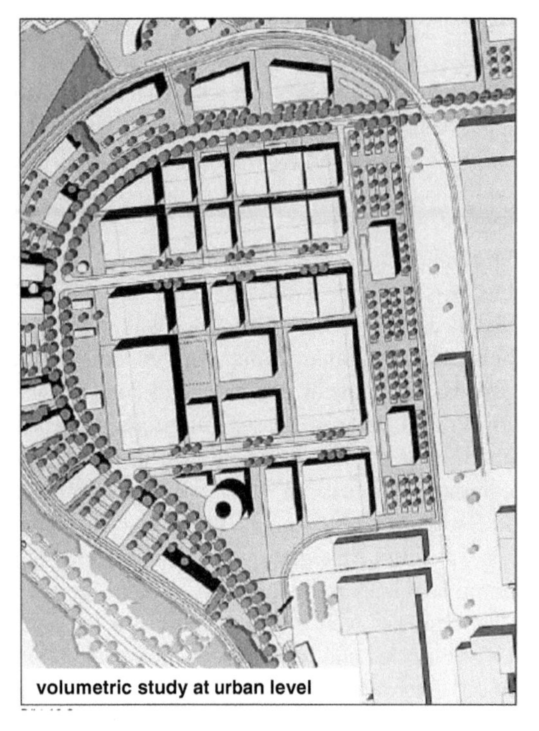

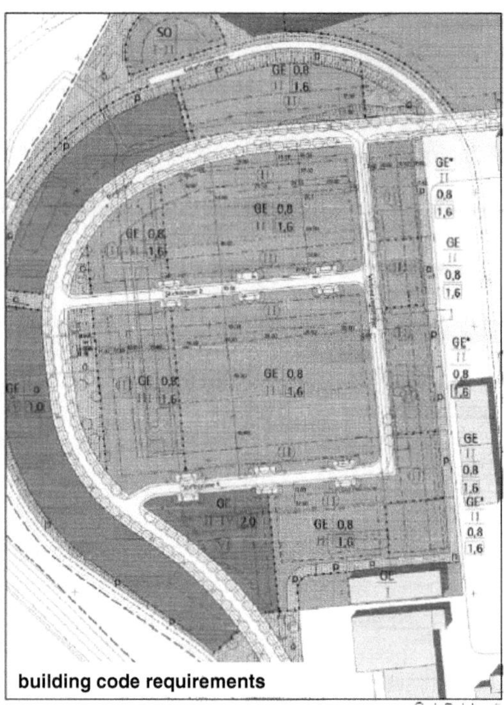

Fig. 13.4 Urban development concept of a business park. © Reichardt 15.230_JR_B

provides obligatory and transparent information for all investors including sanctioning parameters, design manual and visualizations.

Besides the aspects discussed above, the national, regional and state authorities may have a vast number of other legislations and regulations or specific local statutes depending on the place.

13.5 Site Evaluation

The various aspects of development such as supply and disposal, land, environment, expansion possibilities, construction codes, purchase price and municipal taxes and subsidies are aspects to be considered during the evaluation process of the alternative sites. Various case studies of [AGI04, Kin09] help us to identify and list the possible selection factors of location decisions. Figure 13.5 left explains a proven procedure for analyzing the "benefit value" scheme. Starting from the selection of evaluation criteria, the planning team first of all sets a weighting of the criteria, which totals up to 100 %. Then variants for each location included in the shortlist are rated from 0 (unsatisfactory) to 4 (very good) and multiplied by the weighting factor. The individual benefit values thus sum up to the total benefit value. Figure 13.5 right shows an example whose overall utility value is 3.25, which is 81 % of the maximum possible value of 4.0. As a rule of thumb an examined solution with a utility value below 80 % of the maximum possible long term value would not be competitive in the long run.

Figure 13.6 shows an example of the site study of the relocation of a factory dealing with rubber products. The inner city factory plot did not offer sufficient development opportunities for future expansion of the factory. Twelve locations were analyzed and evaluated within a distance of 20 km from the existing location before recommending a particular site. In the left part of Fig. 13.6 the 12 optional sites are listed with their benchmarks and top right are the rating criteria. The sites were further analyzed using so called selection of evaluation criteria → weighting of evaluation criteria → assessment of locations → comparison and choice

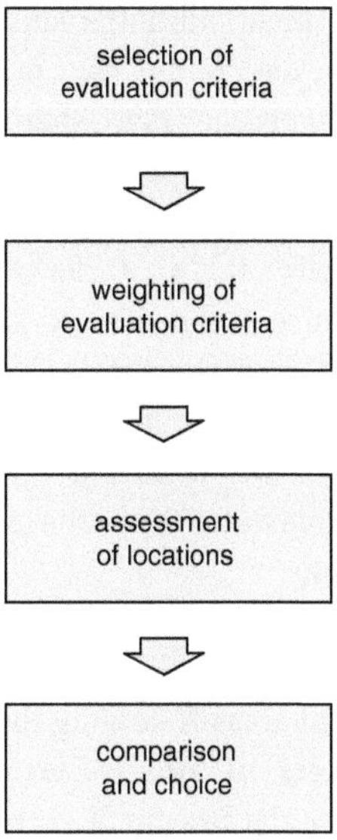

evaluation criterion	weight	assessment	value
• accessibility for the staff	20	3	0.6
• expandability of the property area	15	4	0.6
• purchase price per sqm,	15	2	0.3
• traffic connections for transportation	15	4	0.6
• time for clarification, purchase and regulations	10	3	0.3
• residual wastes or contamination	10	2	0.2
• soil condition	10	2	0.2
• environmentally attractive location	5	3	0.15
sum	100		3.25

assessment : 0 unsatisfactory 1 just still acceptable 2 sufficing 3 good 4 very good (ideal)

assessment steps **benefit calculation**

Fig. 13.5 Site assessment (example). © IFA15.231SW_B

assessment:
0 unsatisfactory 2 sufficing 4 very good (ideal)
1 just still portable 3 good

						1	2	3	4	5	6	7	8	9	10
						k.o.	selection criteria								
						suitability of space division for PTT 2005 = 250 m x 270 m	supply with gas/steam	accessibility for staff (local public transport)	extendibility of the total area	price per square meter (incl. incentives)	traffic connection for transportation	estimate time to clarify purchase planning and regulations	existing residual wastes nad contamination	building ground composition	location activity/environment
						yes/no	yes/no	20	15	15	15	10	10	10	5
location								20%	15%	15%	15%	10%	10%	10%	5%
Address	size in m2	expansion-area in m2	price per m2	price in euro	assessm. location	assessment according to assessment scale									
KH, Wenzendorf, A1 Rade	61000	139,000	77	4697,000	3.25	yes	yes/no	2	4	2	4	4	4	4	3
KH, NW-Schleppelsb. A1 Rade	61000	233,900	77	4697,000	3.10	yes	yes/no	2	4	2	4	3	4	3	4
KH, Beckedorf, A7 Marmsdorf	61000	309,000	77	4697,000	2.95	yes	yes/no	3	4	2	4	3	2	2	3
KH, Mienenbütte, A1 Rade	61000	839,000	77	4697,000	2.95	yes	yes/no	2	4	2	4	1	4	4	3
KH, Tangendorf, A7 Thieshope	61000	363,900	65	3965,000	2.80	yes	yes/no	1	4	3	4	0	4	4	3
KH, Allermöhe, Rungedamm,	61000	4,500	100	6100,000	2.35	yes	yes/no	2	0	1	4	4	4	3	2
KH, Nöldekestr. Bahn	26400	?	35	4880,000	2.35	no	yes/no	4	0	4	2	1	2	3	1
KH, Schlachthofstr. Bahn	61000	9,000	80		2.3	no	yes/no	4	0	2	4	1	2	2	2
KH, Hausbruch, FHH	61000	5,300	100	6100,000	2.25	no	yes/no	2	0	1	4	3	4	3	2
KH, Obergeorgsw. Dyke, FHH	61000	339,000	95	5795,000	1.70	yes	?	0	4	1	1	0	4	3	2
KH, Reiherstieg, DEA -east	50591	86,250	94	4755,554	1.55	no	yes/no	1	3	1	2	1	2	1	1
KH, Reiherstieg, DEA west	61000	75,841	94	5734,000	1.25	yes	yes/no	1	3	1	2	0	0	1	1

Fig. 13.6 Results of a site assessment (example). © IFA 15.232SW_B

"knock out" criteria which could be either non compliance or immediate disqualification. Following further the explained scheme according Fig. 13.5 left, the summing up of the benefit values of the individual locations is given in the lower right side of the table. The sorting by descending values results for the location "Wenzendorf" in a benefit value of 3.25, corresponding to 81 % of the possible maximum value. Crucial to the first rank were in this case, among other in particular the ground figure of the plot and the future extension possibilities.

13.6 Environmental Aspects

Heat gains and losses of the proposed factory building should be evaluated as part of a comprehensive evaluation process by means of energy simulation from the very beginning of the planning stages. In particular, opportunities for power generation using renewable sources like sun, wind, water or geothermal energy should be explored based upon the climate conditions of the specific site, e.g. temperature profiles, sunshine hours, rainfall or prevailing wind directions, etc. An intelligent architectural solution, in most cases, would generate a building typology related to the specific location optimizing construction, building shell and required utilities. Past meteorological data is very useful when finalizing energy performance based preliminary design concepts. As a rule of thumb, different climatic zones require different spatial concepts, building volumes, orientation strategies, generation and use of alternative energies, etc.

Air movement is important for fresh air supply and should be allowed to move freely between the buildings. Construction projects result in conversion of natural landforms into hard surfaces and building volumes—structures of various sizes and shapes. In most cases, such interventions increase the overall air temperature, result in turbulent air-flow, driving rain or unseasonal rainfall or other such changes to the micro-climate. It is advisable that new industrial areas align the position of their buildings such that the prevailing wind directions are not blocked and everybody gets a fair share. Attempts should be made to compensate for built-up areas with a generous amount of green and open spaces, boulevards, etc. in order to maintain the ecological balance of the site and its surroundings. It is an established fact that green cover and planting of trees contribute significantly to the improvement of the micro-climatic and enables infiltration of rainwater for recharging of the ground water table.

In many cases there are opportunities for new industrial and commercial areas to be planned as an extension to existing green areas following the landscape planning principles of Anglo-Saxon industrial parks. Planning guidelines for a comprehensive 2,000,000 m^2 (4942,000 ac) technology park, developed around a green center, are shown in [Mat94]. The shrubs, green spaces and the spacious park landscapes provided much recreation spaces for the employees and the nearby residents. Even during the planning of the buildings, every attempt should be made to enhance the image and brand value of the Company through green and sustainable technologies for a natural and livelier environment. Car parking lots should be generally designed with trees while main access roads could be lined with trees on either one side or both the sides. Further, roof and facade landscaping could contribute significantly to a better quality of ecological environment.

13.7 Summary

The choice of a location is the most important decision in the context of any new development of factories. The long-term aspects of development, supply and disposal of media, utilities, rules, regulations and increasing ecological and environmental concerns influence a company's present and future changeability. The multitude of relevant parameters requires a thorough and systematic assessment of possible options.

The discussions on site and master planning have shown that a synergetic approach to problem solving should help in resolving most of the developmental capabilities of the chosen site in the long term at the very beginning. These aspects are continued in Chap. 14, strategic site development.

Bibliography

[AG104] Arbeitsgemeinschaft Industriebau e.V.: Standortplanung im Industriebau—Ein Leitfaden für Architekten, Ingenieure und Unternehmen (Working Group Industrial Construction: Site Planning in Industrial Buildings—A Guide for Architects, Engineers and Companies). Callwey, Munich (2004)

[Kar88] Karsten, G., Reichardt, J.: Flächensparendes Bauen für Industrie und Gewerbe in Berlin. (Space Saving Construction for Industry in Berlin). Report commissioned to the Berlin Senate in 1988

[Kin09] Kinkel, S.: Erfolgsfaktor Standortplanung (Success Factor Location Planning), 2nd edn. Springer, Heidelberg (2009)

[Mat94] Matzig, G.: Neue Mitte—Technology Park Nürnberg—Fürth—Erlangen (New Center Technology Park Nürnberg—Fürth—Erlangen). City Bauwelt, **36**, S. 1912 f (1994)

[Rei94a] Reichardt, J.: Multi-story industrial buildings—new chances for the city development? In: Lecture Manuscript International Congress—New Technology in Town, p. 36 f. Rotterdam (1994a)

[Rei94b] Reichardt, J.: Arbeiten in der Stadt—neue Lebensräume. Chancen für verdichtete Stadtstrukturen (Work in the City—New Habitat Opportunities for Compacted Urban Structures). DAB, **9**, S. 1328 ff (1994b)

[Rei97] Reichardt, J.: Entwicklung des Gewerbegebietes «M1». (Development of the industrial area "M1"). BAUKULTUR, **2**, S. 36 ff (1997)

[Sch92] Schulitz, H.C.: Industrie contra Städtebau (Industry versus urban development). In: Constructec Price 1990 Industrial Architecture in Europe, pp. 8–41. Ernst and Sohn, Berlin (1992)

14 Strategic Location Planning

14.1 Introduction

Searching for a location from a strategic perspective is not a primary responsibility of the factory planning, but rather of the enterprise's management. Nevertheless, since knowledge about the underlying key assumptions strongly impacts the later features of the factory—in particular the required degree of changeability—they are highly significant for the factory planner when developing an existing or new location. The factory planner should also be familiar with how such strategic decisions are made so that they can estimate the certainty or uncertainty of some of the assumptions especially in regard to the future product spectrum. With the following remarks we can only illustrate the framework and basic steps involved in the search for a strategic location, those wishing more in-depth information are referred to e.g., [Au07, Kin09, Abe08, Heg08].

14.2 Location Planning Triggers

Our discussion about change drivers in Chap. 1 clearly demonstrates that a plethora of external and internal change drivers impact a factory. Not all of them are reason to question the location. Usually, only the simultaneous and sometimes almost imperceptible appearance of several factors triggers an insidious drop in corporate earnings and growing unease in management. This in turn can lead to the realization that the current location perhaps no longer ensures continued success in economic, ecologic and social terms.

Figure 14.1 depicts internal and external drivers that can bring a location as a whole into question. The external influences from technology, market and environment tend to represent constant seeds of change. Only when obvious weaknesses in the logistic performance and cost situation combine with fluctuating quality and ongoing capacity constraints does management begin to recognize that continually improving individual weak points is no longer sufficient. It then becomes clear that a comprehensive approach based on a long term competitive strategy for the production is required. This can mean focusing on one of the targets named in the diagram (cost, quality, technological, logistics or flexibility leadership) or a combination of them. Then at the latest management will have to deliberate the international orientation of the production. Considerations here can come from two key perspectives: internal competitive strategies stemming from strategic impulses or motives for internationalization. Motives for internalization may include significantly reducing costs, opening-up a new sales market, the pressure to follow one or more customers to their international location or the rather seldom motivation of opening-up specific know-how found in another country.

In Fig. 14.2 we can see that these two perspectives have to fit together strategically [Kin09]. Internationalization is not very useful when for example personnel costs represent only

DOI 10.1007/978-3-662-46391-8_14, © Springer-Verlag Berlin Heidelberg 2015

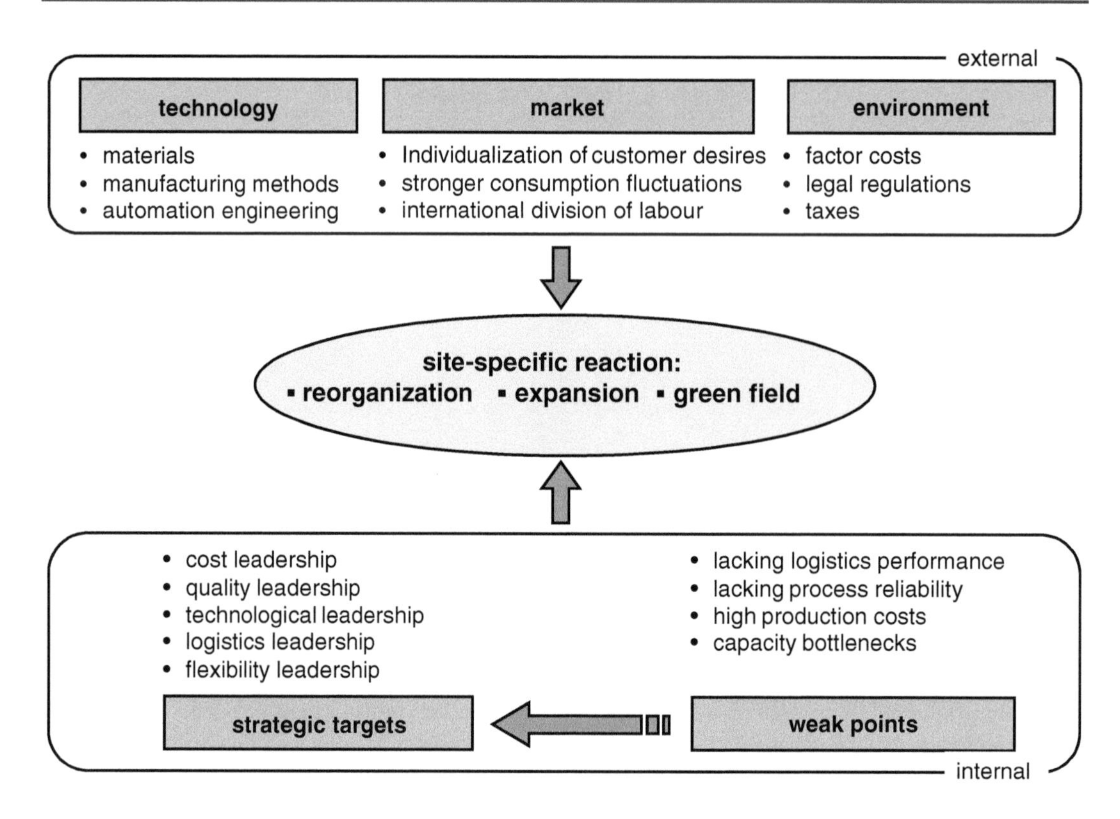

Fig. 14.1 Change drivers and alternative responses on the location level. © IFA 15.252_B

	motives for internationalization			
competitive strategies	costs reduction 1)	opening up markets	following customer	opening up technology / know-how
cost leadership	●	●	●	●
quality leadership	●	●	●	●
technological leadership	●	●	●	●
logistics leadership	●	●	●	●
flexibility leadership	●	●	●	●

● in principle, motive matches the strategy ● motive does not match the strategy

1) also contains raw material and energy costs besides labor costs as well as cost-intensive legal regulations

Fig. 14.2 Strategic fit between motives for internationalization and competitive strategies (to Kinkel et al.). © IFA 15.253_B

20 % of a product's manufacturing costs and the ratio of work costs at a German location to those at an international location is 5:1. In this case, assuming the same material costs, savings would equal 16 % of the manufacturing costs. Additional costs for logistics, start-up and management could however easily erode these savings.

In comparison, regardless of the competitive strategy, opening-up a new market is always agreeable. One the other hand, following one or more big customers to their sales markets is not sustainable from a pure cost perspective. Finally, opening-up new technological know-how only supports strategies focused on quality, technology and flexibility leadership [Kin09].

Depending on the depth and breadth of the analysis, the findings may show that reorganizing the current site (e.g., transforming it into a lean production) suffices. In other cases, needs can be met by expanding locally, possibly combined with consolidating a number of locations. Finally, it might be necessary to construct a new site, either while relinquishing the current location or as an additional location linked to the current site by sharing functions and capacities.

It is critical that this decision is not made impulsively nor due to a single factor (e.g., pressure around costs, proximity to a big customer), but rather is made following a thorough strategic planning. Figure 1.6 in Chap. 1 already presented the argument that failure to pay attention to this rule results in a quarter of all internationalizations being unsuccessful. Moreover, enterprises are then faced with moving the production back to the original country, entailing considerable losses. It is frequently said that such academic exercises take too much time and that prompt action is instead required, however, this ignores the complexity of the task and the thus unreflected risks related to it.

14.3 Suitability of the Current Structure

Before the actual search for a location is initiated, it needs to be ensured that the current and new locations are comparable in regard to the factors critical for success. In accordance with [Jun04] the factors critical for success (which we will refer to here as 'performance factors') are:

- productivity/production costs,
- throughput times,
- capacity for innovation,
- product and production flexibility as well as
- product and process quality.

Obviously it is not appropriate when a location that has grown 'organically' and has not been thoroughly modernized in years is compared with a new location that has been optimized according to the state of the art. Thus, first the latent potentials in the existing production should be explored and evaluated in regard to their possible impact on the five abovementioned performance factors. There are numerous approaches to accomplishing this (e.g., Lean Production, Total Quality Management and Business Reengineering) and are frequently combined together to form an integrated production system.

However, since our concern here is more about estimating the potential and not concretely implementing it, it is recommendable to focus on four fields of modernization, namely technology, organization, personnel and product [Jun04]. Figure 14.3 illustrates the general procedure for doing this. Through a structured analysis of the measures already implemented in these four fields and the search for further improvements, the potential optimization that has already been used as well as that which can still be exploited in regards to the five performance factors can be estimated.

Possibilities for modernizing these four fields are summarized in Fig. 14.4 [Jun04]. In terms of production technology the main focus is on maintaining highly stable processes in the face of internal sources of disruption and flexible automating complete value adding chains. The previously described flexibility of the product and process variants is also prioritized. Ultimately, the aim here is to permanently increase the overall performance.

Based on the example of optimizing the production's performance, Fig. 14.5 provides a detailed overview of how to analyze such

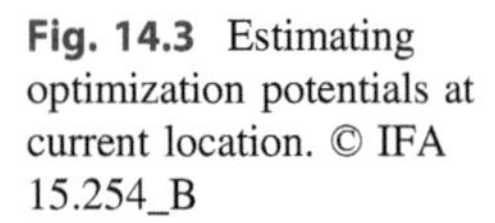

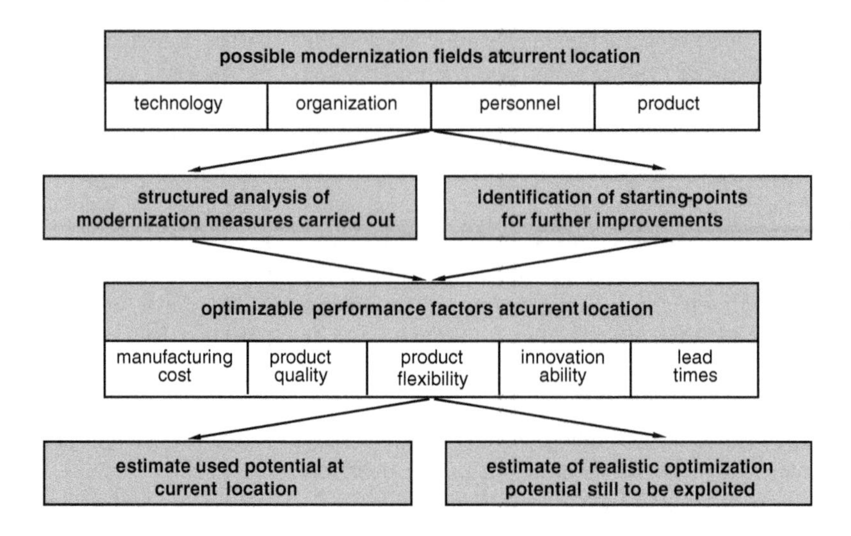

Fig. 14.3 Estimating optimization potentials at current location. © IFA 15.254_B

Fig. 14.4 Potentials for modernizing an existing production. © IFA 15.255_B

modernization fields			
technology	organization	personnel	product
• stability • automation • flexibilization • performance optimization	• organization hierarchy • business processes • vertical integration	• work structures • training • leadership • wage systems	• standardization • modularization • market customization • new products

measures. The measures listed are characterized by the implementation of new machine tool concepts and manufacturing processes which can impact the five performance factors positively, negatively or ambivalently. In this case, the primary focus is on increasing productivity. Since the throughput time represents a relatively small share of the value adding, less focus tends to be placed on influencing it. Although the flexibility is negatively influenced due to the comparably high setup times, the quality can be positively or negatively influenced depending on how controlled the processes are.

Returning to Fig. 14.4, the key possibilities for modernizing the organization of an enterprise are found in hierarchical organizational structures and workflows. It is precisely here that Lean Production strategies come into play, ranging in form from manufacturing cells and segments up to largely independent business units. In view of creating superior products, not only the production but also the product development should be included here e.g., by having development teams working closely with production. Determining the vertical integration and spatial distribution of the value adding poses one of the key starting points for the entire search for a location and is extensively discussed in Sect. 14.4. In general it comes down to ensuring the supply is close to the market while finding the right balance between costs, quality and logistics.

In terms of modernizing personnel performance, new work structures with greater task integration and on-going skill development provide valuable approaches to improvements. Here the aim is to increase employees' understanding of the necessity for continual change, so that it is no longer considered a threat, but rather seen as a

measure	examples	performance factors				
		manufacturing costs/ productivity	lead times	innovation-ability	product-/ production-flexibility	product-/ process-quality
performance optimization	• high speed processing • hard machining • dry processing • machine tools with linear drives • computer integrated manufacturing (CIM) • manufacturing methods for new materials	+	(+)		-	±

+ improvement is in the key focus of the measure
(+) improvement as a side-effect of the measure possible
± ambivalent effect: both improvements and declines are possible
- negative side effect is possible or probable

Fig. 14.5 Impact of an improvement measure (Example: Optimizing performance in the modernization field 'Technology'). acc. to Jung [Jun04] © IFA 15.256_B

challenge and an opportunity. The general approach is characterized by increasing the flexibility of work hours and allocating tasks.

Finally, the potential of integrating standardization, modularization and platform concepts into the product design should be examined. Mastering market-specific product adaptations (e.g., in regards to country specific acceptability and safety standards) and their impact on the production and logistics offers a variety of potential with corresponding repercussions for everything from product structure to packaging design. Developing new products as well as further developing existing products is one of the permanent challenges for an enterprise. Knowledge about the measures intended for this is crucial to the search for a location.

In some case, the impact of selected measures on the performance factors should be estimated not only qualitatively but also quantitatively. According to research conducted by Fraunhofer ISI, segmenting the production, introducing the pull principle, implementing simultaneous engineering as well as task integration and group work in the production have the greatest impact [Jun04, p. 149]. Preliminary studies such as these focus forces on using potential instead of rushing more or less unprepared into a risky relocation abroad.

14.4 Location Factors

There are numerous recommendations for assessing locations e.g., [Au07, Paw08, Abe08, Eve96]. Usually they are differentiated according to global, regional and local aspects. In this chapter, we will only focus on the global, regional and local strategic factors, the local factors will be examined from a spatial perspective later in Chap. 13.

Traditionally, global location assessments focus on production and market factors. According to research conducted by Kinkel though, these should be extended to include the five performance factors introduced in Fig. 14.4. Furthermore, an additional category that arises is the network requirements—in many cases the related expenditures and benefits are not considered sufficiently [Kin09].

Figure 14.6 provides an overview of the resulting location factors with each of the categories further sub-divided into quantitative and qualitative factors. All of the factors should be understood as generic terms for a bundle of parameters which can be further differentiated. Due to the long term binding nature of the decision, the selected time horizon is an important condition and should not be less than 5 years and in special cases 10 years and more. Frequently, one tries to estimate the development of the quantitative factors during this time period.

Starting with the *production factors*, the quantitative parameters include the production factor costs in particular the costs and availability of personnel, followed by capital investments, materials (supplier), energy as well as property and other taxes or fees. Laws, procedures and macro-economic indicators also belong to this list. Qualitative factors are mainly concerned with the quality and long term stability of production factors and infrastructure, but also take into consideration political stability as well as geo-specific and socio-culture factors.

The *market perspective* is primarily characterized by the sales potential and trade hindrances such as the mandatory share of local content or duties. The more qualitative factors include the long term attractiveness of the market and the current/future behavior of competitors.

Performance factors should be assessed in dependence of the pursued production strategy and can be actively influenced by the enterprise.

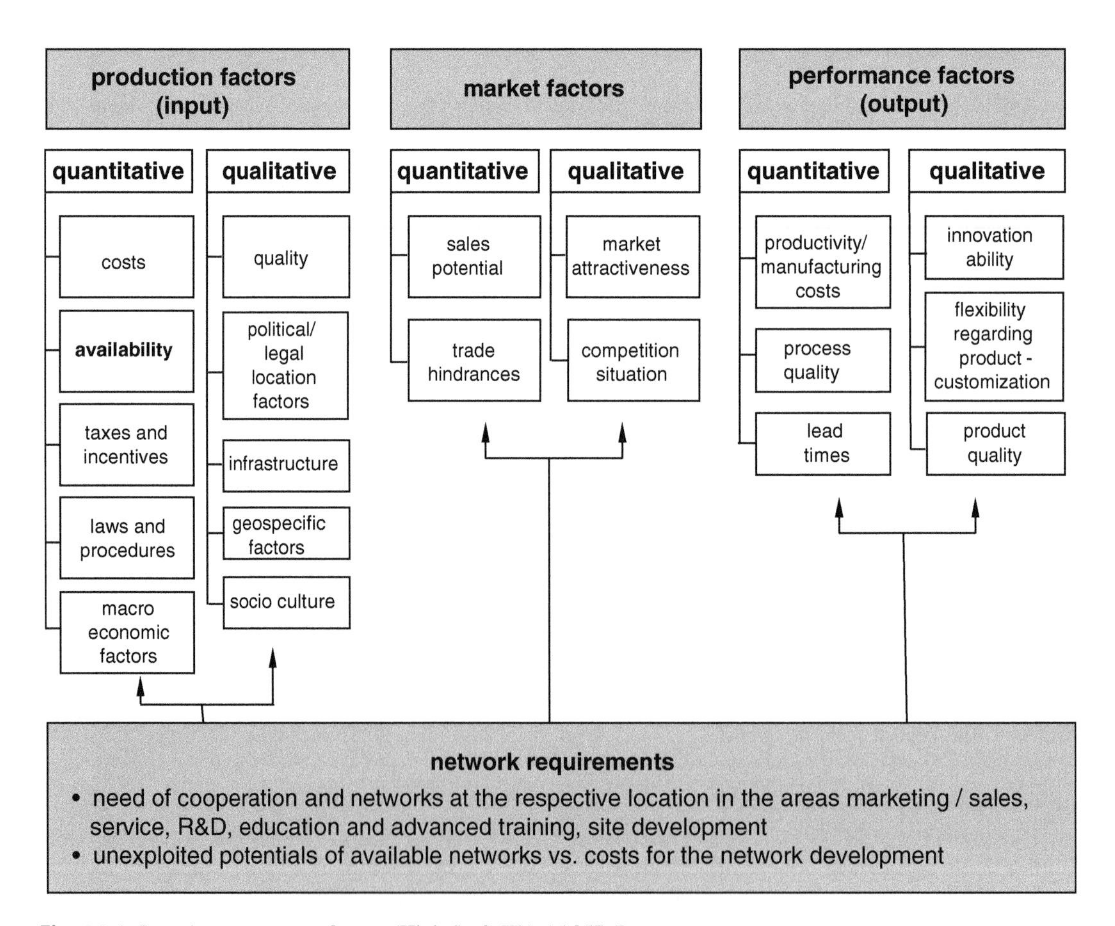

Fig. 14.6 Location assessment factors (Kinkel). © IFA 15.257_B

Thus if the goal is cost leadership, the productivity and process stability ('process quality') are decisive for success, whereas if the goal is high quality products, the product quality, the ability to innovate, and modifying products to meet customers desires are first and foremost. If products are similar to those of rival companies then delivery times and delivery reliability are the keys to distinguishing the enterprise from competitors; in this case a high logistic performance is decisive.

Relatively new factors that need to be considered are those that result from *networking*. The key here is not only the networking of the own production with the other locations, but also the smooth cooperation with suppliers, the domestic marketing and sales, equipment services (IT, maintenance), research and development for adjusting products to local requirements, education and advanced training of existing and locally hired staff as well as development of the location outside of the actual enterprise. Thus there is new potential to be exploited but, among other things, considerable costs are to be expected in order to develop the network.

The assessment itself represents a learning process, which regardless of the general basic steps has to be conducted anew every time by a core team that includes management. Whereas initially positive arguments may be in the foreground of discussions, examining the situation more thoroughly will reveal negative aspects as well.

Depending on the internationalization strategy being pursued, the bundles of factors outlined in Fig. 14.6 have to be further defined and made more concrete (see Fig. 14.7). Although the factors were obtained in the context of a planned internationalization, they are in principle also applicable for the location within the own country, since there can also be appreciable differences within a country in regards to the individual factors.

Based on a business survey conducted by Kinkel, Fig. 14.7 lists the most important factors when developing a new sales market and reducing costs. We will only discuss these briefly here; for a more extensive explanation please see [Kin09].

The most critical aspect when aiming to open new markets is realistically estimating the market potential, followed by analyzing the local competition. Moreover, vital prerequisites include developing a thorough understanding of the market and accessing distribution channels. These in turn allow realistic target prices and margins to be set. Furthermore, it may also be necessary to modify products and/or provide local supports and services in order to access a new market. Obstacles in the form of trade barriers or questions of product liability also need to be considered. One last consideration is monetary benefits in regards to procurement: Since both purchases and sales are made in the same currency, fluctuations in currencies can be compensated for at least be partially.

The management consultancy A.T. Kearney offers a clear presentation of these relationships (see Fig. 14.8) whereby, the relevant product families are examined in view of their positioning in a product portfolio, formed by the product maturity and market complexity [Sch11].

With *product maturity* the aim is to determine where the product is in its lifecycle based on the five criteria mentioned in the lower part of the diagram and whether or not the product is easy to produce remotely. The more mature a product is, the easier it is to produce remotely from a production perspective. Based on the seven criteria listed on the left side of the diagram, *market complexity* describes how closely the product design is tied to customers and which local conditions might be restrictive. The more complex a market is, the more important the proximity to customers is.

This results in four types of plants, each of which is labelled in a way that clearly describes the location. The global lead plant manufactures products that require a strong production competence developed locally and with close proximity to the product research and development. A regional lead plant is located close to the customer, but has developed their own production competence and has a specific R&D capability. The local server plant is also close to the

internationalization strategy

opening-up new markets

1. realistic market potential
2. concentration (number), market power, technology standard (lead or delay) and sunk costs (signal for strategic competition) of the local competitors
3. effort for the development of adequate market knowledge
4. access to well-adjusted distribution channels and networks
5. target prices and margins
6. customization need of products to the market situation
7. need for application advice and service on the spot
8. product liability
9. codified (primarily customs, local content) and not codified trade barriers (e.g. attitude towards "German" products
10. currency advantages on the supply side

cost reduction

all relevant cost types and drivers involved in a total costs analysis:

step 1: direct costs of the production in a broader sense

- wage and salary costs, material - and advance cost, freight (semi finished and final products)
- future development of wage costs and prices at the site

step 2: transition to unit costs

- productivity standard at the location
- availability and fluctuation of workers
- ramp up time and cost for quality and productivity

step 3: consideration of overhead costs

- overhead costs arising at the parent company's headquarters
- training costs

step 4: consideration of

- costs for adapting technology to the qualification standard
- costs for developing local network
- "soft factors" at parent company's headquarters ; confidence, motivation, conflict

not primarily decisive:

subsidies, charges and taxes

Fig. 14.7 Factors critical for success when developing new markets and reducing costs (acc. to Kinkel [Kin09]). © IFA 15.258_B

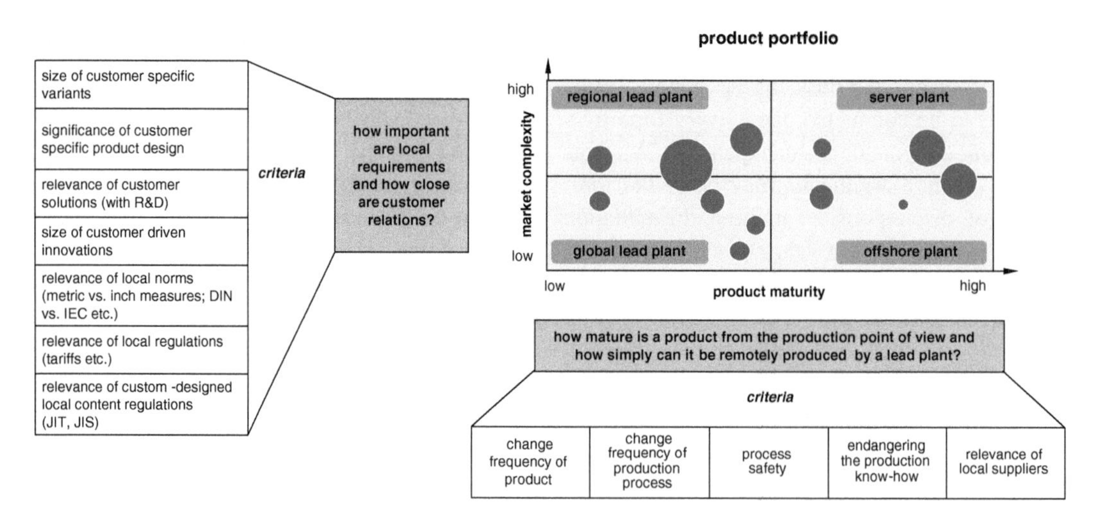

Fig. 14.8 Product portfolio with preferred factory types (B. Schmidt, A.T. Kearney). © IFA 15.259_B

customer, however does not require any specific production competence, while the offshore plant is oriented on standard products which exploit local cost advantages.

The cost reduction strategy requires all of the costs to be taken into account across the direct costs, unit costs and overhead costs as well as so-called sunk costs. The latter refers to costs that have already been incurred at the established location (e.g., manufacturing equipment, training, developing local suppliers and service providers) but are incurred once more at a new location. Moreover, a cost reduction strategy requires for example, that transportation costs be analyzed to demonstrate how far a product can be economically transported in consideration of the labor cost advantages found at the international location.

Figure 14.9 exemplarily depicts the maximum economic transport radius curves for two products (PA and PB) for sea, road and air freight routes [Sch11]. In this case, the transport costs for the affected customer's products do not pose any obstacle for a so-called 'global footprint design'. In accordance with research conducted by Schuh a global footprint design refers to concepts developed according to the long term desired spatial distribution of production capacities, competences, processes and resources, taking into account strategic and economic criteria [Schu08].

The factors for the other two internationalization strategies, "following a customer" and "technology", are summarized together in Fig. 14.10.

Many enterprises in the supply industry are confronted with the demand of one or more customers to follow them to an international location in order to create part of the procurement net there. The location should suffice the usual standards in regard to quality and delivery reliability and increase the share of local added value. The ten most important factors for the "following a customer" strategy are named on the left side of Fig. 14.10. Of these factors, the most critical are:

- the share of sales the customer represents in comparison to the rest of customers,
- the certainty of the purchased quantity as well as the support provided by them during the ramp-up.
- Demands for certification and local content can offer advantages over domestic competitors, just as proximity to the end customer e.g., in further developing the product.
- The new location's potential for development can also create the opportunity to expand sales, whereas the costs of the new plant, the inclination of employees towards changing jobs and the costs related to coordination and quality assurance have to be considered risks.

Seriously weighing all of the advantages and disadvantages including the risks may delay the decision to internationalize or even completely negate it.

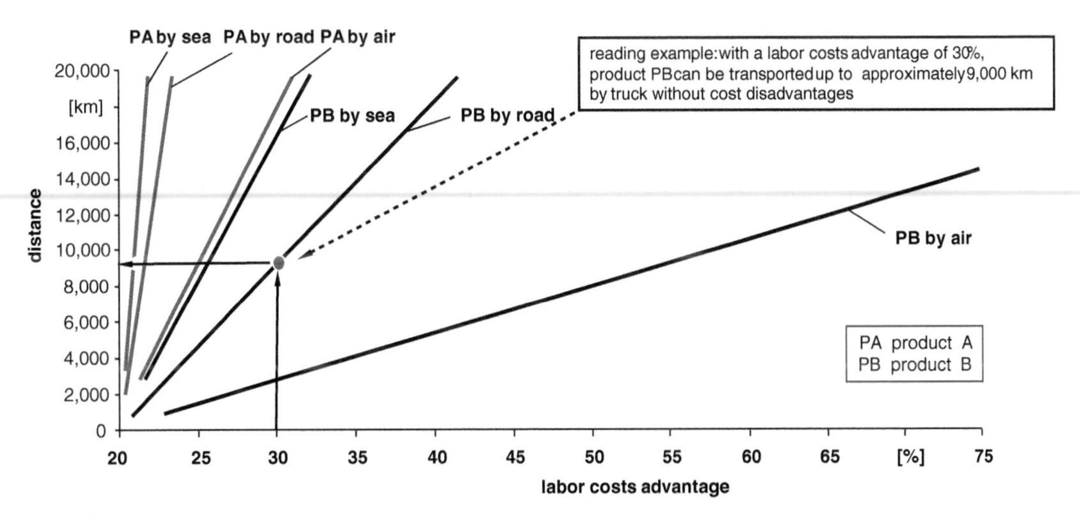

Fig. 14.9 Maximum economic transportation radius (example) (per B. Schmidt, A.T. Kearney). © IFA 15.260_B

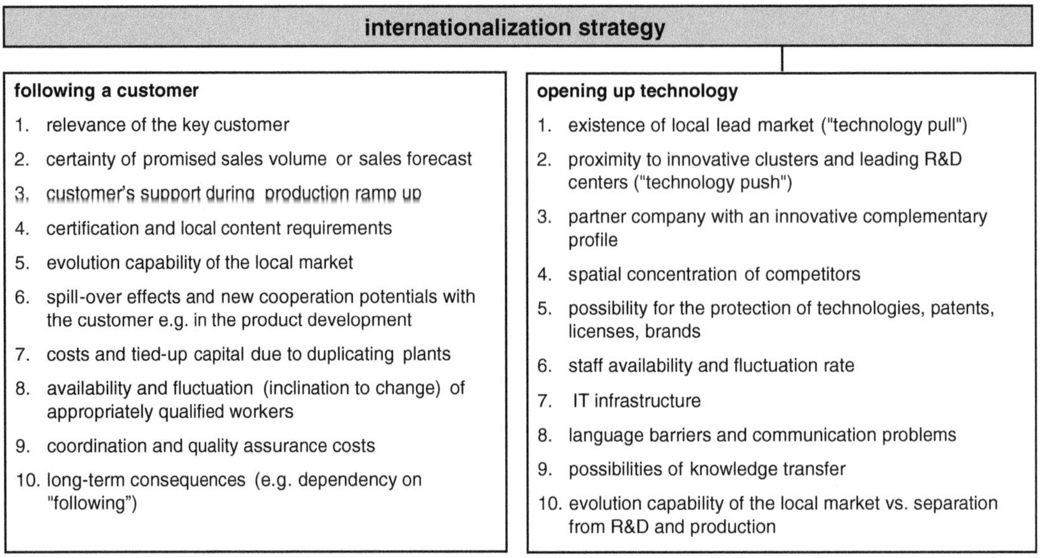

internationalization strategy	
following a customer	**opening up technology**
1. relevance of the key customer	1. existence of local lead market ("technology pull")
2. certainty of promised sales volume or sales forecast	2. proximity to innovative clusters and leading R&D centers ("technology push")
3. customer's support during production ramp up	3. partner company with an innovative complementary profile
4. certification and local content requirements	4. spatial concentration of competitors
5. evolution capability of the local market	5. possibility for the protection of technologies, patents, licenses, brands
6. spill-over effects and new cooperation potentials with the customer e.g. in the product development	6. staff availability and fluctuation rate
7. costs and tied-up capital due to duplicating plants	7. IT infrastructure
8. availability and fluctuation (inclination to change) of appropriately qualified workers	8. language barriers and communication problems
9. coordination and quality assurance costs	9. possibilities of knowledge transfer
10. long-term consequences (e.g. dependency on "following")	10. evolution capability of the local market vs. separation from R&D and production

Fig. 14.10 Factors critical for success with strategies: following a customer and opening-up technology. acc. to Kinkel [Kin09] © IFA 15.261_B

The success factors for the "opening-up technology" strategy (Fig. 14.10, right) start with

- the need for 'lead markets', which are characterized by customers with high demands and conditions requiring innovation e.g., stringent environmental constraints.
- The proximity to leading research institutions, the availability of innovative cooperation partners and the spatial concentration of competitors creates a stimulating atmosphere. Critical factors which need to be considered include: the possible loss of know-how, recruiting staff and the danger of staff leaving, language and cultural barriers and possibly unsatisfactory knowledge transfer to the original location.
- Finally separating the R&D activities from the production can also be disadvantageous. This only applies though when the entire product is being constructed at the remote location not just prototypes.

In summary it can be said that in individual cases it is also possible to combine strategies; opening-up a new market together with reducing costs is the most frequent grouping. Furthermore, it should be taken into account that, for a number of reasons, sometimes only specific components or assembly units are relocated and not always entire products. This strategy combines manufacturing core components that require intensive know-how at the original location with the cost-effective production of labor-intensive components and their completion close to the market. This is examined in more detail in Sect. 14.5 under the heading production stages.

14.5 Procedure for Selecting a Location

The procedure for selecting a location is primarily determined by the production strategy as well as the reason for and extent of the envisaged relocation. The following could be considered typical cases [Mey06]:

- There is a widely defined product and process scope which is relatively minimally networked with other locations and suppliers. This situation is typical of mid-sized enterprises with few locations, customers and suppliers. Generally the motivation for finding a location is reducing costs combined with opening-up a market.

- The enterprise (usually consolidated companies) is active in a large number of business areas, which tend not to overlap as far as the target markets and products are concerned. The focus then is assessing the portfolio areas in which capital returns and operating expenses could be improved the most by a new international location.
- The most complex case is enterprises which offer serial products with many product levels in the automobile and mechanical engineering industries, which operate factories within a globally distributed production network and are directed at a specific purpose s e.g., manufacturing components or final assembly. Here not only do logistic interactions between plants play a significant role, but also the suppliers' flow of goods due to the large number of suppliers. There are frequently repercussions for the selection of technology selection and construction of products. The focus in this situation is on continually adapting the entire production network in view of the total costs.

Despite the differences in these scenarios, it is still possible to recommend a general procedural model which, through a series of gradual restrictions, leads from a large number of possible locations at a national level to a decision for a local site (Fig. 14.11) [Mey06]. We will only discuss it briefly here but a more detailed description can be found in [Abe06].

The process, developed by the consulting firm McKinsey, begins with a pre-selection of countries in which specific products and manufacturing steps seem to be practically producible. Which countries appear to be attractive depends on the previously developed production strategy. The factors presented in Figs. 14.7 and 14.10 as being critical to succeeding with the four typical strategies offer valuable support here. In the simplest case, this pre-selection is made by experienced decision makers, possibly supported by external consultants.

The second step involves determining a not yet more specifically localized site on a country level and integrating it into the market that is to be supplied, the in-house production network and the supplier network. The so-called 'total landed costs' serve as important criterion. The total landed costs are understood as the sum of

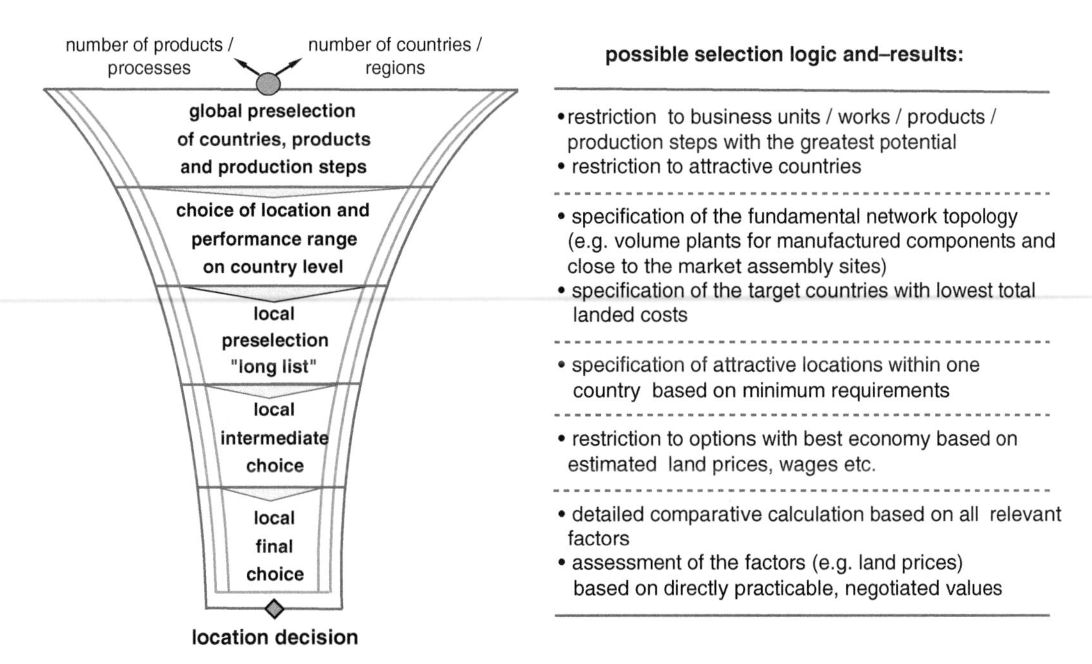

Fig. 14.11 Procedure for selecting a location (acc. to Meyer). © IFA 15.262_B

production costs and all transaction costs for the entire value-adding chain up until it is placed on the market.

The pre-selection is aimed at finding approximately 10–30 attractive locations within a country, which are to be reduced to 3–5 options in the next step. The list of criteria increasingly takes into account local factors in order to finally be able to negotiate (preferably parallel) with the property owners and local authorities.

Selecting the final location requires the preselected sites to be thoroughly evaluated and compared, whereby the most frequent indicators are the amortization period for investments and the share of labor costs. Generally it can be assumed that start-up times and costs are often underestimated. There is a tendency to set the overhead costs for constructing, coordinating and controlling the new location too low and to improperly calculate them. Finally, often dynamic price changes and labor costs at the new location are insufficiently considered.

14.6 Establishing Production Stages

In discussing the process of finding a new production location we have yet to address which components of a product or product group should be manufactured there. Since this data is the starting point for the entire location planning, answering this question is still part of the strategic planning and input information for the factory planning.

Generally the following aspects should be borne in mind:

- Which product components are labor-intensive and which are material-intensive?
- Where is the end product to be completed?
- In view of cost reductions and protecting know-how, which technologies are useful to perform at the original location and which at the new location?
- What are the consequences of a globally divided production on logistic processes and supply chain management?
- How will previous and possibly new suppliers be integrated?
- Does the product structure meet the requirements for a globally distributed production?

As a solution, the so called global variant production system (GVP) has been developed by a consortium of researchers and industrial firms [Nyh08]. With the underlying methodology and corresponding tools, technologically challenging products can be produced at internationally distributed locations manually or automated in variable quantities and a wide array of variants.

The first step in applying this approach is to identify core assemblies that are both know-how intensive and bundle the enterprises' key technological competences. Three production stages result from there which need to be logistically linked: purchasing, in-house production and close to the market completion. Figure 14.12 depicts the GVP system developed for this. The individual steps are extensively described in [Nyh08] and an accompanying CD with all the checklists provides support in using it.

Product structuring here aims to concentrate function related key competences on core components which are then produced in-house in order to protect know-how. Moreover, it focuses on minimizing the internal product diversity in face of the variety demanded externally by the market.

A data model which derives the product structure based on the functional structure of the end product and saves it as a bill of materials (BOM). Every BOM item is stored with the corresponding process structure, whereby a process can be a manufacturing operation, an assembly step or a procurement procedure. Figure 14.13 depicts this in an integrated product model, stored in a relational data bank. The

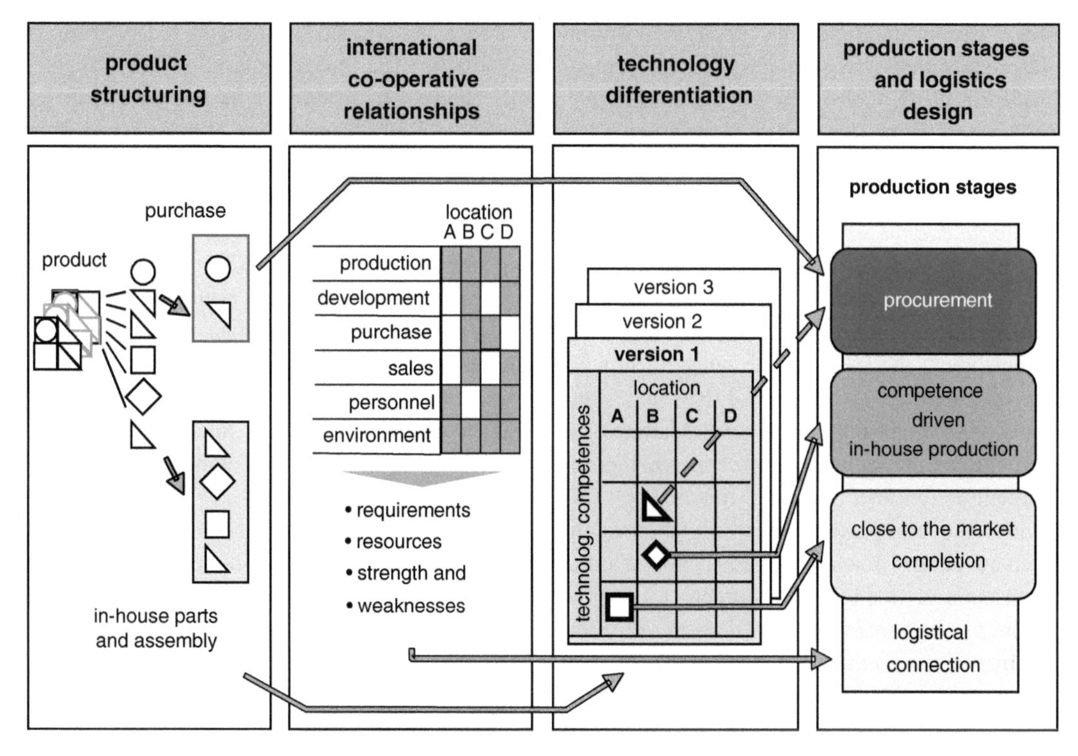

Fig. 14.12 Global variant production system. © IFA 15.263_B

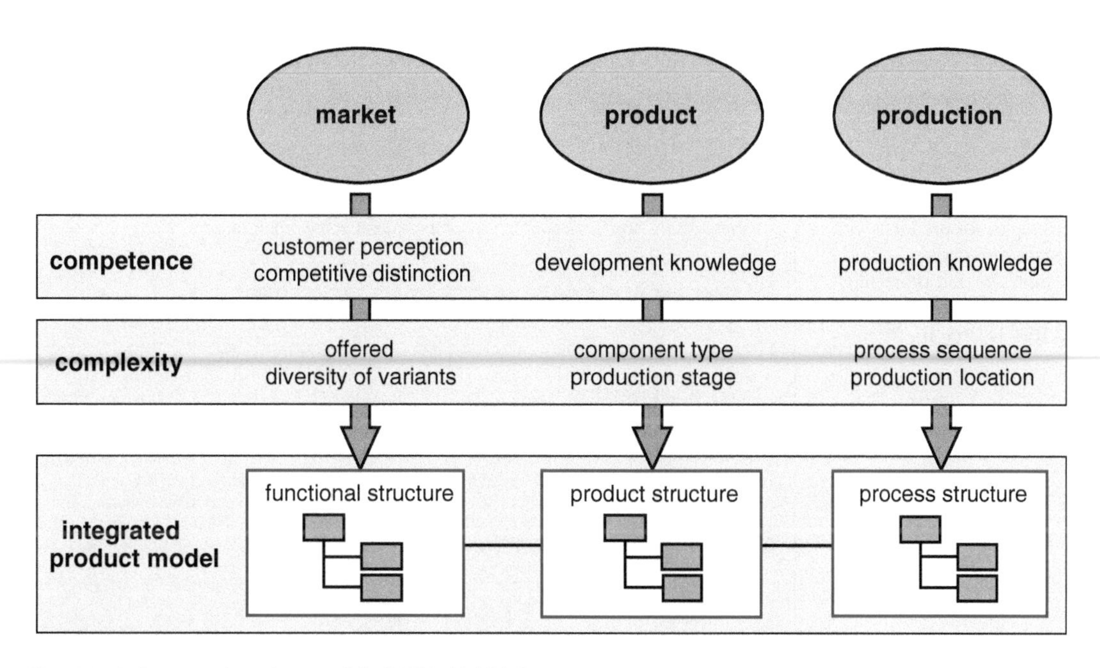

Fig. 14.13 Integrated product model. © IFA 15.264_B

model allows the functions, product positions and process steps to be linked via data systems, thus ensuring the consistency of data. If an element in one of the sub-models is changed, all of the links to corresponding elements in the other two models appear.

The *functional structure* represents the functions the product offers which are classified in regards to the customers' perspective, the competitive differentiation and the complexity. With the customers' perspective, the question is whether the function is a more or less important factor in purchasing decisions, whereas the competitive differentiation represents whether the competitors product has this function or not. Finally, the complexity describes whether the diversity of the function variants can be influenced more or less by the customer.

The *product structure* visualizes the physical structure of the product in a multi-level BOM or 'super BOM' with all of the variants. Depending on the complexity of the product this structure is established down to the individual part level or only down to first and second level assembly units, which are then further analyzed separately depending on their importance.

The *process structure* documents the process type (procure, modify, join, test), the (provisional) location and whether it is conducted in-house or externally. The production knowledge is then classified as to whether it is only available in-house, at a few competitors, at most competitors or is considered state-of-the-art.

By analyzing the available functional, product and process structure starting points for improving the product construction and production processes can be identified. Figure 14.14 lists the nine resulting design principles that are based on the relevant literature and the experiences of the firms participating in the project. Available design approaches include: integration, substitution, elimination and embedding of structural elements.

Before a final decision can be made about distributing the value-adding processes to different locations, requirements for potential cooperative partners as well as their strengths and weaknesses need to be analyzed and evaluated within the frame of international cooperative relationships.

The envisaged partners and their locations form the starting point, followed by the objectives of the cooperation (costs, market, customers, technology) and the object of the cooperation (manufacturing, assembling, development, additional purchases etc.). Requirements regarding

Fig. 14.14 Design principles for products with regard to reducing complexity and protecting key competences. © IFA 15.265_B

1. concentrate product bound core competences in few components
2. concentrate production bound competences into core components
3. concentrate production competence in the in house-production
4. bring external product variety into line with variety of market demand
5. minimize internal product variety considering the external product variety
6. concentrate internal product variety in few components
7. minimize function overlaps between components
8. create external variants late in production
9. create country-specific variants late in production

the partner's resources are derived from there (management, production, procurement, R&D, sales, personnel, capital, reliability etc.) as well as in regard to the local environment (infrastructure, service, suppliers, training centers etc.).

With internal company resources we are referring to the in-house resources affected by the cooperation i.e., which have to work together with the cooperative partner. In addition to the technical resources the personnel requirements need to be critically examined in particular with regard to language skills and cultural knowledge. Finally the cooperation resources concern the shared management of information, knowledge and conflicts.

According to Fig. 14.12 and based on the analysis of the products and co-operations, the technologies can be differentiated next. It serves to analyze the process chain, which spans across the locations, and is necessary for producing the product, based on the competences and attractiveness of technologies. The technology portfolio, originally developed by Pfeiffer [Pfe91] and adapted within the frame of GVP, is created by comparing processes within a matrix (see Fig. 14.15 left). Instead of the original scale of low, medium and high, a cost-benefit value that results from the degree of fulfillment for individual criteria is used here [Sch08].

Technology competences refer to the production technology capabilities that can be influenced by the enterprise as well as the corresponding facilities and personnel-managed expertise for fulfilling a specific process step. In comparison, the technology's attractiveness describes the potential to further develop the respective capabilities.

Positioning the processes in the matrix allows them to be distinguished as core competence, differentiation or standard processes. There are standard strategies for handling each of the process types (Fig. 14.15 right). Core competence processes represent outstanding capabilities with the goal of being a leader in technology. These need to be supported by investment measures and should be applied across as many business areas as possible. Examples here include special coating or joining methods. Standard processes are less strategically important and should be operated as economically efficiently as possible. Differentiation processes cover the area between high competence and low attractiveness to low competence and high attractiveness. Depending on whether or not these processes are applied in core components, they can be invested in so that they become core competences or they can be out-sourced to a strategic partner.

The actual technology differentiation refers to a process in which products are structurally redesigned with the aim to divide its components into different groups. Components containing functional core know-how are produced with a

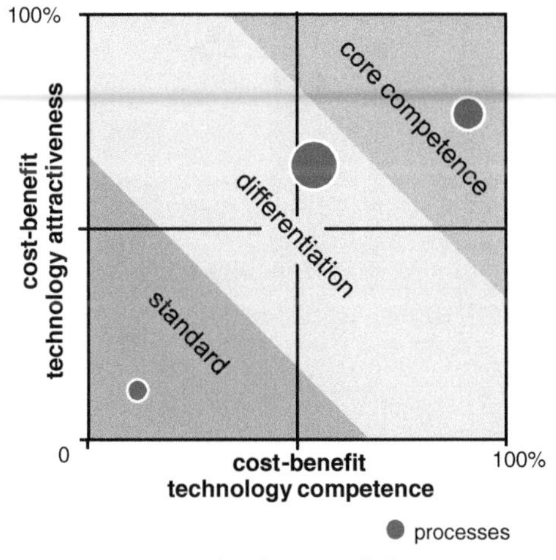

core competence process
- investment
- business field expansion

differentiation process
- investment
- technology substitution
- division in standard or core competence processes

standard process
- reduction measures
- desinvestment
- withdrawal of technologies

standardization strategies

Fig. 14.15 Process classes and standardization strategies in the technology portfolio. © IFA 15.266_B core competence. The remaining parts, which fulfill a standard function, are produced where possible with a standard technology and can thus be produced externally depending on the cost situation.

Finally, within the frame of designing the production stages and logistics, the production can be divided amongst the different locations in view of the competences (Fig. 14.12, right). In doing so, the following key questions are to be answered:

- What is the optimal range of the production at a location?
- Which parts of the value-adding chain should be located where?
- How should the material flow be strategically designed in the global market?

First, an analysis has to provide information about the current value stream and total costs in the production network. Figure 14.16 depicts a fictional example of a value stream diagram for a limited product family. The base material delivered by plane by a supplier is processed in Hannover and then transported by truck to Prague where it is assembled made-to-order for customers and then deposited in the warehouse of the local trade partner. The timeline shows the share of the throughput time that is spent adding value or not adding value. Moreover, the cycle times and inventory values for the location as well as the values for the time, distance and lot sizes for the transportation routes are noted in the data boxes. This provides initial impetus for improving the value flow by for example, eliminating, combining or conducting parallel operations [Wag08].

The total costs of a product in a network are also referred to as the total cost of ownership. In addition to the purchase costs these include handling costs, costs due to poor logistic process certainty (transportation difficulties, tariff changes, losses etc.), control and system costs, transaction costs, transport costs, costs related to currency values, stock and quality costs.

A fictional example for an analysis such as this for the IC programming of an IC chip at a site in Hannover, Germany is depicted in

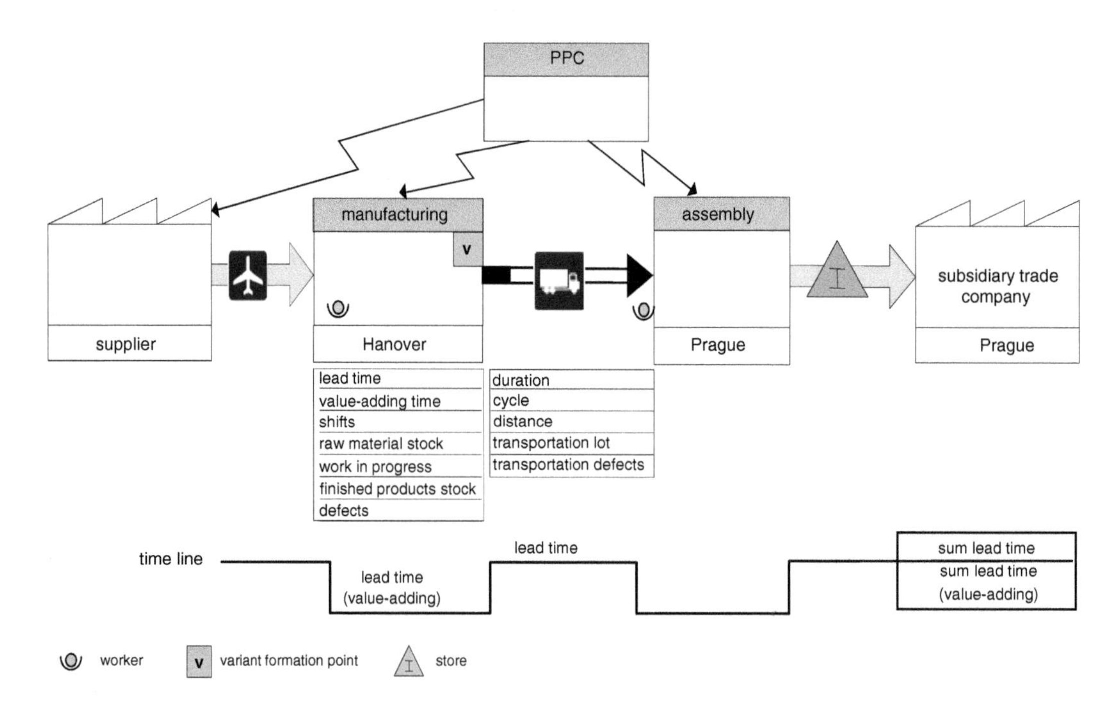

Fig. 14.16 Inter-site value stream analysis (example). © IFA 15.267_B

part name		IC chip
running time (years)		6
annual amount (pieces)		100,000
production stage		**IC programming Hanover**
quantitative assessment		
cost type	description	value [€]
extended unit costs	production costs, cost of materials, discounts, payment conditions, building costs, development overhead costs, administration and sales costs	*5.00*
tool costs	complete tool costs over the run time (incl. repair, substitute)	*0.02*
logistics costs	transportation, packing, insurances, charges, customs, import / export taxes incl.	*0.29*
standard transport	*incl. packing, insurance, charges, custom procedure*	0.09
alternative transport	*incl. packing, insurance, charges, custom procedure*	0.10
customs, taxes	*customs, import / export taxes*	0.10
supplier training	project effort staff + travelling expenses + other costs	*0.01*
series support	series effort staff+ travelling expenses + other costs	*0.00*
inventory	capital bound for goods in transit and safety stock	*0.03*
total costs		*5.34*
additional costs in per cent of total costs		*6.8 %*

Fig. 14.17 Quantitative assessment of a procurement decision (Example). © IFA 15.268_B

Fig. 14.17. Here the additional costs contribute 'only' 6.8 % of the so-called extended unit costs. If the chip was to be programmed in another country, the share of costs would shift accordingly.

Furthermore, it has to be kept in mind that with great distances and many different partners from other cultures there is a risk in regards to the ability to react to sudden changes in demand. This affects for example, the supplier's business orientation, the local services etc. These factors are included in a qualitative assessment which together with the quantitative analysis leads to a final decision.

The subsequent configuration of the sites' production stages is oriented on combining production tasks with equivalent competences. Figure 14.18 depicts the resulting generalized production system with three characteristic production stages.

The procurement stage is concerned with parts for which the enterprise has no competence or cost advantages. Currency related risks can also be reduced by purchasing these in the country with the greatest volume of sales. The in-house production stage is comprised of parts and assembly units which ensure the core competence and production at the location while protecting know-how. Finally, locating the completion stage of core components close to the market allows country-specific variants to be delivered quickly and fulfills requirements for a specific percentage of local value-adding. The thus defined production stages require an adaptable logistic supply chain, whereby priorities include questions about local and central warehousing and linking them as closely as possible to sales regions.

From a factory planning perspective, the GVP approach not only requires local changeable

procurement stage which assemblies are bought from outside?	**competence driven in-house production stage** what is manufactured and assembled in-house?	**close to the market completion stage** what is completed closely to the market?
• opening up of economies of scale • currency conversation compensation • use of suppliers' competence	• safeguarding of core competences • know-how protection • variant reduction at home location • safeguarding home location	• differentiated markets supply • control of countries specific variants • fulfillment of local content requirements

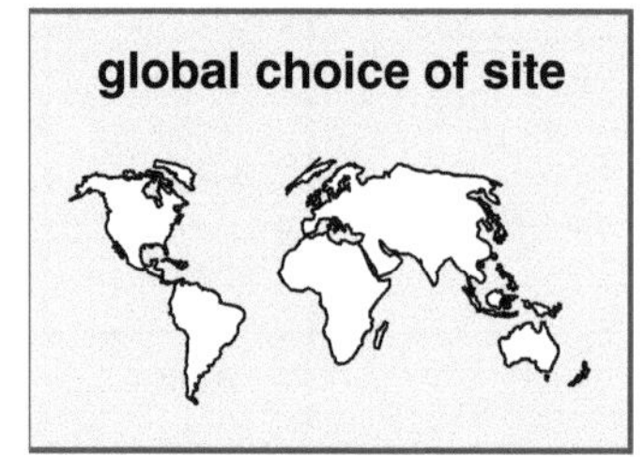

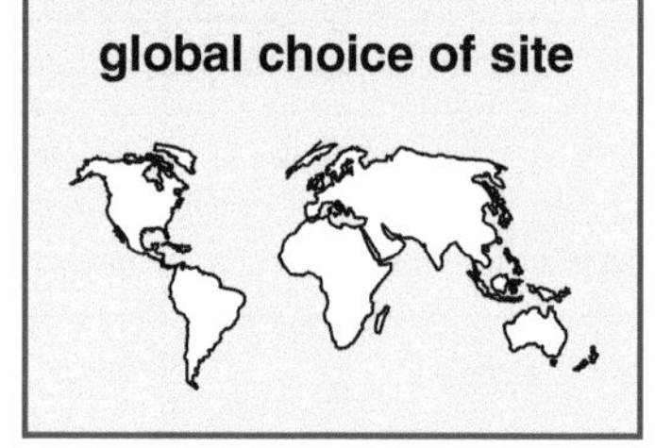

Fig. 14.18 Forming production stages in a network. © IFA 15.269_B

production facilities and buildings, but also an equally changeable logistics distribution network for the physical material flow. Thus with the factory planning, the generally plausible supply and distribution concepts are to be seen as variants whose features are also to be designed with changeability in mind. Available possibilities for designing these are presented in Chap. 9.

This concludes our discussion on all factory levels with regards to their technical, organizational and spatial design. In the next chapters we will consider the factory planning process and the project management required for it.

14.7 Summary

For a long-term successful location of a factory an orientation to the business location planning is indispensable. It is based on a competitive strategy that may relate to a cost, quality, technology, logistics and flexibility leadership. The location assignment is found in steps starting with a global selection at country level, followed by a search on the local level until the determination of the final location. The content of the local production takes place on the product level, the final product assembly and item manufacturing based on the procurement, in-house production and final assembly strategy.

Bibliography

[Abe08] Abele, E., Meyer, T., Näher, U., Strube, G., Sykes, R. (eds.): Global Production: A Handbook for Strategy and Implementation. Springer, Berlin (2008)

[Au07] Aurich, J.C., Wolf, N.U., Fuchs, C.H.: Strukturierte Standortplanung: Konzept zur systematischen Auswahl von Standorten (Structured location planning. Concept for the systematic choice of locations). Industrie Management. **23** (3), 43–46 (2007)

[Eve96] Eversheim, W.: Standortplanung (site location planning). In: Eversheim, W., Schuh, G. (eds.) Produktion und Management "Betriebshütte", 7th edn. pp. 9-40–9-57. Springer, Berlin (1996)

[Heg08] Heragu, S.: Facilities Design, 3rd edn. PWS Publishing Company, Boston (2008)

[Jun04] Jung Ercec, P.: Optimierungspotenzial am deutschen Standort bewerten. Das Konzept des Instruments (Rating the optimization potential at the German site. The concept of the instrument). In: Kinkel, S. (ed.) Erfolgsfaktor Standortplanung. In- und ausländische Standorte richtig bewerten (Success factor location planning. Rating domestic and foreign locations correctly), pp. 131–161. Springer, Berlin (2004)

[Kin09] Kinkel, S. (ed.): Erfolgsfaktor Standortplanung. In- und ausländische Standorte richtig bewerten (Success factor location planning. Rating domestic and foreign locations correctly), 2nd ed., Springer, Berlin (2009)

[Mey06] Meyer, T.: Investitionen in Auslandsstandorte: Bewertung und Auswahl (Investments in foreign locations: evaluation and selection). In: Abele, E., Kluge, J., Näher, U. (eds.) Handbuch Globale Produktion, pp. 102–142. Hanser Verlag, München Wien (2006)

[Nyh08] Nyhuis, P., Nickel, R., Tullius, K. (eds.): Globales Varianten Produktionssystem. Globalisierung mit System (Global variants production system. Globalization with system). PZH Produktionstechnisches Zentrum GmbH, Garbsen (2008)

[Paw08] Pawellek, G.: Ganzheitliche Fabrikplanung. Grundlagen, Vorgehensweise, EDV-Unterstützung (Integrated Factory Planning. Principles, procedures, IT support). Springer, Berlin (2008)

[Pfe91] Pfeiffer, W., Metze, G., Schneider, W.: Technologie-Portfolio zum Management strategischer Zukunftsfelder (Technology portfolio for the management of future strategic fields), 6th edn. Verlag Vandenhoek und Ruprecht, Göttingen (1996)

[Sch08] Schünemann, W.: GVP-Modul Technologiedifferenzierung (Technology differentiation). In: Nyhuis, P., Nickel, R., Tullius, K. (eds.) Globales Varianten Produktionssystem. Globalisierung mit System, pp. 98–123. PZH Produktionstechnisches Zentrum GmbH, Garbsen (2008)

[Sch11] Schmidt, B.C.: Gestaltung Globaler Produktionsstrategien (Design of Global Production Strategies). In: Gausemeier, J., Wiendahl, H.-P. (eds.) Wertschöpfung und Beschäftigung in Deutschland (Value creation and employment in Germany). acatech Workshop Hannover 14.10.2010, Springer, Berlin (2011)

[Schu08] Schuh, G. et al.: Global Footprint Design. In: Manufacturing Systems and Technologies for the New Frontier. Proc. of the 41st CIRP Conference on Manufacturing Systems May 26–28, 2008, Springer, Tokyo, Japan (2008)

[Wag08] Wagner, C., Großhennig, P.: Produktionsstufen- und Logistikgestaltung (Design of production stages and logistics). In: Nyhuis, P., Nickel, R., Tullius, K. (eds.) Globales Varianten Produktionssystem. Globalisierung mit System (Global variants production system. Globalization with system), pp. 124–163. PZH Produktionstechnisches Zentrum GmbH, Garbsen (2008)

15 The Synergetic Factory Planning Process

Due to different products, processes and environmental conditions, every factory planning is unique. Nonetheless, certain procedural steps that are based on a strategic positioning have proven themselves useful for all projects—independent of whether the planning is for a new building, an expansion or a re-organization. Since these aspects are frequently neglected in practice, supposedly due to tight deadlines, we will explain them thoroughly. In a second step, it is critical that the initial data is clearly defined as it provides the basis for the arrangement of the facilities of the factory stage-by-stage with increasing accuracy. Due to the close connection between planning the production and the building, this chapter introduces an approach referred to as 'synergetic factory planning', which merges the procedures put forth in VDI Guidelines 5200 with the construction phases outlined in the German fee schedule for architects and engineers (HOAI).

15.1 Planning Approach

With traditional factory planning the technological and logistical processes, facilities and layout are first structured based on the given tasks and objectives. An architect is then employed to design a—usually as inexpensively as possible—building shell with the necessary building facilities. However, the processes and production systems along with their flows of materials, information and personnel are closely related to the building services (energy, media, ventilation etc.), which are in turn part of the building architecture. Typically, with conventional planning approaches, isolated solutions are found for the location, buildings, building services and processes resulting in a lot of unpleasant symptoms as depicted in Fig. 15.1 [Rei07].

During planning this practice leads to more than just missed deadlines and overrun budgets. It also produces unsatisfactory planning outcomes which are reflected in functional and qualitative defects, the buildings poor performance and insufficient changeability during the operating phase.

The described situation also calls into question the methods used up until now for building planning. A critical look at the typical practice reveals grave differences particularly in comparison to the "digital" methods used in advanced industries and prevents the sub-projects from being viewed as a whole. The location, buildings, building services and process are each planned sequentially compared to a simultaneous engineering approach which has been commonly employed in developing products within the consumer and investment goods industry for some time now.

This brief characterization of traditional factory and project planning shows that a deeper consideration is needed. The synergetic factory planning

DOI 10.1007/978-3-662-46391-8_15,

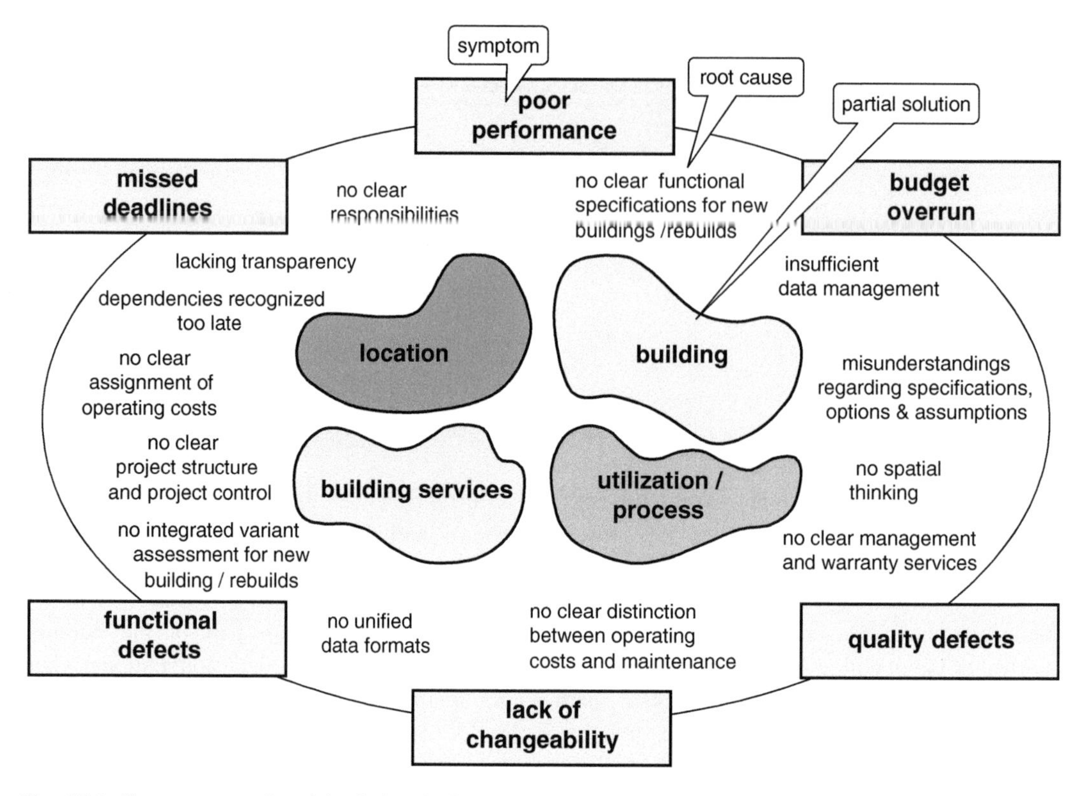

Fig. 15.1 Consequences of partial solutions in factory planning. © Reichardt 15.457

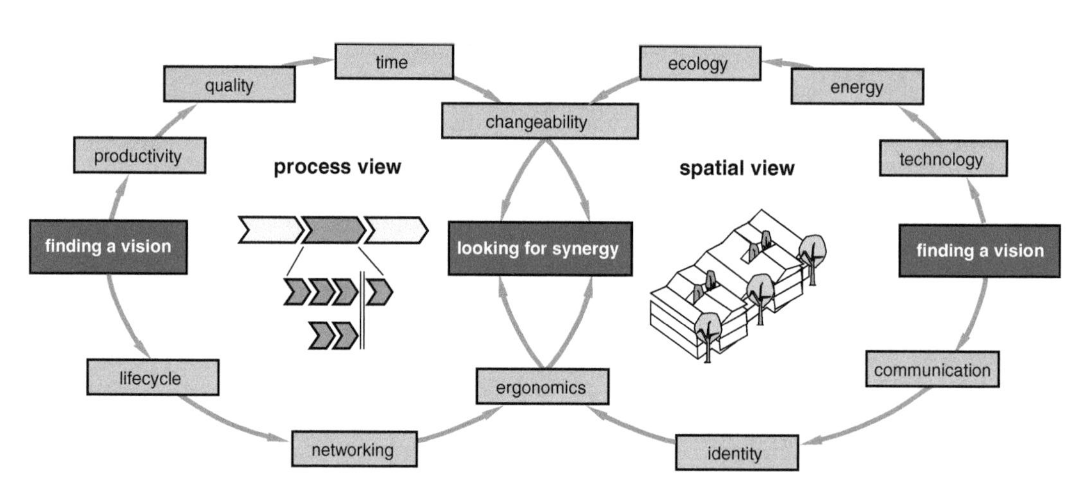

Fig. 15.2 Synergetic approach to factory planning. © Reichardt 15.458

developed by the authors over the course of a decade thus already begins when planning the goals with a combined process and spatial perspective which continues through the planning phase up to the ramp-up phase [Wie96, Wie02, Nyh04, Rei04, Rei07]. Generally, synergetic is understood as the merging of different forces, factors and elements into a coordinated overall performance. The fundamental ideas of a factory planning based on this are clarified in Fig. 15.2.

Each partner first develops a vision from their process or spatial perspective and then with increasing precision establishes the detailed requirements relevant to their expertise over gradual steps. From the *process perspective*, traditional demands for high productivity and quality with short throughput times combined with the relatively new demand for changeability means that an ergonomic design stands in the foreground. Changeability directly impacts the spatial perspective, impacting not only the building structures but also the building services. Moreover, the interactions between the factory and an increasingly networked environment also give rise to requirements including product and facility lifecycles as well as integrating production processes in supply chains and production networks.

Once a vision has been developed the *spatial perspective* begins with harder factors such as building technology and energy consumption. Ecological considerations play a role both during construction (e.g., energy consumption and materials used) as well as with processes (e.g., hazardous substances or waste materials). Soft factors here concern the question of simple personal communication as well as internal and external appearances which help create a corporate identity.

All of the requirements merge into a synergetic approach in which the target projections unite into a solution supported by all of the planning team members. Not only calculable facts are at issue, rather the emotional agreement also matters. The ongoing fascination originating in this conception is crucial for its sustainability.

From a process and spatial perspective, the new quality of a thus defined cooperative planning is found in merging the spatially designed sub-projects (location, buildings, building services and processes) as early on as possible (Fig. 15.3) [Rei07]. The actuating variables here are materials, information and communication, capital and people, which are continually moving within an open system.

Generally, the aim is to create a three-dimensional representation of all objects, which allows the interaction between the planning partners to be supported by data systems. The integrated approach continually refines the sub-projects' 3D-structure as well as the attributive planning data from rough approximations to fixed specifications and evaluates decision-making processes by comprehensively discussing alternatives.

The job specifications can clearly outline the explicit objectives of the sub-projects in view of the target performance features (changeability and investment costs). Subsequently these can be translated into a spatial model and the effects of the entire project can be examined. A continuous updated 3D data model makes use of the potentials of present CAD-CAM database technologies in favor of a comprehensive project optimization and cyclic 3D quality assurance. Chapter 16 will describe this approach in more detail from the process side under the concept of the digital factory and on the spatial side under the concept of BIM (Building Information Modeling).

Moreover, based on the spatial optimization of this synergetic factory, tests can be run to check for model collisions and a quality control can be conducted simultaneously for all of the facilities. In particular, conflicts between the location, buildings, building services and uses/processes which negatively impact costs, time and/or quality can be identified and eliminated early-on instead of in the middle of construction or later once the plant is already operating. One further aspect of the 3D modelling that should not be underestimated is the immediate clarity of the planning results for all participants.

This approach however cannot be limited to just the planning phase, rather in the future it also has to include the entire lifecycle of a building (Fig. 15.4). An integrated facility management data model preferably based on the BIM standard, see Chap. 16) opens up the following possibilities.

Starting with the initial goal projections the *base investigation* can already take into account the target costs. During the *conception phase* and the related feasibility study all of the spatial elements together with the production facilities are three dimensionally modelled and possibly animated. The *rough and detailed planning* is

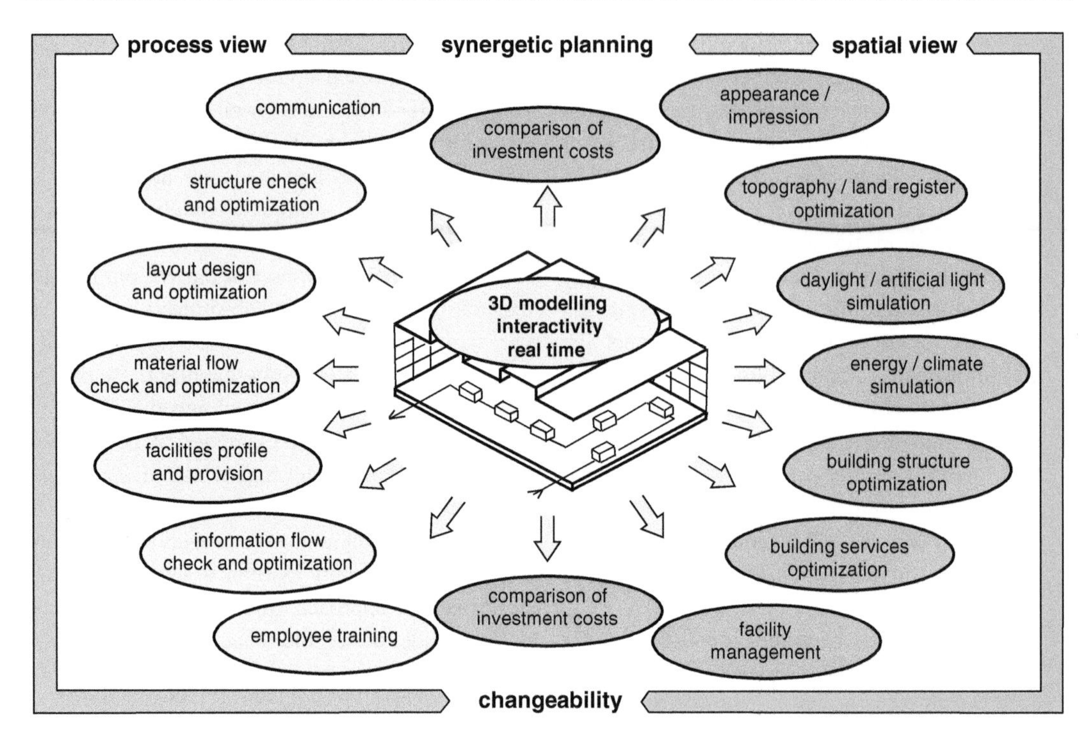

Fig. 15.3 Merging the sub-projects of a factory object. © Reichardt 15.459

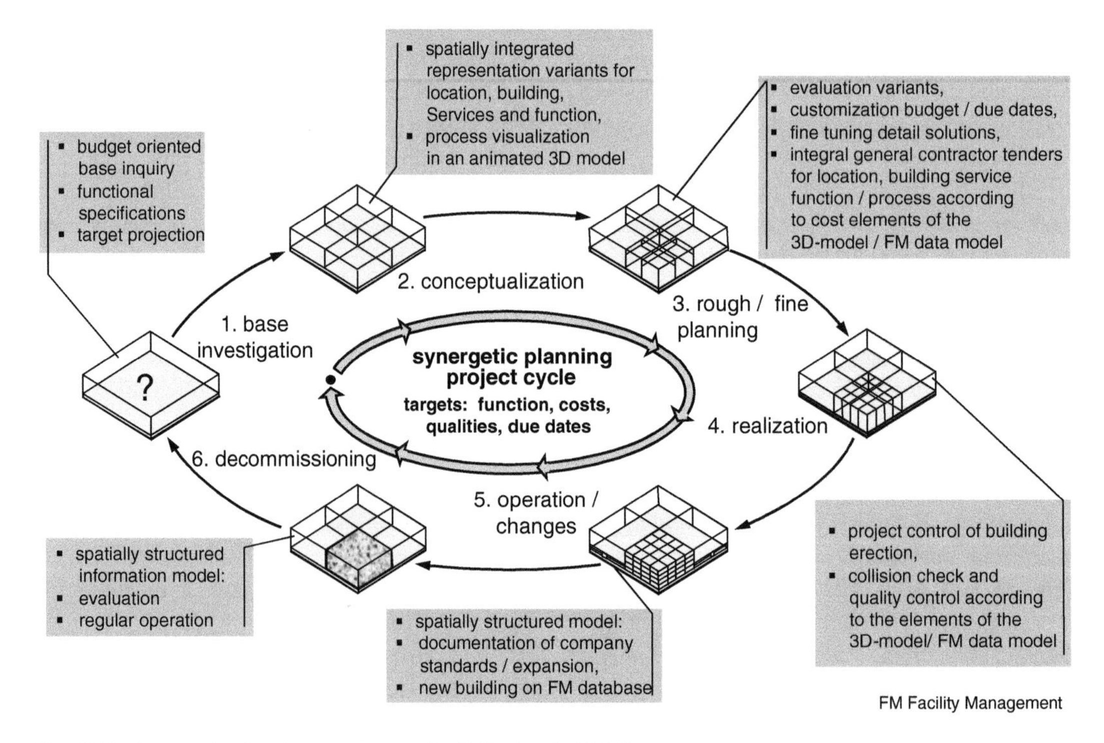

Fig. 15.4 Synergetic planning—project cycle. © Reichardt 15.460

simplified through the model-driven fine planning, tendering and adjustment of the costs and schedule when there are changes and/or variations. The *realization phase* can be supported by the 3D model in regard to collision checks and planning the construction process. During the *operational phase*, a spatially structured model for all documents is useful. Finally, it is also possible to analyze the building operation costs while it is operating.

Using the synergetic factory planning approach however first necessitates a clearly structured process model that can also be applied without a 3D modelling. It must meet the following criteria:

- Establishment of a specification catalogue emanating from a corporate vision and a brand image. Derivation of equally important hard factors (e.g., productivity, material flow, energy consumption) and soft factors (e.g., communication, identity and changeability).
- Structuring of planning tasks into design levels which are developed with increasing precision. These range from embedding the master plan into the local surroundings to the individual buildings and up to sub-dividing them into sections and workplaces.
- Parallelization of sub-tasks as much as possible in order to shorten the planning time.
- Synergetic solutions for the processes, operating facilities and spaces in which these processes take place.
- Development of a data model depicting both processes and spaces three dimensionally.
- Standardization of repetitive tasks in order to reduce planning expenditures and increase planning quality.
- Use of a limited number of standard tools in order to systematize the depiction of results.
- Definition of milestones in the sense of quality gates for improving the quality of results and avoiding undiscovered mistakes from being passed on.
- Installation of a project management according to technical, organizational and atmospheric aspects.

Following we will introduce the structure of this model.

15.2 Process Model

Factory planning projects can differ in their triggers, required precision, organizational conditions, the selected range of consideration, the targeted use of the planning results and how and to whom results are presented. The requirements of the planning process thus differ greatly depending on the type of project: Dependent on whether the project is dealing with a new object, an extension or a re-organization, different objectives result which in turn seem to require separate approaches. Nonetheless, it is possible to identify basic phases of a planning process that have to be conducted in more or less detail.

This approach has already been pointed out by a number of authors (see e.g., [Wie72, Ket84, Agg87, Wie96, Fel98, Gru00, Dan01, Sch10, Paw08, Tom10]) and taken up in a VDI guideline [VDI11]. In regard to a future factory there are currently relatively new aspects that need to be accounted for and which result from their integration in a global production network and the related increasing unpredictability of market behavior. The corresponding approach of the changeable factory was developed in Chap. 5.

Today's factory planning is thus no longer aimed at developing a long term fixed production facility for a mostly stable production program and with an eye to minimizing costs, but rather at designing a space that can provide solutions for distinct production scenarios.

Synergetic factory planning therefore strives to generate the desired planning results within the shortest time possible using strongly networked resources. The form of collaboration shifts from the isolated processing and merging of individual work packages to a creative interdisciplinary dialogue in which the initially only rough vision of the factory is purposefully and gradually transformed into a concrete solution.

This is made possible not only by applying the abovementioned planning principles but also by using the (to some degree) already discussed new media and work techniques such as those used in the digital factory [VDI08] (see Chap. 16), in building information modeling (Chap. 16) and in facility management (Chap. 17) as well as tools such as SharePoint Server or telephone and video conferencing. The increased implementation of tools for simulating the flow of materials, communication and energy as well as for simulating air flow and the distribution of light helps ensure the quality of the results.

The process model we will introduce here provides an organized framework for the mentioned planning situations and is highly adaptable. It results in a modular structure in which the dependencies are precisely described and the connections between the individual processes are designed to be easily understood. Additional information regarding applicable methods and tools and analyzing the quality of the results are also part of such a descriptive model.

The process model for a synergetic factory planning is depicted in Fig. 15.5.

The *production planning* stages (top left) form the starting point. These stages include: analysis, structural design, layout design and implementation of the project once the planning is completed. In increasing levels of detail, this sub-model describes the designing of technological and logistic processes as well as the production facilities and their organization in view of the flows of material, energy, communication etc.

The production planning stages are conducted alongside the *object planning* stages which design the internal and external spaces of a production plant from an architectonic perspective (top right). In accordance with the German fee schedule for architects and engineers (HOAI) there are nine stages starting with the base analysis and extending up to the object support and documentation [HOAI13]. HOAI describes the content of these phases, defines how the fees for architects and engineers planning building and external features should be determined and writes the limits for the fees of the architects' and engineers' services in the schedule (minimum and maximum rates).

The *synergetic factory planning* is a merger of the two above mentioned processes comprised of six demarcated stages in an integrated planning of the factory lifecycle spanning from the preparation of the planning up until the factory is in

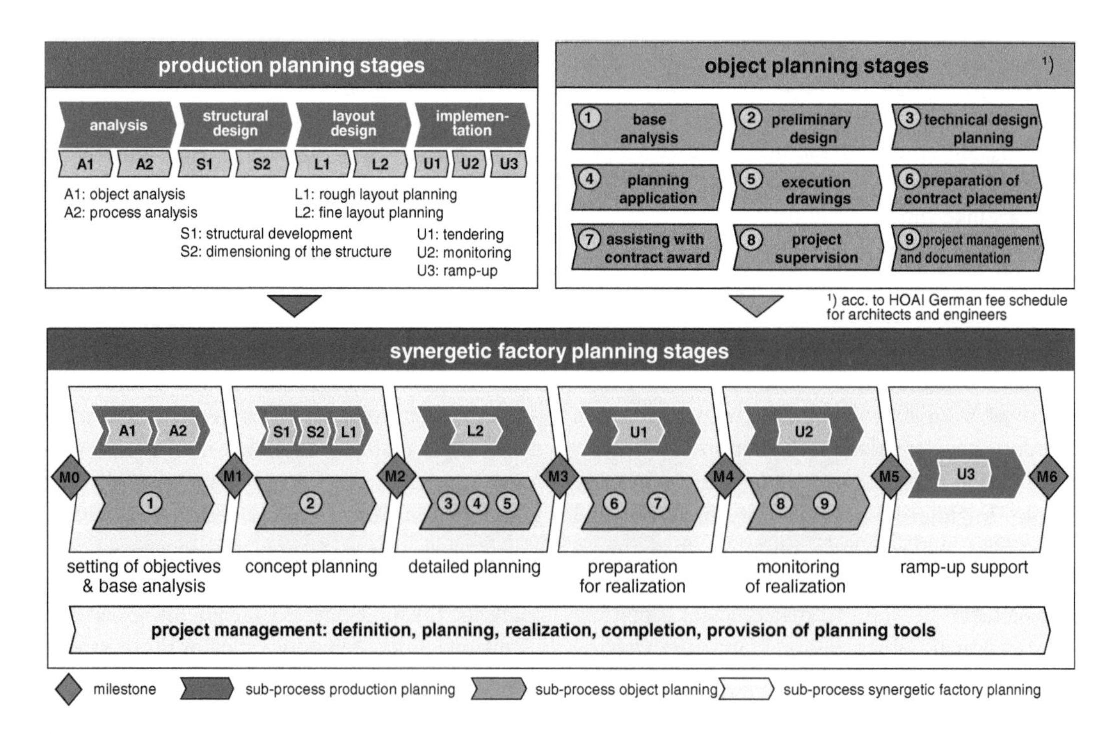

Fig. 15.5 Process model of the synergetic factory planning. © IFA 15.461

operation. The planning process begins at Milestone M0, followed by milestones M1–M6 marking the completion of each stage. These stages are complimented by a *project management* which includes the project definition, planning, execution and completion as well as the selection of tools and communication standards that are to be implemented. These phases are designated pursuant to VDI Guideline 5200 [VDI11].

The synergetic process begins with the Milestone *project resolution* (M0), i.e., the decision to formally start the project, typically resulting from a strategic re-positioning due to a strong catalyst for change. It may, for example, be based on a strategic preliminary study such as that introduced in Chap. 14 Strategic Location Planning. The cornerstones set by management include the production program, the location, the shares of in-house manufacturing and purchases, the networking with other sites as well as the targeted completion date and sometimes the intended capital expenditure.

The following *objectives definition and base analysis* marks the start of the actual factory planning. Based on Milestone 0, the project management and production planning team develop a vision, mission as well strategic objectives for the new factory. At the same time, from the object perspective, the base analysis is conducted. In the next phase of this stage the project conditions are analyzed in two steps from the perspective of the production planning. *Object analysis* (A1) is concerned with the products and their derivatives, variants and the structure of the bill of materials sub-divided into parts manufactured in-house, parts purchased and components. Moreover, an inventory of new or existing operating facilities has to be made, along with the required areas and personnel. This is further extended to include the personnel and office space required for marketing, distribution, research and development, work prep, production control as well as supportive functions (human resources, finance, supervision etc.).

In comparison, *process analysis* step(A2) examines the production processes from a technology perspective based on work plans and organization. Additionally, logistics are analyzed i.e., suppliers' delivery concepts, production control concept, finished goods shipping concept and required storage and transportation facilities. In individual cases further objects and processes have to be included e.g., testing facilities or a school for training customers.

On the object planning side and in accordance with HOAI the task from the building planning perspective is clearly defined in this stage. Particularly with complex industrial projects analyzing the initial data for the building planning is usually difficult. This is due in part to the continually given mixture of hard specifications (e.g., column grids, ceiling heights) with more soft factors (e.g., employee and customer communication) and opinions that vary greatly among participants.

Consequently, at Milestone M1 the following results are available:

- the factory vision and mission,
- a coordinated requirement profile,
- factory strategies and targets as well as the fields of action,
- building specifications according to HOAI,
- a procurement and distribution model,
- a list of machinery with spatial attributes and
- an approximate project schedule.

Along with the corresponding documentation these points form the agenda for the milestone meeting with the steering committee.

In the ensuing *concept stage*, the structural design, detailed dimensioned planning and rough layout planning are conducted from the production planning perspective. In doing so either existing production technology or new technology from a parallel project are adopted.

The *structural design* stage (S1) focuses on determining structural alternatives. In a 2D depiction they describe the relationships between the manufacturing, assembly and logistics sectors and thus define the entire factory concept. These are developed based on specific structural relations such as technologies or product groups.

Subsequently, the *structure dimensioning* (S2) is undertaken. This includes determining the required production equipment along with their spatial requirements, calculating the number of employees and dimensioning the building grids and construction areas in co-ordination with the object planning.

During the *rough layout planning* (L1) the facilities are spatially allocated without details, usually in block form. The structure that is thus found is then converted into an ideal layout in view of more refined organizational criteria such as products, technologies, materials etc.

Parallel to this the object planning concentrates on developing a planning concept for the building construction taking into account the load-bearing capacity, obtaining the necessary permits and approval and the costs according to HOAI's phase 2. The ASTM Uniformat II Classification for Building Elements (E1557-97) could be used as well (see Sect. 16.7.2). This planning stage concludes with a feasibility study of the building. The quality of this feasibility study lies in the synergetic overview of the objectives developed up until this time (for the location, process and organization, building and building services) in a comprehensive spatial model. The model conveys to the steering committee and project team the interplay between the production facilities and the building. Aspects such as communication or the use of daylight are mediated on a visual sensory level. At the same time, realistic characteristics such as the gross floor area and cubature are available as decisive benchmarks for estimating costs. Furthermore, using the data sets saved in the 3D model, video animations of the project can be generated with relatively reasonable expenditure. The persuasiveness of moving images is quite advantageous particularly when comparatively analyzing possible different scenarios. Process animations such as material flow can also be imbedded in the 'storyboard' cooperatively developed in advance by the project team.

Appendix C2 contains a sample video animation generated for a realized project. It shows the internal and external design of the building in its natural surroundings and later extension stages.

Once the concept planning is completed the following results are available at Milestone M2:

- the production concept,
- the building volume and spatial schedule (with room list, gross floor area and gross volume)
- an estimate of costs,
- an integrated 3D spatial and production concept as well as
- an updated timeline.

Following this the *detailed planning* begins. On the production planning side it includes determining the operating facilities for manufacturing, assembly and logistics as well as the detailed layout planning (L2). By taking into account production restrictions and the now known building concept, a realistic detailed plan of the factory is created including the positioning of facilities, routes, utility services etc.

In this stage the object planning produces the object design in accordance with HOAI's phase 3. The necessary permits have to be obtained from the authorities (HOAI phase 4) and the planning for executing the plans (HOAI phase 5) also takes place.

The results at Milestone M3 include:

- the detailed layout with operating facilities positioned precisely,
- approval of the prospective users,
- the approval plan for the authorities,
- the execution plan for contractors
- ensuring the plan's feasibility and cost compliance
- cost calculations as well as
- the room list and
- the detailed project schedule.

Upon reaching Milestone M3, the planning of the factory object as such is concluded.

The ensuing *preparation for realization* focus on implementing the results of the planning. On the production side preparations are made for tendering contracts for the new production facilities and possibly adjusting adopted equipment. The object planner becomes here the essential driver behind the object realization, while the factory planner serves more as a process facilitator.

On the object side, planning the work and awarding contracts is to be realized and facilitated based on the tendering documents and in accordance with HOAI's phase 5, 6 and 7 (the various contracts will be discussed in Sect. 16.5).

At Milestone M4 the following documents are available:

- invitations for bids from potential contractors,
- cost estimates as well as
- the detailed project schedule.

From the synergetic factory planning perspective, the next stage *monitoring the realization* serves to monitor and control the construction progress and the correct installation of operating facilities. The planners of the individual contracts verify the yielded service from both the object side (HOAI phase 8) and production side.

Upon reaching Milestone M5, the factory is ready to move into the new object and the following documents are available:

- performance tests and approvals of facilities and buildings,
- cost statements in accordance with HOAI, and the
- final documentation.

The last stage is the *ramp-up*. During the warranty period, when defects arise and there is an admission of liability, the object planning (HOAI phase 9) arranges for incomplete or faulty constructions to be completed or corrected. Close contact between the spatial and process planners is indispensable here. At the latest now, the collected 'as built' information from the room list (execution phase) and completion documentation for the operating phase should be transferred into a facility management system (see Chap. 17). Nevertheless, this stage is focused on ramp up of the production until it attains the agreed upon production output rate. A more precise view of this is provided in Fig. 15.6 ([Zeug98, Win07]).

In addition to the factory planning perspective that was our primary consideration up to now, the product and production view are to be distinguished. For the product development (Fig. 15.6 top) once the product design is completed, the production ramp-up is part of the series ramp-up, which begins with the prototyping. The product and its properties are of interest here. Automotive companies have their own strictly secured factories for this process, whereas with smaller products, prototypes are usually created in experimental or production facilities. Following that a pilot series, sub-divided into pre-production and pilot series is started on the production facilities which have in the meantime

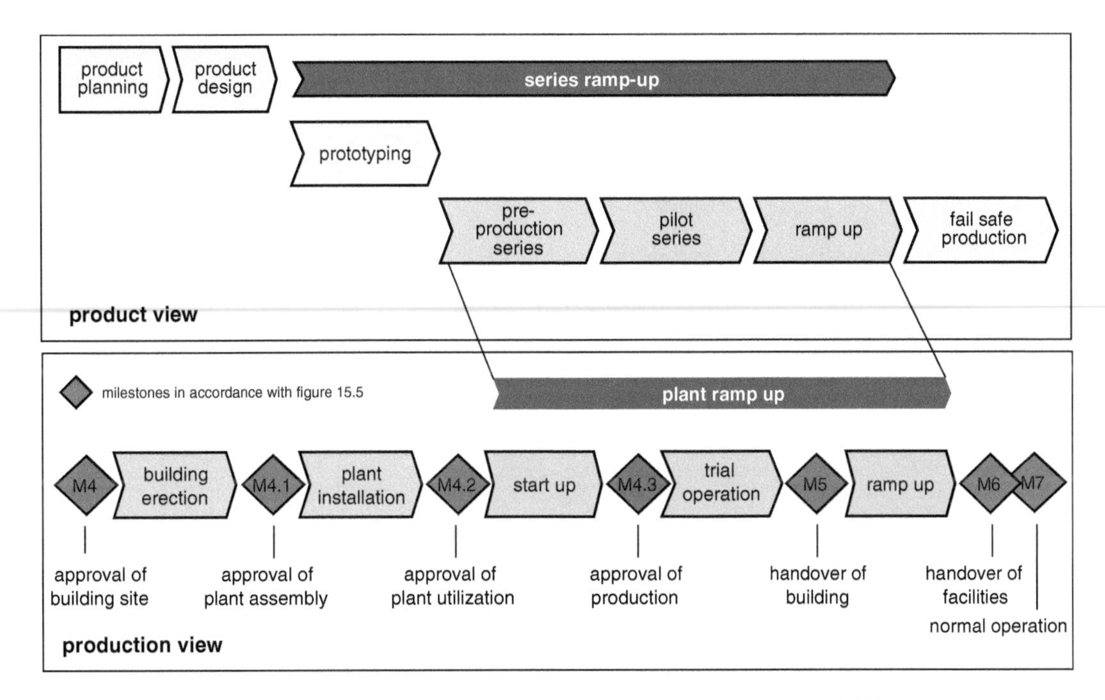

Fig. 15.6 Ramp-up phases of a factory from the product and production view. © IFA 15.523

been constructed. Typically, products then have to be adjusted or equipment has to be changed. When the results are satisfactory, the production is ramped-up until a stable production is attained.

From the production perspective, the realization of the production begins with the start of construction on site, followed by the erection of the building and continues up until the approval to install facilities. Also included here is a thoroughly planned move. From the later users' perspective, the process reliability of the production facilities is of interest. Once the plant is approved for utilization the actual ramp-up can begin. First, facilities are inspected for functionality after installation. In batch manufacturing this is followed by the actual production with pre-production series, pilot series and full load production. In single or small series productions exemplary cases are selected. Steady continuous operation is first achieved with Milestone M7 (not included in Fig. 15.5).

With the completion of this phase (Milestone M6) the factory is in full operation and the project is concluded from the factory planning perspective. Depending on the agreement, the responsibility is now transferred to the user. Both the factory's performance capability as well as the planning project itself should finally be evaluated at this point. Results at M6 include:

- the determination of the full load capacity,
- evaluation of the target attainment (performance control) as well as
- transferring of documentation and balancing of accounts.

The 6 planning stages are facilitated by *project management*. The project management begins with the already described *project definition*, which is concerned with the broadly hedged project objectives. This is followed by the *project planning* which is aimed at defining the project phases and their results as clearly as possible. The *project realization* is concerned with the organization and documentation of the ongoing project and lastly the *project completion* documents the results. Project management is responsible for arranging the tools that are implemented for developing, depicting and documenting results. Details will be presented in Chap. 16 Project Management, in the section on digital tools.

The descriptions of the *milestones* deserve particular attention since they serve the technical, temporal and cost-related coordination as well as the approval through the steering committee. Figure 15.7 schematically depicts the technical coordination process based on the example of the concept planning stage.

In this stage, the planning experts who are oriented on the rooms, facilities and organization

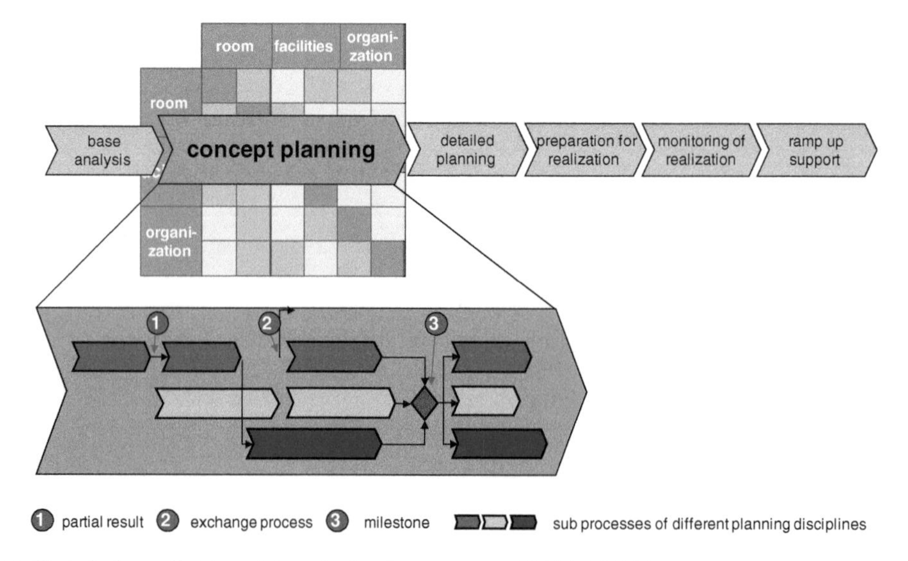

Fig. 15.7 Technical coordination of the sub-planning processes. © IFA 15.462

have to prepare their results and present them to the steering committee at an agreed upon date. The planning objects however functionally interact with one another as is indicated by the matrix shown in the back ground. In order to recommend a building structure, for example, the spatial planners require an organizational model of the factory from which they can identify the basic sections and how they are linked. The partial results are passed on to the affected partners in the form of plans, diagrams, lists etc. (exchange process). At specific points during the conceptualization all of the participating disciplines are coordinated usually via a workshop. This planning alignment represents a milestone for the project management. The result forms the planning basis for the next phase and is subject to approval from the property developer e.g., for the concept and a partial payment.

Figure 15.8 depicts an example of the results created by coordinating the process and spatial planning. During the detailed planning phase the exact position of machinery is determined in step L2. This then defines where in the factory heat or vapors might arise. This information then provides the initial data for the object planning so that the building service planners can take into account the heat sources and can design the technical building systems appropriately.

Differences in implementing synergetic factory planning mainly result from the extent to which the individual stages have to be conducted. Thus for example, if the project focus is on reorganizing a factory more actual data from existing processes is collected than when the project is concerned with planning a new one. Moreover, with a reorganization project the focus of the planning is on optimizing the processes and configuration.

Up to this point, our comments have been oriented on the technical aspects of the planning model and on the whole form the focus of this chapter. Methodical and atmospheric viewpoints, i.e., more soft factors, are also however decisive for a successful planning project. Their importance in solving interest-related target conflicts is usually underestimated and typically only come

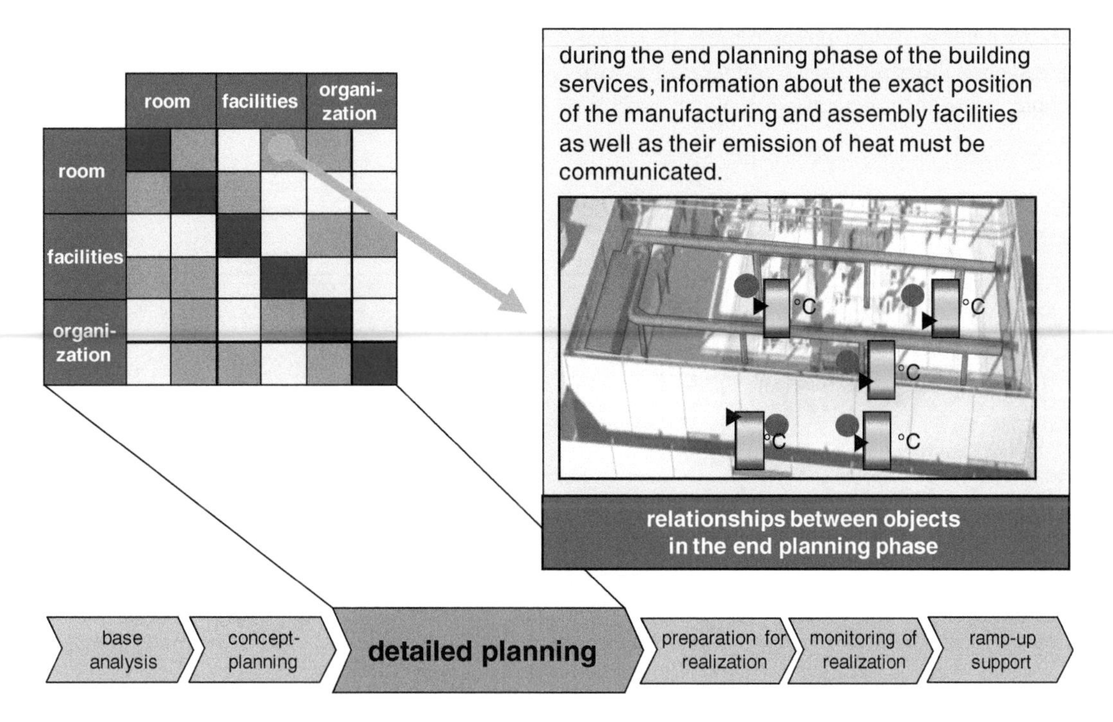

Fig. 15.8 Coordination of the partial planning processes content. © IFA 15.463

to light during the workshop for coordinating the sub-plans and the milestone meetings.

Based on the specific conditions of the project, the technical, methodical and atmospheric aspects of the project work generally need to be distinguished: Fig. 15.9 clarifies these perspectives and the resulting project management responsibilities in consideration of all three aspects based on a suggestion from Wiendahl [WieH11]. Although it was originally developed for designing and implementing a system for order management, due to the comparable complexity of the project, it can be advantageously transferred to a factory planning project.

This socio-technical approach is based on the three-skill approach developed by Katz, which differentiates technical, social and conceptual competences as personnel competences [Kat55, Son07]. In addition it takes into account the systemic analysis, which concerns the general conditions as well as the human, technological and organizational aspects of a system [Uli99, May01]. As a result, it is recommended that a distinction be made between managerial tasks that belong to a more technical project management and those are more along the lines of process facilitation. When possible these two areas of responsibilities should be transferred to two people, who then perceive the three tasks in varying intensities. Figure 15.9 left depicts the relevant aspects from a project perspective:

- The general conditions describe the design object and the actual situation as well as the business strategy pertinent to the production. This step is discussed in Chap. 14 (Strategic Location Planning) but is also addressed in more depth in the Sect. 15.3 on goal setting.
- The technical and special issues result from the factory design fields addressed in Chaps. 6–14. They represent the core of the factory planning and are the responsibility of the project management.
- In comparison, the emotional and social aspects affect the atmosphere and mood in the project and enterprise as well as the informal relationships between the actors. The process facilitator is responsible for ensuring that these are considered appropriately.
- The third aspect is the project organization and methodology. These determine how the project will be approached. The project manager and the process facilitator are responsible together for ensuring that both the technical as well as the social emotional aspects are taken into account. Thus criteria are formulated not only for technical aspects of the project but also pertaining to the methodological and atmospheric quality.

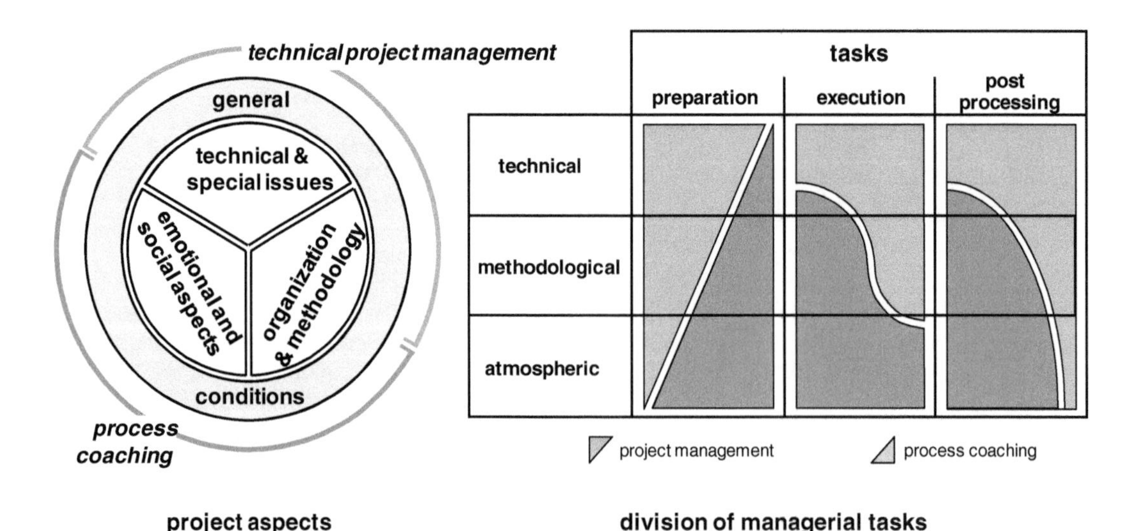

Fig. 15.9 Planning and managerial tasks (Schübel and Wiendahl). © IFA 15.464

Figure 15.9 right details the changeable aspects of both these managerial responsibilities during the sequential tasks: preparations, execution and post-processing.

- During the preparation phase the project manager dominates, whereas the process facilitator or coach gains significance with the methodological and then atmospheric aspects.
- During the execution phase the process facilitator ensures the methodological and atmospheric conditions for a professionally targeted solution finding in order to later pay more attention to compliance with the agreed upon rules.
- The usually unpopular post-processing is divided into two parts: On the methodological side, the project manager has to prepare or be responsible for organizing the results. Attention needs to be paid to ensuring the consistency of the contents, to tailoring results to specific groups and to presenting both consensus and dissent. Whereas on the atmospheric/methodological side, based on participant feedback, the process facilitator reflects on the quality of the results, compliance with rules and the mood of the team with the aim of continually making improvements.

The overall responsibility however remains that of the project manager.

In summary, it can be said in advance that the synergetic factory planning process model has proven to be practicable in many projects and has led to high quality results with fewer coordination problems. The synergy effect is derived from the fact that

- during each planning phase, there is a creative dialogue between the experts who are pertinent for the production and object planning in that phase,
- a mutual understanding of the demands, conditions and possible solutions for each of the specific disciplines is developed,
- interim results which have been agreed upon by both the production and object planners are available at the end of each phase, not just at the end of the factory planning,
- attention is guided early-on and consistently to soft factors such as employee orientation, communication or the sustainability of the factory through the controlled alignment of the production and object planning,
- the planning utilizes 3D representations as much as possible, thus ensuring clarity and consistency of all data in the factory's lifecycle and finally,
- risks during the construction and ramp-up of the factory's processes are reduced.

In the following, we will introduce the synergetic factory planning phases in detail, focusing on the technical content that the production planner brings into the project. Particular attention will be paid to describing the results of the phases and the synergetic linking of the object and production planning. The architectural and engineering services contributed by the object planners and delivered within the frame of HOAI are extensively described in relation to the factory levels in Chaps. 8 (Workplace), 10 (Work Areas), 11 (Building), 12 (Master Plan) and 13 (Location Planning). Whereas, the equally important project management tasks during the construction planning are examined more closely in Chap. 16 (Project Management).

15.3 Goal Setting

15.3.1 Main Steps

As is extensively discussed in Chaps. 1 and 3, it is essential to continually adapt business strategies to market requirements. This is nevertheless only possible when the factory in consideration also has the potential to manage the posed requirements (e.g., variant flexibility, high logistic performance). The basis for creating this potential is set at the beginning of every planning project during a project definition and goal setting workshop. Existing business strategies or a special strategic location planning (see Chap. 14) serves as the foundation for this workshop. Figure 15.10 depicts the three steps involved in such a workshop.

In addition to the planning team, participants in such a workshop include those responsible in

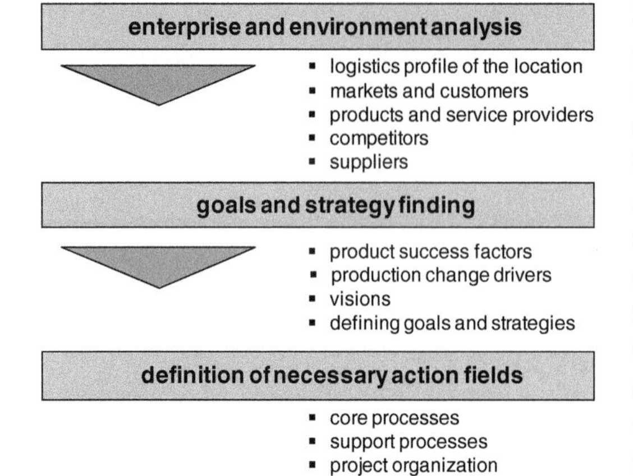

Fig. 15.10 Steps in a project definition and goal setting workshop. © IFA 15.465

management for the process as well as those responsible for the functions of core and support processes. The team is responsible for developing clear goals for the project based on the strategic guidelines provided by management and for finding a common understanding of the task. The first step involves delimiting the location in comparison to other enterprise locations in the form of a logistic profile. Following that the location and its surroundings are analyzed and possible future scenarios are developed. By examining the suitability of existing structures, strengths and potentials of the location can be identified, which will then be divided into concrete sub-goals. The workshop results in a definition of the project which describes the next steps as well as responsibilities. Depending on the project scope, a workshop such as this requires one to two days. Along with the goal setting this planning phase is also responsible for conducting a number of concentrated analyses. So that participants can be suitably prepared for these analyses, the problems to be addressed are provided to participants in advance.

15.3.2 Site Logistic Profile

In order to delineate the considered location and planning tasks that have to be executed, it is necessary to gain a rough overview of all the enterprise's products and value-adding processes that are to take place at the site. Figure 15.11 depicts a simple scheme for this. On the left side, all of the enterprise's product groups and the

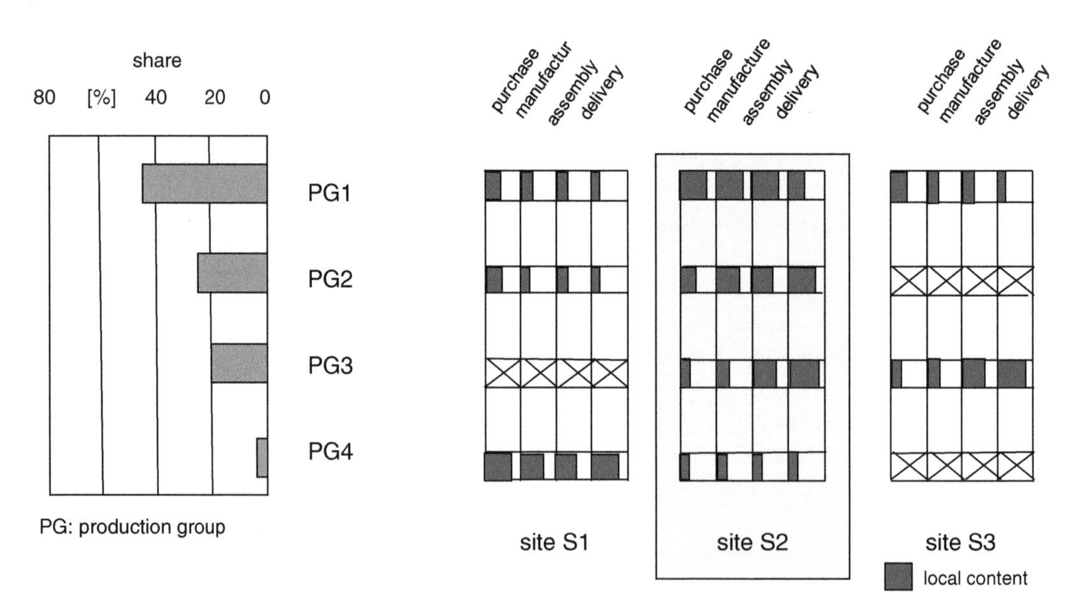

Fig. 15.11 Production program analysis for a site. © IFA 15.466

projected sales turnover for them are displayed in percent of the overall value. On the right side, based on the results of the strategic location planning (Chap. 14) we can see how the product groups are to be divided among the locations and which processes are intended to be executed at what volume in the purchasing, manufacturing, assembly and delivery sections. For the procurement, both the volume and sources (external/internal) are determined for article groups. Manufacturing includes all of the parts produced in-house according to article groups and the related production hours, whereas the assembly can include components for customers and other locations (competence oriented assembly groups), but usually delivers finished products. Depending on the delivery concept, deliveries are made to a sister plant, a logistic provider or directly to an end customer.

From a logistics perspective, topographically visualizing the flow of goods in and out of the site helps provide an idea about the volume and distances that needs to be bridged as well as the network with customers, sister plants and suppliers. Figure 15.12 depicts an extremely simplified diagram for a fictional production network with 3 sites. Due to the usually greater number of products, suppliers and customers, real schemes are naturally much more complex and are effectively separated into a number of layers and depicted according to product groups and actors.

The site's logistics profile derived from this production network is vital to the factory planning (Fig. 15.13). Here the product groups are further divided into customer/market segments, each meaning a specific combination of specific products for specific customers in specific markets which should be produced at this site. As can be seen in the matrixes depicted in the diagram, the functional workplace design (extensively discussed in Chap. 9) determined the procurement, production and delivery concepts for each of the segments.

One might get the impression that there are no other design challenges remaining for the factory planner to solve. However, it cannot be overlooked that the strategic preliminary determinations were made from the perspective of the market and customers and securely supplying

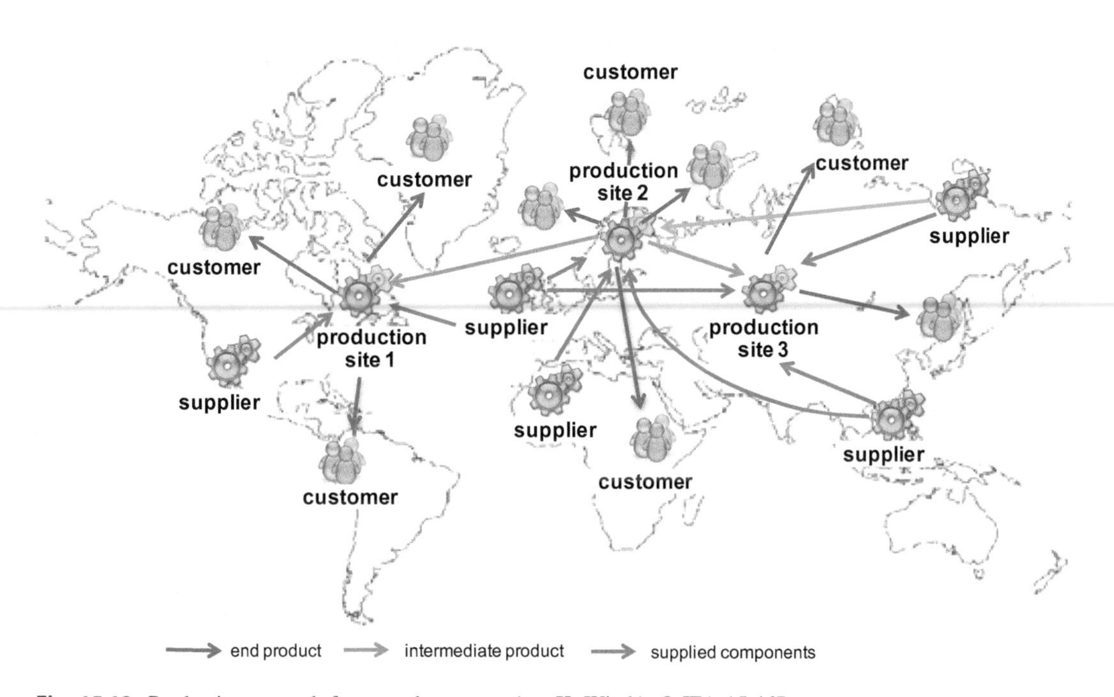

Fig. 15.12 Production network for a product group (per K. Windt). © IFA 15.467

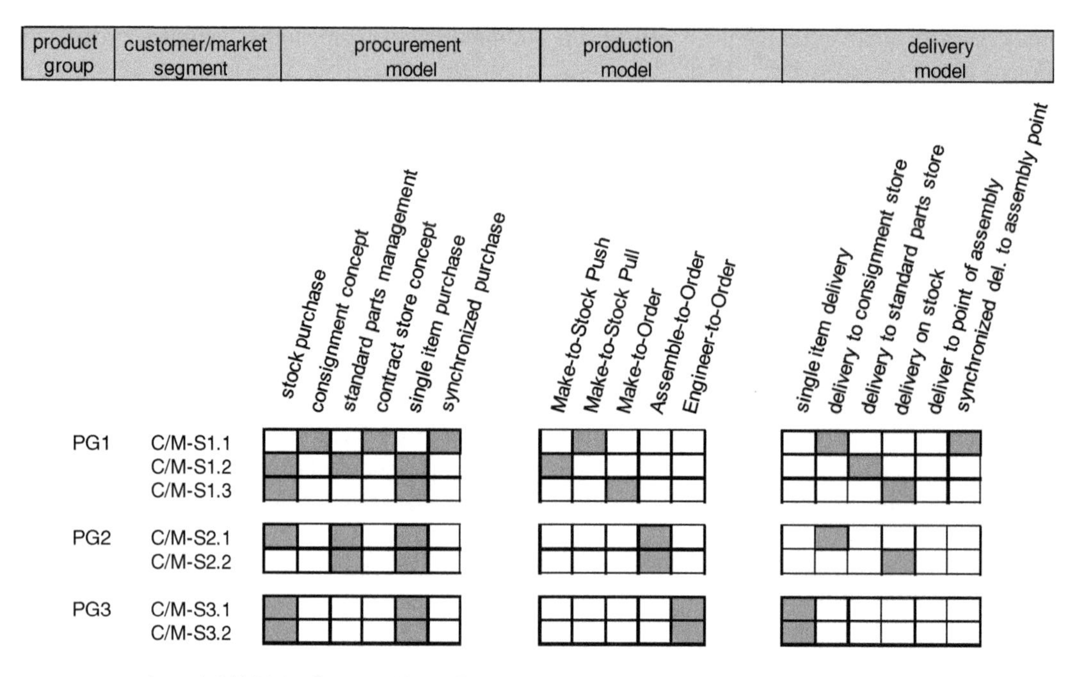

Fig. 15.13 Site logistics profile. © IFA 15.468

them. Technically different products for very different customers may meet-up at the production site though. The challenge thus remains to merge the resulting, to some degree even contradicting requirements, into a changeable attractive local solution in view of economic, ecological and social aspects. This explains why it is indispensable despite all the strategic objectives to first analyze the enterprise and environment in view of the site and afterwards set the contractual goals and strategies for the planning object.

15.3.3 Environment Analysis

The primary goal of the *enterprise and environment analysis* is to gain a compact overview of the enterprise's market, production and competitive situation and its products from the perspective of the considered site. In addition to identifying the current standing of the enterprise, planned future developments as well as emerging trends should be included for a horizon of 5 to a maximum of 10 years. From here the requirements for the new factory are derived. For example, if a company plans to place a new product on the market in the next years, the space required for the necessary production area should already be taken into account in the current planning. Similarly, if new regulations or changes in environmental protection are anticipated, the resulting requirements should also be considered.

The results of such an analysis are depicted in Fig. 15.14 based on the example of a mid-sized supplier of high quality engineering products. This project was triggered by the need to shift the production site in order to respond to capacity demands.

The market is described here through the sales volume and its regional distribution, while the product is described by production relevant characteristics such as base materials, type of functions, dimensions and size of production runs. There are also other sites, but the connection is very weak and thus is not considered further. The competitive situation clarifies the position behind the market leader, while the

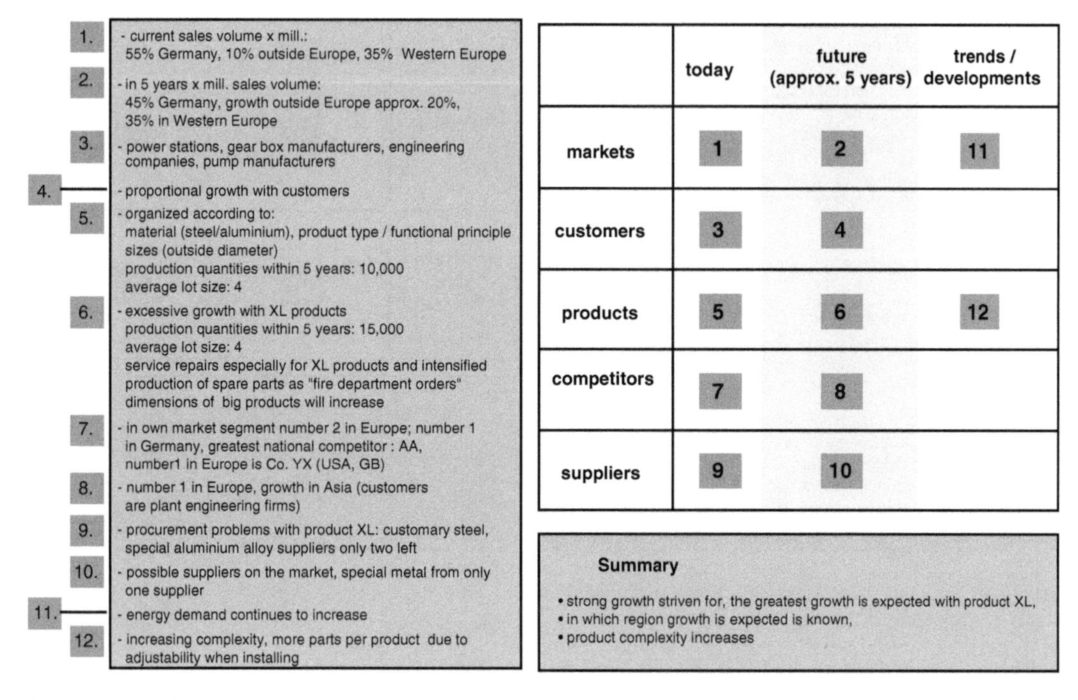

Fig. 15.14 Analysis of the market, product and competitive situation (example). © IFA 15.469

supplier analysis indicates the danger of a sole supplier for a critical material. The summary shows a clear growth strategy based on the growing demand for the customers' products in a specific region and that is ensured by further developing the product's functions. This is nevertheless connected to an increased product complexity.

15.3.4 Factors for Success

One of the basic components of this step in the analysis is identifying the product specific success factors. Here again the current situation is differentiated from future developments in order to derive requirements for the planned factory. If a product on the market distinguishes itself for example, by high delivery reliability, this directly impacts the factory—in this case, particularly the logistics concept. Figure 15.15 presents the results of the distribution and development survey.

In this case, it is clear that the most important factor for success is high product quality, which in the future however needs to be extended to include shorter delivery times, improved service and partial out-sourcing.

15.3.5 Change Drivers

The last step of the analysis is concerned with identifying the internal success factors and the factory's change drivers. Generally, the following questions need to be answered:

- What were the key drivers that lead to changes in the factory previously? How often did these drivers appear?
- What effects did the change drivers have on the factory and which measures were implemented?
- What are the key drivers that will lead to future changes in the production?
- What is the anticipated frequency of their appearance?

Factors considered here include production and logistics costs, added value per employee, floor space utilization, stock-turnover ratio and changeability of the factory concept, but also

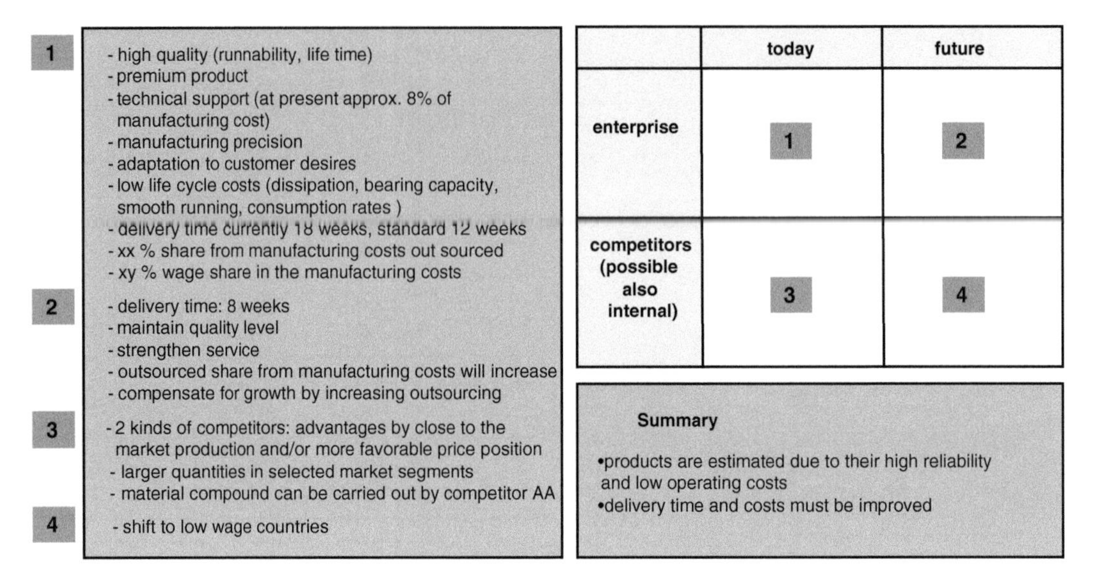

Fig. 15.15 Product success factors (example). © IFA 15.470

employee qualifications and spatial proximity between direct and indirect areas. It is crucial to identify what has worked in the past and what potentials can be created within the frame of the new planning.

Figure 15.16 depicts the results of analyzing the change drivers, which in this case turned out to be relatively short. As already mentioned, the increase in sales volume triggered management's decision to relocate the factory nearby as an independent business unit, whereby the support functions (IT, personnel, finance etc.) would continue to be provided from the central location.

15.3.6 Creating Scenarios

The case selected as an example is comparatively simple. With more complex cases, it is recommended that *scenario management* be implemented. Originally, scenario management was used primarily in product planning, but an abbreviated form can also be applied very well within a workshop. The goal of this method is to determine consistent, readily understandable visions of the future [Gau06].

Scenario management is based on two fundamental principles: (a) the concept of networked

1
- increase in turnover
- shorten delivery time
- markets (segmentation)

2
- Management decision to move

3
- legal independence as own business unit with the following exceptions:
- IT department and support
- human resources management
- finance and accountancy
- purchasing of sub-functions
- maintenance service

4
- creating a closed spatial and organizational unit

	change driver / trigger	consequences / measures
in the past	1	2
in future to be expected	3	4

Fig. 15.16 Production change drivers (example). © IFA 15.471

thinking in which the relationships between numerous influential factors are taken into account and (b) the concept of multiple futures, which states that each factor can develop in different directions in the future (see also [Gau06]).

The process for creating scenarios has five basic steps, whose sequence is clarified in Fig. 15.17. First, the design field that is to be defined is selected (in this case a factory) and described with the aid of 'design field components'. In the case of the factory, these could be the design fields outlined in Fig. 2.8 for example. Subsequently, the influential factors that impact this design field are collected and a network analysis is used to derive the 'key factors' i.e., the factors that are particularly significant for the object of consideration.

Following this, future possibilities for developing each of these key factors (so-called projections) are derived. If, for example, a specific purchased material is a key factor, projections might include, sinking, stable or greatly increasing prices. In a subsequent investigation, the projections for the key factors are combined with one another and examined for their consistency. If, for example, energy and transportation costs are key factors, high energy costs have to correspond to high transportation costs and low energy costs with low transportation costs. Only consistent combinations of projections are entered into the future visions, which are referred to as scenarios. These scenarios are not to be mistaken for typical planning variations such as 'optimistic', 'pessimistic' and 'realistic'.

The final step, the scenario transfer, serves to convert the developed scenarios into a continuous text. The found scenarios span the future providing an indication of possible forthcoming effects on the design fields of the factory.

If the scenario management is applied in a project definition and goal finding workshop, it is recommended to use a strongly abbreviated procedure in comparison to the one described by Gausemeier. The key factors compiled by Hernández are useful as starting information. As outlined in Fig. 15.18, these factors are differentiated as either being from the extended environment (non-steerable) or from the more immediate factory environment (steerable) [Her03].

From this list, a maximum of 4–5 key factors with 2–3 projections each are selected. The participants in the workshop create the scenarios and

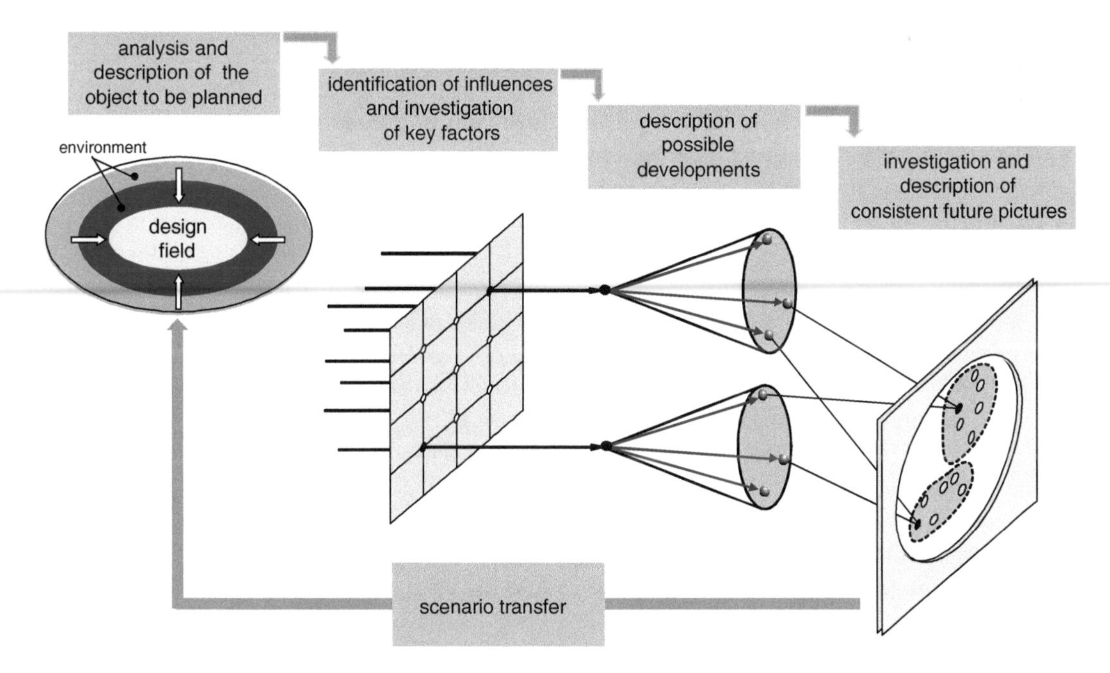

Fig. 15.17 Creating scenarios (per Gausemeier). © IFA 15.472

	non-steerable key factors (global and enterprise environment)
1	market dynamics, laws and development
2	demand trend
3	competition structure / new competitors
4	market structure and segmentation
5	competences of the competitors
6	economic development
7	competences of the partners in the network
8	utilization of the partners in the network
9	production network organization
10	industry-specific standards and norms
11	price requirements
12	innovation speed
13	globalization of the production
14	market strategy of the competitors
15	delivering requirements
16	risk inclination of the capital givers/shareholders
17	supplier structure
18	position of power of the suppliers
19	financial policy
20	technology development
21	ecological policy
22	global research and development intensity
23	raw materials
24	import and export
25	origin and structure of the customers

	steerable key factors (factory environment)
1	product types and types
2	product variants
3	amounts of products
4	degree of specialization
5	product standardization
6	product prices
7	place of the production
8	material development
9	product technology
10	vertical integration
11	production technology and automation
12	product size and weight
13	investment budget
14	current assets and fixed assets
15	sales volume and profits
16	building life cycle
17	cooperation strategy
18	location development
19	distribution strategy
20	market strategy / business field strategy
21	service
22	humane strategies
23	product life cycles
24	enterprise and meta goals
25	logistics strategy

Fig. 15.18 Steerable and non-steerable key factors of a factory (per Hernández). © IFA 15.473

the breadth of their experiences has to ensure the consistency of the projection bundles. The knowledge gained is naturally less than with traditional scenario management, but is sufficient for planning the factory.

Figure 15.19 presents the results from an exemplary factory planning project. Four possible scenarios were developed for the enterprise considered here, each with very different development paths. The 'boom' and 'trend' scenarios both basically assume a continuation of the developments up until now, whereas scenario 3 reflects focusing on a specific product spectrum from the field of exhaust technology. Scenario 4 presents a mixture of a number of development paths, including specialization of a specific product spectrum, orientation as a system supplier and a related decrease in the production volume.

The essential characteristic of these scenarios is that they are not allocated to any specific probability of occurrence. Whether a specific scenario occurs or not, is ultimately dependent on the development of the key factors and their interaction. The only way they can be influenced by internal business decisions is in regard to the steerable factors.

By projecting the individual scenarios on the design fields of a factory, it becomes recognizable which scenarios strongly influence the respective design field and which weakly influence it. The design fields, which require greatly varying solutions for different scenarios, need to have a specific degree of changeability e.g., in the form of mobile or scalable assembly facilities or easily adjustable ventilation systems. This makes it possible to assess whether a planned factory concept is sufficiently changeable to react to all of the development paths or only to selected scenarios. A concept that is only able to cover one possible scenario should be critically examined in view of its future viability. However, if a changeable solution is not used frequently enough, the savings from the related lower changeover costs do not cover the overhead costs of the solution. In this case the changeability has reached its economic limit.

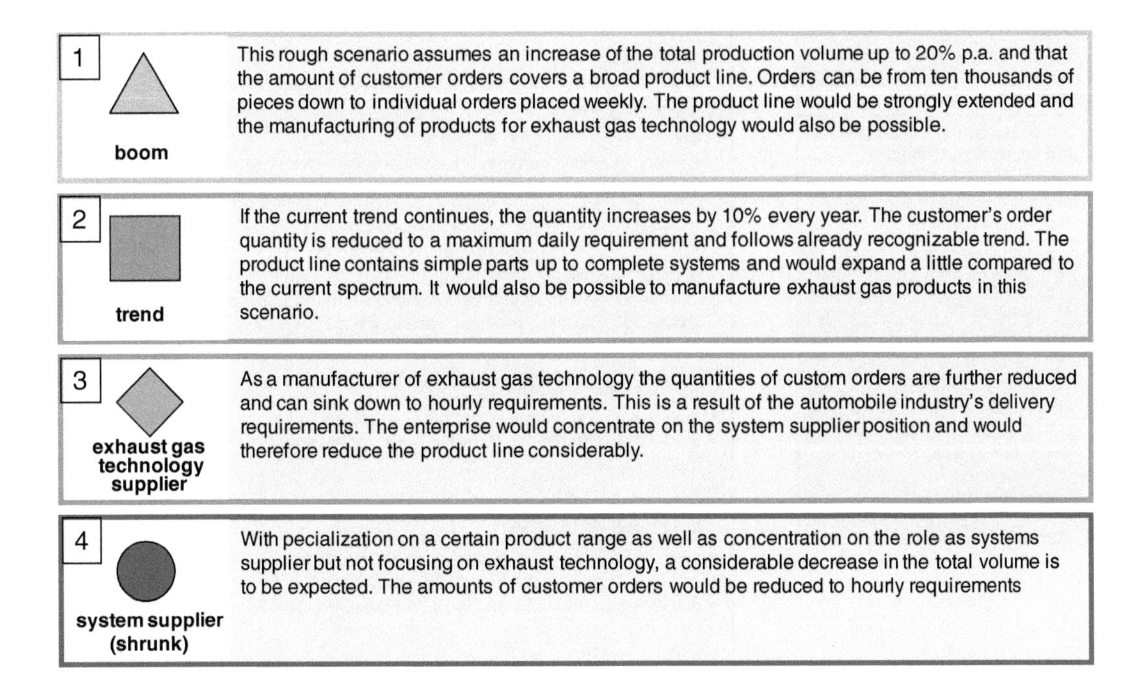

Fig. 15.19 Rough scenarios for an automotive supplier. © IFA 15.474

15.3.7 Finding a Vision

Defining a common vision for the factory involves an intensive discussion process since each of the interest groups (owners, management, employees, etc.) has a very different idea about such a vision. To ensure that a vision is as accepted as much as possible within the project team it should not be determined by the authoritarian rule of management, but rather developed by a team of senior staff. In doing so the following questions need to be answered:

- What is our vision of the enterprise (e.g., market leader, fast follower, best partner, leader in innovation)?
- What are the meta-goals of our location (e.g., sales, returns, delivery times, market share, products, increasing productivity, turnover, minimizing stock, system capability, quality, changeability, volume/output, comparison with competitors, etc.)?
- What strategies do we want to use to reach these goals (e.g., cost leadership, pioneering design, cooperation, imitation, niche strategy, technology leadership, zero-defects, CIP, process orientation)?
- Are emergency strategies required (e.g., through redundancy, expandability in small steps, the ability to re-built)?

Figure 15.20 depicts the results of a vision finding workshop for our case study.

In the left upper part of the figure, the vision is summarized as the result of a collection of idea cards (not shown here), while information for the meta-goals and business strategies is listed on the right side. The meta-goals form the business foundation for the entire factory project and serve as the basis for later controlling the success of the project. In this case, the business strategy for the factory is focused on two aspects of product quality and changeability through modularity. One of the given critical conditions is that the project can be realized within a short time period.

15.3.8 GENEering

As already mentioned a factory is also always an expression of the business culture and reflects the

1. vision of the factory in this project

Our products are reliable precise and economical through our competence, quality and efficiency. For us this means claim and obligation.

2. meta goals of the enterprise for the factory

3. company strategy for the factory

4. general conditions

2. meta goals

- sales volume: increasing from y mill €. to x mill. € within 5 years
- EBIT: min. 10%
- delivery time: from 12 to 8 weeks
  delivery reliability: currently 77% (difference between promised and attained date, exact to the week, delivery ex works); target: 85% in 1st year, 90% within 2 years
- increase in productivity: within two years 6%, as of then 3% p.a.
  inventory: stock turnover frequency from 2.2 to 3 within 2 years
- quality: maintain
- changeability: with regard to increasingly larger product dimensions, arranging all facilities to be as mobile as possible
- establish a service centre in the factory

3. company strategy for the factory

- technological superiority (precision, reliability)
- modularity (grid e.g. 3 x 3 m also advantageous for the move)
- mobility of facilities

4. general conditions

- max. move duration: 3 weeks, plan 50% production volume during move
- move must be completed within 6 months to not endanger other projects
- machine information available for the complete plant that is to be shifted

Fig. 15.20 Vision, meta-goals and factory strategies (example). © IFA 15.475

enterprise's demands on itself and its products. Thus, if a company primarily serves the market for regenerative energy, the technology found there should also be used in the factory. Accordingly, the enterprise's overall vision has to be transformed not only out of the production perspective but also the spatial perspective into a factory vision.

A proven method for achieving this is GENEering [Rei04]. As the term—which linguistically combines 'gene' and 'engineering'—indicates this method is concerned with developing a 'DNA code' from the perspective of the object planning. This sets the structural forming parameters in the factory's future lifecycle and thus the performance of the object's form under the motto "form follows performance".

Figure 15.21 clarifies these 8 factors, which will be visualized in further steps through expressive images from these areas. In the first round these are not directly related to the factory development, but rather are meant to trigger new associations for workshop participants. In the second round, the factors are reviewed in a similar fashion, however this time are oriented on examples from the factory building.

Each individual factor is broken down once more into three sub-concepts and examined from internal and external company perspectives. The results of a corresponding detailed discussion about the factor 'changeability' in a project are depicted in Fig. 15.22. Each of the sub-concepts are designated an importance, whose value can be between 1 (not important) and 10 (extremely important). In this example, modularity and flexibility were assigned a high point value, while mobility was barely pertinent.

The arithmetic mean then flows into an overall image of this object's DNA code as shown in Fig. 15.23. In doing so the factors are divided into hard and soft assessment fields. In the last step, the desired features of the complete object are allocated to the affected design fields, as is indicated on the right side of the diagram. In our example, it can be seen that the soft factors strongly dominate for the site, building and organization.

Through the ensuing group discussion of the target concepts from a production engineering

Fig. 15.21 GENEering. © Reichardt 15.476

and object related perspective, highly original ideas often arise, which are formative for solutions, particularly in the structural planning of the production and building.

GENEering gains particular importance in the discussion on sustainability in industrial construction. Already in 2002, in their highly acclaimed book "Cradle to Cradle" Braungart and McDonough pointed to the—at the time still minimally considered—side effects of industrial societies such as contaminants in products, unresolved waste disposal or irreplaceable resources [Bra09]. The goal of energy efficiency according to this is only one of many aspects of a comprehensive sustainability, especially when planning a factory.

Subsequently, the results are transformed into guidelines for designing the object and can for example look like this:

- Maximize changeability through mobility, modularity, employee qualifications and work-hours' model.
- Use new technology for construction, fire protection and heat recovery.
- Increase efficiency by taking into account limiting values (minimum throughout times, minimum inventories etc.).
- Secure ecological off-set spaces; take into consideration local rainwater seepage.
- Create aesthetics through order, color concept and media routings.
- Promote communication by locating support functions such as work-prep and order processing close to the production.
- Provide a corporate identity through exhibitions, individuality, tradition and myth in a foyer.

15.3.9 Fields of Action

Based on the preceding results, the ensuing 'definition of fields of action' designates concrete sub-projects with responsibilities and due dates.

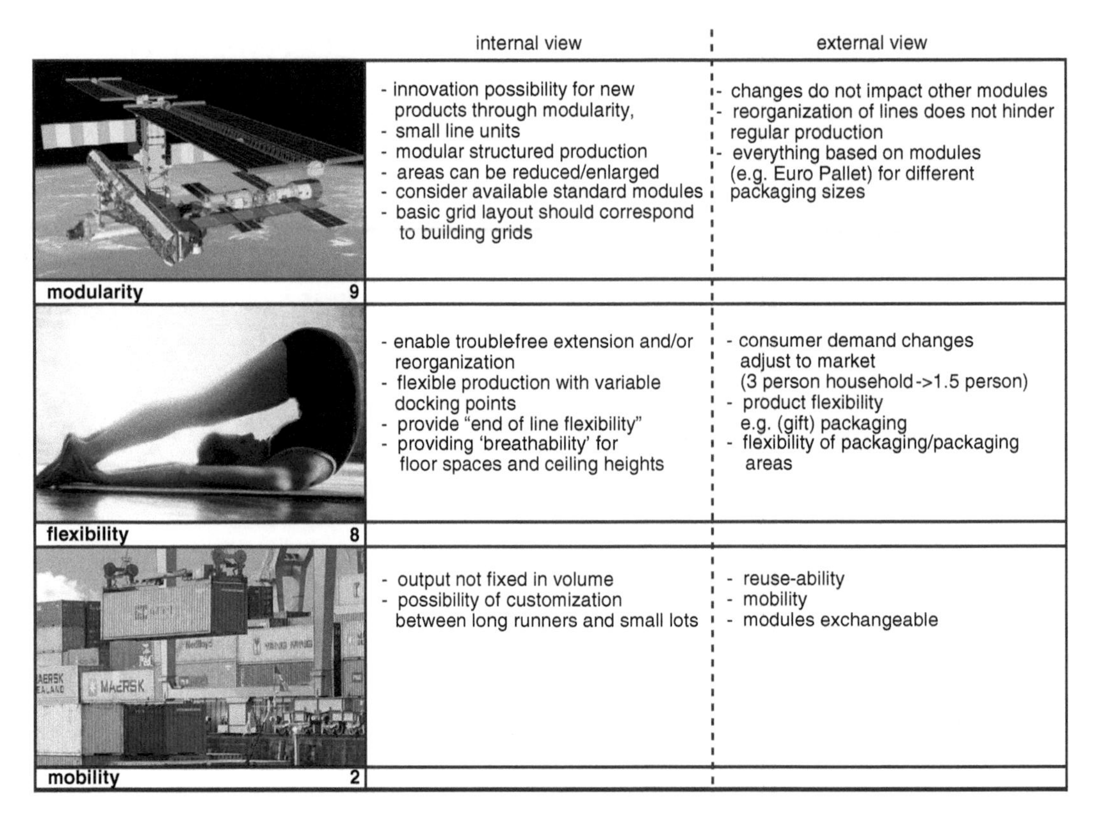

Fig. 15.22 Aspects and assessment of the factor changeability (example). © Reichardt 15.477

This can for example be organized according to a manufacturing company's core and support process. Figure 15.24 depicts a possibility oriented on the factory planning's main processes 'product development' and 'order processing' (see also Fig. 2.7 and [Wie14]).

Marketing and sales search out and define new product ideas, which are then transformed into products by research and development. The production (manufacturing and assembly) along with logistics (procurement, production and distribution logistics) process the order together. The support processes provide services for the core processes. From the perspective of the factory planning, the individual processes are each examined in view of the company's vision, guidelines and location strategy for areas of activity.

Goals in terms of economic aspects and timelines are defined for each process. Moreover, all of the workshop participants have to agree on the identified fields of activity, because this is the only way to ensure a common process. In addition, compliance with the set goals has to be ensured by the steering committee with appropriate authoritative powers.

Once the goals have been defined, the basis for the actual factory planning from the perspective of the process and spatial planning is set. The next step, the base analysis, is discussed below.

15.4 Base Analysis

The base analysis primarily serves to evaluate the product and process data. In case of a re-organization this allows the weakness and potentials of the current state of an existing factory to be determined. With a new planning, it allows reference processes within the planned factory to be

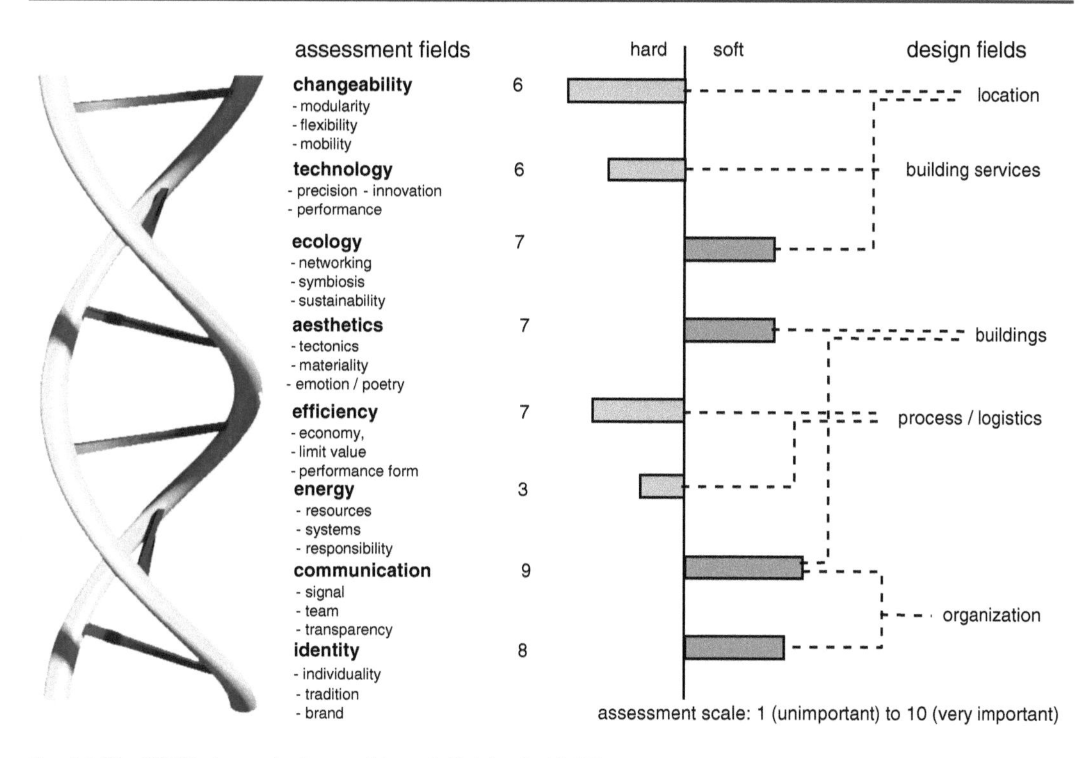

Fig. 15.23 GENEering code for an object. © Reichardt 15.478

described. In order to systematically organize the data collection and documentation it is recommendable to consistently implement a scheme such as the one depicted in Fig. 15.25. Following a brief description of the contents, project management also records the data users, producers and availability along with the state of completion for each data group and saves this information in the project database. Experience has shown that recording and analyzing the data

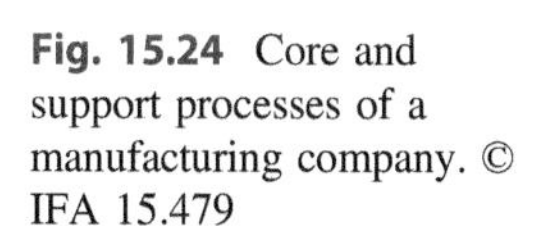

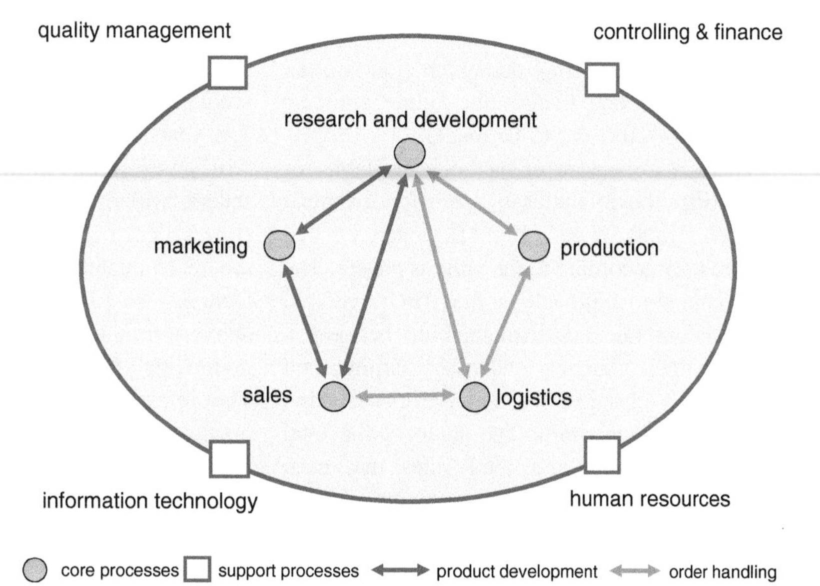

Fig. 15.24 Core and support processes of a manufacturing company. © IFA 15.479

no.	data	description	request from	repon-sible	date	received		
						status	from	when
0	**project management**							
0.1	contact person					3		
0.2	logo of the enterprise	for presentation				3		
1	**enterprise**							
1.1	company structure	group structure, legal form, financial integration				3		
1.2	sales volume					3		
1.3	number of employees	distribution among sections (also indirect employees)				3		
1.4	business calendar	incl. open days, holidays				3		
1.5	shift model					3		
1.6	organization chart	organizational structure (over 3 hierarchy levels)				3		
2	**plant and factory layout**							
2.1	master plan					3		
2.2	current rough layout					3		
2.3	current fine layout					3		
2.4	target rough layout					3		
2.5	target fine layout					3		
2.6	area balance sheets	size of all areas, (incl. unused areas)				3		
2.7	fixed points	layout with fixed points				3		
2.8	3D files	e.g. movies				3		
3	**products**							
3.1	production program	volume development of all variants, changes				3		
3.2	forecasted product program					3		
3.3	name of the products					3		
3.4	product numbers, item number					3		
3.5	variant number					3		

Fig. 15.25 Data requirements list. © IFA 15.480

can take 30 % up to 50 % of the total project expenditures. The complete list is documented in Appendix C2.

The analysis is conducted for the key design aspects of the factory (see Figs. 2.8 and 5.17). In order to limit the expenditures required for this, it should be decided in each project which factory aspects (also referred to as objects) need to be analyzed. It is critical that data is not collected 'blindly' and saved 'in advance', but rather that the project leader together with the sub-process planners determines the scope and depth of required data for each step. It also has to be kept in mind that the existing documentation is often not up-to-date and that in the future strategic assumptions will have to be made.

There is a plethora of methods available for supporting these analyses (see for example [Agg87, Gru00, Wie96, Ket84]), which we will explore later according to the various phases. The most common approach is the *ABC Analysis*, which is suitable for distinguishing between a few important elements and many unimportant ones. It is well suited for example to analyzing the production program. The share of the total production volume and total sales that each product represents is depicted comparatively in a graph, allowing high revenue products to be easily separated from low revenue products. Information such as this is required when designing the factory's structure and makes it possible to divide the production into 'best seller' and 'slow seller' segments.

15.4.1 Object Data

- Production Program

The starting point for analyzing the object data is usually the production program which, based on the overview depicted in Fig. 15.11, summarizes the products to be manufactured at the pertinent location along with their respective production volumes over the course of the next 3–5 years. In the sense of the abovementioned ABC analysis, this shows which products are best sellers, which are steady sellers and which are slow sellers with low production volumes. Figure 15.26 depicts the example of a pump manufacturer, who offers 7 standard pumps with a delivery time of 5 days as well as a special program with 7 types and a delivery time of 3 weeks. This information is primarily required for organizing the production segments later, if the segments are oriented on production volumes.

With serial production it is common to have a BOM (bill of Material) for each product type with parts manufactured in-house separated from

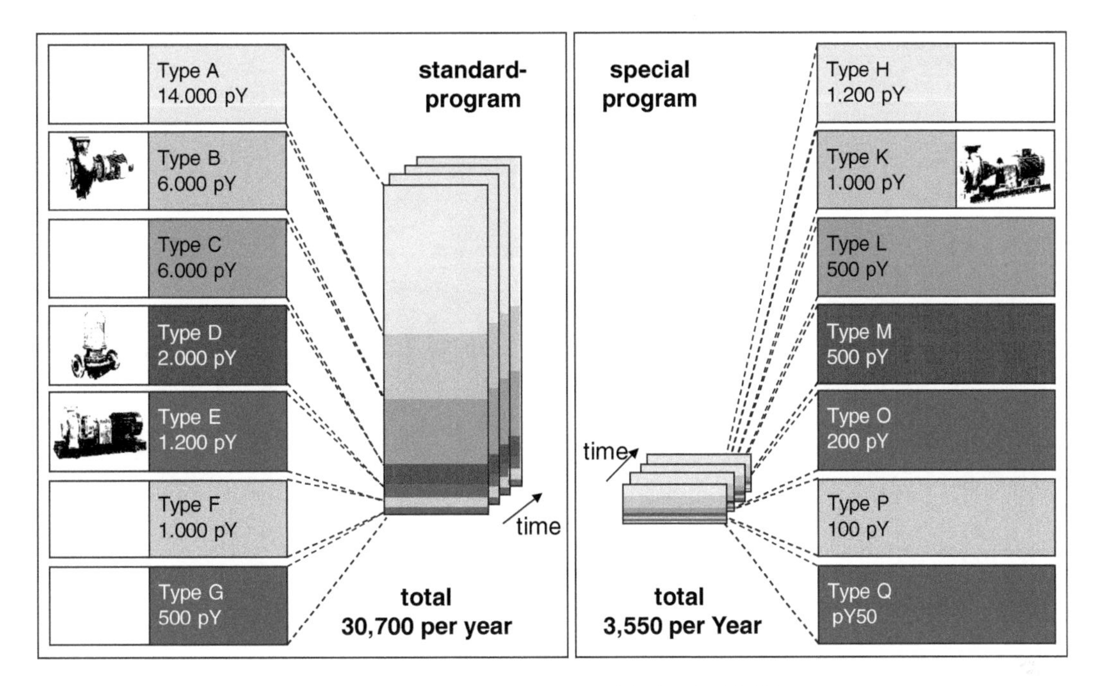

Fig. 15.26 Production program of a pump manufacturer (example). © IFA 15.481

those bought from suppliers. For parts manufactured internally there are work plans which document the individual operations with the corresponding manufacturing methods, setup times and production times per piece. These form the basis for calculating the capacities later. In comparison, job productions with a large variety of products use the load data from invoiced orders and combine it into a representative production program. Section 15.6.2 (structural dimensioning) considers this in more detail.

- Operating Facilities

The operating facilities represent the second largest object group. Here, one starts with an overview of the site, the location of the building, the infrastructure as well as expansion possibilities and existing open spaces. These are documented in the site ground plan; a highly simplified example is depicted in Fig. 15.27. The plan can be supplemented with additional data e.g., the state of the roads and buildings including equipment and facilities (cranes etc.). This information is necessary for assessing locations as well as for planning areas for potential expansions.

Within the building, data about resources that is relevant to the factory planning should be ascertained. Machine specification sheets; which are maintained by the work-prep section and serve as a basis for determining capital assets, are well suited for this. Required floor space, height, weight, energy requirements and media connections are key data for the factory planner. Additional important information includes required ventilation or exhaust vents as well as noise from vibration emissions that may require encapsulation or a vibration absorbing foundation.

The floor space data for the facilities has to be converted into a normal operational state before being applied in further planning, because the current state generally does not correspond to what is required for a proper process flow. To do so, floor space for a local material buffer, service and maintenance areas as well as tools and fixtures need to be included. This can often add up to be a multiple of the initially stated floor space. Section 15.6.2 examines this more closely as well.

Dividing the hall floor space into storage and buffer areas, functional areas, production areas and transportation areas leads to an area balance

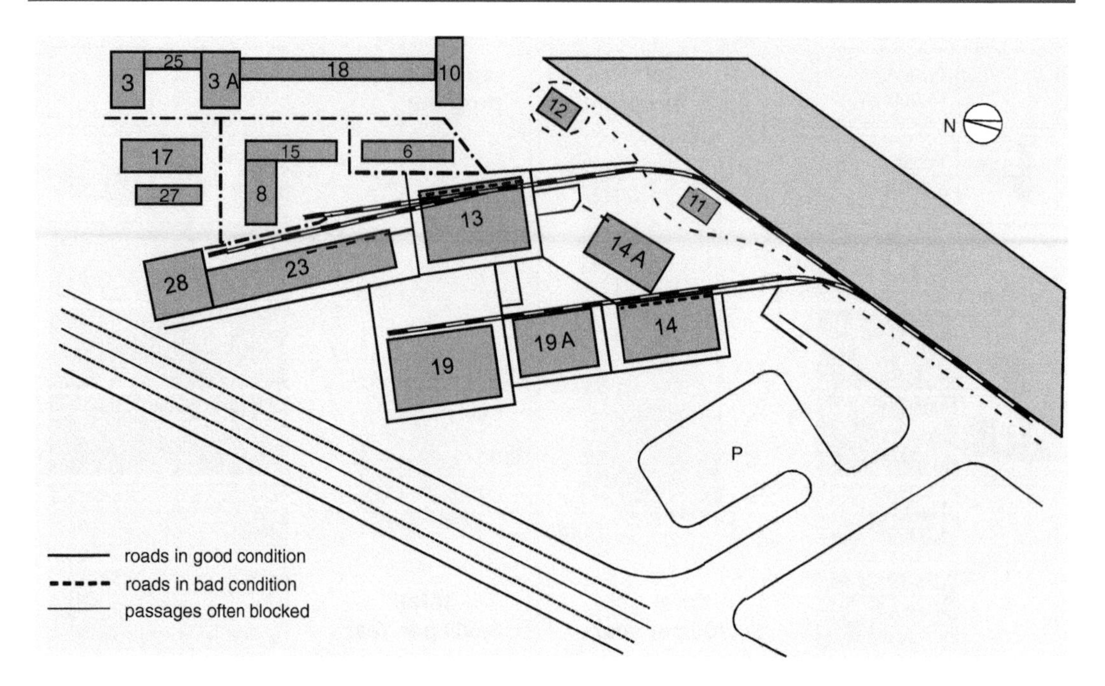

Fig. 15.27 Site ground plan (example). © IFA 15.482

sheet for the entire plant (Fig. 15.28). By comparing the result with standard industry area ratios, it is possible to draw conclusions about the appropriate distribution of individual areas. Frequently an area balance sheet such as this leads to reducing stores and buffers in order to minimize waste.

As a strongly aggregated overview of an object analysis such as this, Fig. 15.29 depicts the progression of the sales volume, total area, the separately indicated office space and the number of employees assumed for an exemplary factory planning. In this case the actual values during the planning year were set at 100 % and

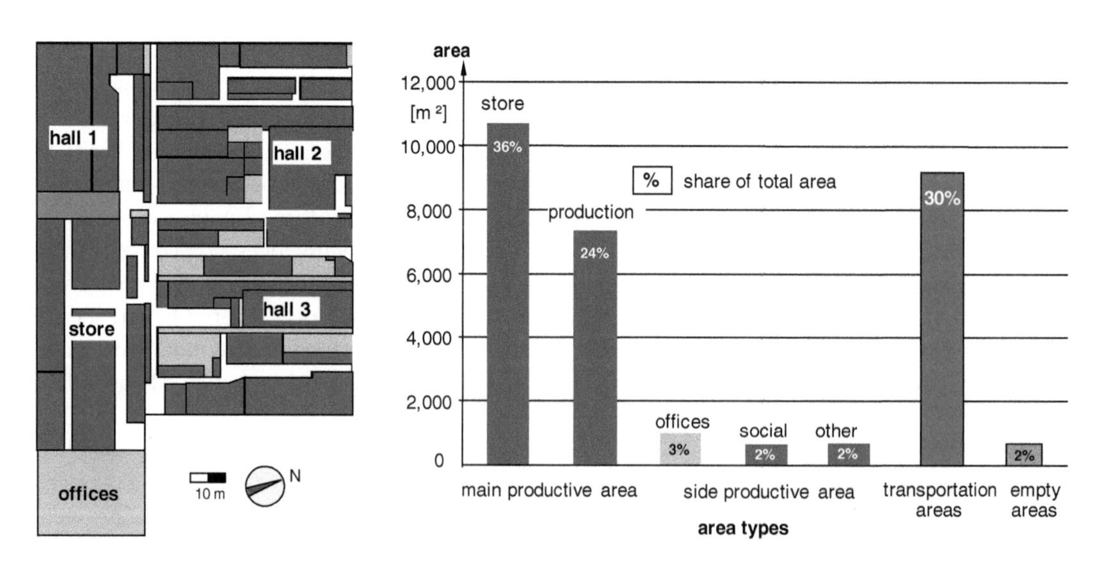

Fig. 15.28 Area balance sheet for an automotive supplier. © IFA 15.483

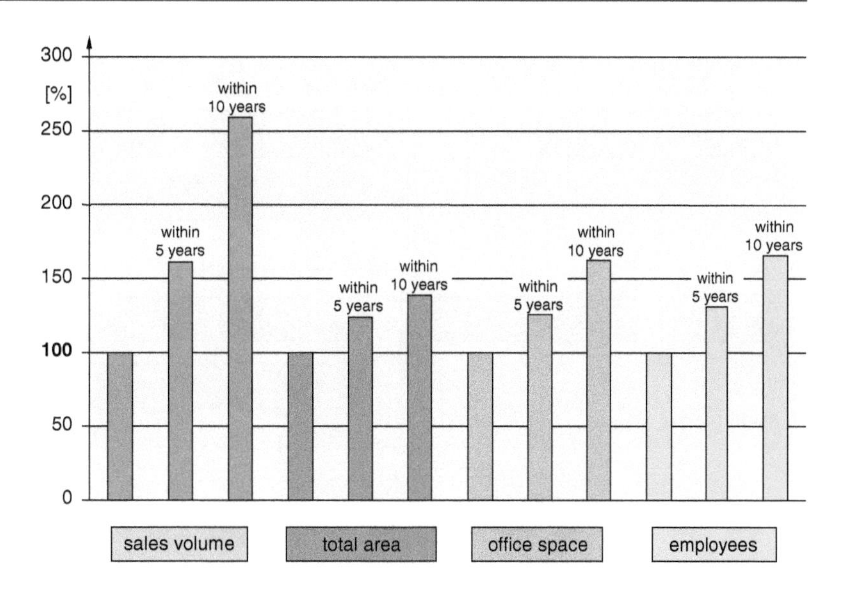

Fig. 15.29 Planning framework for a factory project (example). © IFA 15.484

the planned values were normalized to the respective values. While the sales volume in the next 10 years should increase to 250 %, the growth in total area based on rationalizing and re-structuring is approximately 140 %. The number of employees due to sales related activities rises together with the associated office space to approximately 170 %.

15.4.2 Process Analysis

The goal of analyzing the processes is to describe business processes, materials, communication and value flows. A number of methods and tools have been developed for depicting business processes, many of which originate in corporate information technology e.g., the ARIS system (Architecture Integrated Information System) [Sche01, Sche02]. For business applications REFA Association [Bin11] developed some recommendations, whereas process descriptions focused on logistics are particularly well suited for analyzing factory processes.

Figure 15.30 depicts an example of such the latter, the *Process Chain Analysis*, developed by Kuhn at Fraunhofer Institute for Material Flow and Logistics in Dortmund [Win97]. It uses specific symbols to describe the individual process steps and their connections for a product or product group. So-called connectors produce either a physical or temporal link between a number of process elements e.g., joining a number of parts into a component. In Fig. 15.30 the process chains for two article groups can be seen. Article Group 1 is deliverable within 24 h, while Article Group 2 reaches the customer according to standard service within e.g., 1 week. The second chain has two additional process elements.

Each process element is described by four groups of features, summarized in Fig. 15.31. A process chain element produces a physical or spatial transformation of an object e.g., forming, painting, transporting, storing etc. The processes are described by sources and sinks as well as an inner structure, as depicted in Fig. 15.30 in the lower part as a sub-model. This sub-model describes the topology—that is the layout—of the elements as well as its integration into the hierarchical organization and communication structure of the network.

The elements' resources are classified according to the material and immaterial features, while steering characterizes the planning and control on the normative level (corporate culture

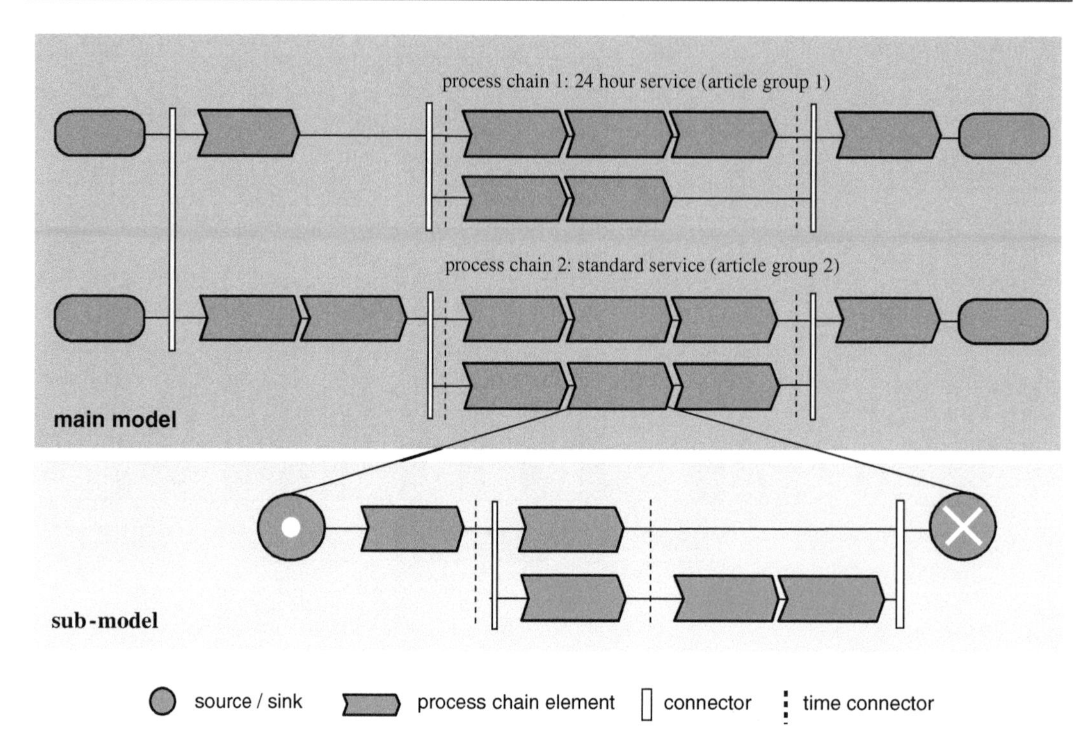

Fig. 15.30 Process chain model (per Kuhn). © IFA 15.485

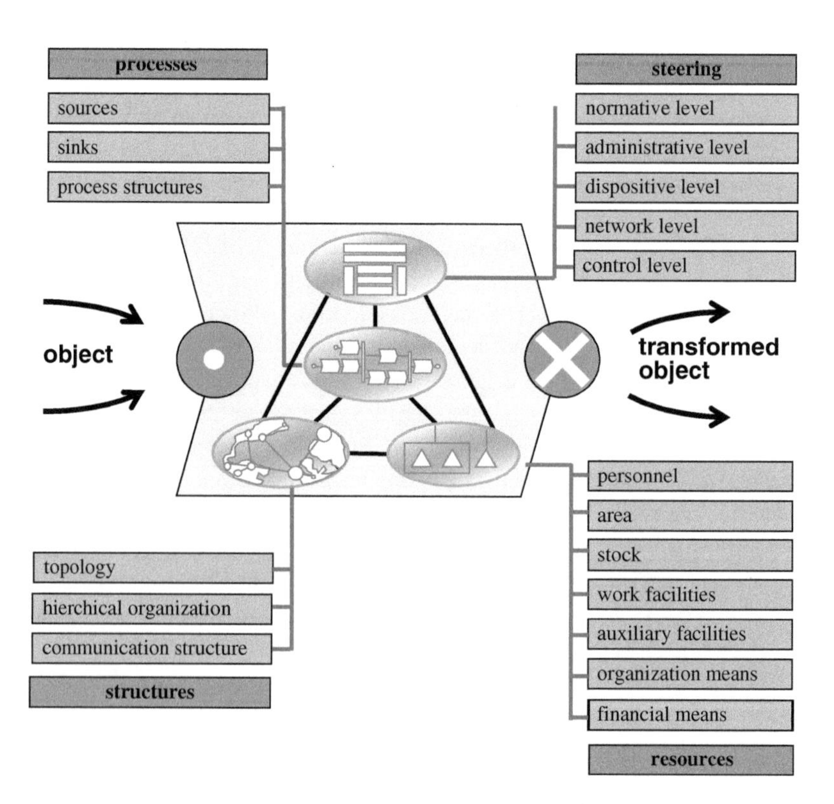

Fig. 15.31 Features of a process chain element (per Kuhn). © IFA 15.486

and rules, administrative, disposition, control and links within the network).

Along with making processes transparent for all participants by representing them in a simple straightforward format, the goal of analyzing a process chain is to uncover and resolve weak points. The recommendations set out in Fig. 15.32 are applicable as rules. Processes are distinguished as either utility processes (these create value—the percentage of these should thus be increased) and support processes (these are necessary due to the selected utility processes, however do not add value e.g., transportation between two workstations—such processes should at least be maintained as a constant). Faulty processes, on the other hand, involve uncertain process designs and need to be constantly reduced, whereas blind processes represent pure waste.

In addition to traditional methods of modelling business processes, *value stream analysis* has proven to be a valuable aid in depicting and analyzing a product family's value stream and its corresponding flows of information and materials. Based on standardized symbols the entire value stream is recorded in the opposite direction of the material flow i.e. from shipping to receipt of goods (see also Fig. 14.16). An extensive introduction to value stream analysis can be found in [Roth03]. Figure 15.33 outlines the basic steps.

The goal of analyzing the value stream is to identify waste in terms of stock, floor space and waiting times in the throughput of products and to thus create bases for developing a flow oriented production. This easy to use method (typically requires a few days including preparation and follow-up) is especially helpful for obtaining a quick overview of the production processes during the analysis phase. The results are carried over as input data into the next phase of the factory planning—the structural design—and can be further refined as needed.

Figure 15.34 depicts an example of such an analysis for the current state of a production process. Following the analysis, the investigation team develops a target state, characterized by a dramatic decrease in the throughput time from 15.1 to 2.9 days (Fig. 15.35). This is attained by discontinuing interim operations and eliminating related buffers.

It should be noted that both the business process analyses and value stream analysis

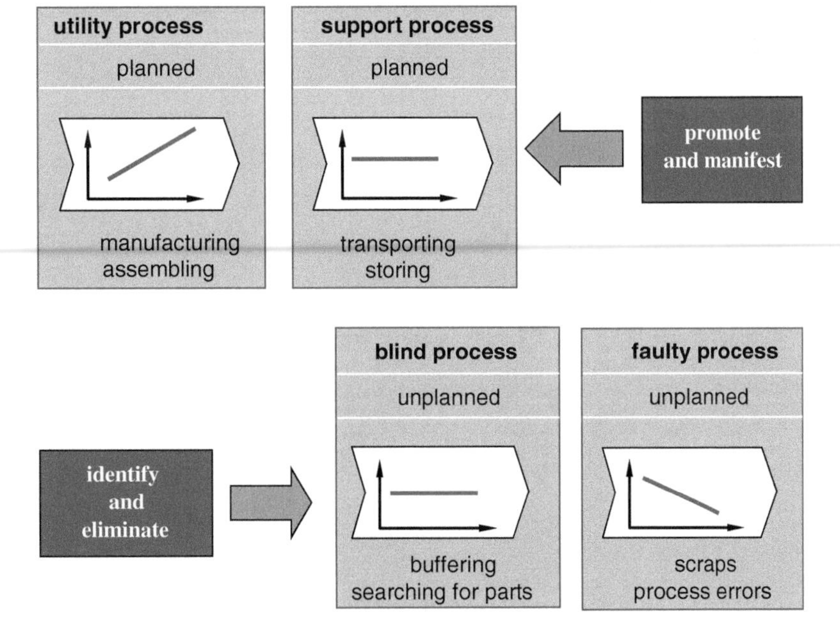

Fig. 15.32 Approaches to improving the process chain. © IFA 15.487

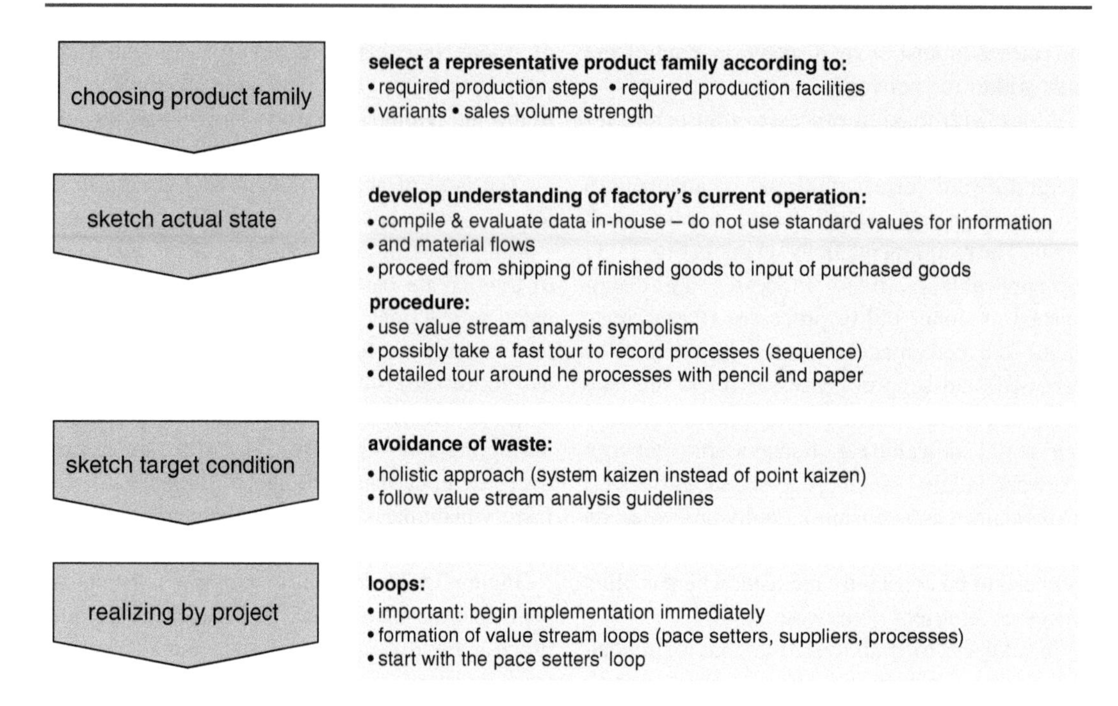

Fig. 15.33 Procedural steps in a value stream analysis. © IFA 15.488

always consider only one product or product group. When a number of products or product families are involved a corresponding number of analyses are then required.

In a factory, several different product groups (typically three to five) with different production volumes are usually manufactured for different purchasers. In addition components for other

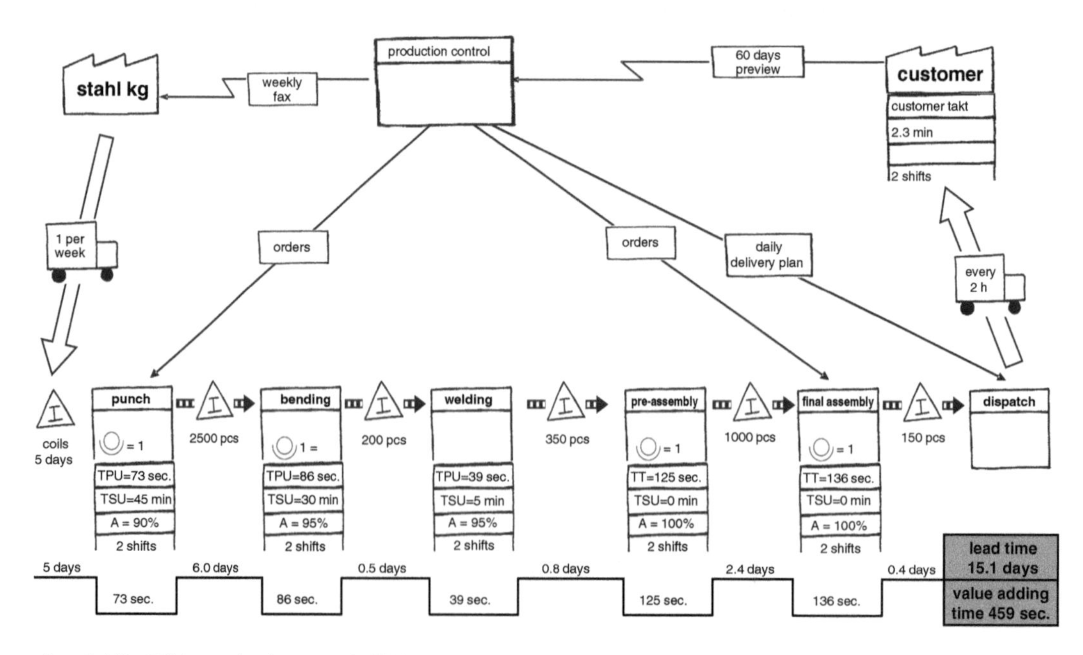

Fig. 15.34 Value stream analysis example—current state. © IFA 15.489

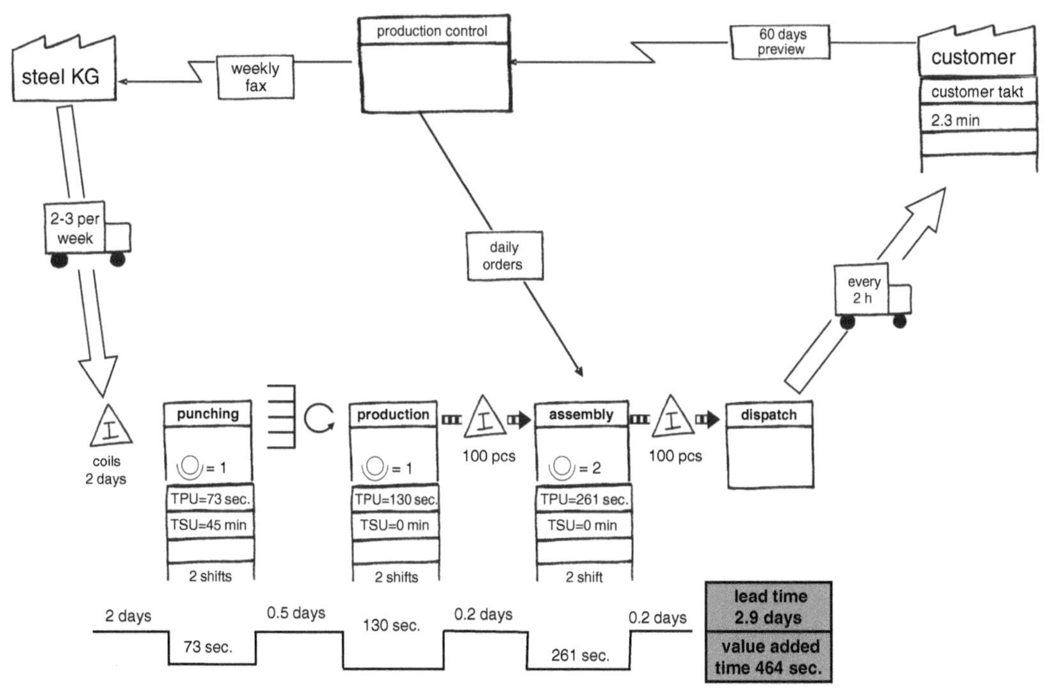

Fig. 15.35 Value stream analysis example—target state. © IFA 15.490

plants as well as replacement parts are also produced. Frequently, due to the underutilization of machines it is not possible to install a separate production for each of the product groups. Therefore all of the products in the entire order stream compete for capacities.

In this situation it is practical to also record the material and information flows from the perspective of the resources. The Sankey diagram (Fig. 15.36 left) is a helpful tool for this. It visualizes the organizational units and flows of e.g., material or information between them. The space required for the units and the distance between them is not considered in the graph. The width of the flow channels is proportional to the transported volume per time unit (here e.g., tons/per month). The flow relations can also be recorded in a 'to and from' matrix also referred to as a *material flow matrix* (Fig. 15.36 right).

The rows represent the transmitting/sending station and the columns the receiving station. The numbers for the flow can refer to the weight, volume or quantity transported per time unit. Deviations between the sum of the inputs and outputs can arise when not all of the sections are documented in the matrix.

As personal communication gains significance, the relationships between the flows of information also continue to take on importance. These can be visualized using the same methods applied for material flows.

The compiled data is prepared and presented in the matrix form. Figure 15.37 depicts an example for three levels: plant manager, section manager and employee. The participants in the survey are allocated to their sections in the same order on both the x and y axis. If the mutual estimation of partners' communication intensity and content is to be compared in both directions, the fields above and below the diagonal should be filled in. The color indicates the intensity, here in three degrees of intensity: low (green), medium (yellow) and high (red). With this visualization we can see the exchange of information on the employee level, but it can also be aggregated

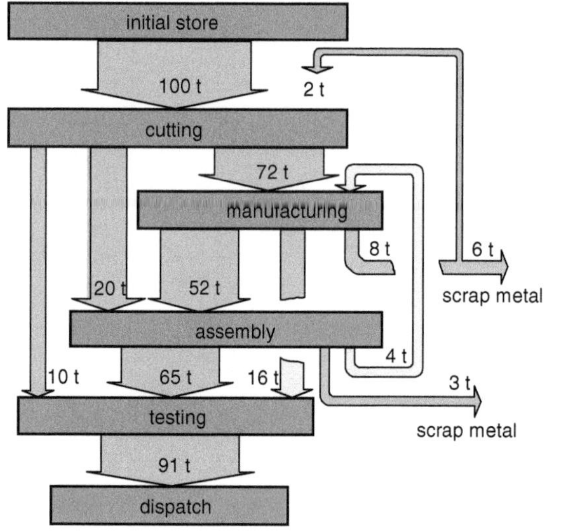

Sankey Diagram

from \ to	initial store	cutting	manufacturing	assembly	testing	dispatch	scrap metal	sum
initial store		100						100
cutting			72	20	10			102
manufacturing		2		52	16		6	76
assembly			4		65		3	72
testing						65		65
dispatch								
scrap metal								
sum		102	76	72	91	65	9	

units in [t/month]

material flow matrix

Fig. 15.36 Visualizations of material flows (examples). © IFA 15.491

to include the relationships between entire sections in the factory.

With a practically designed organizational structure the intensity of communication within a section is typically high (boxes along the main diagonal), whereas the intensity between sections is low. If the basic content of thc communications is added to this information along with the spatial location of the sections from the analysis and then compared with the 'target', the required input data for a communication concept can be generated step by step. The final results provide concrete recommendations for arranging the affected sections (e.g., into an office concept).

Unfortunately currently inefficient processes are often transferred to the new factory with the tacit assumption that everything will then be better somehow. However, without an appropriate program this is never the case. The analysis phase should thus also serve to make improvements possible before moving to the new factory, which can then be implemented step-by-step after the move.

Parallel to analyzing the production facilities, the base analysis of the object planning essentially involves documenting the existing or planned grounds and, where applicable, existing buildings and the relationships of the flows between them. Subsequently, the results of analyzing the production facilities and processes are merged with the analysis of the spatial object from the architectural perspective and in accordance with Fig. 15.5 are adopted in Milestone M1 by the steering committee.

15.5 Concept Planning

According to Fig. 15.5, the concept stage entails the structural development, dimensioning the structure and rough layout planning. The structural design phase is responsible for forming structural units which are technologically, organizationally and economically good functioning. A structural unit is defined here as a module within a complex structure, which fulfills a specific function [Wie96].

The basis for this is system theory [Wil96]. According to it, systems describe a set of interrelated elements with specific properties [Rop99]. The network of the relationships between the elements is referred to as structure. This structure is not random, but rather always oriented on the system's purpose. Elements only represent the smallest units in the observed

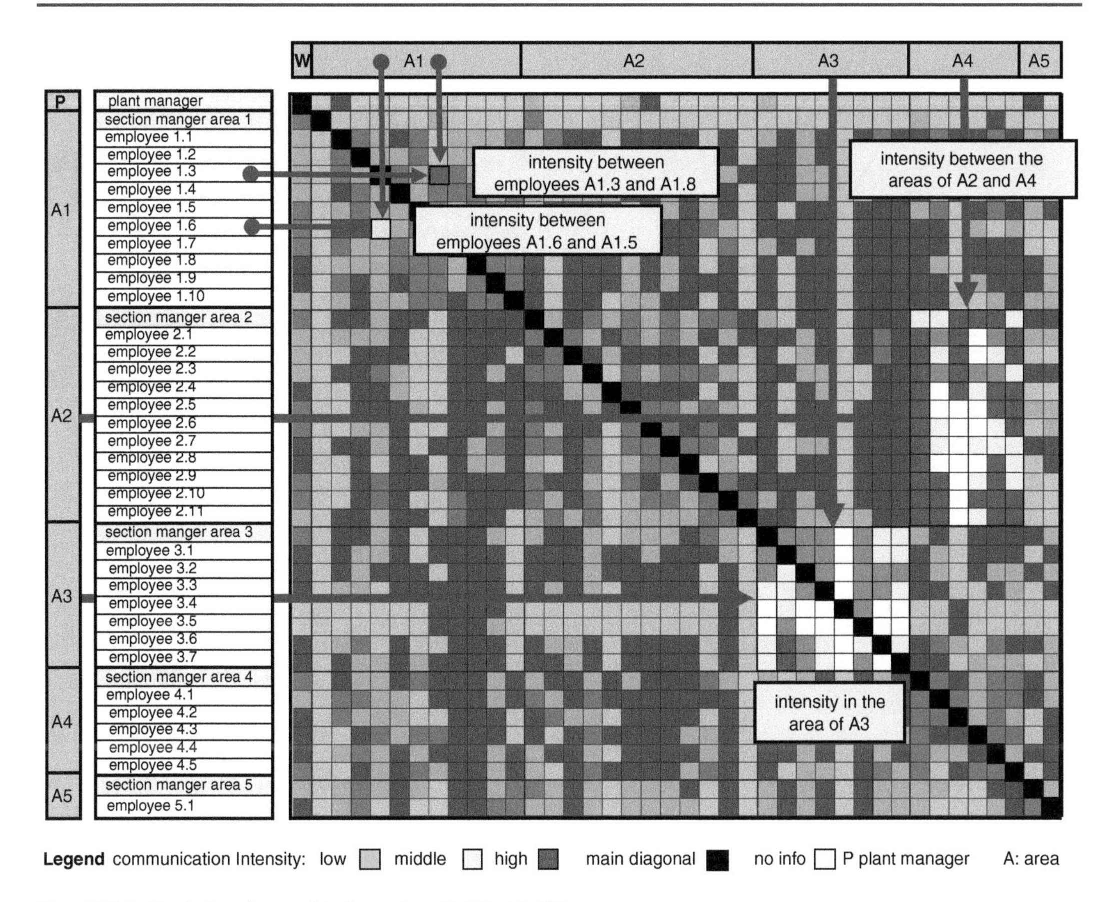

Fig. 15.37 Depicting flows of information. © IFA 15.492

system, but can also be a system themselves. Thus in a factory system, the smallest element is a machine, but it in turn can also be understood as a system (see e.g., Fig. 6.26 Elements of a Machine Tool).

In regard to the factory planning the structure can be seen as the overriding organizing principle of the factory. The structure, however, does not map the spatial characteristics of a system, but rather illustrates the structural units and their relationships.

The structural planning is the most creative and important step in planning a factory. Here the criteria acquired through the goal planning are implemented in a relational structure of the factory objects. From a process and spatial perspective the structure has to be long-lasting, but also allow modifications to changing conditions. Practitioners often find this way of thinking difficult, because they usually have concrete spatial conceptions of factory objects and find depictions of functional relationships very theoretical. Experience has shown though that only this type of abstraction leads to a deeper discussion about the production principles and to fresh approaches.

15.5.1 Structural Design

Structural units are developed by orienting the production based on so-called structuring principles which follow a strategic approach. The starting point for this is the logistic profile of a location (Fig. 15.13). In accordance with Fig. 15.38, the primary focus for the location is the market, products and processed defined during the analysis phase whereas the required

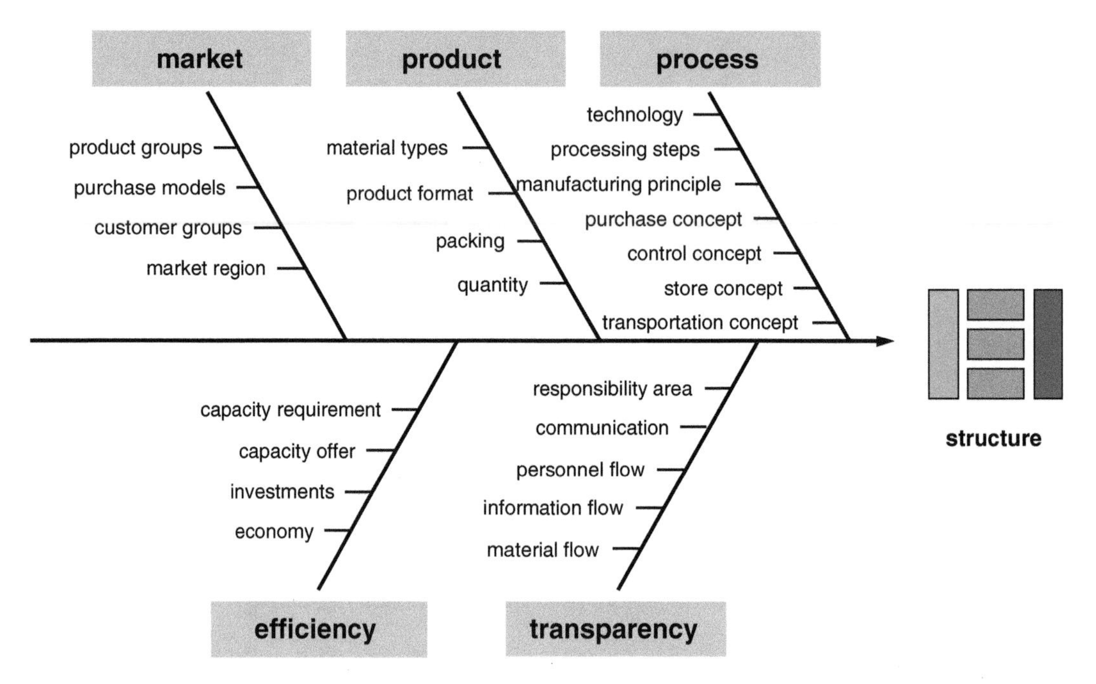

Fig. 15.38 Factors influencing the factory structure (acc. to Harms). © IFA 15.493

efficiency and transparency are considered general conditions. Figure 15.38 breaks these broader terms into further features, which serve as leads for the concrete structuring along with the project specific conditions [Har04].

Generally, it is not possible to find a single consistent structuring principle for an entire factory. It is thus practical to orient on the factory levels introduced in Fig. 5.18. Figure 15.39 illustrates this principle using a fictional example. The upper most level describes the factory structure, for which the individual product groups are decisive for in this case. The structural units are reflected in the organization of the enterprise as well and are often referred to as business units (BU). In addition to the manufacturing section, a business unit is usually assigned to marketing, distribution and product development.

On the section level of the factory, in this example, within each product group specific groups of customers are addressed, which are usually structured into product/customer segments and when possible, combined under one roof. These segments can be further divided logistically on the lowest system level e.g., according to runners (large production volumes), sprinters (average production volumes) and rarities (low production volumes). Runners, sprinters and rarities require different procurement, manufacturing, assembly and supply concepts, as indicated in Fig. 5.13. What is new on this level in comparison to the logistic profile is the decision about the spatial manufacturing and/or assembly structure. The possible characteristics for these have been presented in Figs. 9.7 and 9.8 . For the runners and sprinters, which are in high and regular demand from customers, manufacturing according to the cell or flow principle is well suited. For the more sporadically demanded rarities, a workshop principle is more suitable.

Selecting the appropriate structuring characteristics is illustrated based on the example of a factory for plastic injection molded parts in Fig. 15.40. It can be seen in the upper part of the figure that on the factory level, from left to right there is a raw material store followed by raw material preparation. Attached to that is the production with an integrated store for purchased parts required by the assembly. Subsequently

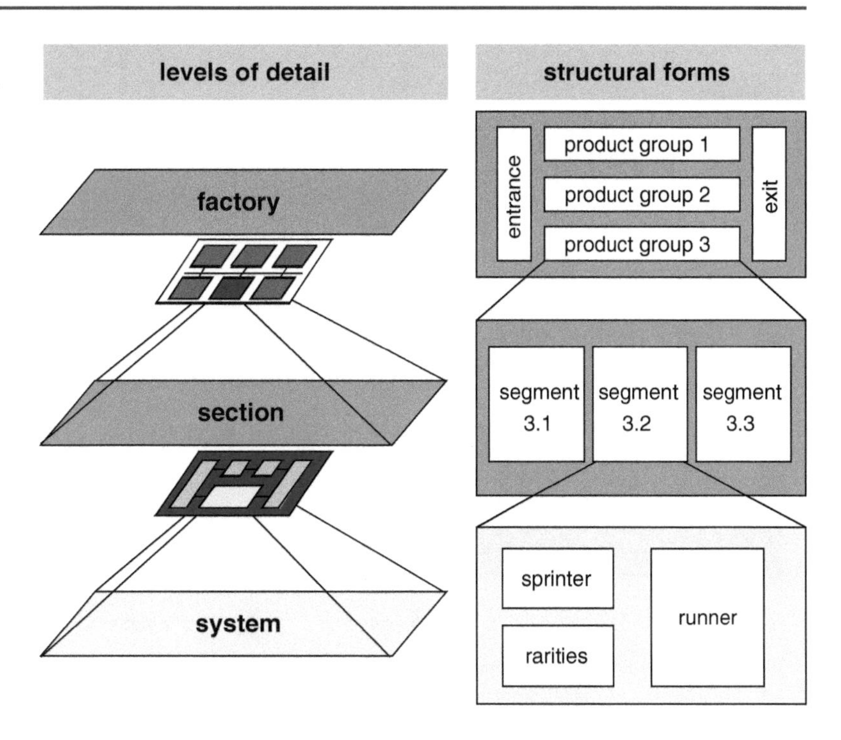

Fig. 15.39 Levels of the structural planning. © IFA 15.494

finished goods are inspected, packaged and transferred to a finished parts store, from where they are commissioned and shipped as needed for order consignments. Ten possible structural options developed for the production in the project are presented in the lower part of the diagram. They are either oriented on the process, customer, technology or on product groups.

With a *process orientation* (principle 1), the processes conducted using similar technologies are combined within a structural unit (e.g., assembly or manufacturing area). One of the advantages of this workshop principle is that equipment and personnel are highly utilized. Moreover, personnel then develop concentrated know-how about the specific technology. If there are changes in the production program they are easily implemented. Capacities within the structural unit are highly adaptable. Nevertheless, there are also associated disadvantages such as high stock levels and long throughput times. If the sections are very large, the workshops can be further sub-divided into machinery size classes (principle 5).

With a *process chain orientation* (principle 10) all of the production steps of a product or product family are combined locally. One of the advantages here is the unit's complete responsibility for a product and the related sub or complete process chains. There is thus a concentration of product related process know-how. The throughput times and stock levels in comparison to principle 1 are clearly faster/lower and the production control is comparatively easy. In return though, not all of the production units can be fully utilized and the technological know-how of individual operations is distributed among several segments.

Nowadays, the process orientation is predominantly oriented on the customer (principle 4), with larger customers on their plants (principle 7), on the products of each of the customers (principle 6) or on the delivery model (principle 8). If the delivery volume is too small for individual segments, the orientation can also be on product families (principle 2).

A selection of possible structural characteristics has to be individually made for each project in order to take into account the project specific conditions. Nevertheless, the possible characteristics of a structure usually range between a process orientation and a process chain

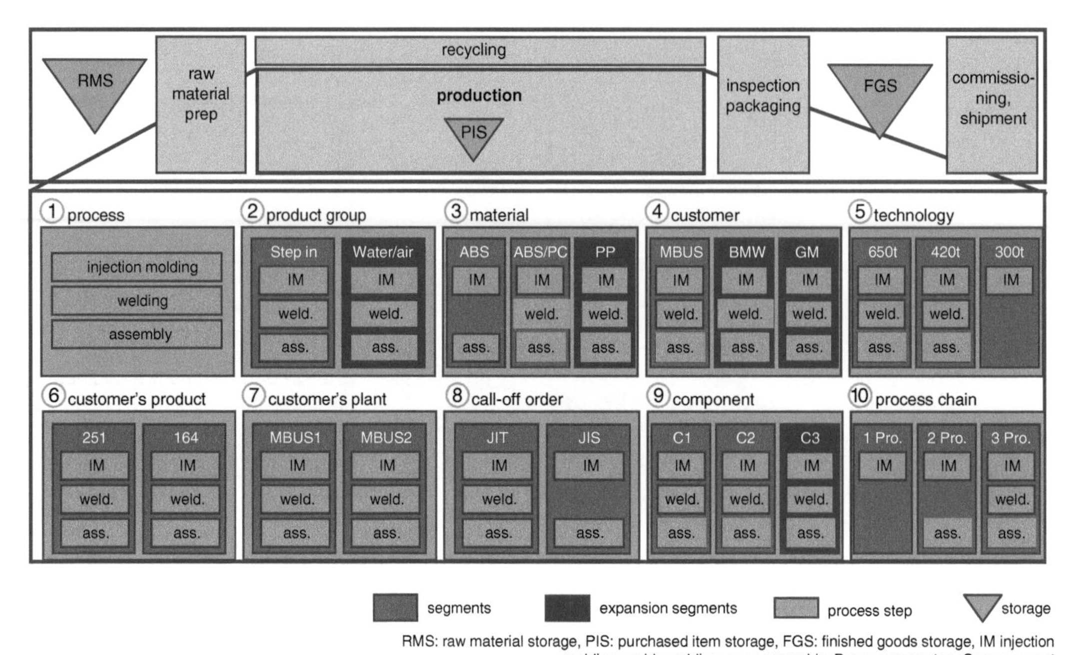

Fig. 15.40 Possible orientations for the structure of a factory (example). © IFA 15.495

orientation. A combination process such as this is depicted in Fig. 15.41 middle. In this case, three process chain oriented segments were maintained, however, the input and output situations were designed together with a functional orientation.

According to the synergetic approach, the object planning should be actively involved in determining the structure. Moreover, variants should be evaluated and selected based on the guidelines set within the goal definition workshop. Once the structure is developed, resources

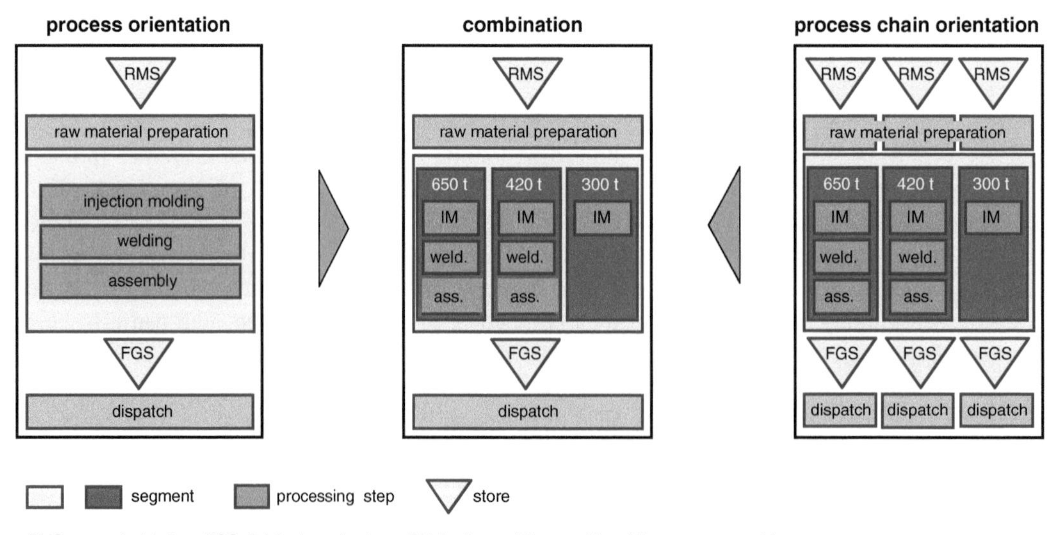

Fig. 15.41 Possible orientation of a factory structure (example). © IFA 15.496

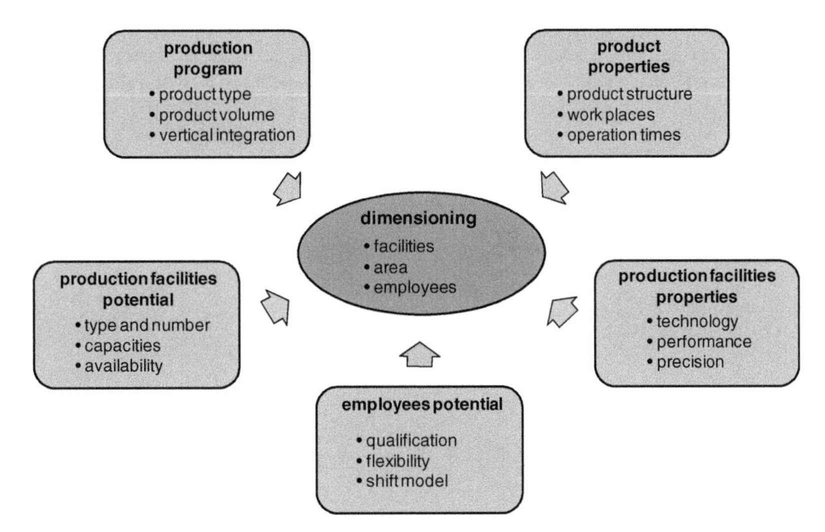

Fig. 15.42 Input variables for dimensioning resources. © IFA 15.497

are then dimensioned before the conceptualization is completed with the rough layout planning.

15.5.2 Dimensioning of the Structure

- Input Variables

The structural dimensioning is responsible for determining the amount of machinery required, the necessary floor space as well as the number of employees required for operation. The input parameters include the production program, the product features, existing production equipment as well as that needed in the future as well as the qualifications and flexibility of the personnel who are to be employed (Fig. 15.42).

The production program is fixed together with the creation of the logistics profile and contains information about the type and quantity of the products. This then serves as the basis for calculating the resources. In addition to current products and their production volumes, information from the analysis phase about the future production volumes, product lifecycles (new, growing and discontinuing) as well as future products has to be available.

- Determining the Operating Facilities

The basic procedure for dimensioning resources is depicted in Fig. 15.43 [Wie72]. The production program typically projected over a 5 year horizon serves as the primary input variable (Fig. 15.43 top left). The product properties, as the second input variable, are determined by the product structure of their individual components, which are documented in the Bill of Material BOM (see Fig. 6.6). The operations sequence along with the required setup times (per lot) and operation times (per piece) can be read from the work plans. The number of end products, components and individual parts that are to be manufactured yearly can be calculated by multiplying the product volume of the production program with the quantities listed in the BOM for each of the components. Next, the yearly demand is to be sub-divided into economic lots. The work content per product can then be determined by multiplying the number of lots by the setup times and the number of individual parts by the operation times per piece. This calculation is clarified in an example from an industry analysis in Fig. 15.44.

In this case, the planning is based on a yearly requirement of 80 machines, which are produced in 8 lots of 10 pieces each. The capacities are broken down according to the facilities classification used in this company as indicated in the work plan per operation. The production volume per individual part per year (many of the individual parts are integrated a number of times in

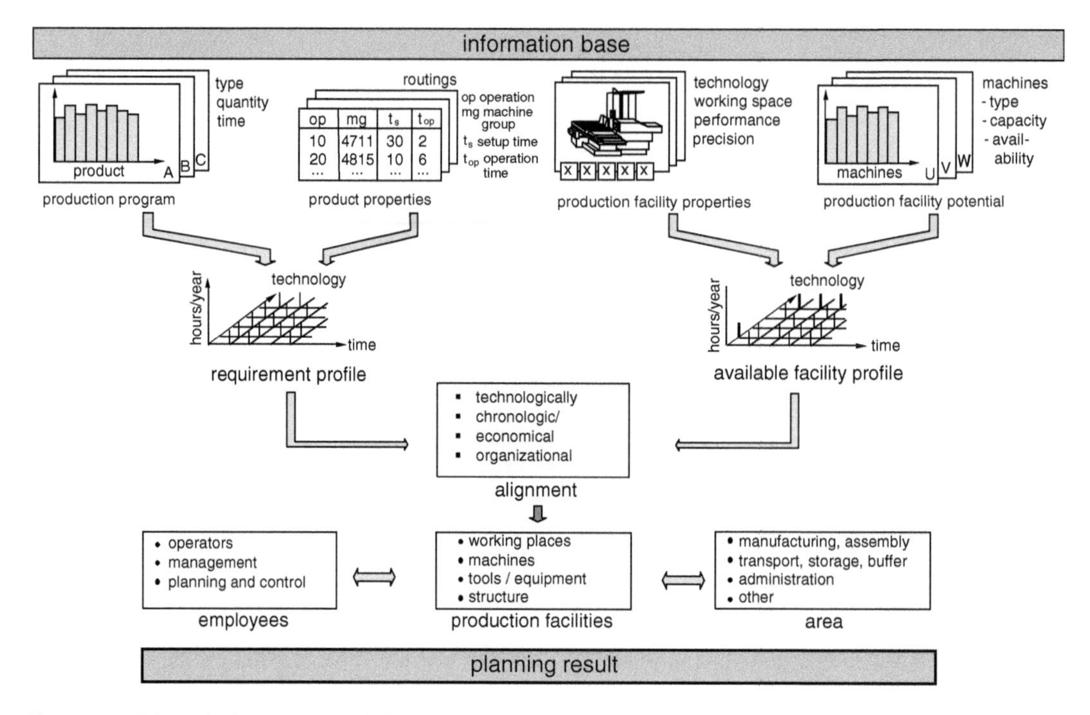

Fig. 15.43 Dimensioning resources. © IFA 15.498

Product PG1

quantity	80	pcs/year
number of orders	8	pcs/year
lot size per order	10	pcs
capacity per machine	1600	hrs/year

no.	capacity group name	quantity [pcs/year]	setup time [hrs/year]	operation time [hrs/year]	work content [hrs/year]	setup time [%]	machines [pcs]
01	center lathes	77,920	555	15,409	15,964	3.5	10.0
03	facing lathes	80	5	60	65	8.2	0.0
03	vertical lathes	800	28	2,764	2,792	1.0	1.7
05	turret lathes	72,320	229	4,385	4,615	5.0	2.9
14	external cylindrical grinder	31,520	82	2,289	2,371	3.5	1.5
16	internal cylindrical grinder	2,960	6	197	203	3.0	0.1
08	milling machines	46,240	215	4,056	4,271	5.0	27
09	large milling machines	560	16	1,203	1,219	1.3	0.8
10	groove cutters	7,600	30	712	742	4.0	0.5
12	inside curve milling machines	880	14	1,295	1,309	1.1	0.8
17	surface grinding machines	8,960	26	888	914	2.9	0.6
20	horizontal drilling machine 1	800	22	540	562	4.0	0.4
21	horizontal drilling machine 2	320	6	583	588	1.0	0.4
22	vertical drilling machine	59,840	184	7,599	7,783	2.4	4.9
23	jig drilling machines	2,480	5	120	125	4.1	0.1
26	shaping machines	6,560	24	248	272	8.7	0.2
27	planing machines	400	12	117	129	9.1	0.1
30	broaching and grooving machine	6,080	58	1,634	1,701	3.4	1.1
31	slotting machines	2,080	28	173	201	13.8	0.1
34	gear milling machines	3,600	57	1,660	1,717	3.3	1.1
	sum ▶	**332,000**	**1,602**	**45,941**	**47,543**	**3.4**	**29.7**

Fig. 15.44 Capacity requirements of a product group (example). © IFA 15.499

this product) is listed in the first column, followed by the setup times, operation times and work contents in the subsequent columns. The work content corresponds to the yearly capacity requirements measured in standard hours.

Sixteen hundred standard hours was accepted as the yearly available capacity per machine. This corresponds to a single shift operation with allowances for breaks, maintenance, disruptions etc. The share of setup time is inserted in order to gain a sense of the lot size's impact e.g., when switching over to smaller lots. The quotient of the work content and capacity results in the number of machines required for single shift operation; usually it is not a whole number. In accordance with Fig. 15.43, adding all of the product groups found in the production program leads to the required capacities profile.

In single-piece productions though it is not practical to derive the requirement profile from the BOMs and work plans of individual orders due to high expenditures and the fact that there are usually no work plans available. In this case, the load values of invoiced orders from the accounting are used. A representative product mix is then determined and extrapolated to the total required capacities based on target sales [Wie74].

The second basic step in the dimensional planning is concerned with determining the available facilities profile. It provides information about the type and quantity of production equipment, their capacities and their availabilities. Considerations about planned investments for replacements or new acquisitions are also included here.

In the third step the two profiles are aligned in three ways. The *technological alignment* verifies whether or not the previously implemented manufacturing technology and machinery should also be used in the future. New technologies capable of a much greater output rate are often being tested (e.g., high speed milling) and require work contents to be corrected accordingly. Independent of new facilities, many enterprises often aim for a yearly productivity increase of 3–6 % depending on the branch. This means that in five years the required capacities would be reduced by 15–30 %.

The *chronological alignment* is concerned with the shift model i.e., whether there should be one, two or three shifts and whether the facilities should be operated on weekends (see Fig. 7.10).

The *organizational alignment* allocates the machinery to the sections set during the structural design, as indicated for example in Fig. 15.39.

Here at the latest the dilemma of economic efficiency versus flexibility comes to light, since usually not all of the machinery in a segment is fully utilized. As a solution, used machinery is often bought for the technologies that do not represent one of the enterprise's core competences (see Fig. 14.15).

Additional resources such as storage and internal conveyor technology result from the number of production facilities and the flows between them. Here, the focus is on the require space, which is discussed in more detail in the next section.

The number of machines and other facilities as well as the shift model determines ultimately the operative personnel and the required qualifications. Possibilities of operating multiple machines (e.g., with fully automated plastic injection mold factories) are considered here. Depending on the structural principle, managerial functions (group leaders, masters, segment supervisors etc.) and services close to production (e.g., NC programming) are also taken into account.

Figure 15.45 shows a graphed comparison of the required and available capacities which should clarify problems concerning increasing uncertainty about capacity requirements in the future. Whereas the ranges for the forecasted values might have seemed exaggerated up until the 1990s, today such values have to be considered normal when planning new factories. This depiction once again emphasizes the significance of a factory and its facilities being flexible and changeable.

- Required Floor Space for Facilities

In the last step of dimensioning the resources, the required floor space is determined from the

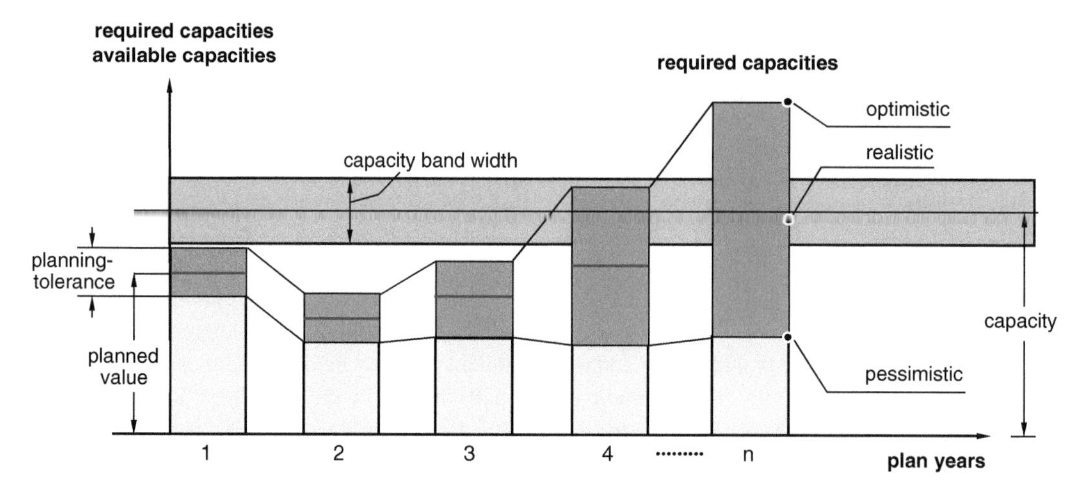

Fig. 15.45 Comparison of required and available capacities. © IFA 15.500

amount of production facilities. Based on the sites total area, Fig. 15.46 provides an overview of the types of spaces and how they are organized for a production facility according to VDI Guideline 3644 [VDI91].

In planning a factory, the following spaces are of importance:

- *production areas*—spaces required for manufacturing, assembling, handling and testing workpieces,
- *storage areas*—spaces intended for receiving and delivering goods as well as supplying material for the production process.
- *office areas*—spaces reserved for administrative segments,

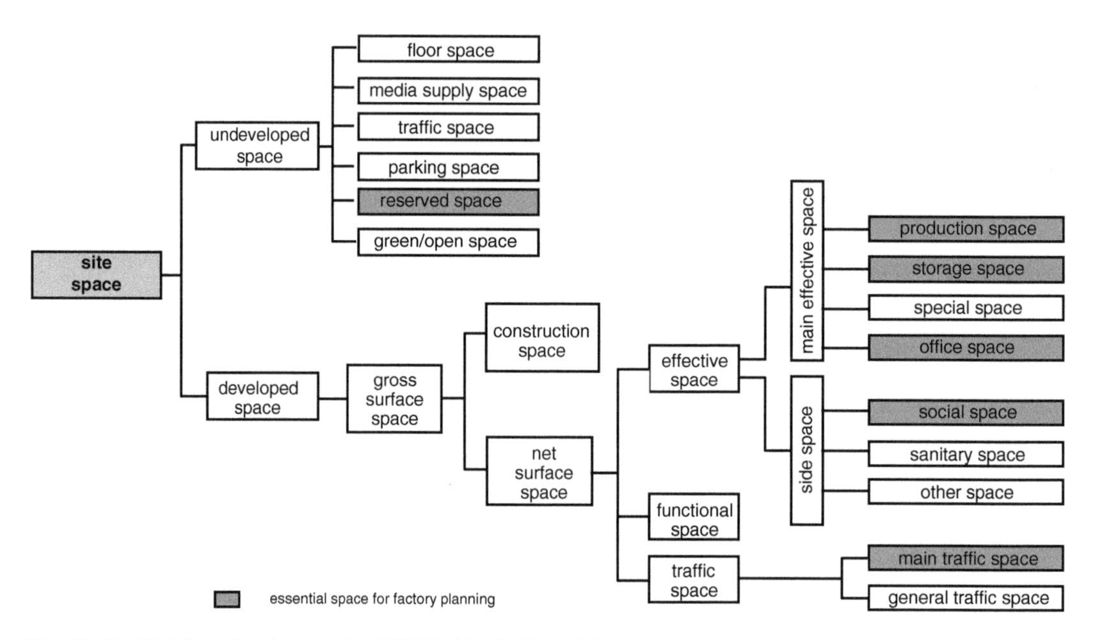

Fig. 15.46 Division of a site area (per VDI 3644). © IFA 15.501

- *social areas*—spaces which predominantly serve the health and care of the workforce,
- *main traffic areas*—spaces kept free and used exclusively for transporting workpieces and personnel in the production units, and
- *reserve areas*—undeveloped spaces reserved for the factory's expansion in case of growth.

In order to determine the space required for production areas there are both key figure based methods and computational techniques. The key figure methods are particularly well suited for estimating in the early planning phases. Spatial requirements are stated either as absolutes (e.g., production areas = 1535 m^2) or relatively (e.g., transport areas = 25 % of production areas). Computational methods include for example, the substitute area calculation and the functional area calculation (see a.o. [Ket84, Gru00]). Both methods start with the machine base area which is then extended with the aid of aggregate factors into the total machine workspace area. The substitute area calculation only adds space for possible transportation units to the machine's base area (typically 25 %) whereas the functional area calculation takes into account individual spatial elements (e.g., for transportation and maintenance). Figure 15.47 helps to visualize the spatial components that can be taken into account with this method.

The aggregate factors in both methods are to some degree centered on workplace regulations e.g., for repairs and maintenance areas (see a.o. Sect. 8.4 Workplace Regulations). However, they can also be set based on experiences, discussions or available operational data.

- Determining Storage and Transportation Areas

In addition to the machine workplaces the other areas mentioned in Fig. 15.46 also need to be dimensioned. The next aspect to consider is the storage and transportation areas allocated to the production. These are primarily determined by the procurement and supply model established in the logistic profile and structural design.

Figure 15.48 depicts the storage areas that are to be taken into account for product group PG1 and that were derived from the logistic sub and elementary processes (see Fig. 6.44). In doing so, the storage areas that are always required are differentiated from those that are dependent on

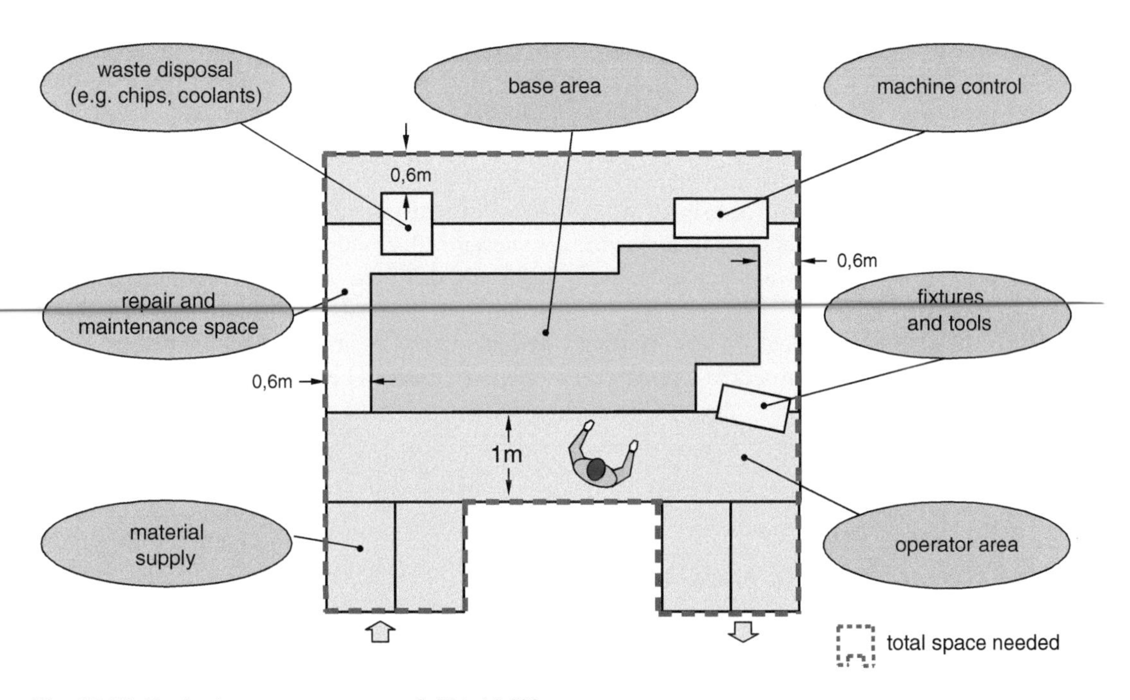

Fig. 15.47 Production area components. © IFA 15.502

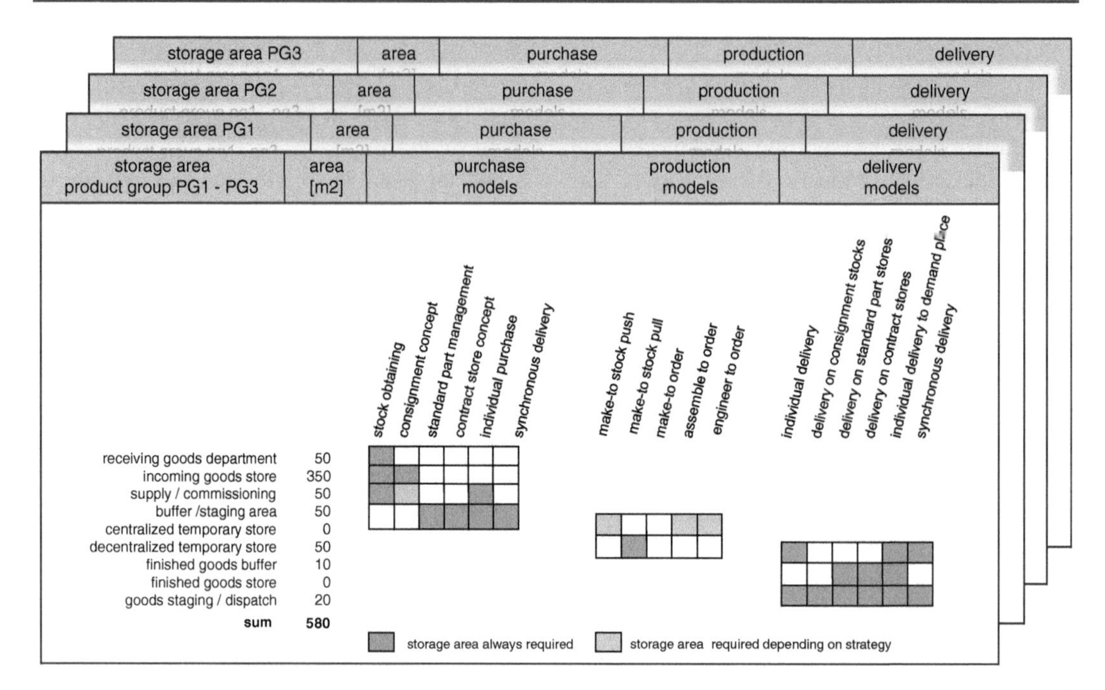

Fig. 15.48 Storage areas to be considered depending on the logistic profile. © IFA 15.503

the logistics concept. The storage area in Fig. 15. 48 is oriented on the material flow, whereby it begins with the receiving goods are and the connected incoming goods store. Once materials have been requested, they are supplied and commissioned when needed with downstream buffer and staging areas. During processing, interim stores for semi-finished goods may also be required. Finished goods are in turn stored in a finished goods buffer or warehouse until they are staged and shipped. The figure also shows which storage areas are required depending on whether this product group pursues a purchasing, production or supply model.

The size of the individual storage areas is dependent on the number of articles needing to be stored or buffered, their size, the number per carrier (stacking container, mesh box, pallet etc.) and from the inventory turnover. The area is often drawn in relation to the dimensions of the Euro pallet (800 × 1200 mm) (31.5 × 47.2 in) and then expressed in parts or multiples of these units.

The resulting areas are then allocated to the respective segments. Whether or not the areas allocated to the individual segments can be combined, first becomes apparent in the rough layout planning. For example a common interim store might be positioned between a push-controlled manufacturing segment and a pull-controlled assembly. In Fig. 15.48 the total required space of 580 m^2 is shown summarized at the bottom of the second column from the left which lists the values from an analysis of all the product groups.

Transportation areas are then taken into account with an aggregate factor and typically represent 20–25 % of the production facility's area. Furthermore, areas for socializing result from the number of employees and legal requirements. As a guideline for offices, areas ranging between 11 and 15 m^2 (118–161.5 ft^2) per person apply for an office with 1–3 occupants.

- Area Modules

When planning a new factory, it is desirable not only to design the production facilities to be changeable, but also the layout itself. *Area modules* are well suited for this. An operating facility can then always occupy a more or less larger module. In cases where a new factory is being planned, facilities can be easily relocated

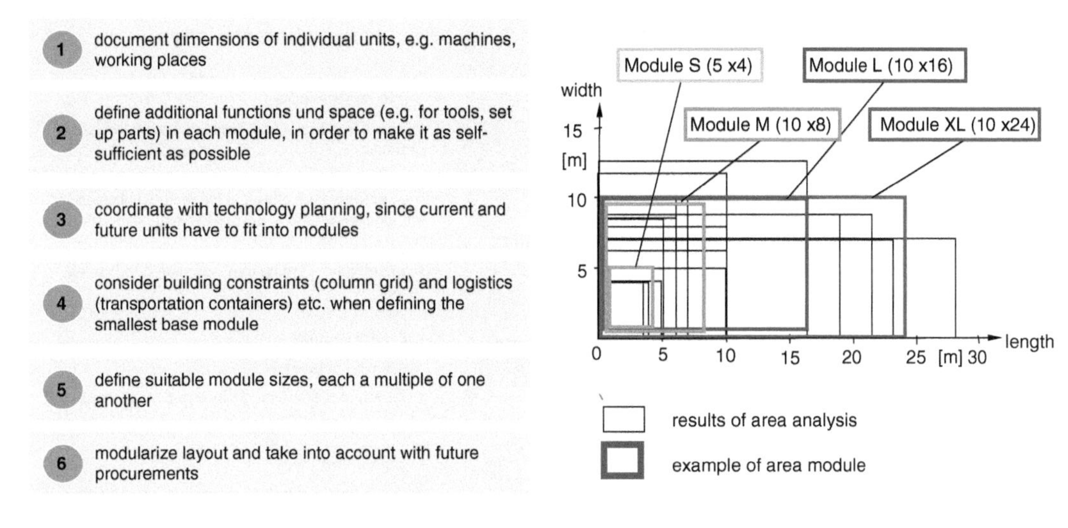

Fig. 15.49 Defining area modules (with example). © IFA 15.504

in a module grid. Nevertheless, one of the disadvantages of this this approach is that the floor space is not optimally used.

As shown in Fig. 15.49, areas are modularized through the following steps: First the required areas for the facilities are determined including the secondary areas indicated in Fig. 15.47 and plotted as a rectangular outline in the coordinate system depicted on the right side of the figure. Following that, areas with similar dimensions are combined into modules. The smallest module should be oriented on the building grid. Modules should be a multiple of the base module in order to allow them to be nested without gaps. Facilities then need to be dimensioned congruently with the modules in order to occupy the areas.

This approach is advantageous in that the layout is highly flexible during further planning processes and subsequent operations. If for example, the positions of machinery needs to be changed due to modifications in the production program this is easily achieved within the individual module's limits, because each machine occupies the same module area size. Within the frame of this concept, vacant areas are identified as conversion areas. The primary disadvantage of modularizing areas is the comparative larger amount of required space. The extra cost of the area therefore has to be weighed against the expected advantages of greater changeability.

The last step in the conceptualization is the rough layout planning, during which both the process and object planners have to find a common design.

15.5.3 Rough Layout Planning

15.5.3.1 Types of Layouts

The spatial positioning of structural units is referred to as the layout (or allocation plan). Depending on the level of the factory, there are different types of layouts with increasing detail, as clarified in Fig. 15.50 according to [Gru00].

The *site layout* is a macro representation that provides a complete overview of all the structural units on a factory site. In the object planning this corresponds to the master building plan (see Sect. 12.4). It shows how the buildings stand in relation to one another on the site and are joined with roadways. A *rough layout* presents the individual production areas within a factory building. The main focus is on the internal logistics, which is why it includes the main transportation and material flow routes. It is important here to take changeability into account.

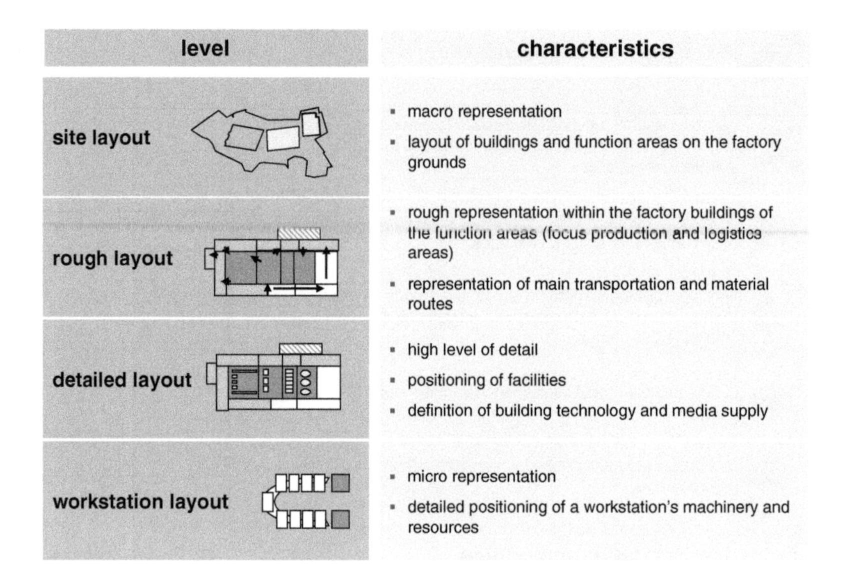

Fig. 15.50 Types of layouts (acc. to Grundig). © IFA 15.505

The layout should thus clearly indicate expansion directions and have material flows which run perpendicular to the expansion direction.

The *detailed layout* presents the exact position and allocation of facilities within an area including building services and media supplies. The highest level of detail though is to be found in the *workstation layout*. In a micro representation it depicts the precise location of all the machinery, tools or materials including energy and media connections.

15.5.3.2 Ideal and Scaled Functional Scheme

A layout is generally based on the so-called *ideal functional scheme*, which is in turn developed from the results of the structural design. An ideal functional scheme, which only represents the area units and the intensity of their material flows, is presented in Fig. 15.51. The area values are adopted from the resource dimensioning, whereas the material flow intensities (expressed in transport units per time unit) stem either from

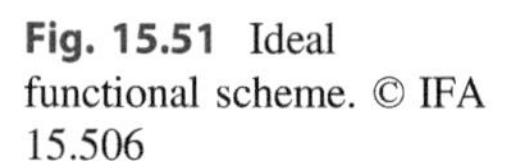

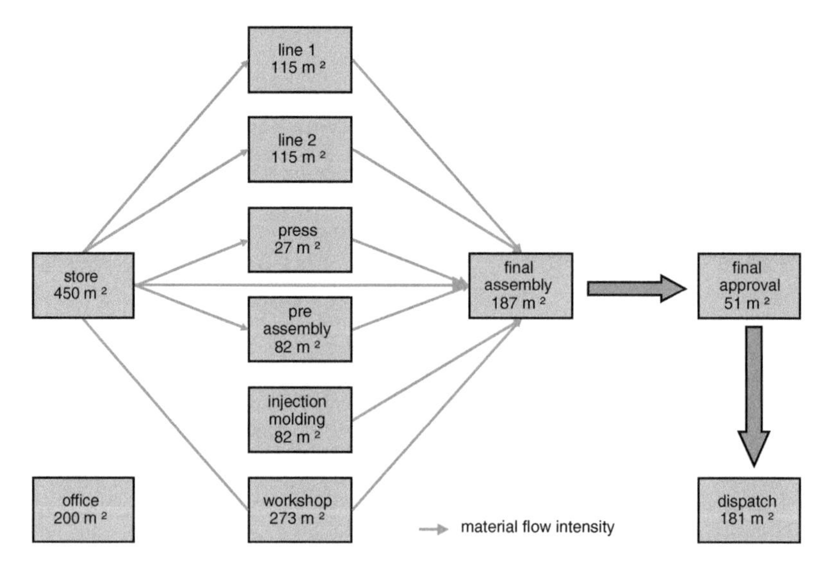

Fig. 15.51 Ideal functional scheme. © IFA 15.506

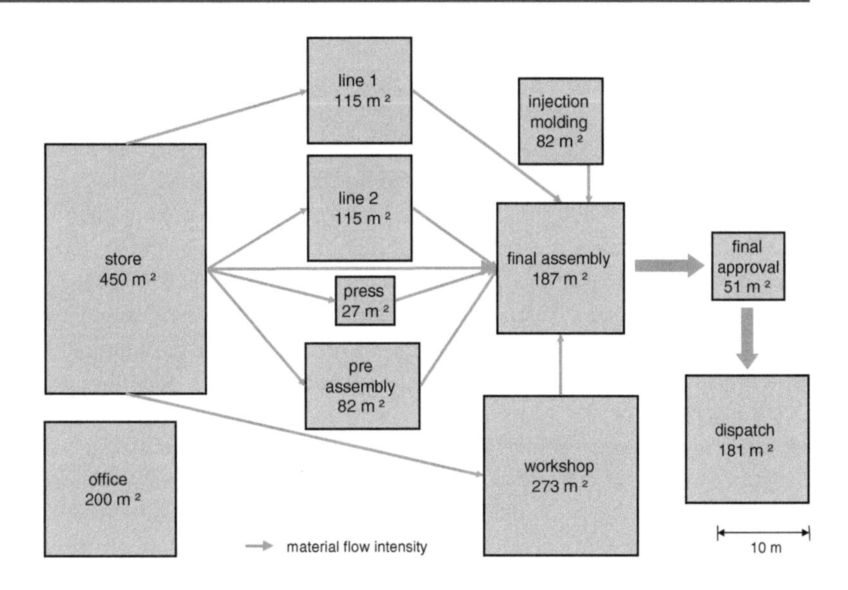

Fig. 15.52 Scaled functional scheme (example). © IFA 15.507

the value stream analysis or a material flow calculation based on work plans. In a first step, the functional units are allocated oriented on the material flow.

In a second step, the functional units are depicted to scale while maintaining the material flow relationships (Fig. 15.52). Since the areas usually vary greatly, the first dimensional distortions occur in relation to the ideal arrangement of Fig. 5.51.

In a third step, the aim is to set the areas in relation to one another according to the target criteria and to obtain a predominantly closed external contour.

15.5.3.3 Ideal 2D and 3D Rough Layout

At the same time this structural planning has been going on, the object planning has developed the initial building framework, which at this point is only described by the span width and columns grid. These two streams of planning are now merged together in a creative process to establish the ideal scaled 2D layout (Fig. 15.53).

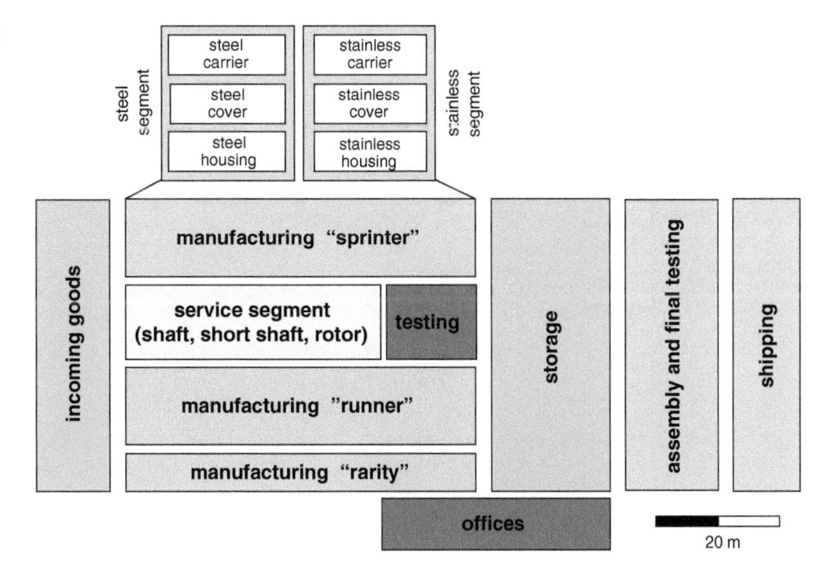

Fig. 15.53 2D Ideal rough layout (example). © IFA 15.508

According to [Sch10], an ideal layout is defined as an "flow oriented, area-related, idealized spatial arrangement of structural units". As can be seen in the diagram, the individual manufacturing segments in this example are oriented on the character of the product groups' production volume (sprinters, runners, rarities), while the interim store and assembly are used by all. In the runners manufacturing segment, the manufacturing of workpieces is divided into those made of normal steel and those made of stainless steel. Due to the risk of flash rust and the need for materials to be separated for swarf removal processes these are not allowed to be processed on the same machine.

In the sense of a 3D oriented synergetic factory planning, the 2D layout can now be converted into a 3D layout, which takes into consideration the ceiling clearance of the individual segments. Figure 15.54 provides an impression of the spatial units and already indicates the room modules derived from the area modules as well as the necessity for overhead cranes in the assembly and test areas. Initial concepts for an ideal building that will envelope the structural units also arise here. Areas required by employees for offices, workplaces and socializing also need to be integrated into the layout considerations; where they are located and how they are developed plays a decisive role in the internal communication.

As already mentioned a number of times, when designing layouts not only the flow of materials but also the flow of personnel, information and communication need to be focused on. Under the primacy of changeability, it should also be possible to easily extend or reduce structural units. Moreover, these changes should be implementable without disrupting the performance capability of the factory. Furthermore, it is also critical to locate indirect segments close to the production in order to promote personal communication, thus increasing the reactive capacity and overcoming the traditional barriers between "white coats" and "blue coats".

The ideal layout planning, which is free from restrictions, aims to release planners from existing conditions and open up their field of vision in order to obtain a benchmark for evaluating the following planning phase [Agg87]. It is advisable to plan an ideal layout both when designing a new building as well as when re-planning an existing factory.

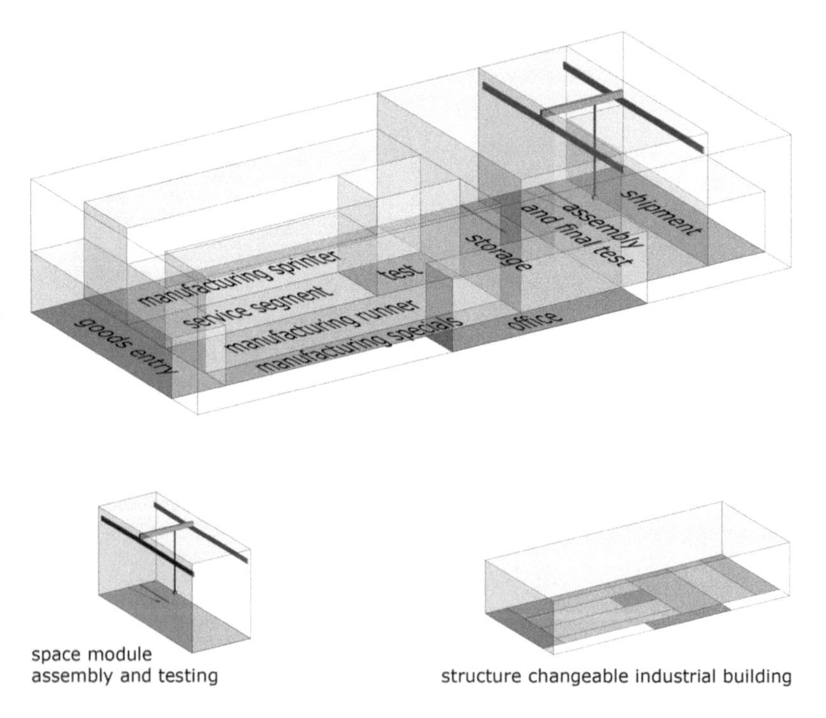

Fig. 15.54 3D ideal rough layout. © Reichardt 15.509

This is recommended because it not only frequently leads to original solutions, but also because it sometimes leads to the realization that the existing building structure simply has no future and the site or building is best to be abandoned.

15.5.3.4 Real Rough Layout

In the next step, the ideal layout is transformed into a rough real layout by adjusting it to the specific operating conditions and restrictions. The term "real layout" is used to describe a spatial arrangement of structural units that can be realistically implemented. It thus also takes into account factors related to the flows, floor spaces, business considerations and regulations that influence the layout [Sch10].

In developing the real layout the various restrictions that originated from the object and process oriented base analyses also require attention (Fig. 15.55) [Sch10]. Local restrictions first pertain to the property and its topography as well as to access rights and contaminated sites. This can mean that the ideal size of a factory has to be changed or adapted. Structural changes may also be required if there are noteworthy differences in elevation. Generally there are also fixed points, which due to large foundations or costly conversions, should not be changed. Further restrictions may arise from the continued use of expensive facilities. In addition, provisions made by the parent organization (e.g., holding or investors) or Good Manufacturing Practices (guidelines for quality assurance in production processes) can also require plans to be adjusted as can limited financial resources. The last factor we will mention here are the legal regulations and requirements (e.g., regional, state, federal) that are non-negotiable and need to be met in order to operate industrial workplaces. These are discussed in-depth in Sect. 8.4 (Occupational Health and Safety Standards), Chap. 10 (Spatial Workplace Design) and Chap. 13 (Location Planning from a Spatial Perspective).

Figure 15.56 illustrates an example of a transition from a rough ideal layout to a rough real layout. The functional units are ideally arranged based on the previously developed structural layout. In refining the 2D layout from Fig. 5.53 additional segments are first formed within the rough structure in view of the specific restrictions for the individual objects. This results in an

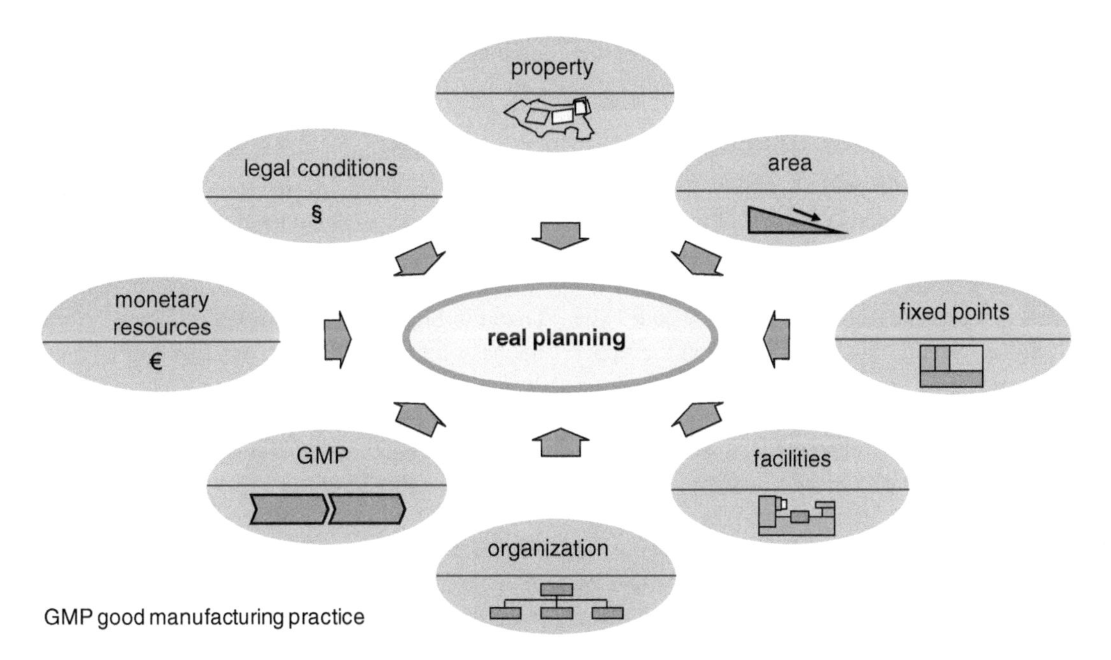

Fig. 15.55 Restrictions in the real layout planning. © IFA 15.510

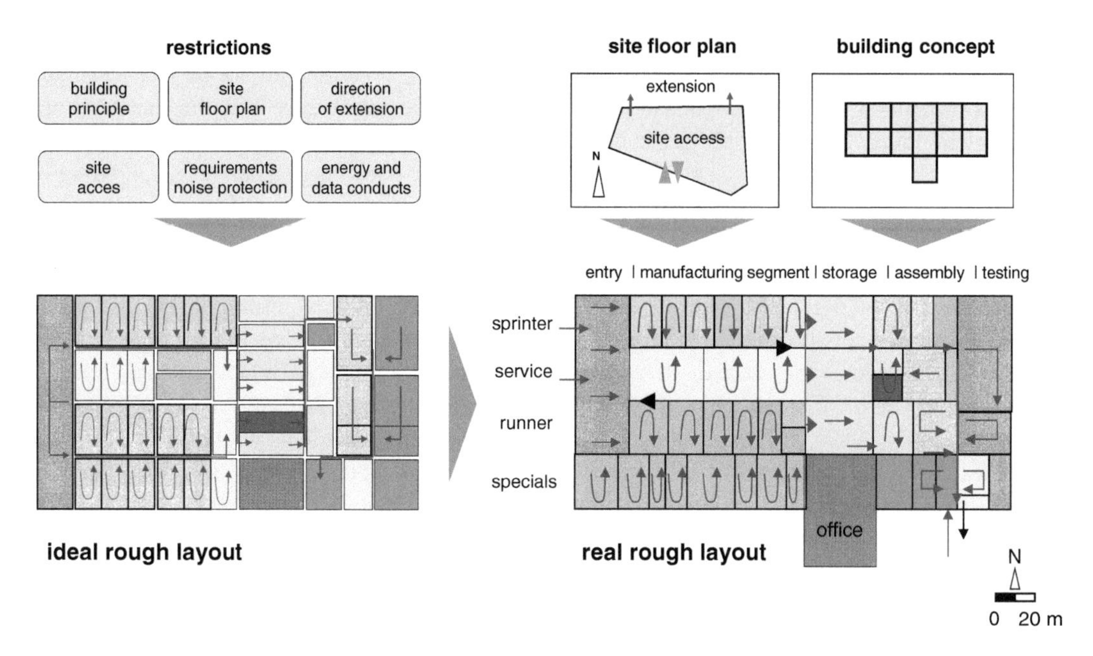

Fig. 15.56 Developing a rough real layout. © IFA 15.511

interim stage of the real rough layout, depicted in the lower left of the diagram.

In this case, in consideration of the property's topography, the idea of an underlying modular structure of the building (previously decided upon during the object planning) leads to the decision to select a building grid of 20 × 20 m (65 x 65 ft). The building can be later extended by offsetting the back longitudinal wall by a module width.

As an important contribution to improving communication, the object planning also recommended accommodating the functions that directly impact external relations (e.g., reception, management, sales, exhibition rooms etc.) on two floors of a module located on the south side. The inward connecting module, located on the upper integrated floor, provides the order processing staff a direct view into the production. These indirect functions essentially include the production control, work preparations incl. NC programming and operative procurement. The ground level floor of this module opens up into the production and is used for example by the maintenance and tool reconditioning divisions. The upper floor thus represents the link between the external and internal functions and ensures the quickest possible way to communicate. The result is the simplified real plan depicted in the lower right corner of the diagram.

At Milestone M2, there is thus a concept available which has been aligned by both the production and object planners. Since the allocation of facilities is also known at this phase, the building design which has been developed parallel to the structural and rough layout planning can also be adopted. In accordance with the synergetic factory planning approach, a rough 3D layout is presented for these purposes (depicted in Fig. 15.57 in analogy to the 2D layout in Fig. 15.56). Whereas both layouts form the basis for the production and logistics oriented evaluation, the 3D model of the external building envelope (Fig. 15.58) primarily facilitates the discussion about (and attainment of) the goals set during the GENEering.

15.5.3.5 Evaluating Variants

In a number of cases, the client requests that several variants of a rough layout and factory building be developed, since e.g., management or

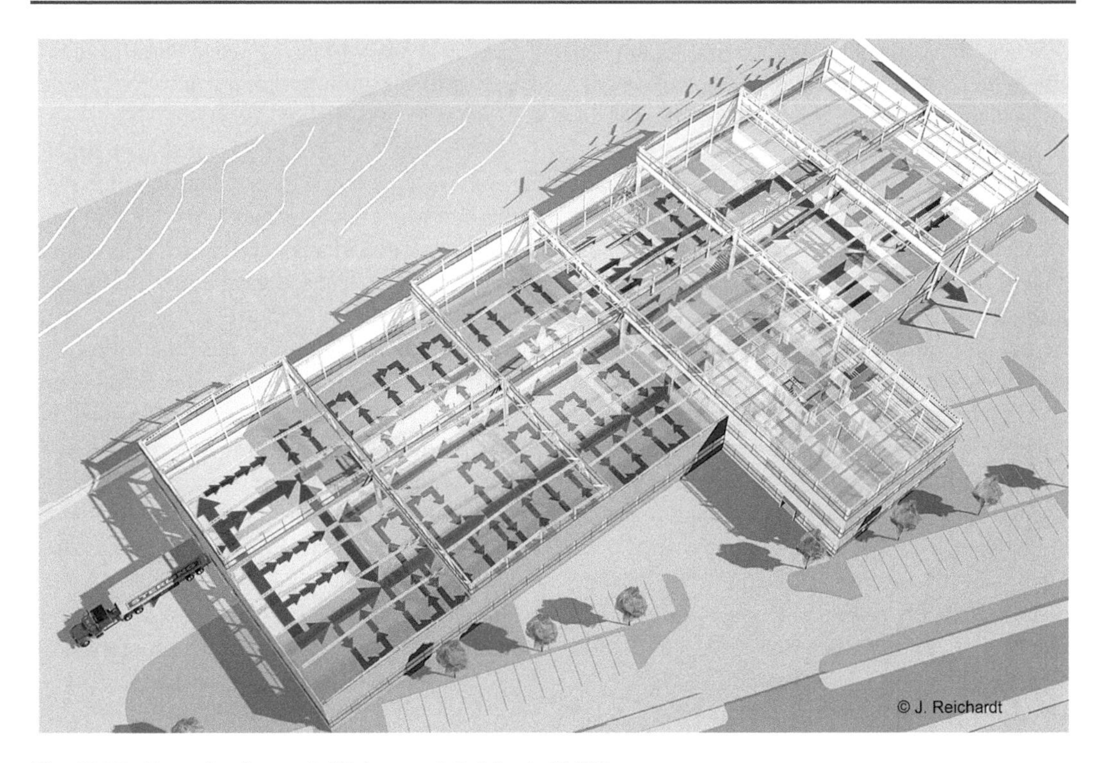

Fig. 15.57 Example of a rough 3D layout. © Reichardt 15.512

Fig. 15.58 3D Representation of an external factory view. © Reichardt 15.513

users already have their own ideas. Variants raise awareness about the quality of a solution and make it easier to obtain a decision by consensus. In the field of architecture, variants are usually generated through a public competition with designs ranked by an independent jury.

When the competition procedure is applied with factory buildings, the concept developed by the production planning is rarely questioned by the architects.

The synergetic factory planning approach however, aims to prevent exactly this effect and to draw on the competence of the object planners as early-on as possible, because the potentials currently available for designing a creative and changeable building are generally not known to production planners. Usually the only selection criterion that counts is the cost of the investment in the building and in seldom cases the building operation costs in the life cycle of the building. Whether or not this preferred approach manages to integrate the object perspective early, is a question of the contract negotiations between the property developer and architects and the firms executing it. This is discussed more thoroughly in Chap. 16 (Project Management).

Independent of the contract negotiations, there are frequently variants available at Milestone M2 which need to be evaluated. The decision made will pose critical points for a future-robust solution. There are countless discussions and proposals about evaluating variants, but with factory planning, these frequently boil down to a simple benefit analysis with weighted criteria [Zan76]. These will be briefly commented on before addressing economic considerations.

The benefit analysis (also called cost benefit analysis) is a method for systematically preparing a decision when selecting from project options. It analyzes a set of variants with the aim of arranging the individual variants according to the preferences of the decision-makers in reference to a multi-dimensional system of objectives [Zan76].

Before presenting the variants the project core team establishes a list of criteria oriented on those developed in the target setting workshop and try to rank them with respect to their relative importance for the project. An example shows Fig. 15.59. The criteria are entered into a matrix twice. In our example here, there are 9 criteria. Next, each of the criterions is examined column for column as to whether it is more or less important than the remaining criteria in the column. This weighting is accomplished most easily by comparing them in pairs. If the criterion being considered is more important than the other, the number of the more important criterion is entered. Conversely, if it is less important the number of the other (more important) criterion is entered in its row.

no.	criteria	1	2	3	4	5	6	7	8	9	weight [%]
1	product handling										2.78
2	volume flexibility	2									13.89
3	traceability	1	3								2.78
4	circulation flexibility	4	2	4							8.33
5	incremental capacity adaption	5	5	5	5						19.44
6	disturbance flexibility	6	2	6	4	5					11.11
7	variant flexibility	7	2	7	7	5	6				11.11
8	pass over flexibility	8	2	8	8	5	6	7			8.33
9	process reliability	9	9	9	9	9	9	9	9		22.22
	further criteria										
		1	5	1	3	7	4	4	3	8	100.00

frequency F_x with which the criterion K_x was judged more important than the other criteria.

$$W_x = \frac{F_x}{\sum_{X=1}^{n} C_x} \cdot 100$$

F: frequency, C: criterion, x: number criterion, W: weighting, n: number of criteria

Fig. 15.59 Weighting criteria in a benefit analysis. © IFA 15.514

For example in the first column, criterion 2 (volume flexibility) is graded as more important than criterion 1 (product handling), thus a 2 is in column 1, row 2. Criterion 3 (traceability) in comparison is assessed as less important than criterion 1 (product handling), thus there is a 1 in column 1, row 3. The value in the bottom row shows how often the respective criterion is rated as more important than the other criteria. In the first column, for example, criterion 1 was only assessed as more important than the other criteria only once, in comparison criterion 9 was prioritized eight times. It is immediately recognizable that in this case criterion 9 and 5 are the most important criteria. The weighting of the individual criterion is then calculated based on a 100 % normed weight using the simple equation inserted in the diagram.

The second step of the benefit analysis is shown in Fig. 15.60. The three concepts available here are evaluated based on the 9 criteria, whereby the scale can range from 1 to 5 or from 1 to 10. A 1 corresponds to the lowest rating, a 5 or 10, the highest. In this case a range from 1 to 5 was chosen. The weighted criteria value results from the product of the criteria weight and the relative rating (which is the quotient of the number rating and the maximum value of the rating scale). The benefit value of each of the variants is equal to the sum of the weighted individual criterion ratings. The value expresses the percentage of the maximum possible rating a variant has attained.

A benefit value of less than 80 % should result in a critical review of the variant. Either: the property or existing building's restrictions prevent a better solution, in which case the site or buildings should be questioned because the solution is not competitive over the long term; or the solution is not yet fully developed and should be sent back to the planning team. In order to facilitate the use of this method Appendix C3 contains a small Excel application.

A second approach to evaluating variants is investment analysis. These methods focus on economic advantages in comparison to other pending investments. According to a study by Brieke, the capital value method has established itself in factory planning [Bri09]. The capital value (also known as the net present value NPV) is the sum of all the discounted or compounded incoming and outgoing payments at a point in time, which are caused by realizing an investment object . The absolute advantage of an investment compared to refraining from it is visible in a value greater than zero, whereas the greater capital value demonstrates the relative advantage in comparison to other variants.

For a capital good such as a factory, whose operating costs over a lifecycle often amount to several times the investment costs, methods such as the total cost of ownership (TCO) and lifecycle analysis are common. This idea is also drawn upon in the construction planning by differentiating between construction costs and operating costs (discussed in more detail in Sect. 16.3).

no.	criteria	criteria-weighting [%]	concept 1		concept 2		concept 3		concept 4	
			assessment	weighted assessment [%]	assessment	weighted assessment [%]	assessment	weighted assessment [%]	assessment	weighted assessment [%]
1	product handling	2.78	1	0.56	1	0.56	5	2.78	5	**2.78**
2	volume flexibility	13.89	4	11.11	4	11.11	3	8.33	4	**11.11**
3	traceability	2.78	4	2.22	4	2.22	4	2.22	4	**2.22**
4	circulation flexibility	8.33	4	6.67	2	3.33	4	6.67	3	**5.00**
5	incremental capacity customization	19.44	4	15.56	2	7.78	4	15.56	3	**11.67**
6	disturbance flexibility	11.11	3	6.67	3	6.67	4	8.89	5	**8.89**
7	variant flexibility	11.11	2	11.11	2	11.11	3	6.67	4	**6.67**
8	pass over flexibility	8.33	1	8.33	1	8.33	5	8.33	5	**8.33**
9	process reliability	22.22	1	22.22	1	22.22	5	22.22	5	**22.22**
	result ▶			**53.33**		**42.22**		**81.67**		**83.33**

assessment scale 1 -5

weighted assessment [%] = criteria weight • (assessment value / maximum assessment value)

Fig. 15.60 Benefit analysis—evaluating variants (example). © IFA 15.515

While it is still possible to determine the incoming and outgoing payments for a factory project with some effort, it is considerably more difficult for non-monetary objectives despite their undeniable economic importance. Brieke thus developed an extended economic analysis which is based on the non-monetary target system depicted in Fig. 15.61 [Bri09]. The listed targets and their exemplary sub-targets have gained significant strategic importance particularly within the frame of a highly responsive factory and have been discussed in a number of cases in the preceding chapters.

Figure 15.62 clarifies the principle of extending the traditional incoming and outgoing payments with the monetary effects of fulfilling the objectives of non-monetary targets. The transformation occurs by converting the degree of objective attainment into a monetary incoming or outgoing value over a time axis. For example, for an assembly segment the target 'employee orientation' is evaluated by whether the selected solution is expected to place high, average or low physical demands on the worker. The transformation into a monetary value occurs via the required additional investment in better ergonomic design. The increase in job satisfaction is on the incoming payment side and is related to a lower fluctuation in employee turnover and less sick days.

On the left side, Fig. 15.62 depicts both the incoming (deposits) and outgoing (payments) cash flows, differentiated according to immediately determinable payments and non-monetary payments that have been transformed into payments. On the right side, Fig. 15.62 illustrates the changes in the capital value for three variants with and without consideration of the non-monetary targets. The strongest changes are visible in Variant C, which with a traditional analysis would have been dismissed in comparison to Variant A and B, but which now dominates due to considering the additional utility.

The extended evaluation is not a quick and simple approach nor will it easily gain a foothold in the practice. Nevertheless, it is important for drawing the attention of decision-makers to the monetary significance of softer targets and for making a rational consideration accessible.

With the final evaluation of the rough layout and the building structure, the concept can be

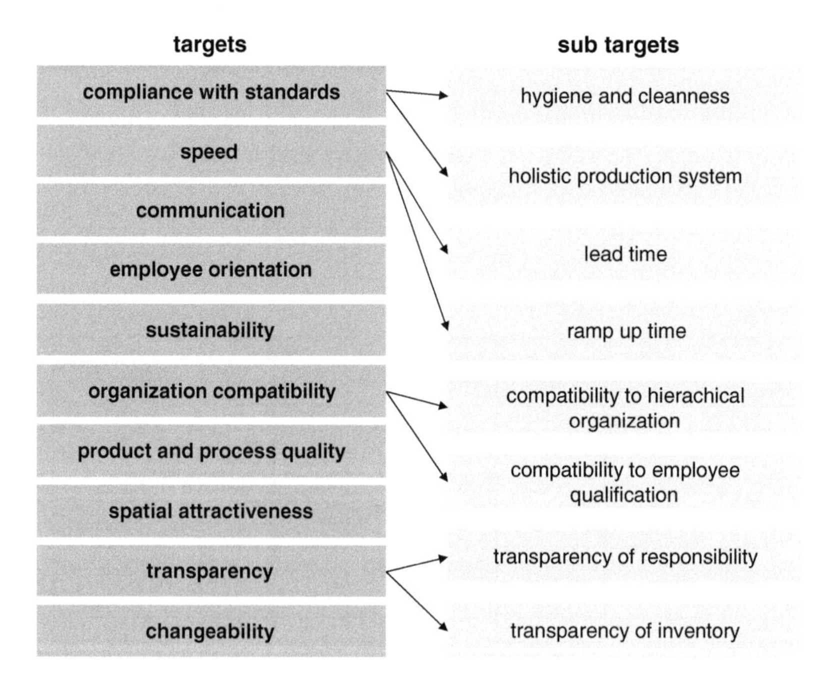

Fig. 15.61 Non-monetary target system for factory assessment. © IFA 15.516

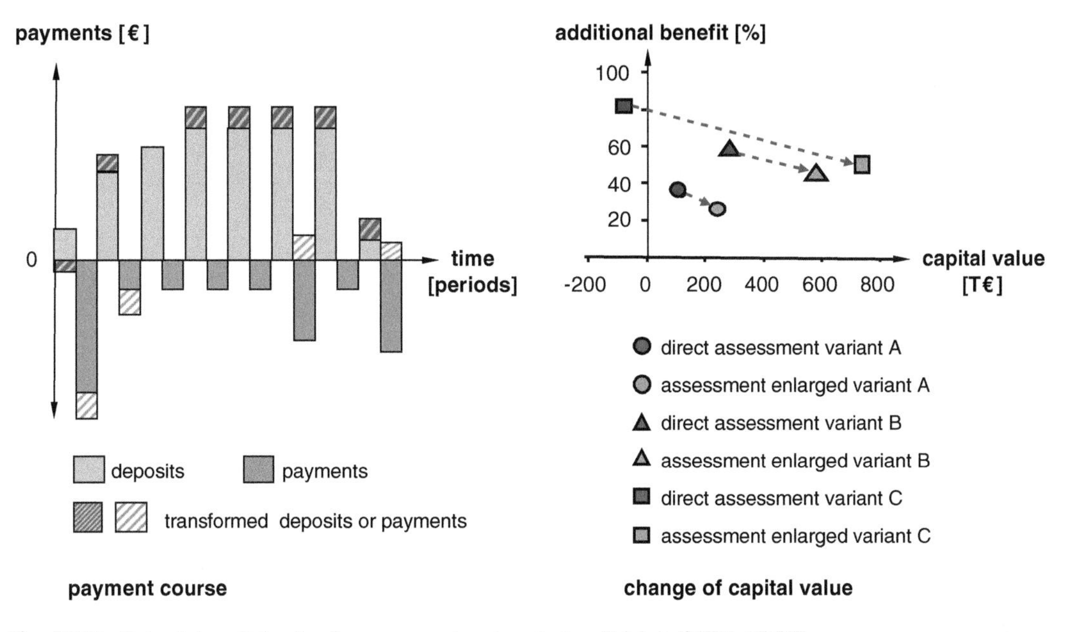

Fig. 15.62 Extended capital value for non-monetary targets (per Brieke). © IFA 15.517

decided upon at the meeting for Milestone M2. The conceptualization phase is thus complete and the next stage 'detailed planning' can begin.

15.6 Detailed Planning

15.6.1 Transportation Route System

An essential part of the detailed planning from the process view is designing the transportation related organization of the individual segments and the transportation route system within the factory. All of the internal material and personnel flows are handled with this system. Depending on the planning tasks, different material flow variations are possible for spatially allocating the routes on varying levels of detail. Within the factory it is important to differentiate between empty and loaded transports, which should not meet or cross one another if possible (Fig. 15.63).

Transportation routes should be arranged so that:

- it is easy to deliver and pick-up goods from structural units,
- the available floor space is used well,
- communication is supported, and
- legal regulations, particularly in regards to evacuation and rescue paths, are adhered to.

This task directly impacts the building design and therefore has to be closely coordinated with the object planning. Within the factory the logistic quality of the layout is determined by the symmetry of the transportation system among other factors.

15.6.2 Fine Layout

In the *fine layout*, the exact position of facilities (within a magnitude of ±2 to 4 inch) can be identified. Moreover, the location and furnishing of indirect segments is fixed, as are in-house offices and transportation routes. In collaboration with the architects, the connections between the facilities and the energy and media supply/removal system are established as well as the location of the necessary access points. In addition, the need for air supply and/or exhaust systems for specific production areas are clarified.

To solve this task many attempts have been made to apply mathematical optimization

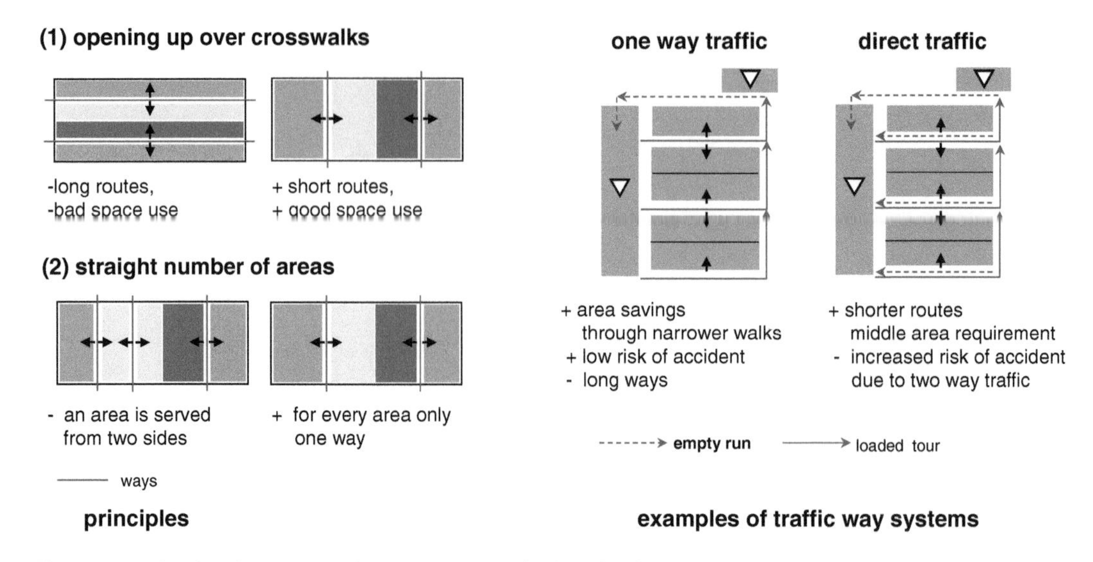

Fig. 15.63 Planning the transportation route system. © IFA 15.518

methods. A brief survey is provided, for example, by Tompkins et al. in [Tom10] Chap. 6: "Layout planning Models and Design Algorithms". In Chap. 7 "Models/Simulation" of [Han04], the authors present 5 programs for the "layout of multiple items". They begin their discussion by noting that "although specialized layout programs are rarely utilized in contemporary design, a brief discussion is presented for historical purposes".

The problem with algorithms to optimize layouts is that almost all of them pursue only one optimization criterion—usually: minimal material flow or greatest space utilization. As we have seen in our discussion of factory targets this is far too simple, especially when it comes to fostering soft targets like changeability and communication. Usually the results of the algorithms are therefore disappointing. In dozens of projects over the course of 3 decades the authors of this book have failed to apply a single layout program. We also surveyed several leading consulting firms about their practice; none of them ever applied layout programs. They all follow the structural concept from structure to rough and to fine layout. For the fine layout in particular, it is almost common practice to develop it in workshops together with the foremen from the shop floor. Figure 15.64 gives an impression of the

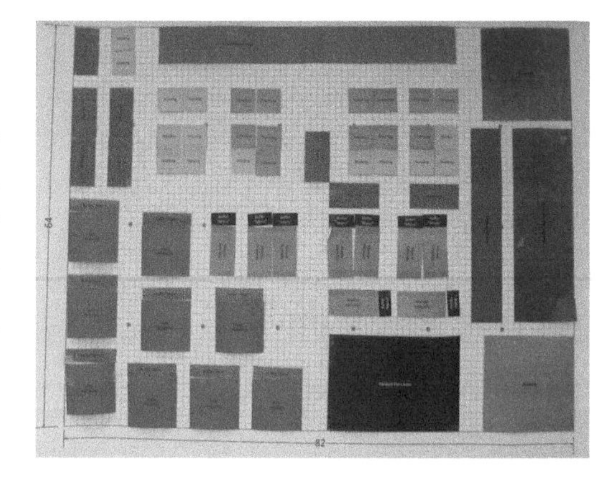

Fig. 15.64 Workshop result fine layout (example)

results of such a workshop. The single machines are modelled as scaled pieces of card board and are inserted into to the grid of the hall.

Instead of algorithmic solutions meaningful results have been achieved however with event driven simulations within the frame work of the digital factory. It has to be recognized though that a simulation does not create a solution; rather it verifies a solution with respect to key performance figures. Details will be discussed in Sect. 16.8.

Based on the fine layout and the building grid, the facilities with their actual dimensions are now adjusted in the area. In doing so a plethora of

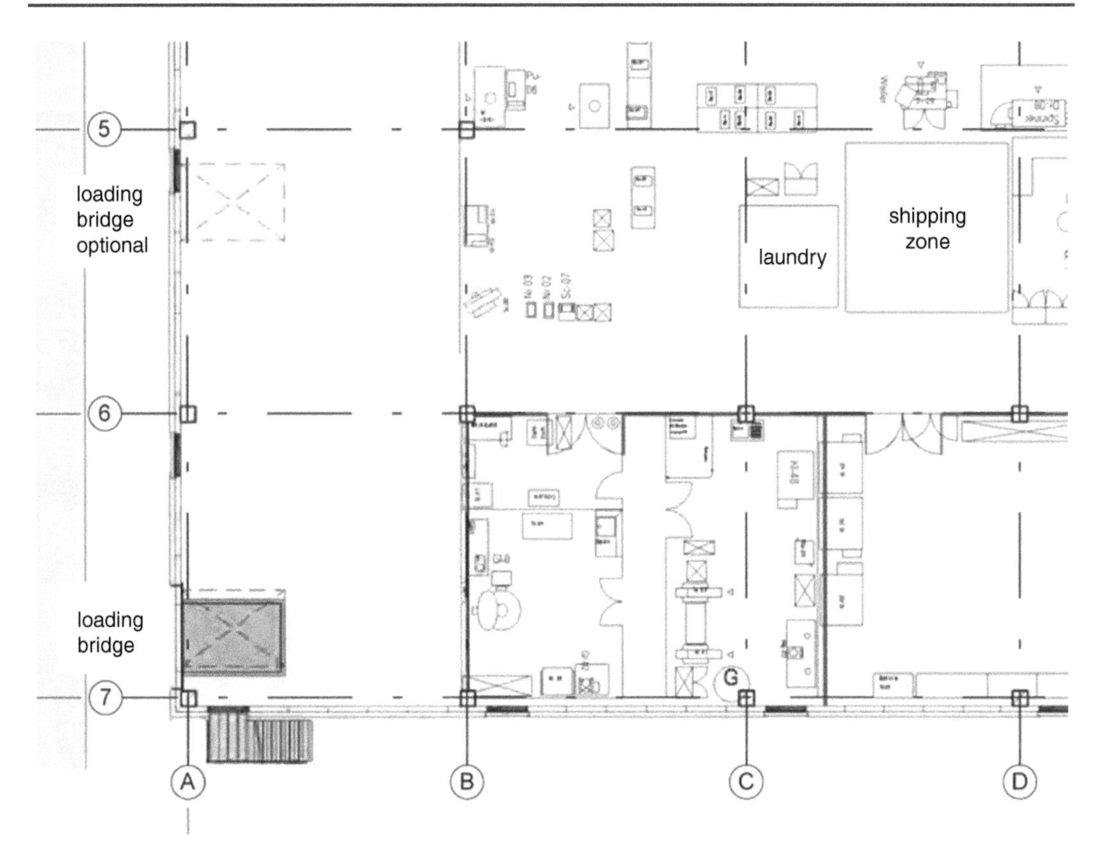

Fig. 15.65 Excerpt from a 2D fine layout. © Reichardt 15.519

small objects have to be taken into account e.g., waste containers, impact protection around columns as well as objects for building services such as distribution cabinets for media, electricity and data. The latter often lead to unpleasant surprises because of their frequently unexpected size and tendency to have to be located at fixed points. At the same time the entire layout also has to suffice the demands for changeability, for a lean production and meet aesthetic requirements. Figure 15.65 depicts an exemplary excerpt of such a plan in 2D.

The fine layout forms the basis for the *technical design planning*. According to HOAI Phase 3, the core of this is the detailed planning of all the technical building systems including the approval of the structural framework plan. Usually a special milestone is negotiated for adopting the technical design.

For the property developers, the technical design, possibly in a number of variants, is paramount to their approving the entire object. At the latest here, the quality of the structural planning becomes evident. As an object becomes more concrete, clients frequently identify desired changes—a good structural plan is able to accommodate these.

According to HOAI Phase 4, the *planning the approval* includes processing the applications for the required permits of the relevant authorities. The contents of a building application are shown in Fig. 15.66. It shows that in addition to the building description, the application also contains an abundance of permit requests. In particular, the plan for the external systems connected to the specific site (e.g., drainage, green spaces) requires a significant amount of effort to coordinate with the authorities and

1 building application forms

a) building application
b) building description for the building application
c) operation description for commercial plants
d) statistical evaluation sheet

2 descriptions

a) system description of the building construction
b) system description of the technical building equipment

3 technical proofs and surveys

a) stability proof (statics)
b) energy proof for commercial buildings according to energy saving act
c) noise protection proof
d) fire prevention survey
e) parking space proof

4 calculations

a) calculation of enclosed space
b) calculation of the gross ground floor
c) calculation of the net ground floor
d) calculation of the highest situated common room above the area surface
e) calculation of the raw building costs

5 official site plans

- extract from the cadastrial map
- qualified site plan

6 plans building

a) graphical representation in ground plans, cross sections, views

7 plans outer plants

a) graphical representation of the outer plants

8 waste water drainage application

a) discharge of dirt water and rain water into the public drainage system

9 water rights permission

a) introduction of rain water into the underground

Fig. 15.66 Contents of a building application. © Reichardt 15.520

possibly with neighbors (e.g., adjusting boundaries, rights of way).

The subsequent *execution drawings* (Phase 5 HOAI) refine the technical design proposal until the solution is mature enough to be realized, including the facilities and their connections. It also frequently contains samples of furnishing details for the walls, ceilings, windows, facades etc. in order to concretize the functional and optical impact. This proposal also has to be approved by the clients.

The exact content of these 3 planning stages (design, approval, execution) is extensively described in HOAI and is negotiated with clients on a project to project basis. These stages represent extensive architectural and technical design services and are the focus of the detailed planning. They far exceed the effort required for the fine layout planning. Numerous details have to be negotiated with the user. For example for the loading bridge indicated in Fig. 16.65 lower left corner the loading area, load-bearing capacity, gate clearance, weather protection, sealant details, gate speed and control have to be determined before sending out an invitation for bids. In this example, with changeability in mind, there is also an optional position planned, which possibly requires constructive precautions, which the call for bids has to indicate.

The fundamentals that are essential for the production planner are extensively discussed according to the different levels of a factory in Chap. 8 (Spatial Workplace Design), Chap. 10 (Spatial Workspace Design), Chap. 11 (Building Design), Chap. 12 (Master Building Plan) and Chap. 13 (Location Planning from a Spatial Perspective). Figure 15.67 depicts an example of merging the sub-plans from the process and spatial perspectives into a 3D model. Here the building already introduced in Fig. 10.6 is once again taken up and the allocation of the facilities on the ground floor is shown as blocks. Furthermore, the main material flow is drawn in: It is fed from the two input areas (marked by the orange rolling gates in the upper and left corner), distributed over the manufacturing area and

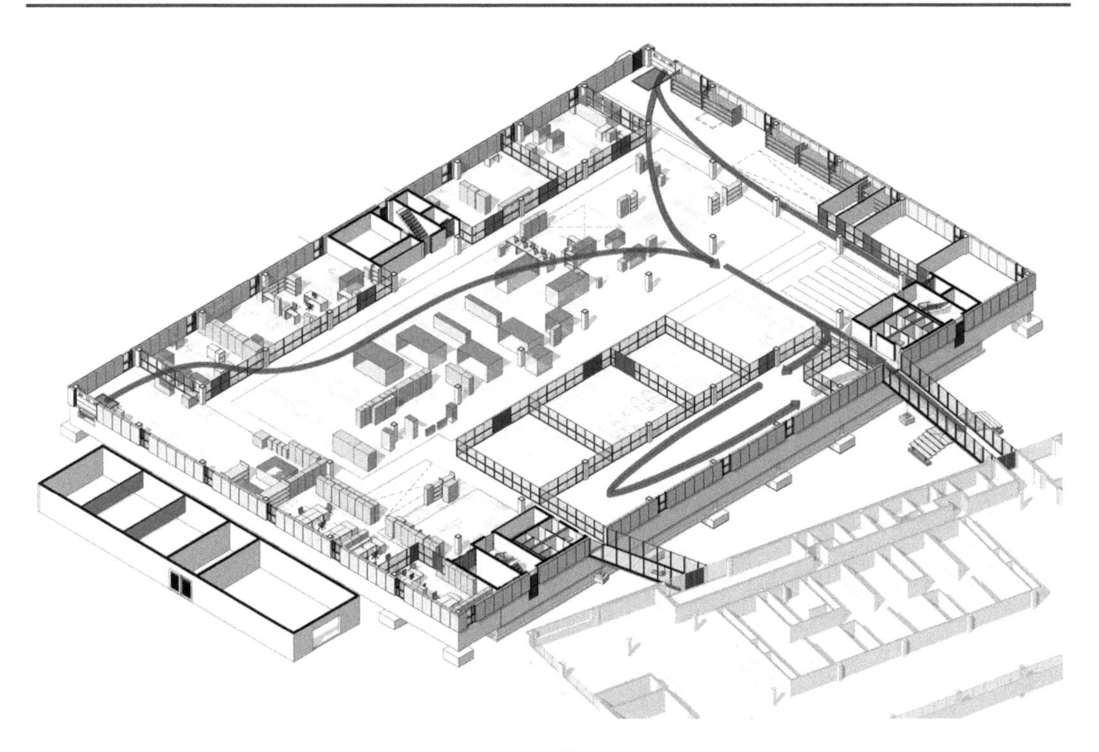

Fig. 15.67 3D detailed layout (example). © Reichardt 15.521

flows over the connecting bridge into the neighboring building to be used further.

In the example, it had to be decided based on the layout at which points incoming or outgoing air vents should be located on the different building levels. The centrals for supplying media and fresh air were already positioned by the architects during the design phase (see Sect. 11.1 Supply and Removal Systems). The 3D design of the building services equipment for the building pictured in Fig. 15.67 is depicted in Fig. 15.68.

The changeability of the building services equipment deserves special attention. Since furnishing a building with utility equipment is generally expensive, changeability is often restricted due to budget reasons. Nevertheless it should be considered whether to equip individual bays with an autonomous technical infrastructure, since this would allow hall segments to operate independently in cases where there are disruptions. Moreover, it also allows the production facilities to be changed or sub-areas of a production to be better controlled depending on the implemented shift model. Milestone M3 is marked by the approval of the execution plan.

15.7 Energy Efficiency

15.7.1 Overview

As was extensively discussed in Sect. 3.9, since the mid-1990s there has been a change in mindset within the industry, in particular due to radical environmental changes and rapidly rising energy prices. Technological leaders try, on the one hand, to manage tighter cost situations with energy efficiency programs and on the other hand, in an increased awareness of their social responsibility, to mitigate the threat of global warning by comprehensively reducing CO_2 emissions. Relevant areas for increasing energy efficiency and reducing CO_2 primarily include industrial production processes, transportation logistics for supply chains, the energetic quality of buildings and the efficiency of the building

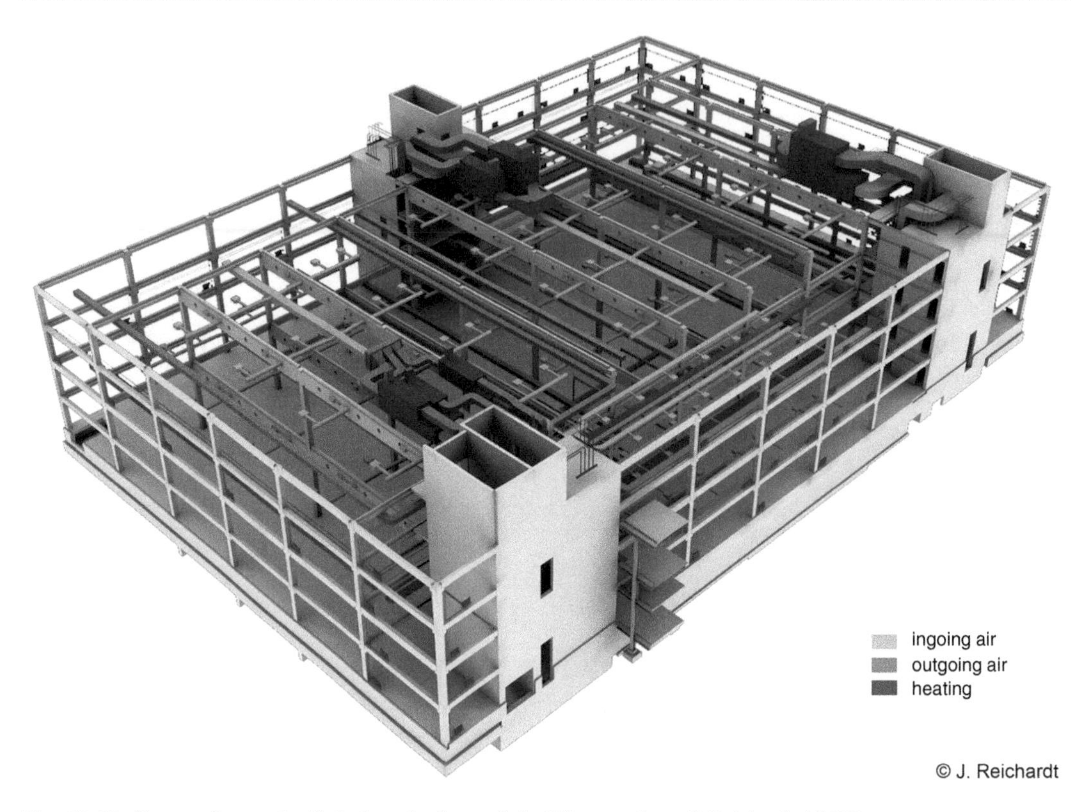

Fig. 15.68 Excerpt from a detailed plan of a factory's building services. © Reichardt 15.522

services. These additional requirements create a relatively new aspect requiring attention when planning a factory. A comprehensive discussion of this topic is found [Neu13]. Since the guiding principles for the project development originate during the base analysis (HOAI stage 1), related considerations already begin there (see also Sect. 3.10 'Sustainability' and Sect. 15.3.8 'GENEering').

Although, the technology related energy savings potential in the industrial sector can be highly rated (e.g., lighting, compressed air generation, pumping systems, cooling systems, heating supply and ventilation systems), up until now only a few measures have been found for actually tapping into it [Eng08, Mül09, Her11, Pet14]. A proposal for codes to design energy saving buildings and power systems is published with the International Energy Conservation Code [IECC12].

In order to clarify the subject, Fig. 15.69 depicts a typical energy flow within a factory [Mül09]. In the middle, the core factory processes can be seen (storage, production, transportation, commissioning). The related materials flows are connected to the outside world via incoming and outgoing deliveries. In terms of the energy balance, a large and significant share is attributed to auxiliary processes such as transforming and distributing electrical energy, generating process heat, process cold and compressed air as well as room heating, cooling and ventilating spaces. These are supplied by solar energy, primary energy sources (oil, gas, coal etc.) as well as district heating.

Against the background of constantly climbing energy prices and an increasing number of legal obligations concerning resource efficiency, it is imperative to more strongly and more comprehensively consider processes both from an economical (e.g., operating costs, lifecycle costs—see Chap. 16) and an ecological perspective (e.g., space consumption, recycling, changeability, CO_2 emissions).

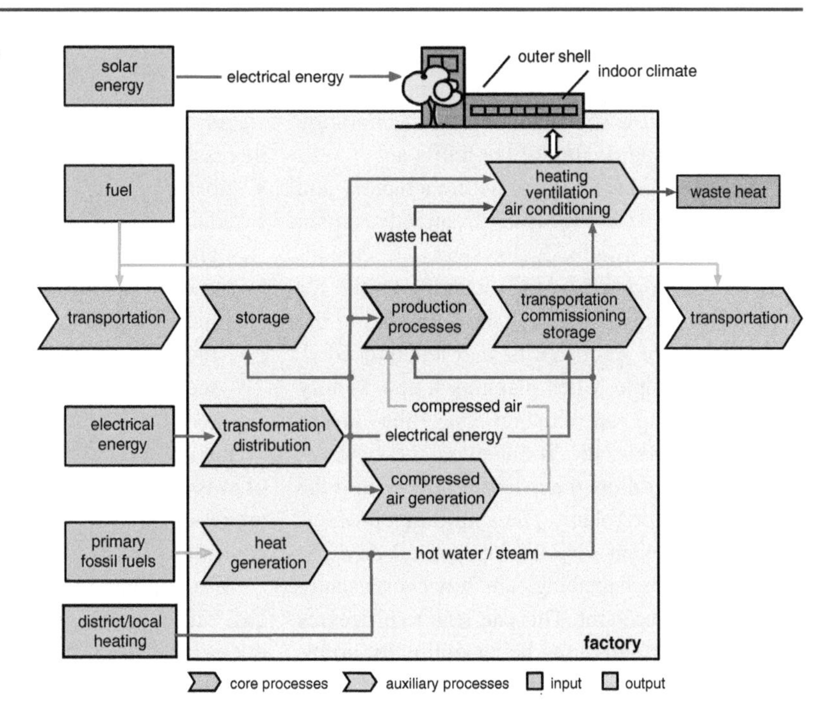

Fig. 15.69 Typical energy flow of a factory (per Müller). © IFA 17.612_B

Currently, enterprises still focus mainly on increasing the efficiency of individual systems. Figure 15.70 depicts a typical analysis which serves as the basis for identifying the main consumers of a factory. This particular example is based on a car body manufacturer [Eng08]. We can see that the laser system for welding the sheet metal parts including the necessary cooling is responsible for 30 % of the total electricity consumption. With such a visual representation though, it is frequently overlooked that every individual system is integrated into the overall system of a factory. Thus using the laser also requires a cooling system, whose consumption is

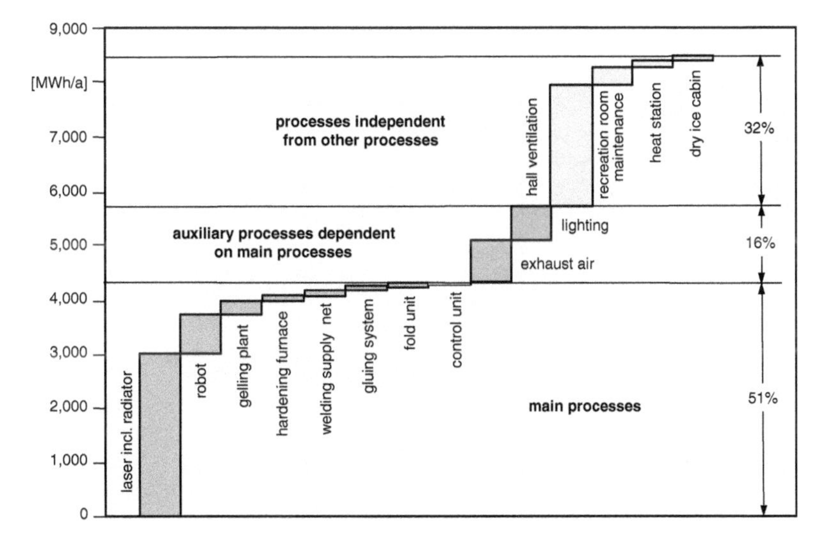

Fig. 15.70 Energy consumption analysis of a car body manufacturer (per Engelmann). © IFA 17.613_B

not separately indicated as well as a dry ice treatment. Moreover, we cannot identify, within this analysis, how implementing this technology impacts the conditioning of the hall's air.

The interactions that arise within a factory and that are indicated in Fig. 15.69 frequently remain overlooked when individual systems are considered. This in turn means the related potential for improving energy efficiency and minimizing the consumption of resources is also left unheeded. Thus, for example, when planning a new factory or reorganizing an existing one, the layout planning and thus the arrangement of systems, typically only follow the material flow within the building's floor plan. The building profiles, which are relevant for utilizing natural daylight and for energy balancing, are however usually not taken into account. This can lead to processes with high heat emissions being optimally arranged in terms of the material flow. Nevertheless, since they then have too low of a ceiling clearance or are subjected to direct sunlight they require cooling. The area therefore ends up increasing the environmentally damaging CO_2.

Müller et al. published a systematic approach to planning and operating an energy efficient factory in the piece goods industry with a focus on mechanical and automotive engineering [Mül08]. The energy relevant functions developed there are summarized in abbreviated form in Fig. 15.71. They refer to a factory building and its building services as well as the production systems. The outside facilities are not considered here.

As fundamental courses of action for increasing energy efficiency the authors mention:

- substituting the energy resources used,
- reducing the required net energy e.g., through:
  - energetically-optimized product-design,
  - energy-saving modes for operating systems,
  - increasing efficiency,
  - reducing waste energy,
  - recovering energy,
  - reusing waste energy as incurred energy

Incurred energy is defined here as "... the sum of waste heat from people, electrical equipment, process heat systems, water heaters etc. that contribute to the heat balance of an area."

In addition to the energy consumed through gas, oil, electricity etc., further environmentally relevant factors such as water consumption and emissions need to be considered, especially the CO_2 induced by using fossil fuels. These resources are primarily a question of the implemented production technology and facilities and have to be addressed early on in the production planning. The interactions with the factory concept are then examined within the frame of the synergetic planning process and influence the contractors' concrete designs on the four factory levels: workplaces (Chaps. 6–8), work areas (Chaps. 9 and 10), buildings (Chap. 11) and location (Chaps. 12, 13 and 15).

1 transmission and use of electrical energy
- electrical energy distribution
- application
- electromechanical drives
- compressed air
- lights
- thermal energy
- other energy types
- information and communication technology
- control technology

2 process heat
- generation of process heat
- heat recovery
- transmission of process heat
- use of process heat

3 process refrigeration
- generation

4 heating, ventilation, air conditioning
- physiological requirements
- technological requirements
- environmental protection requirements
- tasks
- reduce load
- ventilation concept
- heating concept

5 buildings
- layout and orientation of buildings
- building shape
- building zoning
- building construction
- windows, doors, gates

Fig. 15.71 Energy relevant processes and systems in a factory (per Müller et al.). © IFA 17.614_B

15.7.2 Certification Systems

In Sect. 3.9 we discussed the efforts being made world-wide to not only manage energy consumption, but to also protect the environment in general and to develop rules for responsible use. An important example is ISO 14000, which presents a series of international standards for managing energy (http://www.iso.org/). The idea behind EMS is "a management tool that allows organizations of all sizes:

- to identify and control the environmental impact of their activities, products and services,
- to continually improve their environmental performance and
- to introduce a systematic approach to setting and achieving environmental objectives, to attain them and to show that they have been attained."

ISO is working on a new standard for the carbon footprint of products, for quantifying and communicating greenhouse gas emissions (GHG) produced through the work associated with the production of goods and services. A carbon footprint is defined as the total greenhouse gas emissions that are generated by a person, organization event or product (http://www.carbontrust.co.uk/). The new norm is extensively based on ISO 14040/44 and ISO 14025.

For the thus addressed expanded tasks involved in planning a factory, systems have been developed world-wide since the 1990s for certifying the quality of building projects regarding sustainability. Whereas these were first created based on the typology of large building complexes, they are now being transferred to industrial buildings. To do so, evaluation methods in the sense of benchmarks are being established. In the following we will briefly introduce the key systems available globally.

- BREEAM (Building Research Establishment Environmental Assessment Method) was first established in Great Britain in 1990. It provides a point system for the project quality in 8 categories including management processes, energy, water use, ecology, health, transportation, material and pollution (http://www.breeam.org/). Since then, over 200,000 buildings world-wide have been certified according to BREEAM; the assessments span a scale ranging from excellent to very good, good and average.
- LEED (Leadership in Energy and Environmental Design) was designed by the US Green Building Council in 1998 based on BREEAM (http://www.usgbc.org/). Categories that are assessed here include the sustainability of the site and earth, water efficiency, energy and atmosphere, materials and resources, quality of internal spaces, as well as innovation and design processes. Projects assessed are certified as silver, gold or platinum.
- IGBC (India Green Building Council) was developed in India in 2006 based on LEED and is especially designed for buildings in industrial projects. In particular, it takes into consideration aspects of energy efficiency and sustainability which are particular to the hot climate regions of India (http://www.igbc.in/site/igbc/index.jsp). The definition states: "a green building is one which uses less water, optimizes energy efficiency, conserves natural resources, generates less waste and provides healthier spaces for occupants than a conventional building". In 2009 a pilot version was published for evaluating factory buildings.
- DGNB: The German Sustainable Building Council (DGNB) was introduced in Germany in 2007 and has now spread internationally (www.dgbn.de). Its aim is to provide a certification system that can be adapted to local conditions, but which makes buildings in different countries directly comparable. Figure 15.72 depicts an overview of the evaluation system. Six categories are defined with which the ecological, economic, socio-cultural, functional, technical, process and site quality are assessed; weighting is project specific.

Figure 15.73 depicts how the 6 main categories are sub-divided into 13 categories and a total of 51 sub-criteria. In this context, process does not refer to a functional sequence, but e.g., to the transparency of documentation. It is noticeable

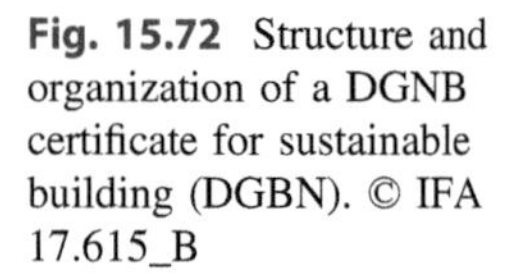
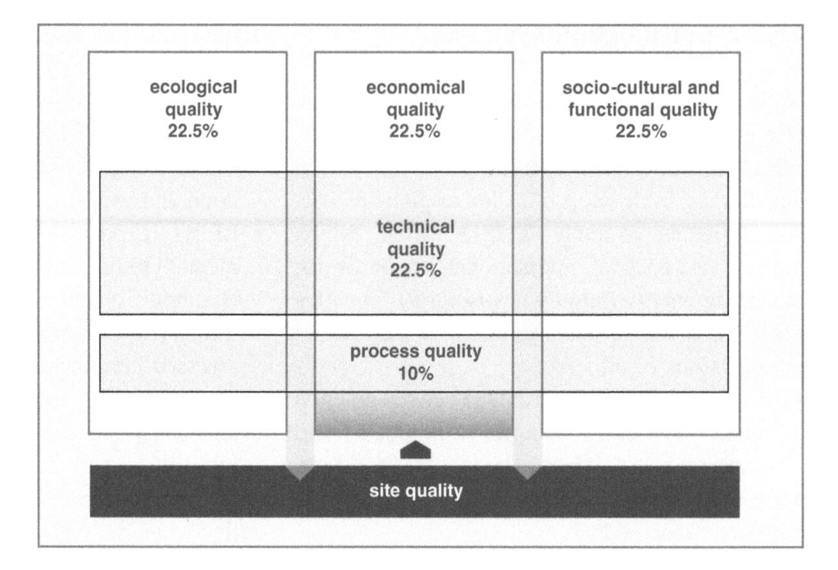

Fig. 15.72 Structure and organization of a DGNB certificate for sustainable building (DGBN). © IFA 17.615_B

that, there are no measureable criteria for the site quality—at least not for the current DGNB certification; information regarding this can be found in Sect. 13.5.

An auditing system moderates the achievement of points during the planning and execution of the project. Depending on the criteria met, ratings of bronze, silver or gold are possible.

Based on the following case studies we will explain how these evaluation systems functions.

15.7.3 Case Studies

Case Study 1 Expansion of a Baked Goods Manufacturer in Germany

Modern bakeries are characterized by the highly technicized use of heating and cooling units. In the conceptualization of the first construction phase for a plant in 1998, the optimal combination of the building structure, technology and energy expenditure was sought using 3D simulation technology [Rei98]. By means of simultaneous engineering, the interdisciplinary planning team was able to attain the goal of an efficient and energy-saving production facility under the banner of sustainability. The planning team participants jointly set the conditions for a holistically optimized new construction including: the bakery specific production processes, the workplace design, the building structure as well as the supply and removal system.

Traditionally, cooling and heating loads are calculated using static models. However, in this project, the building and system simulation program TAS was used to calculate the thermal currents and temperature distributions. TAS is a modular structured dynamic simulation program with which climate, energy and facade concepts can be analyzed (http://www.ifes-frechen.de/en).

The integrated energy and production planning lead to an annual energy requirement of 31 kWh/m^2 for heating and approximately 450 kWh/m^2 for cooling. The, to a large degree recyclable, building structure conceptualized later in Phase 2 was designed as a highly changeable skeleton structure. An energetic and ecological novelty was a completely modular roof and wall construction made of the renewable resource wood. By intelligently networking the process and air conditioning technology as well as realizing the building shell in a passive house standard with an insulation strength of 30 cm (11.8 in), the room air could for the most part be sufficiently conditioned by the process waste heat.

Figure 15.74 depicts the location with Phase 1, Phase 2 of the new project and an existing hall. One can see both the spatial structure with the roof removed as well as a longitudinal section of

Main criteria group	Criteria group	Criteria	Main criteria group	Criteria group	Criteria	Main criteria group	Criteria group	Criteria
Ecological quality	*Life cycle analysis*	Global warming potential			Visual comfort			Ease of dismantling and recycling
		Ozone depletion potential			User control possibilities	**Process quality**	*Planing quality*	Quality of project preparation
		Photochemical ozone creation potential			Quality of outdoor spaces			Integral planning
		Acidification potential			Safety and risk of hazardous incidents			Optimization and complexity of planning method
		Eutrophication potential		*Functionality*	Handicapped accessibility			Evidence of sustainable aspects in call for and awarding of tenders
	Effect on the global and local environment	Risks to the local environment			Space efficiency			Creation of conditions for optimal use and management
		Other effects on the local environment			Suitability for conversion		*Construction quality*	Construction site / construction process
		Sustainable use of resources / wood			Public access			Quality of contractors / prequalification
		Microclimate			Bicycling convenience			Quality assurance for construction
	Ressource consumption and waste generation	Nonrenewable primary energy demand			Social integration			Commissioning
		Total primary energy demand and proportion of renewable primary energy		*Design quality*	Assurance of design and urban development quality in a competition		*Management quality*	Controlling
		Other uses of non-renewable resources			Percent for art			Management
		Waste by category			Quality features of use profile			Systematic inspection, maintenance, and servicing
		Drinking water demand and volume of waste water	**Technical quality**	*Technical performance quality*	Fire prevention			Qualification of operating staff
		Space demand			Sound insulation	**Site quality**	*Site quality*	Risks in the micro-environment
Economic quality	*Life cycle costs*	Building related life-cycle costs			Quality of building envelope with regard to heat and humidity			Conditions in the micro-environment
	Performance	Suitability for third-party use			Building services' backup ability			Public image and condition state of site and neighbourhood
		Marketable			Building services' ease of use			Access to transportation
Sociocultural and functional quality	*Healthiness, comfort and user satisfaction*	Thermal comfort in the winter			Building services' equipment quality			Proximity to use-specific facilities
		Thermal comfort in the summer			Durability			Connections to public services (utilities)
		Interior air hygiene			Ease of cleaning and maintenance			Legal situation for planning
		Acoustic comfort			Resistance to hail, storms, and flooding			Extension options / reserves

Fig. 15.73 Hierarchy of evaluation criteria for the DGBN certification system. © IFA 17.616_B

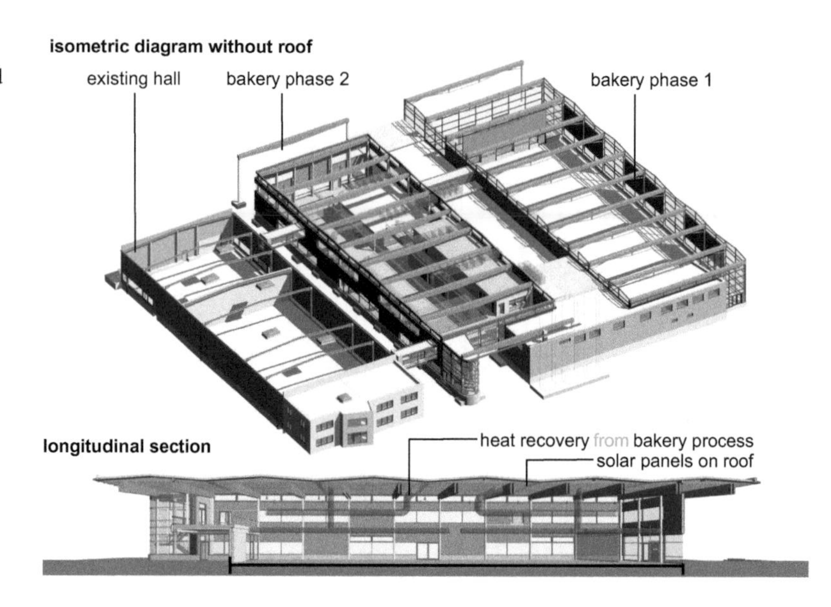

Fig. 15.74 Expansion phases of an energetically and ecologically optimized bakery. © Reichardt 17.617_B

the hall through the wooden structure of Phase 2. In deepening the energetic and building climate strategies from Phase 1, the GENEering process was implemented in the expansion project to meet the client's demands in particular for continuing to reduce CO_2 levels also in the supply chain and increasing the overall energy efficiency. Consequently in the approximately 1500 m^2 (16,150 ft^2) hall, heat recovery technology from baking steam and room air is comprehensively used.

In addition, for the first time in the baking trade, the high global warming potential (GWP) of conventional refrigerant R 4404a was avoided through a two stage cascaded compression refrigeration system. Cascades work with separate refrigerant circuits so that the evaporator temperature decreases considerably from stage to stage. An energy-saving of 45 % compared to Phase 1 was achieved with double the space to cool (i.e. cooling volume). In a future expansion stage, warm water is supplied to the washing machines from the systems' thermal discharge. Furthermore, for the first time in a bakery, an energetic high efficiency and innovative lighting system (in comparison to conventional systems) is developed using LED high bay lighting, which due to the specific design of the prism plates and heat sinks, is suitable for the flour loaded atmosphere in bakeries. Networked together with a daylight dependent control, the system is highly sensitive.

In the final stage, the in total 2200 m^2 (23,700 ft^2) roof area from Phase and Phase 2 is completely equipped with photovoltaic panels. In a first step, on the approximately 1100 m^2 (11,800 ft^2) roof from Phase 2, 12,300 KWh of solar yield is directly fed into the bun baking process as well as in the fueling of the company's zero emission delivery fleet (5 t delivery trucks with solar energy drive electric motors) for the retail outlets. In supplying retail outlets in urban areas, with a mileage of approximately 300,000 km/year and approximately 36,000 l of diesel fuel, approximately 100 t of CO_2 is saved per year. After drawing off the 54,000 kWh/year required for the delivery fleet, approximately 69,000 kWh/year of solar energy can still be directly fed into the bakery's lighting and electrical processes. The approximately 50 outlets distributed around the entire city of Essen thus become "ambassadors" of a forward thinking supply chain strategy.

The planning of the layout and logistics, buildings and building services was facilitated during the entire project duration by a certification according to DGNB. (An overview of which, including the list of criteria, was provided in the previous section.)

MAINCRITERIA GROUP	CRITERIA GROUP	CRITERIA	SCORE MAXIMUM	CRITERIA POINTS ACHIEVED	CRITERIA POINTS MAX. POSSIBLE	WEIGHTING FACTOR	ADAPTATION FACTOR	WEIGHTED POINTS ACHIVED	WEIGHTED POINTS MAX.	PERFORMANCE INDEX	GROUP POINTS ACHIEVED	GROUP POINTS MAX. POSSIBLE	GROUP PERFORMANCE	GROUP WEIGHT	TOTAL PERFORMANCE INDEX
ECOLOGICAL QUALITY	LIFE CYCLE	Global warming potential	10	8,0	10,0	3,0	1	24,0	30	80,0%	173,5	215,0	81%	22,5%	85,2%
		Ozone depletion potential	10	10,0	10,0	1,0	1	10,0	10	100,0%					
		Photochemical ozone creation potential	10	10,0	10,0	1,0	1	10,0	10	100,0%					
		Acidification potential	10	8,0	10,0	1,0	1	8,0	10	80,0%					
		Eutrophication potential	10	9,0	10,0	1,0	1	9,0	10	90,0%					
	EFFECT ON THE GLOBAL AND LOCAL ENVIRONMENT	Local environmental impact	10	7,5	10,0	3,0	1	22,5	30	75,0%					
		Suytainable use of resources/wood	10	10,0	10,0	1,0	1	10,0	10	100,0%					
		Microclimate	10	1,0	10,0	0,5	1	0,5	5	10,0%					
	RESSOURCE CONSUMPTION AND WASTE GENERATION	Nonrenewable primary energy demand	10	7,0	10,0	3,0	1	21,0	30	70,0%					
		Total primary energy demand and share of renewable primary energy	10	10,0	10,0	3,0	1	30,0	30	100,0%					
		Drinking water demand and volume of waste water	10	6,0	10,0	1,0	1	6,0	10	60,0%					
		Land demand	10	7,5	10,0	3,0	1	22,5	30	75,0%					

Fig. 15.75 Criteria for the ecological criteria of a wholesale bakery (per ifes). © IFA 17.618_B

Figure 15.75 lists the criteria for the main category 'ecological quality' for this case study and clarifies the evaluation process. The main category is further divided into three sub-groups with a total of 12 individual criteria. Each criterion can achieve a maximum of 10 points. The achieved number of points is multiplied by a weighting factor and where needed an adjustment factor. This value is then divided by the maximal attainable weighted point value and results in a performance index. The sum of all the weighted point values is set in relationship to the maximal attainable number of points resulting in the group performance indicator, which in this case is 81 % for the ecological quality. The group value has a weight of 22.5 % in the overall evaluation.

Figure 15.76 shows a graphical depiction of a performance profile according to the evaluation criteria radially organized in groups of segments. Already during the planning phase this project received a 'gold' certification with 85.2 points from a maximal 100 possible points.

A more in-depth analysis of the energy potential requires the interconnections between

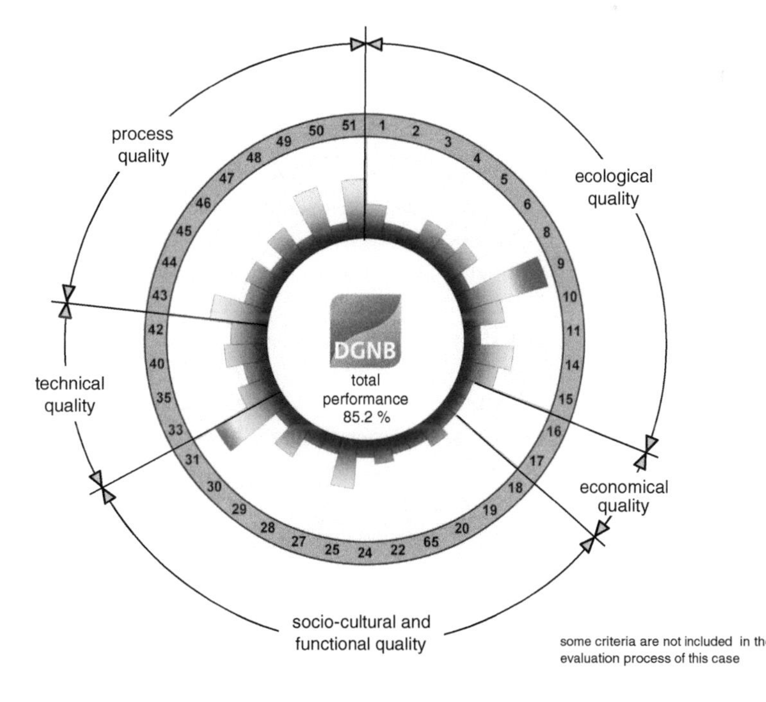

Fig. 15.76 Overall evaluation of a bakery project according to DGBN (per ifes). © IFA 17.619_B

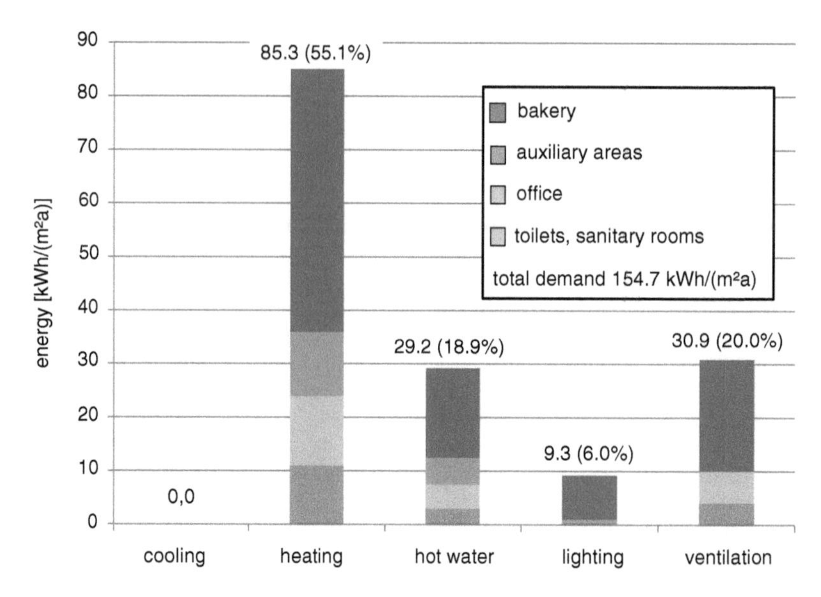

Fig. 15.77 Distribution of energy consumption within a bakery project. © Reichardt 17.620_B

the buildings, processes and building services to be examined in view of self-sufficiency. Here self-sufficient means that the factory is able to operate without the use of external energy e.g., for heating or cooling the building. A detailed breakdown of the building services consumption values for the individual plant segments allows the optimization of the energy currents of the building services to be finely tuned with the building and process loads.

Figure 15.77 shows the energy consumption of the utility areas for the baked goods project.

Case Study 2 Construction of a New Distribution and Assembly Center in India.

This project was aimed at configuring a new factory building of approximately 12,500 m^2 (134,500 ft^2) in Chennai, South India for a global manufacturer and distributor of components for metal fittings for windows, doors and glass walls. Figure 15.78 depicts the master plan in a number of development stages with expansion options for the assembly, storage as well as research and development. The GENEering process for this climatically complex project resulted in the client's conception of a "green building factory". In Chennai temperatures range between 30 and 35 °C (86–95 °F) year round with a humidity of 70–90 %. It is therefore particularly important to ensure the room comfort through an intelligent building structure with minimized operating costs and CO_2 emissions. Based on a comprehensive analysis of the processes, location, climate, buildings and building services, a 3D modelling of the layout, buildings and energy flows was used to examine the possible uses of daylight, the thermal comfort at workplaces as well as the overall energetic efficiency for various alternative configurations.

With the appropriately profiled roof silhouette, the selected cross-section prevents the hot zenithal sun from entering, while at the same time lighting the hall workplaces with an average of 350 lux. In doing so, operating costs for electric lighting and related heat loads as well as CO_2 emissions are avoided. In addition, "soil ducts" (commercial sewage pipes made from concrete), laid under the hall's floor, introduce relatively cool air into the hall. This reduces the temperature inside by up to 4 degrees Celsius (5.4 °F) compared to the outside air, thus increasing the comfort of the occupied area in the hall without expensive ventilation technology and at the same time reduces CO_2 emissions.

In the upper left corner of Fig. 15.79 we can see variants of the air cooling using the soil ducts. In variants 1–3 the number of ducts is

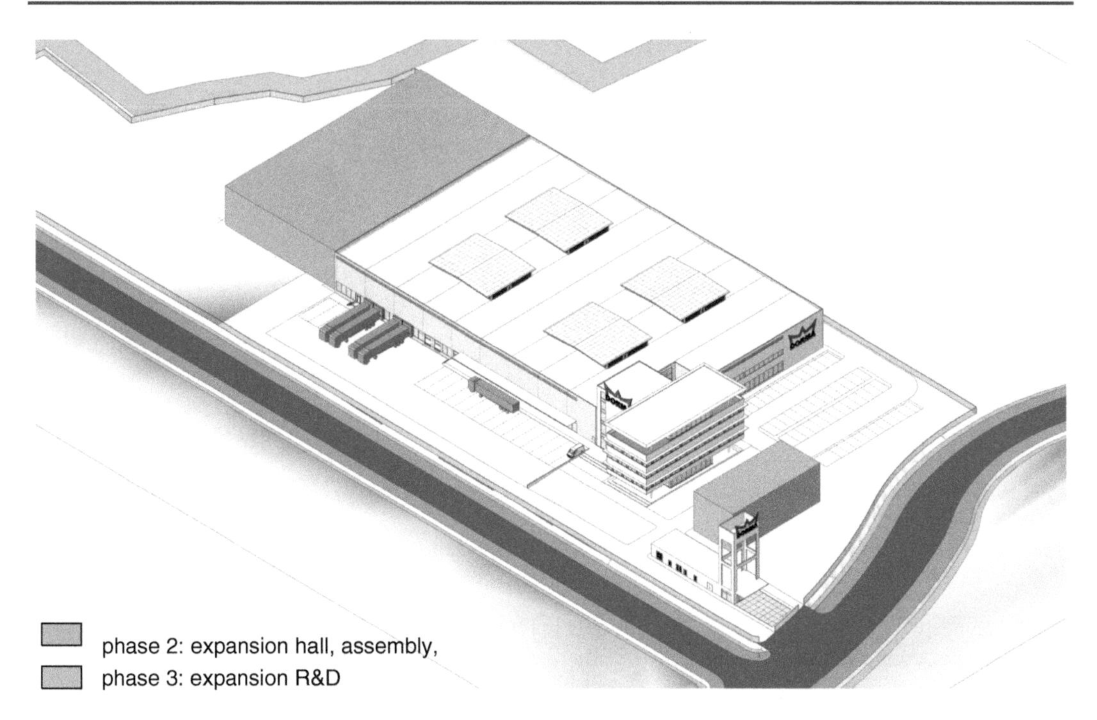

Fig. 15.78 Master plan of an assembly and distribution center (example Chennai, India). © Reichardt 17.621_B

varied (from 0 to 64) and variant 4 includes no openings in the facade. The results of the dynamic simulations in regard to the temperature distribution during the day (Fig. 15.79, upper right) clarify the impact of the underground cooling. In the lower part of Fig. 15.79 we see the temperature layers for variant 4.

"Solar drafts" support the air displacement principle of the halls thermal activity. Solar drafts are created when fans, driven by CO_2 neutral

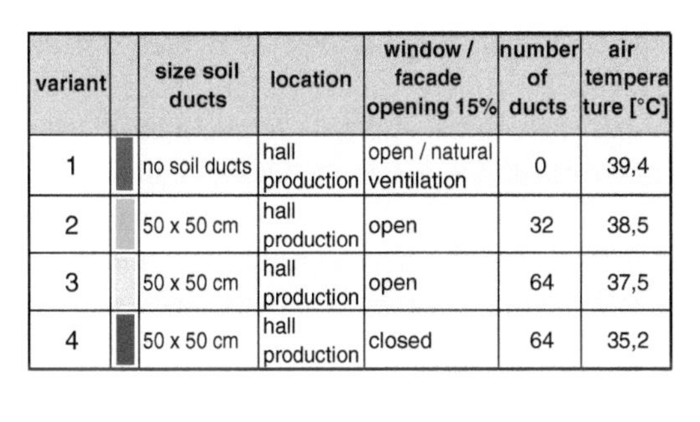

variant		size soil ducts	location	window / facade opening 15%	number of ducts	air tempera ture [°C]
1		no soil ducts	hall production	open / natural ventilation	0	39,4
2		50 x 50 cm	hall production	open	32	38,5
3		50 x 50 cm	hall production	open	64	37,5
4		50 x 50 cm	hall production	closed	64	35,2

variants of cooling ducts

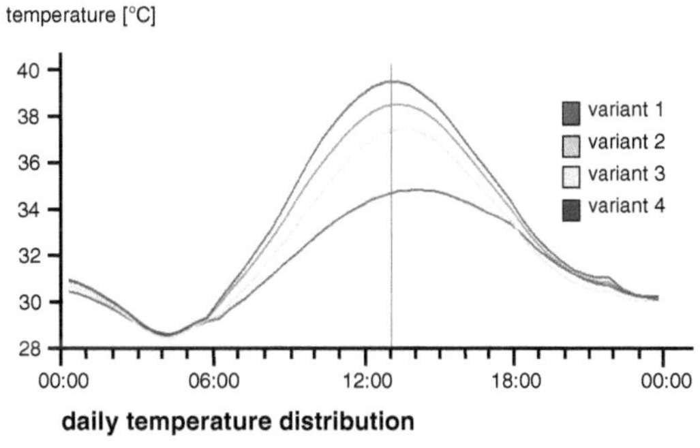

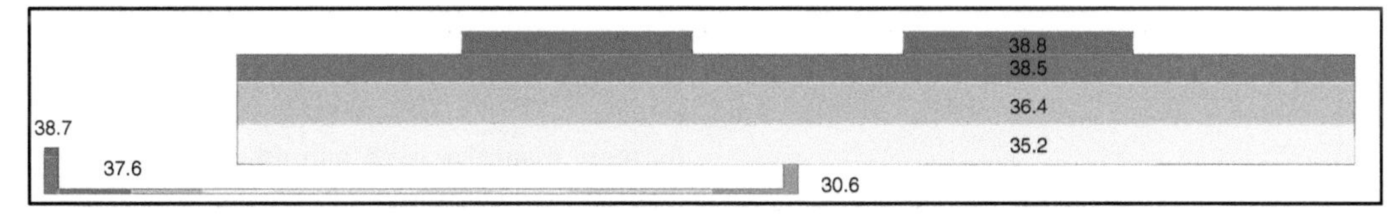

Fig. 15.79 Temperature profile for soil duct variants. © Reichardt 17.622_B

Fig. 15.80 Solar drafts principle for a building. © Reichardt 17.623_B

electricity from the photovoltaic surfaces of the roof skylights, carry the warm air outside when the heat and solar radiation are at their peak at noon.

In a 3D computer graphic, Fig. 15.80 depicts the inflow of outside air through the soil ducts according to the air displacement principle as well as the mechanism behind the solar drafts. The solar drafts carry the upper, hottest layer of air out of the building, so that the soil ducts can conduct fresh air into the hall. Additional measures in the building structure and the offices' building services (e.g., shades for all glass surfaces which prevent solar heat from entering) help maximize the plant's energy efficiency.

The entire project was certified according to the rules set forth by the India Green Building Council (IGBC) including measures for e.g., retaining rainwaters. The preliminary planning was already awarded 'gold'. Despite these 'green building' measures as well as others not mentioned here, the high architectural quality demanded by the client was not compromised in any way.

15.8 Preparations for Realization

The next phase of the synergetic factory planning is the *preparation for contract placement* stage. Here the concretization of the planning from the production and object perspective steps into the foreground. Described as Phase 6 according to HOAI, this stage entails *preparing the requests for bids* for the specific projects and serves to quantify and develop the project outcome specifications in agreement with the expert planners. This applies however, only to the building, its technical equipment and the external systems.

As already mentioned, most of the time when a factory is being planned a significant proportion of the production equipment is taken over from an existing factory. Due to their complexity and value, only large-scale systems (e.g., paper machinery, rolling mills, print shops, press shops, paint shops and final assemblies of automotive plants) dominate the factory buildings. In such cases the building becomes a protective envelope and not a changeable structure with products and production facilities that vary relatively frequently.

From the perspective of the production planning, invitations for bids for the operating facilities defined in the rough and detailed planning are released and awarded. Only the storage and commissioning facilities, as a component of the production logistics, need to be designed based on the new manufacturing and assembly segments; these too are tendered and procured. If automated facilities such as high bay warehouses are concerned, specialized providers are contracted.

The subsequent collaboration in awarding the contracts (HOAI Phase 7) as a conclusion to the preparations for realizing the project means examining and evaluating the bids for their conformity with the specifications of the tender documents, developing a price sheet comparison according to the subservices and assisting in the negotiations with the firms who are awarded the contracts. This phase is strongly shaped by the award form, which is described in more detail in Sect. 16.2 (Award Forms).

15.9 Supervising the Realization

During the realization phase, the production planning has a monitoring function in the sense of ensuring the future production and logistic processes. While the building is being erected there will always be many special questions that need to be clarified such as the accessibility of facilities in case of maintenance or disruptions. The supervising architects are responsible for overseeing the construction work, while the engineers' task is to oversee the installation of the building services. Together with the construction site manager they closely supervise the individual projects ensuring that they are executed according to the contracts and prepare the necessary inspections and approvals.

According to HOAI's Phase 8 (project supervision) numerous control tasks are related from a schedule, cost and legal perspective including identifying and resolving deficiencies. As indicated in Fig. 15.7, with complex production systems, separate inspection trials are conducted once the facilities are installed and brought into operation. The object supervision ends when the individual projects and the entire object have been approved and accepted by the clients and respective authorities.

The last step of monitoring the realization according to HOAI's Phase 9 is the object management. This basically includes developing the maintenance plans, determining defects or deficiencies and compiling the visual documentation. If a computer supported facility

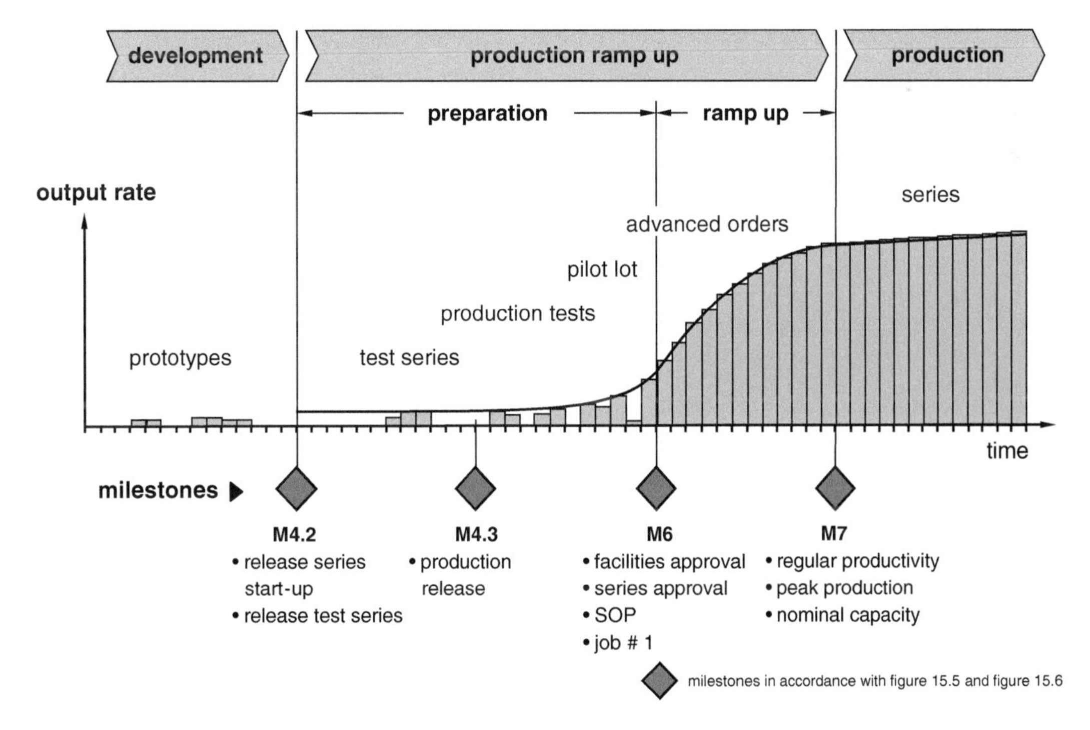

Fig. 15.81 Ramp-up curve of a production. © IFA D 13.966

management is planned, the corresponding documents can be set-up there. Details will be explained in Chap. 17 Facility Management. This step of supervising the realization marks Milestone M5.

15.10 Managing the Ramp-up

The last stage of a synergetic factory planning focuses on the user. The task of planning the move to the new property should not be underestimated. The production should be interrupted as briefly as possible and this disruption should be bridged by producing stock ahead of time. After the move, machinery and systems have to be started and ramped-up to the rated capacity. With complex objects this phase can require a number of weeks. Figure 15.81 provides an idea of a ramp-up curve for a complex production system [Win07].

The stages and milestones clarified in Figs. 15.6 and 15.7 can be identified though the milestones are more precise here. The respective output rate for the prototypes stem from the development requirements, whereas test series and production tests help upgrade the system and prepare it for full production. After the pilot lot, the entry into the market is usually launched and the first advance-orders are processed. Reoccurring disruptions should be minimized though by systematically managing the ramp-up [Ku02, Win07].

Once the facilities have been approved and accepted, Milestone M6 has been attained. The investment account is closed and the final statement is drawn-up. It in turn is the basis for developing the balance sheet for the fixed assets.

With these comments we conclude our discussion of the stages of the synergetic factory planning model from a technical perspective. As described at the start, project management is a supportive service which we will discuss in the following chapter because of its significance for the success of a project.

15.11 Summary

At the beginning of a systematic project course, the question should always be the need for and strategic importance of a factory rebuilding or new construction. The subsequent planning process is divided into 6 phases each of which is terminated with a milestone. Compared to the commonly encountered separate production planning and building design these phases represent the process view and space view in a synergistic and mutually penetrating approach. All phases are planned and controlled by a careful project management.

Bibliography

[Agg87] Aggteleky, B.: Werksentwicklung und Betriebsrationalisierung (Site Development and Operational Rationalization), Vol. 2: Betriebsanalyse und Feasibility-Studie (Operations Analysis and Feasibility Study). Hanser, München (1987)

[Bri09] Brieke, M.: Erweiterte Wirtschaftlichkeitsrechnung in der Fabrikplanung. (*Extended economic calculation in factory planning*). Ph.D. Thesis. Leibniz Universität Hannover 2009. PZH GmbH, Garbsen (2009)

[Bin11] Binner, H.F.: Handbuch der prozessorientierten Arbeitsorganisation. REFA. Methoden und Werkzeuge zur Umsetzung (Handbook of Process-Oriented Work Organization. REFA. Methods and Tools for Implementation), 4th edn. Hanser, München, Wien (2011)

[Bra09] Braungart, M., McDonough, W.: Cradle to Cradle, Remaking the Way We Make Things, 2nd edn. Vintage Books, London (2009)

[Dan01] Dangelmaier, W.: Fertigungsplanung. Planung von Aufbau und Ablauf der Fertigung (Manufacturing Planning. Planning the Construction and Operation of Manufacturing). Springer, Berlin (2001)

[Eng08] Engelmann, J.: Methoden und Werkzeuge zur Planung und Gestaltung energieeffizienter Fabriken (Methods and Tools for Planning and Design of Energy-Efficient Factories). PhD Thesis Univ. Chemnitz (2008)

[Fel98] Felix, H.: Unternehmens- und Fabrikplanung: Planungsprozesse, Leistungen und Beziehungen (Corporate and Factory

Planning: Planning Processes, Services and Relationships). Hanser, München (1998)

[Gau06] Gausemeier, J., Fink, A., Schlake, O.: Szenariomanagement. Planen und Führen mit Szenarien (Scenario Management. Planning and Managing with Scenarios), 2nd edn. Hanser, München (2006)

[Gru00] Grundig, C.-G.: Fabrikplanung: Planungssystematik, Methoden, Anwendungen (Factory planning: Planning Systematics, Methods, Applications). Hanser, München (2000)

[Han04] Hanna, S.R., Konz, S.: Facility Design & Engineering, 3rd edn. Holcomb Hathaway, Scottsdale (2004)

[Har04] Harms, T.: Agentenbasierte Strukturplanung (Agent based structure planning). Ph.D. Thesis, University of Hannover 2004. PZH GmbH, Garbsen (2004)

[Her03] Hernández Morales, R.: Systematik der Wandlungsfähigkeit in der Fabrikplanung (Systematics of changeability in factory planning). Ph.D. Thesis, University of Hannover 2002. Fortschrittberichte VDI, Reihe 16, Nr. 149, Düsseldorf (2003)

[Her11] Herrmann, C. et al.: Energy oriented simulation of manufacturing systems—Concept and application. CIRP Annals—Manufacturing Technology **60**, 45–48 (2011)

[HOAI13] Verordnung über die Honorare für Leistungen der Architekten und der Ingenieure. Bekannt gegeben am 10. Juli 2013 (Regulation on fees for services of architects and engineers), July 10 Federal Law Publication 213 No. 37, pp. 2276–2374 (2013)

[IECC12] International Code Council: 2012 International Energy Conservation Code. International Code Council (2011)

[Kat55] Katz, R.L.: Skills of an effective administrator. Harvard Bus. Rev. **33**(1), 33–42 (1955)

[Ket84] Kettner, H., Schmidt, J., Greim, H.-J.: Leitfaden der systematischen Fabrikplanung (Guide Line of Systematic Factory Planning). Hanser, München (1994)

[Ku02] Kuhn, A., Wiendahl, H.-P., Eversheim, W., Schuh, G., Winkler, H., et al.: Schneller Produktionsanlauf von Serienprodukten (Fast Ramp Up of Series Products). Verlag Praxiswissen, Dortmund (2002)

[May01] Mayrshofer, D., Kröger, H.A.: Prozesskompetenz in der Projektarbeit: Ein Handbuch für Projektleiter, Prozessbegleiter und Berater (Process Skills in Project Work: A Handbook for Project Managers, Facilitators and Consultants). Hamburg, Windmühle (2001)

[Mül09] Müller, E. et al.: Energieeffiziente Fabriken planen und betreiben (Planning and operate energy efficient factories). Springer (2009)

[Neu13] Neugebauer, R.: Handbuch Ressourcenorienterte Produktion (Handbook resource oriented manufacturing. Hanser München (2013)

[Nyh04] Nyhuis, P., Elscher, A., et al.: Prozessmodell der Synergetischen Fabrikplanung: Ganzheitliche Integration von Prozess- und Raumsicht (Process model of synergetic factory planning: holistic integration of process and space view). wt Werkstattstechnik online **94**(4), 95–99

[Paw08] Pawellek, G.: Ganzheitliche Fabrikplanung: Grundlagen, Vorgehensweise, EDV-Unterstützung (Holistic Factory Planning: Principles, Procedure, IT Support). Springer, Berlin (2008)

[Pet14] Petrecca, G.: Energy Conversion and Management: Principles and Applications. Springer, Berlin (2014)

[Rei04] Reichardt, J., Gottswinter, C.: Synergetische Fabrikplanung – Montagewerk mit den Planungstechniken aus dem Automobilbau realisiert (Synergetic Factory Planning—Assembly Plant Realized with the Planning Process from the Automotive Industry), S. 52–55. industrieBAU, 3/2004

[Rei98] Reichardt, J., Drüke, K.: Bäckerei mit mit innovativem Gesamtkonzept (Bakery with innovative overall concept). IndustrieBau 6 34–39 (1998)

[Rei07] Reichardt, J., Pfeifer, I.: Phasenmodell der Synergetischen Fabrikplanung. Stand der Forschung und Praxisbeispiele (Phase model of synergetic factory planning. State of research and practice examples). wt online **97**(4), 218–225 (2007)

[Rop99] Ropohl, G.: Allgemeine Technologie - Eine Systemtheorie der Technik (General Technology—A Systems Theory of Technology). Hanser, München (1999)

[Roth03] Rother, M., Shook, J.: Learning to See: Value Stream Mapping to Add Value and Eliminate MUDA. Lean Enterprise Institute, Cambridge, USA (2003)

[Sche01] Scheer, A.W.: ARIS-Modellierungs-Methoden, Metamodelle, Anwendungen (ARIS Modeling Methods, Meta-Models, Applications), 4th edn. Springer, Berlin (2001)

[Sche02] Scheer, A.W.: ARIS. Vom Geschäftsprozess zum Anwendungssystem (From Business Process to Application System), 4th edn. Springer, Berlin (2002)

[Sch10] Schenk, M., Wirth, S., Müller, E.: Factory Planning Manual. Situation-Driven Production Facility Planning. Springer, Berlin (2010)

[Son07] Sonntag, K.: Kompetenzmodelle im Human Resource (HR-) Management (Competence models in human resources planning). In: Schäfer E. u. a (eds.): Arbeitsleben!

Arbeitsanalyse, Arbeitsgestaltung, Kompetenzentwicklung (Working Life! Job Analysis, Job Design, Competency Development), S. 264–279. Kassel University Press, Kassel (2007)

[Tom10] Tompkins, et al.: Facilities Planning, 4th edn. Wiley, Hoboken (2010)

[Uli99] Ulich, E.: Mensch - Technik - Organisation: Ein europäisches Produktionskonzept (Human—Technology—Organization: A European Production Concept). Betonwerk und Fertigteil-Technik **65**(22), 26–31 (1999)

[VDI91] VDI-Richtlinie 3644: Analyse und Planung von Betriebsflächen. Grundlagen, Anwendung und Beispiele (Analysis and Planning of Operational Areas. Foundations, Application and Examples). Beuth Verlag, Berlin (1991)

[VDI08] VDI-Richtlinie 4499: Digitale Fabrik. Grundlagen (Digital Factory. Basics)

[VDI11] VDI-Richtlinie 5200: Fabrikplanung. Planungsvorgehen (Factory Planning. Planning Procedures) Beuth Verlag, Berlin (2011)

[Wie72] Wiendahl, H.-P.: Technische Struktur- und Investitionsplanung (Technical Structure and Investment Planning). Girardet Verlag, Essen (1972)

[Wie74] Wiendahl, H.-P., Bertram, D.: Die Produktionsplanung als Bestandteil der Unternehmensplanung (The production planning as part of the corporate planning). Ind. Org. **43** (4), 193–198 (1974)

[Wie96] Wiendahl, H.-P.: Grundlagen der Fabrikplanung (Basics of factory planning). In: Eversheim, W., Schuh, G. (Hrsg.) Betriebshütte - Produktion und Management. Springer, Berlin (1996)

[Wie02] Wiendahl, H.-P., Hernández, R.: Fabrikplanung im Blickpunkt. Herausforderung Wandlungsfähigkeit (Focus Factory Planning. Challenge Changeability). wt Werkstattstechnik online, Jhg, **92**(4), 133–138 (2002)

[Wie14] Wiendahl, H.-P.: Betriebsorganisation für Ingenieure (*Operational* Organization for Engineers), 8th edn. Hanser, München (2014)

[WieH11] Wiendahl, H.-H.: Auftragsmanagement der industriellen Produktion. Grundlagen, Konfiguration, Einführung (Order Management of Industrial Production. Basics, Configuration, Implementation). Springer, Berlin (2011)

[Win97] Winz, G., Quint, M., Kuhn, A. (eds.): Prozesskettenmanagement - Leitfaden für die Praxis (Management of Process Chains). Verlag Praxiswissen, Dortmund (1997)

[Win07] Winkler, H.: Modellierung vernetzter Wirkbeziehungen im Produktionsanlauf (Modeling of networked activity relations in production ramp up). Ph.D. Thesis, Leibniz Uni. Hannover 2007. PZH Verlag, Garbsen (2007)

[Wil96] Willke, H.: Systemtheorie (Systems Theory). Fischer Verlag, Stuttgart (1996)

[Zan76] Zangemeister, C.: Nutzwertanalyse in der Systemtechnik: Eine Methodik zur multidimensionalen Bewertung und Auswahl von Projektalternativen (Cost-Benefit Analysis in Systems Engineering: A Methodology for Multi-Dimensional Evaluation and Selection of Project Alternatives), 4th edn. Springer, Berlin (1976)

[Zeug98] Zeugträger, K.: Anlaufmanagement von Großanlagen (Ramp up management of large plants). Ph.D. Thesis, Univ. Hannover. VDI-Verlag, Düsseldorf (1998)

16 Project Management

A professional Project Management Team plays a significant role in the success of a factory planning project, in terms of its functional performance, compliance with the time schedule and cost frame. Many managers underestimate the complexity of this task and wrongly assume that this task can be handled in addition to their daily businesses, based upon their personal experiences in small construction and/or remodeling projects. Such misconceptions may result in dramatic impacts to the extent of project failures and may be to the point of endangering the entire company. For a systematic approach, the proposals discussed in the following chapter are advisable, based on the methodology developed in the Chap. 15, synergetic factory planning. Holistic project management includes skillful organization of a team, processes and planning to manage contract, time, cost and quality using state of the art digital planning tools.

16.1 Tasks of Project Management

16.1.1 Stumbling Blocks

The project management of a factory includes planning, steering, coordination and control tasks, aiming at a reasonably accurate, scheduled and within budget implementation of the project [Rös94, Die04, Rog08, Knu08].

There are many stumbling blocks that must be avoided with professional, organizational and human skills. They are typical of all complex projects including factory design and can be summarized based upon the experiences of Schulte [Schu08] and others as follows:

- The project preparation is often the result of hasty decisions, instead of clarifying mutual expectations of the project team at the very onset.
- A methodological approach to concept generation is missing. Engineers want to "make" instead of spending time discussing targets, communication and workflow.
- There is a lack of common understanding of the solution; moreover many individual solutions are never "optimized".
- The project decisions are often emotional reactions, by positions of power rather than factual arguments.
- Clients lack awareness and underestimate the complexity of a factory project; resulting in unreasonable targets both in time and costs.
- In many cases, it is assumed that a standard procedure exists which automatically guarantees an "optimal" solution. However, each project and its site being unique there is no "right" solution to start with; solutions are usually developed through a conscious team effort and from learning from each other.
- Resistances against certain solutions are not imminently visible creating an imbalance between interests of the planning team and those of the future users of the building.
- Finally in case of conflicts, it is extremely difficult to differentiate between personal

H.-P. Wiendahl et al., *Handbook Factory Planning and Design*,
DOI 10.1007/978-3-662-46391-8_16, © Springer-Verlag Berlin Heidelberg 2015

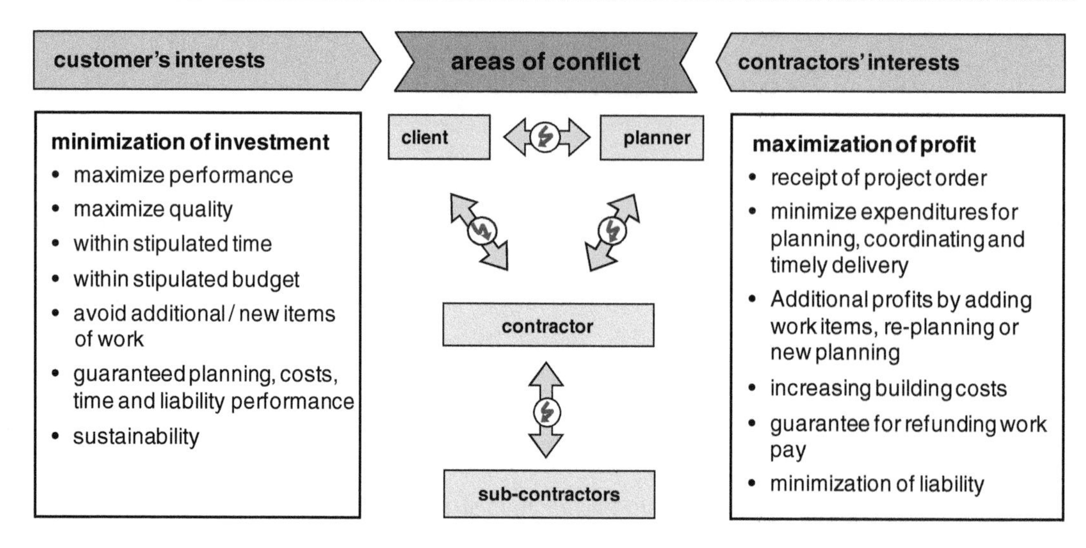

Fig. 16.1 Fields of conflict within a factory planning project (acc. to Schulte). © IFA 15.234_B

interests and relationship issues (values and cultures), property issues (professional views and related interests) and apparent misconceptions (with different interpretations of problems, often unclear communication). Figure 16.1 shows the resulting stress fields between the client, the planners, the contractor and the subcontractors [Schu08].

In general, clients are interested in minimizing investment costs (and/or operational costs); there is also a tendency to minimize risks related to budgeting, delays, change orders, warranties, etc. This applies even to post-contractual cost and time related problems which may be the result due to unclear requirements, belated decision making or too many changes. On the other hand, users would like to take this opportunity to eliminate their long-standing shortcomings through the new project. In some cases, Managements are not keen to coordinate all the requests from the users at the onset and additional costs are often attributed to the planning team.

Building Contractors, on the other hand, are interested in maximizing their profits out of each sub-project and minimizing their efforts and expenses. Generally, planning teams refuse to take overall responsibility or even liability for the client and user requirements, which were not known to them at the time of their contract agreement. It is assumed that the planning teams were contracted in line with the relevant national/international guidelines (e.g. HOAI—guidelines for fee structure for architects and engineers in Germany) and tied to construction costs; thus allowing for adjustments in fees based upon final project costs. The various vendors of building, utilities or process equipment would, in most cases, try to compensate for the discounts (given during final negotiations to win the contract) by skillfully managing claims against changes, additional works, etc. Also contractors, in general, try to minimize their warranty obligations. To avoid conflicts between the various parties, it is advisable to develop basic project objectives in a structured manner, in accordance with the process model of synergetic factory planning (see Chap. 15), keeping in mind process and spatial requirements. This applies to the overall project as well as for the sub-projects.

16.1.2 Task Overview

It appears that an infinite number of solutions are possible guarantying efficient project execution, some of which rely upon the knowledge of

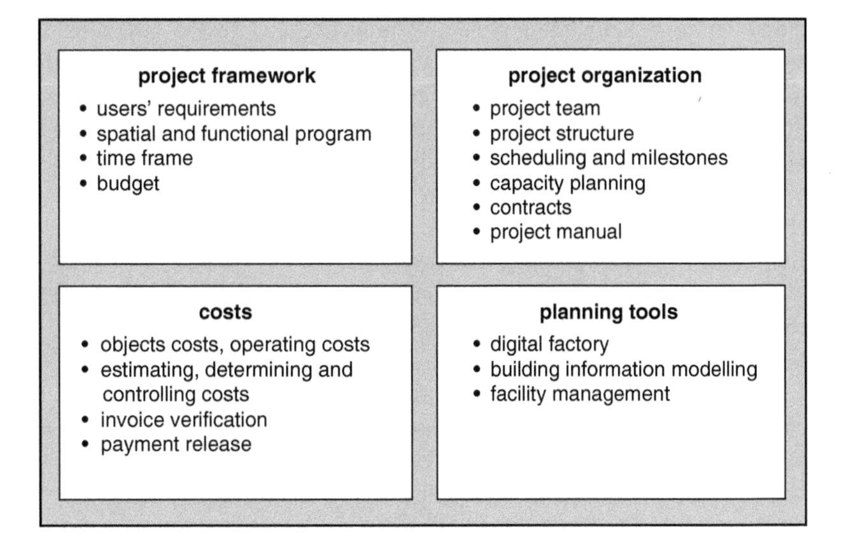

Fig. 16.2 Fields of activity in project management. © IFA 15.235_B

systems engineering. Based upon available literature, for example, [Rös94, Die04, Rog08, Knu08, Mar11] and authors' experience, the main action areas are identified in Fig. 16.2.

The starting point for any project should be based on a project framework which first defines the requirements before Milestone M1 in Fig. 15. 5. Here the users' requirements are developed and used to create a spatial and functional program, referred to as a 'space schedule' or 'space book'. The space book is the counterpart to the specification catalogue for the production facilities and is the basis for successful mediating between the property developers, users and external partners [Rei04].

Space books are generated floor-wise for each building to specify their requirements, utility connections as well as the descriptions of their process facilities. Appendix B shows an example of an appropriately structured Excel Sheet, for the space—function program. Finally, the framework should include a basic understanding of project time and budget, to be derived from the overall strategic considerations of the project.

The second field of action concerns the project organization. It defines the constellation of the project participants for the planning, realization and ramp-up of the factory. Moreover, a project structure has to be determined with 'work packages'. The sequence in which the packages are to be completed is then developed and inserted into the project master schedule with milestones.

Even when the majority of work has been outsourced, the project execution still requires internal capacities that need to be planned. All of the work has to be contractually negotiated, with attention paid in particular to liability issues and resolving deficiencies. It is also critical to agree upon a date after which no further significant changes can be made from the users' side. This time span, known as the "frozen period", allows the project and trades to develop without disruptions. Ultimately, a consistent planning state is ensured by coordinating the information and participation in the project along with documenting the results in the form of a project manual. The project manual should lay down the formats and routines, the process of coordination and the documentation of results.

The third area of action concerns the *costs*. These partly relate to the production facilities and partly to the building project. Usually initial cost estimates, tend to emphasize on the identification and completeness of all cost elements rather than accuracy. It is desirable to distinguish between a one-time or initial object cost and recurring costs due to usage, operation and maintenance. From the above discussions one could derive overall target costs for the building project. A proactive management of all cost items would help in minimizing the life cycle cost of the entire factory including the production facilities and buildings. Finally, with increasing demands for energy efficiency and sustainability and environmental consciousness towards Green Factory Buildings, it is recommended that every attempt should be made to recover energy and conserve resources from the very beginning [Rei98] (see Sect. 15.7, energy efficiency).

These brief and initial remarks should help to clarify the fact that project management plays a key role in balancing the objectives including both—"hard" factors (such as functional objectives, cost, schedule) and more subjective, "soft" factors (such as quality, aesthetics, ecology).

Project management tasks are increasingly influenced by the ever changing digital planning skills and tools. State of the art "Building Information Modeling" (BIM) and its digital counterpart on process side the "Digital Factory" are being increasingly used for precise planning and engineering of buildings. These techniques favour Facility Management during the operational phase of the building, with skillful digital documentation of the entire planning phase. The above process would be elaborated in the final Chap. 17.

In the following sections, three main action areas of project management namely—organization, cost and planning tools will be discussed in detail for an understanding of the basic tasks of a project manager.

16.2 Project Organization

16.2.1 Team Building

Once management has made the decision to start the project a project team has to be established. For this purpose, the responsibilities and powers of the project management, the customers as well as the users, planners and executers are defined within an organizational chart. The principle structure of this chart is depicted in Fig. 16.3. Following common understanding (e.g. [Knu08]) the project management is divided into a Lead Team (decision-making and implementation capabilities) and Project Steering Team (advisory services without decision-making power). Business base for all other partners is the afore mentioned coordinated space—function program.

In most cases, clients underestimate their own management effort for the project, particularly in terms of the required level of professionalism,

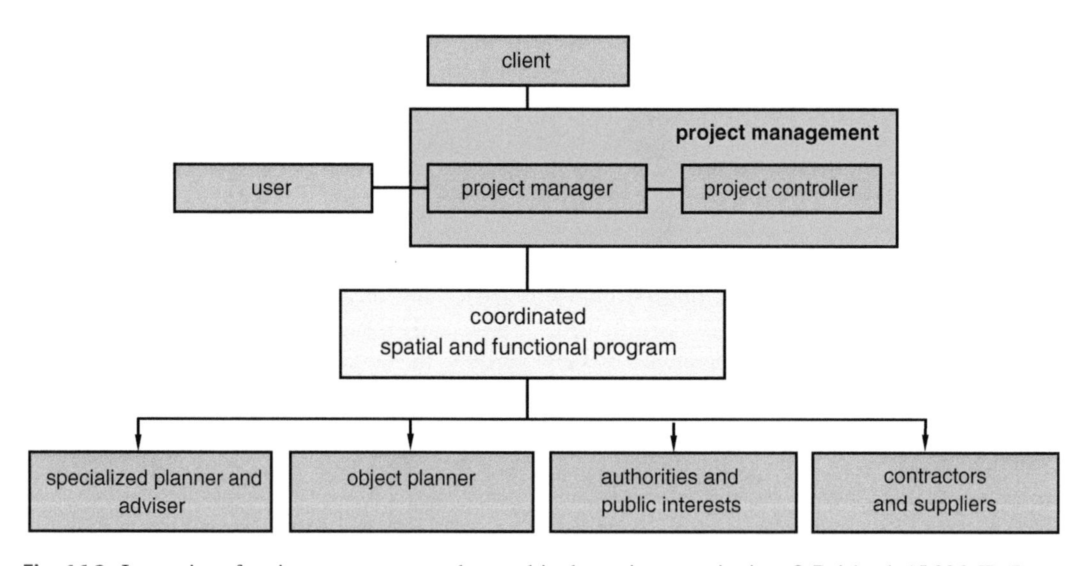

Fig. 16.3 Integration of project management and control in the project organization. © Reichardt 15.236_JR_B

time and leadership skills. There is a tendency to shift all potential project risks to planning and subsequently to the execution teams. But the ultimate responsibility for the project lies in the client's hands, since he is decisive in formulating and controlling the project goals. A responsible client provides all necessary support to the project management team both internally and externally. This is especially important for all internal coordination required on behalf of the client. In case internal management staff of required skills and experience are not available, the client should not hesitate to recruit the services of external management expertise for in-house "client representation". In most cases, by the end of the second step, the project structure is ready. The project should be divided into smaller executable jobs, e.g. building construction, utilities equipment, landscaping, power generation units, waste water treatment plants, etc. to arrive at a pragmatic time schedule.

16.2.2 Example of a Project Organization

Figure 16.4 depicts an example of how the actors in a large factory planning project interact. Together they form the project team [Schu08].

Firstly one needs to identify the external members of the Team. Typically, a *client Steering Committee* consists of relevant members of the management board, representatives of the primary users and—where available—the works Council Chairman. Sometimes an independent external expert may be included too. The *project manager* and, depending on the milestone, the responsible planner or critical equipment suppliers report back to the steering committee. The *users of the project* provide the necessary data during the preparation phase, generated along with the planning team after discussing the various project requirements and adopting required solution/s. Moreover they determine the future organizational structure of

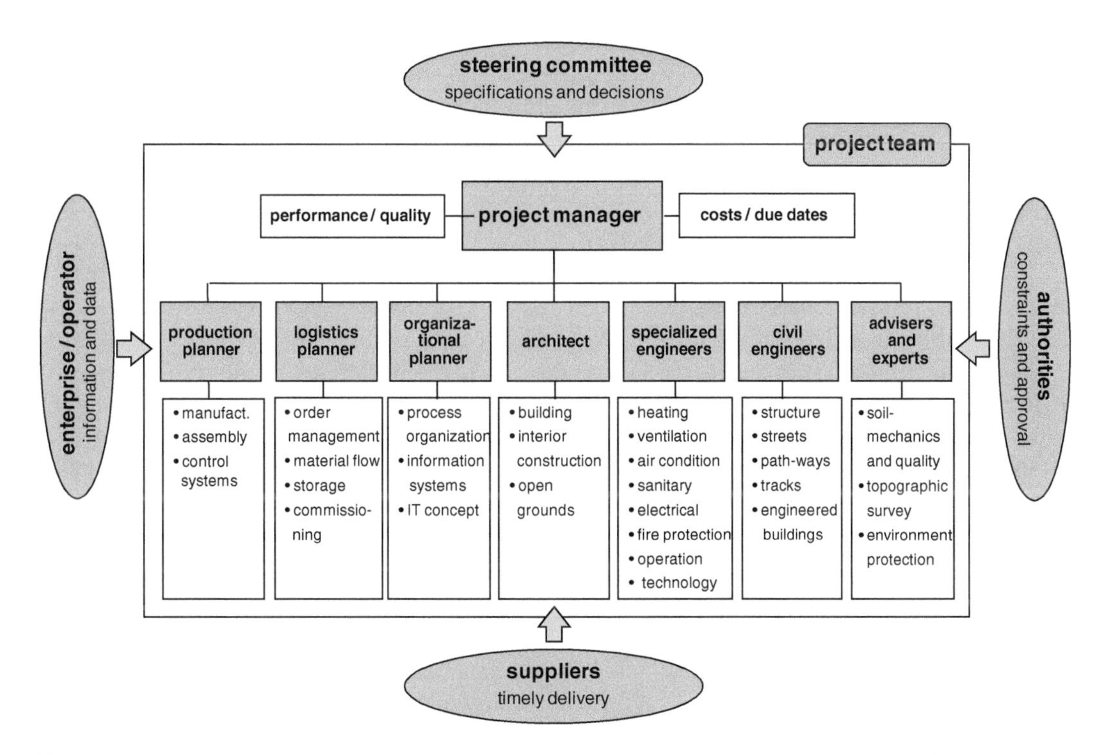

Fig. 16.4 Comprehensive project organization in a complex factory planning project (H. Schulte). © IFA 15.238_B

the works of the new factory and plan for timely recruitment and training of staff. National and international authorities or institutional bodies impose their requirements for occupational safety, health administration and environmental protection. The relevant procedural time for acceptance and licensing to start operation needs to be built into the project time schedule. Finally the suppliers of equipment and services are responsible for timely delivery of their contractual agreements.

The core of the project organization is the project team. It is led by the Project Leader (mostly called Project Manager) who is responsible to the Steering Committee for delivering the agreed performance and quality within the budgeted cost and time schedule. For this purpose, he needs to setup a project management framework methodology for the planning phases after critically examining the qualifications of the people as to whether they fit the requirements of the given planning tasks or not. In setting the rules and agreements around interactions the project manager fundamentally influences the project atmosphere. A project manager's personality should be characterized by their moderation and mediation skills, impartiality towards participants and their methodological competence. It is also critical that the project manager has a direct contact with someone on the executive level who functions as the project sponsor.

The team members in charge of executing the project along with their responsibility can also be identified in Fig. 16.4. Production, logistics and organizational planners are responsible for the production related design fields extensively discussed in the preceding chapters. The architects and the various engineers and consultants coordinated by them are responsible for the buildings' functionality. Together they form the core team, which in some cases may be supplemented temporary with internal or external specialists e.g., landscape planners or fire safety experts.

16.2.3 Rules for Project Team

As already discussed in Chap. 15, Fig. 15.9, the success of a project depends on more than just the sub-projects being functionally processed and the project manager expertly coordinating them: Additional methodological and atmospheric issues that lead to certain rules regarding the interactions of the project team need to be considered. The first of these is that each team member is responsible for their planning object in terms of its function, quality, costs and delivery time. These are coordinated within the project meetings. The meetings serve to prepare, verify, make decisions or exchange information about the project's technical content and processes. Typically these meetings include steering committee sessions, regular fixed jour meetings of the core team (e.g., every Friday from 2 pm to 4 pm) as well as workshops and information meetings.

Further operational rules are useful, based on general problem solving routines. This includes, for example, that for all partial planning there is an unambiguous task description, an attempt to develop and evaluate variants, as well as the overall evaluation of the system on the basis of criteria derived from strategies and general conditions.

In comparison *procedural guidelines* are meant to ensure that members comply with the agreed upon rules and obligations and with the communication of differing opinions. These guidelines might include general specifications such as weekly plans, regulating the core team internally (e.g., "any external communication regarding the state of the project is first agreed upon"), or project related rules such as "additional costs of a sub-project have to be compensated for by savings in another sub-project".

Finally, *documentation rules* define how the course of the project and its results will be documented. This essentially applies to the results of meetings, the chronological progression of work and the technical results, which are saved in the project manual (see Sect. 16.6).

Figure 16.5 summarizes the resulting project management tasks from a process, operational and environmental perspective using the example of a working meeting at the concept stage [WieH11]. The chart assumes that there is a clear distinction between project management and process support (see Fig. 15.9). These tasks are transferred separately to two sets of actors. If the

	tasks		
	preparation	execution	post processing
technical project manager	*technical* • pre-structuring of technical content • brain-storming of possible design options *methodical* • logically sequencing content • didactic processing of technical content	*technical* • clarification of technical content *methodical* • supervising content of technical discussions and interactions between the various experts • identifying logical contradictions within the contents	*technical* • ensuring the consistency of technical concepts *methodical* • suitable presentation of results for the specific groups • presenting agreements and disagreements about contents
process facilitator	*methodical* • developing goals and selecting suitable work procedures • drafting the process plan *atmospheric* • assessing the relationships and motivations of those involved which are critical for success	*methodical* • appropriately implementing the process plan (adapting as needed) • consistent application of moderation methods *atmospheric* • supporting work relationships (group dynamics)	*methodical* • promptly documenting procedural results • understandable presentation of the plan of action *atmospheric* • appropriate interventions (e.g. for conflict resolution)

Fig. 16.5 Management tasks at project meetings (H.-H. Wiendahl). © IFA 15.239_B

teams are experienced both tasks can be independently performed by professional project managers.

The starting point of every meeting are the objectives to be achieved which are generally derived from the work package of the overall project plan. As a preparation for the meeting the project team responsible for the topic prepares contents and proposes professional problem solving, whilst a facilitator moderates the work (e.g. communication with meta plan technique) and the overall scenario, including the terms of reference for the meeting. The targets are checked from a technical and systems point of view for their practicality.

At the beginning of the meeting, the facilitator sets the frame work, taking over the moderation of discussion. The technical project manager pays attention to the possibility and compatibility of the various solutions. Finally, the most viable solution must be agreed upon through discussions and debates.

In follow-up, the project manager documents the work results while the process facilitator reviews the documentation for intelligibility. Depending on the results it can be practical to review the objectives, which also serves as preparation for the next meeting.

Finally, the unavoidable conflicts that arise between participants need to be clarified on the client side according to their main responsibilities. In doing so, other projects including the personal needs of employees involved should be taken into account (e.g., planning vacation and leisure time) as these reduce employee stress and thus serve to prevent poor work results.

16.3 Project Plan Development

After defining the team, structure and rules, the next important task of the project manager is to plan the project time schedule. The main aim is to connect necessary activities of gross work packages with the time axis. In this effort one aims to identify jobs which could be taken up simultaneously in order to minimize the overall project time span. Starting backwards from a wishful completion date, realistic time frames need to be set for planning, approvals, execution of sub-trades and installation. It is important to provide necessary buffers to accommodate unforeseen delays to the extent realistically possible. Figure 16.6 shows a time schedule for a building project, as it is usually presented at

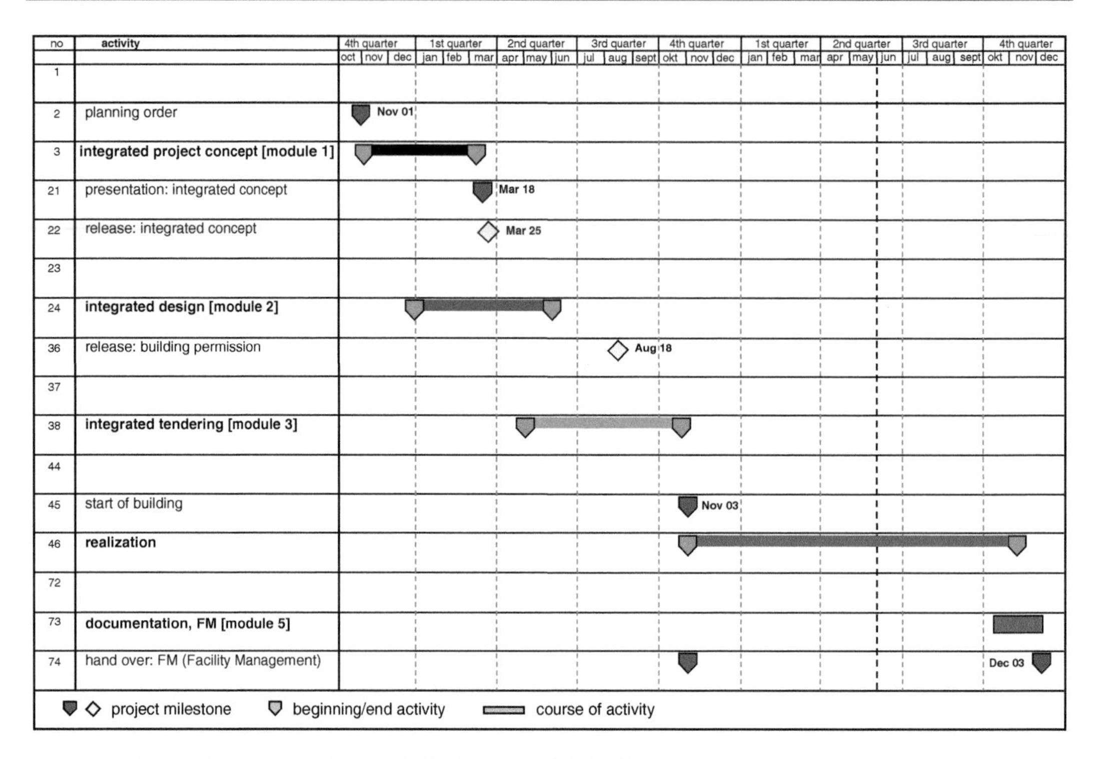

Fig. 16.6 Coarse time schedule (example). © Reichardt 15.240_JR_B

milestone 0. It is based on the work phases of German HOAI fee regulation and is classified under modules—integrated project concept, integrated design, integrated tender, realization and documentation.

The initial time schedules are refined as the planning phase progresses. Figure 16.7 gives a sample of a schedule for a building construction and interior works. The Project Master Schedule is further detailed out based upon dates received from building contractors, equipment vendors, etc.

Apart from the planning stage, there is a corresponding schedule for the planning of production facilities. Figure 16.8 shows a sample schedule, assuming that the operating facilities for the production, assembly and internal logistics are completely new. The schedule needs to be suitably edited to include existing equipment, where necessary.

The milestones, described extensively in Chap. 15, deserve particular attention. They serve to align the interim results according to the principles outlined in Figs. 15.7 and 15.8 and to review the agreed upon interim goals. Milestone reports sometimes also include risk evaluations for so-called due diligence procedures, in which projects are assessed by independent experts at specific points in time e.g., the building shell completion. The assessment often serves to release certain payments for the approved performance of work packages of contractors.

Progressive milestones highlight critical planning phases e.g. approvals of the project (obtaining regulatory sanctioning according to German phase HOAI 4), transfer of detailed design (German phase HOAI 5), tendering (German phase HOAI 6), releasing of the assembly plans to the site (e.g. positioning of delivered machineries according to layout), etc. Finally, project management support deceases (or sometime ceases) on commissioning of buildings and facilities. During this phase, there is a possibility of conflicts about the roles and responsibilities during the defect liability period i.e. facilities, their disposal, guarantee and liability rules [Esc09, Lev09].

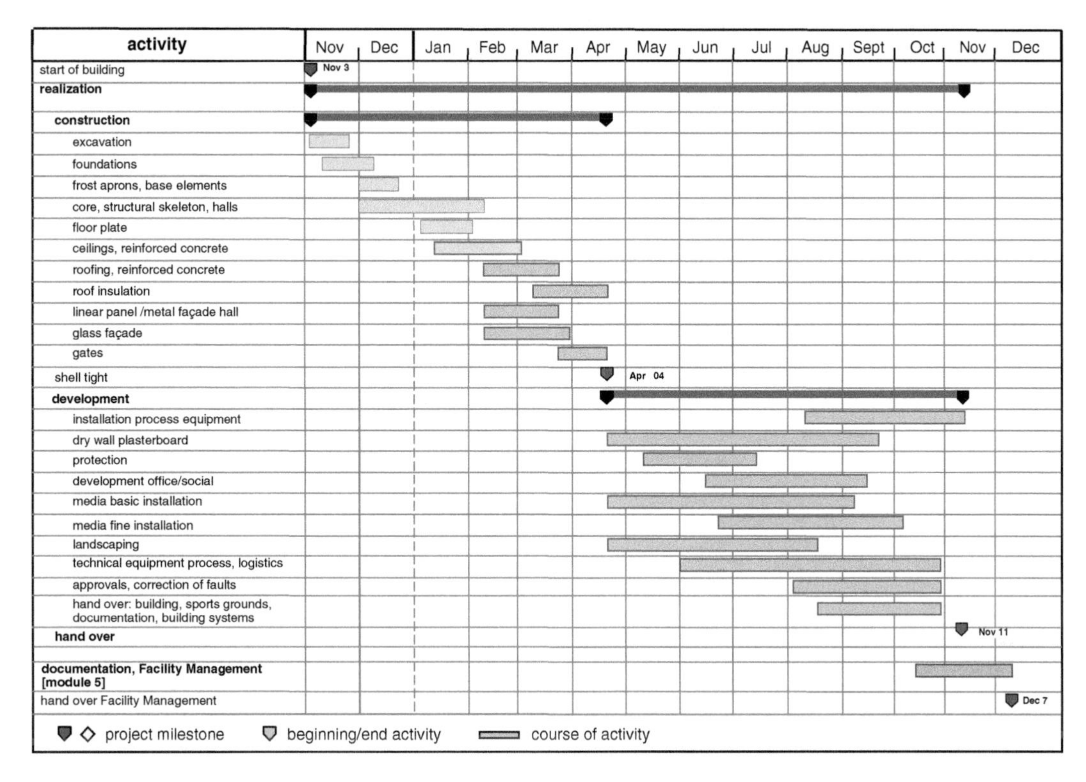

Fig. 16.7 Detailed time schedule (example). © IFA 15.241_JR_B

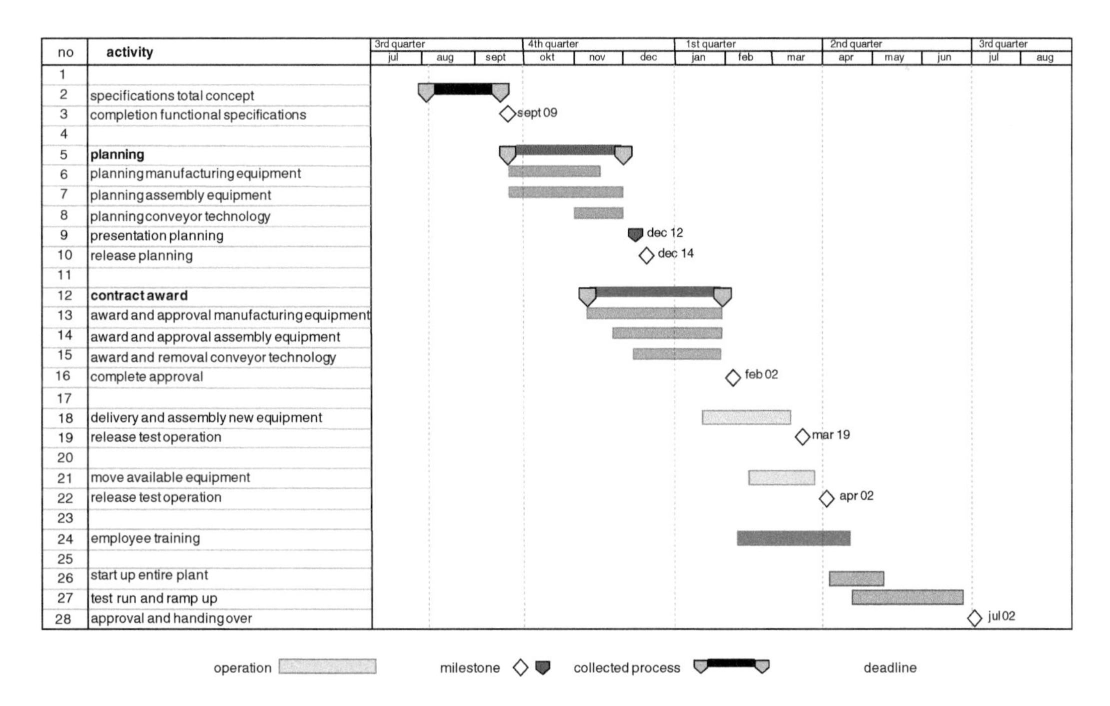

Fig. 16.8 Coarse time schedule for production facilities (example). © IFA 13.969_B

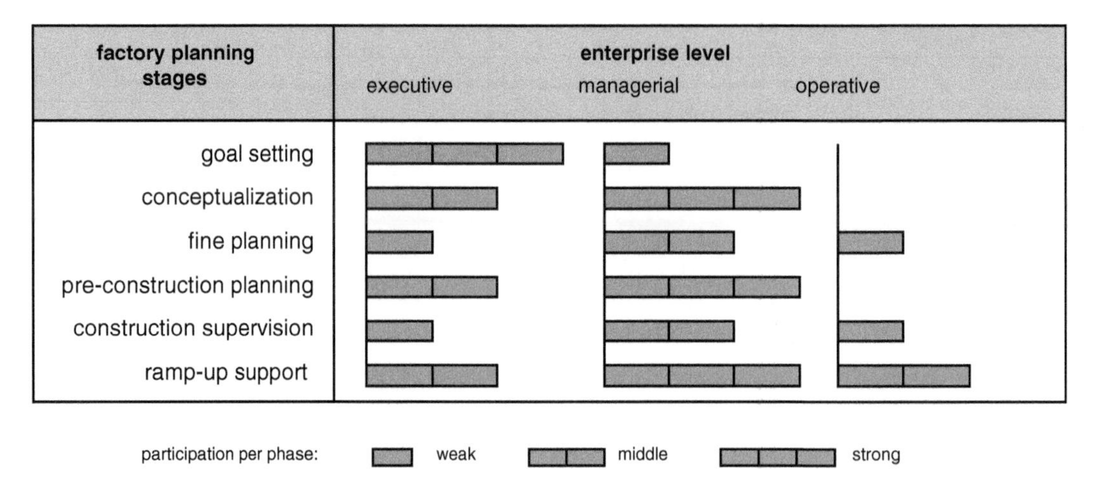

Fig. 16.9 Work load of enterprise levels in a factory planning project. © IFA 16,161 E_B

16.4 Project Capacity Planning

Large projects like 'factory buildings' require the services of not only the external partners but also numerous internal departments. Some of them would be direct users of the new production facility while others would be involved with indirect functions such as process planning, NC programming, production control, etc. There may be others who would be responsible for support functions like quality assurance, information management and communications, personnel management, plant security, etc. In the course of ramping up, the operating employees would be additionally involved. It is advisable to plan for and depute project staff, such that day-to-day business is not hampered and there is clarity between the various levels of participation in each phase of the project (Fig. 16.9).

The management as decision maker is required particularly in the phase of defining overall project targets and in the ramp-up phase. In the intermediate planning and implementation phases the management performs its scheduled tasks as part of the Steering Committee. The mid-level management executives and experts collaborate to resolve production issues, facilities and logistical concerns while the operative employees are involved in training, testing and cooperation during startup.

These broad activities are often underestimated in their scope and should be documented separately to include holidays as well. This would prevent overloading a particular employee wherein he is forced to neglect his normal duties. It is also important to quantify the indirect costs of the above efforts.

16.5 Contract Drafting

16.5.1 General

The next most important task of project management is the preparation and fine-tuning of the contracts with all project stakeholders. Contracts essentially include description of services, timelines and documentation guidelines. Furthermore, it should be clarified between all partners how shortcomings of services would be handled through reworks, revision of prices or others. Other areas of concern are with reference to defect liability guarantees, claims and identification of a justifiable basis for additional costs.

The issues related to contractual liabilities (with reference to German Law) are discussed in detail in Sect. 16.5.4.

16.5.2 Award Variants

Clients of factory projects can take very different organizational and legal forms as can the consultants, planners and executing enterprises hired by them. There are thus four basic plausible constellations and therefore award forms for planning and realizing factory projects. We will briefly identify their specific characteristics as well as a number of advantages and disadvantages next.

- **Individual Planners, Individual Contractors**

  The conventional award form is the trade-specific award of contracts on the basis of itemized specifications and detailed construction drawings. In this case, the client has the maximum control on project schedule, cost and quality since it has direct contractual relationship with concerned construction companies. For the project management, this result in several individual contractor awards and a number of interfaces and coordination efforts between trades with related quality, cost and liability risks. Logically, these advantages and disadvantages also apply during the selection of professional planners for process, space and organization.

- **General Planner, General Contractor**

  As mentioned, the coordination effort required between planning and contracts are considerable. By combining the assignment for overall planning (e.g. for the design of a building with general planning of architecture, structural engineering, building utilities such as heating, ventilation, plumbing, electrical as well as landscaping), from the perspective of interfaces the clients' efforts in time and money, are significantly reduced. Further, reduction of contractual interfaces could be achieved by having a turnkey building contractor with fixed-price contract. However, before entering into such a contract system, all construction work should have defined in detail. Changes, if any, due to ongoing refinements to the process and organizational planning after the award, inevitably leads to pricing adjustments which are beyond the market competition. Such amendments are risky and prove to be costly for the client since it gives the contractor an opportunity to make up for discounts granted during the final negotiations. In addition, the award forms single planners/general contractor exist (rarely) and general planner/single contractors, or general planner/contractor trades packages (rather often).

- **Guaranteed Maximum Price (GMP)**

  In order to overcome the obvious conflicts of interest, a third variant of contracting was developed in the U.S. the so called GMP (guaranteed maximum price) ensures "partnership building" and aims at fixed prices both in the planning as well as the execution phase [Ble00].

  In this case, the client details the tender documents on the basis of the completed preliminary planning. The missing detailed planning, site execution management, as well as overall responsibility for contracted project targets are provided by the general contractor (including those required for services, utilities, landscaping and construction detailing). The negotiated lump sum package for the GMP contract includes a fixed share of fees, risks and profit. A variable component is reserved for subsequent construction to be worked out later by external deliverers, but contracted in advance for ease of co-ordination with fixed profits. This is the principle of "glass pockets"—the client constantly receives "true" information about the state of current costs and schedules [Eit07].

  To what extent this theoretical project transparency of all cost elements is actually maintained in the course of works or skillfully manipulated by the relevant partners, is ultimately a question of trust.

- **Public Private Partnership (PPP)**

  As a result of insufficient investment funds or in order to reduce the investment risk or to exploit tax-saving potential, factories are sometimes built as externally financed objects, which are then rented back to the user according to either a leasing or operator model. With a leasing model, a leasing group takes on the contractual responsibility and the project risk. They secure the financing through borrowing and engage planners and contractors without the collaboration of the client. The leasing contract negotiated with the user, outlines the agreement for the repayment of the loan, other costs as well as the transfer of the ownership at the end of the leasing period. The operator model, in comparison, is based on awarding concessions to operate a building [Bur06].

16.5.3 Advantages and Disadvantages of Award Variants

For the long-term overall success of a project it is important to align the interests of all the parties towards a common goal. Bonus-malus schemes are increasingly being used in the construction industry. Rewards (bonus) are given for meeting the target values while failure to meet the necessary targets usually attracts penalties (malus). Generally lesser responsibilities usually mean lesser freedoms to release orders, especially in terms of the crucial phases of tendering, implementation, acceptance and commissioning. The elimination of trade-specific tender descriptions favoring a functional tender usually helps to stabilize the cost and time schedule at a very early stage. However, these advantages on the other hand usually result in poor or even a loss of quality and sometimes even functional failures.

The strategy to allocate single activities ensures strongest control on the quality of workmanship with best price-performance ratio. The disadvantage lies in the fact that a very high level of coordination is required at a multitude of interfaces and that there is a fear of delay or complete stoppage of work due to insolvency of the single company.

A typical customer prefers a guarantee on the project cost and the delivery time and one agent for the defects liability. Under these circumstances, it is preferable to have a turnkey contractor instead of multiple agencies for various trades. The mentioned advantages are offset by the disadvantages of higher costs and overpriced planning changes after contractual award.

Buying a whole factory as a sum of functional services at a guaranteed maximum price (GMP) is generally possible; however, "the devil is in the details". Especially when the goal is to develop the targeted changeability, the relevant structures and interfaces can only be defined from a process and spatial perspective in a detailed planning much beyond purely functional needs.

When choosing a leasing model, the client waives own investment funds, often in favor of tax effects. A professional leasing company should plan for flexibility in building structure, resulting in a mutable space, which could be easily leased to others at the end of the initial lease period. Therefore, in this model, more emphasis should be given to clear definitions along with careful and detailed planning of building and utilities.

For long-term quality, especially with regard to strategic adaptability and sustainability, building elements should be clearly and unambiguously planned and executed in best possible manner, in line with the award variants discussed before. In addition, the aesthetic grammar of interiors, facades and exteriors demand for careful and farsighted solutions before starting the execution process. These architectural values contribute considerably towards a positive impression of the factory and generate a visual identity for the users and visitors.

16.5.4 Liability Issues

The project management tasks include consulting, coordination and control services. The legal issues arising from this is multi-layered; we can only touch upon the basics in this book. It is important to consider liability issues both for services or works contract in terms of quality, cost and time. According to contracts for work (German Law §§ 631 ff. BGB) the contractor owes the purchaser of a work (e.g. factory projects) the timely execution of the project or sub-project in line with the agreed quality standards along with the client. The client in return owes the contractor the agreed compensation for work. In contrast, a service contract agreement regulates the provision of services by the contracted party (after [Wik14]). It is the content which matters in a contract and not the title and whether the same contributes to the success of the works or not.

In general, planners, such as architects and engineers, are in debt of the overall project success by work contract law. External project managers (as temporary employees of the client), in contrast, merely provide advisory services in accordance with service agreements. Looking at current trends in project management contracts, it appears that project steering tasks are seen as success related and prefer work contracts. However, this is not a general rule and needs to be looked into on a case to case basis.

The liability of the results of project control is dependent on the deficiencies, acceptability of documents created, proposed controls and control mechanisms. The further liability for qualities, costs and time schedules require the transfer of authority for decision-making and setting of decisions from the client to external project steering [Esc01].

In Germany, architects are generally responsible for cost control and in particular the precise calculation of quantities during the planning phases. As stated before, client driven changes in size and qualities of building structures are frequent reasons for cost overruns. Thus an important task of the architect is a correct quantity estimation corresponding to each planning phase for all the building trades. New requirements of the client/customer, such as choice of alternative materials necessitate appropriate cost benefit analysis by the architect on an immediate basis. Failures in warning and advisory duties, e.g. higher costs due to results of soil analysis, can be interpreted later as a breach of duty. The reference to augmenting in cost should always be documented in written form and countersigned by the client.

It is important to define the extent to which planners are liable for their cost estimations depending on their contractual agreements. In some cases the reference to tolerances of cost calculations would not be possible. In reality, contractual agreements to construction cost limits are conceivable for relatively simple planning tasks, e.g. design of a row house; the same may not be applicable to complex factory planning projects.

Claims for damages due to cost overruns by the client may be admissible only after consideration of a previous deadline for rectification. This means, that the client must provide the planners an opportunity to reduce costs by rescheduling or re-planning, as long as this is possible. It is imperative that the client considers a potential increase in the market value of his property, due to higher construction costs.

As already mentioned, the German HOAI (fee structure for architects and engineers) allow for a higher fee in case the construction project costs increases for no fault of the planning team. On its web-site www.HOAI.de a multitude of information on the above topics are available. Among other things, such as case studies, the amended HOAI text version of July 2013 with description of the contents of the HOAI phases as well as fee structure sheets on basis of total project costs for building, structural, utilities and landscaping could be downloaded. As an additional service, readers could also find references to current judgments with reference to cost and time schedule overruns, liability issues and others.

16.6 Project Manual

For meaningful communication in a project, the project manual is a valuable document. It contains addresses of members of planning team, data formats, layer coding of drawings, plan designations and others. Figure 16.10 shows an example of the possible structure of such a project manual. Its structuring is derived from the survey action areas mentioned in Fig. 16.2.

16.7 Costing and Control

When constructing a building, hardly any client would state “no matter what it may cost”. In most cases, the project manager, representing the client, would want to convey to the client the ‘minimum’ project cost, to prove his professional capabilities as quickly as possible.

In reality, this often leads to immature project planning and questionable project budgeting either on the basis of a lump sum or on the basis

1 general	2 organization	3 due dates	4 costs	5 quality
1 planning basis - site 2 planning basis - users 3 space books 4 function program 5 requirement profiles project risks	1 master data/ participants 2 contracts 3 organizational chart 4 codings, layouts 5 drawing lists 6 structural models 2D / 3D 7 protocols 8 charts 9 assessment of options 10 reports and presentations 11 permits & approvals 12 acceptances	1 master schedule 2 milestone schedule 3 detailed timeline 4 realization timeline	1 budget with tolerance ranges 2 cost estimate 3 cost calculation 4 cost quotes 5 final cost statements 6 documentation of changes 7 fee estimates	1 building standard 2 aesthetics 3 systems, brands

Fig. 16.10 Abridged content of a project manual (J. Reichardt, RMA Architects). © Reichardt 15.237_JR_B

of cost per square meter or footage of floor space to be built. If these numbers are known by top management (often before planning has already started), it is almost impossible to change this sum based on developing requirements and dimensions. The same applies to negotiating contracts with the planners; consistent contracts are first possible when they are based on work volume and project sums that have been verified as plausible.

16.7.1 Conditions for Determining Costs

The dilemma of meaningful cost estimations may be clarified through an imaginary example of "cost" of a row house project. A telephone enquiry to a fictitious builder might reveal that the construction costs would be around €250,000 raising a lot of questions. Does the quoted amount include the plot, planning expenses, fees for building permit, fees for surveying, registration costs, civil construction, interiors, fittings like—kitchenette, sauna installation, etc., landscaping, and all the necessary taxes?

So far, only the initial construction costs have been considered. A second, often neglected cost block includes mostly annual recurring costs such as property taxes, garbage collection, street cleaning, maintenance and repair as well as ever increasing costs of monthly operations and maintenance—gas, water and electricity.

This simple case study reveals the difficulties of setting up and communicating reliable cost estimations, often demanded in advance without a clear vision of the proposed project. The cost assessment of a factory project is much more difficult, since data from standardized typologies may not work while data from similar structures may not be available. The specific combination of site, climate, location, processes, organization, building and building utilities and their various changeability requirements results in an unique solution "as no other"; the specifics would have been optimized from all different angles for construction and future maintenance.

Since costs are a direct result of the built volume on one hand and the requirements as well as the quality of workmanship on the other hand, a reasonable cost estimate can be developed based only on a fairly accurate project space and functional programs.

For square meter calculations DIN 277 (industrial norm for Germany, determination of surface areas and volumes of buildings, updated April 2005) is recommended. For square footage (U.S., Britain, Asia) calculation a similar cost structure is recommended. It should be noted that in the construction industry the gross floor area (GFA) and not the net floor area (NFA) is considered as a reference for the construction costs per unit. Since client driven user programs are generally based on net surfaces, these figures need to be suitably enhanced to include circulation spaces, construction systems, utilities, etc. to arrive at a suitable gross area of construction. The same applies to building volumes. Here, too, the gross building volume (BRI) needs to be calculated based upon outer dimensions of building volumes—eaves and ridges of the roof structures and foundations depths.

A significant cost risk is hidden in functional programs related to building utilities, plants and services, fire protection, the computer networking and server spaces, specifications of storage spaces including hazardous substances, etc. They are always underestimated at the beginning due to lack of knowledge and data. In a typical factory project, the cost share of utilities and services (acc. German cost group 400, DIN 276) may vary from 25 to 35 % of the total cost of building construction (cost group 300, DIN 276). It is apparent that risks increase in the absence of clear definitions and guidelines for building utilities and related equipment. This is especially true for ventilation and extraction systems, which not only add to the costs but because of huge duct sizes lead to an aesthetically unsatisfactory roof structure.

Finally, we need to mention the uncertainties related to the location and the approval procedures. In the absence of sound geo-technical and topographical survey reports along with necessary information regarding local planning rules and regulations to be followed, costs estimations would be flawed from the very beginning. For example, local planning rules and regulations may force certain industries to put into place additional measures against noise and fire which, might prove to be a costly oversight in the long run.

Site specific challenges and user specific requirements need to be identified as early as possible through the spatial and functional programs. These needs should be coordinated before starting the planning process and continuously documented in specifications or the space book.

16.7.2 Costs in Building Construction

In order to systemize costing of building construction the US Standard ASTM UNIFORMAT II 276-1 classify construction costs into standardized cost groups and cost elements. Figure 16.11 shows the eight major group elements on the first level followed by second and third level of building and utilities cost elements [Ch99]. The document also describes a 4th level. Similar classifications have been developed in Germany (DIN 276-1), Canada, UK and for Europe. There is no doubt that if the cost structure is consistently followed from the basic to detailed versions a very high degree of cost transparency can be achieved in planning as well as construction phases. It is advisable to estimate costs in this manner for all clients.

A robust costing framework and structure is vital for the success of international projects having multi-cultural or multi-lingual team operating across various boundaries; mistakes or misunderstandings with reference to the building elements, specifications, etc. could pose great risks for the project as a whole. During the preparation of contract documents, the project management team should compulsorily set standards for each part of the project. As a starting point reference values, approximations or statistical cost estimates from a comparable construction projects may be used that needs to be constantly refined as the project progresses. The construction cost information center of German Architects' Chamber (BKI, www.Baukosten.de) produces an annual survey of costs for various completed projects. For example for the year 2013, there are listings under (www.baukosten.de/neubau-2013.html?page=6), classified according to DIN 276 for different building typologies—such as industrial production building, storage buildings,

Level 1 Major Group elements	Level 2 Group elements	Level 3 Individual elements
A SUBSTRUCTURE	A10 Foundations	A1010 Standard Foundations A1020 Special Foundations A1030 Slab on Grade
	A20 Basement Construction	A2010 Basement Excavation A2020 Basement Walls
B SHELL	B10 Superstructure	B1010 Floor Construction B1020 Roof Construction
	B20 Exterior Enclosure	B2010 Exterior Walls B1020 Exterior Windows
	B30 Roofing	B3010 Roof Coverings B3020 Roof Openings
C INTERIORS	C10 Interior Construction	C1010 Partitions C1020 Interior Doors C1030 Specialities
	C20 Stair Cases	C2010 Stair Construction C2020 Stair Finishes
	C30 Interior Finishes	C3010 Wall Finishes C3020 Floor Finishes C3030 Ceiling Finishes
D SERVICES	D10 Conveying Systems	D1010 Elevators D1020 Escalators & movong walks D 1030 Material Handling Systems
	D20 Plumbing	D2010 Plumbing Fixtures D 2020 Domestic Water Distribution D2030 Sanitary Waste D2040 Rain Water Drainage D2050 Special Plumbing Systems
	D30 HVAC	D3010 Energy Supply D3020 Heat Generating Systems D3030 Cooling Generating Systems D3040 Distribution Systems D3050 Terminal & Package Units D3060 Contols & Instrumentation D3070 Special HVAC Systems & Equipment D3080 Systems Testing & Balancing
	D40 Fire Protection	D4010 Fire Protection Sprinkler System D4020 Stand-Pipe & Hose Sstems D4030 Fire Potection Specialities D4040 Special Electric Systems
	D50 Electrical	D5010 Electrical Service & Distribution 05020 Lighting & Branch Wiring D5030 Communication & Security Systems D5040 Special Electrical Systems
E EQUIPMENT & FURNISHINGS	E10 Equipment	E1010 Commercial equipment E1020 Institutionl Equipment E1030 Vehicular Equipment E1040 Other Equipment
F SPECIAL CONSTRUCTION & DEMOLITION	F10 Special Construction	F1010 Special Structures F1020 Integrated Construction F1030 Special Construction Systems F1040 Special Facilities F1050 Special Controls & Intrumenttions
G BUILDING SITEWORK	G10 Site Preparation	G1010 Site Clearing G1020 Site Demolition & Relocations G1030 Site Earthwork G1040 Hazardous Waste Remediation
	G20 Site Improvements	G2010 roadways G2020 Parking lots G2030Pedestrian Paving G2040 Site Development G2050 Landscaping
	G30 Civil / Mechanical Utilities	G3010 Water Supply & Distribution Systems G3020 Sanitary Sewer Systems G3030 Storm Sewer Systems G3040 Heating Distribution G3050 Cooling Distribution G3060 Fuel Distribution G3070 Other Civil / Mechanical Utilities
	G40 Site Electrical Utilities	G4010 Electrical Distribution G4020 Exterior Lighting G4030 Exterior Communication & Security G4040 Other Electrical Utilities
	G50 Other Site Construction	G5010 Service Tunnels G5020 Other Site Systems & Equipment

Fig. 16.11 Cost structure in building construction (ASTM UNIFORMAT II). © Reichardt 15.242_JR_B

offices and workshops. Data from similar sources could be used to arrive at the first cost estimates.

The basic structure of DIN 276 (see Fig. 17.2) as well as ASTM UNIFORMAT II enables three levels of accuracy in costing. During preliminary design a *rough cost estimate* can be easily arrived at based upon the available drawings and the basic quantity estimates. The total cost estimation will be determined by summing up of all cost groups up to the first level of the cost breakdown.

During the design stage there will be further refinement of the cost calculation based upon more detailed drawings and refined quantity estimates. The *total cost estimate* can be determined by the summing up of all cost groups up to the second level of the cost breakdown.

During the tendering phase, a more detailed calculation is possible since all necessary construction drawings and documentation are ready in every aspect including the utilities design and detailing.

On completion of the project, the entire team will have to establish the *actual costs* which were incurred in the project based upon audited financial documents, bills, claims and other statements, "as built" planning documents, drawings and building descriptions.

It is to mention that the tolerance range for cost variances reduce progressively as planning stages progress further from basic scheme drawings to detailed construction drawings and accuracy in quantity estimations increase.

16.7.3 Usage Costs in Building Construction (According to German DIN 18960)

Planning with usage or occupancy costs (also called life-cycle costs) signifies a more comprehensive approach to building construction projects and costing since the focus shifts to the entire life instead of the initial phase. Economic planning in such cases not only looks into compliance with a project construction budget but requires a more holistic approach for cost effectiveness over the total functional life of the project. A logical next step would be to optimize the overall building project costs through all phases of development, planning and construction using materials and systems which suits both construction as well as maintenance. Questions on flexibility aspects of spatial and functional program, choice of building geometry, material

Fig. 16.12 Usage costs—level 1 and 2 according to DIN 18960. © Reichardt 15.243_JR_B

no	usage costs 1st and 2nd level
100	**capital costs**
110	loan capital costs
120	own capital costs
200	**administrative costs**
210	personnel expenditures
220	material costs
290	administrative costs, other
300	**operating costs**
310	supply and disposal of utilities
320	house-keeping and maintenance
330	operation of the technical facilities
340	inspection and preventive maintenance of the building construction
350	inspection and preventive maintenance of the technical facilities
360	control and security services
370	taxes and contributions
390	
400	**repairs**
410	major repairs of the building construction
420	restoration of the technical facilities
430	restoration of the outside facilities
440	repair of the equipment

characteristics and system efficiency needs to be answered at an early stage.

Comprehensive costing of this nature would help in deciding whether it makes economic sense to add value to a specific cost element which would get compensated through simplified operation and maintenance during the usage period of the project or not [Kal02].

A typical example of a holistic financial analysis would be to compare the additional expenses in increasing the insulation value of a building envelope with reference to the energy savings over the life of the given space. In addition, the sustainability ratings of the building would improve through savings in CO_2 emissions in case of increased insulation values. The same applies to discussions of clear spans and column grids in support of building. Large spans offer increased flexibility and thereby the future potentials of changeability and usage; however, large spans directly influence the project costs requiring detailed discussions with clients, about pros and cons, during the planning phase. Figures 11.9 and 11.10 shows relative cost in comparison of different spans, either in wood or steel structural systems.

DIN 18960 (last updated February 2008) usage costs are defined as "all regularly or irregularly recurring costs in buildings and their sites, resulting from the beginning of their use until their break down". According to the increasing accuracy in the planning process of DIN 276, usage costs are differentiated from coarse to fine in usage cost rough and fine estimations, usage cost calculation, and usage costs determination. These calculations would be important particularly in cases of sustainability certifications, as required under specific certification procedures (see Sect. 15.7.3, case studies DGNB-certification). The cost groups' structure includes capital costs, administrative costs, operating costs and maintenance costs. An overview of the various types of usage costs in accordance with DIN 18960 is shown in Fig. 16.12. A similar scheme of the US National Institute of Building Sciences can be found as Life-Cycle Cost Analysis

(LCCA) under http://www.wbdg.org/. As stated in the discussion of investment costs calculation, the German Chambers of Architects (BKI, www.Baukosten.de) generates cost data for the purpose of periodical listings, classified according to DIN 18960 for different building typologies. For the US [Mean14] provides similar data.

Through careful planning, operating costs for particular components of building utilities and services as well as specific parts of the building structure can be reduced. For example, optimizing the air conditioning of spaces in a building drastically reduces the initial investment costs of the equipment leading to substantial savings in energy and maintenance costs; such initiatives are ecologically and environmentally friendly. These cost figures, calculated over the usage life of the equipment, are often multiples of the initial investment. Floors and roofs are further examples of such components of the building. Frequent replacements, costly cleaning of floor coverings or roofing repairs are generally the result of shortsightedly, cheaply purchased materials and solutions.

16.7.4 Cost Management

Ideally, the financial framework needs to be clarified during the basic evaluation phase of the project. As pointed out earlier a carefully structured spatial and functional program would be the minimum requirement for a realistic framework document. It is advisable to include not only the functional aspects, but the qualitative aspects too—e.g. convenience or comfort in this early phase—as they would affect the costs later. In most cases although the data available is incomplete or vague, nevertheless a sound budgeting is expected from the planning team based upon their professional experience. The planning team on their part should carefully document the available data since subsequent changes in data or user requirements may lead to changes in the cost and time schedules. A comparative analysis of spatial and functional programs over the various project phases would help in establishing the project in a clear and transparent manner.

During the preliminary planning stage, when the user's requirements are usually first queried and articulated a plethora of wishes are usually expressed, for example for not previously mentioned air conditioning in manufacturing areas or conference rooms. Often, the client will want to assign the responsibility to the planners for the resulting increase in the cost estimates as compared to the initially allocated budget.

Similarly in comparison to the preliminary plan, in the detailed planing dimensions have sometimes to be increased due to the greater planning depth, but also due to further wishes on the user's side (e.g., additional space for meetings, IT, copiers), the burden of which in turn is frequently left to the planners. The cost calculation which concludes the layout planning should transparently document these changes in the plan in comparison to the previous estimated costs.

In the next phase of approvals or planning application, sometimes unforeseen changes are requested by the relevant authorities for additional fire safety or noise control mechanisms based upon the proposed production planning systems and processes. Such requests may necessitate the services of additional expertise. Usually such services and their related outputs, result in increase in the cost schedule and, at times, even in time delays. In an ideal scenario, these aspects should have been incorporated in the planning phases well in advance in coordination with the various local authorities.

During the tendering phase, ideally one should verify market availability and vendors' delivery schedules before specifying construction materials and system. At times, it may be necessary to circumvent delivery bottlenecks or probable price increases through alternative solutions (e.g. for insulation materials, steel or concrete) before releasing the tender documents for bidding. Informal discussions with potential construction companies before the actual tendering process sometimes help in clarifying the construction drawings, proposed specifications and project budget.

During the tendering phase, when the contract placement is prepared, bid offers are received, evaluated on an equivalent techno-commercial platform and compared with the existing cost estimates. As already mentioned, apparently "cheap" initial solutions can be very expensive in the long run, if maintenance costs and consequences on other building and technical components are taken into account in totality.

The final contract package should include all building, utilities, landscaping as well as process plans and performance texts that identify the project as a whole and including statements about materials, systems and choice of manufacturers. The bidders would prefer a free hand in their choice of sub-contractors or suppliers—a request which needs careful assessment since at times it leads to the installation of inferior components with increased maintenance cost. Also, at times, such scenarios decrease the overall cost transparency and increase the risk of delays through endless negotiations.

Finally, in the project implementation phase, actual construction starts at site. Project monitoring (either periodic or continuous) needs to be arranged to control quality, time and cost with respect to the planned time and cost schedules. For large or complicate projects it can be recommendable to transfer this task to an independent consultant. It is a common practice to report status of work accomplished at site in percentages of the planned time. Similarly, actual cost is monitored element-wise as per the contracted bill of quantities and summed up to arrive at the overall cost status of the project with reference to the approved budget. The final cost then is the fixed determination, in best case meeting the contracted sums.

As discussed, attempts should be made to follow the structured cost framework (DIN 276 and DIN 18960), at every step of the process updating the same with payments released in the intervening period and continuously refining the planned cost schedule on the way. This prevents excess payments and helps to establish a professional method for 'claims' settlement during the implementation phase. Figure 16.13 summarizes an overview of the various tasks of cost management in architectural planning, with reference to the phases 1–8 of German HOAI.

service stage	basic service acc. to HOAI fee regulation	task cost management
1 base analysis	clarifying the objectives	financial constraints / budget as part of base analysis
2 preliminary planning	estimating costs	approximately determine total costs for high-rise buildings according to DIN 276
3 technical design planning	calculating and controlling costs by comparing estimates with cost quotes	approximately determine total costs for high-rise buildings according to DIN 276 with explanations for deviations
4 planning application		cost control with plan changes / additions in the course of the approval planning
5 execution drawings		cost control with plan changes / additions
6 preparation of contract placement		cost control with plan changes / additions
7 assisting with contract award	cost quotes, cost control by comparing quotes with cost calculation	exactly determine expected real costs for high-rise buildings according to DIN 276, comparison with explanations for deviations
8 project supervision	cost control by checking invoices of executing enterprises, comparing contract prices with quotes	proof of actual costs, written cost control, continuous supervision and comparison of planned / actual cost and cost limits

Fig. 16.13 Cost management tasks. © Reichardt 15.244_JR_B

One needs to be cautious about the permissible tolerances during various cost estimation exercises, especially with reference to claims against incorrect cost estimation. The involvement of clients and users in possible budget increase, particularly due to changes in space and function programs has been pointed out in detail earlier. The planning team could be held responsible for their inexperience in assessment of taxes, inability to foresee sudden increase in costs of specific building materials, inability to plan for special deliveries or bottlenecks, increase in fuel costs resulting in higher transportation costs, predict adverse weather conditions, etc. Once again, it is best to use a standardized structure to develop a transparent system from the beginning to avoid such pitfalls.

As a rule of thumb, depending upon the specific situation the following tolerance values are generally acceptable in Germany, [Fes05]:

- Cost estimate: +/−15 to 33 %,
- Cost calculation: +/−10 to 25 %,
- Cost fixed estimate: +/−5 to 15 %.

It may be noted that these deviations are relevant to the budget and represents an exclusion criterion for the planner. The cost of the building with reference to the total investment is comparatively low; in most cases production costs due to the amortization period (say 30 years) would be much more significant.

An example will clarify this statement. Assuming a cost of 20 € per m^2 per month for a production hall, the "rent" for 8 h/day and 25 working days per month works out to be €0.1/h. Typical tool machinery requires an area of about 20 m^2 which translates to €2.00/h. On the other hand current hourly rates for medium-sized tool machineries are at 50–80 €/h, thus the rental cost share is 4–2.5 %. Of course there are also jobs with lower hourly rates, particularly in assembly. But the building cost share in the low single-digit percentage range. Thus although the above mentioned differences in building costs are negligible, they play a significant role in liquidity planning. A professional cost management team should not only aim to determine deviations, refuse unnecessary change requests, but proactively manage the interplay between all parameters of the overall project economy.

As already mentioned in the context of capacity planning and internal management both from the perspective of client's and planning teams, their effort is a significant factor. The hereby generated costs may be significantly reduced through state-of-art working methods, improved levels of planning through digital tools and improvements in communication system between all parties involved. These topics are covered quite extensively in the available literature and the reader may choose to refer to them. However, two important trends need special attention—data integration and technical documentation of the planning stages. Section 16.8 outlines the concept of the digital factory for production planning whereas Sect. 16.9 discusses Building Information Modeling (BIM) for facilities planning.

16.8 Digital Factory

16.8.1 Concept

In the field of factory planning efforts were made since 1980s, to prepare 3 dimensional digital models for production facilities to predict their behavior as realistically as possible. For this approach, which can be regarded as a further development of CIM (Computer Integrated Manufacturing) the term "Digital Factory" has established [Bra11]. The German VDI guideline VDI 4449 defines as follows:

"The Digital Factory is the generic term for a comprehensive network of digital models, methods and tools—among others simulation and three dimensional visualization—which are integrated by an integral data management. Its goal is the holistic planning, evaluation and continuous improvement of all relevant structures, processes and resources of the real factory in connection with the product".

Figure 16.14 illustrates the underlying concept [Küh06]. The goal is the continuous digital processing of product development on one hand and process and factory planning, in their respective life stages, on the other hand. This approach include the product, process and

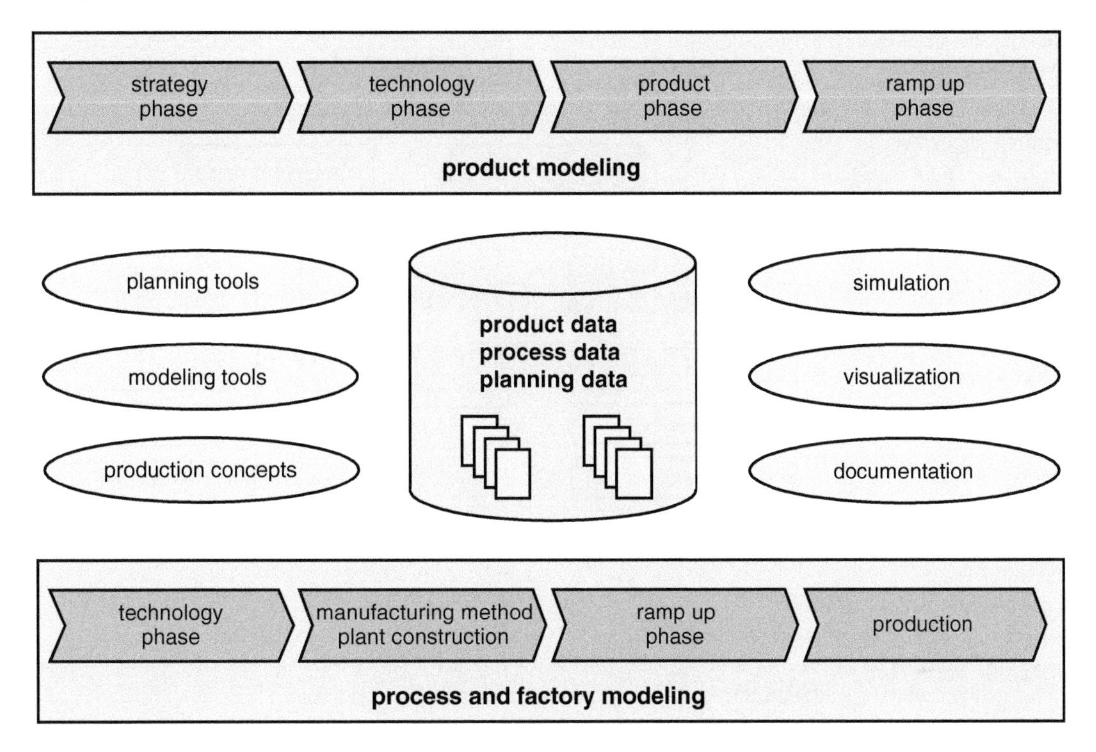

Fig. 16.14 Digital factory concept (Kühn). © IFA 15463

planning data usually leads to holistic production concepts. Furthermore, simulations help to visualize operations on the different levels of the factory including time dependencies. From the perspective of the factory planning, it is important to improve planning quality, shorten timelines and set-up transparent communication systems.

The static and dynamic models of planned objects shown in Fig. 16.15 [Küh06] play a central role. The product description includes the bill of material, 3D models of the parts and construction groups. The production processes are described in work plans, which refer to the equipment used for manufacturing, described in detail in Sect. 6.3.

Not only the products but also the facilities equipment can be detailed out by 2D and 3D models. During the concept stage, production planning is usually carried out through 2D presentations, which are converted in due course of time into 3D models, in order to link them to the 3D building models.

Static models represent a certain status, e.g. a material flow matrix provides a certain understanding of an observation or calculation. Dynamic models are preferred for systems which are dependent on time and space. Discrete-event-models depict interlinked individual events with reference to time, e.g. the movement of transport facilities within a factory. Kinematics (or the study of body movements) can help in understanding the relative movement of a work piece and tool in the working area of a machine tool. FEM (finite element model) splits a body into 3D-elements, based upon certain relations e.g. forces, strains, temperatures, etc. Thus the development or heating processes of a work piece can be simulated to visualize probable issues.

A number of methods and tools are currently available for the Digital Factory; Fig. 16.16 shows an overview as suggested by Bracht [Bra11]. The information and data collected generate a data base for the various models of processes and equipments and is supported by

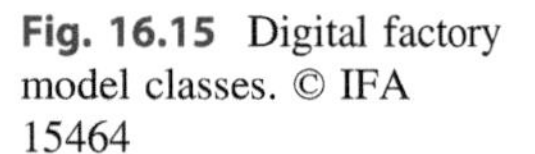

Fig. 16.15 Digital factory model classes. © IFA 15464

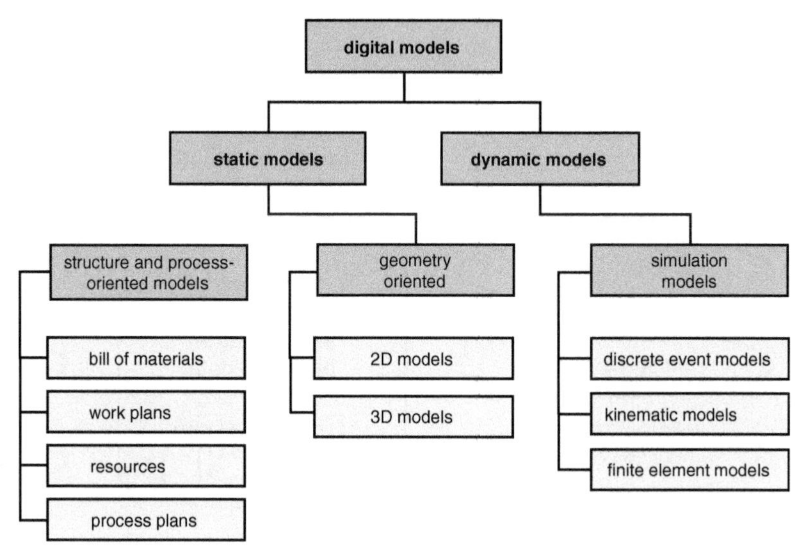

method class	assigned methods
information and data collecting	• survey • personal observation • automatic observation (3D-Lasers canning, Motion-Capturing, object identification) • document analysis
presentation and design	• modelling of processes • information and data modeling • situation modelling
mathematical planning and analysis	• mathematical optimization • methods based on graph theory • statistics and stochastic • comparative quantitative evaluation methods
simulation	• continuous simulation (Finite-Element-method, multi-body simulation, ergonomic simulation) • discrete event simulation
artificial intelligence	• multi agent systems • neural nets
visualization	• static graphical models • dynamic visualization (monitoring, 2D- und 3D-animation) • Virtual Reality • Augmented Reality
collaboration	• technical communication facilities • information space • knowledge management • work-flow management • workgroup computing • project management

Fig. 16.16 Methods of the digital factory (after Bracht). © IFA 15465

automatic laser scanners, motion sensors or even direct questioning. In addition, there are proven methods that serve the object and workflow description of the factory without any problem. Mathematical planning and analyzing methods are rarely used in factory design; they are usually applied to optimization of large scale assembly, manufacturing systems, storage and transport facilities. For simulation, models are of fundamental importance.

It is also important to devise an effective balance between effort and result. For example, the complex dynamic simulation of a system is outdated if in the course of planning changes in the output rate or sequence of processes come up. Here, the value lies in the understanding of the system behavior, for example towards disturbances. Artificial intelligence methods, similar to mathematical methods, are used more for optimization of complex systems and rarely used in factory planning. State of the art visualization methods help to clarify factory objects in 2D or 3D view; they play an important role in understanding the often complex relationships, whether between process and spatial objects or towards processes in the factory. Finally the methods of collaboration are equally important, for cooperating in an IT-enabled environment, allowing for the smooth exchange of data, information and knowledge. The data is available for authorized team members only in so called information rooms. If certain defined processes are maintained at regular intervals and agreed upon, a set of workflows could be drawn upon for documents, information and tasks in a certain order for all stakeholders.

The interaction of these models in the context of factory planning begins with the static scheduling of resources and capacity, as described under Sect. 15.4, concept planning, whilst in the dynamic phase, the refined planning focus points are e.g. buffer and storage sizing under various possible scenarios of production. The level of operational planning is optimized based on dynamic models of the production operation and its control within the realm of quality and logistical challenges. Finally by means of 3D animation, time and space dependent operations can be made visible and by using VR technology. Thus digital factory creates not only a unique factory project, but also the concept of an adaptable factory which could be changed or modified quickly to handle extensive changes in process or spatial needs.

The automotive and aircraft industries are the pioneers of the Digital Factory. An application for a typical automotive factory is described for example in [Loos13]. The remaining industrial sectors, due to their medium sized enterprise structure and because of the considerable preparatory work and investment, only hesitant move in this direction. Another barrier has been the lack of interoperability of models used, complicating the exchange of data at various levels.

Digital Factory consists of data management systems, to which the various users have access, as suggested in Fig. 16.17. It can be seen on the data management level wherein the essential objects such as product, resource, process and project are described by different modules. This data is referred to from side of the indicated function modules, e.g. to create a bar chart, 3D model of a facility or the simulation of a material flow. The operational level is the desktop of the user, on which he recalls the specific tools of the functional level relating to his tasks, which are

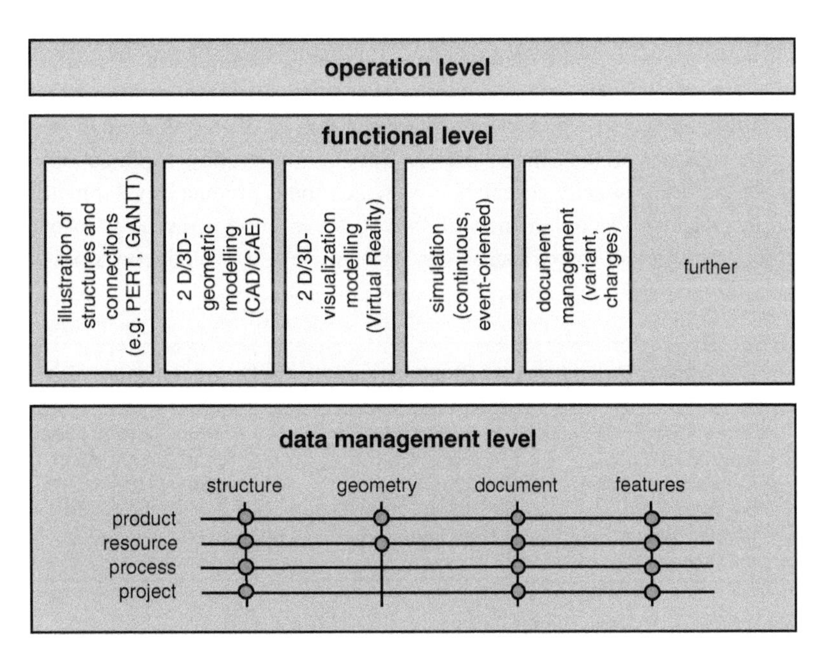

Fig. 16.17 Basic structure of a possible software architecture for the digital factory (VDI 4499). © IFA 15466

derived from the overall process model of factory planning.

It should be noted that at the moment digital factory, concentrates on shaping and operation of the production facilities. An interaction with the surrounding buildings is yet to be made available. A possible interaction of production facilities with surrounding building is shown in three examples in [Thi12].

16.8.2 Digital Tools

Regardless of the comprehensive approach of the digital factory many digital tools are being individually used successfully in practice. They provide valuable support in the planning phases based upon methods mentioned in Fig. 16.16. Figure 16.18 shows some important tool classes and examples.

Conventional tasks such as word processing, spreadsheet calculation, data storage and management, presentation of results and management of tasks and staff are mostly being carried out with programs of the Microsoft Office family. The Internet with appropriate services for conveying of text, graphics, videos and sound files etc. is increasingly playing a dominant role for communication. Telephone and video conferences as well as working on documents from multiple locations are quite common nowadays. For complex projects, it is becoming a common practice to identify and set up a project server, with appropriate back-ups, at the Client's premises or at any other location mutually acceptable, on which the project manual is stored in line with Fig. 16.10. In case, the project participants are located in multiple time zones, a "follow the sun" approach may be implemented wherein the work may be transmitted at the end of the day to a partner firm across a time zone so that the results could be made available to the client the next morning.

As already mentioned, simulations play a pivotal role in the concept of the Digital Factory; however simulation can be used even in the absence of such an environment. Figure 16.19 shows other possible areas of application of simulation in the production planning from process level of the machine to the field level at the factory floor. In case of building design, powerful simulation programs to quantify and simulate airflow—ventilation systems, temperature (see Fig. 11.35) or lighting levels (see Fig. 10.10) are already available.

The visualization of planning results can be generated using different techniques. Traditionally 2D and 3D representations were created using CAD based software. Objects could be moved in space and viewed from all sides. The next level of realism could be achieved with the aforementioned virtual reality technology, an example of which is given in Fig. 16.20.

On the left had side of the figure is the so called 'cave', which leads the visitor into a virtual reality space. By making 'cave walls' transparent, projecting the external environment on to screens and using special 3D glasses the impression of a spatially floating object can be created. Positions in space as well as dimensions of the object can be changed with certain techniques. Applications can be found mainly in product development of automotive and aircraft industries, e.g. to evaluate the design from the perspective of a passenger. Simpler solutions are

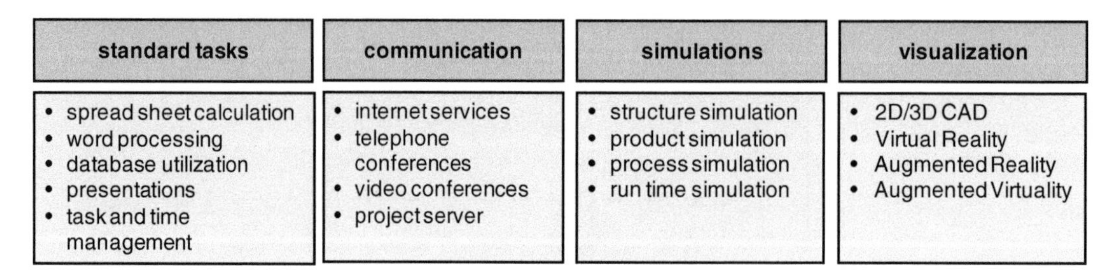

Fig. 16.18 Key classes of digital planning tools. © IFA G9352_Wu_B

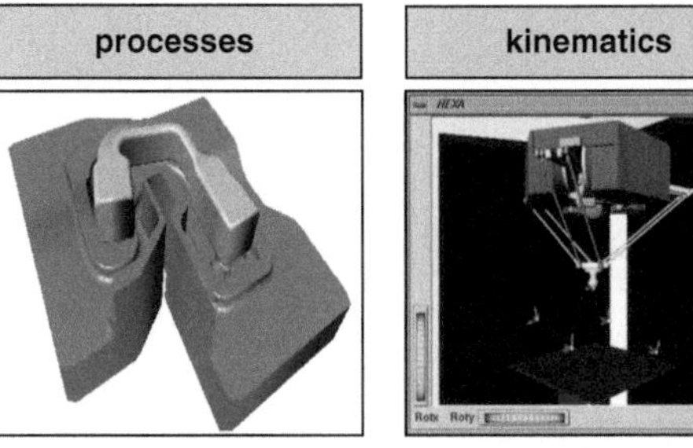
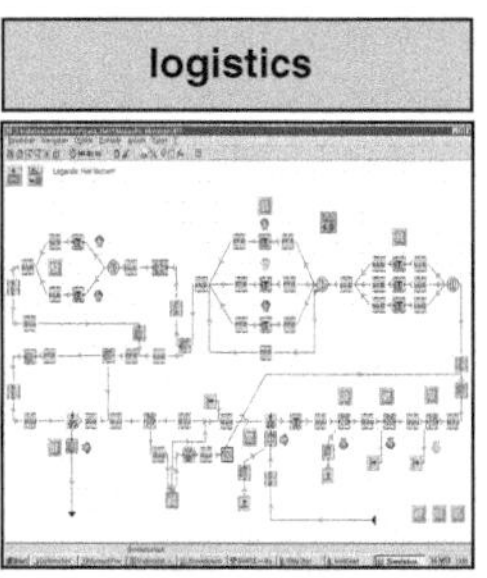

Fig. 16.19 Exemplary simulation applications in production. © IFA 12.360NP_B

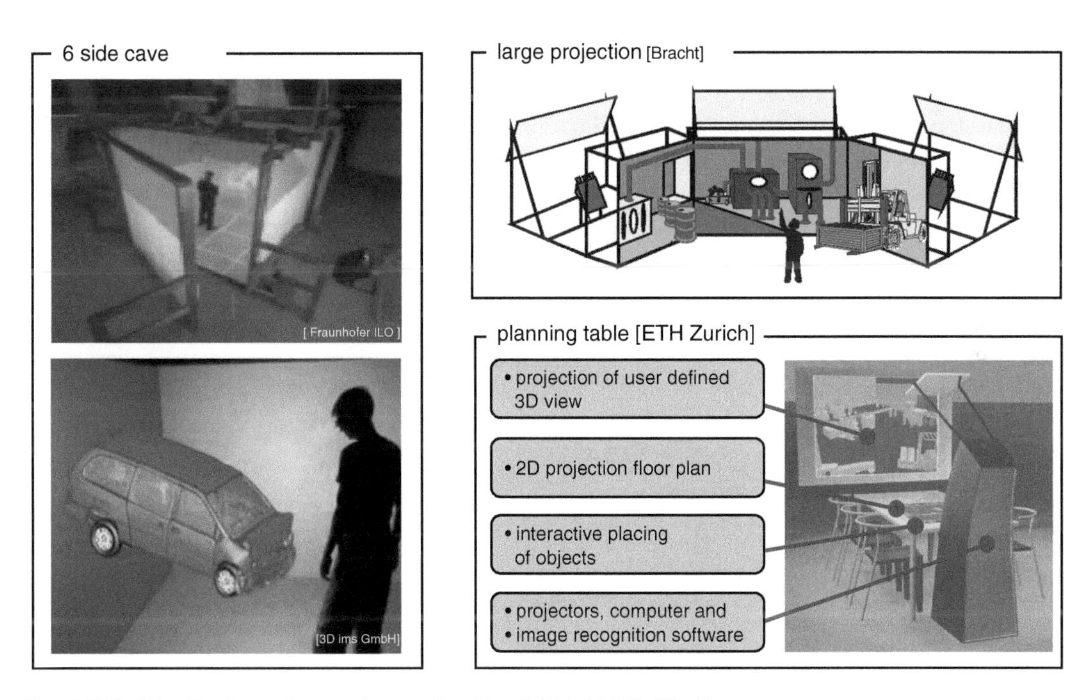

Fig. 16.20 Visualization—planning in virtual reality. © IFA 11.727_Wu_B

large-scale projections that create just a spatial impression, but can be seen more or less only from the outside.

Figure 16.20 bottom right shows a "planning table", as originally invented at the ETH Zurich and developed further by the IPA in Stuttgart and in a slightly different form at the TU Chemnitz. A 2D layout is projected on a table top or a large screen with movable objects, e.g. machines, shelves, transport vehicles, etc. The arrangement is finalized by the participants through planning meetings in line with agreed criteria such as segmentation, etc. The background software database of the objects then calculates the performance figures, such as the use of space or material flow density. For spatial visualization, in addition to 3D, a layout appears on a separate screen.

16.8.3 Simulation Example

From the perspective of planning complex production systems, the dynamic behavior of variants of manufacturing and assembly lines is the most important field of application for discrete-event simulation. The conceptual study of a flexible car body shop may serve as an example [Mei07]. The starting point of the exercise was that three car body models A, B and C were to be produced on one production line in any order. Figure 16.21 shows the main steps of investigation at a glance. During a preliminary study, four structural manufacturing concepts for the body were developed. Then, four scenarios were simulated with a different model mix each. The car variants wander through each of the four life cycle phases of start-up, full load, part load and decline of a vehicle generation. This results in a total of 16 simulation runs, allowing a statement about the output rate behavior of the 16 constellations.

Figure 16.22 shows a scaled layout of the structure variant 1 with the input and output interfaces for the three models A–C, the welding robots and the final welding stations, as well as the connecting conveyor technology. In addition, the targeted output data of the examined scenario 4 is also shown.

Figure 16.23 shows the layout modeled with a simulation program. The target was to deliver 166 completed units per day in two shifts. The back-up processes which are controlling the execution of the production program for each structural concept, the performance of the conveyor system and the availability of practical experience values of the individual aggregates are not visible. The average values of the key figures of the simulation are given in the figure. The output exceeds slightly the target, nevertheless the maximum utilization of the robots reached only about 67 % of the maximum possible rate.

It is equally interesting to note the dynamic behavior of the system. In Fig. 16.24 this is visible for the situation explained under Fig. 16.23. The individual values of the manufacturing systems show considerable variation. By varying the parameters of the control method (the details of which are beyond the scope of the current discussion) this effect can be reduced.

In the example, a slight rating advantage of the structure variant 1 over the other concepts is

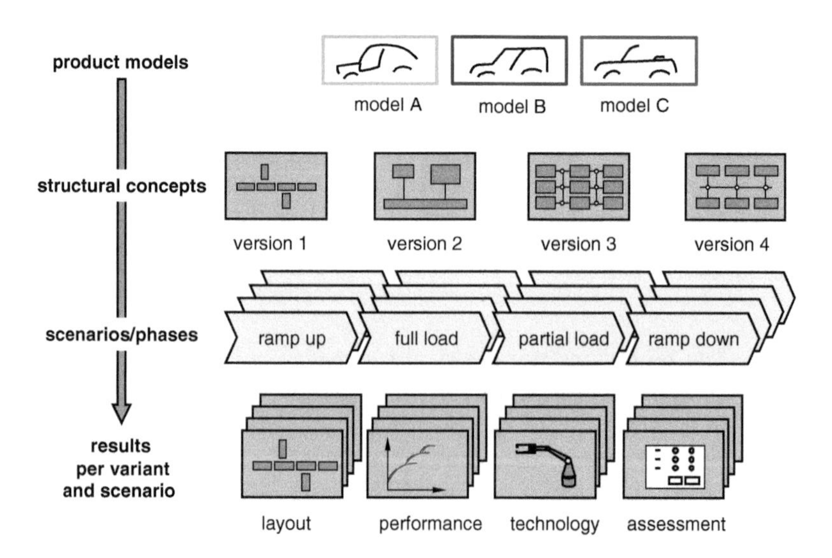

Fig. 16.21 Simulation concept of structural variants for a body shop (Meichsner)

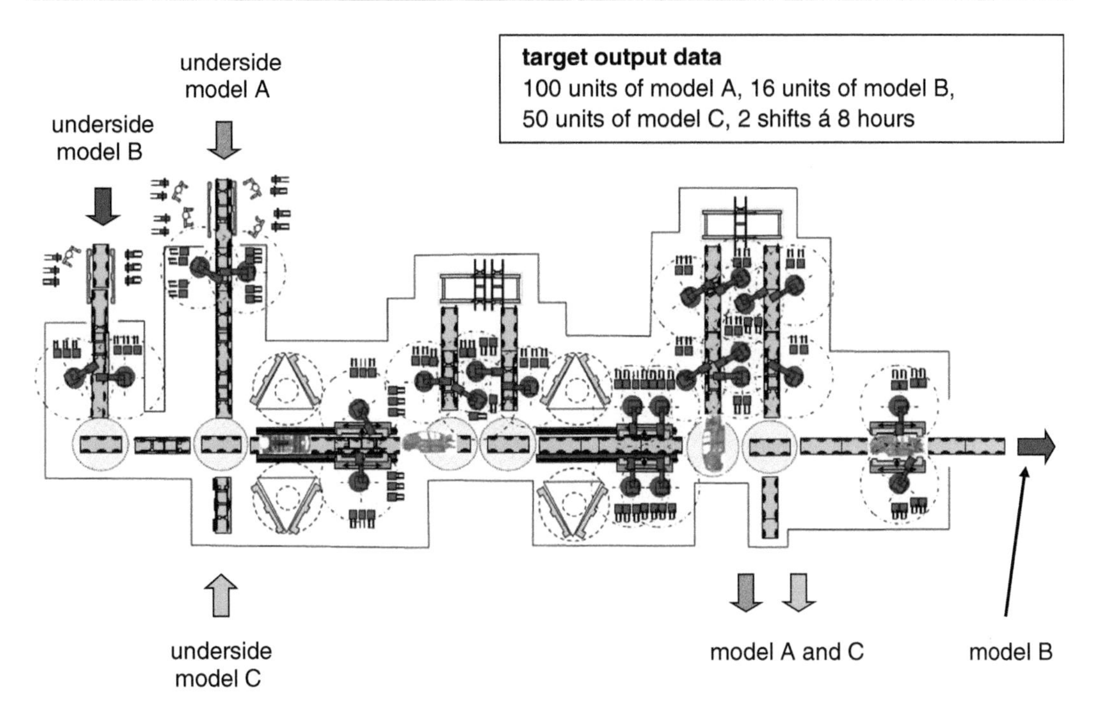

Fig. 16.22 Layout structural variant 1, scenario 4 (Meichsner). © IFA 13984

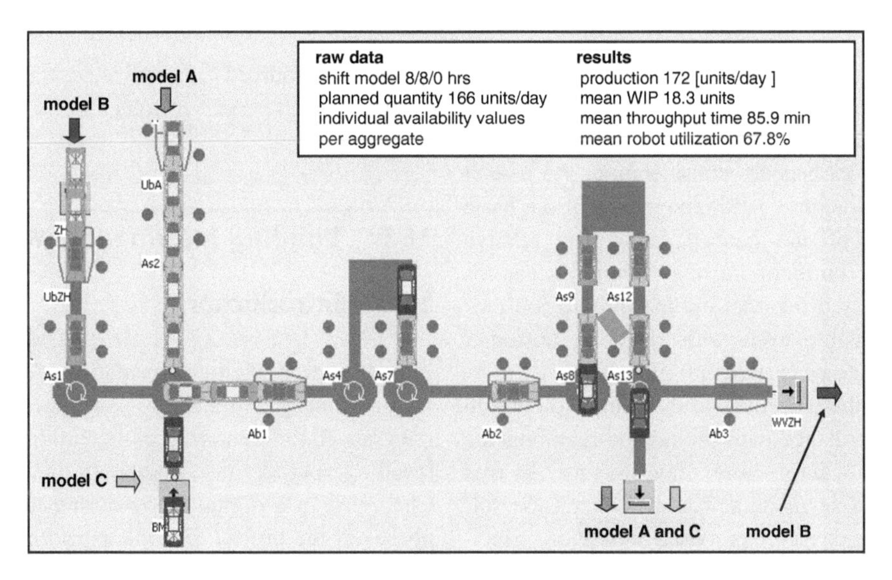

Fig. 16.23 Simulation model for structural variant 1, scenario 4 (Meichsner). © IFA 13985

shown. If the lead is not very large, the two best concepts are usually transferred into refinements, to allow for more accurate statements based on criteria such as ease of maintenance and availability of items.

The relation between efforts and benefits of simulation is clearly apparent from the example. The modeling of structural variants is possible relatively quickly if the performance data is available and the individual components are

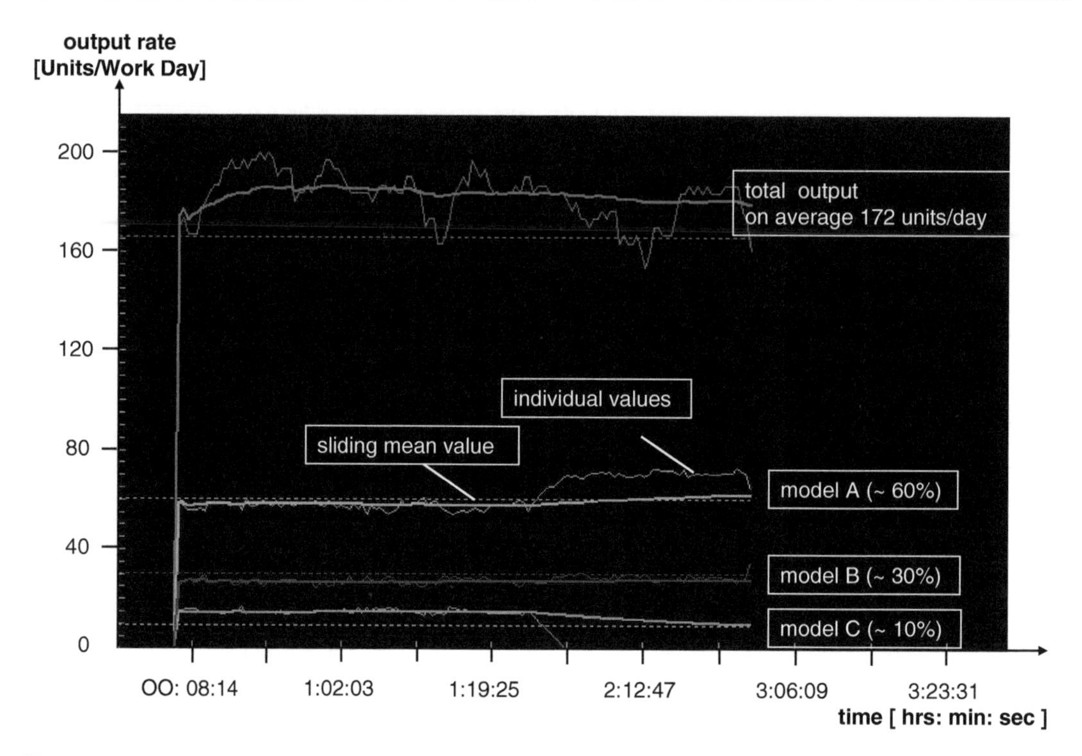

Fig. 16.24 Simulation results structural variant 1, scenario 4 (Meichsner). © IFA 13986

known. The programming of the control method is more complex and requires expertise and intensive discussions.

For the case under consideration, the benefit of the simulation lies in recognition of the basic feasibility of the concept and their relative advantage. Furthermore, the planning scenarios conclusively prove that optimal production output is possible even with changing customer requirements or changes in sales.

One of the objectives of the simulation was to verify compliance with production requirements with significantly lower investments. It was confirmed that modular facilities expansion following Meichsner's proposed migration principle [Mei07] leads to a considerable reduction of the investment volume and a significant increase in efficiency which correlates with the already established approach of quantitative flexible production concepts (see Fig. 3.3).

The implementation of such simulations requires a lot of experience. In the absence of appropriate in-house expertise implementation is usually outsourced to specialized consulting companies or research institutes.

16.9 Building Information Modeling

16.9.1 Introduction

In construction, the implementation of the digital factory concept is relatively new. The goal is to integrate all different parts of the building and the resulting output (as discussed in detail in Chap. 15) in a digital mode during the various phases of the project. For this purpose, the term Building Information Modeling (BIM) was established [Kym08, Eas11, Smi09, Bot13]. The technical basis for this would be the standardized 3D-CAD interface IFC (Industry Foundation Classes), an open standard for digital description of building models (www.ifcwiki.org/). IFC is supported by numerous CAD software manufacturers' for the purpose of interoperability.

Based upon the software system, a detailed and precise three-dimensional building model is created during the design stage. In addition to the object geometry, the central database of BIM collects and networks with the project team to store all data related to production, analysis and optimization as well as subsequent operations. All those participating in the construction—from the users, to the architects and experts in statics and building services technology up to the process and logistics planners and finally those realizing the building refer to the central building information model and could use the information relevant to them. Eastman et al. [Eas11] and Smith and Tardif [Smi09] explain in detail the many advantages of central data model for owners, managers, architects, engineers and entrepreneurs. Changes during the planning phase are automatically incorporated in real time in all the "classical" planning documents such as floor plans, sections, elevations, 3D isometric drawings and schedules thereby avoiding redundancies.

The following example illustrates the possibilities and the interaction of the various professional planners using the BIM approach in more detail:

The architect, in consultation with the factory and logistics planners for the facilities layout, develops a spatial concept for a new building. On completion of the preliminary building planning, the structural skeleton is transferred to the structural engineer, who derives a static analysis model. As the project progresses, the structural engineer refines and details the analysis model and dimension all components of the structure from time to time. As and when further amendments are needed, he refers the same to the architect, who in turn manages the prevailing 3D model without redundancies. The building utilities engineering team now uses the overall three-dimensional building model as generated by architect and monitored by the structural engineer for installing technical plants and services pipe routings. The architect studies all the partial models—structural, building utilities and process facilities—and tries to resolve possible points of conflict thereby preventing errors in overall planning. In the process, conflicts between the technical sections as well as expensive adaptations during the construction period are reduced to a bare minimum.

The production and logistics planners may use the Building Information Model early in the project phases for their own process presentations, as well as monitoring the user coordination of the space programs. In the later phases, basic layouts in the form of simple cubes or more refined 3D visualizations may be added and related to building zones or rooms and assigned to the database automatically. Moreover specific requirements such as weight, utility connections for electricity, compressed air, clean water, effluent, etc. may be integrated when needed. In case there are existing process models made out of other software e.g. Autodesk Inventor, these can be imported into the overall building information model and its technical circuit values may be used for engineering the building utilities services. Increasingly the CAD software suppliers also offer integrated project specific library elements besides the actual construction software packages.

For example, under (www.autodesk.de/suites/factory-design-suite), the user receives a library of 3D elements of process and conveyor technology, enabling quick development of the whole layout facilities. Nevertheless, a specific project solution is usually unique and cannot be entirely derived from standardized library elements. All interactions of the various professional sub-trades are error free based since they are based upon the filtered out sub model or overall central model.

Figure 16.25 shows in a realized automotive factory project the superimposition of the partial models "architecture" (gray), "process facilities equipment" (blue) and "utilities service" (orange).

On completion of the building project, the integrated planning data base can be passed on to a CAFM system (Computer Aided Facility Management) and used for management operation (see Chap. 17). Later conversion steps can be coordinated easily making subsequent efforts and costs for documenting the existing building almost superfluous.

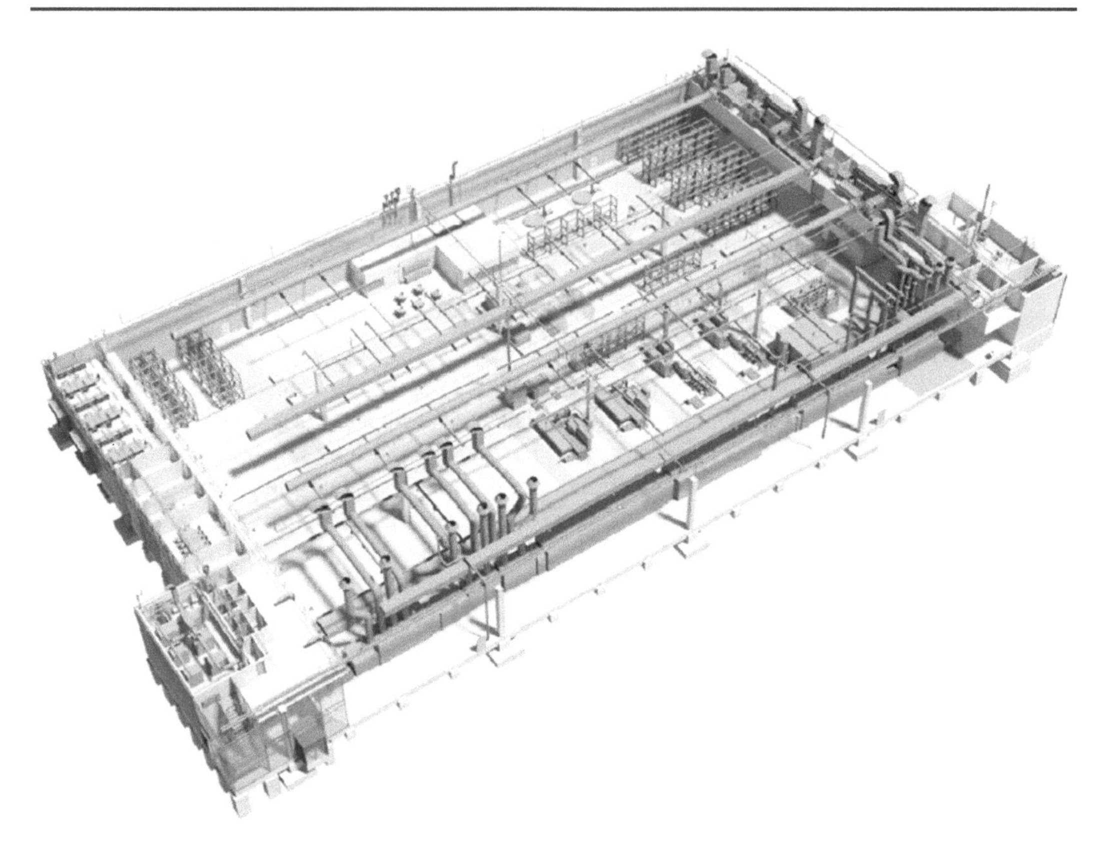

Fig. 16.25 Integration of architecture, building services and process sub-models in a project example (RMA Architekten). © Reichardt 15.245_JR_B

16.9.2 Evaluation of the Building Information Model

The following discussions gives an example of a realized factory as well as details of a "virtual model factory" to explain some of possibilities of the digital building information model.

- **Presentation of design variants**

The preparation and presentation of design variants along with comparative building costs are quite common in the factory planning process. In the conventional design workflow of a project plan, management of variants are often costly, time consuming and leads to redundant plans. Using Building Information Modeling, it is possible to develop, consider and carry all the relevant design options in the same building information model without much loss of time. Each option can be preplanned in the model from basic to the detailed, depending on required depth of representation with or without the relevant information about surfaces, materials, components, etc. Changes that affect the ongoing parts of the project and are not part of the design options are made only once; the same gets automatically included in all planning options. This saves considerable amount of time and keeps all the data concerned in the project consistent.

Figure 16.26 shows the early planning stages of the fictitious model factory project wherein the volumetric expansions of various stages are explained, along with the scheduling options for surfaces and masses. The simultaneous graphical presentation of expansion possibilities as well as abilities to simultaneously calculate gross floor area, gross volume and corresponding indicative costs in the visual model allows for rapid assessment of the possible solutions by the project team.

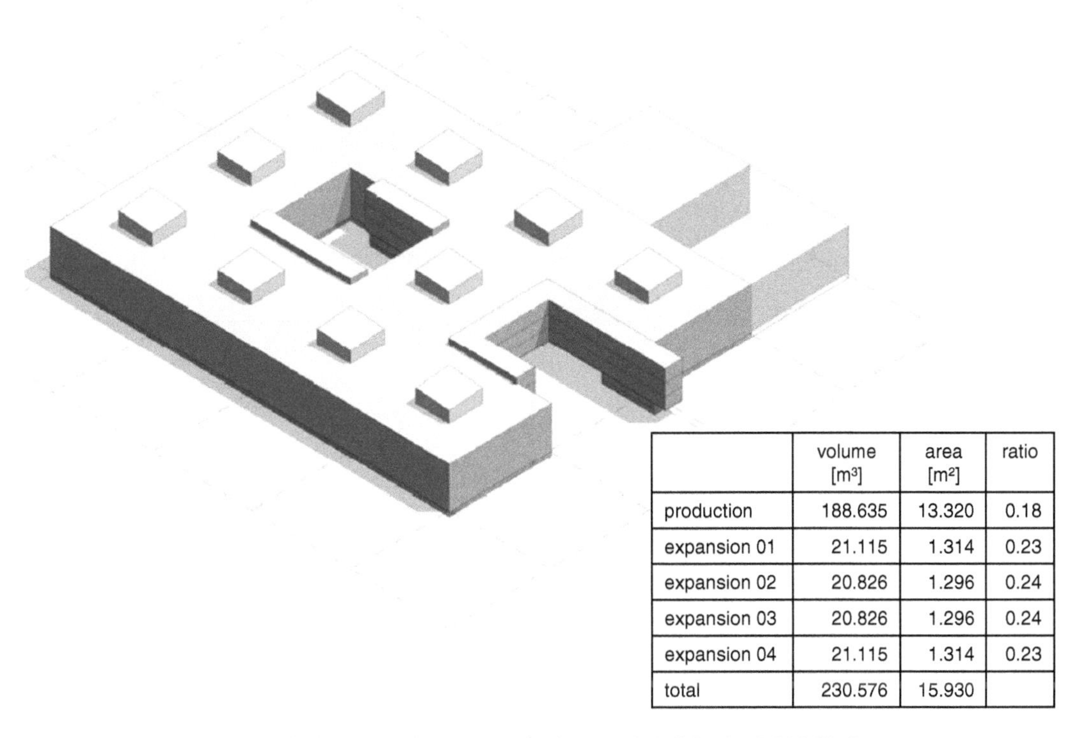

	volume [m³]	area [m²]	ratio
production	188.635	13.320	0.18
expansion 01	21.115	1.314	0.23
expansion 02	20.826	1.296	0.24
expansion 03	20.826	1.296	0.24
expansion 04	21.115	1.314	0.23
total	230.576	15.930	

Fig. 16.26 Modeling and analyzing expansion stages of a factory. © Reichardt 15.246_JR_B

- **Optimization of the topography**

With modeling tools within the BIM software, it is possible to import graphical and numerical information of the topographical features of a site and geotechnical analysis reports relevant for foundation detailing to automatically generate a 3D site model [Maa01].

For sites with a steep slope, earth mounds, trenches, etc. unnecessary cutting and filling of earth can be reduced by balancing the levels thereby saving costs during the construction phase. In addition to the obvious time advantage due to three-dimensional terrain modeling, this method provides a much higher accuracy when compared to the manual calculations with ground sections and interpolations.

Figure 16.27 shows for the phases "inventory", "removal" and "completion" the topographical project based on the three dimensional terrain model as well as calculations of the required earthworks.

- **Parametric**

The virtual building information model consists of parametric elements with digital attachments of a lot of information as well as relationship to other objects. For example a wall is not only allocated the dimension of its absolute height in meters, it is also linked to the edges of the floor slabs. If for some reason the absolute height of the wall is changed all other objects that are bound to the wall, are adjusted automatically to the current scenario. Moreover all related documents such as floor plans, sections and elevations are generated automatically.

- **Surfaces and mass analysis**

The building information model can generate a lot of additional views based upon the available geometric data of the objects. Building element may be grouped to special schedules, offering numerical representation of objects in the form of tables and reports besides classical planning documents such as floor plans [Sei01]. The user

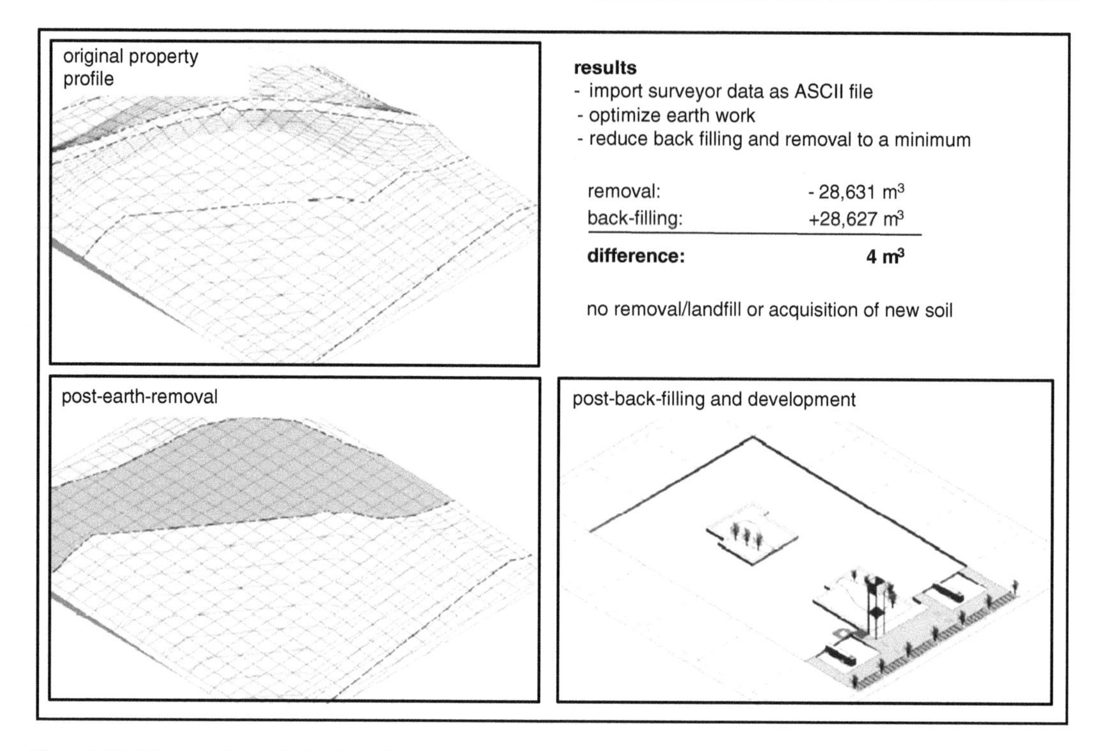

Fig. 16.27 Topography optimization of a site. © Reichardt 15.247_JR_B

has the option to consider, filter and change individual information as and when required. Areas and rooms can for example be represented in the form of a clearly ascertainable space book. As mentioned in Sect. 16.7, surface and volume investigations are the basis for all cost statements, cost estimation and cost control. This fact highlights the necessity of correct investigation of these values based upon the latest 3D design model or its variants. The resulting planning discipline and effort should not be underestimated. Last minute surprises due to faulty manual calculations e.g. in the tendering phase are highly unpleasant and one of the most common reasons for cost increase. In BIM, text attributes e.g. model surfaces, volumes as well as the number and definition of individual construction and cost elements such as doors are generated and affiliated automatically to the object geometry data. Information like room number and room name, not explicitly related to the geometry, can be filled in separately by the user in tables.

In addition to the architect's rooms and areas, component objects of the utilities engineer, structural engineer, as well as the factory and logistics planner are included in the evaluation of listings and schedules, thereby avoiding possible areas of conflict. In addition to structuring of production facilities, corresponding process elements are automatically assigned to rooms or spaces when inserted in the BIM model and can be detected in the combined space books or equipment lists. This is an important feature, for example, in the coordination of production equipment with building utilities services.

- **Construction processes and virtual construction site**

Finally after presentation and optimization of design variants, necessary approvals from the client and for tendering processes, the next step would be a trouble-free construction of the factory project, a very important phase due to the usual time and cost pressure. Both planners and construction companies strain to avoid site driven conflicts and time schedule overlaps. Building

Information Modeling, also, allows for assigning time components to each of the construction relevant object. Thus, not only individual construction phases, but also complex construction processes can be simulated and visualized 3 dimensionally before the groundbreaking.

- **Simulation of daylight and artificial light**

The BIM technology also makes it possible to create daylight and artificial light simulation. Thus from the very beginning of the design, an understanding of the light and shade conditions can be simulated, along with necessary qualitative and quantitative proof. In case of daylight simulation, the 3D building model is transferred into a light simulation software, along with informations about the site, surroundings, global positioning and the sun's location. After defining the site specific weather information and the required time intervals for simulations, evaluations can be made, e.g. animated shadow movements along with color representations of light distribution (see Fig. 10.8). Moreover lighting simulations help to distribute work places throughout the building so as to avoid direct glare on a specific work space or desk of an employee. All possible issues related to sunlight penetrating the building can be detected early and prevented through suitable measures (such as optimization of the transparent building shell glass values or creating solar shading devices, etc.). Figure 16.28 shows a daylight situation in the model factory, on a cloudless sky at 7 o'clock in the morning.

In the case of an artificial light simulation, lights are added with manufacturer-specific lighting information to the 3D building information model using IES format, an internationally accepted data format to describe the light distribution of luminaries at the desired locations. IES stands for Illuminating Engineering Society (http://www.iesna.org/).

By means of a "Light Meter", a user defined grid for virtual measurement of illuminance, the area under discussion is examined from all possible angles using various lighting concepts. This may apply to different requirements of lighting

Fig. 16.28 Day light simulation of a factory model. © Reichardt 15.2488_JR_B

based upon different functional usages, specific areas or entire building complexes. In most cases, time-consuming physical sampling of lighting fixtures on site is no longer required; necessary engineering decisions can be taken virtually. Figure 16.29 shows the interiors of a production hall with artificial lighting levels of 500 lux; the specific hall luminaires were virtually installed directly from the product catalogue of a manufacturer's internet (e.g. ERCO, Zumtobel) into the hall ceiling area.

- **Energy analysis and minimization**

In the past, often low construction investment and the possibility of amortization of the pure construction costs in a few years, were decisive criteria for design selections.

Over the years rising energy prices, the introduction of the Renewable Energy Heat Act 1 (Germany: EEWärmeG) and the introduction and steady tightening of Energy Saving Ordinance for buildings (Energy Saving Ordinance) has substantially changed the thinking of those involved in planning like architects, engineers and especially the investors and users. Even seemingly simple design decisions in the early design stages, such as the geographic positioning of building volumes on a site, can have significant impact on the future energy balance of a building and avoid unnecessarily high operating costs for heating and cooling.

Building information modeling enables architects and engineers to make their virtual 3D models ready for an energy related assessment using simulation tools like IES Virtual Environment, TAS, Autodesk Green Building Calculation, Design Builder, etc. from an early stage. It allows benchmarking the influence of design decisions on the overall energy performance of the structure.

For the preliminary assessment a basic model, with related information on the outer walls, transparent facades, room areas as well as ceilings and inner walls, is sufficient. The building information model uses the special gbXML data interface (energy and green building extensible markup language), supplemented with the

Fig. 16.29 Artificial light simulation of a factory model. © Reichardt 15.249_JR_B

required information for calculating the components along with zoning and technical equipment and exported to the respective simulation applications. Typically such an exercise requires zoning of the 3D model based upon the various temperature requirements for spaces like production, offices or circulation areas.

Figure 16.30 schematically shows examples of such specific requirements for heating, ventilation and lighting of different possible zonings for production, office and traffic areas as well as the resulting schemes for the central utilities plants. A change or manipulation of the variables, parameters of process and building (e.g. waste heat from the process equipment, solar gains of the transparent building shell, insulation values of the building roof and envelope shell) dynamically changes the resultant internal energy, climate model and the related calculation results.

Depending on the complexity of the building model, the results of the simulation are after a short calculation time either displayed through graphical images or detailed as room-wise reports. The findings may range from the primary energy demand, overall performance evaluation of the building envelope, to summer heat load of the office spaces or even to the overall CO_2-emission.

Tools such as TAS (a Dynamic Building Simulation package) or IES Virtual Environment (common in Anglo-Saxon countries) are also capable of more complex CFD simulations (Computational Fluid Dynamics) to provide for even greater accuracy in the simulation results. These methods are currently more expensive than mere dynamic simulations and necessitate sound engineering knowledge. These tools are normally used for complex projects with special requirements; see (for examples of dynamic simulations refer to Sect. 15.7.3, case studies of biscuit factory and an assembly and shipping center, Chennai, India).

- **Visualization**

In addition to quite abstract simulations, analysis and evaluations described above, visualization

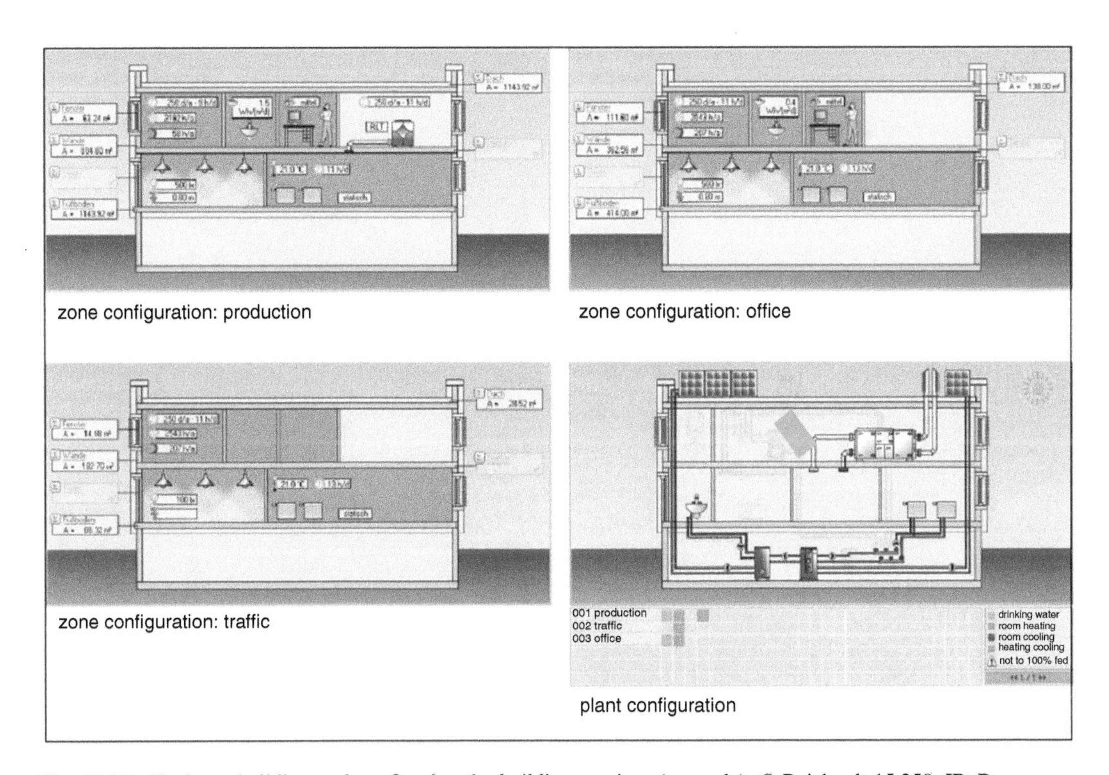

Fig. 16.30 Zoning a building and configuring the building services (example). © Reichardt 15.250_JR_B

is an important tool for conveying design ideas and the operation of technical systems and processes. Materials patching, colour and shade are standard visualization tools of state of the art BIM models, allowing for a realistic impression of the virtual model during the design stage.

In addition, photo-realistic impressions and camera movements in and around the building require minimal additional effort and are widely in use. Such video animations of a manufacturing plant for cosmetics, biscuit goods, automotive cooling systems and tire production as examples are annexed under D1, D2, D3 and D4. The applications provide BIM users with powerful tools for photorealistic visualization of spaces along with settings in a relatively short time. Alternatively, the market also offers other high-end applications such as 3D Studio Max or Cinema 4D. These programs require a sound knowledge in the field of graphics and animation and are used in large planning offices having their own specialized visualization departments. Figure 16.31 gives an example of such an image focusing on the entrance situation of a "Green Model Factory".

16.9.3 Conclusion

Building Information Modeling BIM is a relatively new technology in the field of the planning and building construction industry and offers tremendous potential for increasing efficiency and quality of the entire planning process.

The digital 3D models can comprehensively carry, for example, all related information about their usage, the insulation value of the building envelope, solar heat gain and the structural components [Kry08].

During creation of the model, the sources and consequences of design decisions can be generated directly and attached to the relevant areas. Thus the architect is able to quickly benchmark the decisions, such as the effect of changing the

Fig. 16.31 Rendering of a model factory's entry. © Reichardt 15.251_JR:B

orientation of the building volume on the property or varying the facade structure in line with the energy balance at an early stage of planning. Building Information Modeling is therefore not just a tool, but rather a process that helps users, planners and construction companies to control the ever growing quantum of information and complexity of factory planning projects. The manifold synergies in planning and execution justify the increased effort and in particular the requirement for digital discipline in the overall planning process.

With these thoughts the essential aspects of project management in factory planning are brought to an end. Because of the increasing importance of rational management and documentation of building projects the subject of facility management (as explained in following Chap. 17) will complete this book.

16.10 Summary

Professional project management plays a significant role in the success of factory planning project, in terms of functional performance and compliance with the time and cost frame. The complexity of this task is often underestimated resulting in gross miscalculations which, at times, could be economically disastrous for the client or company. A systematic approach is recommended with complete documentation of assumptions, definitions and decisions. The digital tools currently in use for planning deny individual solutions and advocate a more integrated approach of all subprojects, from the view of process as well as space. State of the art digital technology based upon object oriented 3D models, permits integration of process, construction and further dynamic simulations for evaluation of energy or light. BIM (Building Information Modeling) engineering allows for appropriate know-how, discipline and transparency in planning and in particular, exact space and cost evaluations throughout the project phases thereby enhancing as well as simplifying the functions of project management.

Bibliography

[Ble00] Blecken, U., Schriek, T.: Konzepte für neue Wettbewerbs- und Vertragsformen in der Bauwirtschaft (Concepts for new competition and contract forms in the construction industry). Bautechnik **77**(2), 119–130 (2000)

[Bot13] Both, P., Koch, V., Kindsvater, A.: BIM—Potentiale, Hemmnisse und Handlungsplan. (BIM—Potentials, Barriers and Action Plan). Fraunhofer IRB Verlag, Stuttgart (2013)

[Bra11] Bracht, U., Geckler, D., Wenzel, S.: Digitale Fabrik. Methoden und Fallbeispiele (Digital Factory. Methods and Case Studies). Springer, Heidelberg (2011)

[Bur06] Burghardt, M.: Projektmanagement—Leitfaden für die Planung, Überwachung und Steuerung von Projekten (Project Management—Guidelines for the Planning, Monitoring and Control of Projects), 7th edn. Publicis Corporate Publishing, Erlangen (2006)

[Ch99] Charette, R.P., Marshall, H.E.: ASTM Uniformat II classification for building elements (E1557-97). US National Institute of Standards and Technology (1999)

[Die04] Diederich, C.J.: Knowledge Management for Construction and Real Estate Professionals 1—Basics, 2nd edn. Springer, Berlin, Heidelberg (2004)

[Eas11] Eastman, C., Teichmann, P., Sacks, R., Liston, K.: BIM Handbook: A Guide to Building Information Modeling for Owners, Managers, Designers, Engineers and Contractors, 2nd edn. Wiley, New Jersey (2011)

[Eit07] Eitelhuber, A.: Partnerschaftliche Zusammenarbeit in der Bauwirtschaft—Ansätze zu kooperativem Projektmanagement im Industriebau (Partnership in the Construction industry—Approaches to collaborative Project Management in Construction Building). Kassel University Press, Kassel (2007)

[Esc01] Eschenbruch, K.: Bauverzug: Haftet Baubetreuer/Projektsteuerer? (Construction delays: Is the project manager legally responsible?). In: IBR 2001, H. 1, S. 34

[Esc09] Eschenbruch, K.: Projektsteuerung und Projektmanagement: Leistung, Vergütung, Nachträge, Haftung, Vergabe, Vertragsgestaltung (Project Control and Project Management: Service, Fee, Supplements, Liability, Award, Contract Design), 3rd edn. Werner Verlag (2009)

[Fes05] Feske, I.: Toleranz beim Baukostenlimit? (Tolerance in construction costs limit?) BrBp **2**, 60–64 (2005)

[Kal02] Kalusche, W.: Entscheidung bei der Gebäudeplanung mit Hilfe der Nutzungskostenermittlung (Decision in buildings planning on base of utility costs). Zeitschrift für Immobilienökonomie **1**, 55–63 (2002)

[Kym08] Kymmell, W.: Building Information Modeling: Planning and Managing Construction Projects with 4D CAD and Simulations. McGraw-Hill, New York (2008)

[Kry08] Krygiel, E., Nies, B.: Green BIM: Successful Sustainable Design with Building Information Modeling. Wiley Publishing, Inc., Indianapolis (2008)

[Küh06] Kühn, W.: Digitale Fabrik. Fabriksimulation für Produktionsplaner (Digital Factory. Factory Simulation for Production Planners). Hanser, Munich, Vienna (2006)

[Knu08] Knutson, K., et al.: Construction Management Fundamentals, 2nd edn. McGraw-Hill Series in Civil Engineering (2008)

[Lev09] Levy, S.B.: Legal Project Management: Control Costs, Meet Schedules, Manage Risks, and Maintain Sanity. Steven B. Levy (2009)

[Loos13] Loos, M.N.: Daten- und termingesteuerte Entscheidungsmethodik der Fabrikplanung unter Berücksichtigung der Produktentstehung (Data and time-driven decision-making methodology of factory planning with consideration of product development). Ph.D. Thesis, Karlsruhe Institute of Technology (KIT) (2013)

[Maa01] Maas, B.: Lecture: Synergetic use of BIM on the example of assembly plant MODINE Hungary 2. In: BIM Conference 2008, Berlin (2008)

[Mar11] De Marco, A.: Project Management for Facility Constructions: A Guide for Engineers and Architects. Springer, Berlin (2011)

[Mean14] Means, R.S.: Building Construction Cost Data 2014 Book, 72nd edn. Reed Construction Data Inc. (2014)

[Mei07] Meichsner, T.P.: Migrationskonzept für einen Modell- und variantenflexiblen Karosseriebau (Migration concept for a model and variant-flexible white body shop). Ph.D. Thesis, Leibniz University Hannover. PZH Publisher, Garbsen (2007)

[Rei98] Reichardt, J.: Planungsmanagement mit Pflichtenheft und Energiesimulation (Planning management with specifications and energy simulation). BAUKULTUR **6**, 6–12 (1998)

[Rei04] Reichardt, J., Gottswinter, C.: Synergetische Fabrikplanung—Montagewerk mit den Planungstechniken aus dem Automobilbau realisiert (Synergetic factory planning—assembly site realized with planning techniques of the automotive industry). IndustrieBAU **3**, 52–55 (2004)

[Rog08] Rogers, L.: Basic Construction Management: The Superintendent's Job, 5th edn. Builders Book, Washington (2008)

[Rös94] Rösel, W .: Baumanagement—Grundlagen, Technik, Praxis (Construction Management—Fundamentals, Technology, Practice), 4th edn. Springer, Berlin (1994)

[Schu08] Schulte, H.: Fabrikplanung organisieren und durchführen (Factory planning organize and execute) 8. Deutscher Fabrikplanungskongress. Verlag moderne industrie. Ludwigsburg (2008)

[Sei01] Seifert, A.: BIM in the planning process—Building information model and Kostenermittlung. Thesis, Bauhaus University, Weimar, Jan 2008

[Smi09] Smith, D.A., Tardif, M.: Building Information Modeling: A Strategic Implementation Guide for Architects, Engineers, Constructors, and Real Estate Asset Managers. Wiley, New Jersey (2009)

[Thi12] Thiede, S.: Energy Efficiency in Manufacturing Systems. Springer, Heidelberg (2012)

[WieH11] Wiendahl, H.-H.: Auftragsmanagement in der industriellen Produktion (Order Management in Industrial Production. Basics, Configuration, Implementation). Springer, Berlin (2011)

[Wik14] Wikipedia The Free Encyclopedia: Article "Werksvertrag" (contract of work) and "Dienstvertrag" (contract of employment). Read 8 Nov 2014

17 Facilities Management

As already stated in Chap. 16 'Project Management', a systematic approach is recommended through all phases of factory planning based upon complete documentation of the assumptions, definitions and decisions. The tools used for the information processing should avoid isolated applications; instead they have to enable an integrated approach of all current sub-projects from the process point of view as from the building construction view. Initially facility management was understood as a view of the factory more or less after creating the project on-site, as pure administration of the "as built" operational data. Taking into account the various issues discussed in the previous chapters, it is proven beyond doubt that synergies do exist between initial planning and subsequent functions of Facility Management and cover the entire life cycle of a project including planning, implementation and operation. They are in fact cross-functional in nature. The possibility of integrating different professional views (e.g., production planners, architects, building services engineers) and thematic scales (e.g., coarse, fine) on the overall central data enables the highest level of ongoing active collaboration, enabling a consistent project transparency.

17.1 History and Definition

The term facility management (FM) was first coined in the mid-1950s by the brothers Schnelle from the Quickborner Team, Germany. It is said that the furniture manufacturer Herman Miller, hit upon an idea of improving their productivity and operational interaction through interior-shaping of their offices. This marketing idea was quickly turned into a reality. In 1978, Hermann Miller Corporation, Ann Arbor, Michigan invited its customers to a conference entitled "Facility Impact on Productivity". This initiative led to the establishment of the Facility Management Institute (FMI) in Ann Arbor, Michigan in 1979. In October 1980 the National Facility Management Association was founded by 40 professional facility managers. The rapid growth and expansion to Canada led to the new name of IFMA, International Facility Management Association in 1982. IFMA currently claims to have over 20,000 corporate members in more than 60 countries around the world. In Europe, FM was introduced in the mid 80's. Architect Francis Duffy opened an FM Center in Britain, when he chanced upon the idea. By 1985, AFM or Association of Facility Managers and the Institute of Administrative Management/Facilities Management Group (IAM/LTC) were established.

The term "Facility Management" is often used without a clear definition. With respect to location, building and equipment the GEFMA guide line 100 [GEF96] proposed the following definition: "Facility management is the consideration, analysis and optimization of all cost-and quality-related processes of location, buildings, technical and other utilities".

A concise definition was proposed by Nävy [Näv02]: "Facility management is a strategic

H.-P. Wiendahl et al., *Handbook Factory Planning and Design*,
DOI 10.1007/978-3-662-46391-8_17, © Springer-Verlag Berlin Heidelberg 2015

approach to management, administration and organization of all tangible assets in an enterprise".

The term "Facilities" includes all non-human resources like land, infrastructure, buildings, plant, machineries and equipment, in short all the tangible assets of a company. Facility Management includes all those equipment which have economic value and perform at an optimum level in a building project. The term includes, as already discussed in previous chapters, parameters of a "performance" based building project concept. Facility Management combines consulting, planning, organization, management and control of all processes, structural measures and marketing activities throughout the lifetime of a site include buildings and facilities [Boo09].

Facility management accompanies a construction project from planning through its usage to decommissioning. It should be noted that the construction costs are only a part of the entire life-cycle costs. Depending on the specific project the annual operation and maintenance (O&M) costs of the infrastructure may vary between 10 and 20 % of the initial investments. Therefore, when added to an entire life span of a project these annual maintenance costs far exceed the original construction costs. More and more service-oriented companies are offering off-site internet-based programs, to support basic maintenance routines for technical plants or tendering procedures for technical equipment at competitive pricing. However, in most cases due to a multitude of interfaces, such supports lack a holistic view of the asset.

17.2 Tasks and Delimitation

In today's practice, facility management of a property deal with constant changes to the operational demands of location, buildings and facilities maintenance [Ron06, Rop14, Bra07]. It provides a sound basis for optimum planning, installation, operation, rehabilitation and recovery. According to Fig. 17.1 this complex task can be divided into five sub-areas.

- *Investigating and Providing Current Data*
  This includes the collection and maintenance of actual data concerning land, buildings, building equipment and facilities, the details of their occupancy levels and market value. An effective FM needs a reliable communications network, to ensure quick data collection and maintenance at a competitive pricing. Further, all planning and decision-making data should be easily accessible at the concerned departments.
- *Assessment of Locations, Buildings, Facilities*
  This involves the provision of data for planning, operation and maintenance of a safe, humane and functional work environment. This task relates not only to the actual resources, but also to organizational and personnel development concepts. Other requirements include the identification of follow-up costs due to strategic investment decisions, costs towards maintenance of the external appearance—a mark of identity of the company—and the costs towards ensuring year round security of resources, buildings and data.
- *Space and Occupancy Planning*
  The aim of the space and occupancy planning is to set up, monitor and adjust physical spaces in line with legal, ergonomic, organizational and sociological criteria. In particular, the needs of interaction with other departments/ personnel must be considered.
- *Building Operation and Management*
  The main aim of building operation and management is to analyze the operation and service costs, including building life cycle data and costs related to maintaining an overall efficiency in building operation.
- *Budgeting and Assessment*
  Relates to comparative financial evaluation of individual measures and development of alternatives, taking into account the particular life cycle and ecological consequences.

This new role of facility management incorporates a fundamental shift in the FM functions, extending far beyond the term "planning". Planning now is no longer an exclusive and isolated event constrained by time, which concludes with the construction of a building; it is an

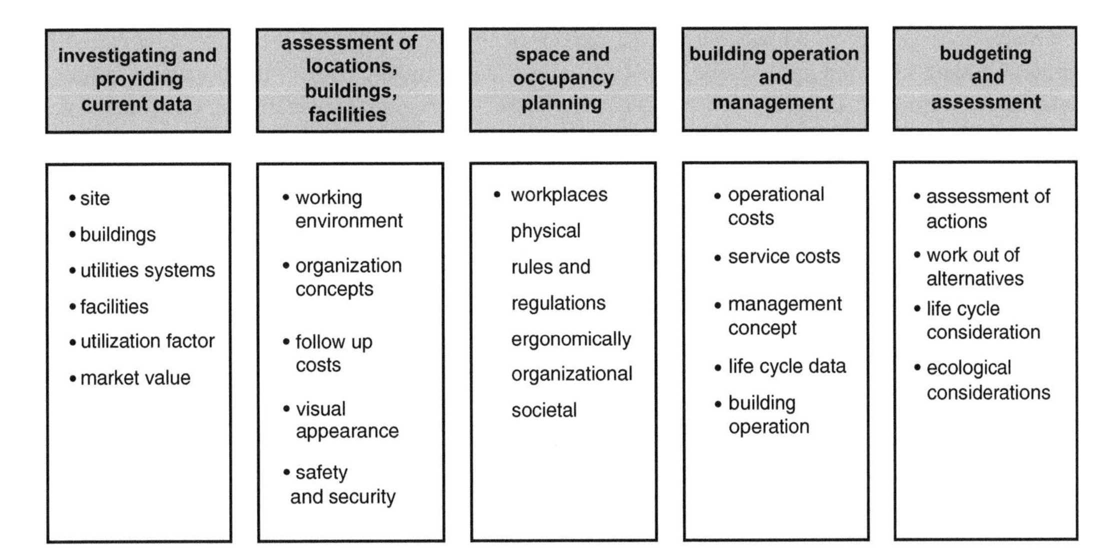

Fig. 17.1 Tasks of facility management. © IFA 15.272ESW_B

all-inclusive process and continues well into the building life-cycle.

17.3 Facilities Management in the Life Cycle of an Object

The earlier notion of independent plans, departments, functions and operational processes is rapidly giving way to a more holistic view of a property over its entire useful life-cycle. Currently data collection and exchanges are coordinated in order to avoid unwanted and repetitive transmissions wasting valuable resources and unwanted errors.

In most cases, the life cycle of a project begins with a consultation service in the form of a workshop. Thereafter current notions of planning, implementation, operation, remodeling, possible repairs, recycling and demolition, by far exceed the original contents of phases of engineering under planning services for architects and engineers (e.g., HOAI, German fee structure for Architects and Engineers). These Phases categorize a building to be a physical object and wherein the service of an architect and engineer ends with its completion and handing over to the owner. Facility management on the other hand, focuses more on the overall usability of a project over its lifetime. As already mentioned, the typical life cycle of a project includes: consulting, planning, realization operating, reconstruction, recycling and demolition. In the following sub chapters the role of facility management in these phases will be looked into in detail.

17.3.1 Development Phase

Usually as part of the Development Phase there are set tasks and targets costs for construction and operation, linked to planning results, to be implemented subsequently. Although the planning phase, when compared to the utilization duration, is very short, basic definitions of the facility management functions are planned and installed thereby establishing and influencing, for example, the heating or cooling demand and thus the future operating costs. Corrections, at a later date either to buildings or utilities or equipment often results in extensive and disruptive renovation activities which are expensive. Targets established early in a project can lead to more holistic view of FM and help generate better solutions.

Without state of the art CAD support and digital database application, complex industrial buildings cannot be realized in the given time. The collection of basic project data should therefore be in a digital format, ideally in the form of "standard" graphics and text files (e.g., defined in a project manual with readymade Excel sheets or facility description cards for the different project requirements) for the facility management procedures. The data exchanges between the project participants within the planning phase via data networks or carriers should be digital and state of the art. The planning phase is closely linked with the execution phase, i.e., the data for the tender documents are usually derived from the basic planning documents. Due to time and cost constraints an integrated approach is suggested, e.g., the tendering process could be executed via internet while building planning could be simplified using 3D BIM (Building Information Modeling).

It is preferable that the structure of tender documents is in line with the cost categories and cost elements of a standard cost matrix systems (e.g., German DIN 276 or ASTM UNIFORM II). Such a system offers the advantage of cost transparency. The system also helps to establish time, cost and payment schedules. German DIN 276 cost standard covers the cost of a building project in accordance with Fig. 17.2 in seven main groups, which are further sub-divided into sub-groups. Each subgroup is divided further detailed out to cover all possible elements of a typical construction project.

Figure 17.3 shows an example of the cost structure of a small industrial project, with systematic sorting of cost groups and further subdivisions of cost elements. The planning and preparation of tenders and the award of construction contracts should be in a holistic digital workflow, including all planning and construction documents. Subsequent adjustments of the property during the operational phase can be based easily on that digital basis.

17.3.2 Implementation Phase

Usually during the implementation phase the building contractor, under the professional

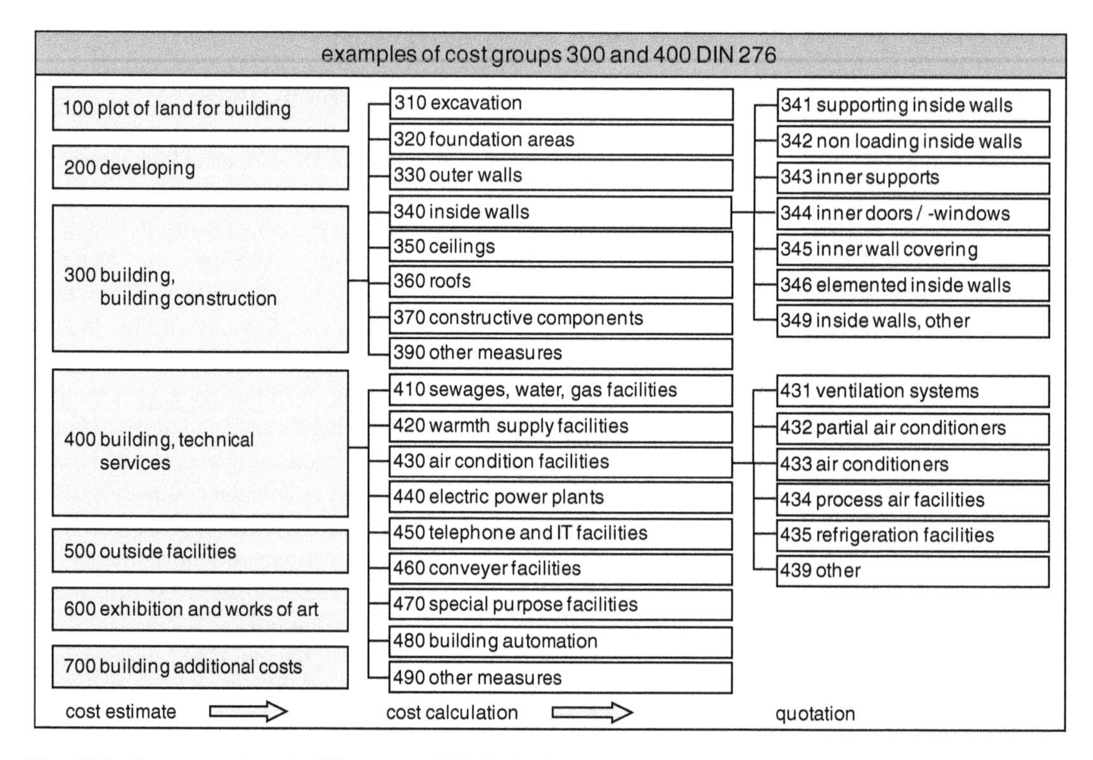

Fig. 17.2 Cost groups for a building acc. to DIN 276 © IFA 15.273ESW_B

Cost Group to DIN 276

100 Property

200 Preparation and Development

300 Building-Constructions

400 Building-Utility Systems

500 Site utilities

600 Equipment and Artworks

700 Additional BuildingCosts

		Areas				Cubature		Reference net values costs per subproject			
Code DIN	Project part new factory	Outside	Areas F1	Areas F2	GFA	F1/GFA	BRI	Construction	House services	Open spaces	Sum
		sqm	sqm	sqm	sqm		m³	Cost Group 300	Cost Group 400	Cot Group 500	Cost Group 3-500
	without reorganizations			6,016.00							
	paved outer area logistics	2,640.00									
	paved outer area by-pass	1,440.00									
	unsealed outer area	0.00									
500	Sum site work	4,080.00									
	Production		4,750.00								
	Dispatch		519.00								
	Adapter/ramp		326.00								
	Workshop		40.50								
	Spare part store		40.50								
	OS laboratory		40.50								
	recreation room		20.00								
	first aid room		20.00								
	extension		42.00								
	fork lift room		44.00								
	toilet men		10.00								
	toilet women		10.00								
300	Sum ground Floor		5,862.50	154.50	6,016.00						
	(only shell construction, gallery, slab, connections wet area)										
	Stairs/shaft 1										
	Stairs/shaft 1										
300	Sum Floor 2		40.00	00.00	40.00						
	Head offices air technology										
	Central compressed air										
	Central heating										
	Maintenance footbridges (optional)										
400	Sum Floor 3		500.00	80.00	580.00						
	Sum areas Proj.		6,402.50	234.00	6,636.00			6,636.00	6,636.00	4,080.00	
	Sum areas Proj.						60,963.00				
	Reference value/m2 open area									55.00	
	Reference value/m2 GFA							405.00	100.00		505.00
	Reference value/m2 NFA							44.09	10.89		54.97
	Total estimated project value							2,687,580.00	663,600.00	224,400.00	3,575,580.00
	add. Reorganization/extension utility centre (estimate)							*50,000.00*	*225,000.00*	*25,000.00*	*300,000.00*

Fig. 17.3 Cost structure for a building according to German DIN 276 (example). © IFA 15.274ESW_B

supervision of architects and project managers, are at site to realize the necessary buildings, utilities and equipment in accordance with the definitions of the contract documents. Amendments and additions or deletions to the contract document during the construction phase are often an inevitable part of any construction process. During the implementation phase the planning status should be tracked regularly with reference to actual data from the construction site thus enabling a transparent project steering in terms of quality, time and cost.

It is preferable to make payments to the building contractors based upon the actual work done against the specific cost elements as against payments at a 'lump sum' rate. During the subsequent management of the building in the operational phase the planning documents may be required again for necessary tendering activities, e.g., to establish the basis for the calculation of an area for cleaning or maintenance work of the technical building equipment. In case "as-built" drawings and documents are not available on completion of the project, the client may be forced to comprehensively re-document the building at additional cost.

This is why facility management is fast developing into a powerful tool for the project management who aims for an inclusive and responsible project control. On completion of the project, all necessary plan documents, data and specifications are normally transferred to the client in line with predefined digital documentation formats. Such documents are usually prepared by the contracted construction companies or architectural/engineering consultants, approved by the project management team and delivered as agreed.

During the implementation phase, possible faults are also reported back to the planning team. This often lead to delays in time and to additional costs. If the same are not updated on a regular basis, they need to be updated on realization of the project.

17.3.3 Operating Phase

The operational phase, as defined from duration of use and resulting cumulative costs, is the most important phase in the entire life cycle of the project. During this phase, there are no significant changes to the material resources other than wear and tear caused by constant usage. In some cases, e.g., in automotive plants, process related machineries and utilities may change due to introduction of new models every three or four years. The task of the Facility Management Team includes keeping a close watch on the daily operations and documenting the wear-tear process along with the associated operating conditions and controls. In addition to issues resulting from non-inclusion of users from the very beginning of the planning and implementation phase, there could be other challenges in the operation phase also. In most cases, the responsibilities of the daily operations of the various resources are split between internal as well as external operators, requiring constant co-ordination.

Another important aspect of the operation phase is an efficient use of space. This helps to reduce operating costs, thereby increasing profit margins. Facility management systems make it easier to track the usage of space in buildings and optimize operating costs by scheduling together or controlling available spaces, workstations and equipment.

17.3.4 Conversion Phase

The conversion phase interrupts the operation phase with the aim of improving the material resources to meet new challenges or requirements (e.g., extension of the production program, introduction of new technology, introduction of new manufacturing and assembly structures, etc.).

Due to different economic and technical life time of the components of a factory, changes could be either more frequent or less frequent. Depending on the business policies, organizational changes could happen as frequent as 2–3 years. IT infrastructure and telecommunications equipment could change every 5 years, furniture every 10 years, lighting systems and fixtures every 10–15 years, household technologies every 5–20 years and interior furniture, fittings and fixtures every 5–30 years. A building structure could be used as a thumb rule over 50–70 years.

In the conversion phase, the same procedure as in the development phase is repeated, with the exception that the status of the existing buildings and facilities needs to be taken into account. In case tender documents of the development phase are available, system solutions already existing might be used during the re-tendering process. Essentially similar operations are repeated, as in the original implementation phase.

After extensive conversion projects, the 'as-built' drawings and documentation should be updated including in-use procedures and structural changes and released for further monitoring. For smaller rescheduling this effort may not require detailed planning. However, there is a clear risk if substantial changes are made without updating the plans and documenting the process. On a daily basis all industrial buildings and plants are characterized by minor changes in their resources; the inability to document the changes might make the 'as-built' plans redundant over a period of time. A stage is quickly reached in which only a comprehensive stock-taking provides the necessary information security, once again. A permanent 'facility' database provides invaluable support at all times during the life cycle of project.

17.3.5 Decommissioning Phase

The decommissioning phase completes the life cycle of objects, the building structure can no longer fulfill its function economically. Also, re-planning or further conversions are no longer useful to the company. The building structure might be sold and continues to be used by the buyer or the new user or the buildings; associated structures could be dismantled, machinery and equipment scrapped. In the case of sale, it is possible to offer the comprehensive facility management database of the building, structures, processes and process equipment to the proposed buyer as an additional selling point. Since the entire life cycle has been virtually documented it provides the new user with additional value. In due course of time when the building structure, processes and process equipment are to be totally scrapped, this information would help to calculate the associated costs.

17.4 Facility Management Systems

17.4.1 Functions

The daily operation and maintenance of a site with its buildings and facilities is generally done by several internal departments partly as a secondary function, or as a primary function of the estate, construction and management department. While the production processes are under higher pressures to optimize operations for higher profit margins, the buildings and their maintenance are considered as "necessary evil". A rethinking of the existing notion of building management is imperative due to the drivers of change outlined in chapter one. This is especially important since maintenance and energy costs are steadily increasing.

This ensures facility management to be an economically viable, efficient, strategic and holistic service, with the following main tasks:

- Real estate management as a proactive task, not only as passive response to individual requests
- Significant decrease of the maintenance costs of the property
- Reduce downtime of equipment and increase safety of use by increasing availability
- Ensuring cost transparency.

An important tool which provides the required information rapidly is CAFM (Computer Aided Facility Management). It is a database supported information system for the storage, maintenance and evaluation of digital object data. It is similar to a PLM (Product Lifecycle Management) Systems for the design documentation of products. (PLM is an IT solution system, consisting of all data generated in the development, production, storage and distribution of a product which when stored safely, can be managed and retrieved easily). Figure 17.4 shows an overview of possible functions of a CAFM system [Näv06].

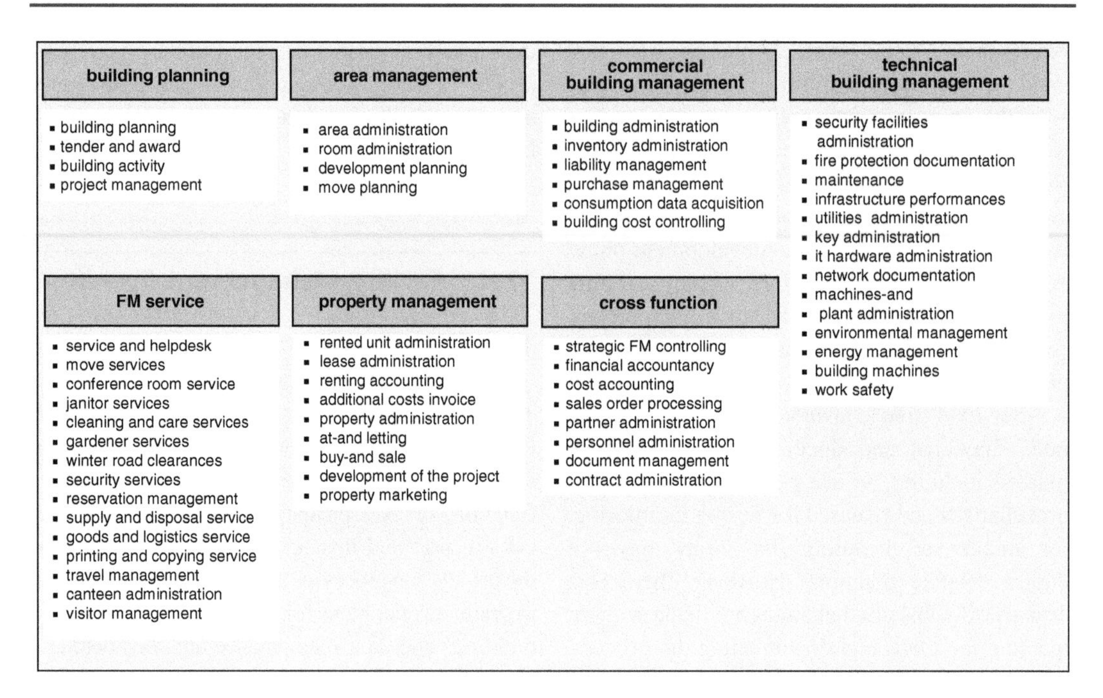

Fig. 17.4 Possible functions of a facility management system (after Nävy). © IFA 15.275BESW_B

For factory buildings key applications of CAFM are the architectural and engineering building design as well as the project management, project documentation and land management (spatial and land management, space and occupancy planning, relocation planning). Other functions may focus e.g., on cleaning and maintenance, security, utilities and waste disposal, particularly in the management of large real estate holdings, commercial and technical building management and service functions. For the operation of complex buildings such as a conference center, services can be managed with such a system and also the entire property management support at rental properties with the essential features of multitude rental aspects managed by such a system. In a specific case, a company may not integrate all conceivable functions, but “tailor” it in coordination with existing commercial systems.

According to studies by Nävy in 2005 the global market is offering about 44 possible FM systems, out of which 80 % are from Germany, 11 % from the rest of Europe and only 9 % of U. S. origin [Näv13].

Facility management systems allow the following applications:

- the analysis of space inventory information based upon departments,
- presenting reports for gross floor space, availability of rooms, service areas,
- the creation of test configurations to compare the space usage of various relocation and occupancy scenarios, and
- seamless integration with drawing programs such as the industry standard AutoCAD, Microstation or Nemetschek to quickly and easily link floor plans with space data.

In addition to the established advantages of transparencies in the planning phase and systematic construction and management of buildings during its lifecycle, it is believed that about 10–30 % of costs can be saved by a holistic FM approach when compared to a conventional approach.

17.4.2 Structure of Data Models

The common basis for all subsequent evaluation during planning, implementation, operation or modification phases are a standard set of CAD, text and image data files. In CAD drawing systems, the graphical information files are usually generated as a function of design software as e.g., DWG (AutoCAD), DGN (Microstation) or DXF file (CAD software drawing exchange file for Autodesk/AutoCAD); for alphanumeric information the industry standard MS Office has prevailed with the programs Word and Excel. Graphics, photos, videos are usually pixel graphical data such as bitmaps. In Computer-aided facility management systems (CAFM), conventional data collections such as drawing cabinets, folders, file boxes, or lists are largely replaced by interdisciplinary databases. CAFM systems allow simplified networking of different information, such as areas, surfaces, costs, personnel, processes and their optimization.

17.4.2.1 Data Structuring

Database systems are used to store and manage large and complex structured data. A major advantage of such systems lies in the generation of different levels through specific queries to the data fields. According to Jedlitzke [Jed02] database models are distinguishable in terms of their task structure; CAFM systems are relational and object-oriented; thus, object-relational database systems are currently in use.

Relational DBS insert data in related tables. Their relationships are clearly marked with key attributes. Object-oriented DBS store object types with attributes that are bound to this. An example of this is the CAD design of a window element with additional textual links e.g., with frame material, glass characteristics or price in an object-oriented CAD design software. Object-relational DBS combine the advantages of different database models, they offer a flexible extension of the data type and are also suitable for the management of multimedia data such as documents, pictures, sound recordings, or films.

The GEFMA guide line 400 [GEF13] classifies according to Fig. 17.5 inventory data, status data, consumption data and other data.

Inventory data of a property are, for example, relevant to land management or building cleaning. Status data report current temperatures, energy flows or disturbances. Consumption data are generated by automatic or manual recording routines through standard building automation or local sensors on site. For organizational as well as financial reason it is not necessary to monitor every possible data, but to collect data which makes sense for presentation, evaluation or ongoing maintenance. Focusing on key operational information enables creation of a solid foundation, which can be enlarged into an open database system during the later stages of the project.

17.4.2.2 Layer Classification

Information in complex CAD drawings needs to be managed in clearly structured layer systems. By grouping functions e.g., all bearing walls, all furniture etc. can be presented. Layers might be compared to drawers, storing drawing files that contain certain information in each of the

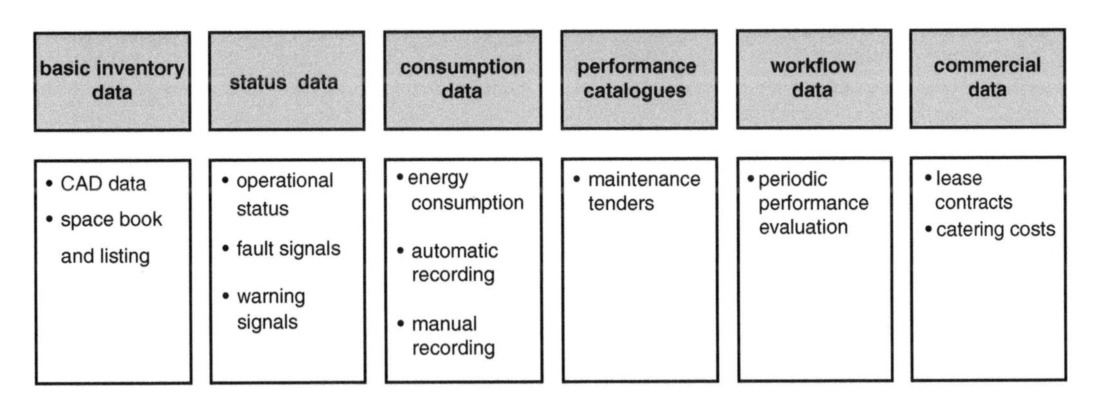

Fig. 17.5 FM data classes, examples (after GEFMA 400). © IFA 15.276ESW_B

represented object, such as contour, color or dimension. By "displaying" or "hiding" layers, the specific graphical representation for a particular application can be adapted.

A project specific layering or referencing structure should be previously established when setting up a CAD drawing, so that the same is consistent through all planning and documentation stages.

Layer assignment helps to classify location, buildings, building services, facility and others. The consistency of a holistic planning requires integrated drawings with pre-determined properties like line thickness and line color. Figure 17.6 shows excerpts of a layer assignment for a small industrial project. In this case, it displays the documentation of the property, the bearing structure and the shell of an industrial building into three main levels, each with several sublevels, which are numbered 11–37. Each sublevel is associated with an object with a short description and color representation.

Another example of a layer layout is shown in Fig. 17.7. Here is the breakdown of the building utilities equipment in focus. Due to the underlying integration of levels for e.g., functions and processes and their spatial association with the building and site levels, it is possible to "overlay" the interaction of processes that take place in the building, with relevant utilities making it immediate visible. This could be, for example, an area in which operating process facilities are located, with superimposed images of compressed air supply systems feeding the process equipment.

	No	Layer Name	Names	Text	Color
Property	11	Land register limits	Land register / building limits	d (documents)	Yellow
	12	Streets, perimeter	Streets / sidewalks/parking lots/ Fence plants	d	Yellow
	13	Height development curves	Height development curves	d	White
	14	Volume control	Stock / new building	d	Cyanogen
	15	Plantations	Trees / bushes	d	White
	16	Parks	Lawn / little plants	d	White
	17	Water	Seas / ponds / rivers / creeks	d	Cyanogen
	18	Tracks	local public transport routes	d	Cyanogen
	19	Signage			
	20				
Building Structure	21	Floor plate foundations	Floor plate / foundations	d (documents)	Yellow
	22	Storey floors	Storey floors	d	Red
	23	Roof insulations	Roof insulations	d	White
	24	Supports	Head-/ façade supports	d	Yellow
	25	Building structure	Head-/ side beams	d	Yellow
	26	External Walls	Masonry / concrete	d	Yellow
	27	Internal Walls	Masonry / concrete	d	Yellow
	28	External stairs			
	29	Signage			
	30				
shell	31	Sky-lights	Smoke exhaust domes / glass fixtures	d (documents)	White
	32	Roof coatings	Water barrier, insulation / green roof	d	White
	33	Façade, solid elements	Sandwich elements / cassettes	d	White
	34	Façade, transparent	Post bolt / cast glass	d	Green
	35	Façade, window	Tilt / open / sliding window	d	Cyanogen
	36	Protection from sun	Louvers / window shades	d	White
	37	Doors	Entrance-/ escape doors	d	Cyanogen

Fig. 17.6 Layer assignment (example industrial project). © IFA 15.277ESW_B

Site

Building

Function/process

CAD construction: Example drawing layer				Building Utilities
	No.	layer name	Names	text
Utilities/heating, ventilation	61	Sprinkler	Sprinkler	d (documents)
	62	Ventilation	Ventilation systems	d
			Input air,	d
			Waste air,	d
			Fresh air,	d
			Stale air	-
	63	Heating	Flow pipe	d
			Return pipe	d
			Radiators	d
	64	Sewage	Waste water	
	65	Sanitary	Supply	d
			Hot water	d
			Cold water	d
			Facilities	d
	66	Compressed air	Compressed air	d
	67	Roof drainage	Rain water	D
	68	Gas	Gas pipes	d
	69	Signage		
	70			

Fig. 17.7 Layer assignment home automation (example). © IFA 15.278ESW_B

The system of layer assignments would be also important for the contractors entrusted with the implementation and monitoring of the specialist functions. On this basis, e.g., the consumption data of all technical centers in the subsequent operation and a database link with the 2D or 3D CAD object modeling can be derived. The current consumption values could be made available by clicking on the respective technical plants either individually or as a whole. For the data structure of a CAFM system an aggregation of similar objects to object classes is advantageous. An example might be simultaneous access to a group of buildings, utilities or process objects in addressing common attributes of all object types. Such classifications are useful according to data type, data format and frequency of change usage [Lut02].

17.4.2.3 3D-Building Model

CAD programs based on a three-dimensional object oriented building model are state of the art. The components of a building such as slabs or columns are stored in their entirety along with their attributes and relations. The CAD software should allow the elements to look at in different views.

One of the biggest challenges in data exchanges between geometric building models and alphanumeric databases is to match the basic logic of the database of the CAD system.

ISO 16739:2013 helped in establishing an open IFC standard (Industry Foundation Classes) for the digital description of building models and standardization of the exchange of data between 3D-systems from different vendors. Due to the complexitiy of 3D-objects, as well as the interests of competing software vendors, the development of these standards was and still is slow. Some CAFM systems such as Archibus/FM combine an open Oracle database structure with the so called "overlay" read-in and read-out functionality of AutoCAD drawing data. This ensures the actuality of the database entries and the drawing data, without redundancies.

17.4.2.4 Space Book

The basis of a CAFM system is a directory of the building under consideration, which is commonly referred to as space book or room book. A space book documents the user requirements from the very beginning and continues to detail out the parameters further during the planning process. Usually space books contain information about the room number, room name, floor, floor area, height of the room as well as more specific information on the building structure (e.g., fire protection). As part of the synergetic planning approach more complex digital space matrix sheets are developed, which can include a variety of additional utility and process features and also allow coupling of these objects with subsequent building interior elements as floor, wall and ceiling. Further, on the basis of consistent data structure it is possible to assign each room with defined utilities and proportionate cost elements for future maintenance or servicing. An example of the planning data of a space book shows in Fig. 17.8

The requirements of each spatial element such as room geometry, building systems for floor, wall and ceiling as well as the building utilities systems and the numbering of specific usage data are merged into matrix sheets. Each spatial element is identified by a serial number in the first column. The spaces are clubbed according to their characteristics, e.g., wet areas. This allows for separate cost evaluations according to building classes as well as consideration e.g., of mixed use gross floor space.

Ideally, the project development of a property should be linked to the starting phase of basic evaluation with a FM oriented data modeling for the detailed planning of location, buildings, building utilities and processes. In particular, the phase of the basic evaluation should be linked to well-coordinated and neatly structured tables for the sake of future planning. In case, precise data are not available for the individual areas, assumptions should be discussed, temporarily recorded and ratified at a later date.

A direct transfer of data from digital space book into the database of the CAFM system is not always possible, since the space books underlying building databases usually are based on individual parameters. Specifically programmed macro tools can help in immediate access to all spatial features by the Oracle database into CAFM system Archibus/FM as explained in Fig. 17.8.

17.4.2.5 Documentation

The various benefits of an early introduction of a CAFM system logic in the planning phase has already been explained in detail. Nevertheless all projects may not be in a position to start from zero e.g., old buildings and existing structures. The extent to which existing inventory documents can be used for the purpose of database generation and integration depends on their condition, accuracy and timely availability. Facilities Management can provide accurate results only if the input data is correct. Clear guidelines for drawings and text data must be provided by the planners addressing potential bidders for the future tendering procedures. File formats, layer orders, line thicknesses, colors, plan formats and numbering system should be defined categorically for all graphical, alphanumeric and pictorial data inputs.

It is advisable to specify the basic data requirements in the form of a separate cost element in the tender documents for the ease of integration by providers at a later date. Figure 17.9 gives an example for the categories that classify the building documentation over an entire life cycle.

According to Morfeld and Potreck [Mor02] the detailed documentation of the building systems is extremely important, particularly in the automotive industry. The research and development department of VW Wolfsburg includes, for example an area of 500,000 m^2. Detailed quantification of tender quantities for maintenance, replacements, and any inquiries about certain components can be transmitted via the CAFM system MORADA. One advantage of this system is its ability to locate complex technical elements according to their specific location in the various buildings of the Wolfsburg plant. In order to achieve this, all components of the supply systems, as well as the utility elements are classified as technical systems with labeling schemes which follow the individual catalogue numbering

		Relation/geometry					Room Equipment: cost group 300					Equipment utilities: cost group 400				
Element	Sub elem.		Floor area	Height	Wall area	Window	Opening Type	Floor	Wall	Ceiling	Facilities	Require. Room	Floor	Wall	Blanket	Facilities
110.01	110.01.1	**Production**	4605.0	6.85			T 1.1	Epoxy raisin	Concrete socket	Steel sheet		Temp:		6 x 230 V/10	Sprinkler	To-and
				3.50			T 3.1	Forklift 1.5 to	Metal sandwich	Trapezoidal		> 19° C		6 x 400 V/16	hall reflector	Waste air plant
							Gate 1.2	Carrier 1.5 to	smooth	changeable		Exchange air		6 x 400 V/32	lights 300lx	
							Gate 1.3		Surfaces	Installation		90.000 m³/h		2 x 400 V/63	Cable stages	E supply for
							F 1.1		window aisles	suspended					escape signs	manuf. units
										ceiling						
								Perimeter 6 cm								
								optional R. B								
								Fine stoneware								
110.02		**Shipment**	458.7	6.85			T 1.1	Epoxy raisin	Concrete socket	Steel girder	4 Thermo-	Temp:		4 x Network	Sprinkler	ceiling air
							Gate 1.1	Forklift1.5 to	Metalsandwich.	Trapezoidal	channelling	> 19° C		4 x 400 V/16	hall reflector	heater
							Gate 1.3	Carriier 1.5 to	smooth	Steel sheeting	2 Thermo-	Exchange air		14 x 230 V/10	lights 300 lx	1 x pressure
							F 1.1	desinfectable	Surfaces		channelling	2.0		8 x 230 V/10		air 1/2 "
								floor	window aisles		Jumbo size	Ventilation		.	Cable stages	
								tilework				(Gates)		2 x 400 V/32	Escape signs	
								Base 3 cm								
								optional R. B								
								Fine stoneware								
110.03		**Store**	90.0	6.85			T 3.1	Epoxy raisin	Concrete socket	Steel girder		Temp:		Electrical outlets	Sprinkler	Air heaters
							Gate 1.3	Act	Metalsandwich	Trapezoidal		> 19° C			hall reflector	Waste air util.
								Forklift 1.5 to	smooth	Steel sheeting					lights 300 lx	.
								Carrier 5 to	Surfaces						Cable stages	E supply for
								desinfectable	window aisles						Escape signs	manuf. units
								floor								
								tilework								
								Base 3 cm								
								Perimeter 6 cm								
								optional R. B								
110.04		**Adapter 1**	322.8	6.85			T 1.1	Reinf. concrete	Concrete socket	Steel girder		Temp:		2 x 230 V/10	Sprinkler	Ceiling air-
							Gate 1.2	smoothed	Metalsandwich.	Trapezoidal		> 15° C		2 x 400 V/10	Light ribbon	heater
							Gate 1.4	Hard layer	.	Steel sheeting		Exchange air		1 x 400 V/10	Emergency	
							F 1.1		Surfaces			by free			lighting	
									window aisles			Ventilation				
												(Gates)				
110.05		**Adapter 2**	280.5	4.00			Gate 1.4	Reinf. concrete	Concrete socket	Steel girder		Temp:		2 x 230 V/10	Sprinkler	Ceiling air-
							Komplettelem.	Wing smooth	Metalsandwich.	Trapezoidal		15° C		2 x 400 V/10	Light ribbon	heater
							T 90	Hard substance	layer	Steel sheeting		Exchange air		1 x 400 V/10	Emergency	High bay store
												by free			lighting	
												Ventilation				

Fig. 17.8 Planning data of a space book (example). © IFA 15.279ESW_B

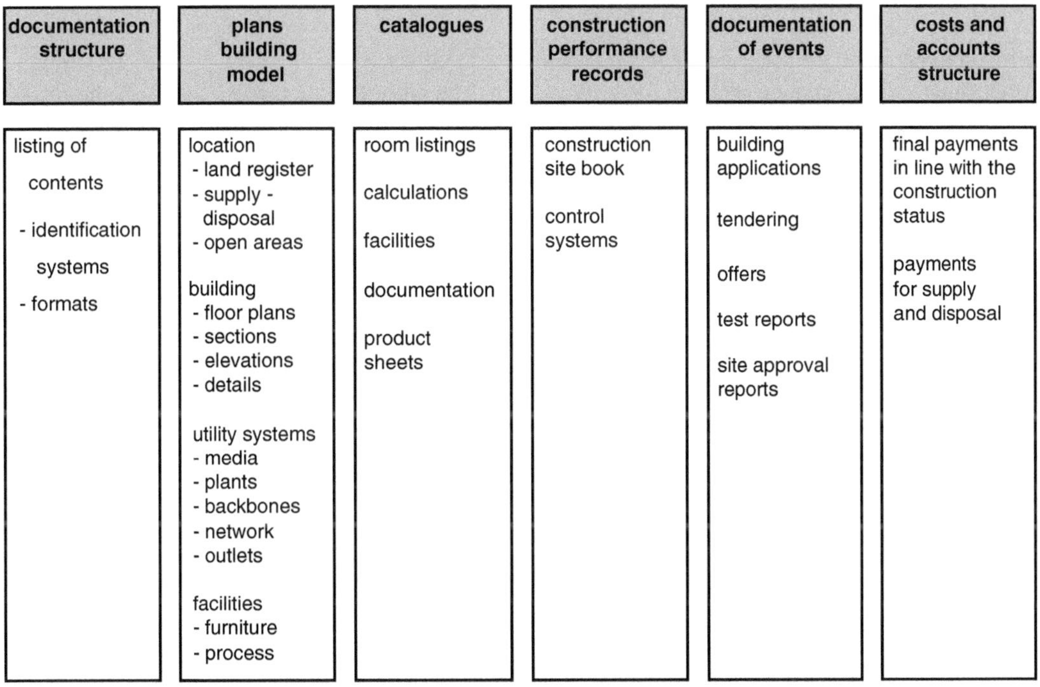

documentation structure	plans building model	catalogues	construction performance records	documentation of events	costs and accounts structure
listing of contents - identification systems - formats	location - land register - supply - disposal - open areas building - floor plans - sections - elevations - details utility systems - media - plants - backbones - network - outlets facilities - furniture - process	room listings calculations facilities documentation product sheets	construction site book control systems	building applications tendering offers test reports site approval reports	final payments in line with the construction status payments for supply and disposal

Fig. 17.9 Classes of a building documentation. © IFA 15.280ESW_B

systems. The schematics of the utilities structure follow the GEFMA guide line 182 while the labeling scheme is modeled on the basis of power plant construction schedules in line with German DIN 6779. In future one could generate a nameplate for each component out of a catalogue. A clear set of instructions enables all contractors to offer their products and data according to these specifications.

17.4.3 Virtual Project Space

For complex projects Internet-based techniques for planning and documentation management are being increasingly used. CAFM systems especially offer World Wide Web communication of Companies that are involved in the management of the property in the planning or execution phase, and provide maintenance and control operations of building facilities online. In this way an external service provider may continuously monitor the plant and discover problems early, without requiring a technician at the local site.

An even higher degree of flexibility and system transparency can be achieved with the establishment of a separate website for a building. Authorized users can then, due to different needs, have access to location, buildings, building services and facility data. In addition, through mobile PDA devices, laptops and mobile phones, data read from devices can be entered online directly. Thus advantages of such a system can be listed down as follows:

- Duplication and errors are reduced, as is ensuring that all work take place on the basis of current plans and documents.
- Plans can be reviewed and commented online, which leads to a considerable saving of time in the assessment and approval processes.
- The risk of losing important files is eliminated, since the current and previous data versions of a drawing or document are stored in a central location.
- An improvement in the flow of communication within the team is achieved through a structured initiating and responding to inquiries.
- A proactive task management facilitates the control of processes.
- Potential defects liability processes are accelerated.

CAFM web based edits are also a cost effective alternative to providing complete hardware and software installation for each user of the CAFM system in the office. Free viewer systems such as volo-view or DWF viewer allow the editing of e.g., AutoCAD drawings without the need of installing the complete software at the individual user.

17.4.4 Navigation

The importance of the visual information for interactive and user friendly systems has already been addressed. Figure 17.10 illustrates the networking of information between location, buildings, utilities systems and function/process with the example of a factory design. Parallel to the planner's view the views of the property owner, operator and user have to be integrated. Their concerns can refer to e.g., the elements of the master plan, components, floor levels, area, room or facilities.

The information merged together in facility management can be visualized as a structured data matrix. A model such as this is depicted in Fig. 17.11.

The upper horizontal bars are views of the system users (planners, owners, operators and object users) with queries on location, buildings, building equipment and function/process while the left vertical bars are used for generating graphical building project information according to different scale levels of the factory from the master plan to the specific facility object. Depending on the query, various possible reporting formats for drawings, texts and images are stored in this information grid; the users may click on the required 'drawers' for their first survey of content. This "navigation system" allows a graphically guided access to the relevant information and easy layering of information. In the "cross-view", for example, one can easily find the arrangement of ventilation layouts for a specific area of a facility layout. Similarly, in the

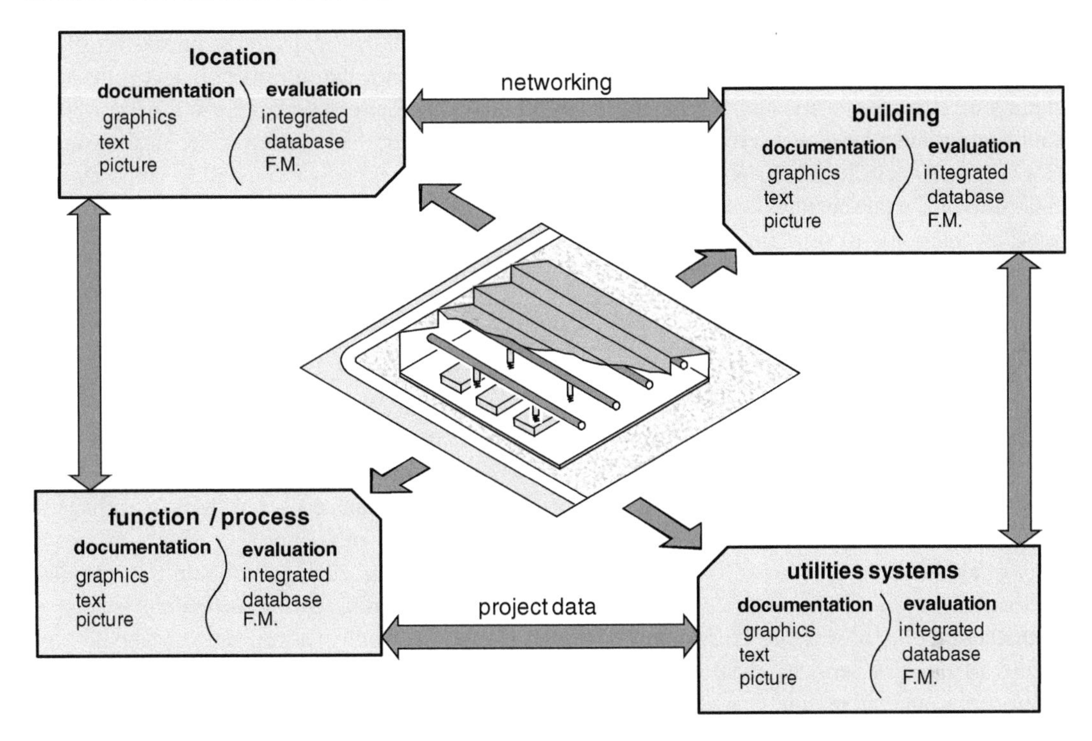

Fig. 17.10 Networking of project data. © IFA 15.281ESW_B

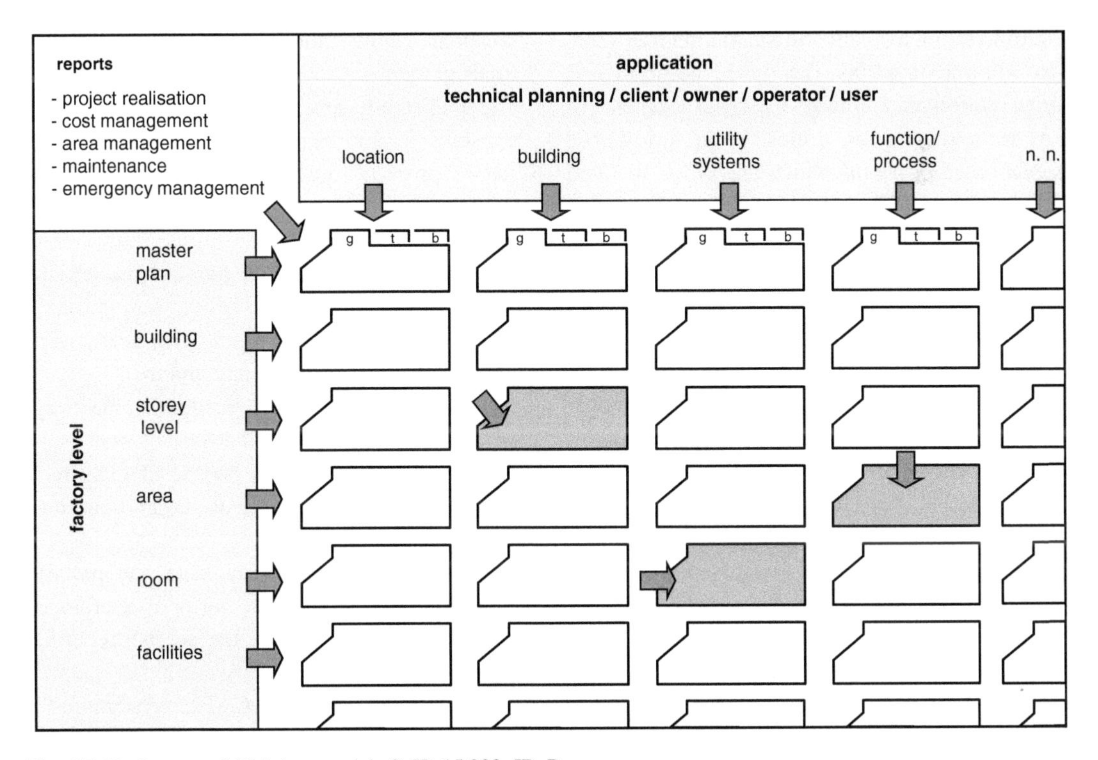

Fig. 17.11 Integrated FM data model. © JR 15.282_JR_B

"longitudinal view", for example, one can find the complete information about a sprinkler system of a large scale technical infrastructure right down to the individual sprinkler head on a given storage rack.

It is conceivable to link classification structures, such as the presented navigation system, to verbal instructions, to develop even more user friendly and interactive modes—from basic graphical inputs to voice command filters. These more intuitive user-specific queries and storages options of information could significantly contribute to the further acceptance of more and more complex facility management systems.

17.4.5 Selection of a CAFM System

Manufacturer of CAD design software, system supplier of building services, planning and consulting firms and increasingly, software manufacturers of more business oriented management programs, such as SAP, provide CAFM software. According to Nävy [Näv06] 43 CAFM systems were available in Germany by 2002; 60 % of which were not available on the market five years ago while 15 systems, developed before 1999, were no longer offered. The development of CAFM systems show a clear trend away from "CAD-based" systems which are to be operated only with special knowledge towards more "neutral" database-oriented systems with variable surfaces for easy operation. The dominance of industry standards such as AutoCAD (77 %) and Oracle (81 %) for the CAD drawing system and central databases is remarkable. In order to ensure the future availability and security of the valuable assets, it is preferable to specify systems that work well with most of the industry standards for graphics and text and having years of experience.

The GEFMA guide line 400 clearly distinguishes between CAFM software and a CAFM system. A CAFM system is understood as a complex and customized software solution that implements the relevant processes in a specific database structure. Before selecting and implementing a CAFM system it is advisable to create a specification that maps the requirements of the user and defines the information levels for variable views. These specifics for a CAFM system depends on the FM processes the users want to monitor and support. Thus it is important to identify the various tasks and queries well in advance by the different user groups (e.g., FM team, owners, managers, internal and external service providers, users, prospects, etc.).

17.4.5.1 CAFM Consulting

According to Warner et al. [War02], and May [May12] the following tasks have to be defined at the onset for any CAFM systems:

- The requirements for a CAFM system have to be structured and detailed.
- A moderating dialogue should be installed between the different "stakeholders".
- A rating and prioritization should be undertaken.
- An agreed process, system and data architecture is to be developed.
- A flowchart of steps and milestones depending on prioritization.
- The success of the implementation must be ensured with an accompanying quality management.

Specialist advisers to the CAFM consulting are useful for moderating the introduction of a CAFM project. Accordingly, the following schedule has been proven to be beneficial in major projects:

- 1st Workshop: spaces, organization, equipment, relocation
- 2nd Workshop: maintenance, technical equipment, energy, building maintenance
- 3rd Workshop: waste disposal, cleaning, security
- 4th Workshop: data sets, license plate system
- 5th Workshop: interfaces to other computer systems.

It is advisable to store the data and process descriptions not only in the form of a protocol but also in digital formats like spreadsheets for ease of transferability.

Selection of a suitable CAFM system for a given purpose needs clarification of the required processes first. Over the years, CAFM systems are clearly moving away from more closed CAD concepts towards more open, flexible database structures enabling the possibility to integrate with widespread CAD suppliers, such as Autodesk (AutoCAD, Revit), providing users with a higher degree of flexibility, availability and security. An alternative is the combination of more business processes oriented software (e.g., SAP) with a suitable graphics visualization tool for the required purposes. This is particularly useful for Companies that already use SAP for their core business processes. An example of a widely used FM system is Archibus/FM, which also allows access to other software operating systems such as ERP or human resources management, in addition to numerous application modules, such as areas, surfaces, furniture, equipment and maintenance management. Figure 17.12 shows an application in which an AutoCAD system can be edited within Archibus/FM by means of the "overlay" function.

17.5 Applications of Facility Management

In the next section we will consider some exemplary fields of application for facility management as far that they are significant for factory planning.

17.5.1 Minimizing Maintenance Costs

In contrast to the conventional building design which is characterized by a fixed construction investment target, facility management starts already in the development phase aiming at optimizing the subsequent maintenance costs or in other words balancing a higher investment cost to lower life cycle operating costs. This is achieved by the fact that at the earliest possible date the decisive factors of maintenance costs, for example energy and cleaning costs are discussed and clearly pointed out. With this support,

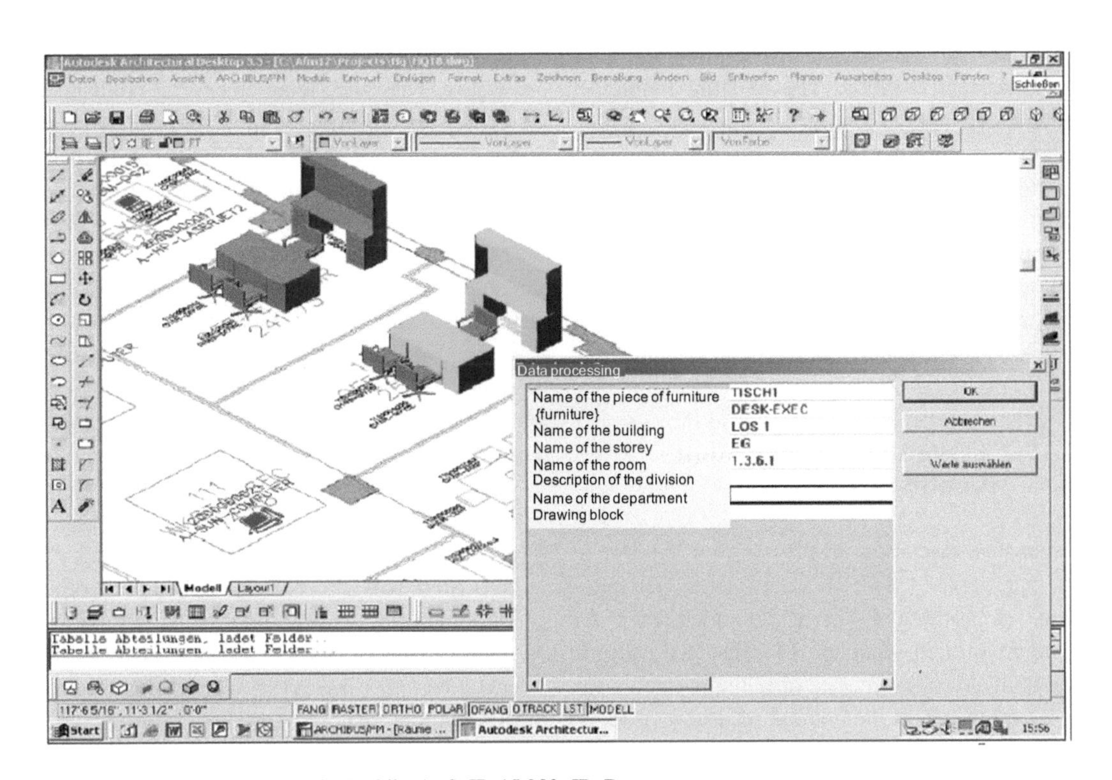

Fig. 17.12 FM system (example Archibus). © JR 15.283_JR_B

the client is not only able to decide on the basis of minimal construction costs but also of minimal maintenance costs. Facilities Management provides for potential savings not only for new construction projects, but also for existing buildings. In these situations FM focuses on optimizing energy, cost transparency, operational processes and the organizational structure, since optimizing a structure or the building services after the structure has been built is generally very expensive.

A variety of service providers, in the business of utilities services, are offering many well-meaning efforts to optimize equipment and processes with "island" or stand-alone solutions and interfaces. Nevertheless, judging on the availability of current professional qualifications, it seems to be appropriate that a trained architect, possessing the most comprehensive knowledge of building as a whole is ideal to merge the many issues related to optimization of the building operation in a holistic manner.

A serious change in the property management discussion about the locations of businesses, as well as in the management of municipalities, communities, countries has been taking place since the early 1990s. The existing municipal budgeting and accounting system of public administration today is focused more on economic optimization and quality of service. In times of tight budgets and funding deficits, municipal property and building management calls for transparent cost element based budgeting of all departmental offices. This is only possible through an FM database of all public property parameters.

17.5.2 Prevention of Allocation Conflicts

A constant discussion in the use of a building is the appropriate carpet area per user or per function, as nothing is more defended than an "acquired" area. Chapter 14 presented various methods used to generate area requirements for working places in the context of a new or renovation works. However, the methods are not suitable for existing ongoing operations and their changes. An FM system offers the possibility to determine the areas actually in use in each department. Implementing an objective method when allocating spaces leads to better and quicker acceptance of the spaces assigned to employees. Reservation plans allow available spaces for new employees to be quickly identified. Depending on the chosen facility management system this allows for:

- the creation of plans and reports with average areas per employee, building and location,
- the allocation of areas by group, space or staff,
- the inclusion of staff symbols in drawings,
- the clearing of the area rates by group, space or an employee within a cost center,
- the possibility to search for rooms with assigned equipment, e.g., display systems and video conferencing.

17.5.3 Spatial Planning

The room and space management function of an FM system includes features that show whether a change of function requires more space or a restructuring of the room occupancy is imminently necessary. The space requirement can be planned according to the number of employees, type of usage, occupancy levels of the premises and logistical costs. The room occupancy data or critical portions of it, can often be transferred into Microsoft Excel or Adobe Acrobat, or even made web-suitable to make this information available to other departments of the organization for the purpose of viewing. Depending on the chosen facility management system, the following functions may be offered:

- the building and infrastructure data can be made accessible to users within their organization,
- the efficiency of space utilization can be compared with the use of indicators to other buildings,
- the allocation of space costs can be effected by occupied space as well as a proportional area of common areas,
- supporting the relocation management and inventory planning.

17.5.4 Lock Management and Key Management

With lock management an unlimited number of locks can be simultaneously managed with unrestricted nesting depth. Groups, individual and central closures are important elements of the lock hierarchy that is to be represented. This clear structure ensures that access permissions for all areas are clearly and comprehensively managed. Moreover, it opens up potential for reporting and analyzing. Plans for individual, central and group locks are often routinely automatically generated and transferred to Microsoft Excel. In addition, it is common to create a powerful generator for individual lists and reports. Integrating lock management in CAD graphic modules supplements the possibility for analyses.

17.5.5 Costs and Building State Control

In order to calculate transparent costing systems for management, it is important that all building costs elements are sufficiently and accurately identified. Moreover all operating costs for owned or leased real estate, as well as taxes need to be registered. Community costs, such as municipal fees and land costs can be factored easily and quickly by means of individual departmental key structures or similar unit allocation routines. The control of any payments can be surveyed easily through the creation of budget and payment plans. This allows for:

- compliance with contractual requirements and deadlines,
- control of the current and upcoming fiscal expenditures,
- assessment of the economical soundness of the properties, and
- status and analysis for each property and real estate.

17.5.6 Reporting

Being able to directly access precise information about floor spaces makes it easier to adapt to external reporting requirements. If an external organization reimburses floor costs (either partially or completely), a considerable difference between the estimated and actual area costs can lead to repayment of thousands or even millions of Euros. Using the appropriate accounting formulas ensures that every division within the organization is responsible for their own usage of space and that the incurred expenses are charged according to the predefined distribution key. Depending on the implemented facility management system it is possible to:

- calculate the room share percentage for partially or temporarily used floor areas,
- summarize the area of each division with the department oriented space analysis reports, and
- provide different accounting methods.

17.5.7 Fire Protection

The basic fire protection standards are outlined in the applicable (e.g., federal or municipal) building codes (see also Sect. 10.5). Facility management in particular, provides outstanding possibilities to comprehensively depict structurally required measures, the operational readiness of the building services as well as the user's responsibility. Especially with changing users and corresponding building renovations, it is indispensable to continually update the documentation concerning the state which is legally agreed upon with the authorities.

Structural fire protection and risk prevention relates primarily to specifics on the following issues:

- distance between the buildings,
- position and arrangement of buildings and structures on the property,
- access roads and areas for the fire brigade,
- building materials, components and overall design,
- partitions,
- positioning and arrangement of emergency routes, and
- location and arrangement of fire department key boxes.

The organizational fire and hazard protection covers constant inputs to the overall fire protection concept on the following issues:

- use of factory-, office- or home fire brigades,
- preparation of fire protection systems,
- provision of object-related application documents,
- limitation of fire loads, and
- timely and effective initiation of security measures.

17.6 Modeling of FM Processes

Thus far, we have only discussed individual facility management tasks. During the operational phase of a building, FM is characterized by a process-oriented approach, typically used in depicting production and logistic functions. These processes are dimensioned by their costs, information and benefit.

The basic processes are the same. It is thus practical to model reference processes and to adapt these to concrete situations. Figure 17.13 depicts the objectives and content of such processes based on the examples of business continuity management, energy monitoring and maintenance and repair orders.

Using the example of a disruption, we will now describe such a process [Kri08]. When there is a defective water faucet in a public building, the process proceeds as follows starting with the trouble alarm and ending with the rectification of the cause: The disruption is registered by the service provider's 24-h emergency service for technical maintenance either via a telephonic trouble indicator or an e-mail. There the report is compiled via a web-based tool for registering disruptions (e.g., with the pit-FM building services software module). Alternatively, the user can also navigate through the room level of the CAFM system and register a "disruption" event in Room '009 Men's Washroom'. According to the level of urgency assigned to it, an operation for awarding the contract is initiated. After the repair, the completion is reported back to the system and the costs that arose are allocated to

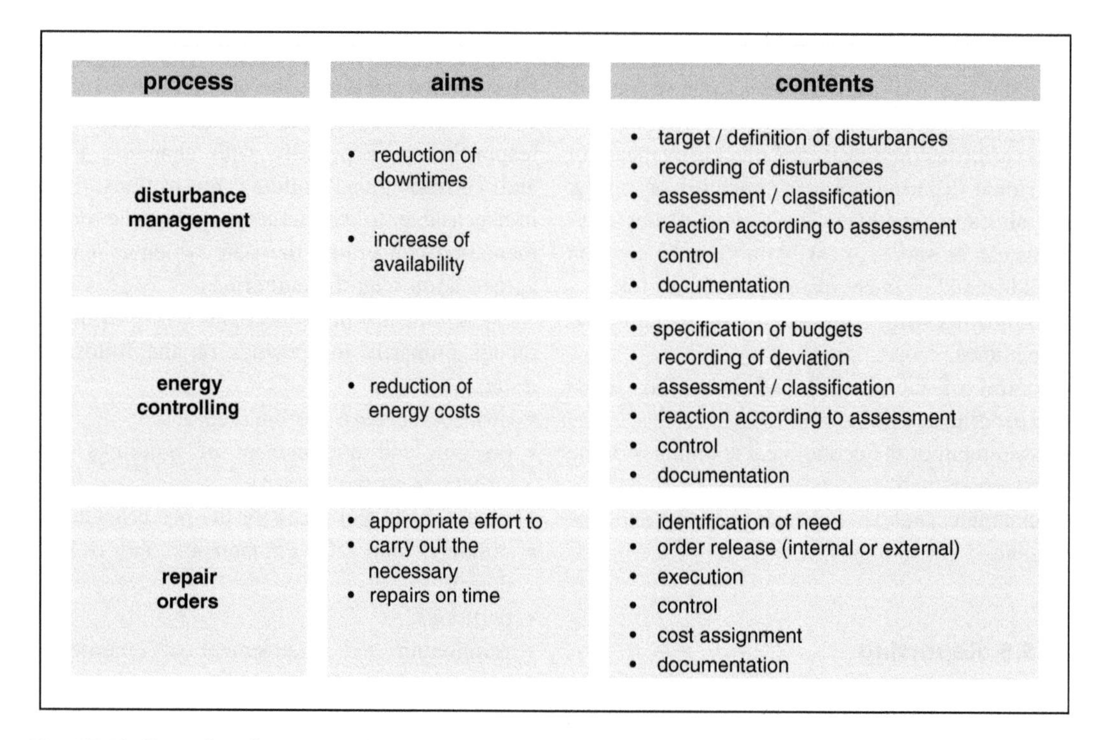

process	aims	contents
disturbance management	• reduction of downtimes • increase of availability	• target / definition of disturbances • recording of disturbances • assessment / classification • reaction according to assessment • control • documentation
energy controlling	• reduction of energy costs	• specification of budgets • recording of deviation • assessment / classification • reaction according to assessment • control • documentation
repair orders	• appropriate effort to • carry out the necessary • repairs on time	• identification of need • order release (internal or external) • execution • control • cost assignment • documentation

Fig. 17.13 Examples of processes of technical building management (after Krimmling). © JR 15.284_JR_B

the respective party (e.g., customer, division etc.).

In order to avoid interfaces in the building services trades and the scarce resources in businesses and communities, "performance contracting" or "energy performance contracting" are becoming quite common. Performance contracting is based upon the fact that all running activities for the technical object operation and optimization of the equipment for measurement, control, building management system, heating, ventilation and air conditioning systems, lighting and power and building service shall be financed through savings in energy costs.

17.7 Case Studies

The Facility Management System of two industrial plants—one for rubber based products and another for hair care applications—are explained hereunder as case studies.

17.7.1 Phoenix AG Hamburg

Phoenix AG is among the leading providers of rubber technology and acoustic systems. Phoenix AG currently belongs to Continental AG. During the 1990s, the Group decided to shift the production facilities increasingly towards Eastern European countries.

The main plant at Hamburg employs around 3100 people and is located opposite the main train station. Unfortunately, the inner city location prevented an extension with new and more efficient factory buildings and the existing industrial structure offered only outdated production facilities based upon increasingly inefficient processes. Meanwhile, the company abandoned this location. Several variants were developed for utilization of existing buildings. Unfortunately, no documentation existed for the approximately 120,000 m^2 (1290,700 ft^2) of gross floor area. The new FM Systems helped to digitally document the inventory for the core areas as well as 3D-construction of premises based upon available old archived plans. Some of the FM tasks were:

- determining the land and gross floor area of the entire area identified in association with nine companies operating there,
- listing of the individual building details under area categories according to German DIN 277,
- generating a complete 3D-building documentation for the entire site (see Fig. 17.14),
- for special elements of the existing construction, such as CAD objects of ceilings, text attributes were added with information about floor loading capacity, structural parameters, utilities status, clear heights as well as general need for building and utilities,
- a digital documentation of all facades of the different buildings.

On this basis, several scenarios/optimization proposals were worked out aiming at answering the following queries:

- correction of floor vacancies by means of better organization of departments,
- improving functional relationships of process, logistics and administration,
- minimize operating costs, and
- identifying possible areas for "outsourcing" for rental or sale.

The 2D/3D-construction, the area and surface determinations were engineered in AutoCAD. The graphics as well as the image data were transferred to the Oracle database of the CAFM system Archibus/FM. The report generator of Archibus/FM allows for merging data as, for example, area calculations, images of different colors for departments and traffic as well as special dependencies of assignment to the desired functions of specific floor space in a simple way (Fig. 17.15).

17.7.2 Londa Rothenkirchen

As an extension to its established plant structure the Londa company wanted to realize a new manufacturing hall with an approximately 6000 m^2 (64,600 ft^2) gross floor area and corresponding service areas at their Rothenkirchen location. Londa is a subsidiary of WELLA AG,

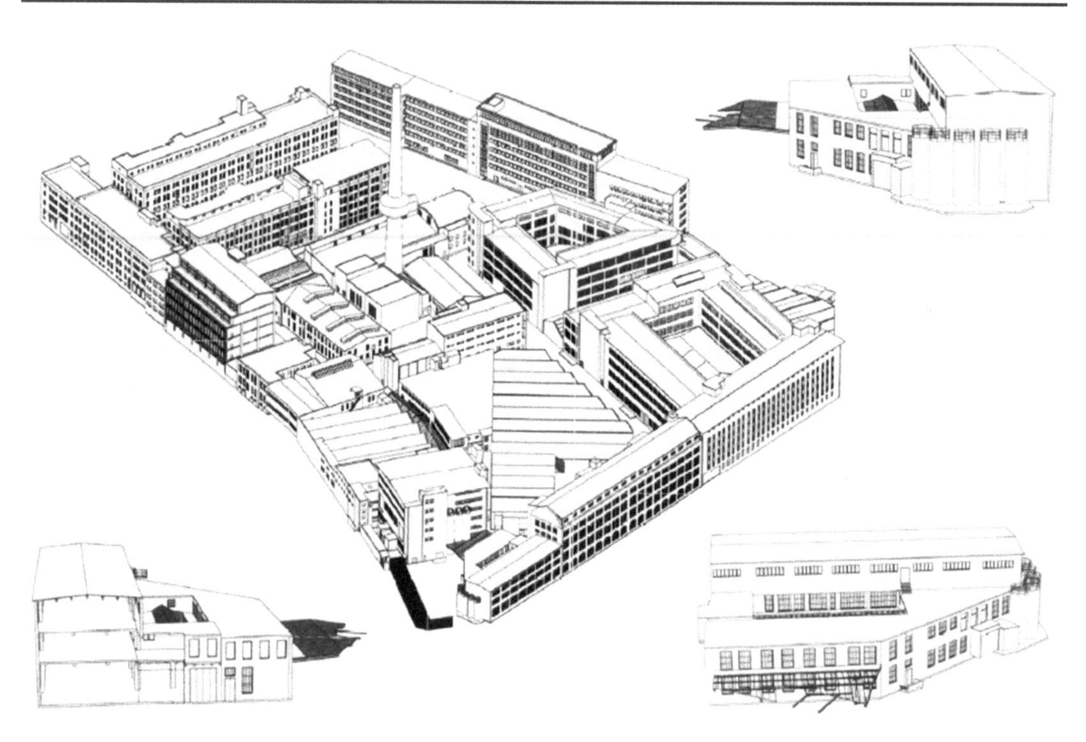

Fig. 17.14 3D building documentation (example Phoenix Hamburg). © JR 15.285_JR_B

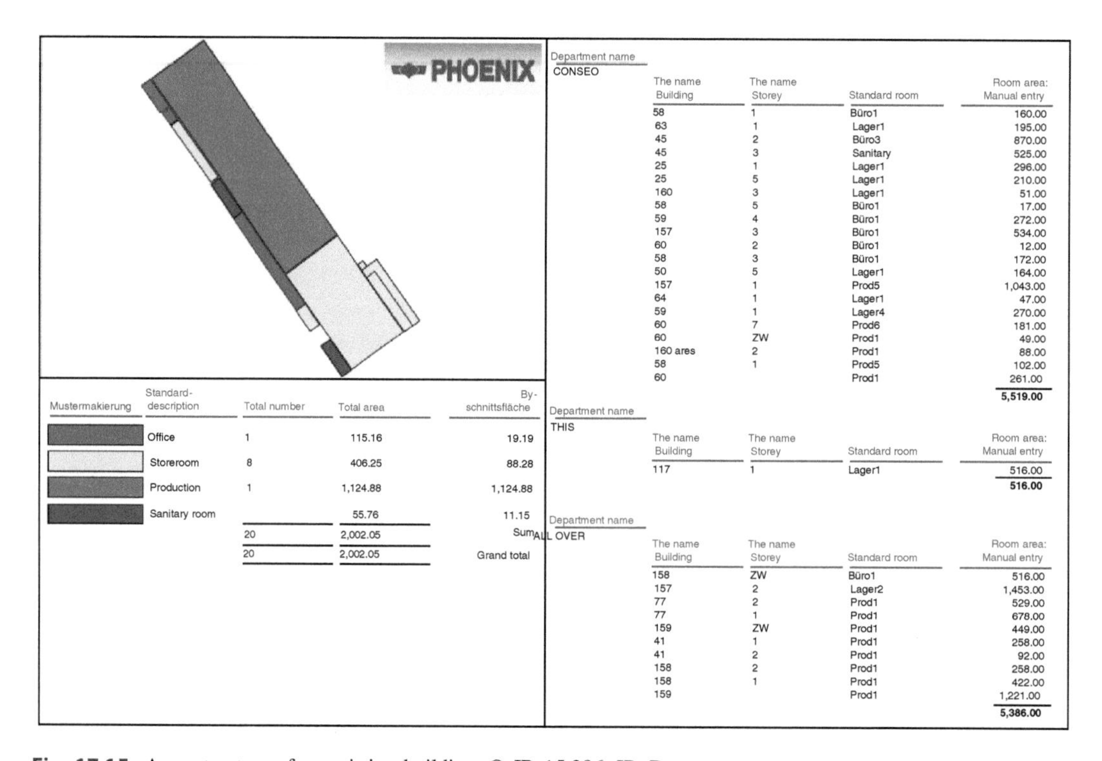

Mustermakierung	Standard-description	Total number	Total area	By-schnittsfläche
	Office	1	115.16	19.19
	Storeroom	8	406.25	88.28
	Production	1	1,124.88	1,124.88
	Sanitary room		55.76	11.15
		20	2,002.05	Sum
		20	2,002.05	Grand total

Department name: CONSEO

The name Building	The name Storey	Standard room	Room area: Manual entry
58	1	Büro1	160.00
63	1	Lager1	195.00
45	2	Büro3	870.00
45	3	Sanitary	525.00
25	1	Lager1	296.00
25	5	Lager1	210.00
160	3	Lager1	51.00
58	5	Büro1	17.00
59	4	Büro1	272.00
157	3	Büro1	534.00
60	2	Büro1	12.00
58	3	Büro1	172.00
50	5	Lager1	164.00
157	1	Prod5	1,043.00
64	1	Lager1	47.00
59	1	Lager4	270.00
60	7	Prod6	181.00
60	ZW	Prod1	49.00
160 ares	2	Prod1	88.00
58	1	Prod5	102.00
60		Prod1	261.00
			5,519.00

Department name: THIS

The name Building	The name Storey	Standard room	Room area: Manual entry
117	1	Lager1	516.00
			516.00

Department name: ALL OVER

The name Building	The name Storey	Standard room	Room area: Manual entry
158	ZW	Büro1	516.00
157	2	Lager2	1,453.00
77	2	Prod1	529.00
77	1	Prod1	678.00
159	ZW	Prod1	449.00
41	1	Prod1	258.00
41	2	Prod1	92.00
158	2	Prod1	258.00
158	1	Prod1	422.00
159		Prod1	1,221.00
			5,386.00

Fig. 17.15 Area structure of an existing building. © JR 15.286_JR_B

Darmstadt which is now part of the Procter & Gamble group.

As part of the IT network of all Proctor & Gamble's properties; the American owner places great value on storing relevant data about the locations, buildings, building services, and facilities in an Oracle database with the Archibus CAFM System. For the plant expansion, it was decided to use the possibility of FM compliant documentation during the planning for all of the disciplines in order to document workshops, rough planning, detailed planning, approval planning, work organization and the calls for tenders.

All professional planners for site, building, utilities, process and logistics collected their 2-D, 3-D, text and engineering data on basis of standard CAD formats and Excel listings which were merged together in line with the methodology described earlier under synergetic factory planning. Digital interface to Archibus/FM was carried out by a macro tool that systematically searched the Excel sheets for new entries and transferred them by means of an "entry", programmed in Access, into the relevant system table of the data fields of the Oracle database of the CAFM system Archibus.

Due to the various reports offered by Archibus/FM P&G management had the ability to constantly access rough or detailed information about areas, process or cost, etc. and could compare these values with other plants in the sense of a global benchmarking. For clarity of the overall architectural project a video simulation had been created, as an integral project simulation model of location, buildings, building equipment and furnishings in the development phase (attached in Annex D1 to this book). These 3D data were further used as basis for a visual information system, following the systems as discussed in Sect. 17.4.4, navigation. The system allowed for generation of a virtual 3D production line for hair cosmetics directly out of a 2D floor plan view and sort out user dependent process information, as line equipment cost or vendor information, by linking drawing information with the CAFM system from the database. Figure 17.16 illustrates the 2D/3D line. For a globally active company such user friendly graphical navigation was really advantageous.

These examples conclude our discussion of facility management. Even though facility management is only spreading slowly, the unstoppable development of the digital factory and the increasing speed of factory changes will continue to reveal the usefulness of FM systems.

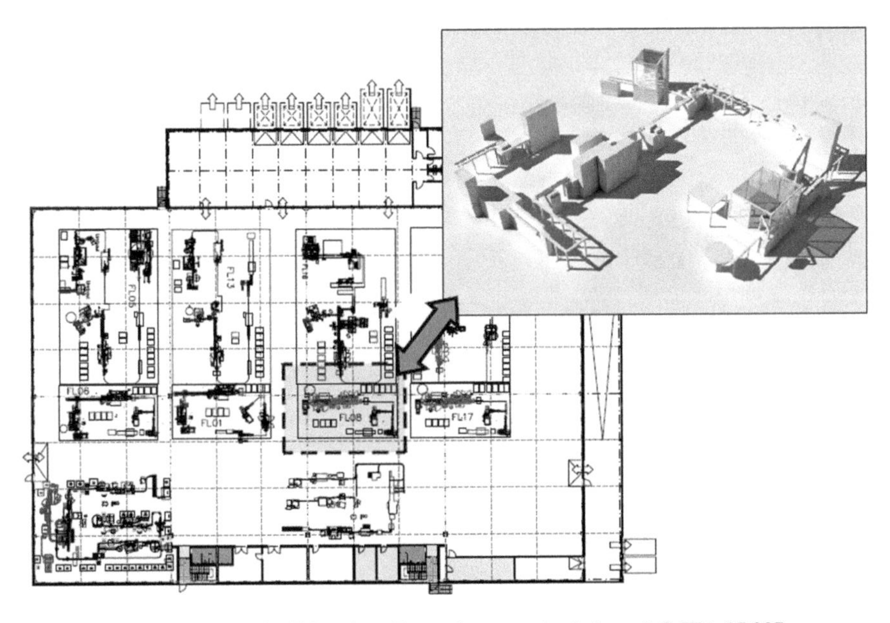

Fig. 17.16 Linking of building and facilities data (Example cosmetics industry) © IFA 15.287

17.8 Summary

Facility Management allows a variety of synergies between space and process. Clear definition of engineering objects in terms of changeability, a conscious structuring and storage of project data of processes and buildings should be documented at the very beginning of every factory planning. It continues during target programming as well as in the development phase along with the documentation of the initial project requirements and workshop results. Project management calls for transparent information gathering across all planning, implementation and operation phases, in particular the latest space and cost evaluations are of major concern. FM definitely helps to fulfill a multitude of tasks related to controlling and quality assurance in all phases of the project. Based upon the available 3D BIM technologies, one can safely assume that more integrated programs would be entering the market in the near future, allowing simulations for thermal comforts, sustainability and energy efficiency. Also, visually aided navigation would help to simplify the current data structures and make the FM Systems more interactive and user friendly.

Bibliography

[Boo09] Booty, F. (ed.): Facilities Management Handbook, 4th edn. Elsevier, New York (2009)

[Bra07] Brown, P., Putter, J., Reents, M., Zahn, P.: Facility Management: Erfolg in der Immobilienbewirtschaftung (Facility Management: Success in the Real Estate Management), 5th edn. Springer, Berlin (2007)

[GEF13] GEFMA Guide Line 400: Computer Aided Facilities Management CAFM. Begriffsbestimmungen, Leistungsmerkmale (CAFM: Definitions, Characteristics). Bonn (2013)

[GEF96] GEFMA Guide Line 100: Facility Management. Begriffe, Strukturen, Inhalte (Concepts, Structures, Contents). Bonn (1996)

[Jed02] Jedlitzke, M.: IT-Grundlagen für Facility Manger (IT basics for facility managers). In: May, M. (ed.) IT im Facility Management erfolgreich einsetzen. Das CAFM Handbuch (Successful use of IT in Facility Management. The CAFM Manual). Springer, Berlin (2002)

[Kri08] Krimmling, J., Preuss, A., Deutschmann, J.U., Renner, E.: Atlas Gebäudetechnik. Grundlagen, Kontruktionen, Details (Atlas Building Techniques. Foundations, Structures, Details). Rudolf Müller Publisher, Köln (2008)

[Lut02] Lutz, U. (ed.): Facility Management Jahrbuch 2002/2003 (Facility Management Year Book 2002/2003). Springer, Berlin (2002)

[May12] May, M., Williams, G. (eds.): The Facility Manager's Guide to Information Technology: An International Collaboration. International Facility Management Association (2012)

[Mor02] Morfeld, E., Potreck, H.: Digitale Bestandsdokumentation zur Optimierung von Wartungs- und Instandhaltungsprozessen. In: Lutz, U. (Hrsg.): Facility Management Jahrbuch 2002/2003 (Digital "as built"- documentation for the optimization of maintenance and repair processes. In: Lutz, U. (ed.): Facility Management Yearbook 2002/2003). Springer, Berlin (2002)

[Näv02] Nävy, J.: Marktübersicht CAFM-Systeme. In: Lutz, U. (Hrsg.): Facility Management Jahrbuch 2002/2003 (Market overview CAFM systems. In: Lutz, U. (ed.): Facility Management Yearbook 2002/2003). Springer, Berlin (2002)

[Näv13] Nävy, J.: Facility Management. Computerunterstützung, Systemeinführung, Anwendungsbeispiele (Facility Management. Computer Support, System Implementation, Application Examples), 4th edn. Springer, Berlin (2013)

[Ron06] Rondeau, E., et al.: Facility Management, 2nd edn. Wiley, New Jersey (2006)

[Rop14] Roper, K., Pyant, R.P.: The Facility Management Handbook, 4th edn. American Management Association, New York (2014)

[War02] Warner, T., et al.: Consulting im CAFM. IT im Facility Management erfolgreich einsetzen. In: May, M. (Hg.): Das CAFM-Handbuch. (Consulting in CAFM. Successful Use of IT in Facility Management. The CAFM Manual). Springer, Berlin (2002)

Erratum to: Handbook Factory Planning and Design

Hans-Peter Wiendahl, Jürgen Reichardt
and Peter Nyhuis

Erratum to: H.-P. Wiendahl et al., *Handbook Factory Planning and Design*, DOI 10.1007/978-3-662-46391-8

This book was published without mentioning the original Copyright. The correct Copyright Information is mentioned below:

Copyright © 2009
Carl Hanser Verlag, München
Alle Rechte vorbehalten.
Autorisierte Übersetzung der deutschen Originalausgabe, erschienen im Carl Hanser Verlag, München.

The online version of the original book can be found under DOI 10.1007/978-3-662-46391-8

H.-P. Wiendahl (✉)
Garbsen, Germany

J. Reichardt
Fachbereich Baukonstruktion und Industriebau, Fachhochschule Münster, Münster, Germany

P. Nyhuis
IFA Institut für Fabrikanlagen und Logistik, Leibniz Universität Hannover, Garbsen, Germany

H.-P. Wiendahl et al., *Handbook Factory Planning and Design*,
DOI 10.1007/978-3-662-46391-8_18, © Springer-Verlag Berlin Heidelberg 2015

Index

H.-P. Wiendahl et al., *Handbook Factory Planning and Design*,
DOI 10.1007/978-3-662-46391-8, © Springer-Verlag Berlin Heidelberg 2015